Second Century of the Skyscraper

The Monograph

Second Century of the Skyscraper documents the future of the high-rise as forecast by tall building experts worldwide. This volume grew out of presentations given at the Third International Conference on Tall Buildings, held in Chicago in January, 1986. It documents the most recent of developments and state-of-the-art of planning and design, the major focus of the Council on Tall Buildings and Urban Habitat.

The material contained in *Second Century of the Skyscraper*, organized to correspond generally with the structure of the Monograph, is outlined below:

Groups and Topics

PLANNING AND ENVIRONMENTAL CRITERIA (PC)

Philosophy of Tall Buildings
History of Tall Buildings
Decision-Making Parameters
Social Effects of the Environment
Sociopolitical Influences
Development and Investment
Economics
Architecture
Interior Design
Landscape Architecture
Urban Planning and Design
External Transportation
Parking
Ownership and Maintenance
Project Management
Rehabilitation, Renovation, and Repair
Design for the Handicapped
Tall Buildings in Developing Countries

SYSTEMS AND CONCEPTS (SC)

Cladding
Partitions, Walls, and Ceilings
Structural Systems
Foundation Design
Construction Systems
High-Rise Housing
Prefabricated Tall Buildings
Tall Buildings Using Local Technology
Robots and Tall Buildings
Application of Systems Methodology

BUILDING SERVICE SYSTEMS (BSS)

Heating, Ventilating, and Air Conditioning
Energy Conservation
Plumbing and Fire Protection
Electrical Systems
High-Tech Buildings
Vertical and Horizontal Transportation
Environmental Design
Urban Services

CRITERIA AND LOADING (CL)

Gravity Loads and Temperature Effects
Earthquake Loading and Response
Wind Loading and Wind Effects
Fire
Accidental Loading
Safety and Quality Assurance
Motion Perception and Tolerance

TALL STEEL BUILDINGS (SB)

Commentary on Structural Standards
Methods of Analysis and Design
Stability
Design Methods Based on Stiffness
Fatigue Assessment and Ductility Assurance
Connections
Cold-Formed Steel
Load and Resistance Factor Design (Limit States Design)
Mixed Construction

TALL CONCRETE AND MASONRY BUILDINGS (CB)

Commentary on Structural Standards
Limit States Design
Selection of Structural Systems
Optimization
Elastic Analysis
Nonlinear Analysis and Limit Design
Stability
Stiffness and Crack Control
Precast Panel Structures
Creep, Shrinkage, and Temperature Effects
Cast-in-Place Concrete
Precast-Prestressed Concrete
Masonry Structures

The original five-volume Monograph on the Planning and Design of Tall Buildings was published between 1978 and 1981 and updated with three volumes in addition to *Second Century of the Skyscraper*. They can be ordered through the Council.

- *Planning and Design of Tall Buildings*, 5 volumes (1978–1981 by American Society of Civil Engineers)
- *Developments in Tall Buildings—1983* (Van Nostrand Reinhold Company)
- *Advances in Tall Buildings* (1986, Van Nostrand Reinhold Company)
- *High-Rise Buildings: Recent Progress* (1986, Council on Tall Buildings and Urban Habitat)
- *Second Century of the Skyscraper* (1988, Van Nostrand Reinhold Company)

Council Headquarters
Lehigh University
Building 13
Bethlehem, Pennsylvania 18015
USA

Second Century of the Skyscraper

Council on Tall Buildings and Urban Habitat

Lynn S. Beedle, Editor-in-Chief

VNR VAN NOSTRAND REINHOLD COMPANY
New York

Library of Congress Catalog Card Number: 88-5577
ISBN: 0-442-22116-9

Printed in the United States of America

Van Nostrand Reinhold Company Inc.
115 Fifth Avenue
New York, New York 10003

Van Nostrand Reinhold Company Limited
Molly Millars Lane
Wokingham, Berkshire RG11 2PY, England

Van Nostrand Reinhold
480 La Trobe Street
Melbourne, Victoria 3000, Australia

Macmillan of Canada
Division of Canada Publishing Corporation
164 Commander Boulevard
Agincourt, Ontario M1S 3C7, Canada

16 15 14 13 12 11 10 9 8 7 6 5 4 3 2 1

Library of Congress Cataloging-in-Publication Data
Second century of the skyscraper.
Bibliography: p.
Includes index.
1. Tall buildings. I. Beedle, Lynn S. II. Council on Tall Buildings and Urban Habitat.
TH845.S38 1988 712′.042 88-5577
ISBN 0-442-22116-9

Council on Tall Buildings and Urban Habitat

Steering Group

Editorial Committee

Sponsoring Societies

International Association for Bridge and Structural Engineering (IABSE)
American Society of Civil Engineers (ASCE)
American Institute of Architects (AIA)
American Planning Association (APA)
International Union of Architects (UIA)
American Society of Interior Designers (ASID)
American Society of Heating, Refrigerating and Air Conditioning (ASHRAE)

The following lists those firms and organizations who provide for the Council's financial support.

Patrons

Jaros, Baum & Bolles, New York
The Turner Corporation, New York

Sponsors

Office of Irwin G. Cantor, P.C., New York
Federal Emergency Management Agency, Washington, D.C.
Kohn, Pedersen, Fox Associates, P. C., New York
Ahmad Moharram, Cairo
Walter P. Moore & Associates, Inc., Houston
National Science Foundation, Washington, D.C.
Nippon Steel, Tokyo
Otis Elevator Co., Farmington
Ove Arup Partnership, London
PSM International, Chicago
Rankine & Hill Pty. Ltd., Sydney
Leslie E. Robertson & Assoc., New York
Schindler Management A.G., Luzerne
Severud-Szegezdy Construction Engineers, New York
Skidmore, Owings & Merrill, Chicago
Swire Properties Ltd., Hong Kong
Thornton-Tomasetti, P.C., New York
Weiskopf & Pickworth, New York
Westinghouse Elevator Co., Morristown
Wing Tai Construction & Engineering Co., Hong Kong

Donors

Bovis (Far East) Ltd., Hong Kong
The Herrick Corp., Hayward
Hyundai Corp., Bellevue
Tishman Realty & Construction Co., Inc., New York
Tishman Research Corporation, New York
Urban Investment Development Company, Chicago

Contributors

Allison, McCormac and Nickolaus, Rockville
American Institute of Steel Construction, Chicago
W. S. Bellows Construction Corp., Houston
Alfred Benesch & Co., Chicago
Bornhorst & Ward Pty. Ltd., Spring Hill
Boundary Layer Wind Tunnel Laboratory (U. Western Ontario), London
Bramalea Ltd., Dallas
Brandow & Johnston Associates, Los Angeles
Brickell Moss, Ltd., Wellington
Cadillac Fairview Corp. Ltd., Toronto
Campeau Corp, Toronto
Capacete-Martin & Associates, San Juan
CBM Engineers Inc., Houston
Cermak/Peterka & Assoc., Inc., Fort Collins
Chin & Hensolt Engineers, Inc., San Francisco
Civil & Civic Pty. Ltd., Sydney
Concrete Constructions Pty. Ltd., Potts Point
Construction Consulting Laboratory, Dallas
Crow Construction Co., New York
DeSimone, Chaplin & Associates, New York
B. M. Dornblatt and Associates, Inc., New Orleans
Englekirk & Hart, Inc., Los Angeles
Flacon Steel Co., Wilmington
F.N.C.S.R.L., Milan
Fujikawa Johnson and Associates, Chicago
George A. Fuller Co., New York
Gutteridge Haskins & Davey, Railway Square
Hayakawa Associates, Los Angeles
Hellmuth, Obata & Kassabaum Inc., San Francisco
Howard Needles Tammen & Bergendoff, New York
The George Hyman Construction Co., Bethesda
Iffland Kavanagh Waterbury, P. C., New York
International Iron & Steel Institute, Brussels
International Masonry Institute, Washington, D.C.
Irwin Johnston Smith Leuchars, Pty., Ltd., North Sydney
Ishihara Construction Co. (M) Sdn. Bhd., Kuala Lumpur
Kadri Consultants, Pvt., Ltd., Bombay
Kajima Corporation, New York
Kerr Construction Inc., Clifton
Ketchum-Konkel-Barrett-Nickel-Austin, Inc., Chicago
Leighton Holdings, Inc., St. Leonards
LeMessurier Associates/SCI, Cambridge
Stanley D. Lindsey & Assoc., Nashville
Low & Hooke (Aust) Pty., Ltd., Five Dock
Albert C. Martin & Associates, Los Angeles
Maunsell Consultants Asia, Hong Kong
McDonnell Douglas, St. Louis

McWilliam & Partners Pty. Ltd., Brisbane
W. L. Meinhardt & Partners Pty. Ltd., Melbourne
Lane N. Meltzer Real Estate, New Orleans
Mueser Rutledge Consulting Engineers, New York
I. A. Naman & Assoc., Houston
New York Concrete Construction Institute, New York
O'Brien-Kreitzberg & Associates, Inc., Merchantville
Ove Arup & Partners, Sydney
Peddle, Thorp & Walker Arch., Sydney
John Portman & Associates, Atlanta
J. Roger Preston & Partners, Hong Kong
Mahendra Raj Cons. Engrs., New Delhi
Ranhill Bersekutu, Kuala Lumpur
Rice & Daubney Architects, Planners, North Sydney
Robert Rosenwasser Associates, P. C., New York
Emery Roth & Sons P. C., Architects, New York
Rowan Williams Davies & Irwin Inc., Guelph
RTKL Associates Inc., Baltimore
Sepakat Setia Perunding (Sdn.) Bhd., Kuala Lumpur
John P. Sheehy/TAC, Cambridge
Shui On Group, Hong Kong
Skilling Ward Magnusson Barkshire, Seattle
South African Institute of Steel Construction, Johannesburg
Steen Consultants Pty. Ltd., Singapore
Syska & Hennessy, Inc., New York
Gary P. TenEyck, Dallas
TenEyck Merritt Barnett Pitt, Dallas
Tishman Speyer Properties, New York
Trade ARBED Inc., New York
Turner Construction Co., New York
B. A. Vavaroutas and Associates, Athens
Werner Voss & Partner, Braunschweig
Harry Weese & Associates, Chicago
Weidlinger Associates, New York
Wiss, Janney, Elstner and Associates, Northbrook
Wong & Ouyang & Associates, Hong Kong
Woodward-Clyde Consultants, New York
Zaldastani Associates, Inc., Boston

Contents

Preface xix
Contributors xxiii

Planning and Environmental Criteria

Introductory Review 1
B. P. Lim

Philosophy of Tall Buildings

The Philosophy and the Future of the Skyscraper 3
Alan Ritchie

History of Tall Buildings

The Two Centuries of Technical Evolution Underlying the Skyscraper 11
Carl W. Condit

The Relative Value of Invention and the History of Tall Buildings 25
Tom F. Peters

Evolution of the Skyscraper: A History of the Tall Building in Chicago 33
C. William Brubaker

Social Effects of the Environment

Time's Arrow: Tall Moving Targets and Social Research 53
David Cooperman

Social and Environmental Factors of High-Rise Living: A Singapore Experience 59
B. P. Lim

Socio-Political Influences

The Urban Ecology of Tall Buildings 79
Leonard I. Ruchelman

Economics

A Preliminary Model for the Economic Analysis of Tall Buildings 87
John P. Wenzelberger
Henry Malcolm Steiner

Architecture

Architecture and Society 101
Paul Goldberger

Tall Buildings as Symbols 117
Bruce J. Graham

Considerations for Urban Architecture and the Tall Building 149
William Pedersen

A Perspective on Architectural Directions 165
Harry Seidler

The Architecture of Large Buildings 183
Fred L. Foote

Impact of European Technical Culture on the Development of Tall Building Architecture 189
Giselher Hartung
Tom F. Peters

Interior Design

Building Design Consultation 201
M. Arthur Gensler, Jr.
Antony Harbour

The Inside Story: How Structure and Services Impact Office Design in Tall Buildings 209
Moira Moser

Urban Planning and Design

Chicago Update 229
Elizabeth L. Hollander

San Francisco Downtown Plan and the Skyscraper 233
George Williams

The Chicago Perspective: Response to the San Francisco Plan 257
David R. Mosena

The San Francisco Downtown Plan: A Tale of Two Cities 263
Jeffrey Heller

Tall Buildings, Tight Streets 275
Tom Lollini

Skylobbies as Interconnecting Links Amid the Cluster of Skyscrapers 287
Valer Mocak

Project Management

Project Management for Tall Buildings and Urban Habitat 297
James J. O'Brien

Scheduling Tall Buildings 309
Denis W. Boyd

Tall Buildings in Developing Countries

High-Rise Development in India 317
Jashwant B. Mehta

Development and Investment

Developing Tall Buildings to Meet Future Needs: Transco Tower Case History 327
Gerald D. Hines

Systems and Concepts

Introductory Review 337
John Rankine

Structural Systems

The Composite High-Rise Building—An Interaction of Planning, Structure, Speed, and Economy 339
John G. Nutt

A Tall Building of Stacked Steel and Concrete Structures 353
R. Shankar Nair

A Tall Building Core-Hanger Interaction System 361
Marek W. Kwieciński
Adam Z. Pawłowski

Transfer Structures 367
Jack Zunz
Chris Wise

Investigation and Prevention of Failures 383
Lev Zetlin

Cladding

Design of Glass Against Breakage 397
Joseph E. Minor

Construction Systems

Innovations in High-Rise Construction 409
John Norris

Times Square Office Center: The Story of a High-Rise Complex 427
L. V. Shute
H. R. Corry

Foundation Design

Recent Improvements in Foundations 433
Edmund M. Burke

High-Rise Housing

High-Rise Public Housing Development in Singapore 443
Khee Poh Lam

Demand for High-Rise Housing in the United States 465
Richard Kateley

A Comparison of High-Rise Housing in India with Industrialized Countries 473
Jashwant B. Mehta

Prefabricated Tall Buildings

Prefabricated Concrete Systems 483
Bohdan Lewicki

INTEGRO: Open-Element Prefabricated Construction System 493
Pavel Cizek

Robots and Tall Buildings

Construction Robotics in Japan 507
Seishi Suzuki
Tetsuji Yoshida
Takatoshi Ueno

Application of Systems Methodology

Computer Integrated Facilities Management 553
Jeffrey E. Harkness

Reduce Operating Expenses by Computerizing Facilities Management 557
Grant S. Farquhar

Perceptions of the Computer in Design 563
Peter Hoyt
Bob Stockdale

Role of Expert Systems in High-Rise Building Design 571
M. L. Maher
S. J. Fenves

Knowledge-Based Systems for Tall Buildings 591
V. Tuncer Akiner

Building Service Systems

Introductory Review 605
Gordon Rigg

Heating, Ventilating, and Air Conditioning (HVAC)

High-Rise Office Buildings: Changes Anticipated by the Year 2000 607
I. A. Naman

Vertical and Horizontal Transportation

Energy Consumption and Power Requirements of Elevators 621
Joris Schroeder

Space-Saving Elevators for the Very Tall Building 629
Joris Schroeder

High-Tech Tall Buildings

Current and Future Trends: The New Age of Buildings 635
Joseph H. Newman

Criteria and Loading

Introductory Review 645
Alan G. Davenport

Earthquake Loading and Response

Behavior of Structures Retrofitted with Diagonal Steel Bracings During the 1985 Mexico City Earthquake 649
Enrique del Valle Calderón

Damage Statistics of the September 19, 1985 Earthquake in Mexico City 657
Gilberto Borja Navarrete
Manuel Díaz Canales
Alejandro Vázquez Vera
Arturo A. Bello Maldonado
Luis B. Rodríguez González
Sergio Escamilla Aguilar
Gerardo Muñoz Gonzalez
Enrique del Valle Calderón

Effects of the September 19, 1985 Earthquake on the Buildings of Mexico City 667
Roberto Meli

Lessons of the Mexico City Earthquake 679
Ignacio Martín

Wind Loading and Wind Effects

Definition of Wind Pressure on Tall Buildings 681
W. H. Melbourne

Air Infiltration and Internal Pressures in Tall Buildings 689
W. Alan Dalgliesh

Effects of Higher Modes on Response of Wind-Excited Tall Buildings 697
B. Samali

The Response of Supertall Buildings to Wind 705
Alan G. Davenport

Fire

Fire Safety Design for Tall Buildings: Recent Developments 727
Margaret Law

Structural Safety and Quality Assurance

Safety, Quality Assurance, and Performance 735
Carl J. Turkstra

Structural Safety: Some Problems of Achievement and Control 741
Graham S. T. Armer

Motion Perception and Tolerance

Motion in Tall Buildings 759
Andy W. Irwin

Damping in Tall Buildings 779
Alan P. Jeary

The Role of Damping Systems 789
Kenneth B. Wiesner

Structural Design of Tall Steel Buildings

Introductory Review 803
Leo Finzi

Commentary on Structural Standards

Structural Standards 807
Geerhard Haaijer

Methods of Analysis and Design

Seismic Design of Tall Steel Buildings 815
Egor P. Popov

Stability

Design for Strength (Stability) 837
Giulio Ballio

Design Methods Based on Stiffness

Adaptive Expert System for Preliminary Design of Wind Bracings in Steel Skeleton Structures 847
Tomasz Arciszewski
Wojciech Ziarko

Connections

Connection Flexibility in Steel Frames 857
W. F. Chen

Load and Resistance Factor Design (Limit States Design)

Serviceability Limits for Tall Buildings 885
Gerhard Sedlacek
Stefan Bild

Structural Design of Tall Concrete and Masonry Buildings

Introductory Review 897
Ignacio Martin

Commentary on Structural Standards

Building Codes 899
James G. MacGregor

Selection of Structural Systems

Concrete Buildings—A Mile High 905
Joseph P. Colaco

Elastic Analysis

Methods of Analysis in the Design of Tall Concrete and Masonry Buildings 921
Alex Coull

Precast Panel Structures

Precast and Prestressed Concrete Directions in Australia and Southeast Asia 945
W. L. Meinhardt

Creep, Shrinkage, and Temperature Effects

Deterioration of Concrete 959
James R. Clifton

Masonry Structures

U. S. Coordinated Program for Masonry Building Research 965
James L. Noland

Project Descriptions

Superstructure of OUB Centre, Singapore 975
S. Sasaki
M. Suko
Y. Atsukawa

The Skyscraper's Base: Architecture, Landscape and Use in the Dallas Arts District 989
Stuart O. Dawson

Appendix

High-Rise Building Data Base 1003

Nomenclature

Glossary 1007
Symbols 1032
Abbreviations 1033
Units 1034

References/Bibliography 1037
Building Index 1079
Name Index 1085
Subject Index 1095

Preface

On the occasion of the one hundredth anniversary of the tall building, the natural question emerges: What will the second century hold?

When one looks at the past, the most telling thing is how quickly we've adjusted to the high-rise life and work habitat. Up until about one hundred years ago, we could not live much higher than we could climb. But all of that has changed. In one hundred years mankind has abandoned the habits learned over hundreds of millions of years. Although some well-publicized problems have developed, it is remarkable that we have adapted as well as we have.

In all likelihood, then, we will continue to adjust to the new environmental conditions that present themselves, if not physically, then sociologically.

So much for preamble. Since the papers in this book were prepared by leaders in the field, one finds a practical flavor in their projections. Although breakthroughs can occur, by and large new schemes are incremental. (The new "world's tallest" is seldom more than 10% taller than the previous world's tallest). Even the most visionary projections in this book have an air of practicality about them, which augurs well for their applicability to professional practice.

A few overriding considerations seem worthy of comment at this point:

We learned in the "first century" that we could not ignore the cultural and social effects of the built environment. So we will see more research on how to build better and more pressure to consider social and cultural factors in planning and design.

As happened in Western Europe, the United States is reaching an urbanization plateau or will reach it in not too many years. Since the correlation is fairly direct, so also will come a plateau in tall building construction. Less and less will tall buildings be designed to meet new needs, and more and more will they be instruments for renewal and the inevitable recycling of the city.

Quality will be strongly emphasized, because of more experience, greater sophistication of the client, and the availability of information in readily accessible form to explain the "how." The computer will become ever more significant, and greater attention will be paid to using it correctly.

The way the users of tall buildings relate to their habitat will also change. People have an opinion about tall buildings and they are expressing it. Tenants are becoming more sophisticated. For office buildings, the one major tenant is looming in importance. The owner is having more influence on the building.

As Gerald D. Hines has said, there are indications that the desire for more discretionary time will lead to more residential high-rises close to or in the midst of downtown office buildings. Downtown living could become the desired alternative.

Tall buildings will be approached increasingly from the standpoint of an urban ecology—that what happens to a part can influence the whole. Providing for public as well as private needs in a tall building project is just one example (facilities for schools, shops, religious, and other needs). More attention will be paid to maintaining streets as lively and interesting places.

Will a new "world's tallest" be built? Will we go a mile high? The answer is probably "yes" to the first, "no" to the second. With the recent spate of super-tall buildings on the drawing boards, going to greater heights was in the back of many people's minds at the Chicago conference. But in the United States, at least, buildings of 70 to 80 stories would appear to provide needed space consistent with economy.

The future, then, is described in depth by papers that go into specific areas. The material is arranged, as it has been in the past for these monographs, according to the group structure of the Council:

PC Planning and Environmental Criteria
SC Systems and Concepts
BSS Building Service Systems
CL Criteria and Loading
SB Structural Design of Tall Steel Buildings
CB Structural Design of Tall Concrete and Masonry Buildings

The contributions of the theme speakers have been incorporated directly into that segment or topical area that appeared to be the closest in order to provide the book with the same kind of coherence that characterized the Third International Conference (Chicago, January, 1986) at which most of this material was presented. We are grateful to the theme speakers for their contributions, both as speakers and authors. They include: Paul Goldberger, Harry Seidler, Bruce Graham, Gerald D. Hines, John Norris, William Pedersen, and Carl Condit.

This book continues the tradition of the Council on Tall Buildings and Urban Habitat to incorporate meaningful research and advanced design information on tall buildings into a comprehensive resource. The Council's original five-volume Monograph on the *Planning and Design of Tall Buildings* was published from 1978 to 1981 by the American Society of Civil Engineers. It was followed by *Developments in Tall Buildings—1983* and *Advances in Tall Buildings* (1986), published by Van Nostrand Reinhold. *High-Rise Buildings:*

Recent Developments was published by the Council in 1986. All of these should be part of the library of any serious practitioner or student of the high-rise.

We first express our appreciation for the dedication of the Council's Conference Committee, Walter P. Moore, chairman, together with Hal Iyengar, and Chandra Jha. We also thank the North American Group Leaders whose editorial guidance facilitated the completion of this volume. They include Leslie E. Robertson (PC), Irwin Cantor (SC), Don Ross (BSS), Alan Davenport (CL), Jerry Iffland (SB), and Ignacio Martin (CB). The committee chairmen, vice chairmen, and editors also deserve commendation for their careful review and editing of the material.

The Chicago Committee on High-Rise Buildings was our host. The person responsible for all coordination was John Zils, who succeeded in establishing an atmosphere into which the future projections could be presented and discussed. Thirty-two countries were represented among the 700 participants. Some presented, some listened, most discussed, and the exchange was rewarding.

We thank the National Science Foundation for its support of research, of the collection of needed documentation, and of technical workshops and meetings that led to this volume. We are equally indebted to the Patrons, Sponsors, and Contributors to the Council and the Federal Emergency Management Agency, which supported a position of the Conference.

The Lehigh staff, under the production editing leadership of Dolores Rice, carried out that important work, both global and detail, that is so essential to the completion of a work such as this. Le-Wu Lu, V. Tuncer Akiner, Ti Huang, and Graham Stewart contributed their technical editing skills to the production of the volume. To all we are gratefully appreciative.

Lynn S. Beedle
Editor-in-Chief

Lehigh University
Bethlehem, Pennsylvania
1988

Contributors

The following is a list of those who have contributed material for this volume. The names, affiliations, and countries of each author are given.

Some contributions were a direct result of the Third International Conference of the Council, and others were received from professionals in the field.

Ackroyd, M., Techtrol Limited, Quebec, Canada
Akiner, V. T., Lehigh University, Bethlehem, Pennsylvania, USA
Arciszewski, T., Wayne State University, Detroit, Michigan, USA
Armer, G. S. T., Building Research Station, Garston, Watford, England
Atsukawa, Y., Nippon Kokan K.K., Yokohama, Japan

Ballio, G., Politecnico di Milan, Milan, Italy
Beedle, L. S., Lehigh University, Bethlehem, Pennsylvania, USA
Bild, S., Engineering Bureau of D. H. Bild, Hagen, West Germany
Bello Maldonado, A. A., Civil Engineers Associated Group, Mexico City
Borja Navarrete, G., Civil Engineers Associated Group, Mexico City
Boyd, D. W., Management Networks Inc., Chicago, Illinois, USA
Brondum-Nielsen, T., Technical University of Denmark, Lyngby, Denmark
Brubaker, C. W., Perkins & Will Architects, Chicago, Illinois, USA
Burke, E. M., Mueser Rutledge Consulting Engineers, New York, New York, USA

Cantor, I. G., The Office of Irwin G. Cantor PC, New York, New York, USA
Chen, W. F., Purdue University, West Lafayette, Indiana, USA
Cizek, P., Engineering and Industrial Concrete Factories, Bratislava, Czechoslovakia
Clifton, J. R., National Bureau of Standards, Washington, DC, USA
Colaco, J. P., CBM Engineers, Houston, Texas, USA
Condit, C. W., Northwestern University, Evanston, Illinois, USA
Cooperman, D., University of Minnesota, Minneapolis, Minnesota USA
Corry, H. R., Turner Construction Company, New York, New York, USA
Coull, A., University of Glasgow, Glasgow, Scotland

Dalgliesh, W. A., National Research Council of Canada, Ottawa, Canada
Davenport, A. G., University of Western Ontario, London, Ontario, Canada
Dawson, S. O., Sasaki Associates, Dallas, Texas, USA
Del Valle Calderón, E., National Autonomus University of Mexico, Ciudad Universitaria, Mexico City, Mexico
Díaz Canales, M., Civil Engineers Associated Group, Mexico City, Mexico

Escamilla Aguilar, S., Civil Engineers Associated Group, Mexico City, Mexico

Farquhar, G. S., Keith Stevens Incorporated, Eden Prairie, Minnesota, USA
Fenves, S. J., Carnegie-Mellon University, Pittsburgh, Pennsylvania, USA
Finzi, L., Politecnico di Milan, Milan, Italy
Foote, F. L., Mitchell/Giurgola Architects, Philadelphia, Pennsylvania, USA

Gaylord, E. H., University of Illinois at Urbana-Champaign, Urbana, Illinois, USA
Gensler, M. A., Gensler & Associates, Architects, San Francisco, California, USA
Goldberger, P., The New York Times, New York, New York, USA
Graham, B. J., Skidmore, Owings & Merrill, Chicago, Illinois, USA

Haaijer, G., American Institute of Steel Construction, Chicago, Illinois, USA
Harbour, A., Gensler & Associates, Architects, San Francisco, California, USA
Harkness, J. E., The Ryan Group, Red Bank, New Jersey, USA
Hartung, G., Landesdenkmalpflege, Frankfurt a.m., West Germany
Heller, J., Heller and Leake Architects, San Francisco, California, USA
Henn, W., Technical Universitaet Braunschweig, Braunschweig, West Germany
Hines, G. D., Gerald D. Hines Interest, Houston, Texas, USA
Hollander, E. L., Department of Planning, City of Chicago, Chicago, Illinois, USA
Hoyt, P., Hellmuth, Obata & Kassabaum, Inc., St. Louis, Missouri, USA

Irwin, A. W., Heriot-Watt University, Edinburgh, Scotland

Jeary, A. P., City Polytechnic of Hong Kong, Hong Kong, Hong Kong

Kateley, R., Real Estate Research Corporation, Chicago, Illinois, USA
Kato, B., University of Tokyo, Bunkyo-ku, Tokyo, Japan
Kwiecinski, M. W., Technical University of Warsaw, Warsaw, Poland

Lam, K. P., National University of Singapore, Singapore
Law, M., Ove Arup Partnership, London, England
Lewicki, B., Building Research Institute, Warsaw, Poland
Lim, B. P., National University of Singapore, Singapore
Lollini, T., Kaplan, McLaughlin, Diaz, San Francisco, California, USA

MacGregor, J. G., University of Alberta, Edmonton, Canada
Maher, M. L., Carnegie-Mellon University, Pittsburgh, Pennsylvania, USA
Martin, I., Capacete-Martin & Associates, San Juan, Puerto Rico
Mehta, J. B., Jashwant Mehta and Associates, Bombay, India
Meinhardt, W. L., W. L. Meinhardt & Partners Pty. Ltd., Melbourne, Australia
Melbourne, W. H., Monash University, Sydney, Australia
Meli, R., Cuidad University, Mexico City, Mexico
Minor, J. E., Texas Tech University, Lubbock, Texas, USA
Mocak, V., Architect and Planner, Yonkers, New York, USA
Mosena, D. R., Department of Planning, City of Chicago, Chicago, Illinois, USA
Moser, M., M. Moser & Associates, Hong Kong, Hong Kong
Muñoz González, G., Civil Engineers Associated Group, Mexico City

Nair, R. S., Consulting Engineer, Chicago, Illinois, USA
Naman, I. A., I. A. Naman and Associates, Houston, Texas, USA
Newman, J. H., Tishman Research Corporation, New York, New York, USA
Noland, J. L., Atkinson-Noland and Associates, Boulder, Colorado, USA
Norris, J., Olympia & York, Toronto, Canada
Nutt, J. G., Ove Arup Partnership, Sydney, Australia

O'Brien, J. J., O'Brien-Kreitzberg & Associates, Inc., Merchantville, New Jersey, USA

Pawlowski, A. Z., Technical University of Warsaw, Warsaw, Poland
Pedersen, W., Kohn Pedersen Fox Associates PC, New York, New York, USA
Peters, T. F., Cornell University, Ithaca, New York, USA
Popov, E. P., University of California, Berkeley, California, USA

Rankine, J., Rankine and Hill Pty. Ltd., Sydney, Australia
Rigg, G., Rankine and Hill Pty. Ltd., Melbourne, Australia
Ritchie, A., John Burgee Architects with Philip Johnson, New York, New York, USA
Rodríguez González, L. B., Civil Engineers Associated Group, Mexico City
Ross, D., Jaros, Baum & Bolles, New York, New York, USA
Ruchelman, L. I., Old Dominion University, Norfolk, Virginia, USA

Samali, B., Maunsell and Partners Pty. Ltd., Sydney, Australia
Sasaki, S., Nippon Kokan K.K., Yokohama, Japan
Schroeder, J., Schindler Management Ltd., Lucerne, Switzerland
Sedlacek, G., Institute of Steel Structures, RWTH Aachen, Aachen, West Germany
Seidler, H., Harry Seidler & Associates, Milsons Point, Australia
Shute, L. V., Turner Construction Company, New York, New York, USA
Steiner, H. M., George Washington University, Washington, DC, USA
Stockdale, R., Hellmuth, Obata & Kassabaum, Inc., St. Louis, Missouri, USA
Suko, M., Nippon Kokan K.K., Yokohama, Japan
Suzuki, S., Shimizu Construction Co., Tokyo, Japan

Turkstra, G. J., Polytechnic University, Brooklyn, New York, USA

Ueno, T., Institute of Technology, Tokyo, Japan

Vázquez Vera, A., Civil Engineers Associated Group, Mexico City

Wakabayashi, M., General Building Research Corp., Osaka-Fu, Japan
Wenzelberger, J. P., George Washington University, Washington, DC, USA
Wiesner, K. B., LeMessurier Consultants Inc., Cambridge, Massachusetts, USA
Williams, G. A., Department of City Planning, San Francisco, California, USA
Wise, C., Ove Arup Partnership, London, England

Yoshida, T., Institute of Technology, Tokyo, Japan

Zetlin, L., Zetlin-Argo Inc., New York, New York, USA
Zia, P., North Carolina State University, Raleigh, North Carolina, USA
Ziarko, W., University of Regina, Regina, Canada
Zunz, J., Ove Arup Partnership, London, England

Second Century of the Skyscraper

Planning and Environmental Criteria

Introductory Review

Bill B. P. Lim

ECONOMICS

The real state of economics of tall building development requires more precise methods of examination of proposals and sensitivities. Studies should be made on the effect of changes of discount rates, building envelopes, land costs, income tax rates, borrowing interest rates, and the influence of inflation and tax strategies as applicable to tall buildings.

TRANSPORTATION

Traffic problems in the proximity of a tall building are affected by the building use. Transportation within the building, both vertical and horizontal, is related to productivity and efficiency of the operation of the building. While trip generation data are generally available, a comprehensive procedure is needed to relate them to the occupancy and use of tall buildings with respect to elevators, sidewalks, crosswalks, and transit arrangements. Special attention should be paid to transportation per unit cost of the area.

THE HANDICAPPED AND THE AGED

Whereas architectural requirements for the handicapped and the aged are well documented in the context of building codes in some countries, there is the need for predesign interviews with the handicapped and the aged so that modifications of the construction detailing may be made with reference to a particular site. There should also be an expert study of time and motion of the handicapped and aged.

INTERNAL ARCHITECTURE

There appears to be an overemphasis by architects in the design of the external appearance of tall buildings and insufficient attention paid to their internal architecture. Airconditioning, lighting, and furniture should be coordinated and integrated with the external design of buildings.

LANDSCAPE ARCHITECTURE

A comprehensive scheme of the entire street scene is more satisfactory than mere piecemealing of spaces around a tall building. The streetscape flanked by tall buildings should be made more human in scale and more suitable for outdoor living than harsh unprotected open spaces as seen in many downtown areas.

INVESTMENT AND DEVELOPMENT

The future of investment and real estate development depends on understanding the needs of customers and the location of sites. Well-designed buildings will not necessarily have full occupancy if the sites are not convenient. From experience, the location of the site may even be more important than the design of the building.

FUTURE TRENDS OF ENVIRONMENTAL CRITERIA

The next generation of tall buildings will be high quality, upscale, flexible, energy-efficient, multiple-use, and adjusted to the environment. The clients are likely to be more enlightened and will be more conscious of the services provided by the consultants. The criteria for planning and environmental control, if formulated by code-makers, should be fully justified by the authorities so that building regulations do not hinder the development of tall buildings, which should be self-regulated.

The Philosophy and The Future of the Skyscraper

Alan Ritchie

It has been stated that the skyscraper and the twentieth century are synonymous and there can be no doubt that the tall building is the landmark of our generation. It is a structural marvel that reaches to the heavens and embodies human goals to build ever higher. The skyscraper is this century's most stunning architectural accomplishment.

But the question of how to design the tall building still continues to taunt, disconcert, and confound practioners. The swing in taste and style is as predictable as night and day, and we are at this very moment busy rewriting the rules of skyscraper design. In the process we are not sure that the right lessons we have learned are not being discarded for the wrong ones.

A successful skyscraper solution and the art of architecture itself depends on how well the structural, utilitarian, environmental, and public roles of the tall building are resolved. Style, any style, must be intrinsic to, and expressive of, these considerations. Architecture is, above all, an expressive art.

The skyscraper has totally changed the scale, appearance, and concept of our cities and the perceptions of people in them. No doubt it will continue to do so. But it is more important today than ever that the builder and architect consider all the factors associated with the design of a tall building and how it is incorporated into its urban setting.

Looking at the whole historical spectrum of skyscraper design, four significant phases can be identified: the functional, the eclectic, the modern, and what is currently called the postmodern, a term coined more by the media, for surely our references to modernism have not changed but have merely broadened.

It is significant that all of the most important structural solutions came early in the development of tall buildings and in a very short space of time. Because these structures were concentrated in Chicago in the two decades at the end of the last century, it was quickly acknowledged and referred to as the *Chicago style.*

The period from 1890 to 1920 was considered the golden age of architecture, and there have been few more masterful and original tall buildings produced than those by the architect Louis Sullivan. Running as counter current to the already emerging eclecticism, Sullivan believed that the design of the skyscraper was the translation of structure and plan into appropriate cladding and ornament and that the answers were not to be found in the rules of the past.

The eclectic phase produced some most remarkable monuments, employing many of the styles and ornamentation from the temples of Greece to the Italian Renaissance. The best examples displayed skilled academic exercises, composed with ingenuity and drama to answer the new needs and aspirations of the twentieth century. These designs so beautifully compiled by architects like Raymond Hood and Cass Gilbert culminated in the famous international competition for the Chicago Tribune Tower in 1922. This competition, which called for "The Most Beautiful and Distinctive Office Building in the World," drew more than 200 entries. The selection of the gothic revival design by Howells and Hood prolonged the eclectic style against the concepts of the modern. For ten years *modernism* as pioneered by a relatively few European architects, paralleled a style that would better be termed *modernistic.* This style was neither pure nor revolutionary, but fused the end of the decorative eclectic style with the modernist theories and has become popularly known today as Art Deco.

The early modern or *international style* skyscrapers are small in number because of lack of courage on the part of the builder and a reluctance to invest in a style not yet accepted. But after the Second World War the descendants of these early modern skyscrapers, such as the McGraw-Hill Building in Manhattan, came to make up the high modern corporate style, the flat top glass boxes that have been the focal point of criticism over the past ten years.

These big buildings have taught us a hard lesson. But it is wrong that so much has been blamed on the esthetics, for such problems owe just as much to investment patterns and social upheavals. Unfortunately the minimalism of the modern esthetics let itself to the cheapest corner cutting. Since this is the most profitable route for the builder to take, it is an elegant and refined vocabulary that was quickly reduced to bottom line banality. Many are already grieving the passing; for it is structure in its purest form, enclosed in a sheer curtain of shaped and shimmering glass, that has produced some of the most innovative designs of our time.

These ideas should not be abandoned in a search for ideal answers. After all, the history of the skyscraper—which is also the history of the century—is a search for identity.

The signs of the modernist movement are many, as anyone associated with design has repeatedly heard; the familiar phenomenon in history of changing values is nothing new. These factors have led to the fourth or current phase of skyscraper design called *postmodern*. Unlike our predecessors the postmodernists want everything back that the modernists had discarded—history, ornament, context, contrast, variety, symbolism, imagery, and metaphor. But above all, it is the references to the past and the incorporation of a new and more permissive architectural vocabulary that make the challenges and potential greater.

The acceptance of the AT&T building designed by Philip Johnson in 1977 made possible the direction architecture is taking today. Some of the buildings now being designed reflect current thinking and show the broadening of the architectural spectrum. It is now acceptable to project many varying tastes and styles, both modern and postmodern, such as has been accomplished in the AT&T Corporate Headquarters, New York (Fig. 1); P.P.G. Headquarters, Pittsburgh (Fig. 2); 190 South LaSalle St. Chicago (Fig. 3); and International Place, Boston (Fig. 4).

Whatever the style, the engineering development of the tall building is one of the truly remarkable chapters in the history of architecture. Structure is the heart of the tall building design. But structural innovation and esthetic preferences can expand the choice of solution only as long as the dollar values work out. Economics is the main contributing factor to large commercial structures today. The modern office building has been standardized as a central service core surrounded by 15,000 to 25,000 ft^2 of space. This standard has been set by business as the optimum work floor area by big corporations. From this standard the shape of the building is dictated and even the uniform building module of four or five feet has evolved out of another economic consideration, the minimum office size. All of these factors have and will continue to have a bearing in the design of large commercial high-rise buildings.

Many of the aspects of commercial tall buildings, such as structure, economics, and floor space, play the same role in the design of tall residential buildings, as do planning and zoning ordinances. The questions of massing, circulation, and integration into urban surroundings will continue to have a significant influence on the development of all tall buildings; and by the very nature of urban building economics it becomes more and more a matter of getting the most building for the least cost.

Some believe that the skyscraper has reached the end of the line—that it has become too large and too destructive. But as long as the cost of land continues to soar and the need of individuals to improve their standard of living is still with us, there will continue to be a need for the high-rise building. But one must seriously question the giant megastructures recently proposed for our inner cities and the need they reflect to build taller and taller. The need for these "supertall" buildings appears to be more the greed and ego of the builder than necessity or the betterment of life.

What must be avoided is the design of tall buildings as monuments. They must be integrated into their urban cityscape, and must relate the horizontal movement at street level with the vertical thrust of the tower. After all, the ingredients for a tall building are the same. It is only the way they are mixed together that produces different end results. But whatever the future appearance the skyscraper may take, it must provide a sense of presence and identity. At the same time it must be for people, giving them a feeling of well being and enjoyment as well as a place to live and be happy. We should endeavor to follow what Sullivan expressed almost a century ago: The tall building must be "A proud and soaring thing."

Fig. 1 AT&T Corporate Headquarters

Fig. 2 PPG Place *(Photo by Brian Rose)*

Fig. 3 190 S. LaSalle St. *(Photo by Hedrich-Blessing)*

Fig. 4 International Place at Fort Hill Square *(Photo by Hedrich-Blessing)*

The Two Centuries of Technical Evolution Underlying the Skyscraper

Carl W. Condit

We can no longer argue that the Home Insurance Building was the first skyscraper (Fig. 1). It was not. Then the question is, what was? Part of my purpose is to demonstrate that there is really no such thing as the first skyscraper, although we can certainly make a case for the emergence of the potential form. My chief argument against the claim for the Home Insurance Building is that it rests on an unacceptably narrow idea of what constitutes a multistory high-rise commercial building. Such a structure is a great deal more complex than what has always been claimed. I am going to use the word *skyscraper* for convenience, but it applies to any large multistory commercial, public, or residential building regardless of its shape or height.

An adequate history of the development of the skyscraper, its urbanistic and economic antecedents, its genesis, its design and construction, and its evolution over the past century, must rest on four fundamental aspects of this particular kind of building—namely, place, structure, utilities, and form. For years historians paid almost exclusive attention to external form, though in recent years a few have become interested in its structure and its history. But that leaves the other two aspects out of the equation, and in the case of utilities our neglect has been unfortunate. An account of the development of building utilities as part of technological history has yet to be written, since it consists

Fig. 1 Home Insurance Building, Chicago (1885)

of a handful of articles and one book, which treats only a very limited aspect of its subject. The purpose of this paper is to describe in a condensed form the historical development of modern building technology over the full period of time in which that evolution took place. In short, the historical growth of all the technical factors that underlie the ultimate skyscraper form and continue to be an organic part of that form will be traced. There are about a dozen parallel and successive lines of evolution, which may be grouped under four primary headings: structure, safety, internal transportation, and habitability. All these characteristics must be inherent in the skyscraper or any other kind of large multistory building that is used by a number of people on a regular, sustained, daily basis. It is impossible to treat these dozen or so aspects in strictly chronological terms. They overlap to such an extent that it will be necessary first to organize them topically, then to give each major area a broadly chronological survey.

STRUCTURE

Iron Framing

Structure is the oldest technico-material aspect of building, without which there would be no building at all. We rightly think of the iron frame—that is, a frame of ferrous metals of various kinds—as an essential characteristic of high-building construction and it is on this basis that buildings like the Home Insurance and the Tacoma in Chicago, for example, are given the place they have in architectural history. There was certainly nothing new about the iron frame, which had been used for more than a century when the Home Insurance Building was placed under construction. The first multistory building in which floors were supported by iron columns was the Calico Mill in Derby, England, built in 1792–1793 by William Strutt, who was a practical builder trained neither in architecture nor engineering. It was late in the nineteenth century before such formal training became the rule rather than the exception. The first iron columns had appeared twenty years before Strutt's mill was constructed, having been introduced initially in St. Anne's Church, Liverpool, in 1772. Iron columns were quickly supplemented by iron floor beams. Charles Bage provided full interior iron framing in the Benyon and Marshall Flax Mill, in 1796–1797, at Shrewsbury, England (where Falstaff played dead on the battle field, *Henry IV,* Part I). Iron columns were introduced into the United States by Benjamin Latrobe, one of the great creative engineer-architects during the formative years of post-colonial building. The particular work was a church in Washington, D.C., constructed in 1808. The first iron roof truss was built by William Murdock for a foundry in Soho, London, in 1810. This truss was also the first metal framework in which there was a precise distinction between wrought iron and cast iron elements on the basis of stress, the wrought iron being used for members subject to tension and the cast iron for those under compression. A long series of experiments carried on by William Fairbairn and Eton Hodgkinson established the basis for this distinction in scientific terms.

Moving rapidly through the chief milestones in the development of iron framing, we might argue that the Crystal Palace in London, 1851, was the best known and the most important. It was followed by the warehouses of the St. Ouen Docks in Paris, 1864–1865, designed by Hippolyte Fontaine, a builder and inventor who helped to develop the first practical electric motor in 1873 in collaboration with Théophile Zenobe Gramme. If our sole criteria for the skyscraper are height and structure, then the warehouses of the St. Ouen Docks would have to take precedence over everything else. They were the first fireproof, iron skeleton, curtain-walled, multistory buildings ever built. The entire dead load, internal floor loads, and live loads are carried entirely on a frame of deep wrought iron girders, smaller cast iron beams, and cast iron columns, together supporting concrete floors designed for a loading factor of 31.12 KPa (650 psf), more than enough for a whiskey warehouse and almost enough for a locomotive. Discussing the United States we turn back a few years in our chronology. Daniel Badger and James Bogardus, both of New York, began respectively in 1846 and 1848 to build multistory structures with cast iron fronts and internal cast and wrought iron frames. By the mid-1860s in Europe and the United States many of the essential features of the skyscraper structure were in place, but it was the United States that first exploited them.

Wind Bracing

A subsidiary but important aspect of structure is one to which an inordinate amount of attention has been paid in the past one hundred years, yet it goes far back into history. Wind bracing was a medieval invention, first introduced for wood construction in the large timber-framed house and barn as they reached maturity in the fourteenth and fifteenth centuries. The common system was knee-bracing, supplemented in larger structures by truss framing of various kinds. The structural system of the mature Gothic cathedral was braced against horizontal forces, the key element being the flying buttress, which is in fact a strut designed to transmit loads across the aisles to the tower buttresses on the periphery of the structure. The scientific investigation of the interrelations between wind velocity and wind pressure and the consequences of this pressure for the behavior of structures began in 1664 with the experiments of Robert Hooke, who worked in all areas of the physical sciences.

Hooke's pioneer work was carried on in the eighteenth century by John Smeaton, creator of the Eddystone Lighthouse and one of the great builder-engineers of his age. Smeaton began his investigations in 1759 and was concerned with the relation between wind velocity and pressure in the action and design of windmills. The French physicist and meteorologist Jean Charles

Borda in 1763 continued these experiments, which went on through the remainder of the eighteenth century and came to a focus in practical building with the design of French lighthouses beginning in 1832. Experiments continued throughout the nineteenth century and into the twentieth, leading in an irregular way to the formula that we use today for the relation of pressure and velocity—$P = 0.00256V^2$—an empirical formula that is not susceptible to dimensional analysis but represents an accurate numerical relationship.

The curious thing is that it was more than 150 years before the scientific investigations began to bear fruit in iron-framed building. First, wind bracing in the form of knee braces was used in all large vaulted structures of wood, most conspicuously in churches. If you explore the space between the vault over the nave and the gabled roof of a large colonial church—Christ Church, Philadelphia; St. Paul's Church, New York; or St. Michael's, Charleston, S.C.; for example—you will find a complete system of bracing often in the form of ship's knees, so-called because of their use in the frames of wooden vessels. It was finally introduced for iron construction in the Hungerford Fish Market, London, in 1835, developed further as a proto-portal as well as double-diagonal bracing in the Crystal Palace of 1851, and as full portal bracing in the Royal Navy Boat Store, Sheerness, England, in 1858–1860.

At this stage we reach a mystery, and I am sorry that I cannot yet unravel it. Henry H. Quimby was the authority at the end of the nineteenth century on windbracing—its history, its applications in building, the accumulated theory, and pioneer uses. Quimby said that bracing was introduced into iron-framed buildings in the United States with the first use of wrought iron columns. These would have been Phoenix columns, the flanged wrought iron column invented in 1861. We are reasonably sure that the first building to be constructed with wrought iron columns was the Brown Brothers Bank in New York, 1863. We know nothing about the bracing in it, or whether it even had any, yet we are compelled to recognize Quimby as the voice of authority (Quimby et al., 1892–93). The ruling view at the time was that if buildings had external bearing walls of masonry, which all of them did except for the two warehouses mentioned, the weight was sufficient to render the internal framework and hence the whole building stable against the wind, so that bracing was regarded as unnecessary. As late as 1893 Adler and Sullivan's Stock Exchange Building in Chicago contained no wind bracing. The Tacoma Building, another candidate for the status of first skyscraper, had a kind of bracing in the form of shear walls of brick extending through the height of the structure. Nevertheless, by the end of the Civil War we know that the necessity for bracing in an iron-framed building was at least recognized.

Foundations

A third category under the heading of structure is that of adequate foundations. Piling goes back to classical antiquity and is described by

Vitruvius, but the watertight caisson of timber sheeting appears to have been a development of the late eighteenth century. There are illustrations in the *Encyclopedie*, the great French compendium of all the arts and sciences of the age. The pneumatic caisson was the invention of Thomas Lord Cochrane of England in 1830. It was introduced in the United States by a builder named L. J. Fleming in 1852 and was developed into its mature form by the pioneer foundation engineer William Sooy Smith. It was given its most conspicuous demonstration by James B. Eads in his St. Louis bridge (1868–1874) and was soon recognized as essential for laying down and supporting the foundations of all large bridges and buildings in unstable, water-bearing soil.

SAFETY

Fireproofing

Under the heading of safety I want exclusively to emphasize fireproofing, although safety factors are involved in all the technical aspects of building. We can follow the history of fireproofing in detail once we come to a decisive and unambiguous starting point. But there was a long antecedent history which began in the late eighteenth century. The overriding reason for tracing these origins to the Age of Enlightenment is that it was also the period of the Industrial Revolution and a new symbiotic union of science and technology. The French builder Ango introduced hollow clay pots into plaster flooring apparently in part to lighten the floor, perhaps the beginning of reinforced concrete because wrought iron beams were incorporated in the plaster work. But there is equally good reason to believe that the aim was to introduce trapped air and a refractory material of low thermal conductivity into the floor. Ango was followed by St. Far, who is credited with building the flooring of an entire house in this way in 1785.

The mature and progressive development of fireproofing began with the construction of the Cooper Union in New York. In 1854 Frederick A. Peterson, its architect, introduced the hollow clay pots into the concrete that leveled up the floor arches spanning between wrought iron beams. The decisive step toward a scientific understanding of fireproofing came with a paper delivered by Peter B. Wight before the New York chapter of the American Institute of Architects on April 6, 1869. In his paper Wight pointed out for the first time that because a mill, bank, or any other kind of commercial building is built with iron columns and beams, concrete floors, and brick walls, it does not follow that the building is proof against destruction by fire (Wight, 1876; 1878). That has been demonstrated again and again, and if you think the lesson has been finally learned you are very much mistaken. The original Metropolitan Fair and Exposition Building in Chicago was destroyed by fire in January 1967, although it was constructed with reinforced concrete walls resting on steel rigid frames. It was supposedly a fireproof building, but it was filled with combustible materials that burned at a temperature as high as 2,500°F. Nobody needs to know very much to explain what happens when the temperature of any ferrous metal is raised to 2,000°F or higher.

A long series of experiments on the relation of loss of strength to rise in temperature was necessary to establish accurate data. What happens to the strength of exposed iron or steel at elevated temperatures? At what point can it no longer be counted on to carry the load imposed upon it? How can it be protected from rapid absorption of heat? Maturity came in fireproofing techniques with the invention of hollow tile cladding for iron columns and beams by George H. Johnson and Balthasar Kreischer in 1871. Their invention was first applied to the iron frame of the Kendall Building in Chicago during the following year. The necessity for this kind of protection was finally recognized, but unfortunately the collapse of the Exposition Building in Chicago was not the last case of the destruction of a supposedly incombustible building. A few years ago a similar case took a very high toll in life as the result of a hotel fire in Las Vegas.

INTERNAL TRANSPORTATION

The third primary heading is internal transportation. When buildings passed five stories in height it was no longer possible to ask people to climb the stairs, especially if it was a prestigious office or department store block. Some other way had to be found for moving from one floor to another, and the solution again came from strictly utilitarian structures. The first power-operated elevators were introduced into English mills about 1835. England was far ahead of any other country in the century from the mid-eighteenth to the mid-nineteenth, but by the latter date it was being rapidly eclipsed by the United States and Germany in technology and industrial development. The first elevator was a primitive device an open wooden platform bounded by rails that was hoisted by ropes in a brick-lined shaft. Hoisting was accomplished by winding the ropes around sheaves connected to the belted shafting of the mill. The apparatus was a homely and unsafe contrivance for the vertical movement of heavy loads and the workers responsible for the task.

Among the obvious defects of early mill elevators was the absence of any means to prevent or slow the free fall of the platform in the event of a cable rupture. The first safety brake, oddly enough, came before the invention of a practical elevator suitable for the movement of passengers. The inventor, Elisha Graves Otis, is perhaps the foremost name in the entire history of elevator technology. He developed an effective though primitive safety brake in 1851, an invention that logically falls in the second category, safety, as well as internal transportation. People were understandably reluctant to rise even a single floor in an elevator if there was a likelihood that it would fall to the basement floor. Otis achieved the second and perhaps decisive step with the invention in 1855 of an elevator moved by a separate steam-powered drum.

He made the first workable installation of an enclosed steam-driven car with safety brake in the Haughwout Building, New York, in 1857. Three years earlier, in the New York Crystal Palace, Otis had given his highly melodramatic demonstration of safety: He raised the elevator four floors and cut the cable from which it was suspended. He didn't come to a gentle stop, but it was better than falling to the bottom of the shaft.

The hydraulic elevator was the invention of two men working independently, Cyrus Baldwin of Boston in 1870 and William E. Hale of Chicago in 1873. We have heard a great deal about builders, developers, architects, and engineers in Chicago, but only one historian has ever mentioned the Hale elevator, and his work has yet to be published. The Hale machines were installed in the Tacoma Building, which has been offered as a better candidate for first skyscraper than the Home Insurance.

The electrically powered elevator had its primitive beginning in 1880 with a demonstration model built by Ernst Werner von Siemens, one of the great creative figures in the pioneer age of electrical technology. A much improved form with more promise for practical use was introduced by Frank Julian Sprague in the Park Row Building, New York, in 1897. Sprague's name ought to be well known: He was the chief creator of electric railroad traction in the United States. But Otis was already at work on a much superior model, having begun his experiments as early as 1890 and having made his first practical installation in 1894. Before the end of the decade he had produced an elevator with all the essential characteristics of the modern machine—the direct-drive electric motor, the cable drum, the counter-weights, the safety brake, and a system of controls that made it possible to provide smooth starting and stopping.

HABITABILITY

The last major category is called habitability, meaning all that makes a building healthy, comfortable, and usable to those who must work or live in it. The importance of this category is underscored with a little more emphasis than previously shown. The overwhelming majority of people who use a building do not care how it was constructed. They do not care how the building was erected, what it was made of, whether it has a riveted-steel or a welded-steel frame, or a reinforced concrete frame, whether it has bearing walls or curtain walls, whether masonry or any other material. But they are vitally concerned for quite understandable, absolutely essential human reasons with whether the building is comfortably heated, whether it has an adequate plumbing and water supply system, whether there are enough plumbing fixtures to serve the users, whether all the factors together guarantee the reliable operation of fixtures, and whether ventilation and air-conditioning provide reasonable comfort for all.

Central Heating

Once more we have to go back to the Industrial Revolution. Central hot air heating was introduced by William Strutt in the Belper textile mill in 1792. Steam heating was another creation of the prolific team of Matthew Boulton and James Watt, who made the first installations in mills placed under construction in 1802. The heating unit was simply a system of parallel pipes fixed to the walls of the mill. Everyone knows the name of James Watt: In a series of patents granted between 1769 and 1790 Watt developed the double-acting reciprocating steam engine with automatic valve and speed control, the first such feedback control or servo-mechanism. Matthew Boulton, primarily a builder, was his entrepreneurial partner. Like the elevator, the first heating systems were crude installations, hardly satisfactory for prestige buildings designed to bring high rents and big returns for their owners. In America the mechanical inventor and mill-builder Oliver Evans first used exhaust steam from boilers to provide the source of heat, around 1811 for his earliest installations. Improvements were slow in coming. Closed-circuit high-pressure hot-water heating was essentially the achievement of Jacob Perkins in England around 1831. He had developed a closed-circuit system in which water under pressure could be raised far beyond the boiling point. Pipes were so hot as to be dangerous to touch, but they gave off plenty of heat. One does not have to know very much about the thermodynamic behavior of fluids under high temperatures to know what might happen if a small fracture occurred in one of the pipes. The water would immediately flash into steam and the whole building would be blown up. It happened to locomotives year after year on the railroads.

A preferable system, a closed-circuit, low-pressure steam-heating system, was primarily the work of Joseph Nason in the United States, and again a series of experiments and innovations took place over a number of years around the mid-century. The chief safety advantage of steam heating is that the pressure may be reduced much below the atmospheric in a closed-circuit system. Nason was awarded the contract for heating the Capitol in Washington, finally completed in 1864 after construction that extended over a period of 72 years. All the essential equipment for a reliable steam heating system—boiler, valves, pumps, piping, controls—could be found described and illustrated in the 1861 catalogue of Morris, Tasker and Company of Philadelphia, the leading suppliers at the time.

Plumbing

After heating, plumbing follows as the most important utility. A building is simply unusable without plumbing equipment, without hot and cold running water, toilets, lavatories, fountains, and if it is a hotel or apartment building, bathtubs and showers. Three factors had to be brought together at the requisite level of maturity for a reliable plumbing system: first, a pressure water supply; second, the necessary fixtures with valves, faucets, seals, and traps; and third, piping. An early and for long an isolated pressure water supply system, with steam-operated pumps, was built in England in 1712. The primitive steam engine was the type invented by Thomas Savery in 1698. To the best of my knowledge the first metropolitan water supply system constructed on a scale adequate to the needs of the new commercial and industrial city of the nineteenth century, was the Croton Aqueduct and Reservoir of New York, 1839–1842, one of the many achievements of John B. Jervis in the formative years of American engineering. The gravity water supply system with its associated siphons and storage facilities provided water at a 30 m (100-ft) head, sufficient for the plumbing equipment that was to come in the city within the next few years.

The flush toilet and the lavatory, with associated valves, seals, and traps, appeared in the latter part of the eighteenth century, but the history of this area of technology has scarcely been touched. One of the important figures in the development of the toilet was Joseph Bramah, the foremost locksmith of England in the full tide of the Industrial Revolution. Benjamin Latrobe is again an important figure in American technology. In the same year in which he introduced the iron column, he built a house with the bathtub, lavatory, and toilet in the same room, making possible the centralized plumbing stacks necessary in a multistory building. He was thus responsible for the curious use of the word *bathroom* in the United States: When you ask your hostess where the bathroom is she does not assume that you plan to take a bath.

With respect to piping, the traditional cast-iron type was clumsy, oversized, and difficult to work, especially in large multiple installations. Wrought iron pipe was seen to be far superior, and once more it was Morris, Tasker and Company who were responsible for its first manufacture. As in the case of heating, all the essential equipment—fixtures, piping, vales, boiler—could be found in the Morris Tasker catalogue of 1858.

Artificial Lighting

The origins of artificial lighting bring us back once again to the eighteenth century. The first coal-gas illuminant to supply multiple fixtures was installed by William Murdock in his own home at Redruth, England, in 1779. The first multiple installation for a large multistory building came with the construction of the Philips and Lee Mill at Salford, England, in 1802. Boulton and Watt were the builders of the mill and William Murdock installed the lighting system. Electric street lighting in its early form depended on Humphrey Davy's invention of arc lighting in 1813, but it was slow in superseding the almost universal gas lighting, which was introduced initially in London in 1814, in Baltimore, 1817, for the first American installation, and in New York,

1825. The proponents of arc lighting, however, were an enthusiastic lot, and by mid-century they were recommending it for universal application, interiors, exteriors, and anything in between. Before very long incandescent lighting was to take the place of all previous forms. Electrical illumination goes back to the experiments and discoveries of Galvani, Volta, and Faraday, whose names are enshrined in the daily vocabulary of electrical technology. After many experiments conducted from 1879 to 1884 practical incandescent lighting was finally realized by Thomas Edison in the United States and Joseph Swan in England.

All large multistory buildings with electric lighting systems incorporated their own generating plant, but as we move toward the end of the nineteenth century these gradually gave way to the purchase of electricity from a central power station. The ancestor of them all was the Edison Station on Pearl Street in New York, opened in 1882, in preparation not only for the incandescent light but the newly invented telephone as well.

Ventilation

The subject of induced-draft ventilation is virtually a closed book. We have only two names, one at the ancestral origins, the other at the beginning of the period of maturity. The French physicist and meteorologist Jean Théophile Desaguliers in 1736 introduced hand-operated centrifugal fans into the holds of naval vessels. The boys who were paid to operate them were called *ventilateurs,* which is the beginning of the technology and the vocabulary of ventilation. Probably the single best known name in the United States was Benjamin Franklin Sturdevant. In a series of experiments initiated in 1855 Sturdevant developed a steam-operated power-driven fan by means of which he could blow air into ducts throughout an entire building. An early installation of great size is in the Cooper Union, New York, where a fan of 3.5-m (12-ft) diameter supplies forced-draft ventilation to the entire building.

CONCLUSION

It is no coincidence that the primary inventions came from England, to some extent from France, and that most of them found their practical expression in the United States. It was in this country that fireproofing, central heating, and reliable plumbing equipment found their first full use. If one is going to choose the buildings that are the milestones, there is no question that the hotel was the decisive building type. Hotel standards in the United States rapidly rose above the level in England and on the Continent except for a few aristocratic spas, where only royalty and the rich might have stayed.

The first hotel to embody the plumbing techniques I have described was the Astor House, 1838–1842, in New York, the building of which coincided with the construction of the aqueduct system that made its plumbing installation possible and workable. An even more advanced work was the Fifth Avenue Hotel, 1857, in New York, marked by the highest standards of utilities and the first hotel elevator equipped with a safety brake. Meanwhile, the builders of the English Houses of Parliament, 1837–1860, had raised the Victoria Tower at one end to a height of 101 m (331 ft) by means of an internal iron frame and external bearing walls of masonry. As I said earlier, if the sole criterion for the first skyscraper is iron skeleton, fireproof, curtain-walled, multi-storied, concrete-floor construction, then the warehouses of the St. Ouen Docks in Paris, 1864–1865, are the obvious candidates. But without utilities the building is uninhabitable for human beings, so that it must be equally obvious that the warehouses do not qualify as skyscrapers, anymore than does the Victoria Tower.

If there is a building in which most of the technical factors I have described—structural system, elevators, pressure plumbing, central steam heating, multiple illuminating fixtures connected to a central supply—are present at the requisite level of maturity and reliability, a high-rent prestige building, it is the Equitable Life Assurance Building of New York, 1868–1870. The architects were Arthur Gilman and Edward Kendall, who, while the building was in the process of design and was expected to cost a fortune, called in George B. Post, a civil engineering graduate of New York University, as a consultant. Post radically revised the structural system: The court walls were treated as true curtain walls supported on an iron skeleton. Other changes followed, and the total cost was reduced by $330,000 as a consequence, but with no sacrifice to the elegance of this Second Empire palace clothed in granite and marble.

To show his gratitude to Post, the Equitable president, Henry B. Hyde, invited the engineer-architect to occupy the office at the highest level of the building, at the highest rent, double the going rate for New York office space. It was height that gave the building its prestige, and the elevator that made it readily accessible. And all this was translated into financial terms for Post: When he needed larger quarters he sold his Equitable lease for $6,000, the equivalent of about $180,000 today. For the first time, first-class commercial building was defined not only by architectural design but by elevators, plumbing, heating, and lighting as well. And we now know, thanks to very recent research (Larson, 1981), that the same company's new building in Chicago, originally known as the Kendall and erected in 1872–73, was in one respect even more advanced than its New York predecessor. It was in this building that Johnson and Kreischer installed the first tile cladding for iron columns and beams. The guarantee of safety against destruction by fire and the easy movement from floor to floor provided by the elevator emboldened Hyde to add two more stories to the height of the company's New York counterpart. If we are tracking down the origins of the skyscraper we have certainly reached the seminal stage in New York and Chicago around the year 1870.

REFERENCES/BIBLIOGRAPHY

Badger, D. D., 1865
ILLUSTRATIONS OF IRON ARCHITECTURE, MADE BY THE ARCHITECTURAL IRON WORKS OF THE CITY OF NEW YORK, Architectural Iron Works, New York.

Bannister, T. C., 1956 and 1957
BOGARDUS REVISITED, Journal of the Society of Architectural Historians, 15:4, December, pp. 11-22; 16:1, March, pp. 11-19.

Bixby, W. H., 1895
WIND PRESSURES IN ENGINEERING CONSTRUCTION, Engineering News, 33:11, March, 175-184.

Bogardus, J., 1856
CAST IRON BUILDINGS: THEIR CONSTRUCTION AND ADVANTAGES, J. W. Harrison, New York.

Brooks, M. A., 1912
REMINISCENCES OF THE EARLY DAYS OF FIREPROOF BUILDING CONSTRUCTION IN NEW YORK CITY, Engineering News, 68:22, November 28, pp. 986-987.

Bruegmann, R., 1978
CENTRAL HEATING AND FORCED VENTILATION: ORIGINS AND EFFECTS ON ARCHITECTURAL DESIGN, Journal of the Society of Architectural Historians, 37:3, pp. 143-160.

Condit, C. W., 1974
THE WIND BRACING OF BUILDINGS, Scientific American, 230:2, February, pp. 92-105.

Ferguson, E., 1976
A HISTORICAL SKETCH OF CENTRAL HEATING, 1800-1860, Charles E. Peterson, ed., Building Early America, Chilton Book Co., Radnor, Pa., pp. 165-185.

Larson, G. E., 1981
FIRE, EARTH AND WIND: TECHNICAL SOURCES OF THE CHICAGO SKYSCRAPER, Inland Architect, 25:7, September, pp. 20-29.

Morris, Tasker and Company, 1858
PRICE LIST, Morris, Tasker and Company, Philadelphia, Pa.

Morris, Tasker and Company, 1861
ILLUSTRATED CATALOGUE, Fourth Edition, Morris, Tasker and Company, Philadelphia.

Otis Elevator Company, 1953
THE FIRST HUNDRED YEARS, Otis Elevator Company, New York.

Quimby, H. H., et al., 1892-1893
WIND BRACING IN HIGH BUILDINGS, Engineering Record, 26:25 (19 Nov 1892), p. 394; 27:5 (31 Dec 1892), p. 99; 27:7 (14 Jan 1893), p. 138; 27:8 (21 Jan 1893), pp. 161-162; 27:9 (28 Jan 1893), p. 180; 27:13 (25 Feb 1893), p. 260; 27:15 (11 Mar 1893), pp. 298-299; 27:16 (18 Mar 1893), p. 320.

Skempton, A. W., 1858-1860
THE BOAT STORE, SHEERNESS (1858-1860), AND ITS PLACE IN STRUCTURAL HISTORY, Transactions of the Newcomen Society, Vol. XXXII, pp. 57-78.

Skempton, A. W., 1959
EVOLUTION OF THE STEEL FRAME BUILDING, Guilds Engineer, Vol. X, pp. 37-51.

Vogel, R. M., 1961
ELEVATOR SYSTEMS OF THE EIFFEL TOWER, 1889, United States National Museum Bulletin 228, Washington, Smithsonian Institution.

Vogel, R. M., 1976
BUILDING IN AN AGE OF STEAM, Charles E. Peterson, ed., Building Early America, Chilton Book Co., Radnor, Pa. pp. 119-134.

Webster, J. C., 1959
THE SKYSCRAPER: LOGICAL AND HISTORICAL CONSIDERATIONS, Journal of the Society of Architectural Historians, 18:4, December, pp. 126-139.

Wight, P. B., 1876
THE FIRE QUESTION, American Architect and Building News, v. 1 (17 June 1876), pp. 195-197; (24 June 1876), pp. 203-205; (1 July 1876), pp. 211-212.

Wight, P. B., 1878
THE FIRE QUESTION, American Architect and Building News, v. 3 (2Mar 1878), p. 76.

History of Tall Buildings

The Relative Value of Invention and the History of Tall Buildings

Tom F. Peters

History is as mutable as the criteria on which it is based. Does tall building history originate with the stiff post-and-beam frame or perhaps with tubular structure? The answer depends entirely on our point of veiw. Is the use of modern steel crucial to the development of the stiff frame connection or is it the abandonment of the cast-iron column? The answer is "yes" to both, which makes it very difficult to posit the beginning of tall building construction as we understand it today. Was or is the curtain wall of any interest in this development at all? Again the question is not an easy one, nor the answer obvious. All the answers depend more on the current interest of the questioner than on any absolute standard of "truth." But that doesn't make them any the less interesting as they determine not only a chronology of invention but also the determination of historical value and our understanding of engineering culture.

WHAT IS HISTORY?

History is a complex subject, ruined for most of us in elementary school where it is taught by grasping at the only testable facts: names and dates.

But history is not names and dates—it is questions aimed at our culture, our problems, and our development, and checked against recorded material.

History is therefore as mutable as the criteria we apply to our questioning. The facts may be incontestible, but the meaning we invest them with and the value we assign them are entirely dependent on how we view our world and how we ask the question. History can be compared to a walk along a river, the flow of time. From any position on the river bank the opposite shore presents a landscape, looking back up the time scale, across at the present, and downstream toward the future. As we, the observers, move along the bank with time, we see the same landscape, past, present, and future, from a different angle each time. Objects that loomed large before are now smaller than those that had appeared insignificant and vice-versa. Objects we had thought to know intimately appear novel when viewed from a different angle and under a different light. Therefore the very nature of history requires that it be rewritten from time to time in order to make it relevant to our current viewpoint. It is a reflexive, analytical tool used to help understand the present world and its problems.

So it is with the history of tall buildings. In order to orient ourselves within such a field it is convenient to determine salient features—in other words, to assign beginnings to a development. But what are the relevant developments and what do we consider to be a beginning? The question is more complex than we would first imagine.

DEVELOPMENT OF TUBULAR CONSTRUCTION

First of all: is it really the stiff post-and-beam frame that determines the tall building? This question would never have been asked if Fazlur Kahn had not begun to design tubular structures for tall buildings in the late 1950's. Today many, if not all of the tallest buildings, are designed in some configuration of tubes. And if we research the history of the tube, we can trace an interesting genealogy, starting for all practical purposes with a bridge of 150 Bavarian feet in span over the Oker River at Brunswick in Germany in 1824, which used tubes of cast iron in compression for the first time. The developer of the system, Georg von Reichenbach, reported having seen a large tubular tripod in Britain in 1792 when he was there on a voyage of industrial espionage (von Dyck, 1912). And we know that the idea of tubular construction originated apparently with John Nash in 1799 (Mehrtens, 1908). Samuel Wyatt (von Dyck, 1912) in 1800 in Britain had been written about by Friedrich Wiebeking in Germany and Emiland Marie Gauthey in France (Mehrtens, 1908) before the Oker Bridge was erected. Nevertheless, we can posit a small beginning here.

The problems of a tube in compression are not the problems of the cantilever, but we can recognize an inkling of the ideas that will later characterize tall buildings. Tubular arch construction was adopted by Antoine Rémy Polonceau in the Pont du Caroussel of 1839 over the Seine in Paris (Mehrtens, 1908), in an identical one erected over the Guadalquivir in Seville in 1851, and finally by James Eads in the celebrated Mississippi Bridge at St. Louis in 1876.

The first large-scale use of a tube in construction, this time in bending, was in the wrought iron Britannia and Conway bridges by Robert Stephenson and William Fairbairn built between 1845 and 1850 in Wales. Then, soon after the building of these bridges, engineers on the European continent began to launch prefabricated tubular bridges, the so-called Red Bridge over the Aare River in Berne in 1858 (Peters, 1981). Now we encounter the *cantilever* in tubular construction for the first time, and the connection with the modern tall building is closer. An essential difference between such bridges and the tall building lies in their foundations, but tower foundations had, of course, long been known in construction.

Masonry towers had also been built as tubes ever since the Middle Ages in Europe, even if the system had not been recognized for what it was. The vault placed at about 1/3 of the height of the brick Asinelli Tower in Bologna (Fig. 1) in 1119, demonstrates that the builders were well aware of the problem of torsion in such slender, tall structures and even knew how to counter it (Council on Tall Buildings, 1981). Iron rods as tension reinforcement for slender bridge piers had occasionally been suggested in order to make the tubular behavior of masonry even more perfect (Stussi, 1943/1944).

It was not until the first quarter of this century that the tube was tried in steel. Ralph Modjeski was the first to build bundled tubes of steel for the towers of his Philadelphia-Camden Suspension Bridge, proposed in his report of 1921 and finished in 1926 (Modjeski, Webster, and Ball, 1927). This bridge was followed by the Oakland Bay Bridge towers and by Joseph Strauss and Clifford Paine's Golden Gate Bridge towers of 1938. These last, at 210 m (690 ft), doubled the height of the Philadephia-Camden towers. Thus we finally have all the elements of the extremely high bundled tube system for tall buildings together in one structural type. The towers of the Golden Gate Bridge were just 10% shorter than the Woolworth Building of 1913 and 55% of the height of the then tallest edifice, the Empire State Building of 1931.

But who is to say in all this development when that part of the concept appeared that specifically constitutes the tall building tube? Which portion of the concept is the crucial evolutionary one and which parts are only contributory? That brings us back to our first question: What do we consider to be a true beginning? The answer, of course, is that it depends on what we regard as the chief element of the concept, and that can change as our preoccupations with this form of structure are modified by our current interests.

SKELETON AND SKIN

The same is true of the development of the post-and-beam frame system itself. Until recently the steel frame was considered, together with the evolution of the curtain wall, to be the crucial element in the evolution of the tall building. The Council on Tall Building's Third International Conference in

Fig. 1 Garisenda and Asinelli Towers in Bologna *(Courtesy: Luigia Binda and Giannantonio Sacchi)*

1986 was timed to coincide with the centenary of the building of the Home Insurance Building by William Jenney, and it was there that the point was closely reexamined.

The structure of the Home Insurance Building was, according to the contractor William Starrett (1928), the first in which the exterior walls were not self-supporting but rather carried by the spandrels of the frame. In part this may be true, but only in part. The building commissioners forbade the placing of columns in the party walls, so that these, at least, were still self-supporting (Randall, 1949). The ground floor was also a self-supporting structure with walls of stone. In 1872, more than a decade before the design of the Home Insurance Building, a remarkable structure had been erected for the Menier Chocolate Factory in Noisel-sur-Marne by an otherwise unknown architect, Jules Saulnier (1817–1881). This building had a frame of wrought iron with riveted connections. Some internal cast-iron columns supported the central part of the girders (Marrey and Chemetov, 1976). But these are conspicuously missing on the ground floor. Their role on the upper floors is therefore of relatively minor import to the frame as a whole. Even before this, the St. Ouen dock warehouse built by Hippolyte Fontaine in 1865 had sported a similar, if not so clearly developed structure (Skempton, 1959/1960).

Whereas the Home Insurance Building had no framing provided for stiffening, the Menier Building was entirely braced by triangulation, and the brick exterior walls were genuine curtain walls which, however, carried their own load over three stories before resting on the lower part of the iron frame. Again it is a value judgement as to which aspects of the independence of frame from skin one chooses to give precedence. However, even though the Menier Factory building may have been the "truer" frame, it had very little immediate influence, whereas some of Jenny's work did.

Jenny had been educated at the Ecole Centrale des Arts et Manufactures in Paris and, although he had long settled in Chicago by the time the Menier building was erected, he may very well have retained personal or perhaps even professional contacts in France and may even have known of it (Turak, 1966). The Menier Building is only the most extreme example of a development that had been going on in France and in Britain for several decades before that. Two glass curtain walls had been erected as part of buildings in Liverpool by the engineer and builder Peter Ellis as early as 1864 and 1866, but the Menier Building is the one of which Jenney might most easily have had knowledge.

I do not wish to say that Jenney was of no consequence in the development of frame construction. On the contrary, it is indubitable that Jenney, trained in the same school as, and graduating a year after, Eiffel in 1856, had an equally profound influence on the evolution of iron frame construction, albeit in an architectural rather than in an engineering sense. In fact, Jenney, not being originally an architect, and yet working within that profession, was uninterested in many, if not all of the formal preoccupations that prevented architects from clearly conceiving a novel formal expression for a new structural type (Turak, 1966), which proved to be a definite advantage, particularly since Jenney trained a whole school of architects who were to become influential in this development. But our interest is focused here on primacy of invention, originality, and clarity of concept in engineering matters.

Another aspect of the preeminence of the Home Insurance Building over all others that were then going up in New York, Chicago and elsewhere, is that Jenney was convinced to use steel beams for the first time in the four top floors of his structure. This decision was courageous on Jenney's part, as the long-term behavior of the material in structures was not yet well known, and it was a decision that Eiffel, for instance, never dared to take. But brave as it was, it was nevertheless of no immediate consequence as far as the development of frame structure was concerned, for two reasons. In the first place, in many cases the steel of the time was little more than wrought iron without the laminated slag inclusions, meaning that the new material contained very little, if any carbon, and that it was therefore extremely ductile and probably quite different from the standard mild structural steel of a few decades later. Then too, whether or not the material of the upper story girders *was* steel in our sense or not, the columns of the building remained cast iron. And it is impossible to effect stiff frame connections between cast-iron and either wrought iron or steel members.

The use of different materials for the columns and girders of high-rise frames remained a serious hindrance to the evolution of a stiff frame for many more years. There were a few exceptions, the most notable being the Rand McNally Building and the Fair Store, which were both completely of steel. The Rand McNally Building of 1890 had a completely riveted frame, except for the ubiquitous party walls, which was designed by the architectural firm of Burnham and Root. The Fair Store was built in 1891 by Jenney and Mundie (Randall, 1949). These were, however, just like the Menier Factory building in France, unique examples, perhaps forerunners, but not yet indicative of a general development. Cast-iron columns continued to be used almost universally until the collapse of the almost completed frame of the Darlington Building in New York on March 2, 1904. This accident appears to have been the chief cause for the change in codes, which thenceforth banned the use of structural cast iron. As far as the evolution of the stiff frame is concerned, or what Freitag (1901) called "cage" as opposed to "frame" construction that required secondary means of stiffening, the banning of cast iron may have been a crucial step in the evolution of the "skin-and-skeleton" building, which has received far too little attention to date.

THE CURTAIN WALL

Given the forty-year period in which most of the current history of high-rise was first examined, from Giedion and Starrett in the late 1920s through Condit in the 1960s, it is clear why so much attention was paid to the development of the independence of the skin from the frame, in other words to the curtain wall. Here is an excellent instance of the decisive influence of criteria on the writing of history discussed at the outset. From the present standpoint, as interest increases toward tubular systems and composite construction, the curtain wall is of less concern today than it was then. In fact it may already seem rather strange to the modern builder that one would even consider disregarding such large and well-positioned contiguous surfaces for the transference of both gravity loads and for stiffening.

Historically, however, the preoccupation with freeing the skin from the structure was not an illogical one. In traditional low-rise construction, masonry walls formed the primary loadbearing structure of buildings, occasionally supplemented secondarily by interior columns. The stiffness of the column-to-beam connection was of little consequence as the primary transference of all horizontal loads transpired between the beams and the walls. Such a system is of little use in construction over 16 floors and is a great encumbrance long before attaining that height, as the few examples that were actually built demonstrated. Relying solely on the columns for gravity loads meant shifting the transference of the horizontal loads from the beam-to-wall connection to the beam-to-column connection. Thus the importance of the wall was reduced and the stiff frame conceived. It was logical then that the interior wall should be reduced to a shear membrane and the outer, for reasons of natural lighting, to a mere skin.

Now that this has been thoroughly achieved, such a clear separation is no longer necessary, and our preoccupation with the history of the curtain wall and with its construction can become less important. Structures once considered of prime importance in the development of the frame may today be placed historically in another context. They do not lose their historic importance, but they do now share the limelight with structures demonstrating other concerns, for instance with those that document the evolution of the bundled tube concept.

CONCLUSION

The object in tracing briefly some aspects of development of the historian's interest in tall buildings is to show the relative nature of the primacy of invention, interesting though it may be. It may go against the grain of those who seek to establish heroes in the history of engineering, but it is a far more complex way of understanding development with more subtle ramifications for contemporary interests. I also intended to show how contemporary the concerns of history really are, to demonstrate how historians attempt to understand the problems of our time using other methods than those of engineers, and what the value of such approaches may be for designers when battling for the clarification of structural concept and for simplicity in building.

REFERENCES/BIBLIOGRAPHY

Bannister, T. C., 1956
BOGARDUS REVISITED. PART 1: THE IRON FRONTS, Vol. 15, No. 4, December, pp. 12-22.

Bannister, T. C., 1957
BOGARDUS REVISITED. PART II: THE TOWERS, Journal of the Society of Architectural Historians, Vol. 16, No. 1, March, pp. 11-19.

Condit, C., 1964
THE CHICAGO SCHOOL OF ARCHITECTURE, The University of Chicago Press, Chicago, Il.

Council on Tall Buildings, 1981
PLANNING AND ENVIRONMENTAL CRITERIA FOR TALL BUILDINGS, Volume PC of Monograph on Planning and Design of Tall Buildings, ASCE, New York

Freitag, J. K., 1901
Architectural Engineering, 1895/1901/1906/1911/1912, John Wiley and Sons, New York.

Giedion, S., 1928
BUILDING IN FRANCE. IRON, REINFORCED CONCRETE (Bauen in Frankenreich. Eisen, Eisenbeton), Second Edition, Klunkhardt & Biermann, Leipzig.

Marrey, B. and Chemetov, P., 1976
FAMILIARLY UNKNOWN . . . ARCHITECTURE, PARIS 1848-1914 (familierement inconnues . . . Architectures, Paris 1848-1914), Catalogue, n.d., n.p. Paris, Bellamy & Martet.

Mehrtens, G. C., 1908
LECTURES IN ENGINEERING SCIENCE, Part II. Iron Bridge Construction, Vol. I (Vorlesungen Uber Ingenieur-Wissenschaften, II, Teil, Eisenbruckenbau, Bd. 1), Wilhelm Engelmann, Leipzig.

Modjeski, R., Webster, G. S., and Ball, L. A., 1927
THE BRIDGE OVER THE DELAWARE RIVER CONNECTING PHILADELPHIA, PA. AND CAMDEN, N.J., Final Report of the Board of Engineers to the Delaware River Bridge Joint Commission of the States of Pennsylvania and New Jersey, n.p.

Peters, T. F., 1981
THE DEVELOPMENT OF LONG-SPAN BRIDGE BUILDING, 3rd ed., Zurich: ETH.

Randall, F. A., 1949
HISTORY OF THE DEVELOPMENT OF BUILDING CONSTRUCTION IN CHICAGO, University of Illinois Press.

Skempton, A. W., 1959-1960
THE BOAT STORE, SHEERNESS (1858-1860), AND ITS PLACE IN STRUCTURAL HISTORY, Transactions Newcomen Society, Vol. 32.

Starrett, Col. W. A., 1928
SKYSCRAPERS AND THE MEN WHO BUILD THEM, C. Scribner's, New York/London.

Stussi, F., 1943/1944
AN UNKNOWN REPORT BY NAVIER (Un rapport inconnu de Navier) in Publications of the International Association of Bridge and Structural Engineering, Vol. VII, 1943/1944, Zurich, pp. 1-13.

Turak, T., 1966
WILLIAM LE BARON JENNEY: 19TH CENTURY ARCHITECT, PhD Dissertation, University of Michigan.

von Dyck, W., 1912
GEORG VON REICHENBACH. DEUTSCHES MUSEUM BIOGRAPHIES AND DOCUMENTS (George von Reichenbach. Deutsches Museum Lebensbeschreibungen und Urkunden), Deutsches Museum, Munich.

Evolution of the Skyscraper: A History of the Tall Building in Chicago

C. William Brubaker

Chicago enjoys a rich architectural heritage. The city is known worldwide for its outstanding buildings. No single style prevails; Chicago thrives on diversity. The city uses a great variety of old and new structures of remarkably high quality.

The tall building evolved here. We celebrate the first hundred years of the skyscraper and plan for the second century.

THE SKYSCRAPER IN THE CITY

The three most famous skyscraper skylines are experienced in Chicago at the edge of Lake Michigan, in New York City on Manhattan Island, and in San Francisco by the Golden Gate.

In the last third of the nineteenth century, tall buildings evolved in Chicago and New York City, and skyscraper became the symbol of those cities. The idea, an American invention, was a great success and spread throughout the country and throughout the world, creating high density centers for commerce that provided convenience, efficiency, excitement, and entirely new concepts of architecture, engineering, and planning.

These high-density, high-rise, activity centers became the focal points for expanding urban areas. Today, one cannot visualize San Francisco without its towers. For better or worse, in Hawaii, Honolulu's Waikiki Beach is a high-rise urban environment. Atlanta, Dallas, Houston, (and Mexico City, Cairo, and Hong Kong) are tall building cities or have tall building downtowns.

THE RACE FOR HEIGHT

For centuries, mankind has given important structures special height. Gothic cathedrals reached upward. Since before the sixteenth century, the campanile at St. Mark's in Venice has been a landmark. Its romantic slender form inspired many American towers in the twentieth century, such as the Ferry Building Tower in San Francisco and a number of skyscrapers in New York City, especially the 1909 Metropolitan Life Tower.

In the 1860s, all of the prerequisites for the skyscraper had developed, including the elevator (Otis's first practical elevator was installed in 1857), available ferrous metal, fire protection systems, and a need for tall buildings. In the design of innovative buildings like the Home Insurance Building in Chicago, completed in 1885, these components came together as an entirely new kind of architecture, the skyscraper office building.

These structures were built higher and higher. In the 1890s some buildings were over 91 m (300 ft) high; in 1913, the Woolworth Building rose to 242 m (792 ft). By 1930, the Chrysler Building in New York was over one thousand feet in height, and in the following years, the Empire State Building was completed at a height of 381 m (1,250 ft)—a record that stood until 1974 when Chicago's 110-story Sears Tower reached 443 m (1,454 ft). That record will be topped. For example, New York City's "Television Tower," publicized in 1986, would have 150 stories and a height of 509 m (1,670 ft). However, the future "world's tallest building" won't necessarily be in Chicago or New York City.

NOT ONLY OFFICE BUILDINGS

Tall buildings were originally designed to be office buildings, but in recent decades, many different functions are accommodated in skyscrapers. Apartments and hotels are conspicuous, but high-rise buildings also serve hospitals and universities.

Different functions are combined into a single structure in the mixed-use development concept. On Chicago's North Michigan Avenue, the Hancock Center, Water Tower Place, and 900 North Michigan, buildings include shops, restaurants, theaters, banks, offices, hotels, apartments, and parking.

EARLY CHICAGO

Marquette and Joliet visited the area in 1673 and noted the strategic location where a short canal could link the Mississippi River to the Great Lakes, but nothing happened until 1803 when Fort Dearborn was built. The small village was incorporated in 1833, and in 1848, the first canal and first railroad served Chicago. The wood and brick community boomed, but in 1871, the Chicago fire destroyed the city. It was rebuilt rapidly and fireproof materials were extensively used in the new buildings.

THE RISE OF THE SKYSCRAPER IN CHICAGO

A few of the buildings built immediately after the fire still survive. The 1874 8-story Delaware Building stands at Randolph and Dearborn, carefully restored (Fig. 1).

Some of the pioneer buildings have not survived. Many were demolished to make way for taller structures. The 10-story Home Insurance Building, designed by William LeBaron Jenney, was completed in 1885, but it was torn down in 1931. Historian Carl Condit, in his book, *The Chicago School of Architecture* called the Home Insurance Building "the decisive step in the evolution of iron and steel framing" (Condit, 1964)

In 1887, the Marshall Field Wholesale Store (demolished in 1930 to create a parking lot) by architect H. H. Richardson, was occupied. Its heavy walls of rusticated granite and arched windows influenced designer Louis Sullivan of Adler and Sullivan, in his design for the Auditorium Building (Fig. 2). This was an early mixed-use development, combining a 4,237-seat theater, hotel, and office building, which today houses Roosevelt University.

On the LaSalle Street canyon at Adams Street, we continue to experience the hundred-year-old Rookery, by Burnham and Root, which continues the Romanesque tradition (Fig. 3). A few years later, in 1891, the same architects completed the Monadnock, at 16 stories, the world's tallest masonry-bearing building. This extremely simple and straightforward, undecorated brick building is an important landmark, but the evolving new concepts of steel framing made its kind of construction obsolete.

THE STEEL FRAME EXPRESSED: THE CHICAGO SCHOOL OF ARCHITECTURE

An entirely new idea evolved in the 1890 decade; thick masonry bearing walls were replaced by steel framing. In Chicago, this efficient new method of construction was not hidden behind eclectic facades but was clearly expressed.

An honest gridiron pattern of columns and beams gave the new Chicago architecture a new appearance. No longer were tall buildings Romanesque or Gothic.

In Chicago's Loop, we still enjoy Holabird and Roche's 16-story 1894 Marquette Building, D.H. Burnham and Company's 1895 Reliance Building (Fig. 4), and Louis Sullivan's 1899 Carson Pirie Scott Store (Fig. 5).

EFFECT OF THE 1893 WORLD'S FAIR

This Chicago School of Architecture, however, was destroyed by new forces of eclecticism, demonstrated forcefully at the 1893 World's Columbian Exposition, where eastern architects educated at the Ecole des Beaux Arts in Paris imposed the old architecture of Europe on the "White City" World's Fair buildings. Americans were impressed, and they wanted Greek temples, Roman banks, Renaissance and Gothic skyscrapers.

For thirty years, historic styles were common. The Wrigley Building of 1921 is a "Renaissance" design, the Continental Illinois Bank of 1924 is "Roman," and the Tribune Tower of 1925 is "Gothic" (Fig. 6).

THE VERTICAL STYLE

In the late 1920s, tall buildings expressed a new idea—the vertical style—with walls of limestone and vertical strips of windows, symmetrical stepped-back forms, and nontraditional streamlined art deco ornament. Also, these buildings were taller. Examples include the Civic Opera, Daily News (Fig. 7), Board of Trade, Palmolive, and the 333 North Michigan Avenue building, which is a vertical style masterpiece designed by Holabird and Root in 1928.

BACK TO BASICS: THE 1950s AND 1960s

Very little new construction occurred in Chicago during the depression years of the 1930s and the war years of the 1940s. Then, in 1952, the twin 860–880 North Lake Shore Drive Apartments, designed by Mies van der Rohe, were completed (Fig. 8). Again, as in the 1890's, the steel frame was clearly expressed. The buildings proved to be a great economic success and inevitably were copied in Chicago, throughout America and the world.

Meanwhile, Skidmore, Owings & Merrill's 1956 Inland Steel Building was built in the Loop, one of the first excellent and innovative office buildings to enhance Chicago's central area (Fig. 9).

Three outstanding structures, each with a plaza, were constructed at the heart of the Loop in the 1960s: (1) the Civic Center by SOM, C. F. Murphy Associates, and Loebl Schlossman and Bennett (with 87 feet long bays and columns, spandrels and mullions of Cor-ten steel); (2) the Federal Center by a team of architects, which included Mies van der Rohe; and (3) The First National Bank of Chicago (Fig. 10) by C. F. Murphy Associates and Perkins & Will (with a lively multi-level plaza that attracts thousands of people on a pleasant day). For the United States Gypsum Building, 1963, Perkins & Will explored new ideas: The Greek Cross plan is turned 45° on the Wacker Drive site, and the gypsum crystal inspired the polyhedral forms (Fig. 11).

Then, Bertrand Goldberg designed Marina City, with two 60-story circular concrete towers plus other elements, to create a popular riverside mixed-use development. Schipporeit-Heinrich designed Lake Point Tower, with 70 stories and a height of 645 feet, making it the world's tallest reinforced concrete building in 1968. Finally, at the end of the 1960s the John Hancock Center, by SOM, was completed not in the Loop, but on North Michigan Avenue (Fig. 12). With wind-bracing expressed, this tapering steel-framed, mixed-use skyscraper is 337 m (1,105 ft) tall.

NEW HEIGHTS AND NEW HUMANISM: THE 1970S AND 1980S

The Amoco Building overlooking Grant Park (Edward D. Stone and Perkins & Will) is Chicago's second tallest, while Sears Tower (SOM), also completed in 1974, is not only Chicago's tallest but also the world's tallest building with 110 stories of offices rising 1,454 ft (Fig. 13). Even higher buildings have been proposed, but one can question the wisdom of some of the proposals.

Water Tower Place, a mixed-use development on the Near North Side, proved to a popular visitor magnet and great economic success. Architects Loebl, Schlossman, Bennett & Dart with C. F. Murphy Associates wisely set the 62-story tower back from North Michigan Avenue.

One Magnificent Mile (SOM) followed in 1983. Meanwhile, in the Loop, the Xerox Center (Murphy-Jahn), State of Illinois Center (Murphy-Jahn), Three First National (SOM), 33 West Monroe (SOM), 30 South Wacker (Fujikawa & Johnson), Associate Center (A. Epstein), 333 West Wacker (KPF and Perkins & Will), and 123 North Wacker (Perkins & Will, Fig. 14) add to Chicago's rich architectural heritage and maintain the city's reputation for innovation, quality, and diversity in the design of tall buildings.

THE SECOND CENTURY OF THE SKYSCRAPER

Some of the early skyscrapers have not only survived but have been restored and adapted successfully to new uses. The hundred-year-old Rookery is being preserved and restored as a first-class office building. The 82-year-old Railway Exchange Building has been beautifully restored as the

Santa Fe Center. The 95-year-old Manhattan office building has been converted into apartments, and the Sears Store (originally the Second Leiter Building of 1891) has been converted into offices.

The lesson is clear: Tall structures will stand and serve more than one hundred years, even though users may change. The skyscraper has proven to be a flexible and adaptable structure.

Future tall buildings will be "intelligent buildings," accommodating changing communications, computer, and other electronic systems. They will be more energy-efficient. Walls will not be identical in all directions, but will be designed in response to the facts of climate and orientation. Walls will be solar energy collectors, putting spandrels and windows to work with photovoltaic systems incorporated into these enclosing components.

Buildings will be lighter weight, more accurately engineered and more efficient, and therefore more economical. (Current planners often forget this lesson from Buckminster Fuller!)

Composite structures will be more common. Steel, concrete, aluminum, and other materials will be combined into more rational, efficient, factory-made parts.

The mixed-use development idea will continue to thrive. The mix will change from time to time. Recreational, educational, cultural, and health care facilities will be a part of the rich mix.

Three dimensional urban planning systems will evolve; upper level sky-lobbies will be linked to the sky-lobbies of neighboring buildings with bridges occasionally (an old idea whose time still has not come).

Will buildings be taller? Technical problems will not limit height, but efficiency and economics, regulations and zoning, safety and human response will temper interest in building taller than 100 stories. However, when all these factors are in balance, we can expect, and plan for, super-tall buildings during the second century of the skyscraper.

REFERENCES/BIBLIOGRAPHY

Condit, C. W., 1964
THE CHICAGO SCHOOL OF ARCHITECTURE, The University of Chicago Press, Chicago, Illinois.

Fig. 1 Delaware Building, built in 1874, survived the Chicago fire *(Photo by Steven A. Herrlin)*

Fig. 2 The Auditorium, built in 1899, designed by Louis Sullivan *(Photo by Hedrich Blessing)*

Fig. 3 The Rookery continues the Romanesque tradition *(Photo by Hedrich Blessing)*

Fig. 4 D. H. Burnham and Company's 1895 Reliance Building *(Photo by Hedrich Blessing)*

Fig. 5 Carson Pirie Scott Store (1899) by Louis Sullivan *(Photo by Hedrich Blessing)*

Fig. 6 Tribune Tower *(Photo by Hedrich Blessing)*

Fig. 7 Vertical style Daily News Building (now Riverside Plaza) *(Photo by Hedrich Blessing)*

Fig. 8 Twin Towers of 860-880 North Lake Shore Drive Apartments *(Photo by Hedrich Blessing)*

Fig. 9 Inland Steel Building *(Photo by Hedrich Blessing)*

Fig. 10 First National Bank of Chicago by C. F. Murphy and Associates and Perkins & Will features multi-level plaza *(Photo by Hedrich Blessing)*

Fig. 11 United States Gypsum Building *(Photo by Hedrich Blessing)*

Fig. 12 John Hancock Center on North Michigan Avenue *(Photo be Hedrich Blessing)*

Fig. 13 Sears Tower, tallest building in the world *(Photo by Hedrich Blessing)*

Fig. 14 123 North Wacker *(Photo by Cabanban)*

Time's Arrow: Tall Moving Targets And Social Research

David Cooperman

SOCIAL EFFECTS TRENDS AND RESEARCH PROBLEMS

A comparison of accounts of interest and inquiry into the social impact of tall buildings over the past two decades with current concerns of the public, planners, architects, and engineers discloses that a problem shift is occurring. Not too long ago the focus was primarily on the problematic aspects of tall buildings such as crime, poverty, family security, environmental pollution, and psychological well-being. Some of these themes continue, but at a much lower level. Environmental psychologists continue to examine microlevel behavior, and public response still focuses on trade-offs between the economic gains of tall building development against potential undesired effects on the city-scape and congestion.

However, there has been a decline in the volume of inquiry into social effects of the environment. The basic reason is that the comparative social uses and types of users have shifted. There have been major increases in the volume of tall building office construction in most large cities, not only in developed Western societies, but in Pacific basin sites; and there has been a demographic shift in residential populations. In the Western world, upwardly mobile single people and retired, comparatively affluent people comprise an ever larger proportion of the tall building residential population. Such people tend to report high levels of satisfaction with quality of life in general, including housing in particular. Hence, there is less social urgency for inquiry in this area. Market research is adequate for the competitive needs of developers of such buildings.

The population characteristics of tall buildings have also been shifting. More mixed-use structures have been built, and developers often propose clustered patterns, which either reflect some quasi-autonomous subcity plan or affect surrounding areas more intensely than single building projects. In effect, tall building habitats have undergone sufficient change to endow the young social science research with a vaguely historic air, as if the old problems shared some of the design features of Modernism of the 1960s as well as the social activism of that decade.

SCALAR SHIFTS AND RESEARCH NEEDS

An imperfect correlation is assumed, not exactly in step over a decade but significant nonetheless, between the size and scope of corporate organizations in a free market and the size and volume of tall office buildings. Even if we averaged out the recent jump in corporate mergers over several years, we are left with some image of very tall building clusters. The phrase *imperfect correlation* means just that. No causal connection is implied; the connection holds best in primary cities, such as New York and Chicago; and the usual caveat, "other things being equal" should be thrown in for good measure. For example, corporate mergers have been just as frequent and significant in Britain and Sweden, yet the scale of very tall building clusters has yet to emerge in London and Stockholm. A rough proposition for North American and some Pacific Basin cities might be, "the larger the size and the greater the complexity of superfirms, the taller and more extensive will be the tall building environment". The Helmut Jahn sketches of Television City for an Upper West Side swath of New York, even if only fiction at present, may be covered by this crude law. And indeed, New York seems to have spawned a fair number of such projects whose architectural characteristics can be quickly gauged and judged but whose social effects are problematic. (Goldberger, 1985*a*).

On a smaller scale, developments in and around the central business districts (CBDs) of regional metropolises, such as St. Louis and the Twin Cities, reflect similar trends. Twenty years ago in San Francisco, the spectre of tall building habitats, which have since been overtopped in Chicago and New York, resulted in the current zoning regulations. Meanwhile, as Manhattan fills vertically, extensions appear horizontally in the shape of tall mixed-use cluster plans for Queens and Brooklyn. The antagonistic responses to various development proposals, such as the Times Square designs, reflect a more encompassing set of interlinked problems than the older antipathies to deprivation of view, light, and increased congestion. A proper understanding of the social effects of alternate very tall building habitats requires a coordinate shift in the scale of research on the subject.

COMPLEX MODELS AND COMPREHENSIVE RESEARCH

A shift in social research does not imply wholesale abandonment of previous work. A review of sections of the monograph volume, *Planning and Environmental Criteria for Tall Buildings* (Council on Tall Buildings, 1981) shows that much research, especially concerning microenvironments and environmental psychology, are both relevant today and can provide the base for the more intricate research projects that should be planned. Other publications on research methods published in the last decade also describe techniques which, with modifications, can provide the basic tools for comprehensive analysis (Michelson, 1975). However, the state-of-the-art methods originate from a variety of disciplines, including environmental psychology, sociology, social impact assessment, and behavioral architecture. Future research on comprehensive very tall building clusters will benefit from logically connected propositions which explain, predict, and model social effects. But what types of research projects can be useful for the analysis of very tall building habitats?

Five years ago a built environment evaluation study was published that won a sociology prize. It was supported by a grant from the National Bureau of Standards, and was a joint venture of sociologists and architects at the University of Michigan. It is a model of comprehensive research and deep evaluation of social-design interface effects of one 4-story federal office building (Marans and Spreckelmeyer, 1981). Micro and macro action are connected, causal design components are empirically identified, and problematics skillfully and rationally described. At first glance it seem implausible and impractical, if not impossible, to project such a study for very tall building clusters in New York or Chicago. Yet, without something like such a venture we will not build with the information we should have. Pruitt-Igoe was not, as some would claim, an architectural failure; comprehensive social foresight was seriously lacking.

SPECIFIC RESEARCH PROJECTIONS FOR VERY TALL BUILDINGS

Each of the proposls in this brief list implies a socially significant aspect of a tall building cluster. Hence it is both a list of information and explanation needs and an attempt to project future social interaction.

1. **Marginal social disutility studies.** Analysis is made of additions to congestion, toward gridlock, by alternate designs. One of the benefits of clustered developments is that optimal social densities can be

obtained with minimum congestion. Multilevel, off-grade skyways and/or tunnels have proven their worth. In cities of secondary rank-size, it is feasible to retrofit roadways in standing CBDs, providing much wider margins for congestion limits while enhancing the social interaction of users. Design alternatives for new developments in major cities can be coordinated with large computer-aided information and simulation of congestion effects under alternative multilevel pathway designs.

2. **Mixed-use Habitats.** The trend to mixed-use buildings is supported by the economics of tall building development at present. But what specific mixes are socially optimal for which populations? Can we increase the range of user-choice options by providing readily accessible services in adjacent, easily accessible structures? What design variations within such buildings can provide people with enhanced choices in such a large, introverted cityscape?

   Maya Lin's recent judgment of Foster Associates' Hong Kong Shanghai Bank Building heaps praises on it for its esthetic/behavioral advances in tall building design, especially with regard to its well-crafted spaces that manage to provide human-scale amenities in large scale contexts (Lin, 1985). Indeed, Lin is most optimistic concerning the future as she contemplates such designs: "As we approach the 21st century, the skyscraper is finally undergoing a marked change in both function and form."

   Regardless of the claim for the particular building, mixed-use habitats enclosing assortments of sculpted spaces should be fashioned with some robust information about their social uses.

3. **Life-course effects.** Traditionally, human population ecologies in urban areas reflect a distribution into locales by standard demographic characteristics, including stage in family life-cycle. Moreover since the beginning of the modern area, from about the eighteenth century in Britain, workplace and domicile have been separate in accord with the division and specialization of labor patterns of societies and cities. Changes in family structure, in household characteristics, and in the nature of work have their effect, through the market, on housing demand. Future tall building habitats, in turn, will affect social interaction patterns. It would certainly be desirable to know what these effects will be. The intense reaction against housing low income families with young children in tall buildings in the recent past might alert planners, developers, and designers alike to the need for valid information and explanations at an early stage of the future.

No doubt, any such proposals require very large bodies of information and complex models to yield the knowledge we need. But the methods of data gathering and the computers needed to do the work are available. Recently, Paul Goldberger (1985*b*) wrote, "The challenge is to make the connections that turn a complex [of tall buildings] into something that possesses genuine urban qualities and is not simply an array of big buildings side by side." If very tall structures are to embrace such human qualities, some very good resource information can be of great help.

REFERENCES/BIBLIOGRAPHY

Council on Tall Buildings, 1981
SOCIAL EFFECTS OF THE ENVIRONMENT, Chapter PC-3, Vol. PC of Monograph on Planning and Design of Tall Buildings, ASCE, New York, Committee 37.

Goldberger, P., 1985*a*
THE PROSPECT OF BIGGER TOWERS CAST A SHADOW, New York Times, December 29.

Goldberger, P., 1985*b*
IS TRUMP'S LATEST PROPOSAL JUST A CASTLE IN THE AIR?, New York Times, December 22.

Lin, M., 1985
BEAUTY AND THE BANK, The New Republic, December 23.

Marans, R., and Spreckelmeyer, K. F., 1981
EVALUATING BUILT ENVIRONMENTS: A BEHAVIORAL APPROACH, Survey Research Center and Architectural Research Laboratory, University of Michigan, Ann Arbor.

Michelson, W., 1975
BEHAVIORAL RESEARCH METHODS IN ENVIRONMENTAL DESIGN, Dowden, Hutchinson & Ross, Inc., Stroudsburg, Pa.

Crystal Court. IDS Building, Minneapolis. Hub of a very large CBD mixed use area. *(Architect: Johnson-Burgee Associates)*

Bottleneck in a critical skyway connector, CBD Minneapolis.

Social and Environmental Factors of High-Rise Living: A Singapore Experience

Bill B. P. Lim

An active housing program in Singapore began in 1927 with the establishment of the Singapore Improvement Trust (SIT) by the British colonial government under the Singapore Improvement Ordinance. However, by 1942 SIT had only completed 2049 houses and 53 shops. The Housing Committee (1947) reported that out of the postwar population of 938,000 persons 68,000 or 72% were housed in the central area and about a third of the population lived in the density of 1000 to an acre. The effect of the SIT in the subsequent 12 years was inadequate to provide housing to the population, which grew to 1.6 million by 1959, and 40,000 units were built by the public and private sectors during this period, accommodating only 300,000 persons.

The present government, which took office in 1959, established the Housing and Development Board (HDB) on February 1, 1960, by the Housing and Development Act. The Board immediately planned two five-year building programs to build 50,000 and 60,000 units respectively. By 1985 when the board celebrated its twenty-fifth anniversary, 81% of the population of approximately 2.6 million lived in HDB flats. The Board manages 508,242 apartments and has rehoused 10,808 resettlement cases. The new town development and the housing estate development are shown in Tables 1 and 2.

Table 1 Current housing estates development

Estate	Total land area (hectares)	Residential area allocated (hectares)	Projected total dwelling units	Dwelling units completed in 1984/1985	Dwelling units completed as of 31 March 1985	Dwelling units under construction as of 31 March 1985
Alexandra Hill	10	9	1,590	–	1,510	80
Bukit Purmei	20	14	2,300	1,190	2,320	–
Delta Estate	10	9	1,420	–	920	500
Eunos	128	37	4,400	–	3,890	180
Geylang East	64	21	3,600	–	3,640	350
Joo Chiat/ Changi Road	3	1	230	–	–	220
Kaki Bukit	196	21	2,910	290	290	4,290
Kampong Ubi	170	21	3,450	1,050	1,050	2,040
Kerbau Road	11	9	640	–	120	520
MacPherson Estate	128	50	10,830	400	10,830	120
Potong Pasir	47	21	3,700	740	3,480	90
Redhill Balance	9	8	2,120	–	1,120	770
Rowell Court	5	5	1,020	590	1,040	–
Simei	105	51	9,400	–	–	2,490
St. George's Estate	14	11	2,440	500	2,550	–
St. Michael's Estate	23	19	3,450	–	3,233	220
Teban Gardens	79	33	6,020	–	3,360	920
Towner Road/ McNair Rd	12	12	2,100	270	900	1,200
Upper East Coast	15	13	1,710	590	590	670

Table 2 New town development *(HDB Annual Report, 1984/1985)*

New town	Estimated population as of 31 March 1985	Total land area (hectares)	Residential area allocated (hectares)
Ang Mo Kio	217,700	671	269
Bedok	209,200	751	277
Bishan	6,100	525	170
Bukit Batok	67,000	733	133
Clementi	105,800	344	143
Hougang	65,100	440	180
Jurong East	56,300	236	82
Jurong West	62,300	709	354
Queenstown	123,200	285	149
Serangoon	35,200	212	110
Tampines	107,600	957	331
Telok Blangah	58,000	96	68
Toa Payoh	161,000	339	167
Woodlands	68,600	1,223	421
Yishun	82,700	903	391
Zhenghua	0	380	211

[a]288 units were reclassified under Bukit Merah Town Centre

THE PLANNING OF HOUSING ESTATES

The location of housing estates follows broadly the master plan of Singapore with reference to transportation and proximity to employment. The expressway system for Singapore now near completion links the major housing estates to the central business district as well as subcenters.

The Mass Rapid Transit System now under construction will eventually provide direct linkage among the housing estates and with the central business district. Thus the housing estates will serve both residential functions and as subcenters of commercial, industrial, and service activities. Figure 1 shows the urbanized areas in the concept plan of Singapore (Yeh, 1975).

In each housing estate the overall layout of the road network is scaled down systematically from the major heavy traffic roads leading to the new towns to the low traffic cul-de-sacs in the precincts.

The precinct design, introduced some 7 years ago, has been further improved. Buildings are grouped within each precinct to define boundaries for easy recognition. Communal facilities are located to promote a stronger sense of community living among residents. Major structures such as town centers and religious buildings are strategically located among the residential buildings to give each neighborhood district landmarks.

In addition land parcels are provided for the construction of the multistory factories, which are leased to industrialists for light manufacturing industry which is usually pollution free.

rojected total welling units	Dwelling units completed in 1984/1985	Dwelling units completed as of 31 March 1985	Dwelling units under construction as of 31 March 1985	Distance from city center (km)
49,500	-	49,480	-	10–14
48,800	160	47,540	-	10–14
24,600	1,380	1,380	6,400	8–10
24,800	11,010	15,230	10,110	14–17
24,500	890	24,050	-	11–13
25,500	4,960	14,800	11,210	9–13
12,800	1,410	12,800	-	14–19
39,500	4,550	14,150	1,810	19–21
28,000	-	28,000	-	5–8
18,000	5,260	7,990	4,460	8–11
45,700	8,240	24,450	6,320	14–18
13,700	-	13,190[a]	-	5–7
36,600	-	36,600	450	6–8
55,000	6,400	15,600	8,020	22–25
40,000	11,390	18,800	17,070	19–21
30,000	-	-	3,740	15–19

APARTMENT DESIGN

The design of HDB apartments is being upgraded continuously to provide better living conditions to the occupants. Recently the design of flats has been streamlined to facilitate mechanized and prefabricated constructions, reducing costs and building time.

Over the years the HDB has introduced many types and designs of apartments for the various personal and family needs and expectations of the people. The HDB offers for sale a choice of nine basic designs comprising three models each for three-room units and four-room units, two models of five-room units and one model for executive apartments. Recently a number of variations have been introduced to give the blocks character and identity.

NEIGHBORHOOD AND TOWN CENTERS

Neighborhood and town centers are established in each housing estate to meet the needs of the residents. Shopping and other communal and recreational facilities are provided. In addition sites are found for the building of schools, homes for the aged, child care centers, community centers and sport complexes to provide the residents with amenities normally available in the downtown areas. The housing estates are self-contained in social activities and have a sense of identity and cohesion.

The architectural treatment of these centers has also been improved pro-

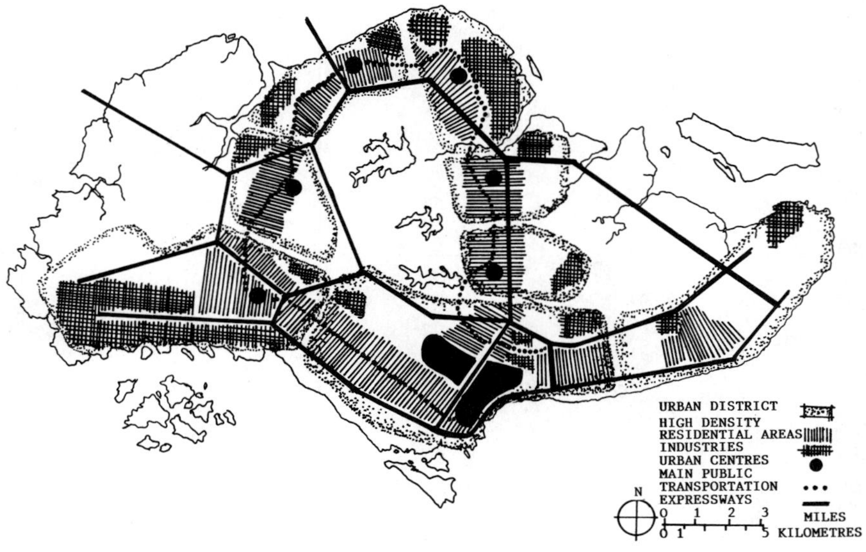

Fig. 1 Urbanized areas in concept plan of Singapore *(Yeh, 1975)*

gressively, and more variety of planning and design is evident. The centers are well patronized by residents especially in the evenings and on weekends.

INDUSTRIAL DEVELOPMENT

As part of the overall industrialization program, clean industry is allowed to be developed near the housing estates to provide employment opportunities for the residents. Siting of factories nearer to housing development will minimize the time required to travel to work, which is particularly advantageous to working mothers, as domestic arrangements may be made for young children to be looked after by neighbors or relatives. In this way the estates do not merely function as dormitories, but become vital communities with their own economic development.

LANDSCAPING

Well-landscaped parks, gardens, and recreation areas are provided in the housing estates to beautify the environment. Children's playgrounds are located near the housing blocks, and gardens and parks are decorated with colorful flowering trees. More recently fruit trees of different varieties have also been planted. The "green city" image of Singapore is enhanced by the foliage, color and fragrance of the vegetation, and the landscaping modifies and softens the otherwise regimental appearance of the estates.

FINANCING LOW-COST HOUSING

A major housing policy in Singapore is the "home ownership for the people" scheme. Purchasers are permitted to use the Central Provident Fund (CPF), which is a social security contribution, as down-payment and monthly installment of the apartments. The combined CPF contributed both by the employee and the employer is 50% of the salary of the employee, divided equally between these two parties with the upper limit of S$1,500 for the employee. It is held by the CPF Board and earns interest. The CPF contribution is primarily for old age pension and hospitalization expenses but also is available for the financing of apartments.

This "save-as-you-live" system, though not too different from pension or superannuation schemes in most countries, is unique in that it provides financial resources for housing through compulsory saving. While take-home pay may be reduced by the large percentage of contribution to the CPF, the financing readily available when the buyer is ready to take over the flat (after a waiting period of about three years) is such that little additional expenses are necessary for occupation. A worker who earns S$800 per month would have enough CPF put away after three years to meet the initial financial purchase requirements without other sources of finances. The monthly installments may then be met by CPF contribution without using take-home pay.

RESALE OF FLAT

Provisions are made for residents to sell their apartments after five years of residence to enable them to upgrade their accommodations. The apartments may be sold to the Board or to private buyers. Owners may also trade their apartments for larger ones within the five-year period, and transfer fees are payable.

Before September 22, 1982, owners who did not apply for a second apartment after reselling the first one did not have to pay the transfer fee, but were barred for 30 months from applying for another apartment. After this date, flat owners are permitted to sell their apartments to private buyers without the 30 month restriction, thus allowing the sellers to apply for another apartment immediately. In return for the concession, sellers are to pay the transfer fee on the resale of the first apartments. The second and subsequent apartment must be resold to the HDB at current posted prices.

The policy is aimed at discouraging flat owners from making excessive profit from the resale of government subsidized housing. The new transfer fees are 10% of the resale price for three-room units, 15% for four-room and 20% for five-room units. Previously the fee was fixed at 5% for all units. The new fees apply to all owners reselling their first apartments to private buyers regardless of whether they are applying for a second apartment. Flat owners would still make a substantial profit even with the higher transfer fee charge. The expected profits of resale of HDB flats are shown in Table 3.

The Board's resale policy is to encourage upward mobility. Owners who

Table 3 Resale of flats

	Ghim Moh		
	3-room (Improved)	4-room	5-room
Selling price			
(average present market price)	$43,000	$75,000	$129,000
Less what is paid for flat (about 6–8 years)	$13,500	$21,500	$35,500
Less what the HDB recovers at			
3-room – 10%	$4,300		
4-room – 15%		$11,250	
5-room – 20%			$25,800
	$25,200	$42,250	$67,700
Less what is paid for renovation/improvement	$4,000	$6,500	$9,000
Net Profit	$21,200	$35,750	$58,700

have occupied their first apartment for 5 years are allowed to sell it to a buyer of their choice at market prices and with the profits buy another subsidized unit from the Board. These factors can continue under the new rule but the requirement that second and subsequent units be sold back to the HDB at current posted prices will ensure that buyers make large profits only once. The HDB may exercise its right to buy back a unit if its owner intentionally under-declares its value when selling to a private buyer to pay less in transfer fee and stamp duty.

Systems such as the one described above are the results of modification of the Board's policy according to changing circumstances without altering the general principle of providing the population with low-cost housing that may be upgraded without giving rise to profiteering. The Board is conscious of the changing needs of the society and amends its *modus operandi* to suit the operation of the society.

HIGH-RISE LIVING

The Board's acceptance of building high-rise apartments from the beginning has structured the living pattern of Singaporeans. In the early 1960s only 10% of the population lived in high-rise buildings; now some 80% or more are housed in public and private high-rise apartments.

Some have criticized the rapid shift from low-rise to high-rise living with the period of only slightly over 2 decades. However before the HDB embarked upon the high-rise programs, the majority of the population lived in substandard quarters, including squatter settlements. The much improved accommodations provided by the Board, were considered far superior to those in existence at the time. It was hoped that the populace would then adjust its lifestyle accordingly.

To answer the question of whether the residents have accepted high-rise

Tao Payoh			Bedok/Chai Choe			Marine Parade		
3-room (Improved)	4-room	5-room	3-room (Improved)	4-room	5-room	3-room (Improved)	4-room	5-room
$37,000	$72,000	$118,000	$42,000	$80,000	$120,000	$48,500	$100,000	$140,000
$7,800	$18,500	$30,000	$11,800	$18,500	$30,000	$13,500	$21,500	$35,500
$3,700			$4,200			$4,850		
	$10,800			$12,000			$15,000	
		$23,600			$24,000			$28,000
$25,500	$42,700	$64,400	$26,000	$49,500	$66,000	$30,150	$63,500	$76,500
$4,000	$6,500	$9,500	$4,000	$6,500	$9,000	$4,000	$6,500	$9,000
$21,500	$36,200	$54,900	$22,000	$43,000	$57,000	$26,150	$57,000	$67,500

living as a way of life, a survey conducted in 1973 indicated that 56.8% of residents were willing to live no higher than the fifth story. In 1981 the percentage dropped to a minority of 46.3%. On the other hand the percentage of residents willing to live on the tenth story and higher has increased from 13.1% in 1973 to 20% in 1981 (Table 4). As to the preferred highest story, 35% of the residents preferred to live no higher than the fifth story. In 1977 this dropped to 29.7% and in 1981 to 22.3%, while the percentage of residents prepared to live above the tenth story increased to 47.3% in 1981 from 35.7% in 1977 and 27.9% in 1973 (Table 5).

MAINTENANCE

Emergency services of the HDB estates are managed by the Essential Maintenance Service Unit (EMSU). The 24-hr service mounted by the unit handles an average of 1,400 telephone calls a day. The performance of the unit in a crisis was tested on February 5, 1983 during the island-wide power failure. Of the 864 elevators checked, the majority were made operational quickly with the help of the automatic rescue device installed in them. Residents are now more confident with the manner by which the Board responds to emergency operations. Except for major blackouts, there is now minimal disturbance to the everyday life of the residents.

PUBLIC SECURITY

A survey made by the HDB in 1981 indicated that the residents found the public security in their neighborhood to be satisfactory, and 92% of the

Table 4 Preferred story of flat (%) *(Housing and Development Board, 1982)*

1973		1981	
1–5th story	56.8	1–5th story	46.3
6–9th story	29.9	6–9th story	33.7
10th and above	13.1	10th and above	20.0
	99.8		100.0

Table 5 Highest story prepared to live on (%) *(Housing and Development Board, 1982)*

	1973	1977	1981
1–5th story	35.1	29.7	22.3
6–9th story	37.0	34.5	30.3
10th and above	27.9	35.7	47.3
	100.0	99.9	99.9

residents perceived their neighborhood as a safe place to live. When asked whether they felt safe to leave their home vacant during the day, 82.8% responded in the positive, 80% felt safe to leave their house empty at night, and 64% felt safe to leave their home empty for several days consecutively.

While the crime rates in the HDB estates are relatively low, the design of the apartments have been modified to minimize breakins by eliminating direct access to apartments from staircase landings. The neighborhood watch scheme organized by the residents supplements police patrol of the housing estate.

SOCIAL INTERACTION

Social exchange among neighbors, including greetings and casual conversation, is considered necessary to foster the neighborhood spirit. A survey conducted in 1981 reported that 65% of the residents had such social exchange with at least five neighboring households, 29% of the residents were acquainted with one to four households, and only 6% said that they did not know any of their neighbors.

Familiarity with neighbors improves considerably with length of stay. 44% of the residents were reported to be acquainted with at least five neighbors, 58% for those who stayed for 2 to 3 years, and about 72% for those who stayed for more than 3 years (Table 6). Also residents with relatives within walking distance of their homes are considerably more likely to know more neighbors than those without relatives nearby. For example 86% of the residents with five or more relatives living nearby were reported to know at least five neighbors compared to 69% of those who did not have any relatives (Table 7). The circle of neighbors is wider if there are relatives of various age groups who live nearby.

The multiracial society of Singapore is integrated well within the HDB estates. Singapore's 2.6 million people are comprised of 77% Chinese, 15% Malay, 6% Indians and 2% other ethnic groups. These proportions are repeated in the estates in which the residents have found interracial familiarity. Nearly 80% of Malay residents were reported to know at least one Chinese

Table 6 Number of neighbors known by length of stay

Number of neighbors known	Length of stay: 1 year or less	2–3 years	4–5 years	More than 5 years
	%	%	%	%
None	10.8	5.7	3.6	6.1
1–4	45.2	36.0	24.2	22.1
5 or more	44.0	58.3	72.2	71.8
Total	100.0	100.0	100.0	100.0

neighbor, and 46% at least one Indian neighbor. Among Chinese residents, 30% were reported to know at least one Malay neighbor and 14% know at least one Indian neighbor. Among Indian residents 74% were reported to know at least one Chinese neighbor and 56% at least one Malay neighbor (Table 8).

The survey helps alleviate the notion that high-rise living tends to isolate neighbors and restrict intercommunal relationship. The HDB's decision to group related households in the same block in cases of resettlement also minimizes isolation.

RESIDENTS COMMITTEES

The effect of residents' committees (RC) is shown also in the 1981 survey. The proportion of the residents knowing at least five neighbors was 65% if they lived in neighborhoods with RC, compared with 56% for those living in neighborhoods without RC (Table 9). The Community Centers (CC) and the RC, sponsored by the Government, and working closely with the HDB, are useful links between the residents and the management of the estates.

Table 7 Percentage distribution of households by number of neighbors known and presence of relatives in the neighborhood

	Presence of relatives in the neighborhood			
Number of neighbors known	None	1–2	3–5	More than 5
None	7.1	5.7	6.0	5.7
1–4	23.9	29.0	21.1	8.2
5 or more	69.0	65.3	72.9	86.1
Total	100.0	100.0	100.0	100.0

Table 8 Percentage distribution of households by inter-ethnic familiarity and ethnic group

	Ethnic Group			
Percentage Distribution of residents	Chinese	Malay	Indian	Others
Percentage of residents knowing at least one Chinese neighbor	NA	79.0	74.3	66.8
Percentage of residents knowing at least one Malay neighbor	30.2	NA	57.3	47.2
Percentage of residents knowing at least one Indian neighbor	14.0	45.7	NA	29.2
Percentage of residents knowing at least one Eurasian neighbor	2.4	10.2	11.4	NA

Readers must bear in mind that the opportunities to know individuals of other ethnic groups are determined by the proportion of ethnic distribution as a whole.

CLASS SEGREGATION

Setting up the Housing and Urban Development Cooperation (HUDC) for middle-class housing resulted in distinct neighborhoods from the lower income housing estates. The distinctiveness of the HUDC estates in design and construction gave the impression of class segregation. The unintended yet undesirable effects were point out by the Prime Minister Mr. Lee Kuan Yew in 1981 (Housing and Development Board, 1982). The Board took measures to rectify the situation immediately. Consequently the middle-class income housing work is carried out by the HDB, which incorporated the operation of the HUDC, and the low-income housing is integrated with the middle-income housing.

While class segregation continues between the population living in public housing estates and those living in the various types of private housing, the price ranges between the more expensive public housing and the less expensive private housing are now merging, giving the public more choice. The large numbers of the population now living in public housing tend to identify themselves with the estates in which they live with a sense of belonging through communal activities.

FAMILY SIZE

The high-rise living practiced in Singapore is not designed for large families, since the units consist of one to three bedrooms only, with limited living and dining spaces. The traditional extended family system is by and large nonexistent, and the nuclear family system similar to that of the West prevails. Several nuclear families living together in the traditional way was as much a matter of culture as a matter of necessity when accommodation for single families was unavailable.

The social support system of the elderly may decline as a result of the

Table 9 Percentage distribution of households by number of neighbors known, presence of residents' committee in the neighborhood and how long ago residents' committee was formed

	Whether there is a residents' commitee				
		Yes, formed			
Number of neighbors known	No	Less than 6 months ago	6–12 months ago	13–24 months ago	More than 24 months ago
None	9.1	4.5	5.3	6.4	4.5
1–4	34.8	31.8	27.5	27.9	27.4
5 or more	56.1	63.7	67.2	65.7	68.1
Total	100.0	100.0	100.0	100.0	100.0

development of the nuclear family system. The government, through its social programs and moral education, encourages filial piety and care of the aged to meet the needs of the social system, which may be impaired by the change of the family structure. At the same time the government has restructured the CPF scheme for medical and welfare services to ensure that senior citizens will be able to afford the medical services they need after retirement, and the aged without family support will have the necessary welfare.

SOCIAL REGULATIONS

The social behavior of the residents is modified by the regulations of the HDB. Antisocial behavior such as vandalism, throwing objects from apartment windows, harboring illegal aliens, or use of the apartment for criminal purposes, could result in the eviction of tenants from rented premises or compulsory acquisition of sold properties. The offenders may also be barred from renting or purchasing HDB apartments again for a period of five years. Thus to some extent the HDB has also become the monitor of social behavior as it affects the community.

THERMAL COMFORT

The moderate-to-high temperature and solar radiation, high humidity, and slight to moderate winds are typical features of the Singapore climate. While the daily temperatures are not as high as those in the hot arid region, discomfort is experienced mainly because of high humidity. Discomfort is further aggravated by solar radiation, which may be intense during certain periods, even in partially cloudy sky conditions. To overcome thermal discomfort the inhabitants of the region traditionally wear light clothing, and cooling is provided by air movement around the body, which loses heat by the evaporation of perspiration.

The study of thermal comfort in the tropics has been carried out in a number of countries including India, Australia, and Papua New Guinea. In Singapore the work began in the 1950s chiefly by C. G. Webb (1960) and Dr. E. P. Ellis (1953). Webb constructed an *Equatorial Comfort Index* (ECI), which could be read from a nomogram for a certain environmental conditions given by the dry and wet bulb temperatures, air movement, and radiation if it is present. A neutral point of ECI of 26°C (78.7°F) was found whereby the observers did not feel warm or cool. Webb's nomogram is shown in Fig. 2.

The work of Webb and Ellis shows that some difference of thermal comfort exists between dwellers in the tropics and in other parts of the world. Webb's ECI differs from the summer optimum effective temperatures normally used in other countries, and Ellis's finding indicates the difference between British and Asians in their acclimatization in Singapore.

Recent work by Lim and Rao (1977) in Singapore followed Webb's ECI and led Lim to recommend that Webb's ECI could be used for Singapore. The neutral ECI may be determined at 25°C (77°F), which is 1°C lower than that derived by Webb (26°C). This small margin may be because Webb used only 14 subjects. It may also be that in the 20 years or so during which Singaporeans

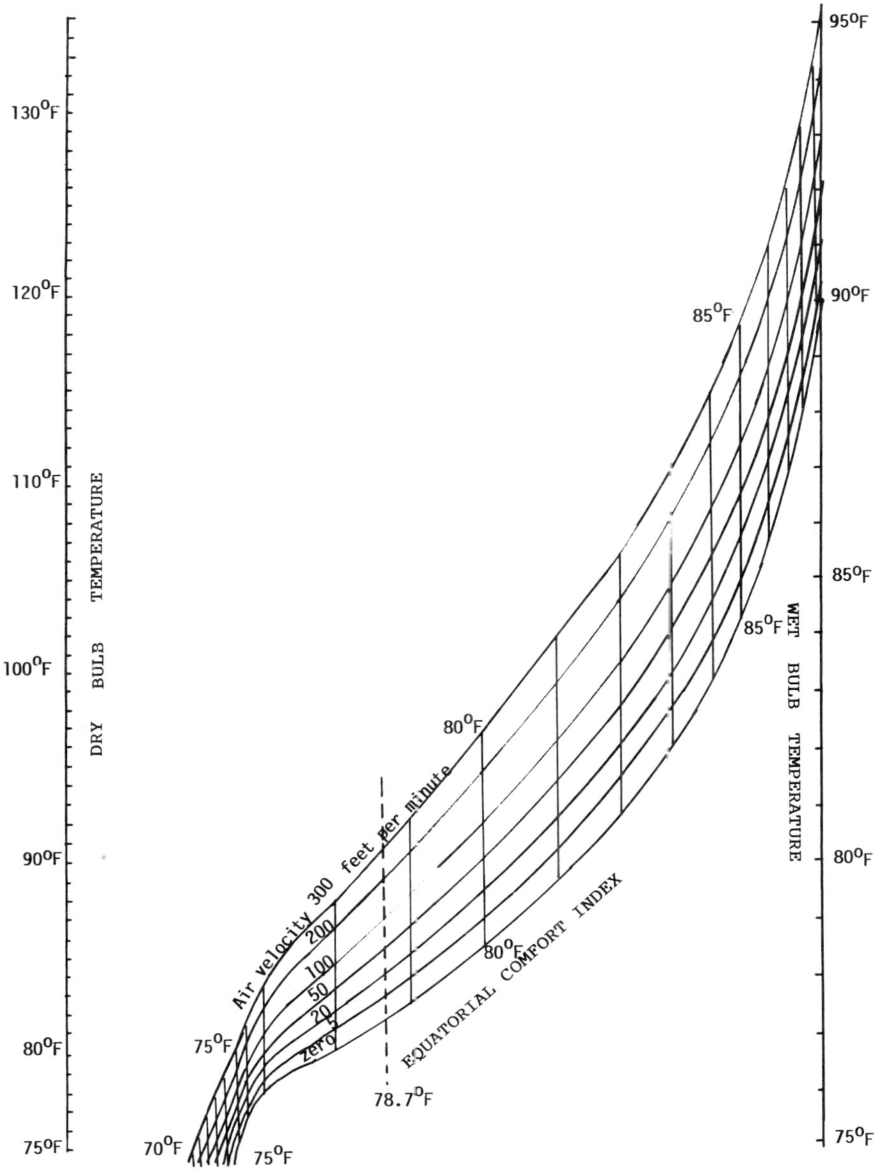

Fig. 2 Equatorial comfort index nomogram

have experienced air-conditioning, there is a slight shift of the neutral point to the cooler side. Whatever the reason, I am of the opinion that for sendentary workers in the equatorial climate with light clothing, an ECI (corrected for radiation if necessary) of 25°C is a possible design optimum.

Experience of inhabitants in Singapore show that, provided that the dwellings are well-ventilated to remove excess heat caused by people and appliances, the residents with light clothing may be kept reasonably comfortable by means of fans.

VENTILATION

It is well known that the wind speed at some height above ground level is normally higher than that near the ground due to the fractional loss to terrains, buildings, and vegetation that reduces the wind flow. Consequently high-rise apartments receive more breezes than low-rise ones. It follows that for a certain density, the land coverage of high-rise apartments will be less than that of the low-rise ones as the spacing between high-rise buildings is wider. An unobstructed air flow due to wide spacing between buildings gives better ventilation through the apartments.

Dutt (1984) demonstrated by wind tunnel tests that typical apartments of the HDB could give wind velocity coefficients between 0.12 and 0.36 at level 1, 0.24 to 0.40 at level 7 when the wind was at right angle to the building, and 0.17 to 0.45 at level 1, and 0.31 to 0.51 at level 7 of the same building with the wind coming at the same direction but at 45° to the building facade. As cooling breezes are necessary at all times in the hot humid region, it is evident that high-rise apartments have better thermal environments (Fig. 3).

Adequate ventilation not only provides a healthy and odorless environment, it also removes heat in a given space due to sensible and latent heat given out by occupants and appliances. In the hot humid tropics the provision of natural cross-ventilation through the living space could keep the indoor temperature rise to a minimum provided that solar radiation is absent from the living space. Lim, Rao, and Rao (1980) found that the increase of ventilation from two to ten air changes per hour could minimize the increase of the indoor temperature to about 1°C above the outdoor air temperature. As ten air changes per hour are normally available in high-rise buildings, the indoor temperature of the living space is not likely to rise above that of the outdoor temperature to more than 1°C or so (Fig. 4).

NOISE IN HOUSING ESTATES

In a recent sample survey conducted by the Department of Mechanical Engineering of the University of Singapore (Lim et al., 1980), sound pressure levels were taken in 2150 flats and distributed in eight housing estates in

Singapore. Table 10 shows the correlation between the type of flats, unit floor area per person, and the average sound pressure levels There is some correlation between the floor areas, the number of residents, and the noise levels. The smaller the flats, the less built-up area per person, the higher the noise levels. Table 11 shows the distribution of sound pressure levels measured.

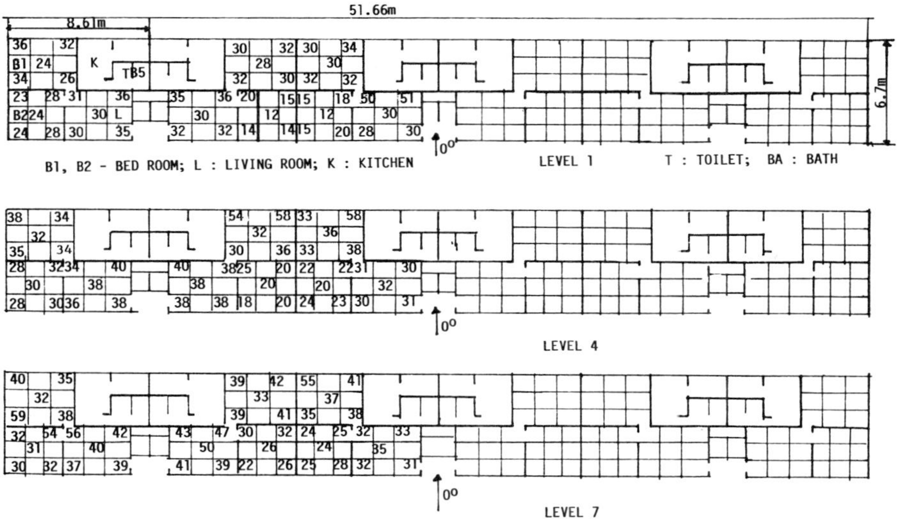

Fig. 3A Wind velocity coefficients in the apartments at level, 1, 4, and 7; 0=0° × 100

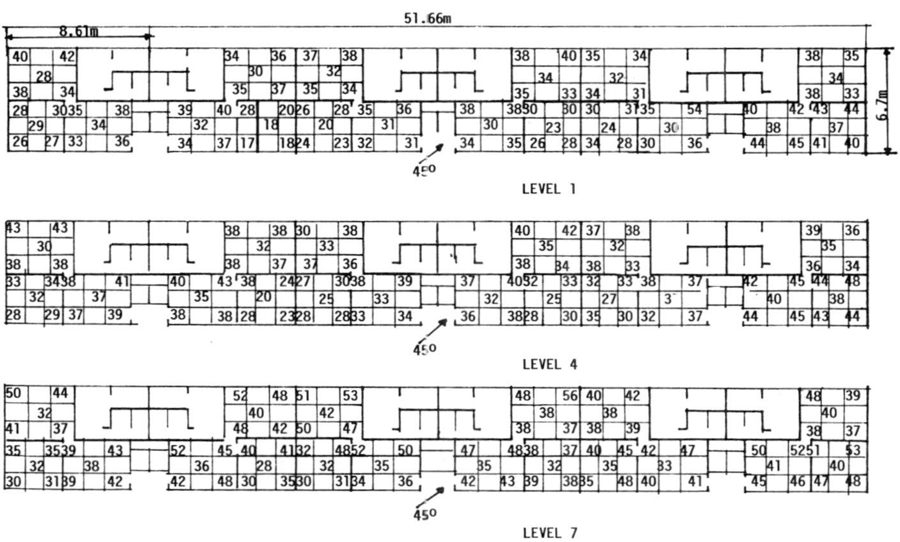

Fig. 3B Wind velocity coefficients in the apartments at level 1, 4 and 7; 0=45°

More than half of the noise levels were found to be between 61 dB and 55 dB. Table 12 shows the rating of the noise by the residents.

From Tables 10 and 12 it may be concluded that the residents, by and large find the flats moderately noisy at about 61db–55db, meaning possibly that some degree of noisiness is recognized. Table 13 shows the types of noise

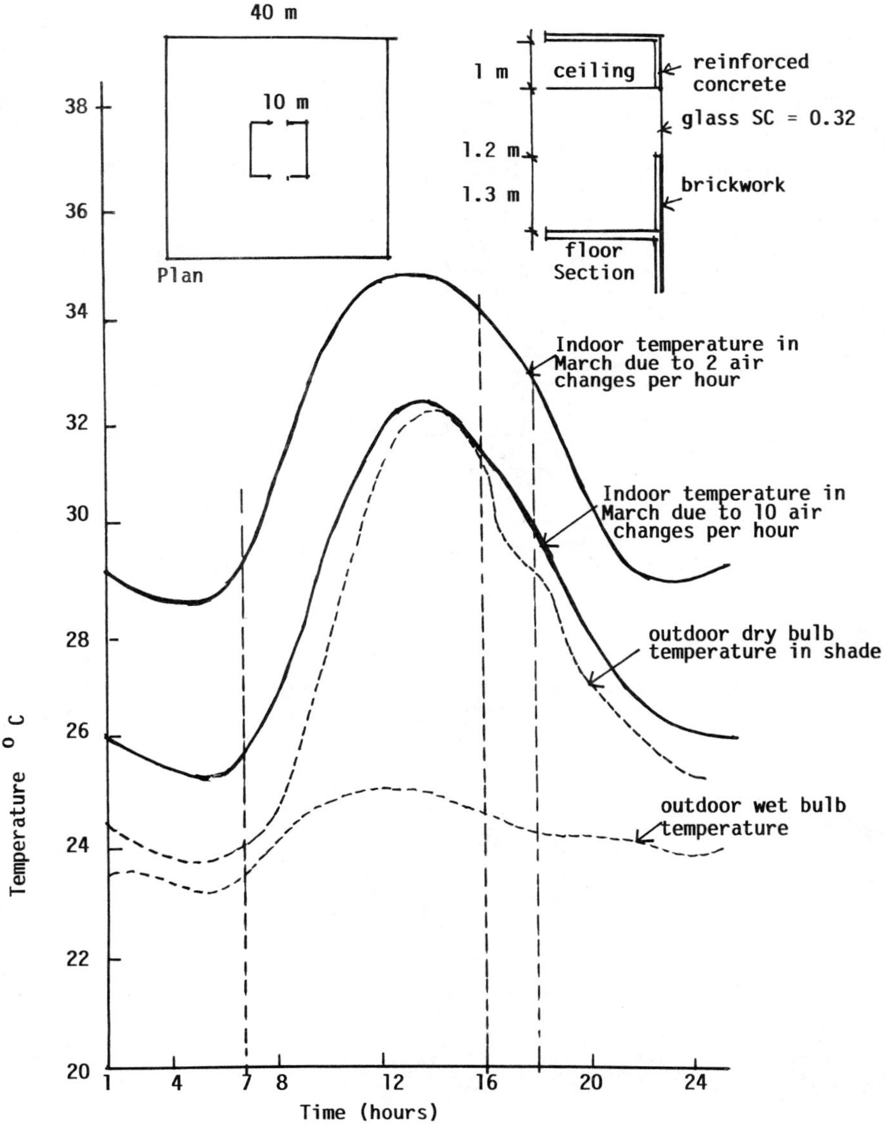

Fig. 4 Indoor temperature variation due to ventilation of typical floor of hypothetical building (not air conditioned)

Table 10 Average sound pressure levels in housing estates

Data	Flat types[a] (No. of living and bed rooms)				
	1	2	3	4	5
Net area per flat m^2	33	45	60 (improved) 69 (new) 65 (average)	83 (improved) 93 (new) 88 (average)	123 (point block) 127 (slab block) 125 (average)
Gross area per flat m^2	42	65	69 (improved) 98 (new) 79 (average)	99 (improved) 110 (new) 105 (average)	140 (point block) 144 (slab block) 142 (average)
Built-up area per person (m^2/person):					
Net area basis	11.0	13.6	15.1	19.6	32.9
Gross area basis	14.0	19.7	18.4	23.3	33.0
Sound Pressure Level* dBA	65	64	63	59	60

[a]These sound pressure levels are averages of SPL's taken at 8 housing estates.

Table 11 Distribution of sound pressure levels dbA

	dbA ≤ 55	dbA 56–60	dbA 61–65	dbA 66–70	dbA 71–75	dbA ≥ 76	Total
No. of readings	106	343	1112	420	115	46	2142
Percentage	4.95	16.01	51.91	19.61	5.37	2.151	100%

Table 12 Degree of noisiness in Flats

Survey	Very Quiet	Quiet	Moderate	Noisy	Very Noisy	Total
No. of Responses	50	330	1038	567	96	2081
Percentage	2.40	15.86	49.88	27.25	4.61	100%

Table 13 Types of noise in housing estate

Type of noise	Traffic	Air-craft	School	Radio or TV	Children at play	Others	Total
No. of readings	825	170	16	216	920	195	2342
Percentage	35.23	7.26	0.68	9.22	39.28	8.33	100%

Table 14 Periods of noisiness in housing estates

Periods	Morning	Afternoon	Evening	Night	Total
No. of responses	510	484	913	436	2343
Percentage	21.77	20.66	38.97	18.61	100%

identified in flats. Traffic and children at play are the main sources of noise in housing estates. The noisy periods are shown in Table 14, showing that generally the most noisy periods are in the evenings when the traffic is at its busiest and the children are at play.

In some flats sound pressure levels L_{10} and L_{90} were measured. (L_{10} is the sound pressure level reached or exceeded for 10% of the time, and L_{90} is the sound pressure level reached or exceeded for 90% the time). It can be seen that if 60dBA is regarded as the average or normal noise level in housing estates (L_{50}), L_{10} would be about 5dBA higher (See Fig. 5).

It may be concluded that the noise level in housing estates should be somewhat reduced. The survey so far does not give any indication of the desirable noise level but it may be assumed the reduction of 5 dBA would improve the situation considerably. These may be done by reducing the length of corridor space outside the flats to minimize external noise and to improve privacy, reducing traffic noise by vehicle control and proper architectural planning, and reducing population density to achieve lower background noise levels.

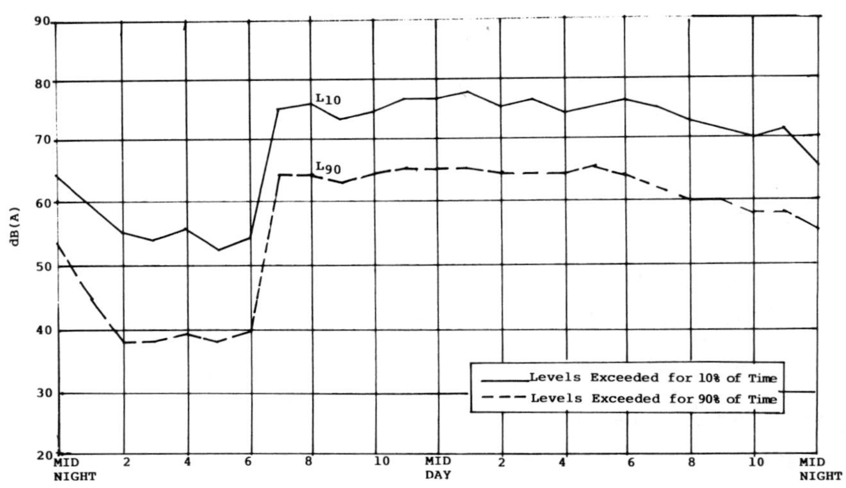

Fig. 5 Sound pressure levels at a typical residence over a 24-hour period

CONCLUSIONS

The success of the HDB is largely due to the national commitment to low cost housing for the population, a dedicated leadership, and a well structured organization. Proper social and economic planning and management provides a framework for low cost housing. The transition from substandard housing to a more orderly and permanent living environment in which modern conveniences are provided has been satisfactory, showing the adaptation of the residents to the changing environment. The commitment to high-rise housing development, while presenting some problems, has been shown to be a reasonable choice in view of the high density requirements in the small island republic. Environmental factors such as ventilation, noise, and thermal comfort are found to be generally acceptable. The HDB will continue to upgrade the planning and design of the housing estates to provide a proper built environment for the population.

NOTES

Singapore	136.8 km north of Equator Area: 618 km^2 Population: 2,471,800 (1983)
HUDC	Housing and Urban Development Corporation in charge of middle-income housing, now included in the activities of HDB.
JTC	Jurong Town Corporation, a statutory organization in the charge of industrial development. It also constructs middle-income housing.
PUB	Public Utilities Board that supplies electricity, gas and water to the households.
Currency	U$ 1 = $2.2

REFERENCES/BIBLIOGRAPHY

Dutt, A. J., 1984
ENVIRONMENTAL EFFECT OF WIND FOR NATURAL VENTILATION IN APARTMENT BLOCK IN SINGAPORE, International Journal of Housing Science, Vol. 8, No. 3, July, pp. 259-267.

Ellis, E. P., 1953
THERMAL COMFORT IN WARM AND HUMID ATMOSPHERES, Journal of Hygiene, Vol. 51.

Housing and Development Board
ANNUAL REPORT 1984/85, Housing and Development Board, National Development Building, Maxwell Road, Singapore 0106, Republic of Singapore.

Housing and Development Board, 1982
OUR HOME, A Housing and Development Board Publication, June.

Lim, B. P. and Rao, K. R., 1977
ENVIRONMENTAL CONTROL OF BUILDING, Journal of the Singapore National Academy of Science, Vol. 6, No. 1, p. 78 ff.

Lim, B. P., Rao, K. R. and Rao, S. P., 1980*a*
AIR-CONDITIONING IN SINGAPORE—IS IT NECESSARY?, Proceedings of the Conference on Energy Conservation and Management in Buildings, Institute of Engineering, Singapore, March, pp. 108-138.

Lim. B. P., et al., 1980*b*
DEVELOPING NOISE CRITERIA FOR SINGAPCRE, Seminar on Noise: Problems and Control, The Science Council of Singapore et al., January, Singapore, pp. 48-74.

Webb, C. G., 1960
THERMAL DISCOMFORT IN AN EQUATORIAL CLIMATE, Journal of the Institute of Heating and Ventilation Engineering, January.

Yeh, S. H. K., 1975
PUBLIC HOUSING IN SINGAPORE, Singapore University Press.

You, P. S. and Lim, C. Y., 1984
SINGAPORE: TWENTY-FIVE YEARS OF DEVELOPMENT Nan Yan Xing Shou Liahne Zaobao, Singapore, p. 323.

The Urban Ecology of Tall Buildings

Leonard I. Ruchelman

Ecology can be defined as the study of the relationships among objects and organisms within an environment. It is the study not of the creatures and objects themselves, but rather of the relationships among them and the functions they perform in their setting (Burgess, 1925; Hoyt, 1939; Harris and Ullman, 1945). Using ecological reasoning as a framework, buildings must be viewed not only as products in their own right, but as integral parts of the larger urban environment. To a greater extent than ever before, buildings are being judged not only in terms of their individual design and utility, but also on the basis of whether or not they are good neighbors.

This poses a special challenge to the tall building industry. As more and more cities become exposed to intensive forms of high-rise development, an important question is how it affects ecological patterns. Because of its height and size, a tall building is likely to have a greater impact on the social and physical environment than would be the case for a low- or mid-rise building. Where a building upsets existing functional relationships, or where it fits awkwardly into the neighborhood setting, it is likely to be a source of controversy. Indeed, a growing number of city halls in America and other countries are being pressured to slow or otherwise restrict high-rise building booms.

The purpose of this paper is to account for the kind of issues being generated by tall building development in the United States. Through a better understanding of ecological factors that come into play, it is believed that tall building decision makers (planners, architects, developers, engineers, and public officials) could better accommodate the urban environment, thus making for more livable cities. In this way, many tall building issues could be moderated or avoided entirely.

TALL BUILDINGS AND THE ECOLOGICAL CHARACTER OF CITIES

In the ecological landscape of most cities, there can usually be found certain features to which residents attach symbolic value. In a study of land use in central Boston, Walter Firey (1945) observed that many acres of valuable land in the central business district had been allowed to remain in uneconomic use such as parks or cemeteries. He theorized that sentiment and symbolism play an important part in determining spatial distributions, pointing out that the 48-acre Common in the heart of downtown Boston had never been developed commercially and that Beacon Hill had largely remained an upper-class residential area in spite of its proximity to the central business district.

Similarly, citizens may react to tall building development on the basis of perceived effects on valued ecological traits such as open space, the skyline, views, transportation patterns, or historical and cultural landmarks. In contrast to European societies, which tend to reject tall buildings in the historical centers of their cities, Americans put their skyscrapers in the central business districts. Thus, the question of how to best fit tall buildings into the surrounding urban environment is usually a more complicated issue in American cities.

This issue was raised in San Francisco where the board of supervisors recently approved a new zoning law designed to limit the height, size, and number of office towers that can be erected in a section of the downtown covering about three quarters of a square mile. The new zoning law will also protect 251 architecturally or historically significant buildings from demolition and offer partial protection for more than two hundred other structures. The plan is intended to quell complaints of San Francisco residents that their city is being "Manhattanized"—in other words, their scenic vista of steep hills and low white buildings is being obscured and replaced by a skyline of look-alike towers.

In contrast to San Francisco, Philadelphia is a city that has decided to break with tradition and allow its skyline to go higher. The Philadelphia City Council has approved construction of two 60-story office structures that will rise 112 m (367 ft) above the statute of William Penn that sits atop the 152-m (500 ft) City Hall. The approval came only after an emotional effort to stop the project by citizens who claimed "Bill Penn's nose" was, by custom and good sense, the highest that buildings there should rise. An unwritten law kept developers from building higher than 150 m (491 ft), leaving Penn's statue a visible monument for the city. By agreeing to establish an eight-block skyscraper zone—where the sky would be the limit—Mayor W. Wilson Good and the city council claimed it would pump billons of new dollars into the downtown area and create thousands of jobs. Thus, the city's need for new business won out over tradition. (In addition, developer Willard Rouse succeeded in giving assurance that the quality of Philadelphia's downtown would be enhanced and that many more people would be attracted into the district.)

Los Angeles represents a more complex case of a city searching for its identity. At the turn of the century, the city passed a law that limited buildings to thirteen stories. The purpose was to encourage development of an urban environment that would be different from the smokey, high-rise dominated cities of the East. This law, however, contributed to a vast sprawling city that did not have a real downtown or a central core. After the law was repealed in 1957, a number of glittering new buildings ranging up to sixty-two stories were constructed downtown. They helped create thousands of office jobs and brought new vitality to the once-decaying sector. Now, however, the booming downtown work force is causing traffic and parking problems. Critics want to impose a moratorium on high-rise buildings to avert traffic gridlock that they believe is bound to develop at the present rate of development. There are esthetic concerns as well. Some public officials want to restore the 13-story limit in the financial district because of concern that proposed skyscrapers will dwarf buildings erected when the limit was in effect.

In comparison to the aforementioned cities, New York has always encouraged skyscraper development as a vital part of the city's economy. City officials, however, have been subject to protests over the enormous size of some of the projects, especially those planned for mid-town Manhattan, in Times Square, and the site of the New York Coliseum. The basic issue is how to preserve the traditional character of that part of Manhattan, which serves as an entertainment center in addition to performing office and retail functions. The critics worry that the theaters and specialty shops may disappear along with the bright lights and active street life that give Manhattan its special quality. So far, plans for the development of Times Square have been revised many times.

TALL BUILDINGS AND THE ECOLOGY OF NEIGHBORHOODS

Neighborhoods can be defined as urban subcommunities the boundaries of which are defined by residents on the basis of shared interests. Though they overlap, the ecology of neighborhoods can be distinguished from the ecology of the city to the extent that neighborhood residents are more inclined to react to the more proximate intrusions of new land uses that are likely to affect them directly. Where a new high-rise structure is proposed, people in the immediate area are likely to respond on the basis of whether they expect the quality of their lives to improve or decline (Downs, 1981).

A basic factor that influences this response is concern over street effects. Outside of the home, streets are the most important part of the neighborhood environment (Appleyard, 1980). Yet, in many instances, modern high-rise buildings have made surrounding streets impersonal and uncomfortable places in which to be. William H. Whyte (1980), who is director of the Street Life Project in New York City, explains that where this happens, social interaction and ultimately a sense of community are seriously impaired. For example, he views Detroit's Renaissance Center and Atlanta's Omni International as being antistreet and antineighborhood. These structures were built to be quite independent of their surroundings; their enclosing walls are blank, windowless, and, to the street, they turn an almost solid face of concrete or brick.

In addition, many high-rise buildings favor the automobile as the prime means of transportation. They ignore the fact that the older neighborhoods of most central cities were originally designed to accommodate pedestrians and are essentially walking neighborhoods. At Houston Center (Fig. 1), an individual can drive in from the freeway to the center's parking garage, walk through a skyway to another tower, work through the day, and then return to the garage and the freeway without ever having set foot on Houston Streets at all. At the street level of Houston Center there are neither stores nor many people. The sole retail activity is a drive-in bank, and the only acknowledgement of pedestrians are flashing lights and signs telling them to watch out for cars (Whyte, 1980; Scully, 1985).

According to Downs (1981), other factors that neighborhood residents are likely to be concerned about are as follows:

- The withdrawal of a key local institution such as a hospital, a school, a church, a park, or a shopping center
- The transition to either higher-income or lower-income occupancy
- Negative effects on the street level such as too much shadow, wind currents, or increased congestion
- An increase in transient uses
- A decline in mixed uses
- An increase in automobile traffic
- The quality of building maintenance and appearance
- The adequacy of public and private services

FUTURE CONSIDERATIONS

In the future, it is likely that tall building development will take on new dimensions in height, scale, and functions. Not only will buildings be made taller and larger, but the surrounding environment will in many instances be entirely reconstituted. An important consequence is that the old distinctions

Fig. 1 Houston Center, Houston, Texas *(Courtesy: Walter P. Moore and Associates)*

between the public and private sectors will be diluted if not nullified entirely, and private developers will be expected to assume greater responsibility for the provision of public needs as well as private needs.

This view is portended by Donald J. Trump's recent proposal to build the world's tallest building—a 150 story obelisk-topped triangle—as part of a mammoth development on the Upper West Side waterfront of Manhattan. Described by Trump as "a tremendous city within a tremendous city," the complex would be constructed on a 100-acre site and would include housing for 20,000 people, 160 thousand m^2 (1.7 million ft^2) of shopping space, and more than 40 acres of parks and public spaces. Another developer, Samuel LeFrak, has submitted a plan for the building of "Newport City" in New Jersey that would include 9,000 apartments, a huge shopping center and more than 370 thousand m^2 (4 million ft^2) of offices on the western bank of the Hudson River. Mr. LeFrak heralded his proposed development, which provides for miniaturized versions of Central Park, Broadway, and Riverside Drive, as "probably the largest job that has ever been built since the pyramids."

Such grandiose forms of development must be sensitive to pertinent ecological factors if they are to promote the health and vitality of community and neighborhood life. The following considerations are offered as guidelines for the development team:

Provide for the public as well as the private needs of the population groups being served by large scale projects. These needs should be identified as part of the overall planning process.

Acknowledge the symbolic attachments of residents in determining spatial distributions and land use, including such ecological factors as open space, views, skyline effects, and historical and cultural landmarks.

Preserve old buildings wherever possible and integrate them with the new buildings. Old buildings not only provide space for new low-cost enterprises, but they also break the visual monotony and perpetuate the historical and architectural character of the community.

Provide for a mixture of neighborhood functions—residence, work, shopping, entertainment, leisure—to the extent that it insures the presence of people and activity in public and private places during different times of the day and night.

Account for street effects in the design plan of any large building, including a consideration of shadows, wind currents, congestion, and the availability of amenities likely to appeal to the pedestrian. Streets should be maintained as lively and interesting places (Fig. 2).

Preserve and possibly improve the range and variety of institutional facilities available to neighborhood residents including schools, hospitals, day-care centers, transportation, and recreation centers.

REFERENCES/BIBLIOGRAPHY

Appleyard, D., 1980
LIVABLE STREETS, University of California Press, Berkeley, California.

Burgess, E. W., 1925
THE GROWTH OF THE CITY, The City, University of Chicago Press, Chicago, Il., p. 47-62.

Downs, A., 1981
NEIGHBORHOODS AND URBAN DEVELOPMENT, The Brookings Institution, Washington, DC.

Firey, W., 1945
SENTIMENT AND SYMBOLISM AS ECOLOGICAL VARIABLES American Sociological Review, Vol. 10.

Harris, C. and Ullman, E., 1945
THE NATURE OF CITIES, Annals of the American Academy of Political and Social Science, Vol. 252.

Hoyt, H., 1939
THE STRUCTURE AND GROWTH OF RESIDENTIAL NEIGHBORHOODS IN AMERICAN CITIES, U.S. Federal Housing Administration, U.S. Government Printing Office, Washington, DC.

Scully, V., 1985
BUILDINGS WITHOUT SOULS, The New York Times Magazine, September 8.

Whyte, W. H., 1980
THE SOCIAL LIFE OF SMALL OPEN SPACES, The Conservative Foundation, Washington, DC.

Fig. 2 Streets should be maintained as lively and interesting places to be *(Courtesy: New York Convention and Visitors Bureau)*

A Preliminary Model for the Economic Analysis of Tall Buildings

John P. Wenzelberger
Henry Malcolm Steiner

In today's dynamic economic environment, tall building designers, developers, and investors are faced with the very difficult problem of making long-term decisions while faced with constantly changing fiscal policy. Considerable sums of money are involved. Congress has swung back and forth from generous tax incentives to tight taxation of business investments, changing the law in this area 15 times in the last 30 years.

The advent of the microcomputer and powerful software spreadsheet programs now provide the economic analyst with the flexibility needed to examine properly alternative proposals and to conduct sensitivity analyses of design variables while including the effects of different tax strategies and the effects of projected inflation rates. This paper describes such a model and provides a few examples of its application to the economic decisions of tall building design.

LIMITATIONS OF BEFORE-TAX ECONOMIC ANALYSIS

The apparent disparity between the tremendous construction boom of tall buildings in recent years and the lackluster rates of return of before-tax economic analysis of these projects is primarily a result of the income tax incentives provided to investors in these projects.

A proper economic analysis of proposed tall building projects must consider the effects of depreciation methods and financing arrangements when evaluating alternatives. In addition, the effects of inflation on the investment must also be considered. A correct economic decision can be made only when all of the economic factors influencing the investment are included in the analytical model, adding a degree of complexity to the model that has deterred the construction and use of such analyses until very recently. The widespread availability and use of microcomputers and spreadsheet software programs, in the past few years, now makes it possible to create a realistic analytical framework for the analysis of tall building designs from a proper economic viewpoint. Such a model is described in this paper. It has been designed to provide maximum flexibility for alternative investment analysis. The effects of depreciation methods and financing arrangements on taxes are included as well as the effects of inflation. Different time horizons for analysis are easily selected by specifying the year in which the property would be resold.

TAX CHANGES AFFECTING REAL ESTATE INVESTORS

At some point in its life, an asset such as a tall building has deteriorated so much that it is worth nothing to its owner and must be replaced. Because that deterioration occurs over a period of years, tax policy makers have consistently agreed that some part of the asset's initial value should be written off annually, just as a bad debt or an employee's salary should be deducted in computing

Table 1 Business tax changes in the past 30 years

1954	Accelerated depreciation enacted
1962	Enactment of 7% investment tax credit
1964	Corporate tax rate cut from 52% to 50%
1965	Corporate tax rate cut from 50% to 48%
1967	Investment tax credit suspended
1968	Income tax surcharge of 10% imposed
1969	Investment tax credit reinstated
1970	Investment tax credit removed, surcharge reduced, accelerated depreciation reduced
1971	Surcharge removed
1972	Investment tax credit reinstated, depreciation periods shortened
1975	Investment tax credit raised to 10%
1979	Corporate tax cut from 48% to 46%
1981	Depreciation shortened and accelerated
1982	Accelerations restricted
1984	Depreciation periods extended

taxable income. They have disagreed, however, on the number of years constituting an asset's economic life and the rate at which an asset's value should be written off.

As previously mentioned, Congress has vacillated back and forth from generous tax incentives to tight controls on business investment, changing the law in this area 15 times in the last 30 years according to recent articles in the Washington Post (1984, 1985*a*, 1985*b*). A brief summary of these changes is listed in Table 1.

Perhaps the most dramatic recent change was the introduction in 1981 of the Accelerated Cost Recovery System (ACRS). Ever since the introduction of ACRS, and in anticipation of the tax "pendulum" swinging the other way, tall building development and construction has been booming.

THE NEED FOR A DYNAMIC MODEL

An analytical model capable of properly analyzing the effects of this rapidly changing economic environment must consider the dynamic situation facing developers and investors who are making long-term decisions involving considerable sums of money, while facing a future that is very difficult to predict.

A preliminary model designed to meet these needs is described in this paper. It is anticipated that the final model will incorporate suggested improvements and the results of "debugging" While the data to date appears mathematically correct, more time is required to assure that all of the features are operating properly.

DESCRIPTION OF THE MODEL

The Analytical Framework

The model is designed to operate on any microcomputer containing at least 256k of memory and capable of running Lotus$_{TM}$ 1–2–3 software. Lotus$_{TM}$ 1–2–3 was chosen as the software framework for the economic analysis model for several reasons. First, it is a widely used software program and is readily available at most universities and industrial organizations. Second, the program is very powerful and is capable of handling the extensive and complex spreadsheet formulas required by this model. Additionally, it has tremendous flexibility for changing input data and modifying cell formulas. One particularly nice feature is the graphical output function, which allows a pictorial presentation of the results. Coupled with the graphical output function is a data table function that facilitates sensitivity analysis by substituting a column of values repetitively into the model for a prescribed input variable and tabulates the output variables selected for analysis.

The model is designed to compare two alternatives referred to as Alternative I and Alternative II. Each alternative is evaluated separately and the difference in the after-tax, after-inflation cash flows is also evaluated.

METHOD OF ANALYSIS AND BASIC ASSUMPTIONS

The model compares the costs and benefits of the two alternatives using the Present Worth method and the Internal Rate of Return method described by Steiner (1980). The two alternatives must be mutually exclusive and should be ranked in order of increasing initial investment. Since incremental analysis is the only correct method of analyzing mutually exclusive alternatives it is incorporated in this model in the form of Alternative (II-I).

As mentioned earlier, the model assumes that the two alternatives being compared are mutually exclusive, meaning that selecting one alternative precludes the choice of the other. Another important assumption, included in the model, is that all transactions during a certain year may be accumulated to a single sum at the end of the year without seriously affecting the calculations (Steiner, 1980).

MODEL INPUT PARAMETERS

The input parameters used to test the model are shown in Table 2. The input data was extracted from an example presented by Swanson (1975).

TESTING THE MODEL

To test the model properly it is necessary to select realistic data for the input parameters. Swanson (1975) presented such data in his example of a 43-story office building under consideration for construction for a nationally known life insurance company.

Detailed data were presented for both a 43-story office building on one site and an above-ground nine-story parking structure on an adjacent site. Alternative I in the test of this model uses the data presented by Swanson for the 43-story office building. Alternative II assumes that both the office building and the nine-story parking structure will be constructed.

The analysis by Swanson is over a 15-year period after the beginning of rental and assumes no resale of the property. The discounted rate of return for this investment, based on a before-tax analysis and with the above assumptions, is only 2.86%, which is considerably less than that expected by a prudent investor.

The model was first tested by analyzing the net present values of Alternative I, Alternative II, and Alternative (II-I) as a function of discount rates

Table 2 Input Parameters

Parameter	Alternative I	Alternative II
Land		
Square Feet	91,440	152,400
Cost per Square Foot	$75	$75
Total Cost of Land	$6,858,000	$11,430,000
Buildings		
Total Square Feet Constructed	1,360,000	1,360,000
Total Square Feet Rentable	1,156,000	1,156,000
ft^2, Office	700,000	700,000
ft^2, Commercial	456,000	456,000
Parking Spaces	0	1,542
ft^2, Other	0	0
Construction Costs		
$\$/ft^2$, Office	$28	$28
$\$/ft^2$, Commercial	$28	$28
$\$/ft^2$, Apartments	0	0
$/Parking Space	0	$4,202.34
$\$/ft^2$ sq ft, Other	0	0
Direct Building Costs	$38,080,000	$44,560,008
Other Costs		
Demolition	$462,000	$1,212,000
Site Work	$90,000	$180,000
A & E Fees	$4,569,000	$5,347,201
Other Miscellaneous Costs	$4,950,400	$5,792,801
Total Investment	$55,010,000	$68,522,010
Investment Schedule		
Year (−2)	$7,410,000	$7,410,000
Year (−1)	$21,420,000	$34,932,010
Year (0)	$26,180,000	$26,180,000
Financing Schedule		
Building Loan, Year (−2)		
% Cash Invested	100	100
Loan Period (Years)	2	2
Interest Rate (%)	0	0
Building Loan, Year (−1)		
% Cash Invested	100	100
Loan Period (Years)	1	1
Interest Rate (%)	0	0
Mortgage Loan, Year (0)		
% Cash Invested	100	100
Loan Period (Years)	16	16
Interest Rate (%)	0	0
Absorption Schedule		
Average % Occupied (Year 1)		
Office	75	75
Commercial	25	25

(continued)

Table 2 Input Parameters (continued)

Parameter	Alternative I	Alternative II
Apartments	0	0
Parking	0	56
Other	0	0
Average % Occupied (Year 2)		
Office	100	100
Commercial	47	47
Apartments	0	0
Parking	0	77
Other	0	0
Average % Occupied (Year 3)		
Office	100	100
Commercial	63	63
Apartments	0	0
Parking	0	87
Other	0	0
Average % Occupied (Year 4)		
Office	100	100
Commercial	79	79
Apartments	0	0
Parking	0	94
Other	0	0
Average % Occupied (Year 5)		
Office	100	100
Commercial	95	95
Apartments	0	0
Parking	0	100
Other	0	0
Revenue Rates		
Rent		
Offices ($/ft^2/yr)	$9.00	$9.00
Commercial ($/ft^2/yr)	$10.00	$10.00
Apartments ($/ft^2/yr)	$0.00	$0.00
Parking ($/space/yr)	$0.00	$360.00
Other ($/ft^2/yr)	$0.00	$0.00
Operating Expenses		
Operating and Maintenance Costs		
Rented Areas ($/ft^2/yr)	$2.75	$2.75
Unrented Areas ($/ft^2/yr)	$1.75	$1.75
Parking Spaces ($/space/yr)	$0.00	$60.00
Real Estate Taxes		
Before Rental (% of Land Value)	4.62%	4.62%
After Rental ($/ft^2/yr)	$2.15	$2.15
Parking ($/space/yr)	$0.00	$90.00
Projected Rates of Increased Cost and Revenue		
Operating Costs (%/yr)	0.00	0.00
Rental Revenue (%/yr)	0.00	0.00

Table 2 Input Parameters (continued)

Parameter	Alternative I	Alternative II
Property Resale Value (%/yr)	0.00	0.00
Overall Inflation Rate (%/yr)	0.00	0.00
Income Tax Considerations		
Tax Rate (%)	0	0
Depreciation Method (Enter 1, 2, 3, or 4) (SL=1, ACRS81=2, ACRS84=3, ACRS85=4)	2	2
If Straight Line, Enter Years	16	16
Value to be Depreciated	$48,152,000	$57,092,010
Resale Value		
Year Sold	16	16
% of Total Investment Before Inflation	0	0

varying from 0 to 4%. The results correlate with the data in Exhibit 2 of Swanson's paper and are shown in Fig. 1.

Next the effects of resale value on the discounted rate of return, also known as internal rate of return, were examined. Resale values varying from 0 to 100% of the initial total investment were examined and the results are presented in Fig. 2.

To test the sensitivity analysis features of the model, the effects of varying construction costs and land costs were examined. These results are presented in Figs. 3 and 4.

The after-tax analysis features were tested by varying the income tax bracket of a hypothetical building owner and evaluating the resultant internal rates of return, which is illustrated in Fig. 5. Figure 6 assumes that only 10% of the investment, required at years (−2), (−1), and (0), is in the form of cash with the remainder borrowed, for 30 years, at the interest rates shown in Fig. 6. ACRS 81 depreciation is used and it is assumed that the taxpayer is in the 45% bracket. Resale value, at year 16, is assumed to be 15% of the original total investment with the remaining mortgage balance being paid off with the resale over 15%. The resultant internal rates of return indicate that this might not be such a bad idea if proper interest rates can be obtained.

SAMPLE OUTPUTS

Sample outputs of the model test results are presented in Figs. 1 to 6. Below each figure is an explanation of the assumptions and a tabular listing of the models numerical output. This data represents only a small fraction of the model's ability to analyze tall building investment alternatives.

CONCLUSIONS

The preliminary economic model developed for the study and analysis of the economics of tall buildings provides a necessary tool to the tall building designers, developers, and investors faced with the difficult decision process in a dynamic economic environment. It has the required flexibility, can handle hundreds of input variables, incorporates sensitivity analysis features, and utilizes a computer graphics output.

Some of the after-tax features involving taxes on the disposition of the property and recapture of depreciation have not yet been activated, although they are integral parts of the model.

The alternatives chosen to demonstrate and test the model provide a convenient check of the mathematics. However when it comes to a realistic choice of alternatives, the input data is somewhat lacking since Alternative II is Alternative I plus a parking garage. Although this does provide two alternatives for comparison, the economics of the parking garage does not merit the additional investment as shown by the incremental analyses.

REFERENCES/BIBLIOGRAPHY

Conlin, Walter F., Sr., 1972
ECONOMICS OF HIGH RISE BUILDINGS, Proceedings of the International Conference on Planning and Design of Tall Buildings, Vol. 1(a), Lehigh University, Bethlehem, PA, August, pp. 119 to 132.

Grannatt, Milton H., III, 1975
THE ECONOMIC IMPACT OF AN OFFICE BUILDING HEIGHT RESTRICTION, Doctoral Dissertation, Lehigh University, April, pp. 43 to 89.

Steiner, H. M., 1980
PUBLIC AND PRIVATE INVESTMENTS: SOCIOECONOMIC ANALYSIS, John Wiley, New York, pp. x to 414.

Steyert, Richard D., 1972
THE ECONOMICS OF HIGH RISE APARTMENT BUILDINGS, Proceedings of the International Conference on Planning and Design of Tall Buildings, Lehigh University, Bethlehem, PA, August, pp. 103 to 118.

Swanson, Vernon E., 1975
THE ECONOMICS OF HIGH RISE OFFICE BUILDINGS, Pan Pacific Tall Building Conference Proceedings. Paper read before the Regional Conference under the auspices of the Joint Committee on Tall Buildings, University of Hawaii, Honolulu, Hawaii, January, pp. 97 to 110.

The Washington Post, 1984
THE WASHINGTON POST, Washington, DC, 14 October.

The Washington Post, 1985a
THE WASHINGTON POST, Washington, DC, 23 February.

The Washington Post, 1985b
THE WASHINGTON POST, Washington, DC, 1 December.

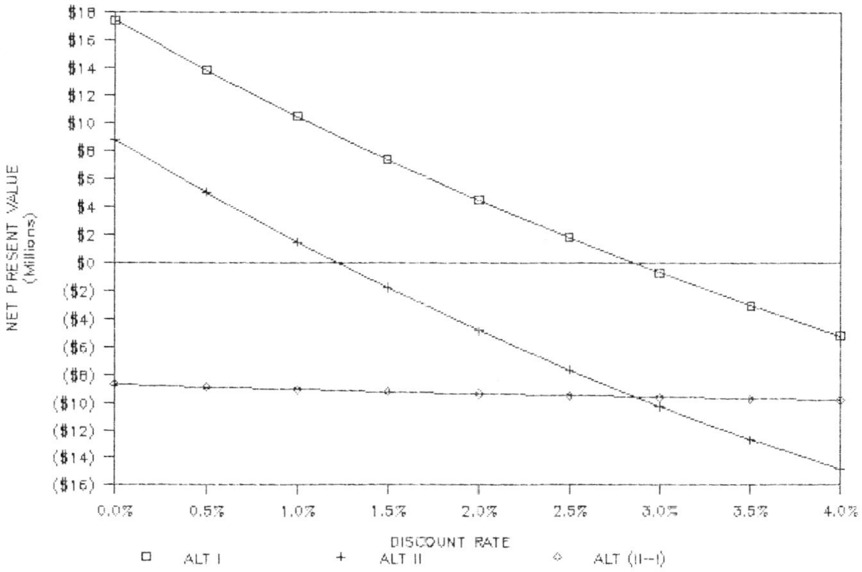

NET PRESENT VALUES ($)

DISCOUNT RATE (%)	ALT I	ALT II	ALT (II-I)
0.0	17,461,794	8,852,418	(8,609,376)
0.5	13,860,862	5,060,858	(8,800,004)
1.0	10,516,754	1,542,340	(8,974,414)
1.5	7,409,812	(1,724,054)	(9,133,866)
2.0	4,522,020	(4,757,496)	(9,279,516)
2.5	1,836,855	(7,575,569)	(9,412,424)
3.0	(660,843)	(10,194,405)	(9,533,562)
3.5	(2,984,996)	(12,628,818)	(9,643,822)
4.0	(5,148,395)	(14,892,420)	(9,744,024)

Fig. 1 Swanson's Scheme 5B

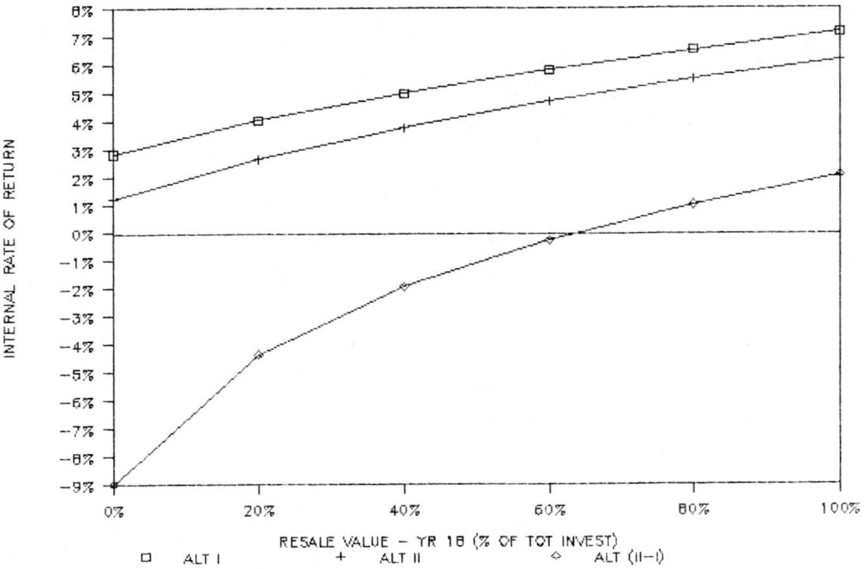

% OF ORIGINAL INVESTMENT	INTERNAL RATE OF RETURN ALT I	ALT II	ALT (II-I)
0	2.86%	1.23%	−8.97%
20	4.06%	2.68%	−4.35%
40	5.02%	3.80%	−1.94%
60	5.84%	4.73%	−0.26%
80	6.54%	5.52%	1.03%
100	7.17%	6.21%	2.09%

Fig. 2 Effect that different resale rates, as a percentage of the initial investment, have on the internal rate of return of the before-tax analysis

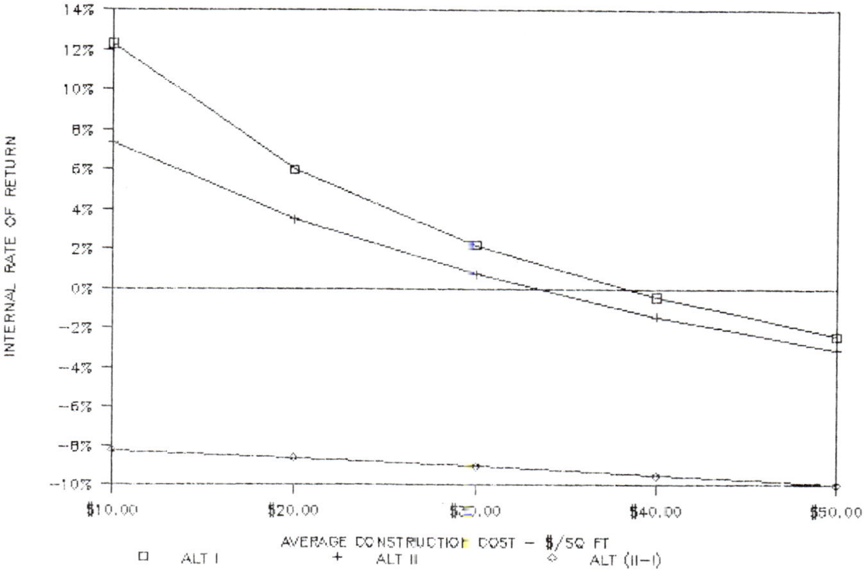

CONSTRUCTION COSTS	INTERNAL RATE OF RETURN ALT I	ALT II	ALT (II-I)
$10.00	12.3[illegible]%	7.37%	−8.22%
$20.00	6.00%	3.51%	−8.63%
$30.00	2.23%	0.74%	−9.06%
$40.00	−0.40%	−1.38%	−9.51%
$50.00	−2.38%	−3.08%	−9.97%

Fig. 3 Effects that different costs of construction have on the before-tax internal rates of return

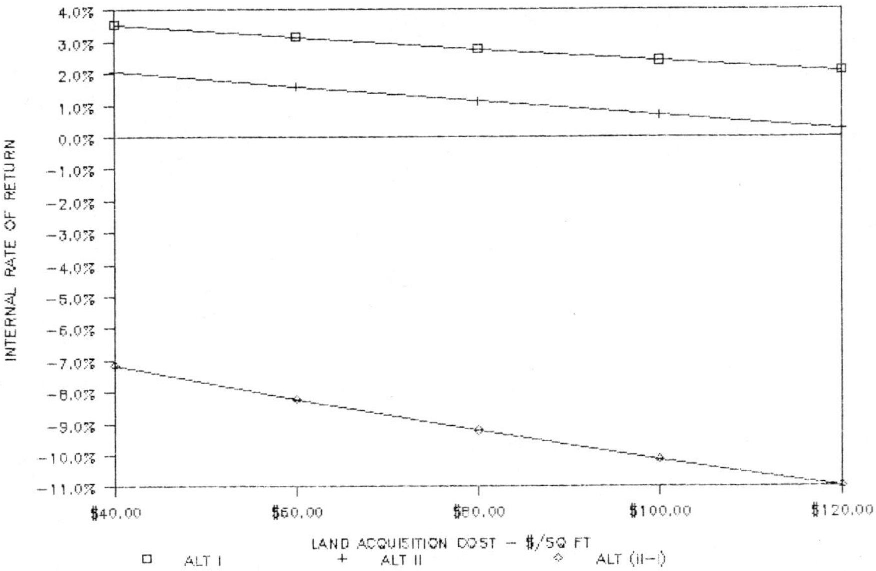

COST OF LAND ($/SQ FT)	INTERNAL RATE OF RETURN ALT I	ALT II	ALT (II-I)
$40.00	3.55%	2.06%	−7.13%
$60.00	3.15%	1.58%	−8.22%
$80.00	2.77%	1.12%	−9.21%
$100.00	2.41%	0.68%	−10.12%
$120.00	2.07%	0.26%	−10.96%

Fig 4 Effects of changes in land costs varying from $40.00/ft^2 to $120.00/ft^2 on before-tax internal rates of return

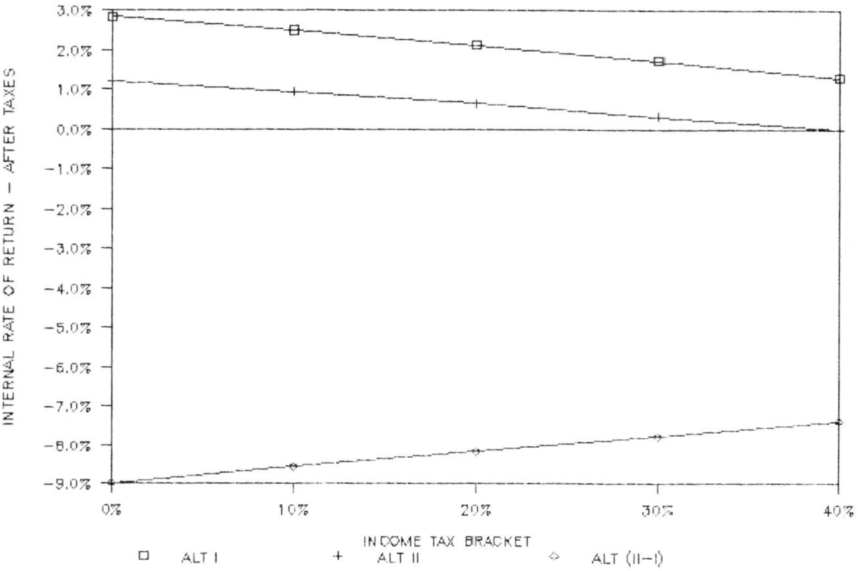

INCOME TAX BRACKET	INTERNAL RATE OF RETURN ALT I	ALT II	ALT (II-I)
0%	2.86%	1.23%	−8.97%
10%	2.51%	0.95%	−8.57%
20%	2.13%	0.65%	−8.17%
30%	1.73%	0.33%	−7.77%
40%	1.30%	−0.01%	−7.36%

Fig. 5 Effects of changes in income tax rates. Assumption is made that the property is held for 16 years with no resale value. Depreciation is by the ACRS 81 method and internal rates of return are calculated for income tax brackets from 0 to 40%.

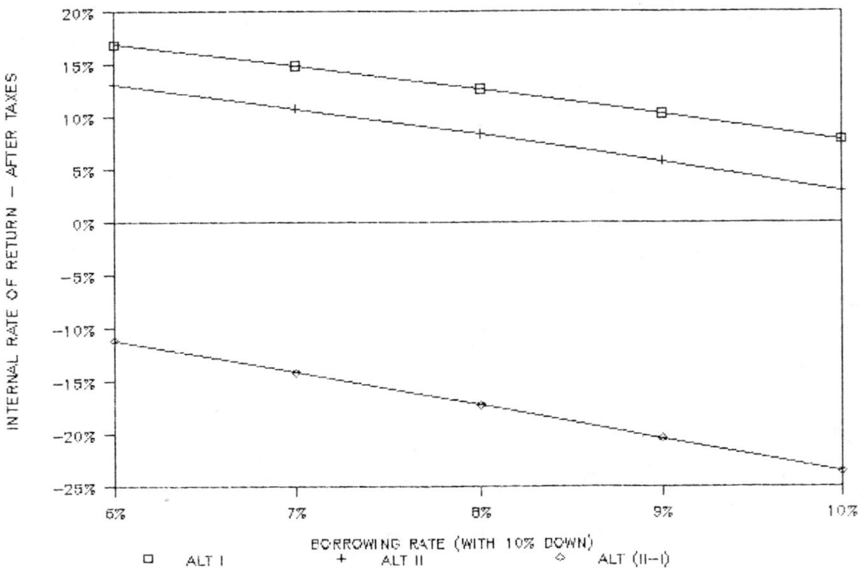

INTEREST RATE	INTERNAL RATE OF RETURN ALT I	ALT II	ALT (II-I)
6%	16.97%	13.15%	−11.06%
7%	14.91%	10.88%	−14.11%
8%	12.71%	8.44%	−17.21%
9%	10.37%	5.84%	−20.35%
10%	7.89%	3.07%	−23.52%

Fig. 6 Effects of borrowing interest rates. Assumption is that only 10% of the total investment is paid in cash and that the remainder is financed, using a 30-year mortgage, at an average interest rate as shown. It is also assumed that the property is held for 16 years with all but 15% of the resale value used to pay off the mortgage balance at year 16. The tax rate is 45%, and depreciation is by the ACRS 81 method.

Architecture and Society

Paul Goldberger

In no arena more than in skyscraper design, Chicago is truly our preeminent city; to hold the Council's Third International Conference in Chicago was as correct and appropriate in every way as it would be to hold a conference on Renaissance painting in Florence. Chicago is the heart of the skyscraper culture—not, to be sure, the only place in which it flourishes, or even the place in which it necessarily flourishes most intensely today. But the city is most intimately associated with the skyscraper's birth, and as we look ahead to a second century, the same perspective is not available from anywhere else.

Not only in Chicago, but now in every American city, the skyscraper is at once the triumphant symbol and the unwelcome intruder. We seem, after a full century of them, still not to be at peace with tall buildings; they shatter scale and steal light, and it is no surprise to hear them denounced as monstrous constructions. Yet we also hold them dear. What brownstone has ever been the symbol of New York that the Empire State Building is (Fig. 1)? What lakefront park the icon of Chicago that Sears Tower has become? To visitors and natives alike, these buildings are these cities, as much as the Cathedral of Notre Dame is Paris or the Houses of Parliament are London. They are absolutely critical to the identities of our cities. Indeed, we might say that in many cities, the skyline is the image—that the body of skyscrapers a city possesses is, collectively, that city's symbol, more than any individual building could be.

Although the skyscraper has taken on an international presence in the last generation, it is still a fundamentally American phenomenon. In the sky-

Fig. 1 Empire State Building *(Courtesy: AISC)*

scraper we see, more clearly and directly than in any other architectural form of our time, the merging of technology, energy, and commerce, that rampant capitalist spirit that seems so particularly American. The skyscraper tends to be spoken of primarily in American terms with full knowledge of the immense spread of skyscraper construction around the world and with great respect for much of it. The fundamental ideas and issues skyscrapers seem to raise are ones that come most clearly into focus on the American stage.

It may no longer be possible to say the same thing now that the extraordinary building by Norman Foster for the Hongkong and Shanghai Bank is complete in Hong Kong or when the I. M. Pei tower, also in Hong Kong, is finished. But for now, given that most of the major towers around the world do not carry either the air or the science of skyscraper design significantly beyond their American counterparts, this paper mainly discusses the issues as they appear in this country.

CHANGING VIEWS

The skyscraper's double-barreled identity—its ability to be at once a triumphant symbol and an unwelcome intruder—is nothing new to our time; it has been the skyscraper's fate for much of its existence. It is not difficult to find objections to early tall buildings, going back even before the turn of the century. Just after this century began, in 1908 the architect David Knickerbocker Boyd wrote in *American Architect and Building News* that "Aside from all the esthetic considerations the continued erection of the so-called 'skyscraper,' the excessively tall building, constitutes a menace to public health and safety and an offence which must be stopped."

Plenty of others held similar sentiments. Architect George B. Post complained even earlier, in 1894, that streets, if lined with tall buildings, would seem like canyons, dark, gloomy and damp. Ernest Flagg, despite his status as architect of the Singer Building, one of the finest of the first generation of Manhattan towers, expressed the hope that some kind of system could be developed in which neighboring owners could buy and sell air rights, thus limiting the actual number of high buildings and assuring that a particular district did not become overcrowded. And Montgomery Schuyler, the distinguished architecture critic, worried frequently in print about the absence of any legal or other stimulus to lead real estate developers to restrict the size and design of skyscrapers.

Schuyler wrote of his concern that "the aspiring dollar-hunter would continue to protrude stark parallelepipeds"—that is his word, and his word alone—"into the empyrean, just as he does now. . . . The parallelepiped is the form which gives him the most space for rental and which can be most cheaply built. To prevent him from building it would seem to him a great outrage."

But this was hardly the only kind of view that was expressed. A few months before, the *New York Times* had written "So swiftly do the wheels of progress revolve in New York that one great achievement may not be finished before another and more wonderful improvement is on the way. It is so with the two tallest skyscrapers ever constructed, and which are in the course of construction here," making reference to the nearly completed Singer Tower by Ernest Flagg and the recent start of construction of the 213-m (700-ft) Metropolitan Life Tower by Napolean LeBrun, which would eclipse Singer as the world's tallest.

The skyscraper has thus always been the source of debate, at least so far as its social and urbanistic implications. We have never made complete peace with it, yet neither has it been consistently an enemy. It is both feared and admired, the source both of dismay and exhilaration.

What, then, is different about skyscrapers today? What makes the question of the skyscraper and its social significance different now, as we begin the second century of tall buildings, from what it was in the first century?

The most significant issue, surely, is quantity. A generation or more ago one had to go to New York or Chicago to see tall buildings in any significant number. They were the sign that one had arrived from the hinterlands and reached the big city. This is no longer the case. Skyscrapers are everywhere, in small and medium-sized cities as well as large ones. And in the large cities, where there were once a few towers of particular height, now dozens stand, cheek-by-jowl.

It is difficult to overestimate the effect of this on one's perception. The tall building is no longer a truly special thing. If there are skyscrapers all over the place, and so densely packed in major downtowns, what do they signify? It is surely something different from the day in which the Woolworth Building (Fig. 2) suggested the new prominence of commerce in national life, standing as a powerful symbol of it, or the time in which the Chrysler Building (Fig. 3) and the Empire State Building stood as perfect symbols of the jazz age. The tall building is now commonplace, and this cannot fail to reduce its effect on public consciousness.

QUESTION OF HEIGHT

Without addressing the related issue that the explosion of skyscraper construction raises (the question of what tower after tower does to the fabric of our cities), another issue more closely connected to the individual skyscraper, the question of height, is important. Here, too, nothing is so startling as it once was. Once it was 152 m (500 ft), then 213 m (700 ft) then 242 m (792 ft)—the Woolworth—then 305 m (1,000 ft) (Chrysler reached that threshhold first), and on and up. The numbers have not continued to mount in recent years the way they did in the 1930s and again in the early 1970s with the World Trade Center and then Sears Tower, although there is again talk of another leap. But

Fig. 2 Woolworth Building *(Courtesy: Library of Congress)*

Fig. 3 Chrysler Building *(Courtesy: Library of Congress)*

in terms of height the issue seems again to be one of quantity, in that there are not many buildings continuing to cross the 100-story barrier, but so many are being built at 60, 70, even 80 stories. Structures of these sizes, which once had the power to stop us in our tracks, now rate barely a glance.

For example, the RCA Building at 30 Rockefeller Plaza in New York, the centerpiece of Rockefeller Center and in some ways the finest commercial office tower of the Twentieth century, for generations held sway over the imaginations of architects, urbanists and, most important, the general public. There is nothing exceptional about its 65-story height, and while its design is remarkable, it is almost lost amidst the chaos of mid-Manhattan, saved only by the breathing space of the low buildings of the Rockefeller Center grouping that surround it.

What do architects do when tall buildings are so commonplace that they can no longer hold the power over our imaginations that they once did? There are two separate ways in which to move.

DESIGN

The first route is one of design—to emphasize the building as an esthetic object, and to make it stand out in a way it would not otherwise do, given the competition on the high-rise front. The beginnings of this go back for at least a decade, since the banal and chilly glass boxes of the modernist generation began to fall out of favor, and architects began to search for more distinctive visual forms.

It is always a bit odd to stand in the city of Mies van der Rohe and speak of the failings of the glass box, of the international style which Mies came to symbolize, but of course the international style was always much more than Mies, and most of what it was after Mies was not very good. It is no disrespect to Mies van der Rohe to speak of Third Avenue in New York or La Defense in Paris; these are testaments to the absence of any real humanistic impulse in a certain kind of large-scale, commercial modernism. They are without sensual meaning and without urbanistic coherence. It is small wonder that by the late 1960s and early 1970s the most sensitive architects were beginning to search for ways to make large-scale, tall buildings that spoke a somewhat different language.

THE SOCIAL SKYSCRAPER

One of the ways reaction began against the banality and austerity of the international style was an attempt to make the tall building a more appealing social presence in the city, to integrate it into the economic and social life of a city in which most international style towers, which were so determined to stand aloof, could not do. The leader of this generation, the model, is a building too often forgotten these days; it is Johnson and Burgee's IDS Center in Minneapolis, which contains a glass-enclosed court, a kind of roofed town square lined with shops and a hotel. Now, with even Citicorp Center in New York almost a decade old, such mixed-use projects are old-hat, but they were an important component of the reaction against the international style.

The social skyscraper—the skyscraper that is part of a mixed-use project, the skyscraper that contains a public plaza or atrium or retail space or a cultural facility—is a significant advance in our time. A great many of these buildings emerged out of zoning laws that granted excessive bonuses in exchange for the provision of these social amenities, thus making the buildings far larger than they should have been. The presence of social amenities has thus become expected in large-scale building in our time. We no longer expect the skyscraper to be an isolated element so far as the living patterns of the city are concerned, existing only for people to live and work in during a set period of hours. Those who neither live nor work in a major tower now expect to have some sort of involvement with it, and this must be considered a good thing, whatever the architectural results.

POSTMODERNISM MOVEMENT

The reaction to orthodox modernism has been manifested most strongly, at least so far as the recent esthetic history of the skyscraper is concerned, in the movement that is known by the term *postmodernism.* It has brought a generation of skyscrapers that rely heavily on historical architectural elements, sometimes taken literally, more often reinterpreted, sometimes put together into an eclectic mixture, sometimes used in a more narrow stylistic framework. Philip Johnson and John Burgee's AT&T Building in New York, the notorious Chippendale skyscraper, stands as a kind of symbolic parent of this generation of buildings, and its notoriety has made it the most important, although it is by no means the best. Johnson and Burgee's own Transco Tower in Houston and Republic Bank Tower in Houston are significantly better, as are many of the buildings by Kohn Pedersen Fox; Skidmore, Owings & Merrill; Cesar Pelli and numerous others who have come in recent years to follow a similar path of allusion to historical form.

Not the least of the benefits of this movement is its restoration of the idea that a skyscraper should have a top, that it deserves a beginning, a middle, and an end. Louis Sullivan, whose buildings were flat-roofed but by virtue of their extraordinary cornices made the same point, understood this, of course, but most modern architects since Sullivan (and, to be fair, Frank Lloyd Wright) did not. Whatever else can be said about post-modernism, it should always be praised for restoring the tops to towers—for recognizing that the way in which a skyscraper meets the sky can be as important as the way in which it meets the ground, that the profile a tall building makes on the skyline can be as important as the impression it makes close up. The Chippendale top of AT&T is hardly the top of the Chrysler Building, but it has served a valuable polemical purpose, and for this it has earned a certain place in history.

ROMANTIC MODERNISM

There are other ways in which architects have attempted to break away from the boredom and banality that turned out to be the sad legacy of architecture that in Mies's hands could yield greatness, but in the hands of so many others yielded much less. Some—and I think here most particularly of Kevin Roche, Cesar Pelli and Edward Larrabee Barnes—have stayed within the modernist vocabulary, but made it less rational, less dogmatic, less rigid, even more picturesque, using the modernist vocabulary of sleek surfaces to what might almost be called postmodern ends, seeking pure visual pleasure above certain rationalist goals. This approach is something I have elsewhere called the *computer esthetic,* for these utterly sleek, smooth buildings seem not so much to have been constructed as to have been whirred out of some microchip. The sense of metal and glass placed one piece onto another, which in Mies's buildings is as clear a sign of the feeling of construction as the sense of stone in a much older building, seems to disappear.

The historicist strain and the computer esthetic, which others, most notably Charles Jencks, have called *late modernism,* have shown signs of coming together in the last few years in numerous buildings that employ sleek, modern materials, but use them to echo historical form. Cesar Pelli's World Financial Center towers at Battery Park City in New York, Helmut Jahn's addition to the Board of Trade in Chicago, perhaps Johnson and Burgee's Transco Tower, and surely much of the work of Kohn Pedersen Fox, all are examples of this phenomenon. It is something I think is best called *romantic modernism,* and as such it seems to express the feeling of this moment best—a time in which we want to be romantic as opposed to rationalist as the international style appears to have been, and yet a time in which we do not want to cut all ties to modernism either. Romantic modernism does not deny the heritage of modernism; indeed, it exploits it through a knowing and willing use of modernist materials, technology, and engineering. Yet it seeks to merge this with at least some of the romantic esthetic that history presents us with.

None of this, of course, is a prescription for how to design. Nothing is worse than seeing style in this fashion, or talking about it as if there could be prescriptions. No one style either guarantees good architecture or prevents it. Much deeper, much more difficult things determine the esthetic success or failure of a tall building. Proportions, scale, texture, materials are entirely specific, and they depend entirely upon how they are used in a particular situation.

A building succeeds or fails esthetically not on the basis of its style, but on the basis of much more fundamental concepts—how good its elevations are, how good its plans are, how well scaled, how well proportioned it is. The success or failure of these elements are what makes Michael Graves's Humana Building in Louisville basically a good building, whatever its problems, and Graves's Portland Building in Oregon, which preceded Humana and in many ways laid the groundwork for it not so good a building. The basics, the fundamentals of architecture, are not handled nearly as well. It all reminds us that architecture is an art of specifics always; it is never an art of generalities, and never a question of rules or formulas.

ESTHETIC STATEMENTS

In each case here the building is mentioned as if it were a single element in the city, disconnected from everything that is around it—a pure sculptural object, as it were. Unfortunately, all too often that is precisely how architects and developers see buildings. Even the buildings that make certain social gestures toward their surroundings in the form of public space tend to be aloof and isolated as formal objects. The reason for this is clear. The one real problem that the resurgence of interest in the skyscraper esthetic has brought us is the tendency to want to make every building a foreground object, to believe that each and every building must stand out in a way that is all its own, to be a kind of prima donna on the landscape. Prima donnas do not go very well next to each other; you cannot make a whole opera out of them. But now, architecture is marketed by real estate developers, who proudly fill their ads with architects' names and talk of their structures as "significant architectural events." When that is being done, there is little care about what is next door. Each man to his own, each block on its own, each building a thing unto itself.

It is an odd price indeed to pay for having a public and a commercial marketplace become interested in architecture. It is not what I would have expected back when so many of us cried out for years for more interesting architecture, as real estate developers tended to produce only the most banal and dreary skyscrapers; now that they are confirmed converts to architecture, we are suffering an unexpected fate, that of having to cope with all kinds of buildings that seem desperate to make an esthetic statement, and which shriek excess at us all the time.

This issue of the building as a kind of prima donna is critical because, contrary to the impression so much of our architecture today gives, no building in a city really does stand alone. Every tall building is but a building block in a larger composition, and that composition is the city. We should be coming to realize as we move into the second century, that the tower is not as envisioned by Le Corbusier and even by Frank Lloyd Wright, as a proud and separate thing; it is part of a larger whole, connected to what is around it both sociologically and visually. It cannot be seen apart.

A COHERENT WHOLE

Buildings did seem to make a coherent whole for the first 50 years or so of the skyscraper's first century, but perhaps this was almost unconscious. It was not out of a real knowledge of the problem. For much of the first half-century of skyscraper construction, there was a highly consistent vocabulary of materials, mainly masonry. While there were significant stylistic differences, most particularly between the more structurally expressive Chicago School and the more decorative, theatrical New York school, the common vocabulary of materials tended to obscure these differences. All can see it clearly in the many instances in which Chicago-like skyscrapers were built in New York, and New York-esque skyscrapers went up in Chicago. In neither case were they a jarring presence.

Beyond common materials was a common sense of scale. Even when towers were permitted to grow very large, as at Cass Gilbert's Woolworth Building, the scale was not overwhelming, and it was able to render the building compatible with much smaller structures adjacent. A third reason the city seemed to be coherent was the utter and complete respect for the street line. Virtually all construction was built out to the street, keeping an even line; think of Park Avenue in New York, of Michigan Avenue in Chicago, of virtually every downtown all around the country.

All of these things began to fall apart in the post World War II era. I do not want to fall into the trap of blaming orthodox modernism for all esthetic and urbanistic problems, but it is difficult not to consider it highly culpable here. The common vocabulary of materials was the first to go. At the beginning its loss was actually quite pleasing, even exhilerating. How dull Park Avenue had begun to look by 1950, and how exciting, how full of promise of a new age, did it look when Lever House's glass slab came to it instead! No one could know at that point how poorly glass worked in terms of making an entire city, how it could not yield the kind of texture and scale that is necessary to make a city of background as well as foreground structures.

And so scale, too, began to slip away, considered less important by orthodox modernism. The signs that relate parts of a building to each other and to the size of the human figure were lost in an onrush of abstraction, in a desperate search for pure, sculptural form. Finally, after the splendid plaza of the Seagram Building opened in 1958, the fallacious belief came that because this plaza worked well, then plazas everywhere were a good idea, and the street wall meant nothing at all. But the mid-1960s, the sense of urban coherence, the kind of unwritten contract that had brought buildings together, had begun to fall apart. Its demise was hastened by the revised zoning code New York adopted in 1961, which specifically encouraged the breakdown of the street wall.

BACKGROUND AND FOREGROUND

We are now in a period of reaction to all of this. Respect for the street wall is coming back, as well as respect for scale and texture, factors that are absolutely critical to the esthetic success of any tall building, and which are vastly more important than style. But we are only beginning to understand that the problem really is one of background and foreground, one of making cities which are wholes and not merely disparate, competing parts. In any good city the whole is something much more than the sum of the parts, but in too many of our cities, the whole is not more at all. It is vastly less.

There is no better example of this than the first scheme for the Upper West Side of Manhattan, the project called Television City, which included six 76-story towers and one 150-story tower which, if built, would have been the tallest skyscraper in the world. It has a certain excitement to it. Who could fail to be moved, even today, by the words "the tallest building in the world"? For the entire history of skyscraper construction, height has had a power over architects, builders, everyone. To build taller seemed, for so long, to be the goal, like winning a race, and not only like winning it, but like winning it better than anyone had won it before. One generation could produce the four-minute mile, the next could produce a miler who could run it in 3:50, and so it would go, from 80 stories to 110 now isn't the logical thing to go on to 150, just as we keep trying to run the mile faster and faster, keep on shooting for the moon, keep on trying to do everything?

GOING TALLER

Every one of you knows that such a building is plausible technologically, and indeed, that even taller buildings than 150 stories could be built. The structure is not the problem. But the whole analogy of the race, of the record, of getting bigger with each generation is false. It gets us far away from architecture, far away from engineering, and into something else altogether. There is a grave problem with a 150-story tower, despite the allure it undoubtedly has, despite its ability to hold sway over our imaginations.

For if architecture and the building of cities mean anything, they have to do with making civilized places for people to live in, use, be inspired by, be uplifted by. The proposed 150-story building for New York does cause the heart to beat faster for a moment, and granted credit is due for that. But I fail to see where building 150 stories worth of condominium apartments in a tower that, by virtue of its vast bulk must contain 2,600 separate apartment units, will be anything other than a Buck Rogers fantasy. And while Buck Rogers may be fun to contemplate for amusement, in real life—which the middle of New York City all too certainly is—it would be more of a science fiction nightmare.

It would seem like a nice leap—a wonderful way, in fact, to commemorate the beginning of the skyscraper's second century—to be able to make this jump in magnitude to an entirely different kind of building. And Helmut Jahn's plan, which would put the building on a large, relatively open site, makes more sense than many earlier schemes for buildings of this great height in denser parts of Manhattan. But these things provide only momentary appeal; in real life, such a building would be otherwise, a case of technology ability completely and entirely outpacing common sense. Because we would build it, I am not convinced that it would deepen and enrich the experience of urban life at all.

If anything, the quest for the 150-story building comes directly out of the numbness as a result of skyscraper glut. The excess of tall buildings has, by now, made it so difficult to become excited by any of them. It is almost as if we need stronger and stronger drugs to stimulate us, so numb have we become to the drama and excitement tall towers can provide. If we build this building, we are conceding a kind of addiction to technological determinism and to the thrill of height. We are allowing these things to become far more important than other things that make up a true city.

My own opposition to this building project, which has since been replaced by a revised design that includes a 130-story tower, has not come without anguish. It would be pleasing to be able to endorse with enthusiasm a chance to push the frontier onward. But that is just the point. I am no longer sure that such a building really does push the frontier onward, despite how it appears at first. For I am less and less sure that height alone really is the frontier anymore, that getting taller and bigger really is the issue. It was the idea for a long time, and it was done.

But now that we can go high, far higher even than this 150-story proposal, perhaps the real issue that must be faced is not all the way up in the sky 488 m (1,600 ft) versus 518 m (1,700 ft) versus 549 m (1,800 ft). Perhaps it is closer to the ground—back to the whole question of making a civilized city, of trying to see the tower not as an isolated object, but as a part of a larger whole, as something that seems to grow organically from everything around it, enriching its surroundings and in turn being enriched by them.

Now is still, for all the imperfections of so much of what is being built and being proposed, a great time for the skyscraper. The most encouraging thing is that we have begun, after years of uncertainty, to settle into a relatively clear esthetic direction, that of romantic modernism, which is an attitude or impulse more than a style, and that is just how it should be. As we move into the late 1980s we are going beyond the excesses of the early years of reaction to modernism; there is no longer a foolish sense on the part of some architects, as there was a few years ago, that modernism was an evil best purged from our culture. We see it now as a great cultural and technological heritage, just not as one that needs to be taken literally, but more as a resource, a language, that we should be reinterpreting and reusing in our own ways.

The passion to be interesting, which has both enlivened skyscraper design in the last 10 or 15 years and turned it into a sad free-for-all, is beginning to settle down, to mature we might say, and this, too, is encouraging. A lot of the esthetic excesses of the last generation were inevitable results of the reaction against modernism's excessive restraint, and as the esthetic pendulum swings more toward the middle, a certain degree of common sense will prevail. We see it in the best of the romantic modernist buildings now under construction or proposed, the buildings that make strong esthetic statements yet do not seem frivolous, tired one-line jokes, the buildings that relate to the greater stream of architectural history without being directly or simplistically imitative.

The best architecture comes always out of specific circumstances, not out of ideological predisposition. We are looking to advance the art of skyscraper design by looking not only at the tallest and most technologically advanced, but also at the buildings that seem to emerge out of the cities of which they are a part and, in turn, enrich those cities. It is encouraging that Rockefeller Center is turned to constantly as a model for admiration by architects today; so is Carrere & Hastings's splendid 26 Broadway in lower Manhattan, or Holabird & Root's Board of Trade in Chicago, or Van Alen's Chrysler or McKim, Mead & White's Municipal Building or Hood's Chicago Tribune.

These are all buildings of strong personality, of strong image and character, yet they are all buildings that exist to make a statement about the life of the cities of which they are a part, and they are not isolated objects. Some connect to their surroundings more than others, but it is impossible to imagine any of them existing anywhere except precisely where they are—on pieces of land in the midst of cities with which they have come, by now, to have a deeply symbiotic relationship.

And so it should be with every tall building. The skyscraper has, in the end, a special responsibility. Its image is powerful, and if handled well, it can be among the most compelling visual experiences architecture can provide us with. The Monadnock, the Wainwright Building, the Woolworth Building, the Chrysler Building, Rockefeller Center, Seagram—these greatest of tall buildings belong on any list of the greatest of all American buildings. But as we have lived with tall buildings for a century, we by now should know that they alone, for all their glory and power, do not in and of themselves make a city.

In King's Dream, the celebrated drawing of a tower-filled New York by Harry M. Pettit published in King's Views of New York in 1908, the vision is one of bigger and bigger buildings, all one connected to the next, with bridges and arcades between and airships above, and the promise is of a more and more glorious city. Now that the image of King's Dream has become, at least in part, a reality, we have learned that it is not so easy. The magical city, the Jerusalem of towers, does not come by itself; even the greatest of skyscrapers do not automatically make a city a civilized place. If there is any urgent mission for the second century of the skyscraper, it is not, then, to go bigger—it is to turn back, inward in a sense, and to struggle to find ways to make of the towers the great city we were promised long ago, the coherent urban world that was always the dream of every skyscraper architect, the civilized city that, so far, has eluded our grasp.

Architecture

Tall Buildings as Symbols

Bruce J. Graham

In the last fifteen years it has become fashionable to identify large-scale with "Evil" and small-scale with "Good." This is hardly rational. We are destroyed more by viruses than by cosmic explosions, but this is still a common feeling among those who fear the world, who fear the universe, and who fear eternity.

Recently, many have begun to identify towers with bigness and, therefore, with Evil. This attitude implies that cities such as New York and London should disappear and only those on the scale of Wheaton, Illinois or Des Moines, Iowa should remain. When it comes to the quality of life, however, scale is not the issue. What *is* important is how poetic and beautiful is the large or the small. Responsible architects must respond to the totality of human experience, from the most basic human needs to the highest human aspirations.

In all of their elements, both large and small, beautiful cities should reflect the richness of human activity. Man's ideas range from macro- to micro-scale and his intelligent responses to complexity distinguish him from all other animals inhabiting the earth.

HISTORY OF GRANDEUR

Evidence of human traditional reverence for grand structures still remains in Egypt, Mesopotamia, and the Far East, where towers have functioned as stairways to God in a literal sense. In western culture, the mythical Tower of Babel (Fig. 1) is a familiar symbol of both man's innate power over nature and God's innate power over man.

Through the centuries, forts and towers have been built as symbols of power and cultural achievement, such as the twelfth to fourteenth century towers that grace the Italian hilltown of San Gimignano (Fig. 2). Nobelmen constructed elegant structures to signify individual prestige and family status. Thirteen of the original 56 towers remain as testaments to the pride and ingenuity of the ancient families they represent.

Foreshadowing larger cities of the future, these towers enjoy particularly pleasant relationships through accidental arrangements around the battlements of a proud village (Fig. 3). Their size, shape, and placement were not influenced by zoning boards or planning commissions, but the group is unified by the natural materials and local worksmanship that constitute their architectural vocabulary.

Tower-building became the great creative outlet for the Gothic age, as Chartres Cathedral testifies, where two soaring spires were developed and refined over a period of 300 years (Fig. 4). Architects were well-supported then. The people of Chartres harnessed themselves to carts and dragged the enormous limestone blocks for their new cathedral from the quarry to the site. The magnificence of Chartres was most apparent when the town was still a village; the great towers celebrated civilization rising above the farmland of northern France. The Gothic architects ended by performing feats of extraordinary virtuosity to the point of occasionally over-reaching themselves. Dramatic evidence of the exuberance of the age is found at Beauvais, where pointed arches were pushed beyond the power of stone, causing the cathedral to collapse in 1284.

In another civilization, the delicate towers for minarets became the symbols of Islam, as in Cairo with its hundreds of mosques (Fig. 5). This may create a pollution of sound but does not contribute to the pollution of the surrounding low-rise slums. Source of the call to prayer, Cairo's mosques both reach to Allah and communicate with the people.

The language of the twelfth century Bab Zowayla gate (Fig. 6) covered the Muslim empire from Spain to India. Towers were not symbols of hostility, they were symbols of welcome. In Europe, as well, their popularity continued. It is difficult to imagine the grand piazza of San Marco in Venice without the companile (Fig. 7). Earlier low-rise buildings of exquisite quality, which would be landmarks in any European nation, were destroyed to create a tower of greater value. The campanile is an essential part of the entire composition. It did, in fact, collapse at one time and was completely rebuilt. It is a tower in the right place, at the right time, and it has given us a beautiful landmark that epitomizes the achievements of its culture.

THE AGE OF INDUSTRIALIZATION

In the nineteenth century the industrialization of the Western World was well underway, and tall towers became an expression of the age. During the great Paris Exposition, a tower was built in spite of the violent objections of those who feared the future as demeaning the past. Over time, however, the Eiffel Tower (Fig. 8) has become not only the symbol of Paris, but the symbol of France. Located with great care, and in dynamic relationship to the whole of Napoleanic Paris, it is not a symbol of an empire, but a symbol of the high level of human achievement given expression by a great French engineer. It marks an epoch. The tower is beloved, not only by all Frenchmen, but also by many foreigners.

In America, towers took on a new significance in the city of Chicago. Chicago has reason today to celebrate towers, for it was here that, most say, the skyscraper was born. The Reliance Building of 1895 (Fig. 9), a product of the engineer-architect, was a new language of architecture built from the ashes of a great fire. In bold strokes, and without fear modern architecture was born. Towers in the plain were created with an old language given new dimensions. Towers became useful tools for human work and habitation, not just symbols celebrating human achievements or the power of God.

THE ARCHITECTS

There was a man who was searching for a vocabulary with which to express the skyscraper, both on the plains and in New York. After frustrating attempts with Neo-Gothic architecture, Bertram Goodhue traveled west and during his involvement at the Panama-California Exposition in San Diego discovered the simple language of California's colonial architecture. Returning to New York, he won a competition for the capitol building in Lincoln, Nebraska (Fig. 10). This great tower in the middle of magnificent plains celebrated the power of the people and their relationship to the land and, in turn, had a great impact on New York architects and changed their vision of New York City.

The visionary drawings of Hugh Ferris (Fig. 11) are a reflection, however insensitive, of the lessons of Goodhue. They were serious attempts to search for the relationship of tower-to-place and tower-to-sun on the densest island in the world. Rockefeller Center, produced by students of Goodhue in 1932, was the realization of that idea (Fig. 12). With disregard for simplistic zoning ordinances, it stands as one of the greatest American urban spaces ever created. The lower frontal buildings relate to Fifth Avenue through an ingenius space opening to a great tower beyond. It seems that such examples are not easily understood or reproduced. New York went on in its cannibalistic way, attempting to solve architectural and urban problems with mathematical formulas (Fig. 13). It is impossible to comply with the current zoning for sunlight without extensive computer analysis. The typical New York "ziggurat building" (Fig. 14) is a travesty that confuses architectural and esthetic problems with politics. Land values, favoritism, corruption, and greed cannot be controlled by simplistic policy-maker decisions. The proper locations of buildings in cities is an art, but it is less difficult than flower arrangement. It is not that architects cannot deal with these issues; it is, rather, that they are not allowed.

Towers have been built even by those, such as Frank Lloyd Wright, who maintained that successful building solutions could all occur within two to three stories. Once given the commission, however, he immediately created the Price Tower in Bartlesville, Oklahoma (Fig. 15). It is a tower that is inefficient but still successful as a symbol for Mr. Price and Bartlesville. San Gimignano's signoria rejoiced in her symbols, so why not a western oil man?

Once given the temptation, the great master succumbed again and produced a sketch for a mile-high building, even more inefficient than his first attempt and, certainly, unbuildable (Fig. 16). But the excitement it generated in was marvelous indeed. In 1922, the architect Mies van der Rohe was experimenting with another technique, a technique that had not yet been put into practical use (Fig. 17). It did away with symbols of antiquity since they represented the Evil that culminated in World War I. A search for buildings based on the machinery of industrial labor, on the products of the proletariat, commenced. At the same time, the barons of Chicago were seeking those very same obsolete symbols to express the might of their enterprises. The Neo-Gothic expression of the Tribune Tower (Fig. 18) was chosen over the free-thinking concepts set forth by Walter Gropius and Eliel Saarinen.

By and large, the direction taken by Goodhue's followers in New York found its way to Chicago in the Board of Trade building (Fig. 19), a building deliberately located to enclose LaSalle Street and to form a room which was, in fact, the financial center of the city. There did not exist a zoning ordinance but a gentleman's agreement to build the city beautiful.

CITY SPACE

These great cliff-like spaces have an exciting human quality best expressed by Wall Street, that beehive of human activity that contains an explosion of people from Monday through Friday and during times of popular celebration (Fig. 20). I am offended when such spaces are defined as inhuman since I cannot conceive them being created by monkeys, dogs, or cows. Those who fear such spaces are those who fear humanity and, therefore, life. But, errors are committed with high-rise buildings, as well as with low-rise, buildings. Surely one of the greatest errors committed by competent architects is the Pan Am building and its destruction, not only of Park Avenue, but of Grand Central Station (Fig. 21). This building is another example of the inability of government to govern itself. The air rights owned by the government should not succumb to the greed of those governing. Air rights belong to the people and not to their elected servants. What was once a grand space in New York was fouled by the inability of government officials to plan cities with vision or to repress speculation and greed.

Placement, therefore, is all-important. The Inland Steel building, for example, was well-situated when first built. However, the placement is even better today because other citizens opened up squares where sun, light, and space could penetrate the Loop (Fig. 22). While plazas are unpopular today, taking up space that could be used to produce private revenue, it is still appreciated that the Federal Center, the First National Bank, and the Civic Center were constructed to allow our business centers room to breathe, creating spaces for demonstration, celebration, and relaxation.

RELATIONSHIPS WITH PEOPLE AND PLACES

Every tower has to relate to its people, and to the unique character and psyche of each city. Houston was, at one time, a city that had a degree of unity in both color and form. It was essentially buff or light brown with green trees. The One Shell Plaza building (Fig. 23) for Gerry Hines related to these colors and forms, clad in travertine instead of black steel. Today, Houston is a jungle of self-expression with very few redeeming qualities and can only be said to gratify unrestrained ego.

A 100-story mixed-use building in Chicago's Near North Side, the John Hancock Center, was a challenge to express the brashness and structure that is Chicago's personality and character (Fig. 24). The balance of uses has proved prophetic, minimizing office space and creating residential and commercial spaces, predicting the growth of what is now the most powerful shopping avenue in America.

The Sears Tower, at 115 stories, 465 thousand m^2 (5 million ft^2), was another experiment resulting in a very large steel building but reducing its impact by shaping the towers into logical steps that reduced the scale in relationship to its neighbors (Fig. 25). Height was never a problem technically. The building's color and texture were derived from Chicago's own color and texture.

With my engineering partners Fazlur Khan and Hal Iyengar, we explored the tube structure, the bundled tube, and finally the megastructure. This particularly interesting study was for a 160-story tower where loads would be transferred to exterior corners of the building, freeing the interior portion of the tower for grand spaces and multiple uses (Fig. 26). The times now demand of the architect more than simple boxes to be sold by the square foot. Spaces are required now that incorporate varied human experiences both day and night. This 160-story tower could have housed hotels, office buildings, and shopping and exhibition spaces. The flexible program allowed atria and other spaces that would give each user a sense of place while still maintaining an awareness of the whole and a relationship to the overall city.

Recently, during the New York Coliseum competition, I was reminded of the casualness with which New York deals with urban spaces. The project at Columbus Circle is on property owned by the city (Fig. 27). No vision of the relationship of uses to the totality of the city was contemplated. The only guiding force was the amount of money to be collected by the city, later to be so frivolously spent as to still leave uncollected garbage and decay. However, solving the problems of separating the structure from the skin of the building developed an idea of how to articulate a structure by varying scale and proportion according to use. This concept was initiated in the late 1950s with the Business Men's Assurance building in Kansas City (Fig. 28), where the delicate steel structure creates an important tower of middle scale that expresses both the quality of spaces and the relationship to the base—an understanding of structure not dissimilar to that achieved by Eiffel in the grand tower of Paris. The location on top of a hill surrounded by parks was ideal as a counterpoint to the center of the city.

This concept is also useful for small towers such as those at Perimeter Center in Atlanta, where the scale and benign climate allows a sense of playfulness and frivolity with structure (Fig. 29). While clear in concept, it entertains both observer and user. Structure here is not only functional but communicates an understanding of scale and stability that results in serenity.

Onterie Center in Chicago is an elaboration of concepts dating back to the John Hancock Tower and an unexecuted project in New Orleans which Fazlur and this author designed in 1965 (Fig. 30). The exterior structure, as convincing as that of the Hancock building in the background, expresses the scale and vitality of a very large building.

REPUBLICBANK CENTER

The RepublicBank Center in Dallas is a 60-story tower that combines many of the lessons learned in other projects (Fig. 31). The materials and texture combine to form a design responding to the city of Dallas. Light stone will appear as a native material. Blue glass is utilized to counter the driving sun. Complex spaces and forms, however, are now much easier to achieve.

The structure had to respond in form and space to confusing clues from neighboring buildings (Fig. 32). Harry Cobb's rhombus-like building was difficult at best, as were the hard-edged, treeless streets, but an enthusiastic owner made the project quite rewarding.

Computer-aided design, in which we have invested more than twenty years, is finally giving us the tools to deal with and respond to complex problems. Sections in three dimensions are easily generated by the computer system. Figure 33 shows the RepublicBank ground floor plan, generated by computer. The technology allows for a greater study of detail and proportion, and enables the in-depth exploration of alternatives at both large- and small-scale.

The tower, which creates a large domed void in the center, is a result of the work we did on the 160-story tower in Chicago (Fig. 34). Loads are pushed toward the exterior, freeing the core for flexible spaces, following the tradition of the grand buildings of the beginning of the century. Sophisticated engineering now allows us again to achieve grand ideas economically. The low buildings and arcades around the tower are very pragmatic, but the towers often generate wind problems. Exciting solutions to these problems can be found by using computer-generated tower models and micro-scale wind tunnel analysis (Fig. 35). Unlike the early days of towers, developers responding to public demand now require complex ground floor uses. The first three levels of the RepublicBank building include shopping areas, landscaped courtyards, and sculpture gardens. Atria rise over the first third of the building, creating a large stable footprint which then reduces to a slender tower, much like the Hancock or the Sears, with reduced windsails at the top. The express system of elevators, an idea taken from the Hancock building, reduces the inefficiency normally associated with a building of over 2 million ft^2. The exterior wall of glass, aluminum, and granite is installed with systems learned over the last 20 years in the manufacture and fabrication of stone by machines, even more efficient than the metal walls of the 1960s. The texture of the building responds to the symbolic quality of the rising towers and to an interpretation of the character of Dallas.

Varying floor plans that respond to a variety of needs and problems can be solved by teams of sophisticated engineers, architects, and planners using one drawing board, the computer, and with one objective: a grand work! The tower floors can have the elegance of smaller towers while enjoying the spectacular views provided by very tall buildings. The ground floor is a transition from the exuberance of height to the calm of a Texas courtyard (Fig. 36).

CANARY WHARF, ENGLAND

While Chicago wallows in the misery of a new imaginative tax, the commercial lease tax, which was created in a great flash of wisdom by the city legislature, England did exactly the opposite. She created an enterprise zone where every pound of construction is deductible, pound for pound from taxes. The result is a 1.2 million m^2 (12.5 million ft^2) project creating 40,000 new jobs with an average income of £70,000 per person.

This financial center, located on Canary Wharf, will be less than 2 miles from downtown London (Fig. 37). The project will also generate another 120,000 jobs in and around southern England. Unlike zoning in Chicago, New York City, or San Francisco, building heights, location of towers, plazas, roads, and railroad stations are not being determined by greedy speculation, public or private, but by architectural decisions carefully considering the past and anticipating the legacy for the next millenium.

All capital costs for Canary Wharf, including subway systems, will be borne by the private sector (Fig. 38). This, in a nation that some Americans think is dead.

We like to think that the landfall on the Thames with its new squares will leave memories like the Etoiles in Paris. We would like to think that Founders Court will measure well against Barclay Square. We would like to think that the towers, like the Eiffel, will become the kind of symbol that Goodhue left in Lincoln, Nebraska. Finally, we would like to think that the community of architects and engineers will measure up to those of previous civilizations. Looking from Greenwich (Fig. 39), we would like to think that Christopher Wren would like what he saw.

Fig. 1 Tower of Babel (painting by Pieter Bruegel, c. 1563)

Fig. 2 San Gimignano

Fig. 3 San Gimignano towers

Fig. 4 Chartres Cathedral

Fig. 5 Cairo cityscape

Fig. 6 Bab Zowayla (southern gate to Cairo, twelfth century)

Fig. 7 San Marco (eleventh to thirteenth centuries)

Fig. 8 Eiffel Tower (1889)

Fig. 9 Reliance Building (1895)

Fig. 10 Capitol in Lincoln, Nebraska (1920-1932)

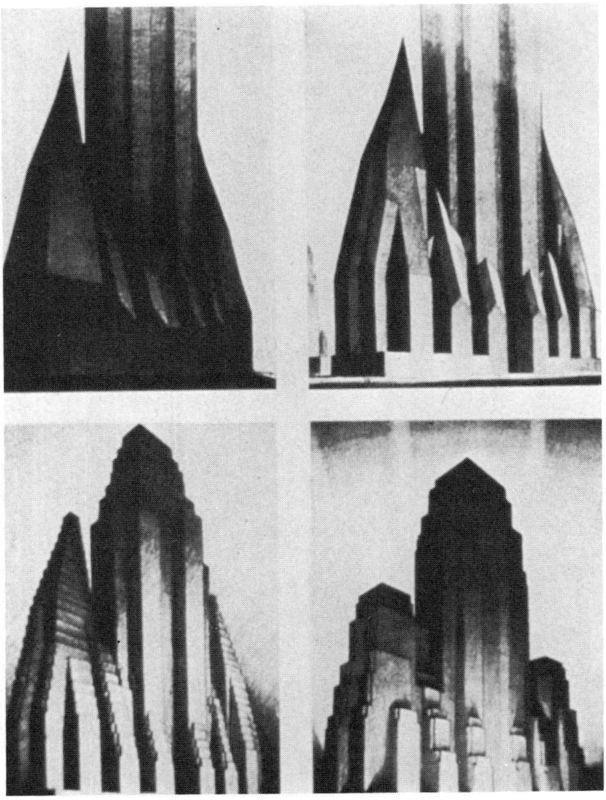

Fig. 11 Hugh Ferris renderings

Fig. 12 Rockefeller Center

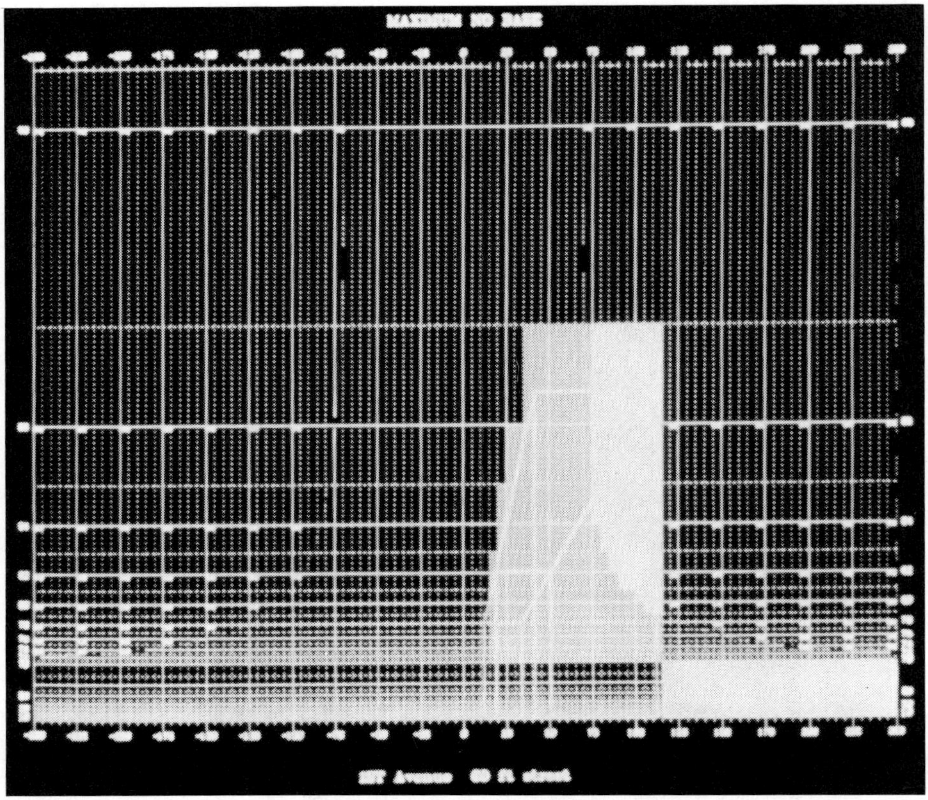

Fig. 13 Computer calculation for New York Zoning

Fig. 14 "Ziggurat" Building

Fig. 15 Price Tower, Bartlesville, Oklahoma (1952)

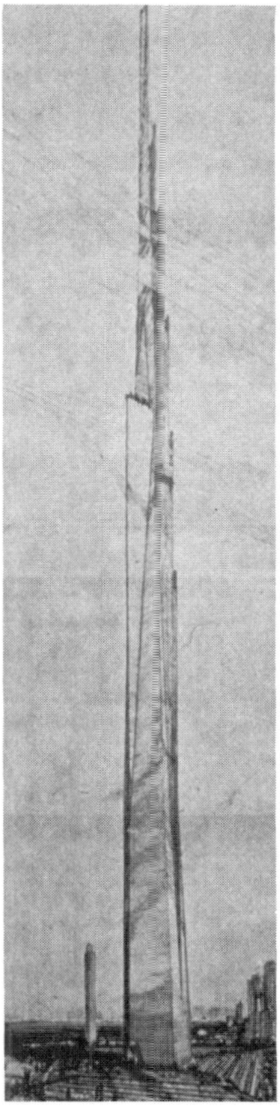

Fig. 16 Frank Lloyd Wright's Mile High Building

Fig. 17 Mies van der Rohe's high-rise in Germany (1922)

Fig. 18 Tribune Tower (1925)

Fig. 19 Chicago Board of Trade building (1930)

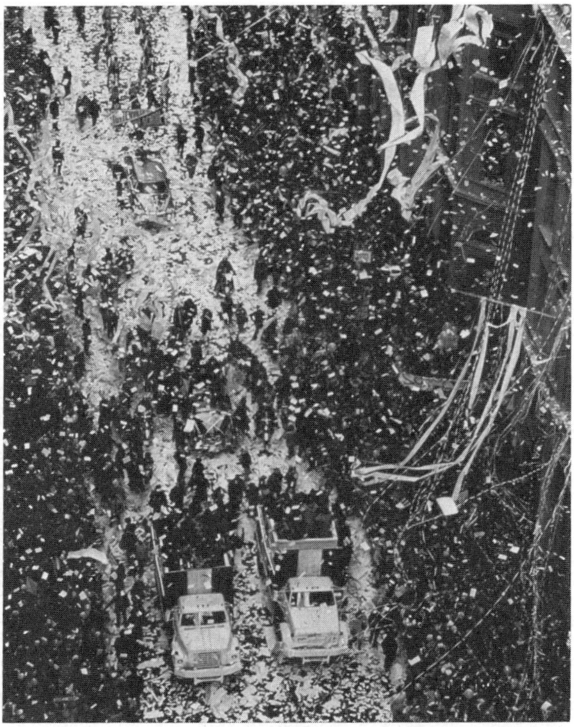

Fig. 20 Wall Street

Fig. 21 Park Avenue vista, facing Pan Am Building

Fig. 22 Inland Steel building *(Photo by Hedrich-Blessing)*

Fig. 23 One Shell Plaza *(Photo by Hedrich-Blessing)*

Fig. 24 John Hancock Center *(Photo by Hedrich-Blessing)*

Fig. 25 Sears Tower *(Photo by Timothy Hursley)*

Fig. 26 160-story building *(Photo by Hedrich-Blessing)*

Fig. 27 Model of proposed New York Coliseum
(Photo by Hedrich-Blessing)

Fig. 28 Business Men's Assurance building
(Photo by ESTC)

Fig. 29 Terraces at Perimeter Center *(Rendering by Carlos Diniz)*

Fig. 30 Onterie Center *(Rendering by Rael Slutsky)*

Fig. 31 RepublicBank Center *(Photo by Hedrich-Blessing)*

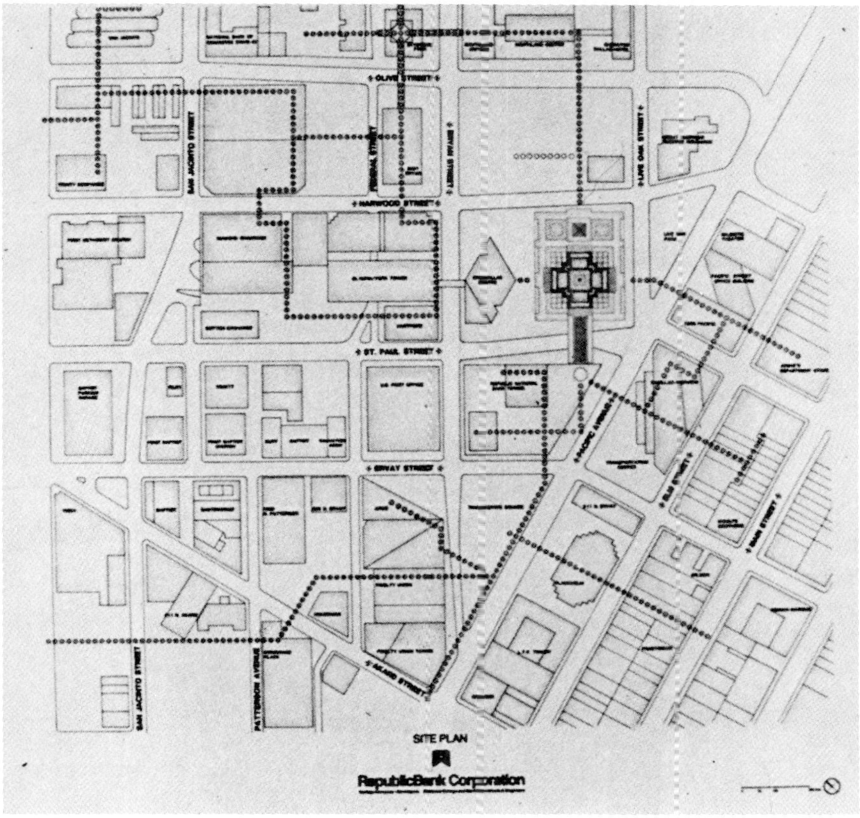

Fig. 32 RepublicBank site plan

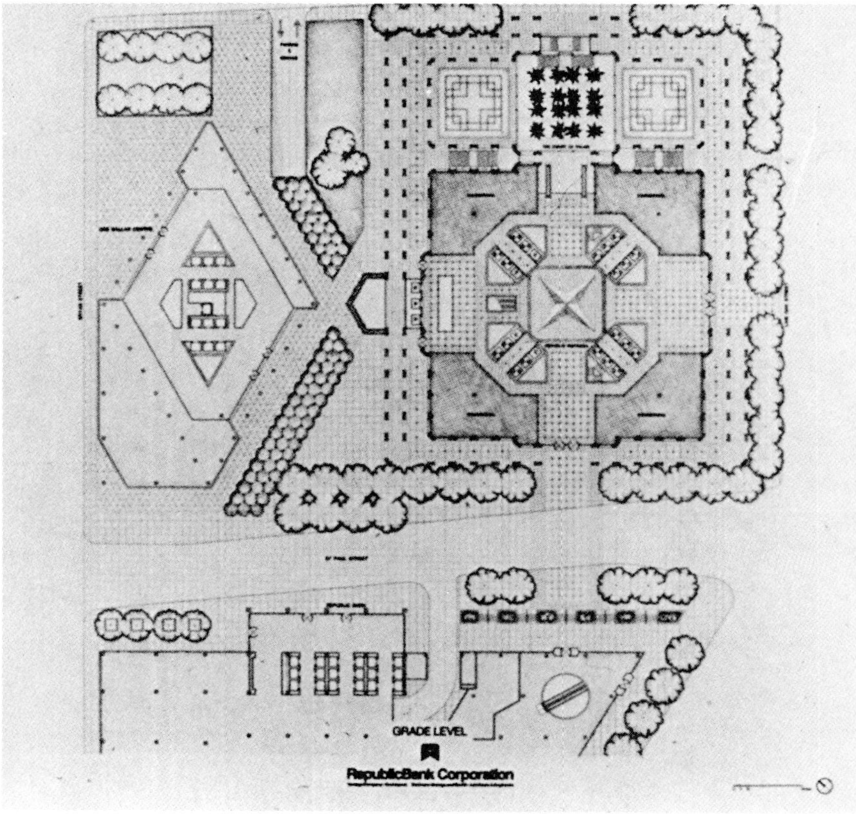

Fig. 33 RepublicBank ground floor plan

Fig. 34 RepublicBank rotunda section

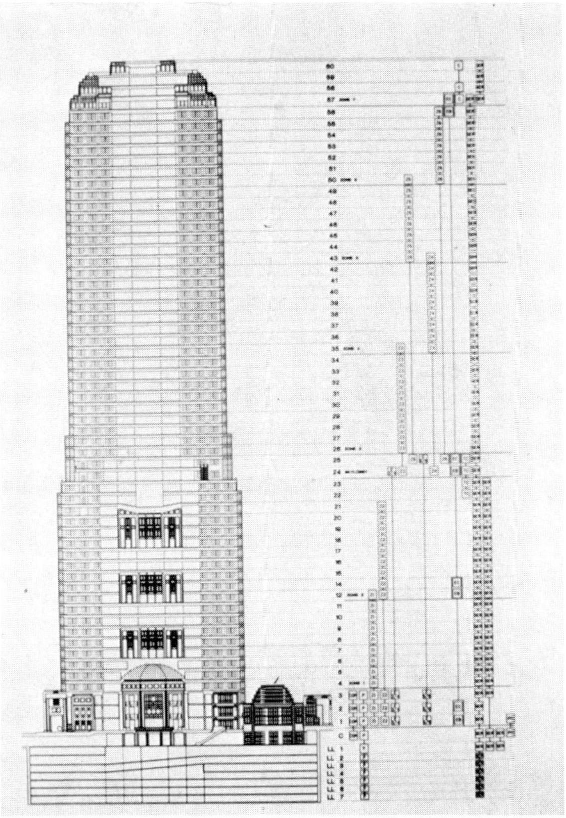

Fig. 35 RepublicBank computer section

Fig. 36 RepublicBank, landscaped courtyard *(Rendering by Rael Slutsky)*

Fig. 37 Canary Wharf model *(Photo by Hedrich-Blessing)*

Fig. 38 Canary Wharf rendering *(Rendering by Hanna/Olin Ltd.)*

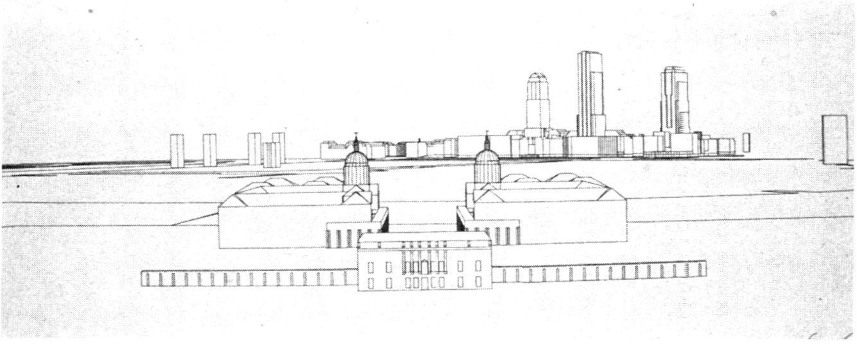

Fig. 39 Canary Wharf computer drawing, looking from Greenwich

Considerations for Urban Architecture and the Tall Building

William Pedersen

The intent of this paper is two-fold: to offer a strategy for design of the very tall urban office building; and to act in defense of the design for a 96-story tower submitted to the Bank of the Southwest in Houston, Texas. The very tall office building as a single structure—albeit of great size—is but one piece in the fabric of a large city. Thus, its capability to detract from or contribute to the milieu into which it will be placed is limited. Nevertheless, a city is formed, ultimately, of many such individual units. Their additive influence is great or small depending upon the attitude each has for the other and for the whole. It is our responsibility, as architects of individual buildings, to search for an approach to design that will allow this additive potential to achieve the highest objectives of the society that these buildings ultimately serve. Beyond that, no more can be asked or achieved. But before one can proceed with the parts—the individual buildings—an examination of the whole is in order. This examination must include the nature of the present urban condition, both as it exists in general, and as it exists in Houston specifically.

THE URBAN CONDITION

Modern cities, particularly those in America, for the most part lack those physical qualities necessary to enrich the spiritual lives of their inhabitants.

The economic, social, and political reasons for this are too complex to enumerate here. Yet, it is fair to say that our cities suffer most from the absence of collective (public) realm in shaping their physical forms; the social and physical implications inherent in the act of "gathering"—of human congregation—are simply neglected.

Consider the argument presented so brilliantly by Colin Rowe; an argument by now well known but, with remarkably few exceptions, little acted upon. Mr. Rowe has stated that modern architecture has exhibited a preoccupation with the external angle that creates form as "object," as opposed to the internal angle that creates form as embracing space. Certainly the *city of modern architecture,* as envisioned by Le Corbusier and others, is conceived as an environment in which the buildings, as figural objects, reside in ubiquitously undifferentiated and unformed space; a city where one's visual perception is dominated by buildings as freestanding objects, rather than by the spaces into which these buildings are placed. These buildings, or objects, are for the most part, of the private realm and the spaces within which they are positioned are of the public realm. Thus, one is left with a condition in which the private realm is dominant over the collective realm, a condition that can be defined as antiurban.

Although the ideal city of modern architecture has not, and cannot, be fully realized, a city such as Houston possesses most of the least desirable characteristics of that model with few of its intended amenities. In Houston the needs of the collective realm, traditionally represented in physical form by the *street* and the *square,* have been unrecognized and abandoned. Each

South Ferry Plaza

building stands as an autonomous figural object without recognition of its urban responsibility to the street it addresses, or to the other buildings with which it must join to create the larger urban context. Each building is autonomously conceived, insular and discrete.

In a *traditional* city, on the other hand (Colin Rowe uses Rome as a model), the public spaces are dominant and, hence, figural; conceived as a substance that can be shaped and formed as clearly as buildings are shaped in the modern city. These public spaces are defined by buildings that act, behind their facades, as a form of inhabitable poché. The facades join to define the walls of enclosure for purposefully constructed *exterior rooms.* The public or collective realm, as represented by the now triumphant spaces, is dominant over the private realm: demonstrably, at least in spatial terms, this is the ideal condition of urbanism, long admired but, in modern cities, long rejected.

There are those, such as Leon Krier, who would have us return to the traditional city, formed of streets and squares; and, in so doing, presumably would have us use traditional building types to achieve this objective. The pragmatics of modern urbanism do not, however, allow this objective to be easily achieved. The "solids" which are our modern building types are not nearly so accommodating as the inhabitable poché of the traditional city. Even the "long skinny building game" called for by Colin Rowe is problematic. The modern building types that comprise the basic texture of our urban fabric are larger, more demanding and less maleable pieces than those that confronted designers of the traditional city.

Chief among the more demanding pieces that comprise our modern city is

South Ferry Plaza

the very tall office building. More than any other building type, it dominates our urban fabric. Before we can hope to regain the traditional qualities of urban life, we must deal with this building type: deal with it in such a way that its primitive character (freestanding, autonomous, insular, and uncommunicative) can be civilized into a more social state; a state that will allow it to be a successful participant in forming collective spaces that can support and enhance the public realm. This, it seems, is a central task facing architecture today.

FACADE

Historically, architecture that has best supported the urban environment has dealt both with the external demands placed upon it by its context, and with the internal forces basic to its nature. One thinks immediately of Baroque Rome where the facade, which represents the division between the

Allegheny International

internal and the external, is concretized as a pulsating membrane upon which the interaction of these two sets of forces play out.

The making of "public urban rooms," defined as street or square, requires that the individual building act, in part, as facade to the street or square. Generally, this means that the lower portion, depending upon the height required of the enclosing facade, must be drawn to the property line so that it can be linked visually with adjacent facades which, hopefully, make similar gestures. This is the first step, admittedly only spatial, in generating the enclosing urban wall, and it ensures two of the basic properties of manmade place: concentration and enclosure. With this gesture accomplished, the very tall building can proclaim its fundamental internal nature: that of being a figural object free to rise above its base.

Early designers of the very tall office building (Cass Gilbert and Louis Sullivan being the most notable) were, as a result of their Beaux Arts training, thoroughly imbued with a concern for facade. Presumably it was quite natural for them, when faced with this new building type, to have addressed issues of facade while allowing the nature of the very tall building to proclaim itself as object. In formal terms, they shaped the mass of this new fellow (for this building type was immediately recognized as a "he"), as a combination of street defining mass on the lower levels, and a tower rising, as object, from this mass. The tower terminated with a triumphant *fioratura* on the sky.

Allegheny International

However, it was not until the completion of Rockefeller Center in New York City that significant urban spaces, formed by buildings acting together in a controlled composition, were achieved with very tall office buildings. Here figural solid and figural void were drawn into a fluctuating state of coexistence. A new urban model was constructed, and its example was clear. But this example was never followed, at least until the present day. Concurrent with the completion of Rockefeller Center the *new architecture* had arrived, sweeping before it all notion of shaping street, square and, hence, collective space. The street was, in fact, anathema to the new architecture, proclaiming, as it did, the building as object purely conceived with all sides acting freely in space. It has taken almost fifty years for these attitudes to run their course.

BOUNDARY

At this point, one must introduce the concept of boundary. Boundary separates that which is public (street) from that which is private (building). More specifically, urban boundary in the horizontal plane is defined by the property line. Given the basic geometry of the city (organized by grid or labyrinth or a combination of the two), the property line is the only common element shared by both the public and the private realm. This line defines the edge of the public realm, and it is at this line that the walls of our

383 Madison Avenue

collective external spaces feel their presence. The most elementary determinant in the spatial evolution of the city is a consistent attitude toward the property line as boundary. One has only to look at the Nolli plan of Rome and compare it to the present plan of Houston to understand the issue at stake.

Traditional cities are characterized by their pragmatic, yet profound respect for the sanctity of the boundary between public and private realm. All collective void is shaped by this boundary. Conversely, modern cities randomly disregard the controlling presence of property lines because the concept of boundary inhibits the self-proclamation of each structure as autonomous and discrete. The concept of boundary demands respect for facade and denies the building as object. If the evolution of the very tall office building is toward contextual linkage with its neighbors, toward defining collective space, then respect for the property line is a fundamental point of departure. Perhaps the concept of boundary would be more palatable to the democratic mind if it were looked upon, not as an end or limit, but as a beginning: "A boundary is not that at which something stops but, as the Greeks recognized, the boundary is that from which something begins its presencing" (Martin Heidegger).

383 Madison Avenue

Once respect for the property line is reestablished, other more complex issues surface. Respect for the property line implies that buildings or, in the case of very tall structures, the bases of buildings, will join together to form the lateral boundaries or walls that enclose and define external space. Unless the facades of individual buildings can join together in such a way that the wall they create conveys, in its totality, some coherent meaning; and unless that meaning is appropriate to the buildings' context, the presence that the wall generates will dissipate. This need for coherent and appropriate meaning raises the issue of visual linkage.

VISUAL LINKAGE

If one accepts the proposition that no object has meaning except in its relationship to other objects, then one begins to recognize that cities can convey meaning to their inhabitants only when the objects that compose them have meaning in their relationships to one another. At one time buildings in the traditional city generated meaning through the shared language of Classical architecture: columns, plinths, cornices, entablatures, pediments, windows and portals all spoke a common language that generated the meaning of the wall and, hence, of urban architecture. Early modern architecture, obsessed with the building as object, purified its language by eliminating these Classical elements. Buildings were no longer intended, nor allowed, to establish relationships, or visual linkages, with one another.

Visual linkage between buildings is made possible when buildings are *composed* of elements derived from common concerns. *Composition* implies gathering various parts into one. In composing visual linkages among buildings, a greater complexity of elements allows for richer possibilities of combination. By reducing a building to only one part (as did the new architecture, further reduced to absurdity by the introduction of mirrored glass), one allows for visual linkage with adjacent structures only as related wholes, and with no secondary or tertiary relationships.

Perhaps if one applies the concept of boundary to the vertical plane, the argument will be more complete. A building as facade is bounded as it rests on the ground; its relationship with the earth is a fundamental condition of its existence. A building is also bounded as it meets the sky. Both the meeting of earth and the meeting of sky imply conditions of boundary, but they are profoundly different, and different architectural elements are needed to satisfy them. Christian Norberg Schultz has said that "when a town pleases us because of its distinct character, it is usually because a majority of its buildings are related to the earth and to the sky in the same way. They seem to express a common way of being on the earth. Thus they constitute a *genius loci* which allows for human identification."

The expectation that all buildings address the ground and the sky in similar ways is unrealistic, even undesirable in American cities. Yet it is not unrealistic or undesirable to expect that buildings address these two fundamental conditions in specific ways. Modern buildings have lost their ability to convey meaning largely because they have ignored the imperatives of boundary. The boundary between earth and building requires architectural elements specific to its needs. One thinks of the rusticated wall of the Renaissance palazzo as mediating between a work of man and a work of nature. Continued pursuit of the nature of boundary will lead to the need for renewed and reinterpreted elements of architecture to satisfy its demands. Indeed, one might say that the elements of architecture evolve to satisfy these issues of boundary. Boundary between portal and wall, between window and wall, between internal and external—all require elements of resolution. These elements can and should come from many sources, yet they must always be present because they convey meaning as they contribute to that architectural language necessary for the dialogue, or visual linkage, between buildings.

TEMPORAL LINKAGE

Can the concept of boundary be extended to time to define past, present, and future? All buildings, either consciously or unconsciously, speak of their

333 Wacker Drive

time. Can they speak also across the boundaries of time by establishing linkage to the past as well as pointing to possibilities of the future? Colin Rowe would call this the dialogue of the "retrospective and the prophetic." In America we find ourselves presently confronted with conflicting objectives and philosophies. We are not ruled by an autocratic form of government and our cities are not composed of citizens of similar heritage. Our technology leads us into space, while our arts often return us to the ancients. Our lives and our cities are swayed by the juxtaposition of conflicting forces. At times they threaten to tear apart our culture, yet they also strengthen and give vitality to our lives. Why is it not possible, then, that within individual buildings the juxtaposition of opposites might also be felt, so long as they fuse to create an ultimate unity of form? Perhaps our buildings can cross temporal boundaries by combining elements conscripted from history with those of the present and (dare one say?) the future. The ancient Greeks created Janus, the god of beginnings and endings, out of their concern for such temporal juxtaposition. Is it not conceivable that our buildings might express similar concerns?

The concepts of linkage and boundary have been introduced in this argument for a specific purpose: to find ways that a building and its facades, when joined with other buildings and facades, can convey meaning. The meaning a building transmits should be drawn from the place it inhabits. The character and spirit created with architecture must be "of the place," not universal to all places.

Norberg Schultz has said that a building can gather its meaning from the place in which it resides. We understand this to mean that a building should draw references from specific pieces that are strongest in creating the *genius loci* of the site. These references may be several and they may even seem to be visually conflicting. To carry sufficient meaning, however, they must be distillations of the spirit of the place.

Gathered meaning implies that a building can be created as an assemblage of pieces specific to the building's context. It also implies a stylistic and temporal juxtaposition of pieces to create the whole. The fusion of these pieces must, in the end, create a unity: specifically that "difficult unity" of which Robert Venturi speaks; a unity consistent with and expressive of the variety and juxtaposition of forces within our modern culture.

Having established the urban responsibilities of the very tall building, let us turn to a specific strategy that calls for the design of the building to emerge as the resolution of two sets of forces acting upon it. One set of forces, which can be called internal, arise from the building's nature regardless of its contextural position. These are forces that generate a building's natural or ideal form. The other set of forces can be called external, and these forces are generated exclusively by context. These external forces, however, must act on something, and that something is understood to be the building's natural or ideal form.

IDEAL FORM

The notion of *ideal* form in architecture has obvious connections to Plato's theory of forms which, in very general terms, states that for each form that exists in an imperfect earthly state there exists an absolute and perfect form of which the earthly one is only an approximation. Louis Kahn certainly hints at Platonic theory, without specifically acknowledging his debt, when he talks of a building "wanting to be." No building type is more immediately applicable or susceptible to the yearning after the ideal than is the very tall office building. This building type is, historically speaking, relatively new. When it was first explored it brought rapturous verse from the pens of its early creators. Cass Gilbert called for it to "grow more inspired the higher it rises" and Louis Sullivan (1924), characteristically eloquent, said:

> It demands of us, what is the chief characteristic of the tall office building? And at once we answer, it is lofty. This loftiness is to the artist-nature its

70 East 55th Street

> thrilling aspect. It is the very open organ-tone in its appeal. It must be in turn the dominant chord in his expression of it, the true excitant of his imagination. It must be tall, every inch of it tall. The force and power of attitude must be in it, the glory and pride of exhaltation must be in it. It must be every inch a proud and soaring thing, rising in sheer exhaltation that from bottom to top it is a unit without a single dissenting line—that it is the new, the unexpected, the eloquent peroration most bold, most sinister, most forbidding conditions.

This initial enthusiasm bore fruit as philosophical desire and physical means were joined. The great achievements of the early 1900s, culminating in such fine examples as the Chrysler Building, the Empire State Building and the very model of the contemporary urban complex, Rockefeller Center, demonstrated this synthesis. Today, with the revival of interest in the possibilities of expression for the tall office structure, it falls to those of us who are exploring anew these possibilities to demonstrate the logic of our approach to its design.

This logic is best structured by first striving after ideal form, in the Platonic sense, and then subjecting that ideal form to the specific demands of the building's context. This confrontation will force the ideal form to be altered and distorted as it adjusts itself to those demands. One can offer, by way of example, the classic demonstration of placing iron filings under a magnetic field. The formal patterns that the magnetized filings assume vary continuously according to the different fields of force. *Ideal form* is then represented by the filings in an unmagnetized state, while *contextual form* is represented by those same filings under the influence of a specific magnetic field.

Before beginning the search for the ideal, however, it may be helpful to define precisely what is meant by a *very tall office building*. The definition is one of slenderness rather than absolute height. Slenderness is determined by the ratio of a building's height to its width and is generally agreed to be in the range of five to one. The ratio of height to width is the point at which the lateral loads created by wind forces exceed in importance, for determining the building's structural support, those forces which are generated purely by loads of gravitation. Years ago it was common to design tall buildings with small floor areas. Consequently, a structure of thirty to forty stories was considered to be a *very tall building*. Over time, however, floor sizes have increased substantially, and today a building must exceed sixty stories to qualify as *very tall*. With this definition behind us, we can explore the ideal form of the very tall building, which is determined by four fundamental concerns: the structural, the solar, the functional and the esthetic.

The Structural

With the very tall office building defined as one for which the lateral forces of wind loading play the dominant role, it follows that an ideal form, structurally speaking, would direct itself at accommodating these forces with minimum penalty, which can be accomplished only by designing a mass that is smaller on the top that it is on the bottom. By reducing the mass on the upper levels, the overturning moment of the structure due to lateral wind loads is also reduced; by increasing the mass at the base of the structure, the structure's natural capability to resist the overturning moment is increased. The full potential of this shape will be realized only if the lateral loads are resisted directly by the enclosing structural envelope of the mass. What results, in essence, is a perimeter wall that forms a structural tube, and an internal core that takes only its share of remaining gravitational loads.

Wind forces are, for design purposes, equal on all sides of a building. The ideal cross section, or plan, of the mass at any point would, therefore, be a circle. The circle, however, makes a difficult shape within which to lay out office functions. The most practical equilateral shape, in functional terms, is

712 Fifth Avenue

the square. Consequently, we are left with an ideal building which, from a structural point of view, has a square plan and a greater mass at its base than at its top.

The Solar

The sun strikes the surface of a building differently on each of its faces. As the human eye adjusts to differing levels of brightness, so must a building be capable of controlled response to its solar loading. Since the north facade receives little or no sun, it requires greater fenestration to receive what light is present. The east and west sides, somewhat solarly symmetrical, receive a low direct sun and, for each, fenestration must be minimal to control heat gain. Similarly, the south side of a building, which in this hemisphere receives the highest sun, requires minimal fenestration. (These requirements assume that the building's systems are not designed to capture and store solar energy.)

Recent developments in reflective tinted glasses have rendered them most efficient in regard to sun control; they are far cheaper and more effective than any exterior shading device used in combination with clear glass. Since it is not technically possible to use an exterior shading device in combination with tinted reflective glass (shading devices cause extreme temperature differentials that can cause reflective glass to crack), the state of the art requires that solar heat gain be controlled by varying the apertures of vision on each of the building's sides. The ideal building, from a solar point of view, would therefore provide maximum fenestration area on the north side, with significantly reduced fenestration area on the east, west, and south sides. In this way not only is the solar heat gain reduced, but also the levels of natural illumination within the building are more equally distributed on all of its sides.

The Functional

Functional considerations for the very tall modern office building are so consistent and well known that they need not be described; indeed, they can be thought of in the same category as other prototypical conditions that determine ideal form. The greatest of these functional concerns is the building's ability to accommodate the most general needs of office planning on a typical office floor. What is the ideal floor plan in respect to both dimension and shape?

At the center of each office floor is a core that houses building services: vertical transportation, toilets, stairs, shafts, electrical closets, fan rooms, and the like. These functions serve the office space and the office space, itself, surrounds the core. Today's rental market calls for a consistent dimension of between 10 to 14 m (35 to 45 ft) from the core to the enclosing outside wall of the structure. This same rental market also encourages a floor plan size of between 1800 to 2800 m² (20,000 to 30,000 ft²). Hence the ideal building, from a functional point of view, calls, again, for a square plan with each of its sides in the range of approximately 44 to 53 m (145 to 175 ft).

The Esthetic

The philosophical notion of what best represents the true nature of the very tall office building has taken many physical forms since the inception of the building type. However, the many verbal expressions of this notion have all stood on common ground. Some have called for the tall building to follow organic design, inspired, no doubt, by the plant kingdom: the tree with its massive root system, its slender trunk and its noble crown is one such model. Others have called for the tall building to follow the model of the Classical column, which consists of a base, shaft and capital. Still others look to anthropomorphic inspiration and envision the foot, the body and the head.

What is consistent among all these models is that they call for dividing the structure into three parts: a beginning, a middle, and an end. Again, to quote Louis Sullivan, these three parts "aver the beauty of prime numbers, the mysticism of the number three, the beauty of all things that are in three parts."

The tall building esthetically considered is seen by the viewer from three vantage points. The first is the distant viewpoint, from which the entire form may be perceived. The second is the middleground. The third is the immediate foreground from which only the bottom of the building is seen. The ideal building must speak in form and detail to each of these separate perspectives. Let us consider, then, the esthetic demands of the three parts: the base, the middle, and the top and the relationship of each to the other.

The bottom of a building must allow a relationship to be established between the human being and the building as a whole. It is at the bottom of the building that the detail and refinement of the mass is most evident. Since it has already been concluded that, from a structural point of view, the bottom of the building is necessarily the widest portion of the mass, it seems logical that this width can be articulated into smaller volumes embellished with rich detail and ornamented to entertain the eye while affording the scale transition from building to street. The bottom of the building also provides entry. The significance of entry and procession offer possibilities to enrich and strengthen a building's place in the urban fabric that surrounds it.

The middle of the building houses floor upon floor of repetitive office space. The nature of this repetition is fundamental to the expression of the very tall office building. The various sides of the middle, or shaft, of the building should respond, in their surface treatment, to both the needs of the solar orientation and to the esthetic nature of the structural system of the building.

The top of the building gives to the mass its distant reading. It establishes the building as a personality in a community of structures: It is the building's signature on the skyline. As such, it represents the final culmination of what is one of architecture's potentially most noble creations: the *skyscraper.* Whereas demands for a functional usage of the top place limitations on its poetic expression, they also present the kind of challenge that is the basis of true accomplishment.

It is the interrelationship between the base, the middle, and the top that gives a building its appeal, its oneness. To quote Louis Sullivan (1924) again, "It must be every inch a proud and soaring thing, rising in shear exhaltation that from bottom to top it is a unit without a single dissenting line . . . ." The ideal building rests finally in the hands of the artist and not the logician. Whereas true art must rest on a logical foundation, its ultimate realization transcends academic impulse.

Having considered the four primary concerns that determine the Platonic ideal for the very tall office building (structural, solar, functional and esthetic) it must be made clear that ideal form is not the objective. The ideal must be subjected to the real demands of context so that it may give to, and take from, that context elements that allow it to be a meaningful participant in a society of buildings. Ideal form must be affected by the external forces of this context; it must bend and adjust to them while always retaining the unique characteristics that define its personality.

A Perspective on Architectural Directions

Harry Seidler

In the area of inner urban development, both in what is allowed to be built and what architects choose to build, we are in a time of great conflict—in a cultural dilemma that has become alarmingly evident in most developed Western countries. Given immense thrust through the media, the emerging new laws and images make us believe that the direction of development in the last 80 years or so has been totally ill-orientated, that it has created nothing but environmental and visual chaos. We are told forcefully that the time has come for a complete turn-about. We should abandon all past notions on city planning, discard theories on architecture developed in our time and change direction totally. I find the propositions and the ensuing visual results thrust at our eyes appalling and quite unbelievable. They constitute a violent reaction, an irrational turning back of the clock from gradual logical and consequential development in our time—a totally antiintellectural stance.

Demonstrating the speed and effectiveness with which the media today disseminate these dissenting notions is the evidence offered by authorities in Melbourne, Australia, in their objections to a large city building. They quote, verbatim, the recently implemented San Francisco plan and insist that this incredibly reactionary set of new rules imported from the United States be adopted, such as, prohibiting buildings with flat roofs or any blank walls, and calling for "a generous use of decorative embellishments." To demonstrate what benefits are offered in return, these decorations are even allowed

Grosvenor Place, Development, under Construction, 1984

to protrude outside the zoning envelope! The San Francisco plan requires buildings to be "shaped to appear delicate and of complex visual imagery." Worst of all there is a dictate to "retain the street wall", in other words the construction of street fronting bases for tall buildings, which are to have distinctive tops and tapering shafts. The rules in fact outlaw any towers reaching the ground with limited site cover. (Author's note: After subsequent negotiations concerning the building in question, the Melbourne City Council's planner was over-ruled, and the project proceeded.)

To enforce such rules of this or any other persuasion I find illogical and contrary to fundamental freedom of action, freedom for the advancement and development of architecture. To stifle creativeness by law is intolerable. We should want no part of a system in which bureaucrats become powerful arbiters of taste.

The irrationality of insisting that urban development built to an index of 14 to 1 shall have 100% site cover is obvious. To allow an increase in the population on a city block of that extent and then strangulate pedestrian circulation by restricting it to 30-m (10-ft) wide footpaths is inhuman and unworkable. And it is all apparently done for no other reason than a misplaced and misguided romanticism trying to recreate eighteenth and nineteenth

Grosvenor Place, Development, Sydney, 1982

Grosvenor Place, Development, Sydney, 1982

century urban patterns with long gone low population densities, when buildings were rarely more than three or four stories high.

The fashion toward solid bases (for towers) "street architecture" is highly questionable. One must reject it for practical and esthetic reasons because it forces architects to design huge, deep, windowless, commercially unviable podium spaces that are structurally and constructionally unworkable.

It is socially irresponsible to build to high indexes of 12 or 14 unless there is a limit on site cover of no more than 25% to 35%. This limit should be so not only for the sake of the health and clarity of the inevitable huge structure that results, but also to generate some breathing space for the additional thousands of people that work in such buildings to create genuinely useful, new, open or sheltered urban spaces—places of repose and recreation—much needed open public space on private land.

The design professions must, as a matter of principle, fight against governments being given rights to codify and thereby dictate design in such detail, otherwise we will find ourselves all living in environments such as the Stalinallee in East Berlin.

By all means let there be enlightened, in other words flexible, three dimensional control strategies that protect the community from excesses, making the intent understood, amending it with time, but pointing toward a viable totality without imposing a dictatorship over the language of form.

On the second issue, the matter of taste and architectural design, much wordy journalism tells us to go back to the 1920s and other fragmentary sources in history for inspiration. Bulking together and labelling everything built in our time as being in the much maligned "international style," the manipulators of media power distort historic facts with great abandon. First of all the term is a misnomer. It was an anathema to the methodology that was expounded by the pioneers of modern architecture. There was, of course, Johnson and Hitchcock's (1932) term "International Style" of the 1930s. No one took their book seriously and, it is rather lowbrow; the term was coined by Johnson, the dilettante, instant "historian" then, rather than the "enfant terrible" of architecture he became later.

Gropius (1946) expressed his contempt for the term. To him the only such style around meant quite rightly, "Those classic colonnades borrowed from the Greeks placed in front of important buildings anywhere from Chicago to Moscow to Tokyo". According to Gropius, by today's standards those who perpetuate and now practice the international style are the "Rats, Posts and Other Pests" as Aldo van Eyck (1981) aptly referred to them in the RIBA Annual Discourse in 1981. Who else but those he so pointedly describes would proceed from doing parodies, caricatures of Le Corbusier for years, to reaching real depths of depravity by turning to emulate Alber Speer's Reichskanzlei mixed with Mussolini's visions and dish these up in Portland, Oregon?

By the definitions of the approach applied, modern architecture could never be a style as it must remain in constant flux, responding not only to inevitable regional differences and social demands, but reflecting the changing visual responses expressed in the art of the time and the ever-expanding wealth of technological means.

What did unlie the methodology was "a clear approach which allows one to tackle a problem according to its peculiar conditions—not by ready-made dogma, or stylistic formula, but an attitude towards the problems of our generation which is unbiased, original and elastic" (Gropius, 1946). In solving such problems "there are three elements which have to be dealt with simultaneously and merged: the bringing into unison considerations of social needs and usage, expressive and to the designer valid art forms of the time, with the most technology can offer to help mould and materialize the results" (Gropius, 1946). The successful marriage of these diverse aspects formed the basis of the pioneers' work and their teaching.

To me, this guiding methodology has never been but a framework on which to hang very different and potentially changing images, the opposite to frozen stylistic molds. It is an approach that grows and mutates in that it is based on the cultural essentials of the time and place. Neither the clarity of this concept nor the specifics of especially the esthetic components (built on the study of visual fundamentals) have ever become the basis of what has generally been built since the last war. Misunderstood, unskilled, superficial images with hideous clichés have covered Western cities and resulted in the present media war on the so-called International Style.

What had originally started as a fight against traditional "style" was utterly misunderstood, and was imitated insensitively to become so banal as to be termed a style itself. Journalists and opportunistic writers capitalized on the

Riverside Centre, Brisbane, 1983

inevitable public distaste for the ubiquitous result, and proceeded deliberately to misrepresent the facts, rewrite history and discredit the dead pioneering initiators and blame those whose work originated development away from the superficial "art for art's sake" architecture of the fin-de-siècle era. I find the results of the alternatives they promote degenerate and unworthy of our time. They make me feel ashamed that I live in an era that can give rise to such an appalling cultural decline.

What is being proposed, seemingly unchallenged, is that we remove from the record and reverse the entire course of the theoretic structure of rational response to environmental needs, expunge and abandon more than half a

Riverside Centre, Brisbane, 1983

Hong Club and Office Building, 1980-1984

MLC Centre, Sydney, 1972-1978

MLC Centre, Sydney, 1972-1978

MLC Centre, Sydney, 1972-1978

century of the evolutionary course of architectural development and change course drastically. The suggested directions are the very antithesis of the visual and technical concerns of our time: they show us ponderous, earthbound, heavy pyramidal compositions standing there flatfootedly and exposing their childish broken pediment "metaphors" to make us feel closer to "history". Ignoring and defying all constructional, let alone structural logic, they are the tantrums of a rich spoiled child, delighting in being contrary, shocking us with corny stylistic idioms, not to say ludicrous bad taste. It could all be ignored if there were not the danger, caused by all the wordy journalism supporting and surrounding it, of being taken literally and seriously by the young and uninitiated, blown up and catapulted into the significance of a new design philosophy. The labels abound: Adhocism, Pluralism, Contextualism, Post modernism, Inclusivism, Late-Modern, Post modern Classicism, and others. They write about all the tastelessness these terms imply with obtuse verbiage, heaping insult on history by parodying the past.

A remark Marcel Breuer made in the 1950s when discussing his reaction to the then fashionable Classicism—that kind of sugar-coated misunderstood Miesian mode prevalent in America then—puts these things in perspective. "Nur abwarten," ["Just wait patiently"] he said in German. Who remembers or takes these fads seriously now? Or the Brutalists in England with their poor imitations of Corb's rough concrete of the 1940s; pathetic in retrospect? With that record, what lasting validity is to be put on the "metaphors" so verbosely elaborated to describe the present reversions to licentious decorative caprices?

Degeneration has gone full circle when one remembers the Western architectural world's outrage at the East's cultural inferiority evidenced by Berlin's Stalinallee at the time Corb built Marseilles after the war. A complete reversal of roles now makes communist East Germany rebuild the Bauhaus buildings better than new and declare them national shrines at the time when Bofill, in the West, builds a public housing scheme that boasts new concrete classic orders and giant fluted Roman columns for fire stairs—the kind of architecture that totalitarian regimes of both left and right have always favored at various times.

The present schizophrenia oscillates in adulation between Post modernism and "Modernistic" stylism, that painful fashion of the 1930s to exhibitionistic displays of technological acrobatics, just for its own sake. Rather than serving any constructional needs this mode exposes its vulnerable arteries, which ensures anything but a permanent life of the building. These, as any fashions, cloy the appetite. They are transient and self-extinguishing, becoming tediously tiresome in the insidious way they grate and annoy the senses in the end. They are antiintellectual modes, regressive, defying reason, art and technics. They are not a worthy product of our time, whose creed is one of restraint and disdain for willful waste, or physical or visual extravagance.

To me, there is a discernible visual direction in our age. The essence of this, which has manifested itself throughout our immediate history, is best defined by the painter, Josef Albers: "Where the discrepancy between physical fact and psychic effect is maximized, there lies the threshold of art," and "One plus one is three—in art (Albers, 1961)."

This credo of getting the most esthetically and physically for the least in effort and material is directly applicable to architecture. Not only is it valid for economic reasons, but it will heighten the value of that which, by a short-cut of the mind and with penetrating insight, finds Gordian Knot solutions to esthetic, planning, and constructional problems The indulgently capricious will fall by the wayside and be seen as a hollow victory indeed.

The simplistic way in which this essential element has been misinterpreted is the cause for much of the harm that has been done in the name of modern architecture. To do the minimal only leads to stagnation and rejection, but to do little so that riches result, both visually and tangibly, that is where our direction lies.

From the earliest days of the Modern movement, emphasis has been on the study of visual fundamentals, of just how our eyes respond in predictable ways to visual phenomena. To understand these will make us realize that our eyes change with changes in other areas. What is valid in 1930 can no longer be actively so today because our senses will respond differently to altered social conditions and advances in technology.

However much we may admire Le Corbusier's buildings that have $2m^2$ (20 ft) span structures (which was all they could do economically then) they are superseded today, as much as their planning, plumbing and everything else about them is. We may still find his spatial flow poetic, enticing and valid, however, even if achieving it meant the use of excessive hand labor or constructional devices no longer realistically plausible.

We live in a world of vastly varying social and economic climates. I have built on four different continents and what is possible, and in fact desirable, in one country with ample, willing and undemanding labor but poor technology, will be unthinkable in a high labor cost location with advanced industrial potential. These considerations will inevitably produce regional differences even if there may be a common denominator or aim in exploiting a subtle orchestration of spatial intricacies.

Twentieth-century human eyes and senses crave space in a new and changing way as only our advancing technology can muster. Instead of assemblies of connected finite volumes, we seek a sense of the infinite and yet simultaneously intimate, a sense of the beyond. In the same way, I believe that visual tension, rather than the phlegmatic earthbound, arrested images of the past, speaks our language: the channelling of space and surfaces in opposition; curve against counter-curve; sun and shadow; the juxtaposition and sequencing of compressive low to the surprise of high. Even if the expression may be

City Mutual Headquarters, Sydney, 1985

City Mutual Headquarters, Sydney, 1985

exuberant or flamboyant, that economy of visual means will heighten the value of the result. Once a strong form element is evolved, it must find its reuse, its echo throughout the work, even if in mutated form (as against the arbitrary assemblage of unrelated geometries).

Our horizons in the choice of appropriate form have broadened with time. The initial restrictive puritanical rigidity has been allowed to widen into an embraceable, all-encompassing search yielding a wealth of new expression. We have learned not to exclude history. By this I do not mean the puerile adaptation of decorative paraphernalia, but looking at the essential forces behind the images. The subtly brilliant geometric systems that came into being in the seventeenth and eighteenth centuries, for instance, can have some validity in our approach to developing system-oriented methods of construction. A new freedom of visual discoveries contributes to the shaping of our elements.

Free reign must be given to the expression of the laws of nature—not what is "imagined" to be so by many structurally naive architects, but the unassailable physical truth of statics. Richness of expression can result from such search, which will have that irreplaceable quality of longevity, of remaining valid, being born of the immutable and irrevocable truth of nature.

The approach to constructional systems has been far too simplistic, accepting any dull repetitiveness to be economically valid. Just as the revivalist architecture at the end of the last century was out of tune with the emerging industrial means, so I believe architects are not taking the lead in today's technological and manpower conditions with its new construction methodologies. That is why we are losing the grip on vital decision-making and are being replaced by hustling technicians. To design a tall building today that simply takes too long to build is a self-arresting, hollow victory that remains on paper.

It is our task to maximize systems of mechanization appropriate and "in tune" with the particular task. Even though these must vary in different socioeconomic and industrial climates, they must not stop at considerations of structure and covering only, as is so often the limit of prevalent thought, but also encompass simultaneously integral solutions to the problems posed by all services without the usual nightmarish afterthought complications of most modern buildings.

True modern architecture is not dead, as some will have us believe. We have hardly started to explore the potential of its methodology. The high principles and clear moral consequentiality of the pioneers needs to be constantly interpreted anew. They demanded basic integrity and an intrinsic honesty of approach. Only by making these part of our work will frontiers of development be pushed forward.

There is a need to return to reason, sensitivity and skill or as Mies van der Rohe has said in Chicago: "I don't want to be interesting—I want to be good".

REFERENCES/BIBLIOGRAPHY

Albers, J., 1961
DESPITE STRAIGHT LINES, Yale University Press, New Haven, Conn.

Gropius, W., 1946
DESIGN, Vol. 47, No. 8, Living Architecture (or INTERNATIONAL STYLE), April.

Hitchcock, H. R. and Johnson, P., 1932
THE INTERNATIONAL STYLE: ARCHITECTURE SINCE 1922, W. W. Norton & Co.

van Eyck, A., 1981
1981 ANNUAL DISCOURSE for the Royal Institute of British Architects, London.

Architecture

The Architecture of Large Buildings

Fred L. Foote

The architecture of large buildings is influenced by many factors: market forces, economics, conditions of a particular site, structural expression, zoning, image, energy considerations, and construction technologies, to name only a few.

This discussion focuses on the *context* of large buildings in the urban setting.

Adapting large buildings into a proper fit with the urban fabric of our cities is a major challenge. (Large buildings are beginning to be built in suburban areas, but this question will not be dealt with here.) A particular building, built for whatever purpose, is a momentary act in the continuum of time and the development of the urban place. However it is generally considered a permanent one. In recent years we have seen the emergence of a new force in the conception of large buildings. This force, which consumes architects, clients, owners, engineers, and tenants, may be described as the mating of ego and economics, resulting in what might be called *"egonomics."*

This force is not particularly new, given projects such as the Chrysler Building, but it seems to be reaching new heights in this high-technology age. Although architecture as *image* may be welcomed by architects, it may be too seductive in reality. Individual large buildings can be very beautiful and can be jewels on our skylines. The Empire State Building in New York and the Sears Tower in Chicago are testimony to the individual building as impressive and beautiful objects in a city. The design by I. M. Pei and Les Robertson for the Bank of China office tower in Hong Kong is very beautiful and may symbolize a new era of the expression of economic power.

Not everyone, however, believes that a proliferation and concentration of these individual tall buildings is best for our urban environment. In a growing number of cities, developers and architects are faced with the possibility of limitations to tall building construction. Remember the extreme reaction when the Club of Rome published the *Limits to Growth* statement? San Francisco has recently adopted its Downtown Plan, and Philadelphia, after spirited public debate, has permitted buildings taller than Philadelphia City Hall for the first time, but in a very limited zone. Washington has had very stringent limits on height for many years. These limits are not only being championed by environmentalists, preservationists, or reactionaries, but by the public at large.

Context may be the key to a partial resolution of the conflict between an apparent need for very large projects in our cities and the desire for a more sensitive impact on the built environment. Responding to the focus on context requires a balance of many influences, but with a conscious priority for the contextual idea, usually not otherwise supported in the economic picture, much less in an ego-driven project. Difficult though it may be to focus on context in tall buildings, it is possible and necessary. It also may be desirable to consider alternative approaches when accommodating large amounts of floor space on limited site areas. Mitchell/Giurgola Architects has, in their modest efforts to date, attempted to speak to context while also responding to the usual issues and pressures of large projects.

The Penn Mutual Tower (Fig. 1), a 21-story addition to the existing company headquarters, represents the traditional high-rise response to creating new space in the city. In this case, the context is a key factor since the project site was located at the south end of Independence Mall, directly behind Independence Hall and the Liberty Bell Pavilion. The north facade was designed to provide a context for the historical buildings while responding to the existing Penn Mutual structure at the cornice and to the street level with the reconstructed 1835 facade by John Haviland (Fig. 2). In addition, the east facade features a sun-shade wall and the north facade is clad in glass curtain wall, both in response to the environmental context.

An example of an alternative to the high-rise solution to projects requiring extensive floor space may be seen in the Strawberry Square development in Harrisburg, Pennsylvania. A mixed-use project, containing two large office buildings totalling over 93 thousand m^2 (1 million ft^2), was designed with a combined screen wall that relates in material, color, and scale to the urban setting (Fig. 3). This design approach also accomplishes the goal of integrating the 12- and 16-story office buildings into the urban design for the block while allowing freedom at the street level to meet pedestrian needs through an arcade and retail storefronts.

Another alternative large-building solution is the National Place project

Fig. 1 Penn Mutual Tower with Independence Hall and the Liberty Bell Pavilion in the foreground *(Photo by Rollin LaFrance)*

on Pennsylvania Avenue in Washington, D.C., featuring the 776-room J.W. Marriott hotel, two office buildings, a retail mall and a below-grade parking garage (Fig. 4). This approach, dictated partially by the limitations imposed in Washington, responds in a major way to the conditions of context. The facade of the Pennsylvania Avenue office building is responsive to the form and material of the historic National Theater. The hotel is intended to be a

Fig. 2 Penn Mutual Tower incorporating the 1835 Haviland facade *(Photo by Rollin LaFrance)*

Fig. 3 View of the Harrisburg Strawberry Square project from the Pennsylvania Capitol *(Photo by Rollin LaFrance)*

Fig. 4 The National Place project on Pennsylvania Avenue *(Photo by Tom Crane)*

dignified structure on Pennsylvania Avenue, enhanced by and enhancing the Willard Hotel to the west. The F Street office building completes the block with a sensitive response to the adjacent National Press Club and League of Cities buildings. The demanding functional needs have been met, as well as considering the orientation of each element, the creation of courtyard amenities and, not least, the special need related to the urban context. These examples are but a brief indication of how the issue of context may be balanced with the other more powerful influences on any project.

In the future, the urban environment will be more threatened but more critical to civilized life in every respect. We must resolve the conflicts and discover the balance between the pressure of time, expediency, and *egonomics* with the concern for the built environment, concern for the national environment (which still exists in rare instances), and awareness of the impact on the many city systems—in short, the urban context. In looking to the future and the potential for limitations on the development of large projects, good judgment will always provide a better result than legislation. With good judgment and a proper concern and respect for context in the process, we may look forward to a vitality of growth that reflects urban and humane values.

Architecture

Impact of European Technical Culture on the Development of Tall Building Architecture

Giselher Hartung
Tom F. Peters

This paper explores the idea that the European development of the iron frame led to the separation of skin and structure as an architectural concept that was then transplanted to Chicago through both architectural and engineering education in the latter half of the nineteenth century.

EVOLUTION OF THE FRAME

The evolution of the steel-frame building began in a modest group of textile mills in the Midlands and north of England built between 1792 and 1804. The development of these structures has been treated by Bannister, Hamilton and Skempton (Bannister, 1951; Hamilton, 1941; Skempton and Johnson, 1962).

The chief criterion for the use of iron framing in mills was the risk of fire. In 1792, William Strutt, industrialist, practical scientist, inventor and engineer, introduced iron into the structure of one of his factories at Millford and into that of a warehouse at Derby. The floors were brick arches and sprang from heavy timber beams spanning from wall to wall. These beams were supported internally by two rows of cruciform cast-iron columns. A more basic change followed in 1796 with the erection of the Castle Foregate Flax Mill in Shrewsbury by Charles Bage. Bage used cast iron for both the columns and the continuous beams over them. He understood the existence of a maximum bending moment at mid-span as he made his beams deepest at that point. Bage did not, however, appreciate the existence of negative moments over the supports. Nonetheless, the mill is still in good condition today (1986) and documents the appearance of the problem of the post-to-beam connection, but not of the rigid connection, because the surrounding masonry walls continued to support all horizontal loads.

The zenith of this first period of development was reached in buildings like the Quadrangle Store, a multistory warehouse in the dockyards at Sheerness, Great Britain, built in 1830 (Fig. 1). Every part of this building was made of

Fig. 1 Quadrangle Store, 1830

nonflammable material. The doors, window frames, roof structure (Fig. 2), hollow cylindrical columns and the inverted T-beams were all cast iron.

Structural frames and mechanical installations are connected from the outset, a novel approach to building evidently derived from machine construction in iron. The Armley Mill built at Leeds in 1804 by Benjamin Gott used hollow cylindrical columns in a dual role: for support and as conduits for a centralized steam heating system. The foundry hall of the Sayner Hütte, West Germany, also built in 1830, is a hitherto disregarded missing link in the evolution of iron structures and also illustrates the connection between structure and mechanical installation (Fig. 3). The architect, Ludwig Althans, worked for the Berlin Public Works Department, the head of which was one of the most famous architects of the nineteenth century, Karl Friedrich Schinkel. Althans, like Schinkel, had visited England before the foundry hall was designed. We may assume that he had seen the famous Ironbridge at Coalbrookdale, particularly since the cross-section of the foundry hall is so similar to the structure of the bridge. The frame rests on 12 outsize, hollow cast-iron columns, 6.50 m (21.3 ft) high and .65 m (2 ft) in diameter. Arched

Fig. 2 Quadrangle Store roof

proto-vierendeel trusses span the aisles with a truss of more orthodox Gothic derivation over the nave. The trussed members across the nave and the longitudinal ones serve as quasi portal bracing. Curiously designed fishbelly trusses, far in advance of their presumed invention by Georg Ludwig Laves in 1838, carry the travelling crane above the blast furnace door. The parabolic bottom chords are probably made of spring steel. Carriage springs are called to mind here, and that is possibly the source of their invention. Robert Stephenson previously had attempted to use cast iron for the bottom chords of an analogous lenticular truss on a less successful railway bridge in West Auckland. The swivelling jib cranes on the cast-iron columns are another invention in the Sayner Hütte Foundry. They turn on the earliest known gigantic cast-iron ball bearings. Again, mechanical installation and structure are inextricably interconnected.

Then followed the development of the cast-iron front. Previously, all iron

Fig. 3 Foundry Hall, Sayner Hütte, West Germany, 1830

structures had been hidden behind masonry walls. Now they began to push through the surface and "decompose' the massive outer walls. The frame became a determining formal element, and with the degeneration of the wall arose the need for bracing systems. Iron fronts made their appearance in the 1840s. Although not the first to build such fronts, James Bogardus and Daniel Badger of New York were instrumental in the breakthrough of cast iron as an architectural material. On a journey to England, Bogardus had studied the versatility of the new structural material. After his return, he became the main proponent of iron architecture in the United States. To demonstrate his ideas, he designed his own factory in 1847, the first building in New York built completely of cast iron, and took out a patent 3 years later based on its design. The frame was made of cast I-beam girders. These were framed into channel-shaped entablatures and supported by Doric semicolumns. For the floors and roofs Bogardus used interlocking tongue-and-groove rolled iron plates that lay on timber beams framed into the girders. The spacing of the exterior bays was determined by the size of the window openings, not by the interior framing. Thus, from the very beginning, the North American cast-iron front was divorced from the construction behind it, a pragmatic, but not a conceptual separation.

Considering the enthusiasm for the structural use of iron in Victorian England and the extent of the export trade in prefabricated iron buildings, it is curious how few such commercial buildings were erected in British cities. Cast-iron fronts, comparable to those that lined the business streets of North American cities from the 1860s on, remained rare.

Glasgow and Liverpool, the major centers of the iron trade in the nineteenth century, were exceptions: In Glasgow, the Gardener Building of 1855 had a exterior completely of iron and glass (Fig. 4). Internally, framed girders made of flat bars of what is probably wrought iron were used together with underspanned cast-iron frames. Although such frames were known at the beginning of the century, these had recently been patented by the iron founder Robert McConnel who was probably also the contractor on this building. The structural grid is more widely spaced than in contemporary North American buildings and corresponds to the glazed bays of the exterior, doubtless due to McConnel's framed girders. Iron buildings in the United States that did have internal iron structures, such as Bogardus's factory of 1847 or the Harper's Building of 1854, allowed no formal expression of the internal structural bay on the exterior.

The first iron-framed building to create an international sensation was, without doubt, the Crystal Palace of 1851 in London. It had a formal impact on architecture in America, whereas in Europe its influence was both formal and technical. Another missing link very closely connected to the construction of the Crystal Palace was the Building for the Exhibition of Industry in Munich of 1854, the so-called *Glaspalast* (Hutsch, 1980). This structure was built under similar limiting conditions as the Crystal Palace: lack of time and the criterion of demountability. Therefore the building commission decided

Fig. 4 Gardner Building, 1855

to use iron and glass as building materials and awarded the contract to the Cramer-Klett Iron and Steel Works in Nürnberg. The project engineer, Ludwig Werder, was as important to the conception and erection of this building as Charles Fox was to the Crystal Palace. Werder had already built an open market hall, supported on cast-iron columns with a corrugated iron roof, the *Schrannenhalle* in Munich in 1853. The architect Voit had worked in iron once before together with Cramer-Klett and Werder in the erection of a conservatory for King Max II of Bavaria. The official report preceeding the erection of the *Glaspalast* expressly requested use be made of the knowledge gained from the building of the London Crystal Palace.

One of the important changes Werder made was to improve the stiffness of the frame. First, the ends of the square tubular cast-iron columns were connected by 12 bolts to the baseplates instead of only four as in the Crystal Palace. Then, to decrease the concentration of stress in the critical connections between the trusses and the columns, Werder separated the column-to-column connection vertically from the column-to-truss connection. Furthermore, he contrived clamps, of which no depiction has survived, to avoid fracture caused by bending stresses in the columns as had most probably occurred in the exterior one-story columns of the Crystal Palace. These clamps helped secure the columns to the beams and apparently formed a semirigid connection. Thus Werder managed to stiffen the building without having recourse to crossbracing. He was perhaps the first constructor ever to erect a truly stiff frame using the heterogenous materials cast- and wrought iron. Werder's solution predates the well-known Boat Store of 1860 at Sheerness, which will be discussed presently.

The stiffness of the column-to-beam connection is a complex chapter in the history of the frame. Many attempted to render such connections rigid using patented "iron cements" or "rust cements" as binders. Several other buildings must be mentioned in connection with this problem. The first is the former Museum of Science and Art in London, built by D. Young of Edinburgh in 1856 under the supervision of Sir William Cubitt. Cubitt had supervised the construction of the Crystal Palace. The museum has three aisles, each 13 m (42 ft) wide, resting on cast-iron posts 8 m (26 ft) high. The inner circular ones carry galleries over the side aisles. The outer ones are H-shaped, which was novel and permitted a more rigid connection between post and beam. The skin of the building was of corrugated iron panelled with wood on the inside. Knee-braces ensured the longitudinal stiffness of the structure, and although we do not know just how the skin was attached to the frame, it may have contributed to the stiffness of the whole. In his critique, the editor of the *Builder* snidely termed it the "Brompton Boilers," thus expressing a distaste of the use of a "non-architectural" building as a museum. Today, hidden behind brick instead of corrugated iron, it graces London's East End as the Bethnal Green Museum.

The Boat Store, built by Colonel Geoffry T. Greene in 1860 at Sheerness Dockyard, was directly affiliated with the Museum of Science and Art (Fig. 5). And it was just as astonishing a structure as the Munich *Glaspalast.* Skempton first described it in an aritcle in 1959 (Skempton, 1960). The structure has

Fig. 5 Boat Store, Sheerness, England, 1860

three bays, is 64 m (210 ft) long, 41 m (135 ft) wide and has an overall height of 16 m (53 ft). The central bay (Fig. 6), open the full height of the building, is spanned by three travelling cranes. In the two four-story bays, four rows of cast-iron posts, spaced on a 5 × 9 m (16 × 30 ft) grid, are bolted to riveted wrought iron plate girders and these to cast-iron joists. Stability is ensured by the bolted connections. Skempton points out that the H-form of the double-story-high posts contributes greatly to the fixity of the connections, much as it had in the Museum of Science and Art. The corrugated iron skin may also contribute to the stiffness. The outermost square corner posts are bolted together in four sections to a total height of 12 m (40 ft). They are hollow and act as downspouts. The decisive factor in the development of framing construction is the design of the rigid post-to-beam connection between 1854 in Munich and 1860 in London.

Fig. 6 Boat Store, Sheerness

THE SKIN

There are two curious buildings built in Liverpool by engineer Peter Ellis that document the evolution of the curtain wall. The Oriel Chambers of 1864 owes its name to the glazed cast-iron oriels or bay windows suspended between extremely thin stone pillars. The two street facades are, in themselves, remarkable designs, presaging the work of the Chicago school and particularly that of Daniel Burnham. On the inside, cast-iron frames made of H-shaped posts and inverted T-girders form the loadbearing structure. The courtyard skin is suspended in front of the cast-iron posts forming a true glass-and-iron curtain wall. As far as is known, this was the first such wall ever built. The skin is set back at each floor level, providing skylights that give excellent additional light even in the narrow courtyard. Ellis adopted the same solution at the rear of his building on Cook Street in 1866, which has an even narrower courtyard. A fascinating feature of this courtyard is the spiral, cast-iron staircase standing free of the facade, totally clad with glazing and iron plates (Fig. 7). After the savage reviews that appeared again in the

Fig. 7 16 Cook Street courtyard facade, Liverpool, England, 1866

Builder, Ellis never designed another building. He worked solely as a civil engineer for another eighteen years.

CONCLUSION

The separation between skin and structure is incomplete in the Crystal Palace of London, in the cast-iron fronts of Badger and Bogardus, and, it appears, in the Museum of Science and Art and the Sheerness Boat House. The Munich *Glaspalast* presents the first clear instance of the separation of the loadbearing structure from the nonloadbearing parts of the building, which then reappears in the work of Ellis. This marks a clear departure in the development of conceptual thinking in Europe from that in the United States and was to become influential in the development of European architectural theory and philosophy in the first half of the twentieth century. This eventually lead to the building of the Fagus Factory by Gropius in Germany in 1911, then to the skyscraper projects of Mies van der Rohe and their first realization in the United States in 1952 and 1956. In France, Le Corbusier's studies culminated in the United Nations Secretariat Building in New York, also in 1952.

As far as the development of the unique indigenous North American case of separation between skin and structure in the Chicago school is concerned, it would be of interest to examine the following possibility: Jenney, who trained many of the early Chicago architects, was trained as an engineer in Paris. Charles Strobel, an influential early representative of the steel industry and structural consultant in Chicago, the author of the first Carnegie steel tables of 1881, was trained in Stuttgart. The conceptual separation between skin and structure that developed in high-rise building in Chicago was in part caused by the pragmatic need to solve the problem of differential expansion between tall walls and iron frames, but only in part. The loadbearing party walls of the 1889 Tacoma Building, for instance, stood in spite of the fact that they were not hung on the frame as the facades were.

The builders of the Chicago structures were trained in European theoretical approaches to design as were many other designers in the United States. But, in contrast to most of the others, they concerned themselves early with high-rise structures and therefore with aspects of structure that determined architectural form. They followed the same conceptual logic of distinction between skin and structure that characterized the European frame. This attitude shows in the technical detailing of the Chicago structures and in their architectural implementation, as expressed for instance in the theories of Louis Sullivan, and is quite distinct from the indigenous high-rise development in New York, which only began to differentiate between skin and skeleton under the influence of Chicago designers such as Burnham and contractors such as the Starrett brothers.

REFERENCES/BIBLIOGRAPHY

Bannister, T. C., 1951
THE FIRST IRON FRAMED BUILDINGS, Architectural Review No. 107.

Hamilton, H. B., 1941
THE USE OF CAST IRON IN BUILDING, Transactions of The Newcomen Society, Vol. 21.

Hutsch, V., 1980
THE MUNICH GLASS PALACE 1854-1931 (Der Munchner Glaspalast 1854-1931), Munchen.

Skempton, A. W., 1960
THE BOAT STORE, SHEERNESS (1858-1860), paper, read at the Science Museum, London on February 3.

Skempton, A. W. and Johnson, H. R., 1962
THE FIRST IRON FRAMES, Architectural Review No. 119.

Building Design Consultation

M. Arthur Gensler Jr.
Antony Harbour

Office building design is much more than the exterior appearance, however attractive or impressive it may be. Whether dealing with a corporate headquarters or a speculative office building, the most successful office building projects are those that complement and enhance the surrounding area, provide optimum space utilization, and serve the needs of the user.

The building design consultant's contributions to a successful office building project will vary according to the client's requirements and the stage of development of the building design. Because of escalating construction and maintenance expenses and increased user emphasis on space planning efficiency, it makes sense to test a building's design to allow for early modifications that will result in an economically sound product. This service can benefit two types of clients: the tenant/user and the owner/developer.

The main emphasis of building design consultation services for the tenant/user is toward tailoring an otherwise speculative office building to the specific requirements of the user. For the owner/developer, the emphasis is to achieve a highly efficient building—one that is flexible to meet a variety of tenant needs and is competitive and leasable in the marketplace.

BENEFITS OF GOOD DESIGN

Good design can benefit both the user and the owner in the following:

a planning module that meets anticipated tenant needs;

a building configuration that optimizes space utilization;

flexible systems that avoid expensive future alterations;

maximum energy conservation techniques;

calculations of usable and rentable areas to determine accurate tenant charge per square foot;

basic marketing data for the leasing agent's use;

project controls that expedite project implementation and tenant move-in.

Whether these services are provided for the tenant/user or the owner/developer, several major areas of investigation and analysis typically occur. This paper addresses some of the questions that relate to this analysis, including how rentable space and usable space are measured and their importance, and how the exterior of the building, the core, the planning module, and the building systems affect efficiency and space utilization.

Basically, every square foot of floor area, except vertical shafts, is included in the rentable area. In some cities, the space housing mechanical equipment, the lobbies, and loading docks are included. Each community and developer uses slightly different criteria. The efficiency of a building for the owner is determined by comparing the rentable space the tenant actually pays for versus the total gross office area. The higher this percentage, the more efficient the plan. Obviously, the potential tenant is looking for the building that offers the most square footage of actual usable space for employees, whereas the building owner is seeking maximum rental space (Figs. 1 and 2).

MAXIMIZING EFFICIENCY AND ECONOMY

The building design consultant can assist in maximizing efficiency for the tenant while increasing the economical benefits for the owner. The size and efficiency of the building core, which contains elevators, stairs, washrooms, and mechanical, electrical, and service facilities, are a basic factor in determining the ratio of rentable/usable space. Ideally, it should be the smallest possible size while still effectively accommodating the necessary functions. Frequently, by rearranging the core elements, space efficiency can be increased for the user. If only a few feet of additional rentable space can be generated through rearrangement of the core elements, the economic benefits realized over a potential 20-year building life span can be considerable. Consider an increase of 465 m^2 (5,000 ft^2) of rentable space, representing a 1% increase of a 46,500-m^2 (500,000-ft^2) building. At $161.50 per m^2 ($15 per ft^2), over the 20-year span, the additional income to the building owner would be $1,500,000.

Location and design of the core strongly impact the utilization of the space. If the core is offset to one side, the resultant large open area can accommodate more efficiently open office planning. A centralized core with a consistent distance from core wall to exterior more easily accommodates private offices

Fig. 1 The impact of the structural system on space planning. In this case, the columns slope through several floors creating odd shaped spaces that are ineff cient and difficult to utilize.

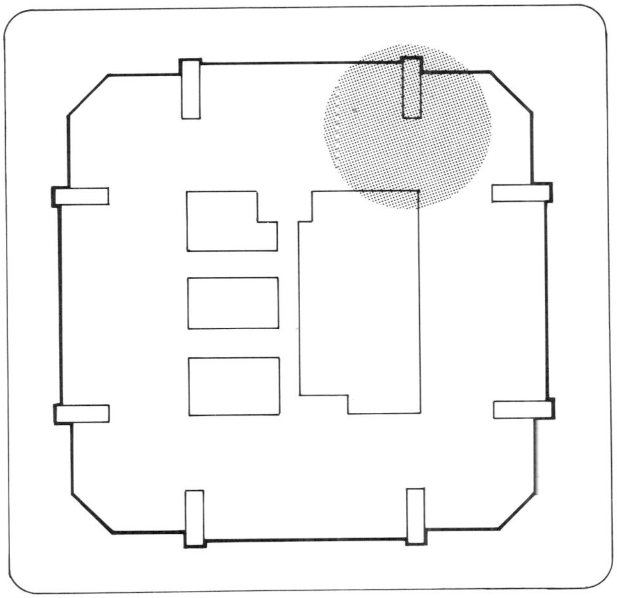

Fig 2. Floor space as affected by the structural system.

around the perimeter. A knowledge of anticipated tenant needs is therefore critical.

The design of facilities within the core also has a direct impact on plan development. Because the core is the origin of all major circulation patterns on the floor, an efficient layout of elevators, stairs, and restrooms ultimately determines the efficiency of circulation throughout the floor. As noted above, the location and number of penetrations into the core directly affect the configuration and length of corridors adjacent to it.

Unfortunately, this critical aspect of building design is often an afterthought. All too often a designer, unconcerned with the long-range impact of building code exit requirements, will develop a core plan that requires extensive "loop" corridors for the plan to function legally (Figs. 3 and 4). The tenant pays for this inefficiency year after year in rental of unusable corridor floor area.

IMPACT OF EXTERIOR ON INTERIOR

The exterior design of the building impacts the interior spaces of the building. Some of the buildings now being designed have unusual shapes

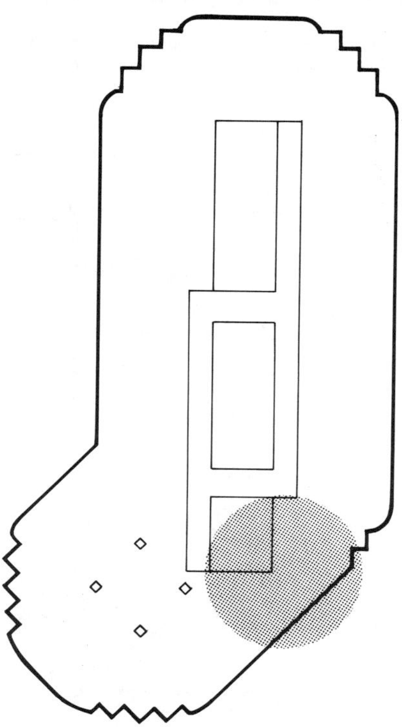

Fig. 3 The original building design had a corridor that was excessively long with multiple entries into the core.

that may have inherent inefficiencies that compromise the use of interior space. While the best plan shape is not necessarily a square or a rectangle, careful analysis of plan shape and floor size is critical to a successful, efficient plan.

The building design consultant can guide both the tenant/user and the owner/developer through trial layouts of how the spaces can be developed. This process assists in weighing the external design esthetic against potential problems in interior planning. The amount of perimeter window glass, the quality and quantity of light transmitted through the type of glazing selected, and the presence or potential presence of tall, shading neighbor buildings all affect the availability of natural light to the interior spaces. Not only will the amount of energy consumed in artificial lighting be a factor but also the approach to some interior planning concepts.

PLANNING MODULE

Most building design is based on some unit of least dimension, usually the window mullion spacing or some multiple of it, and is called the "module." It

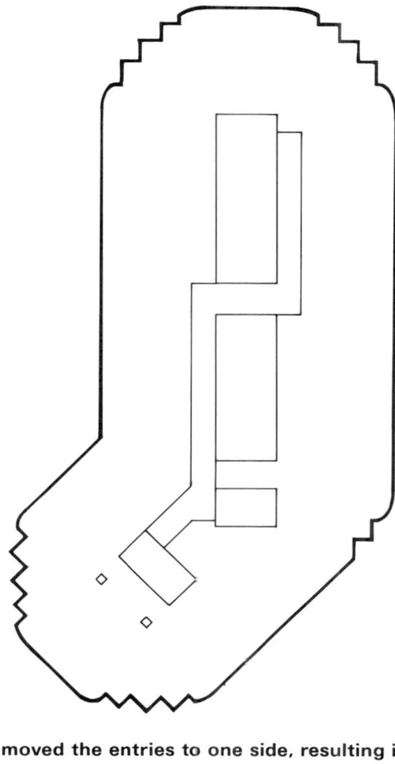

Fig. 4 The revised plan moved the entries to one side, resulting in an increase in rentable space. In addition, the stair portion of the core was rotated to increase the core-to-window-wall dimension.

is established early in the design of a project and is usually a dimension between 1 and 2 m (4 and 6 ft). The module dimension is important because it has a significant impact on the planning and systems for interior spaces. The goal is to establish the most flexible building module to accommodate user requirements. Thus the module will vary depending upon the user's standards.

Different window-mullion spacings will dictate different office widths (Fig. 5). For example, a 1-m (4-ft) module dictates perimeter offices of 3 m (8 ft), 4 m (12 ft), and 5 m (16 ft) and may be more appropriate for tenants with a large operational staff. A 1.5 m (5-ft) module allows 3m (10 ft), 4.5m (15 ft), and even 6 m (20 ft) and thus may be more suitable for enclosed senior management functions. A knowledge of how the spaces will later be used allows the building design consultant to evaluate the impact of this basic dimension on the planning of a new facility (Fig. 6).

Many of the basic systems of a building, such as lighting, ceiling, HVAC, sprinklers, and electric and telephone distribution are based on the planning module. Further, many interior planning systems, such as partitions, floor covering, and furniture, adapt more easily to one module than another. A thorough knowledge of these systems and how they interrelate is critical to module evaluation.

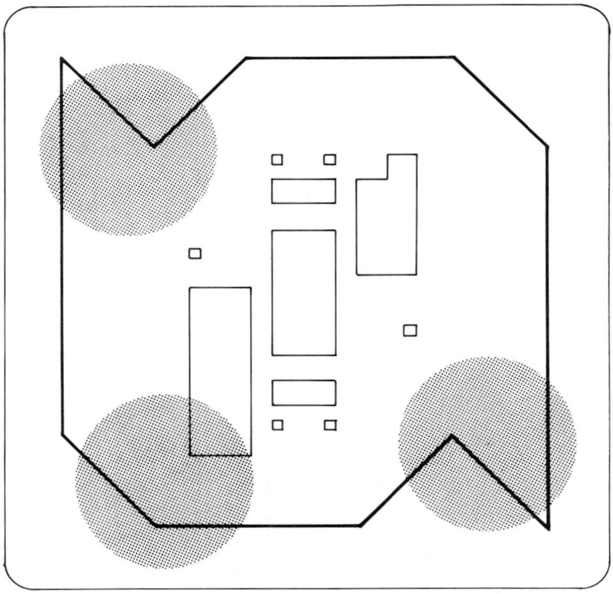

Fig. 5 This floor plan has too many elements interrupting the space. Cut-ins and projections create fluctuating core-to-window-wall dimensions while awkward column placement prevents the efficient use of these areas.

IMPACT OF BUILDING SERVICE SYSTEMS

The electrical, mechanical, communication, and lighting systems impact operational efficiency and flexibility. The building design consultant can study the cost and future use implications of each building system based on anticipated future tenant needs. Electrical, mechanical, communication, and lighting systems account for over 35% of the total cost of a modern office building. Whereas the bulk of this first cost is absorbed by the building owner, day-to-day operating costs, which are usually passed on with the rent, affect the tenant/user. The impact also comes in the costs of subsequent alterations during the life of the lease.

In value engineering, costs of alternative systems can be clearly compared against long-range operation expenses. An understanding of how these systems will be used is key to our analysis and recommendation. User-centered factors such as rate of relocation, personnel growth projections, and hours of operation are essential to the final recommendations. For example, the rate of relocation in the building is approximately 20% per year. In this case, a more expensive but movable partition system that might quickly pay for itself in lower costs for relocation might be considered.

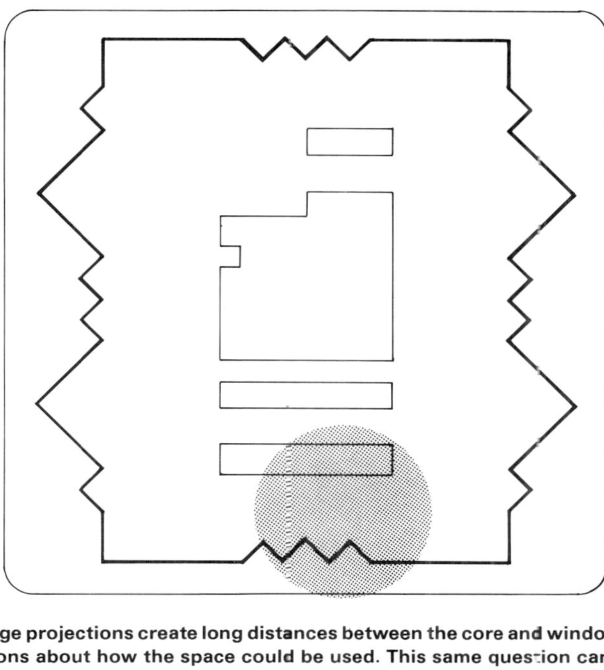

Fig. 6 The large projections create long distances between the core and window wall, raising questions about how the space could be used. This same question can be raised for the smaller projections, which are too small to be used for anything other than a plant stand.

The examination of building systems to anticipate future tenant needs is another factor for the owner/developer to consider. An underfloor duct, preset box system or flat-wiring system for electrical, telephone, and communication distribution may cost more initially than conventional overhead or punch-through installation, but may provide major savings when modifications are made to accommodate change. A systematic approach to the placement of air conditioning and heating ducts and, where necessary, structural penetrations, will allow for future flexibility and simplify renovation of these systems. The number of heating/cooling zones per floor should be planned to accommodate both private offices and open-plan requirements. The selection of light fixtures and their layout should correlate with the building module and accommodate a variety of users with suitable light levels while carefully evaluating the effect on energy usage.

LIFE SAFETY CODE REGULATION AND HANDICAPPED REQUIREMENTS

Every municipality has building codes that are intended to protect the safety of building occupants by regulating the methods, materials, and planning of the building. Many of these regulations are directly related to the preservation of life safety and accommodation of the handicapped. Salient features of life safety codes include sprinklers, smoke and fire dampers, stand-by power, emergency lighting, fireman communication and fire control systems, smoke detectors, control and ventilation systems, elevator control and safety systems, alarm systems, smokeproof stair towers, exit requirements, and compartmentalization of floors into safety zones. How these solutions to the requirements will be implemented is critical to tenant improvement costs and tenant layouts.

CONCLUSION

The tenant/user is often unable to gauge accurately the quality of finishes, materials, and equipment offered in a building against the costs of these appointments in rental dollars. The levels of quality should be assessed against other competitive buildings. This knowledge also benefits the owner/developer by helping him select a palette of standard finishes, materials, and equipment that is both appropriate in the marketplace and appealing to a wide range of tenants.

Building design consultation, then, when done early in project planning, can accommodate the economic needs of both the user and the owner/developer. The success of the project can then be judged by more than the esthetics of the building design.

The Inside Story: How Structure and Services Impact Office Design in Tall Buildings

Moira Moser

An architect knows only too well the desire to create from every high-rise project a structure with a sense of place, with an impact on the skyline, perhaps a visual symbol of a city's growth or a corporation's success. But the architect is also a space planner who creates offices inside other architects' tall buildings and is sometimes confronted with clumsy interior spaces that are the result of this emphasis on the exterior. Architects too seldom design high-rise office buildings around the end-user's requirements.

In the "one-off" type of building where the owner is the user, creation of good interior spaces is considered as much a part of the architecture as the form and detail of the structure itself. But when it comes to high-rise office towers, where the end user may be anonymous, too many architects concentrate on the exterior at the expense of the interior spaces. A number of examples of high-rise buildings in Hong Kong are presented here. They illustrate the impact that structure and building services have on the design of interior office spaces.

The view of Hong Kong's central business district covers the range: The Hongkong Bank Building, a "user-specific" office building; Connaught Centre, one of Hong Kong's early high-rises; and Exchange Square, one of the most recent. Hongkong Bank Building, by Foster Associates, is an owner-user, tenant-specific high-rise, reputed to be the most costly office building in the world, with the entire structure, both inside and out, designed to meet the owner-end user's specific requirements (Fig. 1).

Connaught Centre, by contrast, is one of the central district's older high-rise general office buildings (Fig. 2). The structural engineer said it was designed on the principle of a piece of bamboo with the round windows being like the holes in a bamboo flute. This rather unusual fenestration affects the harmony of the interior spaces.

Not everyone wants an office dominated by a large round window. The spacing determines whether one will have a one-window or two-window office. To make it less obtrusive, the very unusual shape demands a unique treatment, such as the method of framing the windows and the view, with a Southeast Asian motif shown in Fig. 3. The low cabinet along the wall actually houses the air supply units, a situation in which the mechanical system of the building physically limits what can be done with the interior spaces. However, on the plus side, the ceiling space is not a vast return air plenum. Return air is ducted away from the ceiling grills and this makes full sound isolation of individual offices a simple matter of carrying partitions all the way up through the ceiling void.

Exchange Square is the newest premier office high-rise in the central district (Fig. 4). The two towers, with fabulous views, are each a series of

Fig. 1 Hongkong Bank Building

Fig. 2 Connaught Centre

Fig. 3 Exxon Chemical Asia Pacific Limited

Fig. 4 Exchange Square

Fig. 5 Example of a divided corner office in Exchange Square

interlocking circles and squares in plan. However, both floor trunking and ceiling systems are laid out on a strict rectilinear grid. This grid makes enclosed offices difficult to lay out so that, even though Exchange Square is Hong Kong's most prestigious office address, some call it "Strange Square". The center section of each tower is a shallow arc, creating a symmetrical space that no architect would want to cut up. This space lends itself most easily to a symmetrical conference room. The corners of Exchange Square have the best view, the problem being that the sleek exterior skin was only achievable by pulling massive columns just inside the skin. What happens then when two partners both want the corner office? You give each one half a corner, divided on the diagonal, and hide the column in the wall between them (Fig. 5).

A recurrent problem on these floors is a lack of acoustical privacy because although the ceiling tiles are state-of-the-art, the ceiling space is one vast return air plenum. Even partitions above the ceiling with transfer ducts do not solve the sound problem. Any changes in the lighting of this integrated ceiling must be reinstated by the tenant at the end of the lease (usually 3 to 5 years) so the building's uniform lighting is supposed to serve all users, from computer operators to chief executives.

Ceilings in this prestigious building are only 8 ft high. And space above ceilings was squeezed by a few inches on each floor so that one more floor could be added to the tower within the total height. Squeezing the ceiling space of course meant not enough room for ductwork, creating a sound problem.

How did Hong Kong's office buildings develop in the last decade from Connaught Centre to Exchange Square? A look across the harbor to Kowloon will show where some very large mixed-use developments were taking place in these years. For Kowloon, these 12-story buildings really are high-rises because they push the upper limits of what is allowed in the airport flight path. New World Centre combines two hotels, transient apartments, a major shopping center and two office blocks all compliments of Skidmore, Owings & Merrill (Fig. 6). The building's unusual configuration provides the opportunity for many corner offices, as shown in plan (Fig. 7), but the irregular shape is a real challenge when planning a tight series of enclosed spaces. In this particular office, 50% of staff were executives and managers, each wanting a regular shaped, enclosed office with view and with maximum sound and visual privacy (Fig. 8). (Here the architect made a very common design flaw: window sills are about 2 in. lower than average desk height so there is a constant conflict between furniture and fenestration.) The strong angularity of the building became a planning theme in the open clerical areas (Fig. 9). The ducted air conditioning system, with no return air plenum, allowed the soundproofing of enclosed offices. Transfer air ducts in the ceiling above each door maintain the balanced air flow.

Hong Kong was a developer's paradise from 1978 to 1980. In Kowloon, Tsimshatsui East, a district of office blocks, was quickly erected and sold floor-by-floor to speculators who in turn leased them out—mostly to offices

Fig. 6 New World Centre

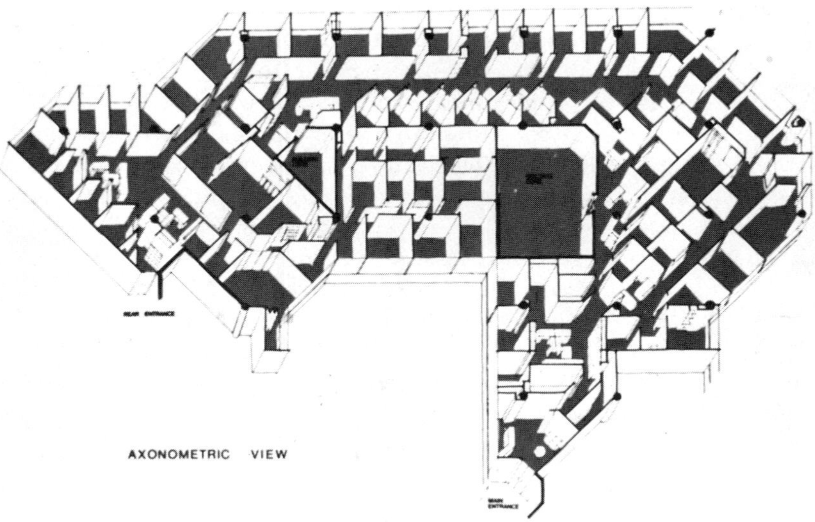

Fig. 7 Du Pont Asia Pacific Limited floor plan

Fig. 8 Du Pont Asia Pacific Limited executive office

Fig. 9 Du Pont Asia Pacific Limited clerical areas

representing commercial enterprises. Peninsula Centre is an example (Fig. 10). It is an extremely long building, with two elevator lobbies and over 20,000 ft^2 of deep space on each floor, appealing to businesses such as garment industries, which are large space users. A reasonable depth from core to exterior wall allows efficient use of the space for clerical—intensive businesses. Figure 11 shows half a floor for a toy company.

The overall space allocation of less than 100 ft^2 per person for this type of business is not uncommon to the crowded conditions of urban Hong Kong. This intensive use of space is one good reason why exterior walls that are half glazed rather than fully glazed are preferred by many tenants.

These businesses often include testing laboratories in their office premises (Fig. 12). This particular toy company needed a 20-ft long toy car test track, and it was provided. The tricky part, however, was adding plumbing for the washing machines and exhaust for the flame test oven to the building's systems without overloading them. Although they attracted a tenancy with special requirements, these buildings are designed only for typical offices.

On Hong Kong Island, the developer boom led to many speculative office buildings with sleek skins. Far East Finance Centre (jokingly called the "Maid's Gold Tooth") has columns just inside the skin, invariably set just inside the corners where the best offices should be (Fig. 13). Another example is Sunning Plaza, I. M. Pei's first building in Hong Kong, with a sleek skin, an unusual corner geometry, and fat, round concrete columns just inside the curtain wall breaking up the interior space (Fig. 14).

Fig. 10 Peninsula Centre

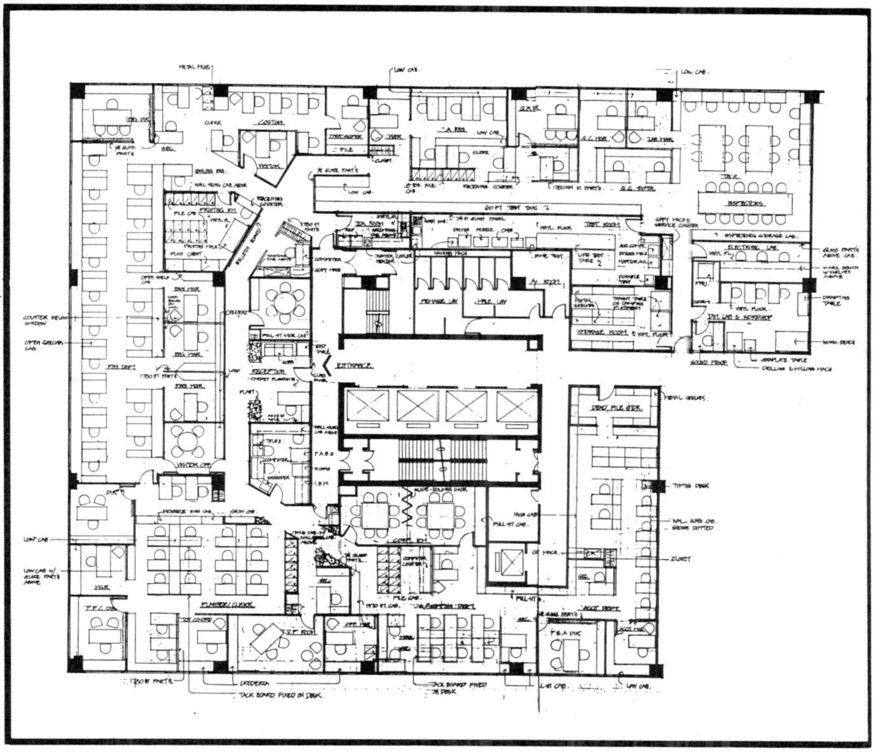

Fig. 11 General Mills floor layout

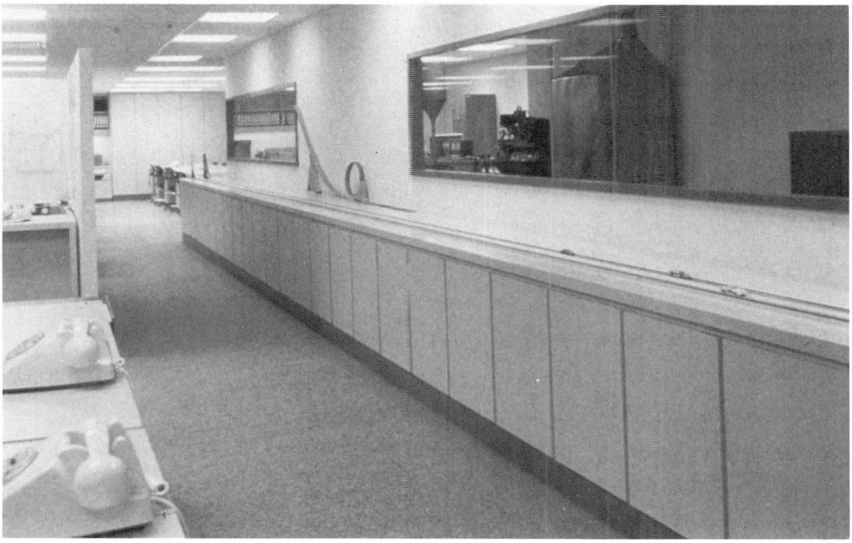

Fig. 12 General Mills testing laboratory

The unusual corner geometry is somewhat difficult to work with inside because the core-to-exterior-wall dimension is only 23 ft (Fig. 15). Here the sills are again 2 in. too low. The narrow dimension typically allows for one row of clerical stations, one circulation path, and one narrow band of offices—again working around the columns (Fig. 16).

In the central district Alexandra House is one of the best office buildings on the island in terms of planning flexibility (Fig. 17). It is not the newest, but it works well. Structural columns, though bulky, are part of the exterior wall and can be more easily dealt with. Spaces are deep and ceiling and trunking grids coincide; window sills are at the right height to accommodate desks against the wall. The architect and structural engineer had the foresight to design a knock-out slab all the way through the building, which we took advantage of to connect five floors for a major law firm (Fig. 18). The interior sill height is correct in this building and easily accommodates a desk or a standard two-drawer file cabinet.

Fig. 13 Far East Finance Centre

The core to exterior wall depth allows ample general office space plus a row of private offices. Floor trunking is well spaced and the floor slab is thick enough to allow additional chasing for electrical conduit. (This method is still cheaper in Hong Kong than undercarpet flat wire.) Sadly, a ceiling space designed as a return air plenum means that fully soundproofing individual offices becomes a near impossibility.

Edinburgh Tower is another excellent building in the central district where the structural grid dictates a reasonable module for interior spaces (Fig. 19). With a very regular 12,000 net ft^2 per floor and full-height exterior glazing, it is well designed for the mainly financial and professional firms who tenant the district (Fig. 20). Surprisingly, landlord-provided lighting was not only dully uniform but was too low. Here tubes have been added to ceiling fixtures for increased illumination and desk lamps have been added for ambience. But in plan, the coordinated window modules, floor trunking and ceiling grid lay out as well for the open plan of a banking platform (Fig.

Fig. 14 Sunning Plaza

Fig. 15 United States Lines

Fig. 16 United States Lines clerical stations

21) as for smaller enclosed spaces such as conference and meeting rooms. And a corner executive office (Fig. 22) can be laid out most generously. The structural window module, in fact, automatically determines one-, two-, three-, or four-bay offices.

Not far outside the central area is China Resources Centre, mainland China's first venture into developing and managing a commercial office building in Hong Kong (Fig. 23). Here again, the column-free depth from core to exterior wall makes it an excellent building for clerical intensive offices (Fig. 24). Punched windows instead of curtain wall allow flexibility for more efficient space allocation with an average of 98 ft^2 per person (Fig. 25). However, the building was not quite designed to meet the high level of electronic requirements of many firms, and the thin floor screed plus inade-

Fig. 17 Alexandra House

quate size and spacing of floor trunking for wiring has been a problem, as has been inadequate air conditioning for the greater heat load of electronic equipment (Fig. 26).

CONCLUSIONS

Generally the architects of high-rise office buildings tend to forget how much the structural shell and the building services systems impact the final interior arrangements. The structural shell affects the interior planning in terms of core-to-external-wall depth, window modules, floor-to-ceiling height, and window sill height. The services primarily affect the interior in terms of air conditioning versus individual soundproofing, lighting, and sufficient provision for the ever-growing number of electronic devices office tenants require. But from round windows to unorthodox shapes, Hong Kong's architects, like architects everywhere, are in love with the exterior. Only a very few know how to design a high-rise office building from the inside out.

Fig. 18 DEACONS

Fig. 19 Edinburgh Tower

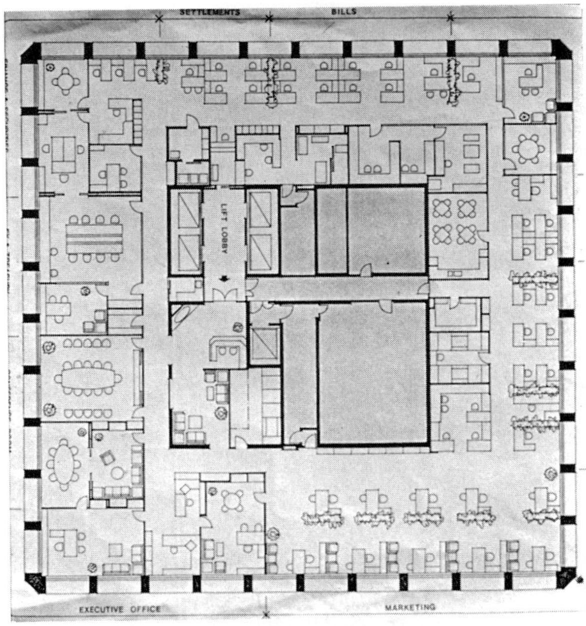

Fig. 20 Amsterdam Rotterdam Bank layout

Fig. 21 Amsterdam Rotterdam Bank

Fig. 22 Mellon Bank

Fig. 23 China Resources Centre

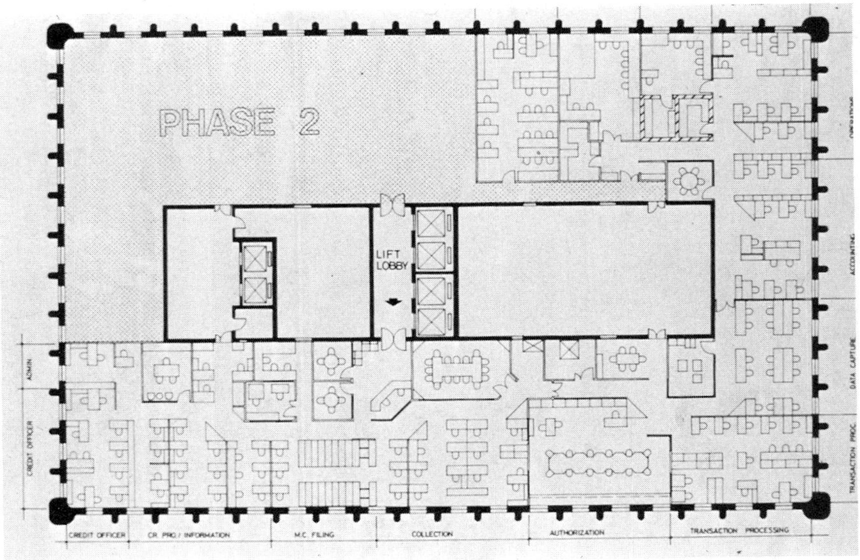

Fig. 24 East Asia Bank Card Centre layout

Fig. 25 East Asia Bank Card Centre allowed 98 ft^2 of space per person

Fig. 26 East Asia Bank Card Centre

Chicago Update

Elizabeth L. Hollander

Chicago, city of the first skyscraper, the Home Insurance Building, celebrated that building and many others at an important exhibit—150 years of Chicago Architecture—at the Museum of Science and Industry during the Third International Conference on Tall Buildings, held in Chicago the week of January 6–10, 1986. This important conference brought together professionals from every aspect of skyscraper development, creating a vital forum of discussion of the future of the world's major cities. Those who attended represented the talents and resources necessary to take a view of what is happening and what needs to happen in this industry.

This work is important. Municipal governments and planning departments across this nation and world need to hear from these professionals.

From the Chicago perspective, the development rush is fantastic. We've completed two million square feet of office space a year, every year for the last ten years, and we are absorbing space at nearly the same pace.

Chicago in 1986 was a headquarters city, a major capital on a number of levels. A recent study undertaken by the Chicago Department of Planning and Northwestern University fixed numbers to this recent growth; we have experienced construction investment of $5.7 billion in the past eight years. Projects valued at $2.5 billion will be completed over the next three years.

New construction, rehab and adaptive reuse are taking place in every corner of the city's central business district. On the north side of the river at Dearborn and Wacker the Hotel NIKKO is going up adjacent to the international headquarters for Quaker Oats. The 48-acre Dock and Canal development immediately east is a $3 billion 20-year project and is the largest downtown renewal project in the United States. Construction for the first phase, headquarters for NBC's Chicago operation, began in October, 1986.

The North Loop Redevelopment Project is also moving ahead in the same general area. This $1 billion development involves office, residential, and recreational uses. These projects will help keep Chicago's downtown a vital urban center 18 hours a day.

We are broadening our development concerns to make sure that each new tall building serves its users well by respecting the urban environment, taking into consideration such things as streetscape, sunlight, wind effects, and vehicular traffic. Buildings must be judged in a context of the whole city and its workplace.

We are convinced that the future will bring better tall buildings that meet the needs of a changing economy—one affected by a growing central city service sector and changing family demographics. The Urban Land Institute just reported that the number of one-person housing units has increased by 84% since 1970 compared to an overall increase of 35%.

We believe that Chicago's future will be as bright as our past, and that it will continue to be a center for innovative building design—inside and out.

REFERENCES/BIBLIOGRAPHY

Ludgin, M. K. and Masoti, L. H., 1985
DOWNTOWN DEVELOPMENT, CHICAGO: 1979-1984, Evanston, IL; Center for Urban Affairs and Policy Research, Northwestern Univ.

Ludgin, M. K. and Masoti, L. H., 1986
DOWNTOWN DEVELOPMENT, CHICAGO: 1985-1986, joint publication of Northwestern University Center for Urban Affairs and the city of Chicago, Department of Planning.

Cityscape view with Grant Park and Burnham Harbor in foreground. Left to right are Sears Tower (the world's tallest building); Prudential Building (the first major tall building erected after the depression in the early 1950s); Standard Oil of Indiana Building (white marble sheathed, narrow windowed for energy efficiency); John Hancock Building; and Lake Point Towers (residential).

Aerial view of North Lake Shore Drive, "Gold Coast" area of city with Mies Van Der Rohe apartments in foreground. John Hancock Building (mixed-use, residential and commercial) dominates photo right. Standard Oil and Sears Tower to the rear and left of Hancock.

Bird's eye view looking north up Michigan Avenue, an elegant commercial street with fine shops, offices, and condominium residences. The Neo-Gothic Tribune Tower in foreground; The Hancock Building and Water Tower Place at the top of the photo.

San Francisco Downtown Plan and the Skyscraper

George Williams

In essence, the Downtown Plan is about how to come to terms with the skyscraper or, to use the more contemporary but less evocative term, the high-rise.

San Francisco's "development wars" have caused that term to be used by some as an epithet. These wars have come about because the face and character of San Francisco have been dramatically changed by the explosion of high-rise office construction that has occurred in the city over the last 20 years. In that time the amount of office space has more than doubled from 2.4 million m^2 (26 million ft^2) to over 5.6 million m^2 (60 million ft^2). Most of that development has occurred in the 140-acre downtown office district. Downtown development—and the high-rise—has been a raging political debate for over a decade. With increasing frequency opponents of downtown development have gone to the ballot with a voter initiative to slow down or stop the construction of high-rise office buildings. They have gotten from 40 to 49.8% of the vote each time.

It was in this environment that the Downtown Plan was born. In technical terms the Downtown Plan consists of an area plan that was adopted as part of the city's master plan and a series of amendments to the city's zoning ordinance which became effective on October 17, 1985.

The plan deals primarily with issues of growth management and urban design. It includes a series of exactions for downtown support infrastructure:

$54/m^2 ($5/ft^2) of office space must be contributed to support expansion of public transit, $57.50/m^2 ($5.25/ft^2) must go to expansion of the housing supply, $35/m^2 ($3.25/ft^2) must be contributed for a park downtown park, and $10.75/m^2 ($1.00/ft^2) must go to child care facilities. In addition 1% of the hard construction cost of the building must be spent for on site public art. All these exactions total about $156/m^2 ($14.50/ft^2).

The plan also establishes a limit on the amount of office space that can be approved city wide over a three-year period. The average is 88,250 m^2 (950,000 ft^2) a year, which is one half of the average construction rate over the past 20 years and 1/3 the average construction rate of the last five years. This limitation was not part of the original proposal. It was added by the mayor and the board of supervisors (our local city council) because of their political sensitivity to their constituents concern about the pace of growth. Under the limitation, the amount of office space is to be allocated competitively. In January, 1986 there were 34 projects, totalling 0.5 million m^2 (5.5 million ft^2) under formal review in the department, which must compete for about 93 thousand m^2 (1 million ft^2) of space. The decisions are to be made by the city planning commission, an appointive body, acting on the recommendation of the director of planning. The criteria to be used in making the decision are stated in the ordinance in very general terms. The include: location advantages and disadvantages of the project; the demand for the kind of office space to be provided; whether the plan furthers or detracts from the policies of the master plan; and most significantly, the architectural quality of the building. It is because of the last feature that the process is being called the "beauty contest" by the press.

Projects will be batched for comparative evaluation and decisions will be made every six months. The first formal review period was begun in 1986. Development decisions traditionally are made by the market place, so this system for rationing office space appears to be unique.

With respect to urban design, zoning was used in a very precise and fine-grained way to implement urban design objectives. Through zoning the plan deals with the location, size and shape of buildings, skyline appearance, streetscape, and preservation. In a single legislative act all of the landmark quality buildings in the downtown were designated and five preservation districts were created.

The plan deals with the pedestrian environment, mandating ground floor retail uses in the retail districts, and providing strong incentives to ground floor retail elsewhere, particularly in the financial district. It requires the provision of open space, principally outdoors but allowing some indoor spaces. It also deals with microclimate issues such as shadows and wind. Close work was done with the College of Environmental Design at the University of California at Berkeley. Using an old WPA model of the city, which was lent to the university, a simulation of how downtown would look if built by these rules was made.

SAN FRANCISCO—THE CITY

San Francisco has always been an urban place. Streets are precisely defined by buildings standing shoulder to shoulder at their edge (Fig. 1). From the beginning its urban qualities were so pronounced that Californians from one end of the state to the other referred to San Francisco simply as "The City" (Figs. 2 & 3).

The 1906 earthquake and fire, however, virtually leveled the city, and a more restrained architecture based on the Beaux Arts style replaced much of the previous downtown. A conscious attempt was made to coordinate design efforts to achieve an overall effect (Fig. 4). Belt courses, for example, were adjusted to reinforce the lines set by the few good surviving buildings (Fig. 5). Bold projecting cornices, bold enough to be seen in our imperfect peripheral vision, framed and defined a human scaled streetscape (Fig. 6). And although quite different styles were often employed—because of shared traits—the buildings worked together to a common effect. The net result was a city well-suited to the needs of the pedestrian, the daily user of downtown. Most important, the early architects were not just creating buildings, they were assembling strong, richly-endowed street spaces and streetscapes.

Streets of this kind offer new rewards with every visit. Neither individual buildings nor their combined effect can be fully absorbed in a glance, but invite repeated study and examination. There are always features or qualities not observed before (Fig.7).

Up until the 1950s cupolas, domes, finials, ballustrades, and slender building tops softened the presence of large bulky commercial buildings on the skyline (Fig. 8). The net effect of adding these small scale features was to make these larger buildings fit harmoniously with their diminutive neighbors. The skyline of downtown thus shared many of the qualities of other parts of the cityscape. And in a city of hills where downtown buildings are seen against a fine-scaled backdrop of residential buildings on nearby hills, that is important (Fig. 9).

San Francisco retained its delicate skyline and cohesive streetscape through the 1950s; then the forces for change began to mount (Fig. 10). The old skyline was quickly hidden by variations of the slab and the standard box-top office tower. Without small scale features providing a visual linkage with the scale of the surrounding city and imparting a human liveliness to their forms, the aggregate effect was often contrary to the character of the surrounding city (Fig. 11). At the street level the new buildings ignored their neighbors and offered little of the qualities that give scale and visual richness to the street (Fig. 12).

Citizens discovered that their cherished views of the bay and city were not guaranteed. The whole form and character of the city seemed under attack. Then a series of threatening mega-projects along the northeast waterfront surfaced, triggering a citizen revolt that helped secure adoption of the Urban Design Plan in 1971 and place limits on the height and bulk of buildings.

THE PLAN

The height and bulk controls succeeded in stopping the proliferation of view-blocking slab towers and super scaled projects but were inadequate by themselves to secure building forms sensitive to the context (Fig. 13). In many other ways new development ignored existing patterns and characteristics of their surroundings. Stronger measures were clearly needed. To project the qualities that make San Francisco's downtown unique, to moderate the rate and ultimate amount of growth, and to assure that new development will enhance rather than detract from the city, the Department of City Planning is proposing new rules and standards for future development.

A fundamental premise of these new rules is that great cities are not built by tearing down fine old buildings (which in San Francisco were being lost at the rate of seven or eight a year) only to replace them with less distinguished efforts. Henceforth 250 buildings of the highest architectural importance will be protected from demolition except in the most extreme situations (Fig. 14). But the city's new preservation controls mean much more than the protection of fine old buildings—they mean the conservation of a unique urban environment. These buildings are conserved not as lifeless museum pieces, but as a vital functioning part of the city. Almost the entire retail district and five smaller areas in the financial district will be made conservation districts. Here there would be incentives to preserve 183 buildings of somewhat lesser quality—called contributory buildings—which, together with the more significant buildings, create areas of unique quality and character. Here new development would be closely controlled to make certain the new buildings "fit" and do not detract from the scale and character of the area.

Transferable Development Rights

The plan's preservation policies are made workable by allowing the transfer of the unused development potential from the sites of architecturally significant and contributory buildings to other sites within downtown where new development is more appropriate (Fig. 15). This concept of transferable development rights (TDR) is a keystone of the plan. It helps achieve the preservation of buildings. At the same time it shifts the focus of new development from the already dense North of Market area where the architecturally important buildings are concentrated , to an underdeveloped area, immediately south of the high-rise core where higher heights and densities are appropriate. Under the new rules these higher heights and densities may be achieved only through the purchase of TDR.

Working with economists and the Environmental Simulation Lab at the University of California Berkeley, we have analyzed where the next 930 thousand m^2 (10 million ft^2) of office development would be likely to occur under the old and new rules. In Fig. 16, under the old rules, the vertical line in the middle is Market Street with the highly developed North of Market area to the right and South of Market to the left. Under the new rules, new development is dramatically shifted to South of Market (Fig. 17). Project proposals that are being reviewed indicate that this shift will occur and that the TDR scheme will work.

Open Space

The manmade canyons of a high-rise center can be exciting vital places to be, as long as one can easily get to a sunny comfortable open space. The plan is premised on the principle that no one should have to walk more than a couple of blocks to find some kind of outdoor space. The accessibility of diverse forms of recreation and settings contributes to the richness and satisfaction of urban living. On a warm San Francisco day a sunny park can be delightful, but on a windy, foggy or rainy day availability of a weather protected arcade, greenhouse, or galleria would be more enjoyable for many

Open space does not always have to be on the ground. A sunny terrace high on an office tower with spectacular bay views can offer an exciting alternative to more traditional open space. At less exalted elevations open space is also potentially exciting and successful, provided public access is obvious and it possesses the qualities that will make it attractive to the public.

Because adequate open space is important for a liveable downtown, the new plan makes its provisions mandatory on site, if possible, offsite if not. The open space requirements, stated as a ratio of gross square feet of open space to gross square feet of development are found in Table 1. The criteria for acceptable open space include minimum standards for seating, sun exposure, public access, wind protection, and food service. The city sees no compelling reason to accept substandard open space in the future.

The largest amount of open space in any downtown is in its streets, and when these streets are attractive and inviting they too lend themselves to recreational use. San Francisco's retail district is nestled close by the office core offering many workers the alternative of a shopping trip or window

Table 1 Plan for open space

C-3 District	Ratio of Open Space to Development (in ft^2)
C-3-O	1:50
C-3-R	1:100
C-3-G	1:50
C-3-S	1:50

In mixed commercial-residential projects, some open spaces could serve both uses.

gazing during the lunch hour. Streets in the retail area are relatively sunny and bright because of the many small buildings mixed in with somewhat larger ones. The presence of sunlight and the comfortable scale of the buildings is considered an important ingredient to the continuing success of the area. It assures that the many buildings with the area designated for preservation will be saved within a context that respects their scale.

Height Limits

The reduction of height limits in the retail district and setting its eastern boundary at Kearny Street is part of a larger pattern of reductions between the office core and a much smaller cluster of towers near the intersection of Van Ness and Market. The intent is to retain this visual separation and restrict the growth of the office core along Market Street.

The 1971 height limits that ranged from 43 to 122 m (140 to 400 ft) in the retail district—with heights in excess of 91 m (300 ft) permitted in most of the area (Fig. 18)—are reduced to a base height of 24 m (80 ft) with the potential for an additional 15 m (50 ft) provided it did not block required sun access on specified streets or contribute to significant shadowing of the sidewalk during the important part of the day, and the added height respects the scale and pattern of the street wall (Fig. 19). Height limits covering the northern half of the office district are generally lowered, in effect tilting the opportunity to build large buildings toward the south. To avoid the visual "benching" of buildings that can occur when there are many buildings of similar height in an area, the size of the individual zones have been reduced.

Not as graphically apparent but dramatically affecting the potential for future development are the reductions in the Floor Area Ratios (FAR) governing the amount of development that may be placed on a given site. The permitted ratio of development to land area has been reduced in all parts of the downtown (Fig. 20).

Bulk Controls

While the 1971 bulk controls succeeded in preventing further proliferation of awkward view-blocking slab-shaped towers, early on it was clear that, as formulated and applied, the controls were inadequate to prevent high-rises (widely perceived as being too large) from overwhelming or manhattanizing the city. It was not so much their height as the intrusive and uncompromizing form of the box shaped towers (Fig. 21). As skyline features they formed harsh compositions. Their flat tops fought the vertical thrust of the hills and the delicate scaled tops of the older buildings (Fig. 22).

Typical box style buildings, however fine architecturally, make no gesture to street scale or the contextual needs of meritorious older buildings. Their bland repetitive facades (Fig. 23) do little for the definition and enrichment of streetspace, nor do their shapes respond to concerns about sunlight and air. The inflexibility of the box-top format telegraphs a message of indifference to the context and the needs of people.

Bulk controls cannot address all these problems directly, but to the degree that they can lessen the stranglehold of the box top tower formula it facilitates a more responsive architecture (Fig. 24). To break the box form and generate an upper tower form more in scale with the cityscape, the top 37% of towers over 61 m (200 ft) high are restricted in mass to an equivalent to a modestly scaled building typical of many older apartment and office buildings in the city. The portion of the upper tower so limited diminishes as the height of the tower drops below 67 m (220 ft). No upper tower reduction is required of buildings less than 49 m (160 ft) in height. To encourage even more slender tops, a 10% increase in building height above the height limit is offered in exchange for a decrease in the volume of the upper tower. In effect the developer transforms square footage into smaller but more numerous floors (Fig. 25).

A side setback is also required starting at one and one-quarter times the width of the abutting street (Fig. 26). The intent is to assure a reasonable spacing of towers and to allow sunlight and air to adjacent buildings and to the street. The requirement affects both the bulk and placement of buildings.

The bulk control formula is not in itself a formula for acceptable architecture, merely the starting point. It is the designer's task to massage the volume about, within the permitted averages and come up with a coherent, graceful design. There are an infinite number of ways this design can be accomplished (Fig. 27).

Building Form

San Francisco's temperate, coastal climate makes two possible impacts of building form—shadow and wind creation—of particular concern since the loss of sunshine or the creation of ground level wind currents can turn a pleasant pedestrian environment into a very unpleasant one. The plan addresses these issues in a number of ways (Fig. 28). In some cases requirements that buildings not encroach on a solar access angle have been imposed. In other cases shadow studies will be mandated to establish a building form that will reduce the amount of shadow the building will create. City parks and squares will receive even greater protection as a result of a recent public referendum. New development above 12 m (40 ft) cannot be allowed if its shadow would have a significant impact on the use of a city park or square anytime between an hour after sunrise and an hour before sunset, year round. An example is the so-called solar fan for Union Square. A solar fan indicates the heights above which a shadow will be cast on the square. The solar fan is created by all of the city parks and squares in the retail and financial districts taking into account the shadows cast by existing buildings.

With respect to wind, performance standards have been set and wind tunnel tests will be performed on each new building to determine whether a proposed building form will meet the standards. Changes will be required if it does not.

CONCLUSION

The plan recognizes that its prescriptive rules may in certain circumstances be too inflexible and unnecessarily rigid. To provide flexibility, each rule has carefully stated grounds for granting exceptions, provided the general objective of the rule is still met. Thus, limited exceptions to the bulk and setback rules can be granted if the result is a superior design and the exception does not have adverse consequences such as loss of light and air to adjacent buildings or public spaces. Similarly, if meeting some of the shadow and wind rules results in an ungainly building form and substantially reduces the development potential of the site, exceptions may be granted if the added shadow or wind impacts are not substantial.

The results of all of the plan's proposals have been carefully tested first, by the number of projects being worked on by developers and architects in anticipation of the new rules, and secondly by simulating likely development at the UC Berkeley Simulation Lab. Figure 29 shows, in model form, San Francisco as it is today. Figure 30 is how it might look in the year 2000 with the completed Yerba Buena Gardens and already approved projects and development under the downtown plan, viewed from high above Twin Peaks.

Fig. 1 Historic San Francisco (nineteenth century)

Fig. 2 Historic street scene (nineteenth century)

Fig. 3 Historic view of Market Street (nineteenth century)

Fig. 4 Example of relationship of buildings at the street level

Fig. 5 Example of downtown belt courses

Fig. 6 Downtown cornices

Fig. 7 Floral ornamentation

Fig. 8 Example of a dormer

Fig. 9 Examples of building tops downtown

Fig. 10 1940s view from Bay Bridge

Fig. 11 Embarcadero Center

Fig. 12 Schematic of building under 1971 bulk controls

Fig. 13 California Street

Fig. 14 Phelan Building

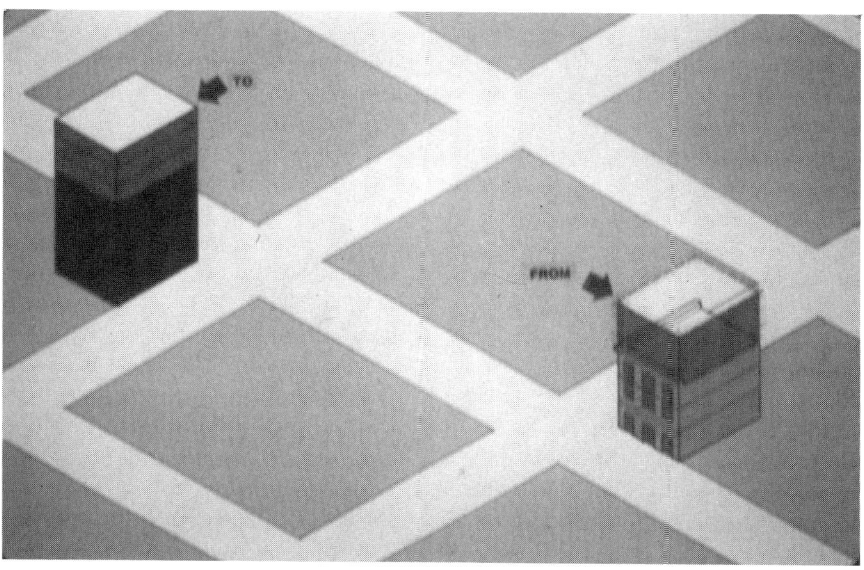

Fig. 15 Diagram of TDR

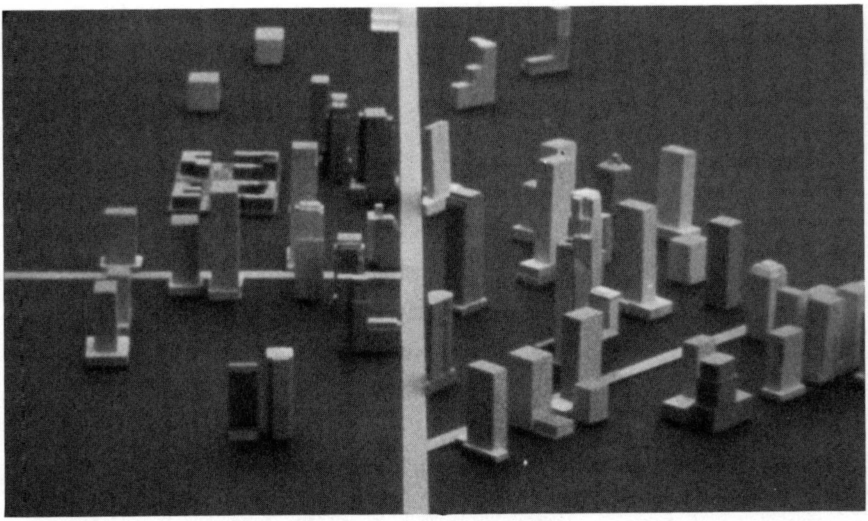

Fig. 16 Model of San Francisco developed by UC Berkeley's Environmental Simulation Lab. Development possible under pre-Downtown Plan rules.

Fig. 17 Model of San Francisco developed by UC Berkeley's Environmental Simulation Lab. Development possible under Downtown Plan rules.

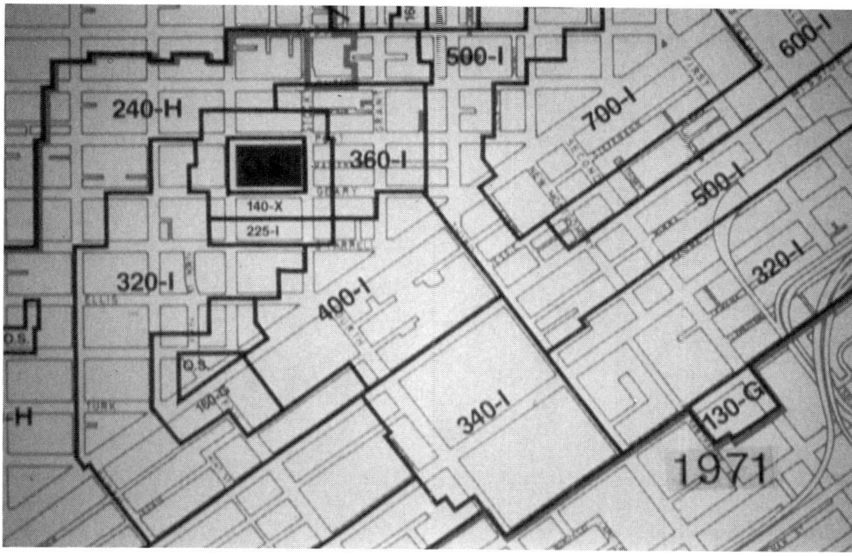

Fig. 18 1971 height map

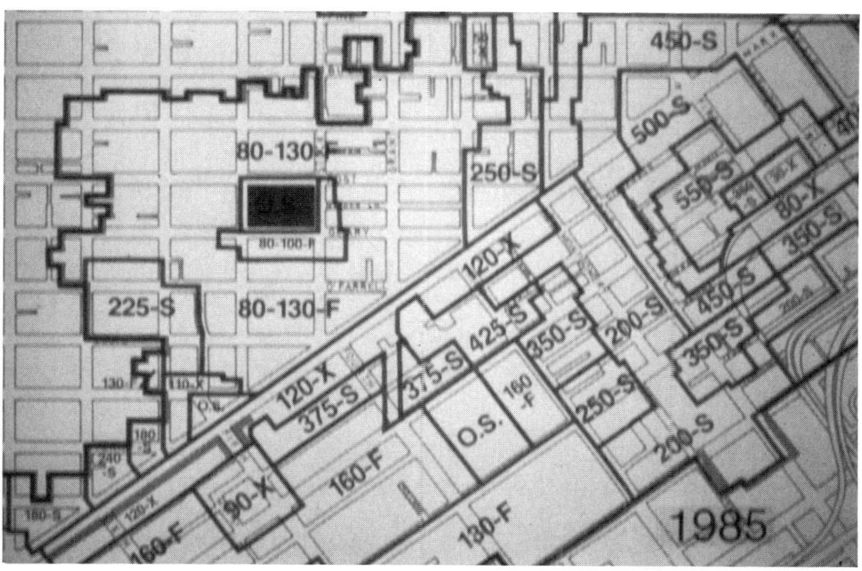

Fig. 19 Downtown Plan height map

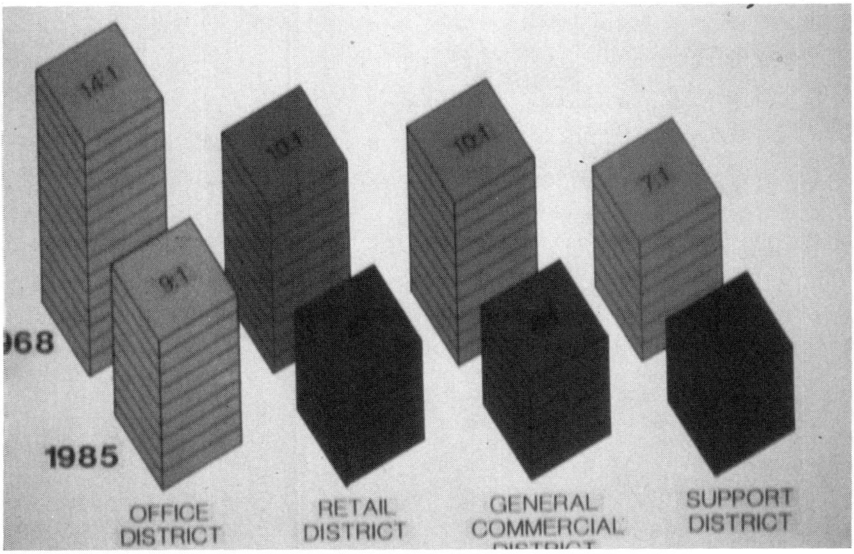

Fig. 20 Diagram of FAR limits under the Downtown Plan and pre-Downtown Plan rules

Fig. 21 Example of existing building tops

Fig. 22 Downtown skyline—example of benching

Fig. 23 Embarcadero Center

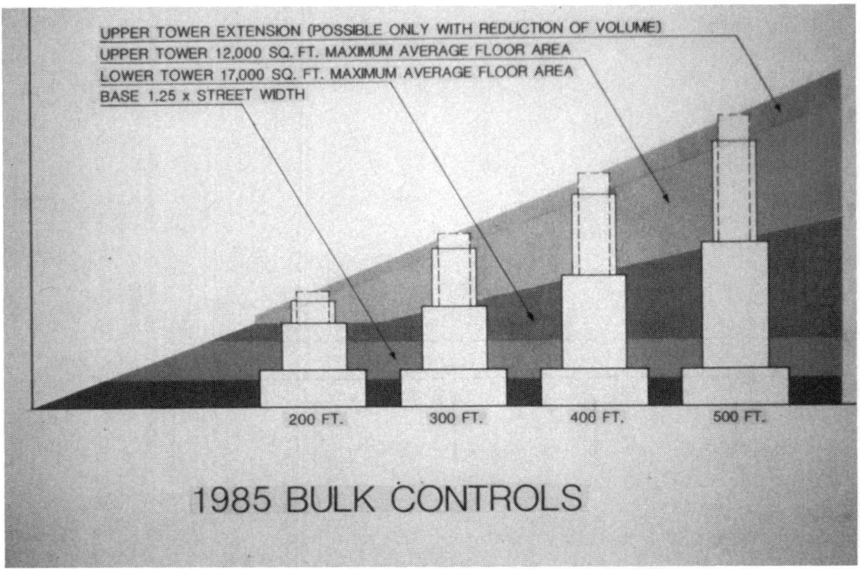

Fig. 24 Downtown Plan bulk diagram

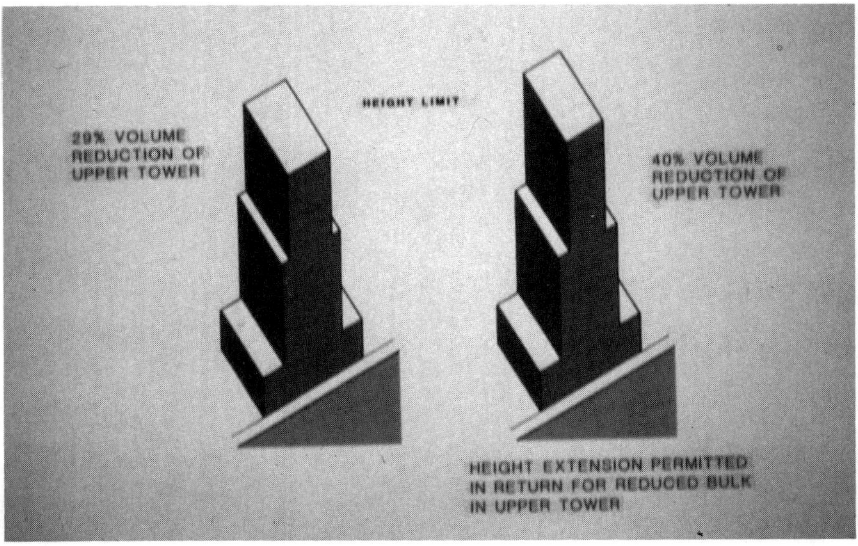

Fig. 25 Downtown Plan bulk diagram—height extensions

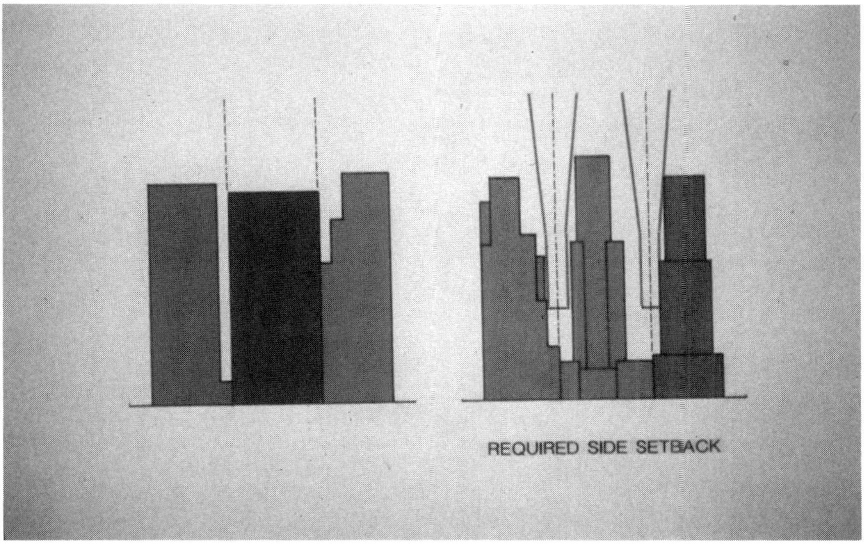

Fig. 26 Diagram of Downtown Plan side setback requirements

Fig. 27 Schematic diagrams of examples of building forms permitted under the Downtown Plan

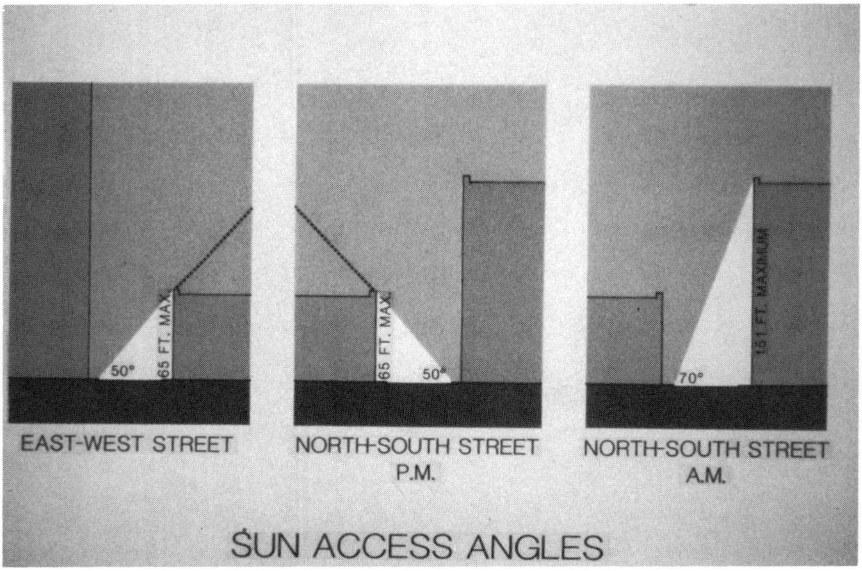

Fig. 28 Diagram of sun access angles

Fig. 29 Model of possible downtown skyline in year 2000 with Downtown Plan rules

Fig. 30 Aerial from Twin Peaks of model of possible downtown skyline in year 2000 with the Downtown Plan

The Chicago Perspective: Response to the San Francisco Plan

David R. Mosena

San Francisco's new Downtown Plan is innovative, visionary, far reaching, and tough—probably the toughest development plan of any city in the country. It is also experimental in the sense that it attempts to do things that no city has tried to do on this scale before.

STRENGTHS

The plan's emphasis on contextualism is important. It demonstrates a strong concern for the collective urban fabric. Architects focus primarily on individual buildings. City plans should address the collective, overall effects of development. And San Francisco's plan does this. Chicago is often praised for its architecture—its individual buildings—but is sometimes criticized for a lack of overall urban design—the broader context.

The San Francisco plan attempts to achieve a compact office core, which is a good idea. Chicago's central area is expanding in three directions, leaving some soft spots in the center. Given an expected contraction in the office building market, the lack of a compact core could present redevelopment problems here in the future.

San Francisco's plan shows great concern for the human scale—the pedestrian environment, the wind tunnel effects of tall buildings, and solar access. These concerns are important to make and keep a city liveable, and they are especially important in San Francisco.

Finally, the plan's efforts to expand housing opportunities in the downtown are laudable. Many cities are making efforts to do this. Chicago has been particularly successful at rimming the Loop with inner city housing (Fig. 1). In San Francisco, where the housing market is one of tightest in the country, it is a particularly important issue.

RESERVATIONS

San Francisco's Downtown Plan is, in the author's opinion, experimental. It attempts to do some controversial things, such as design control, on a far reaching basis. How successful they will be is yet to be determined.

In its attempts to control building design, it goes farther than any other major city ever has in trying to regulate esthetics. But here we get into matters of taste. Reasonable minds can disagree. There often is no *right* answer. Stopping the proliferation of office building *box tops* is one thing; creating great architecture is another. When the Trans-America Building was built in the early 1970s it created much consternation in San Francisco. Today it is one of the more treasured features of an increasingly bland skyline.

Fig. 1 River City is a mixed-use predominantly residential project on the south edge of Chicago's office core *(Photo by Stephen Roman; Courtesy: City of Chicago)*

Of course, San Francisco's attempt to improve its architectural heritage should be encouraged. The trick will be to encourage creative design of lasting value that enhances the city rather than substituting one currently popular style for another.

Chicago's Planning Department is placing increasing emphasis on design review. But most of the focus is on height, bulk, traffic flow, and the street level effects of buildings. They play a minor role in architectural detail, except in historic districts or designated urban design corridors such as Michigan Avenue and the LaSalle Street canyon (Fig. 2). But Chicago has been more fortunate than San Francisco by having a rich architectural heritage supported by a host of some of the world's leading architects.

Restrictions on height limits are another esthetic feature that can be difficult to regulate well. The real issue in Chicago is not height per se, but the effects of massing and bulk (Fig. 3). As developers attempt to achieve their full potential of floor area ratios under the new laws in San Francisco, care must be taken not to let height restrictions alone produce other negative effects resulting from greater bulk and massing. In some situations slender height can be preferable to shorter bulk. Again, context and overall effect is the key.

Transferable development rights, a major element of the San Francisco plan, is also a technique that can be difficult to implement. Success depends to a considerable extent on local market forces. The experience of other cities in using this technique has not been all that promising.

CONTRASTS WITH CHICAGO

The San Francisco Downtown Plan is unique to that city. It contains elements and techniques that are interesting and merit close watching by others, but much of it will not be directly transferable to other cities.

Chicago, for example, is a much different city. Whereas many of the same concerns are shared in theory, they are put into practice differently. As mentioned before, building height in and of itself is less of a concern in Chicago, the home of the world's tallest building, than is building bulk, massing, and the negative effects of too much density in some areas.

San Francisco appears to want to limit office growth (and office workers, for that matter). Chicago, on the contrary, is going through a shift from a manufacturing-based economy to a service-based economy. Chicago is encouraging office development, and still has much room to grow.

San Francisco has one of the tightest housing markets of any city in the country and is taking steps to relieve it. Chicago has housing problems too, but of a different character. Chicago's problem is a shortage of low and moderate income housing in the face of cutbacks in federal housing subsidies. Average housing costs are about half of what they are in San Francisco.

Floor area ratios, a big concern in San Francisco, are less of a concern in

Fig. 2 Manufacturer's Hanover Plaza employs contextual design in enhancing its place in the LaSalle Street financial corridor *(Photo by Fidinem USA; Courtesy: City of Chicago)*

Fig. 3 123 North Wacker Drive Building is typical of the quarter-block-sized office development projects *(Photo by Rubloff Co.; Courtesy: City of Chicago)*

Chicago. Some buildings with the highest FARs have less negative impact than smaller ones, because of the manner in which buildings are designed on sites, how setbacks are used, and other design considerations.

Chicago shares San Francisco's concern for urban design and context. Chicago is now in the process of implementing a number of review measures that will improve the urban design review process and the end products of it. The Planning Department also intends to take a new look at the bonus system in the zoning ordinance to make it more relevant to both developers and the broader concerns of the public at large. The Department is currently in the process of preparing urban design guidelines for a series of downtown districts—the Near South Side, River North, and Streeterville.

CONCLUSION

San Francisco's Downtown Plan contains many good principles supported by most cities, as well as by good architects and developers too. Although it should serve as an inspiration to other cities attempting to address these valid public concerns that make cities liveable, the mechanisms of transferring these techniques must be sensitive to local conditions and needs.

The San Francisco Downtown Plan: A Tale of Two Cities

Jeffrey Heller

On February 27, 1913, The Board of Estimate and Apportionment authorized the nomination of a "Heights of Buildings Committee" on the grounds that the

> "time has come when effort should be made to regulate the height, size, and arrangement of buildings erected within the limits of the city of New York, in order to arrest the seriously increasing evil of the shutting off of light and air from other buildings and from the public streets, to prevent unwholesome and dangerous congestion both in living conditions and in street and transit traffic and to reduce the hazards of fire and peril to life." (Mujica, 1929)

Seventy years later, for some of the same reasons, San Francisco put forward its comprehensive Downtown Plan. Architecture is a social art, not a pure art. Buildings must serve our physical, cultural, and social needs. This is especially true in tall buildings. The bigger the building, the greater the social responsibility. The more sensitive the building to those responsibilities, the greater its public value (Fig. 1).

Buildings are designed to shelter and to fulfill both social and symbolic needs. Aside from the historically documented congestion of older cities, humans had never before faced the substantially negative environmental impacts that tall buildings can create. Principally, these are the unpleasant side effects of shadow and wind. The tall building can become our enemy instead of our servant.

New York's zoning law of 1916 was the first set of serious regulations in the United States to acknowledge the need to manage those problems. (Boston had established regulations earlier, but they were not as comprehensive.) That law produced the marvelous New York skyline of the twenties, thirties, and forties. The Seagram Building honored that law with a brilliant modern approach; the great plaza was created where setbacks were necessary and the tower rose free and clear on the central portion of the site. After the Seagram Building, a series of changes allowed international buildings without the plaza, building out to the property lines, which radically altered the character of our modern cities. San Francisco has suffered from the same conceptual lapse, and the row of boxes along Market Street prove it (Fig. 2).

A series of projects in San Francisco over the last decade has begun to vary by degrees from that situation. This is principally due to San Francisco's inherent sensitivity to the quality of its environment. Levi's Plaza responded to the context of Telegraph Hill and the need for openness in that area. Skidmore, Owings & Merrill's 444 Market Street and 5 Fremont Plaza were early attempts to shape and erode the upper tower mass before any written

Fig. 1 Downtown district, San Francisco *(Photo by Carl Wilmington)*

guidelines emerged. For the previous five years an even more radical approach was taken with the Washington/Montgomery Tower, First and Market Street, and 71 Stevenson Street. The latter design actually conformed to the emerging downtown controls.

The San Francisco Downtown Plan was first conceived around 1980. In 1981 *Guiding Downtown Development,* the first official document, was released (San Francisco Department of City Planning, 1981). Studies continued (Fig. 3), and in 1983 the *Downtown Plan* was published and a building moratorium was instituted for a one-year period. In 1985 the plan was approved amidst continuing political controversy, which has created several uncomfortable conflicts that threaten the Downtown Plan's integrity.

The plan itself is most comprehensive in its requirements and ultimately could be good for architects (Fig. 4). It is the strictest in the country, and this has probably contributed to the stability of the San Francisco market. The plan requires developer contributions for housing, transportation, preservation, open space, and art. Additionally (and for the purpose of this discussion most importantly), it also controls aspects of massing and design. The controls, some of which are remarkably consistent with those in the New York Zoning Ordinances of 1916 as well as that city's newest laws, include regulations for height and bulk (Fig. 5), setback, street relationship, density, surface and

Fig. 2 Model of downtown San Francisco, with existing and proposed projects. Architects: Heller & Leake

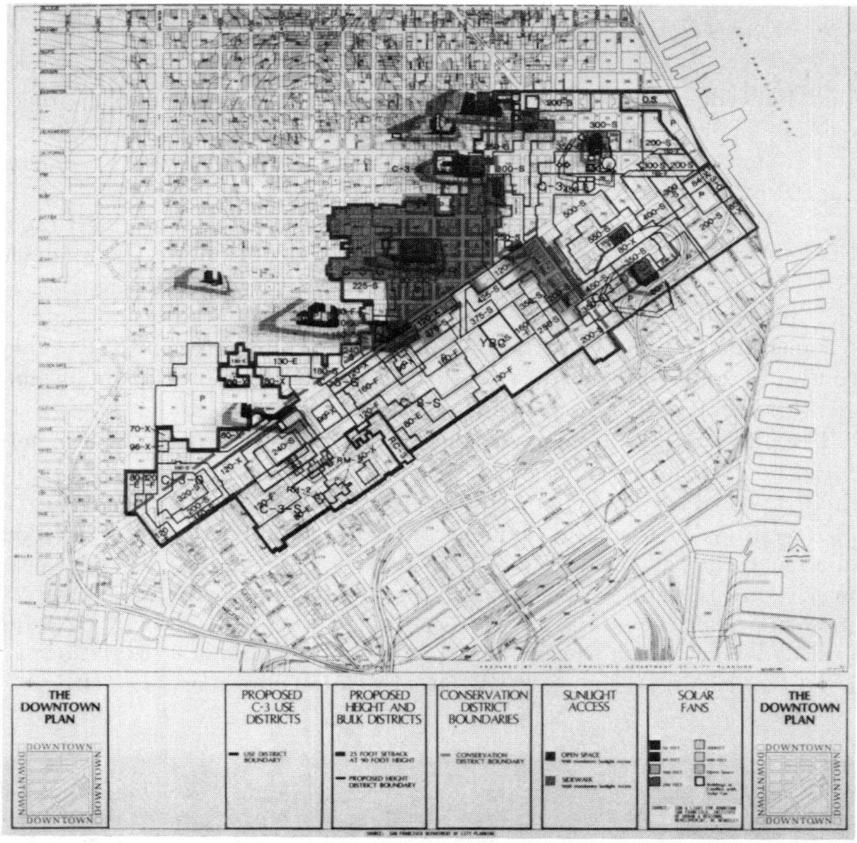

Fig. 3 Composite overlay controls of San Francisco Downtown Plan

Fig. 4 Skyline analysis of San Francisco Downtown Plan

texture, and wind. Wind is included because the stepping and terracing of buildings has much to do with the mitigation of wind impacts at street level.

Two major results for professionals have been obtained: (1) The job of design for architects has become more interesting. The various architectural groups working with the Department of City Planning have increased the flexibility of the Downtown Plan rules to allow substantial latitude in building shape. The architect firm of Heller & Leake has been instrumental in specific changes regarding setbacks (Figs. 6 and 7). (2) The demands of the Plan force developers to accommodate better design, so that the architect does not have to fight that battle alone.

The political "No Growth forces" in San Francisco have persuaded members of the board of supervisors and the mayor to institute an annual growth limitation measure. This measure has once again thrown the development process in San Francisco into uncertainty. Some of the problems that arise are as follows:

1. Unbuilt, pre–Downtown Plan, approved buildings may be built in the vacuum created by a lack of buildings under the Downtown Plan.
2. Politics may prevail over good design in the selection process.
3. Competition is created between central core buildings and outlying

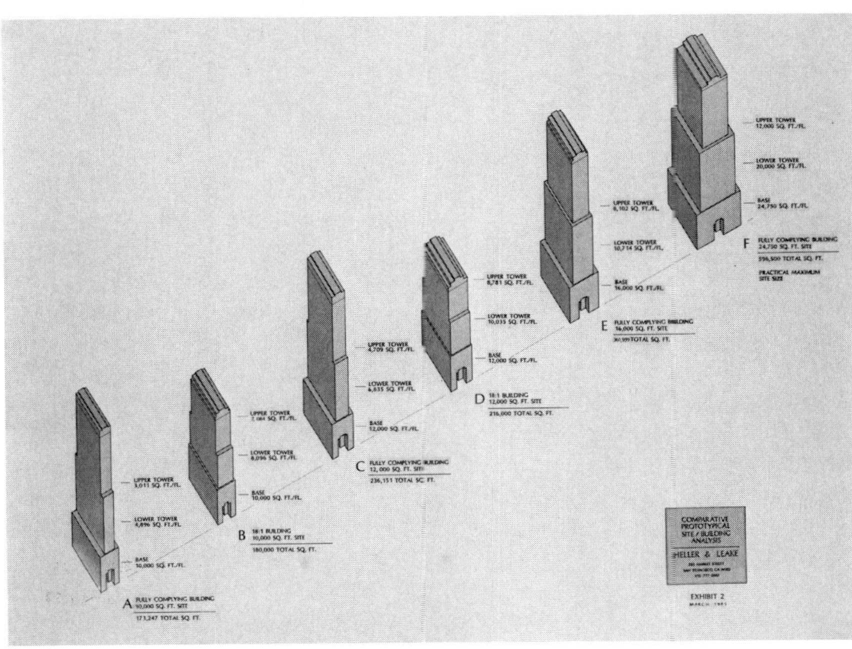

Fig. 5 Comparative prototypical site/building analysis subject to Downtown Plan controls

back-office buildings because the annual limit is citywide, confusing the Downtown Plan and non–Downtown Plan projects.

4. The limited amount of space available on an annual basis in concert with other politically hot issues, such as sunlight access, could produce a leveling effect wherein only short, squat, stepped, small buildings are approved in order to spread the available square

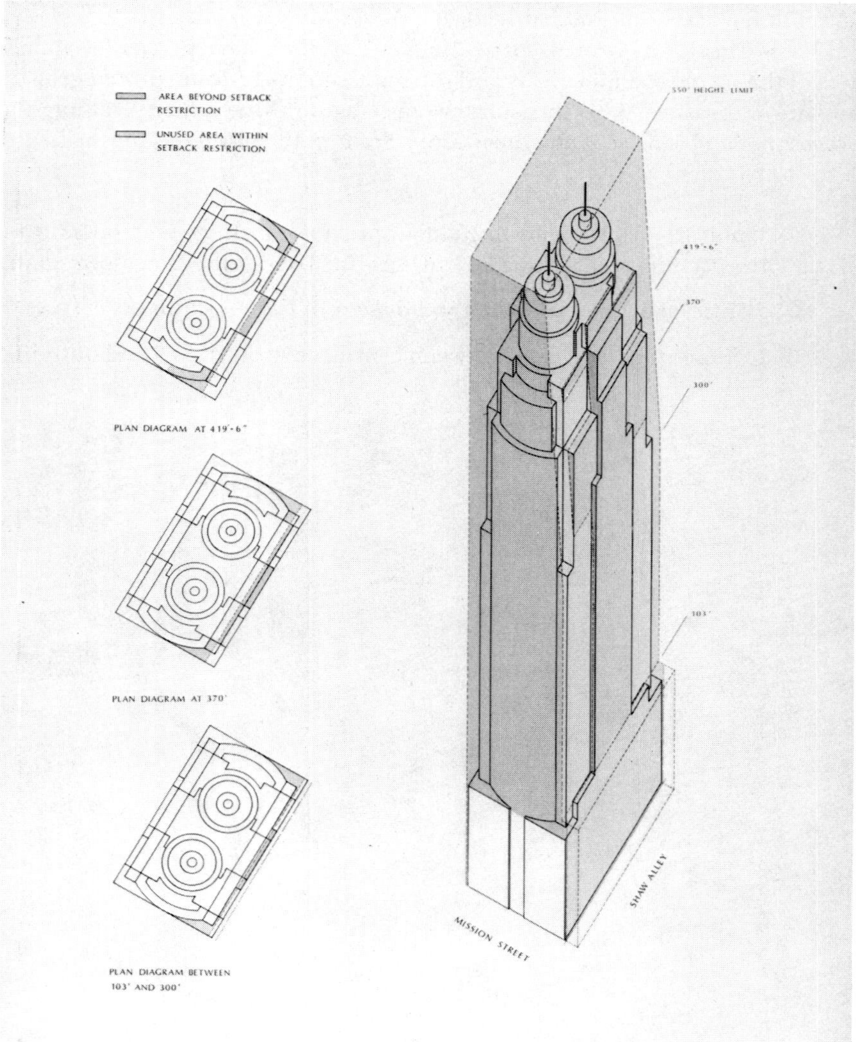

Fig. 6 Setback analysis subject to Downtown Plan controls

footage among as many applicants as possible. This would conflict with urban design goals of the plan for creating shape and transition in the skyline (Fig. 8).

The Downtown Plan, as a document, is an excellent tool. It is comprehensive and sensitive to urban issues and pioneering in its ambition to synthesize and control the forces at work in our modern cities. The darker side of the

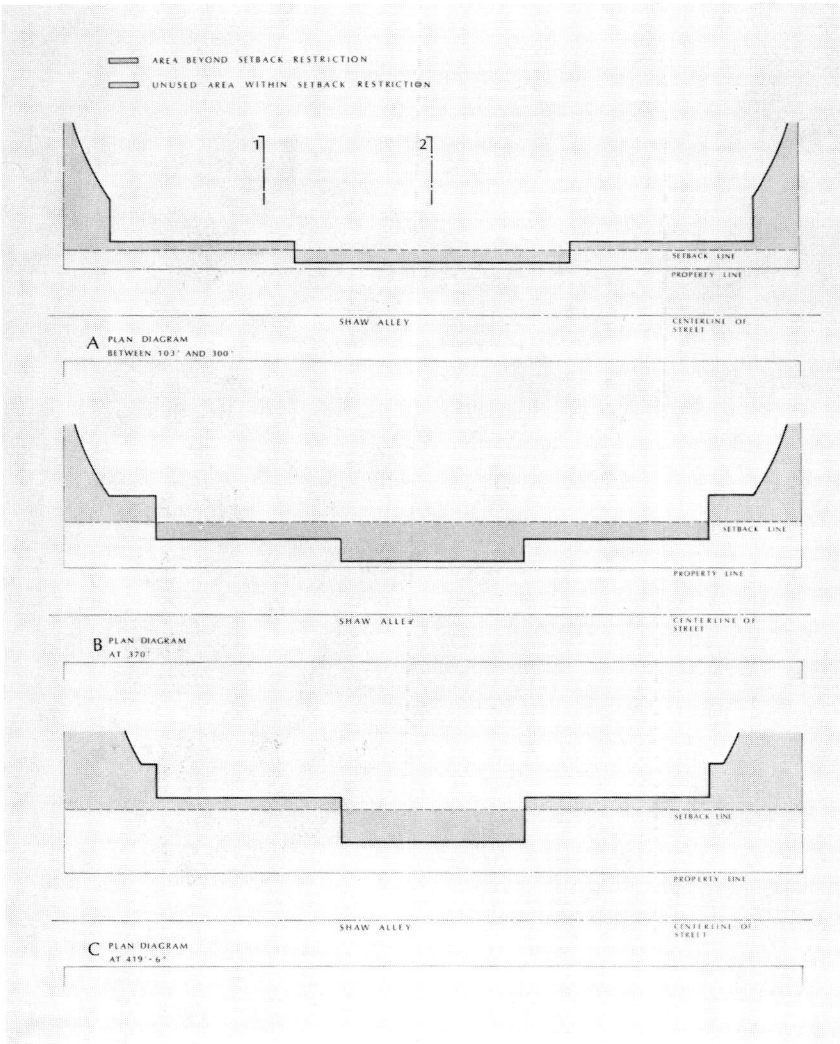

Fig. 7 Setback analysis subject to Downtown Plan controls

picture, however, is one of political uncertainty reasserted in combination with an extremely heavy financial burden at approval. These burdens amount to a development fee of about \$18.00/ft^2 of net new construction. What this will bring in the way of new city buildings or a continued shift of companies and projects to the suburbs remains to be seen. Let us hope the Downtown Plan will provide enough buildings, so that we may see some benefit from all that work in San Francisco.

Fig. 8 Hypothetical solar fan analysis for downtown San Francisco *(Rendering by Peter Bosselman)*

APPENDIX

ANALYSIS OF 18:1 FLOOR AREA RATIO (FAR) CAP IN THE SOUTH OF MARKET C-3-0 AREA

Summary of Conclusions

A careful evaluation of "soft-sites" and development regulations as proposed in the October, 1984 draft of the Downtown Plan (ignoring, for the moment the proposed 18:1 FAR cap) indicates that there are 22 sites in the South of Market C-3-0 district that could accommodate FARs in excess of 17:1. All sites are in the 350 feet or higher height district (shorter buildings are unlikely to achieve FARs in excess of 17:1).

The proposed cap is likely to lead to two major impacts on site assemblage and building form:

1. *Assemblage of large sites.* In those areas where heights are allowed in excess of 400 feet, only one of the 17 currently identified sites is large enough for a development that can achieve the allowed height within an 18:1 cap. With an assumption of further site assemblage, seven additional sites are possible.

   A building in the 550 foot height zone would need a site in excess of 40,000 ft^2 to accommodate what is considered to be the most efficient building (20,000 ft^2 floors in the lower tower and 12,000 ft^2 floors in the upper).

   The impact of the cap, then, would be to lead to a few tall buildings on large sites, with all or most open space provided in plazas on-site (very similar to the impact of pre–Downtown Plan development controls). A greater tendency toward open space provision on-site would severely complicate sunlight preservation efforts.

2. *Shorter, stockier buildings on small sites.* Of the 17 sites identified in the over 400 foot height district, seven have no potential for further site assemblage. For these sites, buildings will very likely be considerably shorter and squatter than the type of building possible under height/bulk regulations alone.

   For example, in the 550 foot height district, a 20,000 ft^2 site could accommodate a 360,000 ft^2 building at an FAR of 18:1. Assuming less than optimal floor plates of 15,000 ft^2 on the lower tower (and base) and 12,000 in the upper tower, the building would rise to 325 ft.

Soft-Site Analysis (by Dan Marks)

Table 1 indicates those sites in the C-3-0 area of downtown San Francisco that could accommodate buildings in excess of an 17:1 FAR in the C-3-0. The

Table 1 Soft-sites with envelope potential over 17:1

Block Lots[a]	Site Area	Conv.	Max Envelope	FAR	Ht
3708					
6	11,819	.65	231,650	19.6	550[b]
7, 8, 9, 10, 11, 12	25,000	.60	645,000	25.8	550
55, 3	21,200	.65	592,500	28.0	550
15, 17, 18, 20	41,250	.45	798,200	19.4	550
3709					
8	14,672	.7	441,600	30	550
3710					
17	18,906	.65	528,400	28	550
3719					
1	16,000	.65	364,000	22.8	400
17	32,441	.55	767,200	23.6	550
5, 6	21,917	.7	429,600	19.6	350
(7–11	28,144	.60	472,800	16.8	350)[a]
3720					
7	24,750	.65	514,800	20.8	400
5, 6	20,625	.65	375,400	18.2	350
3721					
1, 2, 3, 4, 5, 87	25,400		449,150	17.5	350[c]
68, 83	16,400		425,400	26.0	550[c]
70, 73, 74, 75	23,520	.65	611,500	26	500
78, 79, 80, 81, 82	32,160	.60	829,700	25.8	550
47, 48, 49, 52, 53	27,950	.60	670,800	24.0	500
13	12,275	.65	287,200	18.2	450[b]
10, 11, 88	31,870	.55	560,700	17.6	400
14, 15, 16	19,995	.65	467,900	23.4	450
21, 22	20,250	.60	437,400	21.6	450
3722					
(58, 67, 68, 69	35,160	.60	590,700	16.8	350)[a]
3735					
1, 4	20,212	.70	396,200	19.6	350
3736					
92, 93, 94, 98, 99, 100, 101	21,450	.70	470,400	19.6	350
TOTAL	563,666		12,456,450[d]		

[a]Sites with FAR envelope close to 17:1.
[b]Because of small site size, these sites are unlikely to accommodate the building height allowed (each is assumed to accommodate a 350 foot high building).
[c]Proposed buildings designed under the limitations of the downtown plan. Not included in the calculation of FAR is a proposed "open space" on lot 84 (14,500 ft^2) between the two buldings that would satisfy the open space requirement for both. Overall FAR for both buildings, including the proposed open space, would be 15.5:1.
[d]Open space requirement: 249,130 ft^2 (about 6 acres). Office space allowed as of right: 4,942,830 ft^2. TDR needed: about 7,500,000 ft^2.

table is based on soft-site analysis conducted by Roger Boyer, Bill Gary, Clark Manus, and Dan Marks updated to reflect the October, 1984 proposed Downtown Plan (without consideration of the proposed 18:1 cap).

The maximum envelope potential is calculated based on several factors, including

1. *Transferable development rights (TDR) transfer.* It is assumed that enough TDR can be purchased to allow for the maximum build-out potential of a site. Maximum build out is defined as the allowable envelope under height, bulk, and set-back requirements of the Downtown Plan (without reference to a FAR cap).

2. *Prototypical buildings.* Several buildings that have been designed under the guidelines of the Downtown Plan were reviewed to determine the actual impact of the plan on buildings.

3. *Open space requirements.* It was assumed that on smaller sites (under 20, 000 ft^2), almost all open space requirements would be met through some off-site means. As site size increases, it is assumed that proportionately larger amounts of the open space requirement will be met on-site. This reflects actual building proposals.

The "conversion" factor element in the table is an estimate of the reduction in building volume caused by the downtown plan. In other words, it is the reduction in volume relative to a slab covering the whole site up to the

Table 2 Footprint and site FAR, existing buildings

Building	Height	Site Area	Floor Area	Ft Print	Ft Print FAR	Site FAR
Bank of America	780	103,567	1,771,000	34,057	50	17.7
Transamerica	850	54,080	530,000	30,625	17.3	9.8
525 Market	550	45,260	1,041,000	27,395	37.9	23.0
101 California	660	71,300	1,305,800	46,550	28.0	18.0
Crocker Bank						
Tower	500		600,000	20,160	29.8	
Block		110,625				11.6
535 Mission		16,320	425,000	15,200	26,04	
100 First	385	40,500	445,000	21,695	20.5	10.99
Embarcadero						
One	590	1,076,000			35.0	14.0
Two		913,953				12.0
Three		919,318				12.0
Four		1,307,878				14.0
Russ Building	410	40,963	475,000		±20	
Shell Building	380	15,000	200,000		13.3	
Humboldt Building		8,670	105,000	8,670	12.1	
Crown Zellerbach	280	47,060	279,769	16,425	19.0	6.0

maximum allowed height. The conversion factor decreases (leading to smaller volumes), as site size increases, reflecting the relative inefficiency of large sites under the plan and the probable increase in provision of on-site open space on these large sites. The conversion factor was also estimated based on site location and configuration.

Table 2 indicates the likely impact of an 18:1 FAR cap. The cap is likely to lead to a greater tendency toward larger site assemblage in order to accommodate taller buildings. These larger sites will, in turn, lead to more on-site open space provision. On those sites that have no potential for further site assemblage, an 18:1 cap will lead to shorter, squatter buildings than those allowed under height and bulk regulations alone.

REFERENCES/BIBLIOGRAPHY

Goldberger, P., 1981
THE SKYSCRAPER, Alfred A. Knopf, Publisher, New York, created by Media Pojects, Inc.

Mujica, F., 1929
HISTORY OF THE SKYSCRAPER (Unabridged Re-publication by De Capo Press, New York in 1977).

New York Planning Commission, 1982
MIDTOWN ZONING, New York Planning Commission, Department of City Planning, Library of Congress, Catalogue Card #82-81294.

San Francisco Department of City Planning, 1981
GUIDING DOWNTOWN DEVELOPMENT, Department of City Planning, City and County of San Francisco, May.

Tall Buildings, Tight Streets

Tom Lollini

Tall buildings have stirred controversy since they first appeared in American cities toward the end of the last century. Visually, they were exciting. The skyscraper was a new form of building and a uniquely American one. But as tall buildings began to proliferate, changing the shape, the pace, and the patterns of the city, concerns were raised about their impact on the quality of life. Collectively, they created a series of problems that still confront those who plan, design, and build them, as well as those who must live with them. Over a billion square feet of office space has been added to American downtowns since 1950, and this growth continues to alter the character of the urban core faster than planners can develop and evaluate guidelines to cope with it.

IMPACTS

Higher densities and taller buildings have their greatest impact at the street level. While changing the scale and form of the street, they also cast it in shadow and intensify the congestion of pedestrian and vehicular traffic: movement is slow, noisy, unpleasant, and increasingly difficult. Pedestrians are subjected not only to this noisy gloom, but also to the discomfort of winds channeled down these rising canyons (Fig. 1). Finally, downtowns are suffering a loss of character and uniqueness. Important public spaces, historic landmarks, and the older, finer-grained urban fabric are being eclipsed by the volume and scale of modern expansion. In many cases these elements have been eliminated, their multitude of functions absorbed into the relative anonymity of the skyscraper form.

OPPORTUNITIES LOST

Most efforts to rescue the pedestrian environments of our cities have focused on creating outdoor public spaces, such as plazas and parks, while the

Fig. 1 Wall Street District, New York

designs of the buildings associated with these open spaces have largely been ignored. This focus derives from the nearly universal American perception of the classical (Greco-Roman) urban form elements of the forum and the agora, as the ideal model for a sophisticated, civilized culture. This viewpoint has been reinforced by the millions of Americans who have experienced the plazas of modern day Italy and Greece.

This perceptual orientation presents several problems however. First, North America does not have a Mediterranean climate. For most cities the climate discourages use of outdoor public spaces. It is either too cold or too hot, too wet, too windy, too humid, or too cloudy to sit outside comfortably for eight or more months of the year. Second, unlike their European counterparts, the majority of American cities do not have a significant residential population within the urban core to populate these "places for people" beyond the business lunch hour. Third, a comparison of crime statistics justifies the American aversion to using parks and plazas in the evening hours, after the sun has set and the commuters have headed home. Finally, historic zoning patterns and the concentric nature of land values in North American cities have led to the clustering of high-rise districts. This concentration has in many cases devalued outdoor open spaces within these districts by increasing the negative effects of shade and wind patterns.

Critical opportunities for enhancing the quality of the pedestrian environment are being lost as buildings and "public" spaces are developed without a clear understanding of the relative values of the amenities they can provide. The experience of hundreds of lifeless plazas and little used urban parks created in the last 20 years has shown that a more comprehensive approach is necessary.

OPPORTUNITIES FOUND

Amid the development of the San Francisco Downtown Plan, the architectural firm of Kaplan/McLaughlin/Diaz initiated a research project entitled *Tall Buildings, Tight Streets.* The goal of this research was to examine and explore those aspects of public spaces and high-rise buildings that can breathe life into our cities' streets, rather than suffocate them. The study began with the premise that high densities and tall buildings need *not* be detrimental to the quality of urban life; that, far from being liabilities, they can embody some of the elements essential to creating a positive urban environment from a social, esthetic, and economic point of view.

The research consisted of an analysis and evaluation of historical and current trends in the design and regulation of tall buildings. Two surveys were also used: one of the lay public, users of both indoor and outdoor pedestrian spaces; the other of professionals involved in the planning, design, management, and evaluation of them. The open spaces examined included traditional outdoor spaces such as plazas and parks, and dramatic indoor, publicly accessible spaces, such as arcades, atriums, and gallerias.

Certain indoor spaces intimately integrated with the surrounding development were found to be extremely valuable. Searching for a name that differentiated them from less successful spaces, we selected the term *agora.* This term, traditionally used to describe the marketplaces of ancient Greece, implies a public area given over to a multiplicity of communal functions, both formal and informal. The original agora functioned not only as a focus for business and commerce, but also as a place to meet friends and exchange information, to observe the affairs of one's community, or simply to while away some leisure time, reading, watching people, or sitting in the sun. These human needs have not diminished over time, but have, perhaps, grown ever more precious as the stress of urban life mounts. Moreover, they are needs that the modern city must support and nurture.

The survey of public attitudes examined the comparative ability of indoor and outdoor space to provide for these needs. Findings from more than 600 interviews showed that both outdoor and indoor spaces provide places to meet friends, eat lunch, or spend time reading or watching people. Because of their differences, however, each type of space is slightly more useful than the other for certain activities. Most obviously, indoor spaces usually contain a concentration of retail activities, thus attracting a higher percentage of shoppers, while outdoor spaces are appreciated for being outdoors and for being an alternative to the intense activity of the street. Less obviously, outdoor spaces are more frequently used as shortcuts, and indoor spaces are used more often by people who work nearby and enjoy their convenience.

The frequent and varied use of indoor pedestrian spaces demonstrates that they offer an exciting alternative to traditional outdoor open spaces. Indoor spaces can provide most of the amenities that outdoor spaces provide, and more. They serve the public in cool, windy, or rainy weather and during evening hours when outdoor spaces do not; they can be designed to provide alternative pedestrian routes; and they can provide a variety of spatial forms that serve as vital landmarks, a role that becomes increasingly important as our downtowns continue to grow.

The modern agora, like its historical counterpoint, can provide a place for all types of activities from shopping, strolling, people watching or concert-going to eating, reading, relaxing, or simply taking a short cut. They thus enhance the use of public spaces by drawing people to them to take advantage of the amenities they offer.

SUCCESSFUL PROTOTYPES

Hundreds of high-rise developments in the last twenty years have sought to create grand public spaces. For the majority, the results have been little more than windswept, oversized, stone doormats or vaulted lobbies that have more in common with mausoleums than the vital link between the horizontal and vertical cells of the city that they could be. The truly successful developments can be counted on one's hands. Beginning with Rockefeller Center in New York (Fig. 2), a generation passed before projects like Embarcadero Center in San Francisco (Fig. 3) or Citicorp Center in New York began to capture the imagination of the public as truly urban places. The IDS Center in Minneapolis successfully created an indoor town square in a city known for its harsh climate (Fig. 4). The renovated Cleveland Arcade has had similar success (Fig. 5). Observational studies of this space and a well designed park nearby found five times as many people in the arcade as in the park even on a beautiful day. Along 57th Street in New York the combined public atriums of Trump Tower and the IBM Building have also succeeded in creating a pedestrian gathering point where none had existed before. Finally, the Crocker Galleria, though detailed in a somewhat suburban parti, has become an integral part of San Francisco's downtown pedestrian environment.

GUIDELINES

How can our cities foster the development of the modern agora so that each new project adds to the quality of urban life rather than taking from it? The design and regulatory principles that can guide and encourage such developments are several (Fig. 6), and apply to large and small projects alike.

The primary axiom is that the new development must be integrated with the physical aspects of the urban fabric. These spaces must not simply be in

Fig. 2 Rockefeller Center, New York

Fig. 3 Embarcadero Center, San Francisco, California

the city, they must be designed so that they become part of the city. Guiding principles that will achieve this effect are:

Pedestrian access points to public routes and spaces within buildings should be grand and inviting. They should correspond to existing or planned travel routes and view corridors of the neighborhood and in this way become a part of the public pathway system.

Fig. 4 IDS Center, Minneapolis, Minnesota *(Courtesy: E. F. Baker)*

Fig. 5 Cleveland Arcade

Interior spaces including tenant, building circulation, and interior open spaces should be frequently linked with outdoor public spaces such as parks, plazas, and sidewalks. This will provide a more intimate relationship between public and private domains.

Cul-de-sacs in the flow patterns of publicly accessible spaces within the building should be no deeper than can be seen from a public route. This avoids the dead-end syndrome that can cause security problems for users and tenants and exposure problems for retail spaces.

Passage through these buildings and the public spaces within them should be a "public experience". Using the common architectural and environmental language of the street infers that these areas are extensions of the public realm. Some elements of this "public" vocabulary are:

- Building materials and furnishings that are exterior in character
- Skylights or illumination by high levels of reflected or simulated daylight
- Liberal landscaping that also provides a counterpoint to the built environment
- A fountain or some form of moving water, which can be a focal point or simply a pleasant background element
- Bold signage and graphics
- Other "street" elements such as lampposts, clocks, banners

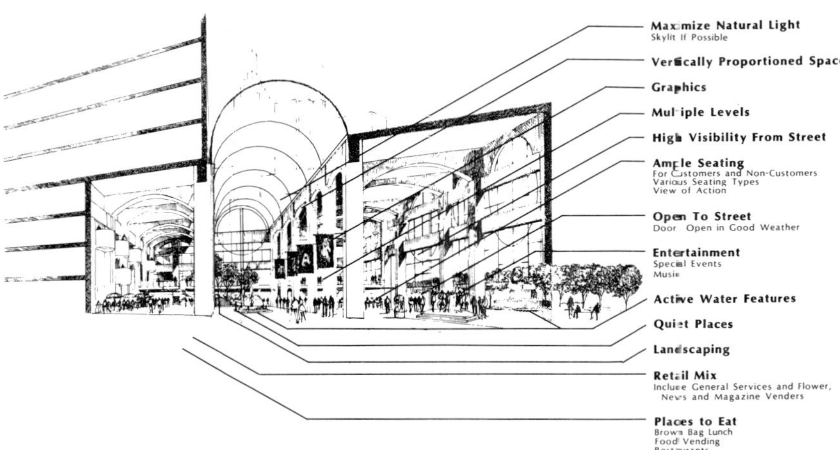

Fig. 6 Design guidelines

- Public art, such as sculptures, mosaics, murals or frescos
- Where and when practical, an outdoor climate. (If it is not practical, the interior climate should be moderated only enough so that people using the space will be comfortable.)

The lower floors of tall buildings should be spatially dramatic with floor heights in excess of 4.5 m (15 ft) and at least two floors of activity generating uses visible from indoor and outdoor public areas.

Public interior spaces and adjacent interior building uses should be highly visible from the street. Transparency of street level facades providing a view of the spaces and activity within can more effectively enrich the character of the street than most forms of traditional or contemporary decoration.

The landscaping of these spaces can take on an infinite variety of forms, textures, and colors, but these elements must invite the pedestrian and not screen or exclude him. Ornamental trees are valuable amenities in themselves, while other types of trees can help define a sense of scale, enclosure, or canopy within larger spaces.

Seating cannot be too strongly emphasized as an essential component. A plentiful supply of comfortable seating in a variety of forms and configurations offering a wide range of choices to the user should be provided. Benches, chairs, steps, walls, ledges, and railings are all possibilities. These seats should be arranged to provide for all the possibilities of social interaction such as sitting alone or in groups, reading, people watching, facing a public route or a focal point like a fountain. One seat for every 20 ft^2 of space is a good rule of thumb. While some tables and seats may be designated for restaurant use, the majority should be available to all.

Sidewalks, shortcuts, and public spaces should be flanked by publicly oriented uses. These uses should be very visible from the public path. Possibilities include, shops; entertainment or drinking establishments and food service such as restaurants and vendors; service retail, such as, cleaners, hairdressers, shoe repair, electronics repair, or financial services; ticket outlets; health clubs; cultural facilities such as galleries, museums, aviaries, aquariums; or potential settings for organized or spontaneous events such as parties, exhibits, musical performances, or the entertainment. Since lunchtime is the period of peak pedestrian activity, food service is the most essential activity-generating use. Commercial services allow people to combine errand running with lunch and/or shopping.

More practical aspects of these planning principles are that security services keep a very low profile. A man behind a desk implies that not everyone is welcome. An obvious message to one person is a subtle message to all. While some spaces must be closed at certain times for safety or security reasons, the goal should be maximum public accessibility.

THE FUTURE

These guidelines point the way toward the next generation of tall buildings. Together these innovative building forms can create a new set of urban spatial dynamics. The incorporation of dramatic, publicly accessible spaces within the lower levels of high-rise structures will expand the public realm and reinforce the vitality of urban life. From the developer's point of view, these civic-minded designs can add significantly to the marketability of a development and provide an economic return through the activity they generate.

These design principles can also act as guidelines for renovating or refurbishing existing, underutilized open spaces. The addition of food service pavilions, retail services, seats, weather protection, or other amenities and activity generating uses can go a long way toward bringing people to places that have served as little more than doormats for an entire generation of modern high-rises. As urban behavioral patterns evolve, so too should open spaces. Who in America could have imagined a wedding party or dinner dance in a public space thirty years ago when Seagram's Plaza or the Ford Foundation were conceived?

The pedestrian environment need not be a simple series of sidewalks and streets, punctuated by the occasional plaza or park. By linking voluminous building interiors with the open air, softening the threshold between inside and out, the city can evolve into a series of inviting public spaces ranging from fully open to fully enclosed volumes, providing an infinite range of opportunities to the public in all kinds of weather and at all times of the day. These new places can become landmarks in their own right, adding character to old downtowns where the historic fabric has been diminished, or complementing those that remain intact. They will be distinguished by their benevolence, ambition, and public spirit, and will thus enhance the urban character of the cities in which they are built. Since we are all pedestrians, we all stand to gain.

Skylobbies and Interconnecting Links Amid the Cluster of Skyscrapers

Valer Mocak

INTRODUCTION

As the recent trends show, more and more skyscrapers are being built, even in cities with no tradition for this type of building. The rising cost of land results in its maximum utilization. Corporate clients are pushing to build tall downtowns as well, with the promise that their identities can be very closely associated with that of a well-designed building. Architecture then becomes a monument of either achievement or failure, to itself as well as to the client who built it.

The ensuing hunt for new shapes, ideas, and concepts takes place as individual identity is expressed by unique means, such as unusual application of architectural vocabulary, new materials, breathtaking structural experiments, innovative stylistic coordinates, unexpected composition of the facade or of the entire bulk of the building group, and revised building typology, to mention a few.

If the postmodern era tries to rectify the Function-Over-Form-type dictate of the International Style with this new approach, the task seems to be easy:

the stagnating, box-like look of past skyscrapers left plenty of room for new design experiments in shapes and forms. Many architects saw opportunities in the departure from the prevailing cliches, boosted by enlightened critics. Changes toward the new were detected also in theater, art, dance, and music. When the client demanded buildings that they would identify with, the new paths for skyscraper designs were secured.

In the process of searching for new uniqueness, at least three basic trends are detectable:

—a refined continuation of the mentioned stereotype of the late, mature period of the International Style, elegant and nonprovocative;
—buildings where the designers are heavily borrowing forms from the past styles: ornaments, historical furniture details, and the like (Many experts are lost in determining whether this trend is original or eclectic;
—the most controversial trend, combining all of the above: inspiration from the past mixed with contemporary elements. There is no hesitation in using an unusual color scheme or in employing large and small windows next to each other, often just for the sake of the image itself. This group is iconoclastically shocking to some and innovatively distinguished to others.

Exploring further the capabilities of the skyscraper design in the next few years, and ultimately into the twenty-first century, attention has been turning to one concept: the cluster of two or more skyscrapers, interconnected in the sky by bridges, or so called *skylobbies* (Fig. 1.).

Such a cluster of interconnected tall buildings would be individual and unique. It would have the potential to satisfy the identity requirement of any developer, corporate client, or a joint venture of more owners. It would introduce a new typology: a group of interconnected skyscrapers as one unit. It would also enrich the silhouettes of the cities by offering a great number of possibilities for its formal grouping, composition, treatment, design, and social impact.

SKYLOBBY CONCEPT

Two or more skyscrapers would be interconnected by bridgelike constructions, approximately four stories high, every fifty or so stories (Fig. 2). The structural aspect of this cross-bracing of several buildings is expected to be beneficial, although more studies are needed to make valid conclusions. The structural elements of the bridges would be covered by the same skin panels as on the adjacent skyscrapers, or they might have a totally different surface, depending on the design solution.

The verticality of the bridges would continue inside into the skyscrapers, creating the mechanical rooms, transfer lobbies from express to local elevators, and public spaces (malls) to disperse people who have no desire to travel to the street level during lunch and coffee breaks, or into congestion during rush hours. By centrally monitoring all public spaces, the security and safety of people can be enforced very effectively.

The skylobbies would house shopping facilities, restaurants, newsstands, boutiques, computer terminals, printing shops, emergency medical stations, libraries, banking services, even chapels, kindergartens, movie theaters, and swimming pools. Figure 3 shows an example using Times Square in New York for a design concept.

ADVANTAGES

Elevators could be more economically utilized, since the total capacity of local elevators and shuttles can be distributed with greater coordination. Since until recently the maximum height of the skyscraper has been limited by the technological level and ability of the elevators to service only so many floors, transfer areas in skylobbies might eliminate this obstacle. The skylobby will become an elevated urban plaza where one can transfer from an upper

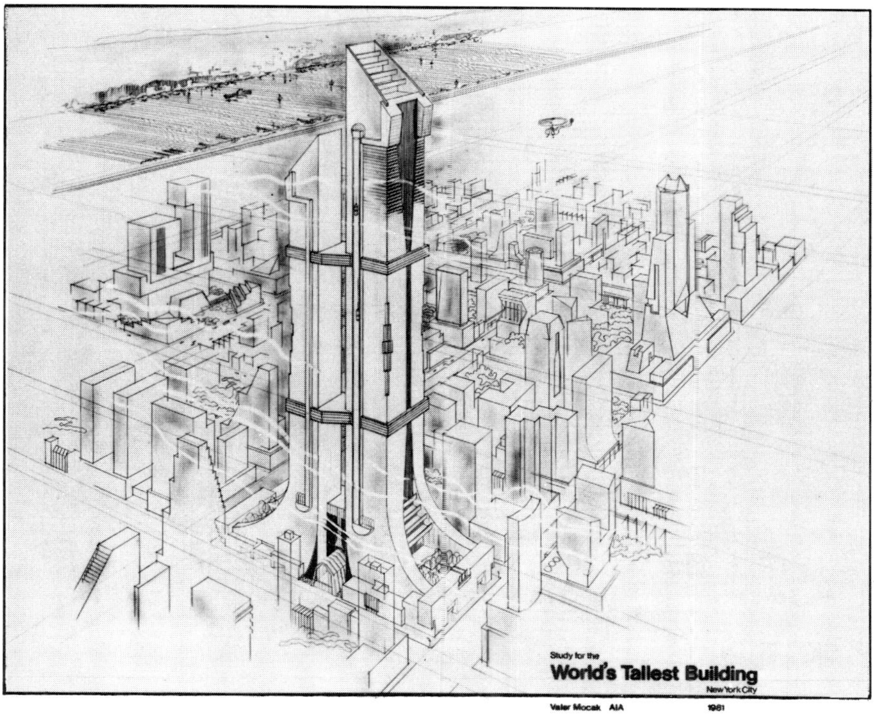

Fig. 1 Study for the "World's Tallest Building" suggests two bridge-like skylobbies to connect two skyscrapers

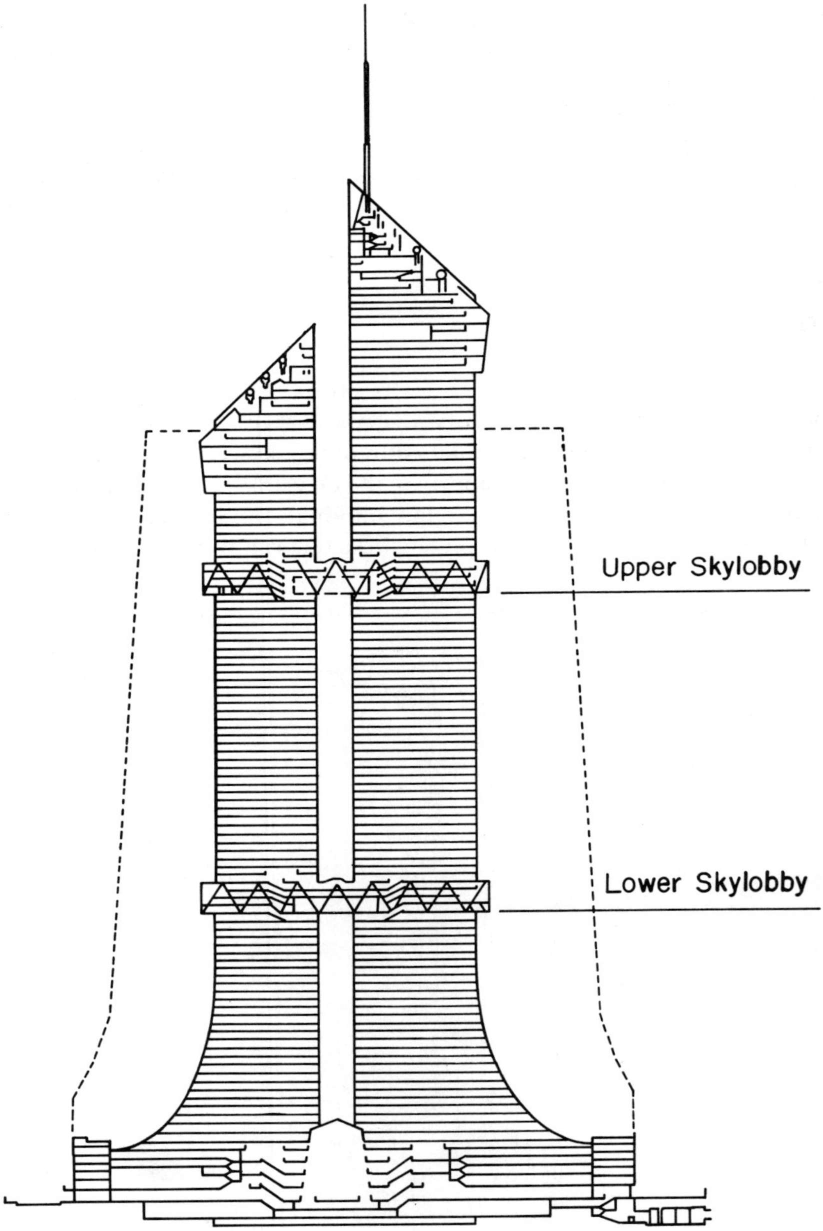

Fig. 2 Section through the "World's Tallest Building" shows the location of skylobbies

low-rise to lower mid-rise group of elevators, or from express to local elevators. While in this public space, one could be shopping, having lunch in a restaurant, or enjoying the park or a swimming pool 80 floors above the street level, and still have the good feeling that all this is backed by reliable mechanical facilities. In case of emergency one could escape through any one of the skyscrapers which, as part of the cluster, is connected to the same skylobby.

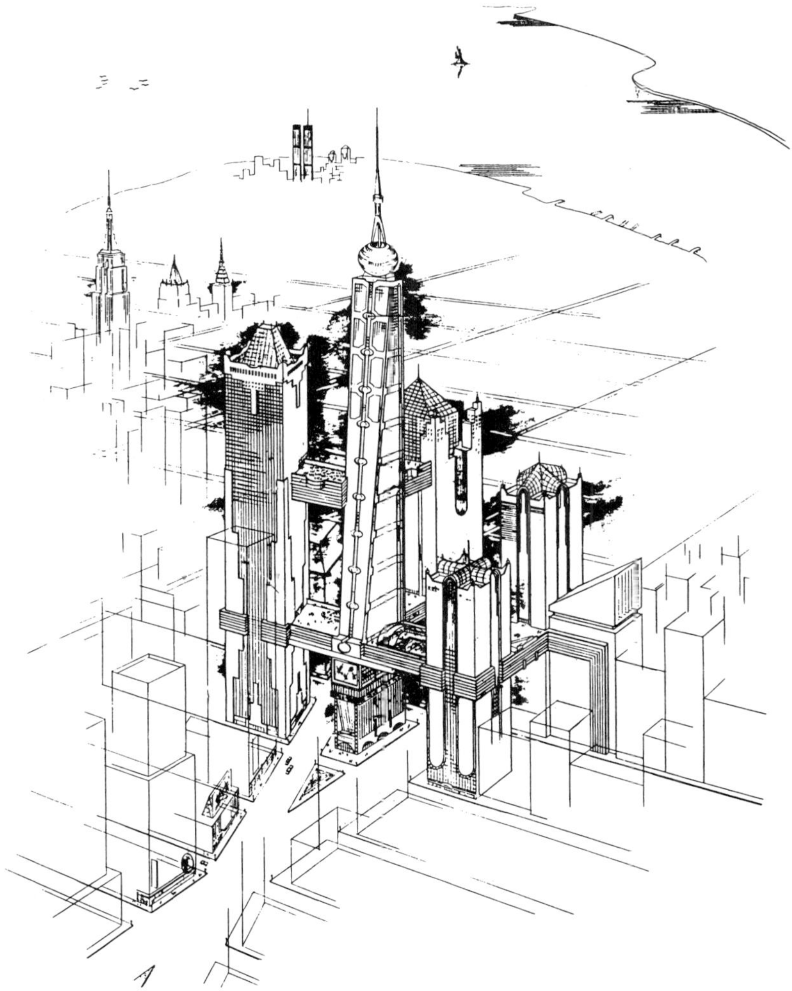

Fig. 3 Concept of the "Times Square" cluster interconnects five towers by skylobbies that are elevated public plazas and malls with restaurants, swimming pool, gardens, shops and other amenities

Structurally speaking, the quality and strength of the steel frame will ultimately become the determining factor of the new height limits.

One of the most obvious advantages of building a skylobby is the elimination of the *Towering Inferno Syndrome:* in case of fire in any part of any building there is at least one other escape route through another building.

Incredible though it may seem, traveling a mile in a vertical direction can take the same time or less as traveling a mile during rush hour through a busy avenue in Manhattan. For some, it is frustrating; others can take it comfortably. As for fears, it is known that riding in an elevator can evoke claustrophobia—an abnormal fear of being in an enclosed place. On the other hand, some people have an abnormal fear of being in an open space, huge halls, public places, plazas—so-called agoraphobia. The state of mind in the condition of extreme height might trigger the fear of heights (acrophobia) which seems not as irrational as other fears. It is often magnified by anticipating the other dangers: the fear of failure of electrical, heating and cooling systems; bomb threats; fire; and the like. A single skyscraper might not give the expected feeling of safety as much as would skyscrapers interconnected with other high-rise buildings at different levels. One would then, logically, have a choice to escape through the egress routes of the neighboring towers.

The movement of the high-rise building due to wind can also contribute to fears and uncertainties. There are not enough data available to show how much the skyscraper can sway, because so many variables are in play, such as column spacing, bracing, horizontal and lateral reinforcing. During the big storms that usually occur once or twice a year, high winds force a steel structure to sway a little. One can hear the squeaks that occasionally accompany such motion, but it is predictable and after several squeaks one becomes used to it. Satellite photos and the use of computer technology can give the occupants of the skyscraper the exact length of these events and so impose the limitations on possible short-lived fears caused by weather conditions.

Thermal expansion usually is not accompanied by rhythmical movement. It can be assumed that in a cluster of several interconnected and braced skyscrapers, the sway could be almost nil; one would feel it less than a vibration on Broadway's sidewalk generated by the movement of a passing subway train below.

The super-high density land area can influence some side effects on the normal social and behavioral pattern of the skyscraper's occupants, but it is very difficult to make any firm conclusions. Nearly all the predictions are based on speculations. The only plausible way to foresee difficulties is to employ the methods similar to those used in futurological disciplines. Today's knowledge of the problem, enriched with the expected new data and projected into some probable developments, can help create alternative models and scenarios.

For instance, the impact of high population density in New York can be different from Houston or Toronto. A distinguished scholar and author on environmental problems, the late Rene Dubos, in his book *Celebrations of Life,*

(1981) made an interesting observation: Hong Kong, the metropolis with an extremely low average income and the highest population density in the world, has been described in surveys and reports as a city with excellent physical and mental health, where "crime or other forms of disorderly conduct were rare". Dubos concluded that high population density becomes a problem "only when other social conditions have seriously deteriorated."

The fear of high population density is not the prerogative of our time only. With the advent of the first 10–12 story skyscrapers a century ago, doomsday predictions for overcrowded cities were rampant. The opposite was true: higher density contributed to a more economical exploitation of utilities and often pushed the population into a higher affluence group. The traditional methods of establishing the optimal degree of density are no longer objective. We used to say that 50,000 people were compacted on a few acres of land, but we failed to add that they would be spread vertically.

FUTURE DEVELOPMENT

The structural and architectural features and possibilities of the concept of an interconnected group of skyscrapers are not describable in their overall complexity. Let's face it: the land surface of our Earth is constant. With the population increase that some predict, we will be forced to go high, to create multilevel cities, at least in certain centers.

The multiple skylobbies will bring a new infrastructure element into existence. Because of skyrocketing land prices, it will be used more and more, thus contributing to the new multilevel type of city, elevating part of its public space into the sky.

Existing zoning laws are not ready to accept these challenging new concepts. In general, existing regulations limiting the heights of buildings are detrimental to progress. Some cities are enforcing the requirement in their zoning law not to exceed a certain floor area ratio (FAR). That means, for example, that to build on the site of 61 m × 244 m (200 ft × 800 ft) with the requirement of FAR 18, one can build a building which has a maximum 268 thousand m^2 (2.88 million ft^2) of total floor space. In reality, it's not so simple, of course. Other factors like setbacks, air rights and formulas for light exposure curves are also influencing the height and shape of the high-rise building.

Here, we are immediately approaching a full series of questions. Why FAR 18 and not 23.8 or more? Who can scientifically prove that a shadow 370 m (400 yards) long thrown by the skyscraper at a certain time of day, damages the environment more than a 270-m (300-yard) long shadow, which represents, say, 25% less floors? How does it apply to all the ecological concerns in the deep forest where most of life thrives in the shadow? Esthetic criteria in general, as well as in relation to building heights, are unmeasurable and can mean different things for different people. To apply height restrictions on the cluster of skyscrapers without considering the new, emerging concepts, as in the idea of multiple interconnecting skylobbies, is ignoring the future. In spite of limitations, the evolution of societies in the past shows that the adaptability of the human race to new conditions should not be underestimated.

We can envision that new FAR formulas will develop, taking into account the floor areas of the public spaces in skylobbies. Air rights will increasingly be borrowed, traded, or bought; light exposures will be revised; and if the need brings enough pressure and is balanced by sufficient advantages, we will witness the creation of more flexible types of special zoning districts.

In the last decade, many developer and investment groups grew to such a size that they now handle multibillion dollar projects. But even some of these mega-developers would have difficulties managing and financing the building of a cluster of skyscrapers with several interconnecting skylobbies. A consortium of public and private sectors, or other investment groups using either domestic or foreign capital (or a combination of both, for that matter), may have to be organized to take part in a project of this magnitude.

CONCLUSION

Skylobbies in the urban design hierarchy can be rated as more than a plaza, more than a shopping mall, more than a traffic junction. Together with a population of the cluster of skyscrapers, they are a sort of minicity, a city within the city. In this sense, they should be viewed as important elements of infrastructure in the emerging new type of multilevel cities.

REFERENCES/BIBLIOGRAPHY

City of New York, 1984
NEW YORK CITY ZONING RESOLUTION, Department of City Planning, New York.

Davis, K., 1973
CITIES: THEIR ORIGIN, GROWTH AND HUMAN IMPACT (Readings from Scientific American), W. H. Freeman and Co., San Francisco.

Dubos, R., 1981
CELEBRATIONS OF LIFE, McGraw-Hill Book Company, New York, St. Louis, San Francisco, Hamburg, Toronto, Mexico.

Ferriss, H., 1986
METROPOLIS, exhibition in the Whitney Museum of American Art at Equitable Center, New York, June 6-July 30.

Giedion, S., 1967
SPACE, TIME AND ARCHITECTURE: THE GROWTH OF A NEW TRADITION, Harvard University Press, Cambridge, Massachusetts.

Goldman, J., 1980
THE EMPIRE STATE BUILDING, St. Martin's Press, New York.

Hruska, E., 1985
TOWARD THE CREATION OF URBAN ENVIRONMENT, (in Slovak: K tvorbe urbanistickeho prostredia), Zvaz slovenskych architektov, Bratislava, Czechoslovakia.

Jacobs, J., 1984
PRINCIPLES OF ECONOMIC LIFE, Random House, New York.

Mocak, V., 1979
SECURITY PROVISIONS IN CONCENTRATED HOUSING DEVELOPMENT, Proceedings of International Conference on Housing Planning, Financing and Construction, pp. 650-661, Florida International University, Pergamon Press, New York, Oxford, Toronto, Sydney, Frankfurt, Paris.

Mocak, V., 1983
SKYLOBBIES: THE EMERGING NEW INFRASTRUCTURE ELEMENTS, Lecture at Fifth International Conference on Urban Design, Washington, DC, October 26-29.

Mocak, V., 1984
SKYSCRAPERS AT WATERFRONT LANDS, Lecture at School of Architecture, Pratt Institute, New York, August 1.

Mocak, V., 1986
ZONING OF SKYSCRAPERS AT WATERFRONT LANDS, Lecture at Skidmore, Owings & Merrill, New York, January 15.

Vernez-Moudon, A., 1983
CITY FORM AND TALL BUILDINGS: CATHEDRALS, PALAZZI, TALL DOWNTOWNS, AND TALL CITIES, Developments in Tall Buildings 1983, Council on Tall Buildings and Urban Habitat, Hutchinson Ross Publishing Company, Stroudsburg, PA.

Wilsher, P., and Richter, R., 1974
THE EXPLODING CITIES, Quadrangle, The New York Times Book Co., New York.

Project Management for Tall Buildings and Urban Habitat

James J. O'Brien

PROJECT MANAGEMENT OVERVIEW

Time, cost, and quality are the three main ingredients that determine the successful completion of a construction project. The total time and final cost are not known until the project is substantially completed. Yet, the owner must make innumerable decisions related to interdependent design and construction problems. At every decision-making point, the owner's primary concern is whether a decision will meet the main objectives of completing the project within the specified time and within budget. The challenge for the owner, then, is to employ effectively as management tools those principles and techniques that minimize the risk of not meeting those objectives.

Until the mid-1950s, there was no generally accepted formal procedure to aid in the management of projects. There were no project management methodologies that would integrate people, resources, and tasks of many organizations and monitor their performance against predetermined plans from conception to completion. The architect/engineer designed the building in accordance with the owner's predetermined requirements, and the general contractor, with the help of specialty contractors, built the project. As technology in the various fields advanced, projects became more and more complex and required a new management concept that would put a better "handle" on the financial cost, schedule, and efficient use of the restricted resources.

A project is always a well-defined task to be accomplished within a definite time period. The project objectives and its overall accomplishment plans are known in advance. Project management is a methodology to achieve project objectives through implementation of the accomplishment plans. Project management functions can be performed either by in-house expertise of the owner (if available) or by outside agencies, specially equipped to handle those functions.

Project management methodology is more suited to the design and construction phases of the project development and delivery process. The key to successful management is the effective phasing of the design and construction through project management methodology.

The complexity of high-rise building construction, quality considerations, and the long time required to complete the project before a return on the investment can be realized—all these factors complicate the problem and demand the expertise of competent management in the design, construction, and operation of the building.

THE GOOD NEWS

The good news is that, as project managers, we have good project management tools to work with. These tools have been developing as an indirect result of the availability of computers. The first use of computers for project management was in the concurrent development of Critical Path Method (CPM) for the civilian sector, and PERT for the Polaris Program. Often referred to as *computerized scheduling* these network techniques introduced a new way of recording planning and thinking on paper, relieving the planners and schedulers from the burdensome task of remembering the interrelationships and sequences. The computer is used for little more than an adding machine in this process, and hence, the "computerized" adjective is really misleading. The important aspect of the first computerized scheduling was the ability to use paper as a recording device, giving the project managers a chance to work out problems on paper rather than in the field.

In the 1960s, project management evolved even further, based principally on the network techniques. Other variations such as resource planning and cash flow planning based on networks were also further developed. A limitation in the 1960s was found both in computer hardware and in computer software.

By the end of the 1960s, project managers began to recognize that they were doing something really different. This consensus recognition occurred a little more than a decade after the first published results of CPM in 1958. In 1969, a group of project managers met together in Atlanta to discuss the results of the prior decade and to determine whether or not project managers were becoming unique enough to identify themselves as a practicing group. This was the first meeting of what has become Project Management Institute, an organization of over 5,000 with members around the globe.

In the 1970s, computers became more capable, and software more comprehensive. MSCS by McDonnell Automation, PMS by IBM, and Project 2 by Project Software & Development, Inc. were all major programs introduced early in the 1970s. Each required a mainframe computer such as the IBM 360 or IBM 370 to carry out its functions. In the 1970s and into the 1980s, this mainframe approach has indirectly performed an interesting result—focusing of information. Using networking as a basis, a major project was able to introduce information on cost and resources as well as time. At KKMC in Saudi Arabia, a network was used with 25,000 work items that not only identified status as to schedule, but also produced information such as: tons of equipment at dockside, USA; tons of equipment due at dockside Saudi Arabia; progress payments; need for construction rooms/beds at the job site; and other similar information. In effect, basic scheduling had grown in the 1970s into Management Information Systems.

As the 1980s arrived, so did minicomputers, microcomputers, and personal/professional computers. A proliferation of project management software is now available, making it possible for the individual project managers to have hands on involvement with scheduling. While adding an interesting new dimension, the capabilities of individual project managers running their own schedules results in a diversity of product, a diffusion of focus, and sometimes ineffectual results.

Nevertheless, the important message is that, sometimes for the wrong reasons, the advent of computers and their introduction to the construction field in 1956–1958 has produced a whole generation of project management tools that have been proven useful in many instances.

VERTICAL PRODUCTION METHOD

The construction industry, by its very nature, differs vastly from other economic systems primarily because of its nonrepetitive nature. Excepting large single-family housing projects, there is no mass production of buildings. And, unlike other industries, there are few possibilities to influence demand to create a better market. Although similar projects may have been successfully completed previously, building projects are seldom repeated in an identical manner. This is particularly true in the construction of a tall building.

In construction, each project has to be treated as a one-time effort to satisfy the individual needs of investors and users. Unique geographic location, creative work of architect/engineers, and unforeseen conditions during planning and construction all create a dynamic project environment. Consequently, in order to accomplish project tasks efficiently, a specific organization must be created with each project to meet those dynamic design and construction requirements of that particular project.

Ironically, the use of computerized scheduling has led to a manual technique that is very effective for high-rise work. Early network scheduling considered all projects to be unique, and it was virtually heresy to suggest that planning factors to be incorporated in networks were repetitive. However, the observant CPM scheduler noted that there were many familiar steps inherent when scheduling similar projects. For the typical floors in a high-rise project, the use of basic CPM can be laborious, and it would be obvious to the CPM scheduler that the input for work on a typical floor was duplicative and tedious. Further, once the scheduler had reached the typical floor, it was possible to predict an end result through basic arithmetic without the use of a computer, suggesting that there were ways of graphing the result by something other than network presentation. This realization resulted in the development of the Vertical Production Method (VPM) for use in high-rise work. This approach is also useful for highway projects and other projects where the basic element of work repeats itself, and the key to a schedule is the establishment of lead times.

Once the "typical floor" (Fig. 1) had been reached in a high-rise project, the high-rise construction became a series of production cycles. No trade can progress faster than the various controlling trades, such as structure. Any following trades that produce more rapidly will catch up and therefore must stop the rhythm of its work, which can be an expensive situation.

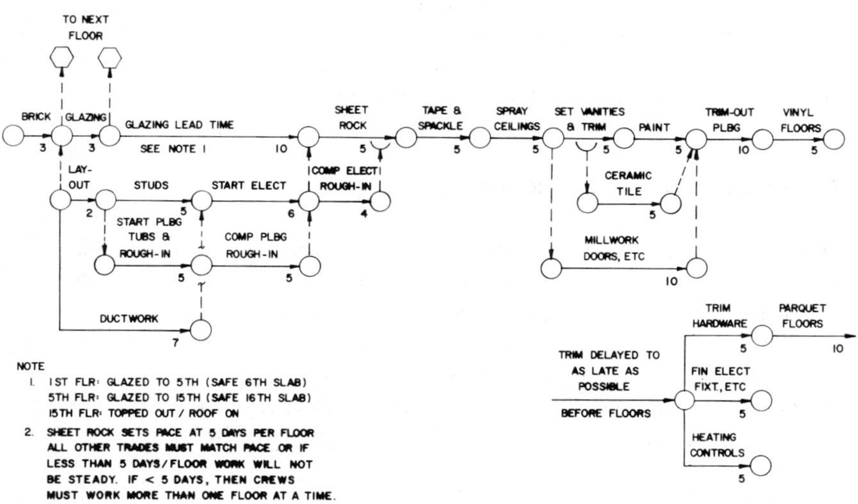

Fig. 1 Typical floor, network format

The schedule is depicted as a rate line on which the vertical axis is the floors by number, and the horizontal is increments of time. The first line to the left is the rate of progress required according to the schedule for start and completion of structure on each floor (Fig. 2). This can identify optimum points for activities such as placing of outriggers (Fig. 3), for brickwork (Figs. 4–6), and where to safe off interim floors. The overall VPM diagram is built up by adding one production line at a time (Figs. 7 and 8). The slower rate curve of the later production lines indicates a typical situation in a high-rise, which is that the mechanical, electrical, and finishing trades usually do not move at as fast a rate as the structure can. Conversely, these slower trades cannot begin until the structure has actually been erected (Figs. 9 and 10).

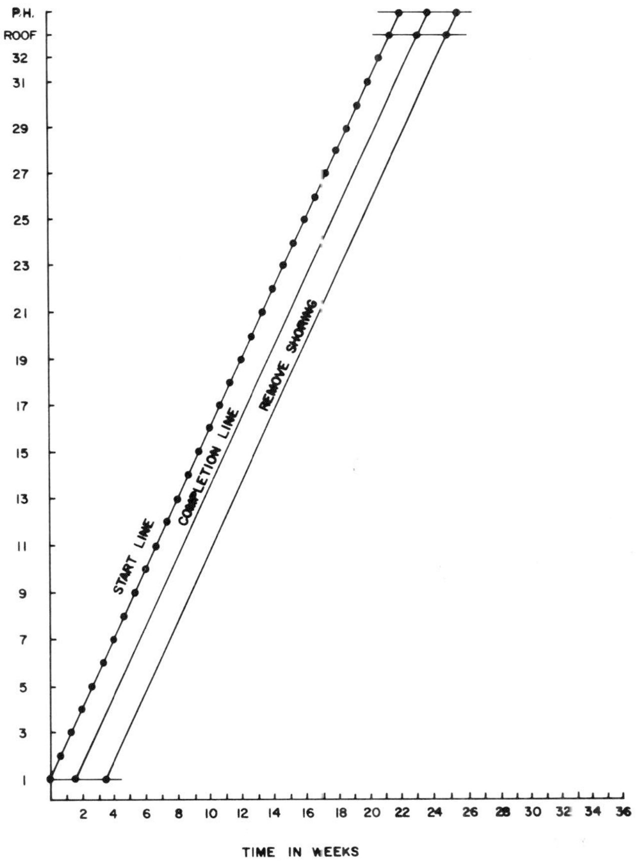

Fig. 2 Structural frame rate: 3 floors every two weeks

THE BAD NEWS

The tall building, with its unusual requirements, provided the opportunity for developing new materials and new methods of construction. High-strength steel, high-strength concrete, composite design, and the use of sophisticated welding techniques in both on-site and off-site fabrication are examples of the use of design and construction know-how that has had a tremendous impact on the construction industry. Also, time saving by fast-tracking (phasing between design and construction) has become a necessity for the owner of a tall building in order to continue to remain competitive in the market. Both new methods and materials have heightened the need for good project management.

The bad news is that owners are not taking full advantage of project management. Part of this failure is due to the lack of understanding on the

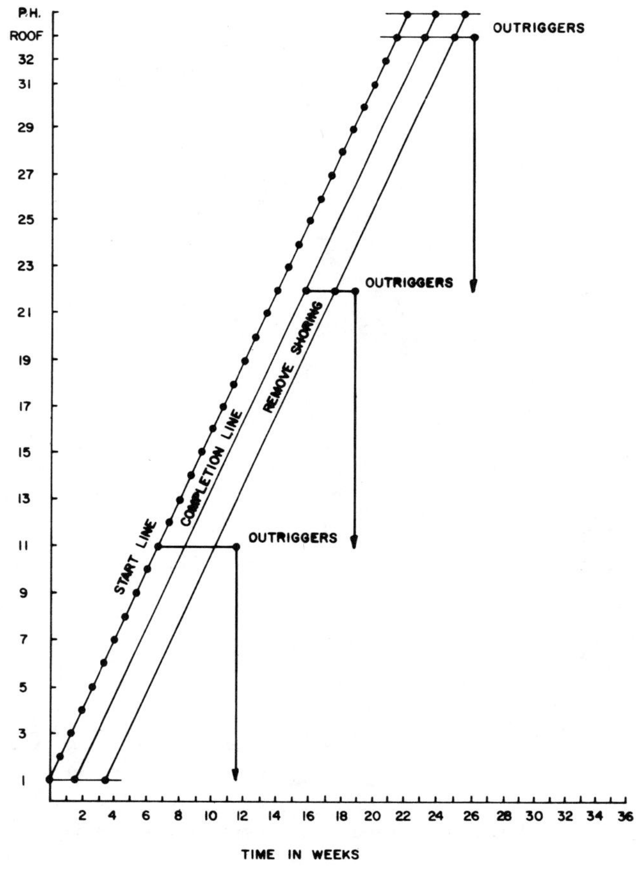

Fig. 3 Structural frame and outriggers

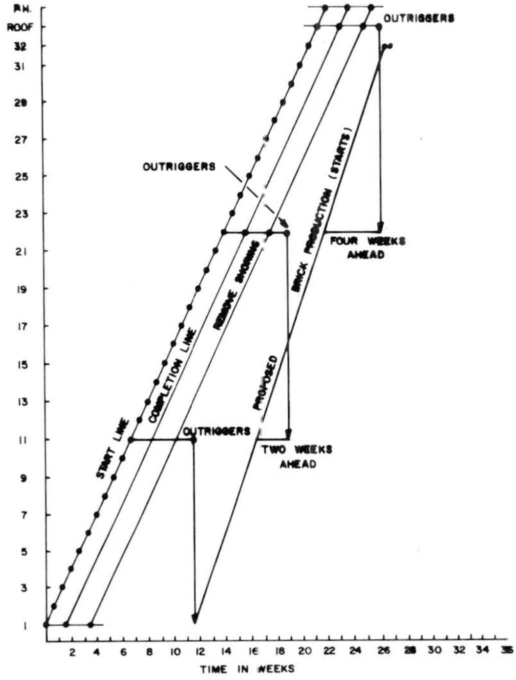

Fig. 4 Proposed brick progress vs. outriggers

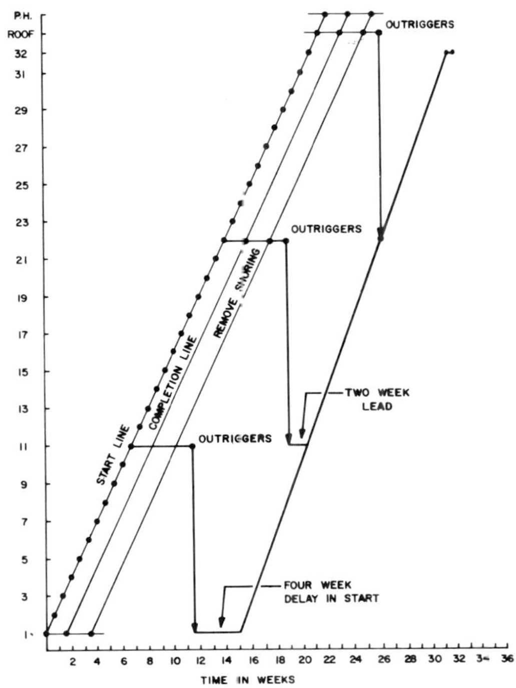

Fig. 5 Brick production—late start

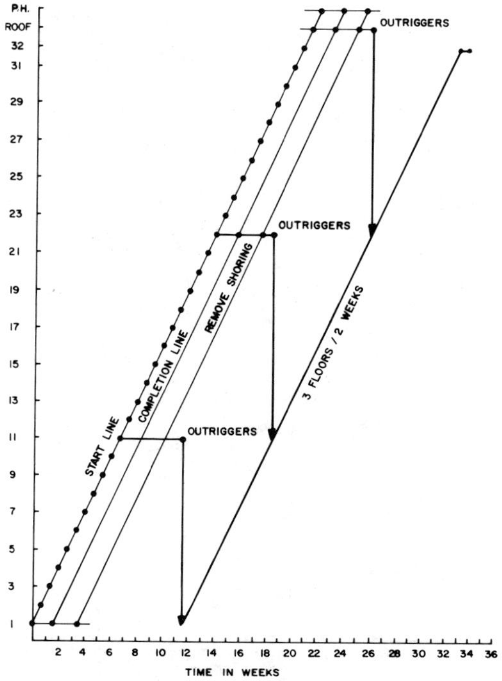

Fig. 6 Brick paced by concrete production (floors 1 to 22), 3 floors per two weeks

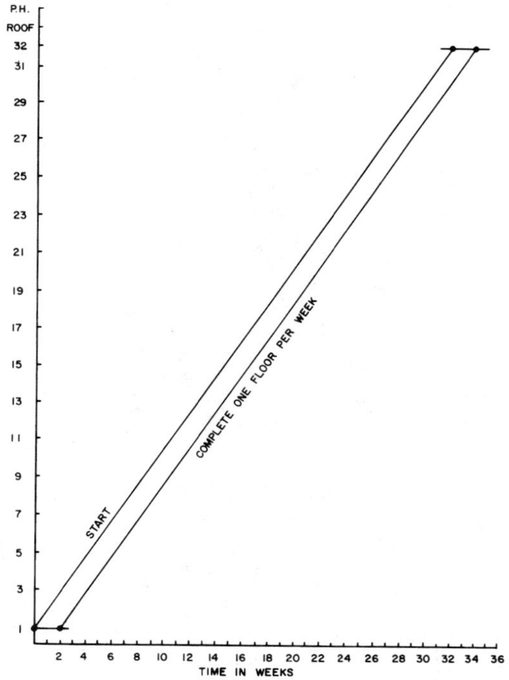

Fig. 7 Proposed glazing production

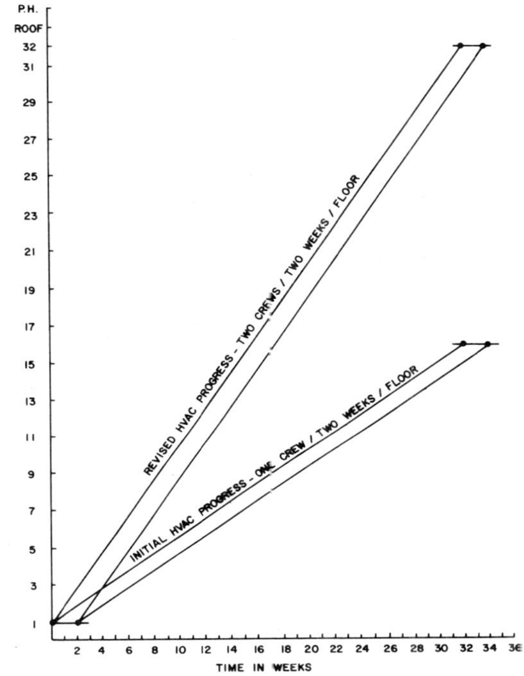

Fig. 8 Proposed rate of HVAC progress

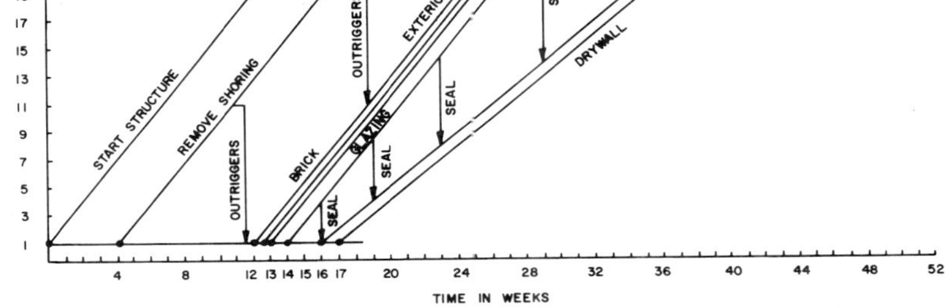

Fig. 9 VPM-drywall-glazing interface, revised

part of many owners and developers that they have a definite input. Their role is analogous to that of a producer in a film or theater production. If their actors and directors and other parties are actively carrying out their roles, the producer's life can be an easy one. Conversely, the producer (owner/developer) has a responsibility of assuring that all of the required tasks are, in fact, being carried out appropriately.

As the buildings have become more complex, it has become apparent that the traditional general contractor approach was incapable of meeting the owner's objectives of completing the project within a specified time and at total maximum cost. Total cost of the project is minimized, not merely by minimizing the time of the construction or design cycle, but through numerous interactions between design, material alternates, competitive bidding processes, contract negotiations, and effective coordination of all construction and design activities. All of this requires dedicated project managers.

The architect/engineer, although highly qualified in planning and designing buildings, is rarely an expert on pricing, construction scheduling, coordinating, and expediting of construction operations. Consequently, it has become necessary for those owners who lacked in-house construction experience to bring in construction specialists from the outside who had the necessary expertise in both preconstruction planning and construction for providing inputs to the architect/engineer in the early planning and design phase. This led into the evolution of the construction management concept enabling the architect, the specialist consultants, the owner, and the construction manager to come together at the beginning as a project team. The construction man-

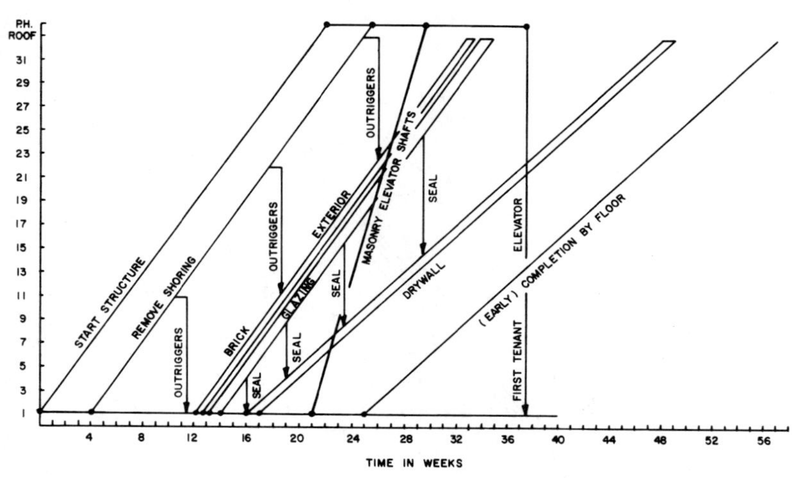

Fig. 10 Overall condominium-apartment VPM

ager in reality must become the leader of the team for encompassing design and construction as a single entity for decision making purposes.

SUMMARY

The selection and assignment of the project manager is a decision of great importance to the tall building/urban habitat project. Further, the authority delegated to the project manager and the manager's appropriate use of it is a significant factor in the success (or failure) of the project management effort. Even here, the owner/developer can be a very positive factor as the project manager carries out his assigned role. Various means can be utilized to maintain active contact with the project manager as the project proceeds. In turn, the project manager needs to find appropriate ways to expeditiously keep the owner/developer advised as to the status of the project.

Graphical summary approaches to demonstrate status (and projections) of both schedule and cost can readily be provided so that the owner/developer can be briefed in a relatively short time. This summary information, if it suggests problems, should also have appropriate hierarchial detailed back-up so that the owner/developer can perform an audit of any topic at any time.

None of the above comes without a price. Project management (and/or construction management) will cost on the order of 5% of the construction cost. However, this is 5% of a moving target. Effective project management can reduce the overall cost, and its value should be viewed in the total picture rather than in the specific expense. "Penny wise and pound foolish" is a sage saying that applies to the underinvestment often evident in project management in tall buildings.

Project Management

Scheduling Tall Buildings

Denis W. Boyd

Today when speaking of construction scheduling we are normally referring to the *critical path method.* The *Critical Path* is the longest path through the construction activities to complete the project. Because of the complexity of most projects the computations are performed on computers using programs especially designed for this task. The critical path method provides information on the earliest and latest start and finish dates for each activity. For dates on the critical path the early and late dates are the same. We say that activities on the critical path have zero float, which means that they cannot slip without delaying the completion of the project.

Tall buildings are a natural application for the critical path method. Because the work can proceed in only one direction, from the bottom to the top, it requires a minimum number of changes in logic to keep track of all the items on the project.

Properly used, the scheduling system can provide the owner/developer with information on the status of the project from its conception to occupancy, information that he will need to make sound business decisions. It furnishes the architect-engineer with required dates for design, construction documents, and approvals. It gives the construction manager or contractor a list of items that need to be completed, their timing, and criticality. To get the most out of the system the scheduler should start about 12 months before the start of construction so that he can provide a project schedule for the preconstruction, as well as the construction periods. A more detailed construction schedule should be developed at the actual start of the work, which will be used as the basis for measuring progress throughout construction. Management of the project can then proceed in an orderly fashion (rather than by the principle of crisis management) and this should save everyone concerned both time and money.

Computerized construction scheduling systems using the critical path method have been available for more than twenty years (Council on Tall Buildings, 1981). Despite their potential value, they have had little impact on the construction industry. Until now they have too often had little real effect beyond producing a chart to hang on the wall or voluminous reports to put in a drawer.

One of the major problems has been the lack of a firm commitment by top management. A computerized scheduling system represents a considerable change from the usual "seat of the pants" method of doing business and will initially be resisted by many managers. The best scheduling system will be wasted if the people running the project simply refuse to use it. Either top management must enforce its use, or the scheduling manager must be given the authority to see that it is being used. Regular meetings with the project managers and superintendents to review the schedule in detail are essential to insure its success.

Another problem is the reporting system. Volumes of paper are often generated and distributed to everyone regardless of their rank or job, while reports should be tailor-made to the user. For example, a report to the owner/developer who has no direct interest in the construction should not exceed two pages, whereas reports to a manager directly involved in field operations may need to be several pages in length. The report formats often need to be improved. They are frequently forbidding to look at and difficult to read. A good report should provide only the information needed by the user in a logical, clear, and concise manner. Even the increasingly popular graphic plotter charts are normally poorly organized and hard to decipher. If used at all they should only be given to a select few.

The hardware that the system is run on has also been part of the problem. In the past it was necessary, because of the size of the system, to use large computers at a central location often far removed from the scheduler or project, making it cumbersome and costly. With the latest generation of microcomputers, however, it has become possible to run even sophisticated main frame scheduling programs. The number of activities permitted may be less than the main frame version, but are still usually adequate for most tall building projects, and far less costly.

Now that the hardware and software are relatively inexpensive, more companies will use the critical path method. As the number of users increases, more systems are being offered. With increasing competition between systems the once cumbersome main frame programs will improve and become easier and more feasible to use.

Now the question is, who will use the system? Historically the scheduling department has been staffed by new entrants into the construction field, often recent graduates. As soon as they gain experience they are promoted into project management so that schedules are always being done by inexperienced people. This turnover leads to unrealistic activity in the field and cannot be used by field personnel. The commitment by top management must provide sufficient compensation to those interested and qualified to make them want to stay in scheduling. For people not interested in or needed for scheduling, the scheduling department still can provide an excellent training ground for future project managers. Companies that do not have the constant volume of work to justify in-house scheduling can always use an outside scheduling service that has the capabilities we are discussing.

Although the computer has come a long way in the past few years, its abilities are still largely limited to addition and subtraction, together with some limited decision-making capability. All these operations are performed at very great speed. The job of the programmer is to take advantage of these capabilities in a creative manner to enable the computer to perform useful and sometimes very complicated tasks. For a scheduling system the computer must add and subtract given activity durations using specified relationships between the activities, then compare the results with a built-in calendar to arrive at dates for those activities, and then be able to report the results in a useful manner.

There are two types of programs or systems: *arrow diagramming* and *precedence diagramming* (Fig. 1). Arrow diagramming has been popular in the past largely because it is simpler to use. Mainly for this reason contractors have been required to use it on government projects. Normally the only relationship allowed between activities is the completion of a preceeding activity and then the start of a succeeding activity. This severely limits this system, because in construction we almost never complete one work item before starting the next item. For example, in a concrete building we never complete one floor of concrete before starting the next floor, and we never complete a floor of curtainwall before starting the next floor. Therefore, the use of arrow diagramming results in artificial schedules that do not describe the work as it is actually performed in the field.

Precedence diagramming systems permit the use of start-to-start relationships, finish-to-finish relationships, as well as the finish-to-start relationship used in arrow diagramming. Some of the more sophisticated systems also permit other relationships giving the scheduler the ability to describe the overlap of activities as it actually takes place in construction. This also makes precedence diagramming more complex than arrow diagramming, requiring an experienced scheduler to get the most out of it. Some precedence diagramming programs may be less expensive or easier to use than others. However, these systems can limit the number of activities, types of reports, types of relationships, number of calendars, and so forth, which limits their usefulness. When scheduling a tall building, which certainly must be considered a major project, the benefits of the use of a good precedence diagramming system by an experienced scheduler will far outweigh the nominal cost.

The schedule should be organized and arranged in a logical manner according to the way the construction actually proceeds, starting with the substructure and ending with the finishes. A suggested outline of major systems and subsystems for a typical tall building is shown in Fig. 2. Some of the systems should be broken down by occupancy zones so that the occupancy date for each zone can be determined. Activities are put into the appropriate subsystem and numbered accordingly. The same breakdown and numbering is used throughout the project, starting with the project schedule that may show only the major systems, through the construction schedule containing systems and subsystems, and into the detailed computerized schedule where the individual activities are input.

Now we have an experienced scheduler using an excellent construction scheduling system on a computer near the jobsite with the full support of management. Still the results are often incorrect, so that the system cannot be

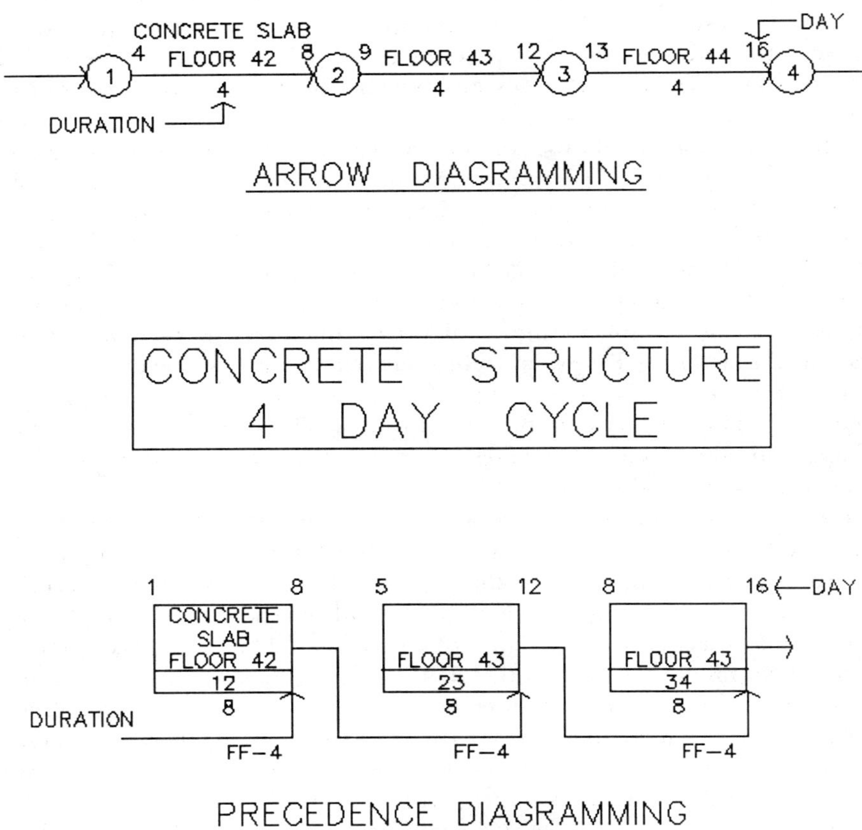

Fig. 1 Precedence diagramming

14. SUBSTRUCTURE
  A. Demolition
  B. Caissons
  C. Site retention
  D. Excavation
  E. Caisson caps
  F. Foundation walls
  G. Waterproofing
  H. Underfloor plumbing/electric
  I. Slabs on grade
  J. Concrete structure
  K. Embedded items

18. SUPERSTRUCTURE
  A. Structural steel
  B. Steel stairs
  C. Metal deck and studs
  D. Embedded items
  E. Concrete slabs
  F. Fireproofing
  G. Miscellaneous

24. BUILDING ENCLOSURE
  A. Masonry
  B. Granite
  C. Curtainwall
  D. Storefront
  E. Exterior doors
  F. Roofing
  G. Waterproofing

32. RISERS AND DISTRIBUTION
  A. Electrical
  B. Power company
  C. Plumbing
  D. Fire protection
  E. Liquid heat transfer piping
  F. Ductwork
  G. Temperature control
  H. Insulation

42. BUILDING EQUIPMENT
  A. Equipment pads/supports
  B. Electric switchgear, lower
  C. Electric switchgear, mid
  D. Electric switchgear, penthouse
  E. Power company vault, lower
  F. Power company vault, mid
  G. Power company vault, penthouse
  H. Emergency generator room
  I. Pump room lower
  J. HVAC equipment room, lower
  K. HVAC equipment room, mid
  L. HVAC equipment room, penthouse

52. ELEVATORS
  A. Brackets and rails
  B. Entrance frames
  C. Machine room
  D. Hoistway
  E. Doors
  F. Cabs
  G. Miscellaneous finishes
  H. Adjustment and inspection

59. ESCALATORS
  A. Trusses
  B. Rough-in
  C. Finishes

62. PARTITIONS
  A. Hollow metal frames
  B. Masonry
  C. Elevator shaftwall
  D. Electric room gypsum board
  E. Core drywall
  F. Perimeter gypsum board
  G. Handrails

72. BUILDING FINISHES
  A. Ceramic tile
  B. Ceilings/lighting
  C. Doors and hardware
  D. Toilet fixtures, partitions, accessories
  E. Electric heaters and baseboard
  F. Painting
  G. Flooring and base

80. LOBBY FINISHES
  A. Granite
  B. Interior storefront
  C. Ceilings/lighting
  D. Ornamental metals
  E. Painting
  F. Baseboard heat
  G. Miscellaneous finishes

82. PLAZA
  A. Concrete
  B. Soffit
  C. Waterproofing
  D. Paving
  E. Miscellaneous finishes

Fig. 2 Outline of major systems and subsystems for tall buildings

used effectively. The remaining problem that needs to be solved is that no one is scheduling the submittals (shop drawings). Material and equipment often arrive at the jobsite late, which does not come the scheduler's attention until after the scheduled date for the start of installation has not been met, when it is too late to take corrective action. Fortunately systems have recently appeared that are designed to schedule the procurement process and interface with the construction scheduling system (Fig. 3). They provide information on the required dates for the submittal, approval, release, and manufacture/delivery of each item in the construction schedule.

The next logical step, once we have a good system that schedules both the construction work and the submittals, is to add costs. By putting costs on the construction and submittal activities we can generate more accurate cash flow reports, trade payment breakdown reports, and other cost reports. Because payments may be made for material, manufacture, and delivery as well as installation, it is necessary to spread the costs over both the submittal and construction scheduling systems. Activities must be selected and broken down with the cost reports in mind so that the reported completed cost, based on percent complete, agrees with the actual cost expended.

With tenants often taking ten or more floors of space in tall buildings, and budgets in excess of $10 million not uncommon, it is very important that a professional scheduler be used to schedule the tenant work. The scheduler must start well in advance of the start of construction to develop an overall project schedule that includes design and construction documents, and long

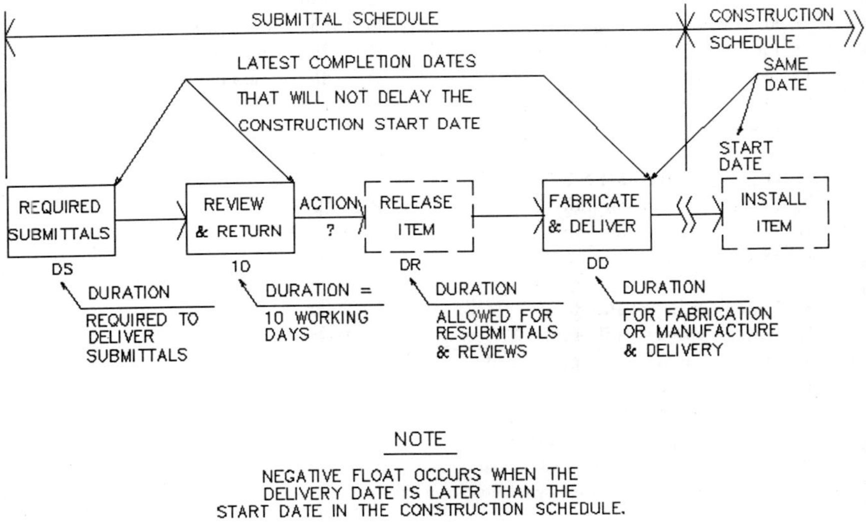

Fig. 3 Submittal scheduling system

lead time items, as well as the construction work. At the beginning of construction a more detailed construction schedule can be drawn for monitoring the construction work. Depending on the size and complexity of the project, the computerized scheduling systems described previously may also be used.

In the future we will see construction scheduling systems more widely and more effectively used on tall buildings as the software and hardware are improved. Such systems can greatly contribute to the success of a tall building project from both the owner and construction manager/contractor standpoint, not only by insuring its timely completion, but by providing the information necessary to get there with the fewest number of problems. An effective system can provide the competetive edge in the highly competitive tall building market.

REFERENCES/BIBLIOGRAPHY

Council on Tall Buildings, 1981
PLANNING AND ENVIRONMENTAL CRITERIA FOR TALL BUILDINGS, Volume PC of the Monograph on the Planning and Design of Tall Buildings, ASCE, New York.

High-Rise Development In India

Jashwant B. Mehta

Ever since the 14 story Life Insurance Corporation Building was constructed in Madras in 1956, a fairly large number of high-rise buildings have been constructed in all the major cities of India including Bombay, Calcutta, Delhi, and Madras. The height of these tall buildings ranges from 8 to 30 stories. In addition to these metropolitan cities a number of high-rise buildings (in the range of 8 to 15 stories) have also come up in smaller cities such as Ahmadabad, Hyderabad, Bangalore, and Pune. An analysis is given of certain aspects of this development with special reference to Bombay, where the largest number of tall buildings are constructed.

FLOOR AREA RATIO

The floor area ratio (FAR) is the ratio of built up area permitted to the area of a given plot. In most Indian cities, the current building regulations have adopted FAR (sometimes also called Floor Space Index or FSI) as the main criterion for permitting allowable floor area for a particular plot. The FAR permitted in major Indian cities generally varies between 1.00 and 3.00, irrespective of the height or number of stories of the building. As compared to other countries, this FAR is rather low. To most Indian architects and engineers, the (maximum) permissible FAR in American cities (Chicago-40,

New York-15, San Francisco-25, Boston-14 etc.) sounds like a fairy tale. Even the figures of some major Asian cities like Tokyo (12) or Singapore (6) appear to them to be too high to be true. Although the FAR permitted in India is so low, many tall buildings are being constructed in our country. An observation of high-rise buildings in Bombay has revealed that wherever the FAR permitted was more than 2.00, tall buildings in the range of 10–20 stories have been preferred by the planners. Even where the FAR has been as low as 1.00 in suburban areas, a number of tall buildings have been built (Fig. 1). Analysis reveals the following reasons:

Availability of large open spaces on the ground. For a 15-story building to

Fig. 1 A typical suburban high-rise. There are numerous such slim high-rise buildings all over Bombay, almost all of which have a nominal built up area of 250-400 m^2 (2700-4300 ft^2) per floor.

consume a FAR of about 2.00 means a plinth area equal to 13.33% of the total plot area, which would mean leaving almost 86% open space on the ground floor. This large availability of open spaces on the ground is used for several purposes such as parking, play areas for the children, and providing amenities such as a swimming pool, tennis courts or other recreation facilities. While basements are popular for car parking, the upper floor car parking is generally not a common feature. Because of the higher unit cost of providing parking on upper floors, parking at ground level is generally preferred. Ground floor open space also facilitates planning for larger entrance halls, providing underground water tank (necessary due to limited availability of municipal water supply), pump room, electric meter room, substation, and the like.

Scenic Views. In a number of instances, scenic views from upper floors are a major attraction for preferring a taller building.

Creating Space for Semiprivate Bungalows. In a number of developments, it was found that owners preferred to construct small row-houses or semiprivate bungalows in the same premises (commanding a higher premium) on part of the plot. These bungalows command a higher price as the occupants get the advantage of an exclusive use of land. The remaining plot area is too small for consumption of the balance of the FAR, unless a tall building is constructed (Fig. 2). Sometimes the owners want to retain their existing structure, which is either a bungalow or a low-rise building, and utilize the balance total potential of the plot. In such cases, a high-rise building is preferred on the remaining area to use the floor area available.

Status Symbol and Sense of Achievement. In a developing country such as India, a tall building is still a fascinating sight. It is a status symbol for the owner and the fulfillment of technological achievement for the builder, engineer, or architect (Fig. 3).

LIGHT AND VENTILATION

A large majority of buildings are dependent on natural light and ventilation. For residential buildings, the ceiling fan is still the most popular means of providing ventilation. Air conditioning, a luxury, is only resorted to during the hot summer months of April, May, and June, mainly in the bedrooms. Window air conditioners are still found convenient and popular (Fig. 4). Very few of the office buildings are fully centrally air conditioned. Unless the entire building is owned by a single large corporate body, it is usual to find window air conditioners or small package units by individual office owners. Most high-rise buildings are condominiums (where individual coowners form a cooperative society) to look after the management.

Because of the dependency on natural light and ventilation the building codes generally restrict the depth of the room to 8.3 m (25 ft) from the external face of the building. The width of the buildings therefore does not exceed

16.6 m (50 ft) to 18.3 m (55 ft). Open space regulations are also quite stringent. As a result, the overall floor areas are much less than elsewhere. In contrast to the massive tall buildings in the United States and elsewhere, with builtup areas per floor as high as 3,000 to 4,000 m^2 (32,000 to 43,000 ft^2), the residential and commercial buildings in India have an average of a mere 300 to 400 m^2 (3,280 to 4,300 ft^2). The builtup area of one of the mighty skyscrapers in the United States such as the Pan American or Empire State buildings would probably be equal to the builtup areas of all the skyscrapers of Nariman Point (a mini-Manhattan of Bombay) put together. It is difficult to conceive this from appearances because height is so deceptive. A building photograph generally reveals more of height than of the other dimensions.

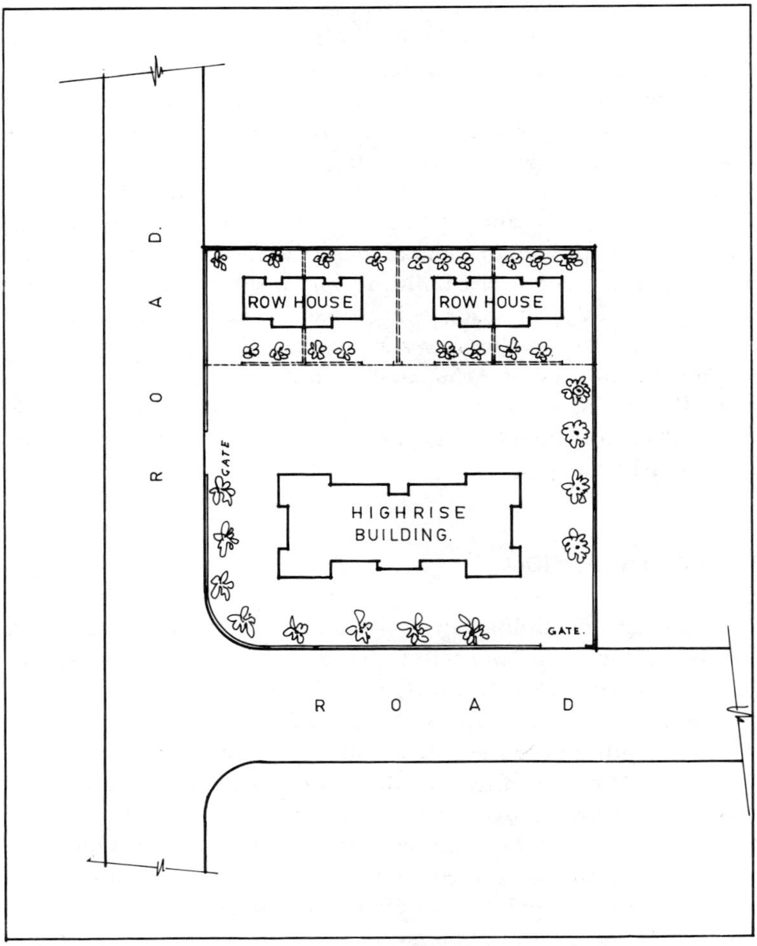

Fig. 2 Typical development site in India featuring a high-rise building with small row-houses on the same plot of land.

Fig. 3 Tall buildings are a status symbol for owners in India.

Fig. 4 A commercial high-rise in South Bombay. The elevation treatment relies heavily on sun-breaking devices such as continuous horizontal overhangs above windows intercepted by vertical fins. Numerous window air-conditioners can be seen.

OPEN SPACE REGULATIONS

Compared with other countries, open space regulations in India are stringent. The requirements of the Bombay Municipal Corporation are given in Table 1.

Table 1 Open Spaces For Different Height Of Buildings For Light and Ventilation

Building height (in m)	Minimum required open space, sides and rear (in m)
10	3
15	5
20	7
30	10
40	12
50	14
53 and above	16

For buildings above 24 m, a minimum front open space of 6 m is required.

ACCOMMODATION FOR SERVICE STAFF AND HOUSEHOLD SERVANTS

An earlier paper (Mehta, 1978) drew attention to inadequate accommodations for household servants and service staff in tall buildings. The following activities are usually carried out manually: cleaning of staircases, corridors and compounds; collection of garbage (both in commercial and residential buildings); security arrangements; and operation of elevators.

On enquiry as a trial case, it was found that a 17-story condominium office building with a total floor area of 27,500 m^2 (296,000 ft^2) employed a total staff of 48 people, consisting of 15 elevator men, 8 sweepers and 25 security men. This does not include the sweepers engaged by individual office owners. All the elevator and security men stayed within the basement and in the outside compound, sharing two toilets provided for common use at ground level. In some cases, some of them stayed in the nearby encampment of huts surrounding these skyscrapers.

The problem of accommodation for household servants is equally acute in most high-rise buildings. Household appliances such as washing machines or dishwashers, so commonly found in developed countries, are still a luxury here. Thus relatively inexpensive manual labor is used. Almost every apartment owner employs at least one fulltime servant. In the upper income groups it is common to employ two or three servants per family. While the bigger flat owners can afford to provide a separate servant's toilet and a sleeping facility with an independent entrance in the same apartment, it is difficult for a smaller flat owner to provide the same facilities. In the absence of adequate sanitary and sleeping arrangements, the plight of servants becomes miserable and the overall standard of sanitation in the building declines.

A workable solution to the problem of facilities for servants, especially for smaller apartments, would be the provision of a common toilet facility on every floor. The ideal location would be at midlanding where it would be of least inconvenience to the flat owners. The author has successfully tried this and it has been fairly well accepted. The facility including its maintenance is shared by the flat occupants on every floor (Fig. 5).

Common passages and corridors are regularly used as sleeping places by servants. While ideal solutions such as a common room on every floor or an

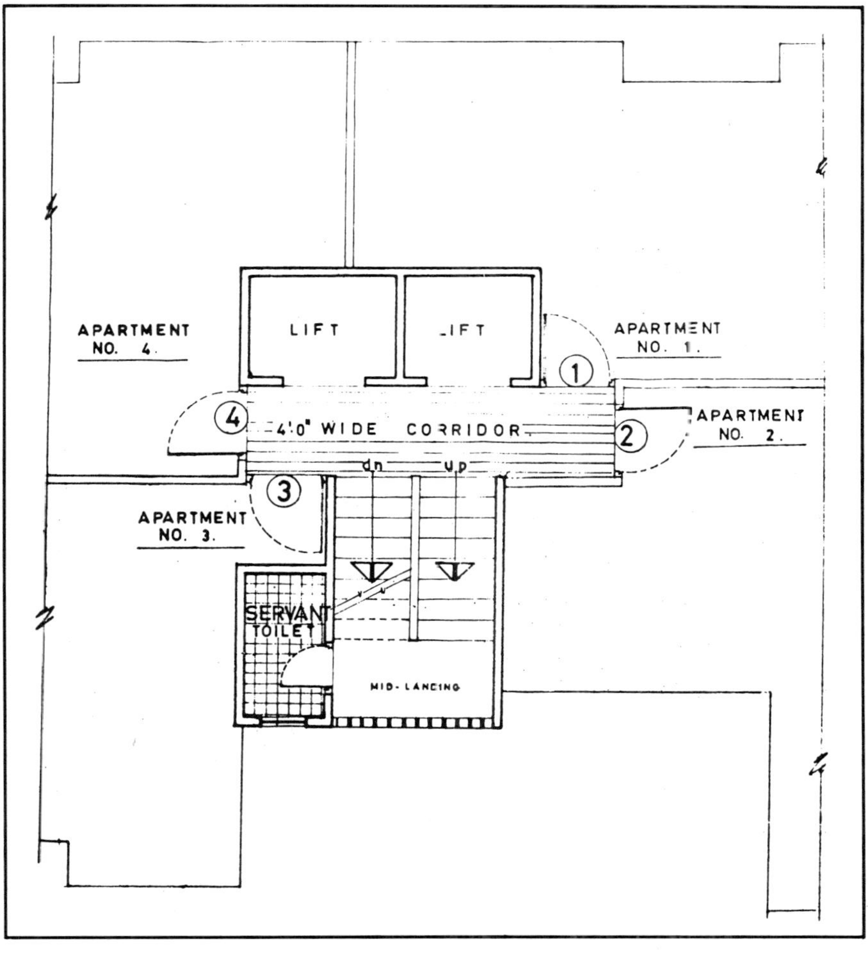

Fig. 5 Floor plan at midlanding level with common toilet facility.

outhouse type dormitory are rather difficult to provide, even the provision of wider passages and corridors could ease the problem and prevent crowding during the nights.

INADEQUATE FIRE-FIGHTING ARRANGEMENTS

A large majority of the tall buildings constructed until now lack the basic fire-fighting appliances required to be provided within the building. While Bombay was among the first cities in India to initiate codes for fire protection and fire-fighting requirements for tall buildings, most other cities had little fire regulation until the early 1980s, in spite of the fact that several tall buildings ranging from 8 to 20 stories had been constructed in these cities. The situation is now slowly changing and fire-fighting regulations are being introduced in all the cities.

In Bombay a survey was conducted to find out the condition of fire-fighting equipment and the presence of fire-fighting appliances within the high-rise buildings after they were occupied. It was observed that out of 150 high-rise buildings (more than 27m (80 ft) in height), only 11 buildings had the necessary equipment in good condition. This means that 92% of the buildings were not in a safe condition. Either the hydrants had missing valves or the fire-fighting pumps and other equipment such as fire-alarm systems and the like were out of order. In Delhi it was reported that as many as 200 skyscrapers constructed before June 1983 were without adequate fire-fighting facilities.

Those buildings that were of single ownership type and were partially or fully occupied by the owners were generally better maintained. However, in those buildings that were fully rented or where ownership was divided, there was an utter negligence of fire-fighting equipment and appliances provided. Awareness of the situation among the occupants and the Co-operative Society of flat-owners (in India this means the Association of flat owners) that manages the building was lacking.

Several fires that have taken place from time to time prove this point again and again. Thus in Delhi, for example, during a 1983 fire in the 13-story Gopala Towers 513 people miraculously escaped only when the fire brigade used an 11 m (36 ft) extension ladder as a bridge from an adjoining building.

In another hotel fire in New Delhi (1985) 37 guests were trapped in the blaze and lost their lives. The situation in most Indian high-rise buildings could be similar to the fires that took place in Sao Paulo skyscrapers (Andraus Building, 1972 and Joelma Building, 1974).

In the event of a fire it would be difficult to rely on the equipment provided within the building. The Bombay Fire Brigade had proposed a special tax on all the high-rise buildings and a routine inspection. However, so far the proposal has not been implemented. In the author's opinion, to minimize the reliance on firefighting equipment provided within the building, the ultimate solution may lie in charging premiums on high-rise buildings, and utilizing these premiums for procuring additional and more sophisticated equipment for the Fire Brigade such as mobile diesel pumps.

More research should be undertaken on the performance of the Suspended Maneuvering System (SMS), which is a helicopter-hung fire-fighting platform fully equipped to tackle the fire and rescue the occupants. Its use for fighting high-rise fires was tried successfully in 1982 under experimental conditions. If the actual field performance is satisfactory, these helicopter platforms could be used by the fire-brigades of large metropolitan areas including those of the developing countries.

REFERENCES/BIBLIOGRAPHY

Government of Maharashtra, 1984
DEVELOPMENT CONTROL RULES FOR GREATER BOMBAY, as amended up to 1984, Urban Development and Public Health Department, Bombay.

Government of Maharashtra, 1974
FIRE PRECAUTIONS IN HIGH RISE BUILDINGS IN BOMBAY, Urban Development and Public Health Department, Bombay.

Mehta, J. B., 1977
AMENITIES IN TALLER BUILDINGS IN DEVELOPING COUNTRIES, 2001, Urban Space for Life and Work, (Proceedings of Conference held in Paris, 1977), Vol. II, CTICM, p. 63-66.

Mehta, J. B., 1978
HIGH RISE BUILDINGS, published by the author.

Purandare, D. D., 1982
FIGHTING FIRE IN THE SKY, The Illustrated Weekly of India, June.

Developing Tall Buildings to Meet Future Needs: Transco Tower Case History

Gerald D. Hines

The recent years have been something of an orgy for those of us who love the romance of tall buildings. The future, on the other hand, may not resemble this past. While the author is in no sense a "doom and gloom" man, it appears that here in the United States, it will be difficult during the next 20 years to replicate the widespread growth of tall buildings that we have seen in the past 20 years. The reasons for the skyscraper still exist, but they are simply not as collectively powerful as they once were.

REASONS TO BUILD TALL

In the beginning, the typical rationale for building tall was to house—primarily, at least—the personnel and equipment of a single company. There were, of course, notable exceptions to this, among them the Empire State Building, completed in 1931. (However, it must be remembered that this exception, a speculative venture, was known for years as the "empty state building.") There was great efficiency in centralized, consolidated operations, and that efficiency fostered the concept of the tall, tall building. Communication, especially, was enhanced when a corporation gathered all or most of its people under one high-rise roof.

That was before the days of computers, satellites, and instantaneous telecommunications. Today, consolidation is increasingly unnecessary and sometimes financially undesirable. Take, for example, such instances as American Express, a New York company that moved its huge credit card facility to Phoenix.

It probably makes little difference where that American Express facility is located, operationally speaking. Modern telecommunications have made such decentralization practical. Contemporary economics have made it attractive. One does not typically get optimal economy for high-density users in a 70-story building with 1800-m^2 (20,000 ft^2) floors, as you know; instead, it comes in two-, three-, or four-story buildings with 4600-m^2 (50,000-ft^2) floors. Of course, various states and cities offer some excellent incentives to businesses who are shopping for advantageous locations. These are hard to resist, because they track straight to the proverbial bottom line.

In short, smart economics are enticing some large companies to decentralize today, and this leaves the potential skyscraper with fewer and fewer potential base tenants.

Another factor that has historically fueled the development of skyscrapers has been the desire of companies, and sometimes chief executives, to use building stature to contribute to their corporate images. And the record suggests that prominence on the skyline does, in fact, have a marketing value. "Bigger is better" remains an important principle when packaging a product or sounding an advertising message.

However, if we have learned a lesson from buildings such as the Transamerica Tower in San Francisco, it is that height is only one of several ways to enhance one's image and to capture public attention. Color and style can do it, too. Unusual designs and unusually sited buildings can achieve the same end. Thirty stories may be plenty. Transamerica, whose design quality is a matter of some debate, has not merely overshadowed the extraordinary Bank of America building in the same city, it has all but replaced the Golden Gate Bridge as the symbol of the Golden Gate City!

So, height, we now know, is not adequate in itself in corporate image-making, further decreasing the number of companies that might be driven to build tall.

Another obvious influence on the construction of tall buildings is the office market. From a national perspective, the market today is in a new phase. Vacancy rates are high and climbing. Some prognosticators see the rate surpassing 20% by 1991. Demographic experts say that it is unreasonable to expect the growth rate of office employment that occurred in the 1970s and early 1980s to continue through this decade. Therefore this, too, cools the fire in the hearts of many who desire to build tall.

The advent of real estate developers as a motivating force has also altered the typical skyscraper scenario. This has special implications for the years ahead, in light of the projections of modest needs for new office space in the remainder of the 1980s. Developers are often willing to take risks on what rents are going to be; they are much less often willing to risk the amount of space they put on the market in a single location. We all know the adage about putting our eggs in a single basket. This wisdom is not lost on entrepreneurial developers, let alone conservative 14-member corporate boards, which would on the whole prefer egg substitutes, because of the breakage factor.

So, uncertainty about the future is further dampening the American urge to build tall.

Put these four influences together with others of more or less importance depending on where you wish to build—factors such as environmental controls, size restrictions, civic pride in older buildings, the density of some of our central business districts, and so on—and you are left with only one factor that spurs skyscraper development as much now as ever: the romance.

ROMANCE OF TALL BUILDINGS

Romance, in the case of the skyscraper, is certainly understandable. Few purely human achievements can compare with it. So the lure of having the company name engraved above a granite arch, topped by six, seven, or eight hundred feet of steel, sculptured stone, and glass is nothing to take lightly. Nor is it a passing fancy, as evidenced by our U.S. cityscapes (Fig. 1).

But how are the intangible values of romance to be made practical? In the years ahead, how will all the needs of major base tenants, contemporary white-collar workers, developers, sophisticated investors, and the citizens of a city be met in a building 40 stories or higher?

The answer is redundant: The needs will be met when the requirements of these groups are satisfied. More than ever, the role of the developer will be to coordinate these ingredients into a pleasing product to all. Our teams of experts must continue to build buildings that make sense architecturally, economically, technically, and romantically. Most importantly, they must address the needs of our *largest* users—companies that alone are able to occupy more than 50% of a proposed tall building. To succeed, tomorrow's skyscrapers must represent unique—almost irresistible—opportunities for these large companies.

One can foresee a period of skyscraper construction that is more market-sensitive, more major-occupant driven, than ever before. And several development considerations will be key to achieving success: customized provisions for these major occupants' privacy (which may also be termed "security"), quality of life offerings, access and parking, and flexibility. Fine architecture will be increasingly demanded. A crucial component in the formula will be public participation, which one can think of as a sharing of the romance.

TRANSCO TOWER: A CASE HISTORY

How these factors can blend into a skyscraper to make it feasible—and successful—now and in the future, is illustrated by the case history of Houston's Transco Tower (Fig. 2). Transco Energy Company typifies the major user of skyscrapers in the years ahead. Tenants such as Transco will separate the dreamers from the visionaries.

The importance of the major tenant could hardly be more clearly illustrated than in the case of the 64-story Transco Tower. The vision of architects Philip Johnson and John Burgee bore fruit because of the mind of Transco president Jack Bowen, and because of the special moment in his company's distinguished history.

Fig. 1 Pennzoil Place and Republic Bank Center *(Photo by Richard Payne)*

Transco's rationale for a new building of its own was abundant. The company had grown considerably until it was squeezed in and scattered throughout five buildings in the Galleria/Post Oak area of near-southwest Houston. The company's first thought was to consolidate downtown. But employees voted that they wished to remain in the suburban location, near major retail centers and where their commute was more agreeable.

That alone justified a building of approximately 40 stories. So, obviously,

Fig. 2 Transco Tower, Houston, Texas *(Photo by Richard Payne)*

there was another dimension to creating new headquarters: creating "the most beautiful building in the world," because of pride in the company and the city—and because the two were not well enough acquainted.

Transco, then, needed something distinctly different from a skyscraper to meet its consolidated operational requirements. On the other hand, it needed something akin to a skyscraper to meet its desire for dramatic new prominence. This was the problem, or rather the unique opportunity, that produced Transco Tower.

Ultimately, the Tower became "a building within a building." Location, public access, parking, vertical transportation systems, security provisions, land use, lower-level architectural design, and other features were made necessary by Transco's operating needs. The remainder of the building was, financial considerations aside, made desirable almost completely by Transco's reach for a new image.

While today the ranking of Transco Tower among the world's beautiful buildings is perhaps open for debate, little question remains about whether the spirit of Transco's request was fulfilled. Moreover, this was accomplished in a way that precisely answered Transco's desire for its own building. Consider a few notable aspects to illustrate what I mean when I say that tall buildings, to meet future needs, will be profoundly occupant driven.

Transco employees wished to remain in the same geographic area. So, Transco Tower places them there. They are today directly across the street from Transco's previous headquarters. Their new home is a landmark in City Post Oak, a mature urban center in the heart of their city's prime labor market, and six miles from downtown Houston.

For employee convenience, and to lessen the impact of the Tower's proposed density on existing traffic, a carefully planned network of entrance and exit points, serviced in part by new streets, was created for the building. The Tower thus may be approached from any of four public streets, including one-way streets bordering the site on the north and south. No building close to this size in Houston is so accessible. This ease of egress and ingress could never have been matched in the central business district.

Because having one's own building also means having a dedicated parking garage, Transco does. The Tower's parking facility is actually two facilities, one of which is open exclusively to Transco employees and visitors. A separate ramp system permits these people independent ingress and egress.

At street level, a covered walkway leads from the public portion of the garage to the Tower's lower lobby. On the level above, an enclosed skywalk connects the garage with a distinct second-level lobby, created expressly for the building's major tenant.

From this second-level lobby, only the lower lobby and floors 3 through 40, Transco's spaces, are accessible. Transco's floors are served by four banks of 22 direct elevators that leave from here. Transco employees and visitors arrive at this lobby from the garage or by escalators from ground level, and thereafter avoid all interaction with non-Transco business and personnel in the Tower. A separate freight elevator and truck dock are also provided.

From the lower-level lobby, five high-speed express elevators travel at 427 m (1,400 ft) per minute to carry non-Transco personnel to a Sky Lobby on the 51st floor. From there, two banks of five elevators each serve floors 41 through 64, the non-Transco spaces.

Prior to the design of this transportation system and the buildings interiors, the incidence of travel and working relationships among Transco employees was studied and documented in various locations. Elevator programming subsequently addressed the problems of vertical separation. In the final plan, interrelated Transco departments were kept in proximity to one another and within the same elevator banks in the new Transco Tower.

To assure that certain quality-of-life offerings near the Tower could be translated into special benefits for Transco employees—whatever the weather—the building boasts a second fully enclosed Skywalk. This one leads not to the Tower but from the garage across West Alabama (to the north), directly into The Galleria. This shopping and entertainment mall offers a celebrated collection of fashion stores, specialty shops, entertainment and service centers, and a broad spectrum of fine restaurants. Atop lies the University Club, an athletic and social club for men and women. Two major Houston hotels also reside within the Galleria complex.

Because of Mr. Bowen's abiding concern for employees' style of life and work, other special requests arose. The company wanted a health/fitness facility, medical facility, and cafeteria, for example. They were provided. Ultimately, an outdoor jogging track and park was also created. An artist himself, Mr. Bowen wished also to have a unique art gallery incorporated into the building. And so it was. Transco now holds 10 to 12 art exhibits annually in the Tower's first-floor gallery. These are enjoyed by all visitors to the Tower, and comprise another distinctive dimension of this truly tailor-made corporate home.

The design of the Tower, from top to bottom, reflects Transco's needs and desires. The base of the building was developed to meet Transco's need for certain larger floors. Twice as large as the higher floors, these lower levels house a major computer and telecommunications facility, as well as other functions shared by all company divisions.

In order that these and all other spaces in the Tower accommodate future advances in technology, which invariably require numerous new, hidden cables, extra space measuring 150 mm (6 in.) was allowed between the floors of the building.

As implied earlier, the upper floors were of value to Transco for, primarily, nonoperations reasons. But that statement was actually only partially correct. In fact, on the top of the Tower today, hidden from view by the peak of the building, rests an array of telecommunications antennae and other equipment. These take full advantage of the building's ultimate height and provide some

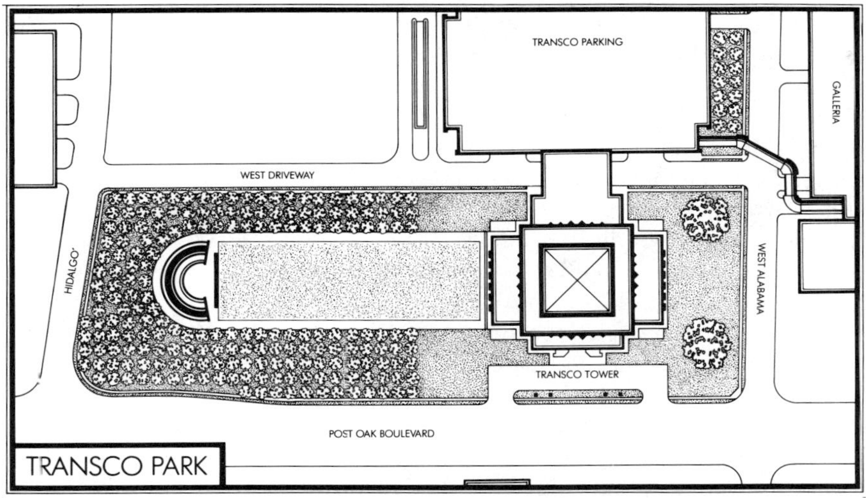

Fig. 3 Site plan for Transco Tower and Transco Park

Fig. 4 New Year's Eve fireworks at Houston's Transco Tower *(Photo by Chris Kuhlman)*

solid rationale for it. From these sky-high antennae, Transco communicates via microwave relays with a multitude of control stations that punctuate its many miles of gas transmission pipeline in the United States. Indeed, the company monitors and controls many of its pipelines' operations—as far away as New York—from right here.

If this brief overview suggests how Transco Tower was developed as a response to one major tenant's essential business requirements, a glimpse at a separate phenomenon of the Tower's development will show how the skyscraper met—and exceeded—the company's desire for greater public awareness.

The city of Houston has embraced this building as it has no other, even sending fan mail to the building. A beacon that rests atop the tower and shines brightly as it rotates slowly each and every night has become a guiding light for the city's travelers. The building and the park at the base of the Tower have become first-class local tourist attractions.

In April, 1985, a 19.5-m (64-ft)-tall water sculpture was dedicated within Transco's three-acre park, 90 m (300 ft) south of the Tower (Fig. 3). This semicircular wall, introduced to the viewer by Romanesque archways and Gothic columns, has become a magnet for Houstonians. More than a hundred requests are received monthly from people and groups who wish to use the fountain and park for private functions—from weddings and parties to religious ceremonies. In December of 1985, ceremonies in the park officially started the celebration by Houstonians of the Texas Sesquicentennial.

A once "invisible" company has become most visible in Houston, day and night. The possible objections to the creation of a skyscraper miles away from downtown never seriously surfaced. Instead, the public has been invited by everyone involved in this Tower to participate in the romance of it (Fig. 4). Abe Lincoln once said that "With public sentiment, nothing can fail; without it, nothing can succeed." This building is an excellent example. The Tower's romance is irresistible.

CONCLUSION

There are, then, no secrets about what will make tall buildings work well in the years ahead. The buildings must be developed to meet the unique needs of major tenants in new ways. They must allow these tenants to control their own destinies in a tall building in precisely the same ways they would control them in smaller, fully dedicated structures. And, the buildings must invite the public to share in the unbridled excitement of a tall building, the romance that is at the very heart of what we deem a "skyscraper."

Will tall buildings be required in the United States in the years ahead? Yes, of course, though not as often as in years past. But let us not be afraid of being moved by the attitude that gave birth to the first of these buildings 100 years ago. Build tall and build beautiful, and you can capture the attention and applause of all the world. Focus attention on meeting the expressed needs of that one very special tenant, and developers will be able to build tall with confidence for years.

The skyscraper represents for us today the same blend of energy and confidence, strength, and intelligence that it did a century ago. The sky remains a frontier. And those who venture into it, in fresh styles and for the betterment of life, express the hopefulness of us all.

Systems and Concepts

Introductory Review

John Rankine

SYSTEM METHODOLOGY

With the advent of the computer in all aspects of human endeavor, it is no wonder that the computer is having a huge impact on our industry. At first, it was used to do mathematical calculations that were originally done by hand and also to assist engineers to perform their tasks by access into data based systems. But now research is leading into knowledge-based and intelligence-based systems, and this is going to have a big impact. Coming from this will be comprehensive three-dimensional design systems that will encompass the whole of the building from design, construction, maintenance, occupation, and change in occupation. We are in for a great future with computers in the building industry.

HIGH-RISE HOUSING

In many countries, high-rise housing is supposedly not greatly successful. All too often we hear the phrase, "Building slums of the future." However, Singapore is one city where high-rise housing is a success. High-rises there have been well planned, they are properly landscaped, and made a very pleasant area in which to live. The area under the buildings is open and allows the tenants to meet, to hold gatherings, to hold parties, wedding receptions, and so forth. Grass and trees are planted around the buildings with ample space between the buildings, and ample car parking is kept well away from the building areas. Housing is in great demand. Apartments are sold to the tenants, and the tenants usually further spend just as much money again to make changes to the units.

PREFABRICATION OF TALL BUILDINGS

Prefabrication of tall buildings is usually not the most economic or viable structural system. However, factors come into play in certain countries that change this situation.

In Singapore, one factor is demand. Some 200,000 units are needed in five years. Although some contracts are in ordinary construction, many are in prefabricated precast. Whole floor systems are manufactured in one day, poured in the morning, steam cured for three hours and put straight onto the vehicles and taken to the site for erection the next day. After an initial two-year period, units are handed over at a rate of 14–15 per day.

In New Zealand, the lack of carpenters and the sudden upturn in the amount of work has forced builders to look to precasting because they cannot get the workers on the site to complete the construction in the time allowed.

In Czechoslovakia, a major factor is control. Most construction in major high-rise buildings there is precast with grid spans of 10–15 m (30–45 ft). The buildings are larger, of unique design and with quality controlled by the factory.

Structural Systems

The Composite High-Rise Building—An Interaction of Planning, Structure, Speed, and Economy

John G. Nutt

Important high-rise buildings are designed by a team of specialized professionals. The selection of a structural system requires knowledge of the appropriate factors of materials and construction together with an awareness of the requirements of architecture and services. As an example, this paper deals with composite construction—the mixed use of large scale steel and concrete units, and the combined use of both in individual elements. It shows how the designer addresses material properties and costs to advantage. Specific reference is made to buildings in the 50-story range. Future trends in the design and construction process are discussed.

COMPOSITE CONSTRUCTION

Two factors in the design of high-rise buildings set them apart from other structures. They are required to resist large lateral loads, and the repetitive nature of the construction requires that the design is refined to enhance the speed of construction and usable areas. Composite construction in high-rise buildings refers to the mixed use of concrete and structural steel in major load supporting elements. In recent years, it has come to have the more restrictive definition of concrete and steel acting in combination within the same elements to share load. The broader definition is adopted here.

Composite construction is not new, but the widespread use of steel in high rise buildings outside North America is undergoing a renaissance. (Firkins, 1984; Haryott and Glover, 1984). Determined publicity and research campaigns by national steel organizations (AISC, CONSTRADO, KOZAI) have promoted the benefits of composite construction. The reliability of the steel supply and the successful construction of important "pathfinder" high-rise projects now result in a composite scheme being studied among other alternatives for all new major high-rise projects.

Composite construction endeavors to use the best properties of concrete and structural steel in the most appropriate way. Briefly these can be categorized as:

Concrete: Cheapness, local production, unskilled labor, plasticity of shape, good compressive strength, built-in fire protection, corrosion protection, short lead time, reuse of molds, alternative hoisting methods.

Structural Steel: Prefabrication, off-site labor, high strength/weight ratio,

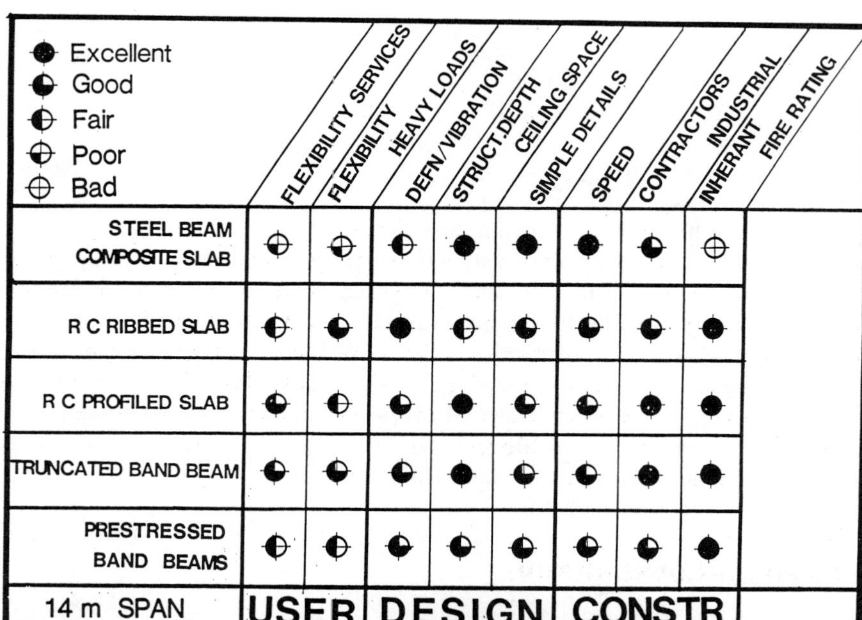

Fig. 1 Floor system—qualitative matrix

high stiffness, high tolerances, smaller member sizes, speed of construction, reduced hoisting, reduced site labor, flexibility of alterations.

The dominant elements where composite construction has been adopted in high-rise buildings are floor systems, columns, and transfer structures. Examples of these are well documented (Pascoe, 1984). In general terms, composite construction must be compared with other forms of construction to be rated. Early comparisons will be qualitative, such as the matrix shown in Fig. 1 becoming more specific and quantitative as planning progresses.

THE DESIGN APPROACH

The approach to the planning of high-rise buildings, as in most situations, is that the important decisions are those that deal with broad issues. Get those right and in balance and the rest follows. The available time is determined by the program and is frequently short. It is therefore necessary to draw upon experience from previous projects.

The major structural zones of a high-rise building are the foundations, basement/podium levels, typical floor systems and the wind resisting structure. To be cost effective, structures must be simple. For composite construction this means repetition and simple details, joints, connections, and penetrations.

WIND RESISTING STRUCTURE

In nonearthquake zones, it is generally possible to use the service core as the main element to resist lateral forces up to about 50 stories, the core cantilevering from the foundations. This also has the advantage of simplifying the beam-column connections on the perimeter.

The core configuration will be agreed upon early in the planning program and sizes held with only minor modifications. Nevertheless the structure must be tightly designed for it can be readily shown that a small loss of floor area around the core can result in the equivalent of a great increase in cost due to capitalization of rentals.

The two buildings shown in Figs. 2 and 3 have adopted a reinforced concrete core, designed for slipforming although that has not necessarily been adopted, with frequent changes of wall configuration and thicknesses.

Wind tunnel testing of buildings upwards of 30 stories invariably pays off because of better definition of moments, forces, and accelerations (Nutt, 1984). Although in a short program member sizes are fixed prior to the results being available, the savings in reinforcement alone will be three to four times the cost of the testing.

FLOOR STRUCTURES

The main influences on the structural design of the typical floors of high-rise buildings are spans, floor-to-floor height, shape, floor area, and configuration of services. High quality commercial office buildings require large areas of column-free space on the typical floor. Spans of 12 m to 14 m (39 ft to 46 ft) are common. Every effort is made to reduce the floor-to-floor height which, for the buildings described here, is in the range of 3.4 to 3.6 m (11 ft 2 in. to 11 ft 10 in.). This reduction has great impact on the overall height of the building and on the cost of the facade, vertical service runs, the vertical structure, and the wind resisting structure. The floor/ceiling height adopted is of the order of 2.65 m (8 ft 9 in.) and the reduction is taken up in the ceiling space.

The runs of services can be established early in the planning process, being

Fig. 2 Grosvenor Place development, Sydney, Australia

fixed by the locations of vertical riser ducts and the zones of floor they service. In the design, the structure is "poured" around these services using a minimum depth of structure where they occur. Beams are stopped off or penetrated, or slabs shaped and reduced in thickness as appropriate. Figure 4 shows three examples of the relationship between services and floor struc-

Fig. 3 Westpac Centre (formerly CBA Centre), Sydney, Australia

ture so as to achieve minimum floor-to-floor height, one using composite construction.

If composite floor construction (steel beams/steel pans/concrete slab) is to be economic, joints and penetrations through the steel beams must be kept simple. Considerable study has been undertaken on standardization of joints, on production of shop drawings (Alves and Berica, 1984), and penetration details (Clawson and Darwin, 1982; Thompson and Ainsworth, 1985).

SPECIAL SITUATIONS—TRANSFER STRUCTURES

Transfer structures, by their nature, are designed to support heavy concentrations of load with limited movements. Composite construction can assist and control of deflections can be achieved by prestressed tendons progressively jacked against the composite action of structural steel and concrete as the load is applied. Thus the properties of the very high strength steel tendons (1790 MPa or 260,000 psi) can be used to advantage. Two examples, transfer girder and *Piloti* transfer structure, are as follows:

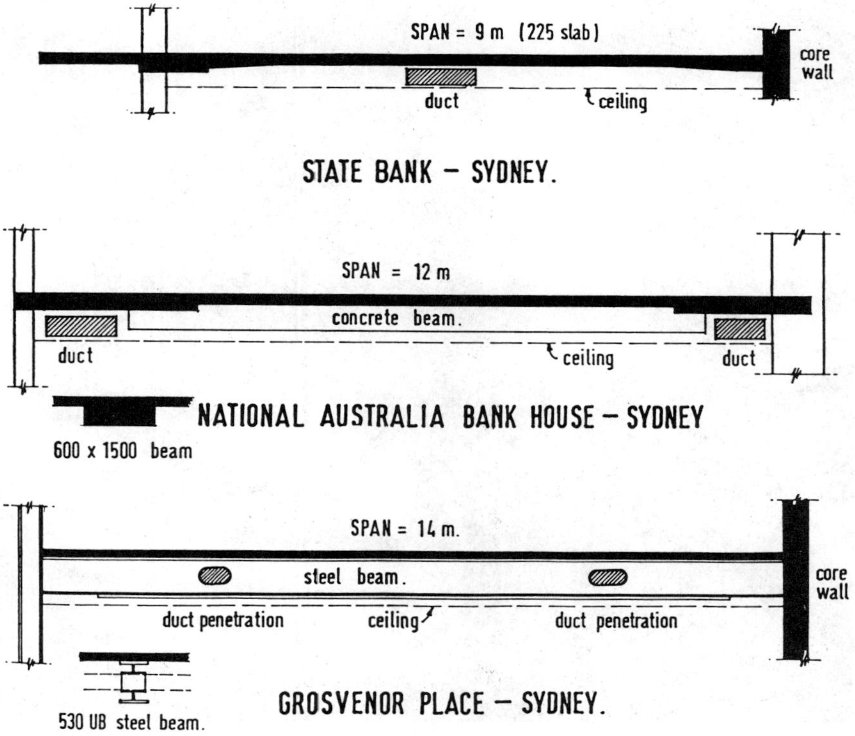

Fig. 4 Floor systems for reduced floor to floor height

1. The OCBC Building Singapore (Thompson, et al., 1976) shown in Fig. 5 is a 44-story building built using an expedited system with the construction of typical floors in each of three banks of floors proceeding simultaneously. Each transfer girder that supports a bank of 14 floors spans 35 m (115 ft) and is designed to carry a superimposed load of 71,000 kN (7,980 tons). It comprises

Fig. 5 OCBC Bank, Singapore—transfer girder shown during construction

four 50 tonne (55 ton) high tensile steel trusses connected together after lifting up the face of the building (76 m (250 ft) for one set, 137 m (450 ft) for another) and subsequently encased and strengthened in the bottom chord with 15 tendons of 79 wires of 7 mm (0.275 in.) diameter of the BBRV prestressing system, which were progressively tensioned to 56,000 kN (6300 tons). Details are shown in Figs. 6 and 7.

2. In the 44-story Grosvenor Place, three facade columns are gathered together at the first floor and supported on a single concrete plinth at ground level. The Italian term *Piloti* is used for such a structure (Fig. 8). The structural elements comprise vertical and inclined steel columns, concrete encased and tied across the top by a post-tensioned composite tie. The post tensioning is provided by 16 high tensile (1080 MPa or 156,600 psi) 38 mm (1.4 in.) diameter VSL/Macalloy bars stressed to 15,000 kN (1700 tons). Details are shown in Fig. 9. The steel alone supports about 20% of the total imposed loading. To control steel stresses throughout the loading history of the elements, encasement is required when 11 floors are constructed.

PROGRAM ASPECTS

Much has been written about the speed advantages of steel and composite construction. However, each locality and building industry has its own characteristics and the rate of progress in construction achieved in one city may not be matched in another. Care must be exercised in examining the claims of various materials. Theoretical cycle times are changed on site

Fig. 6 Erection of transfer girders, OCBC Bank, Singapore

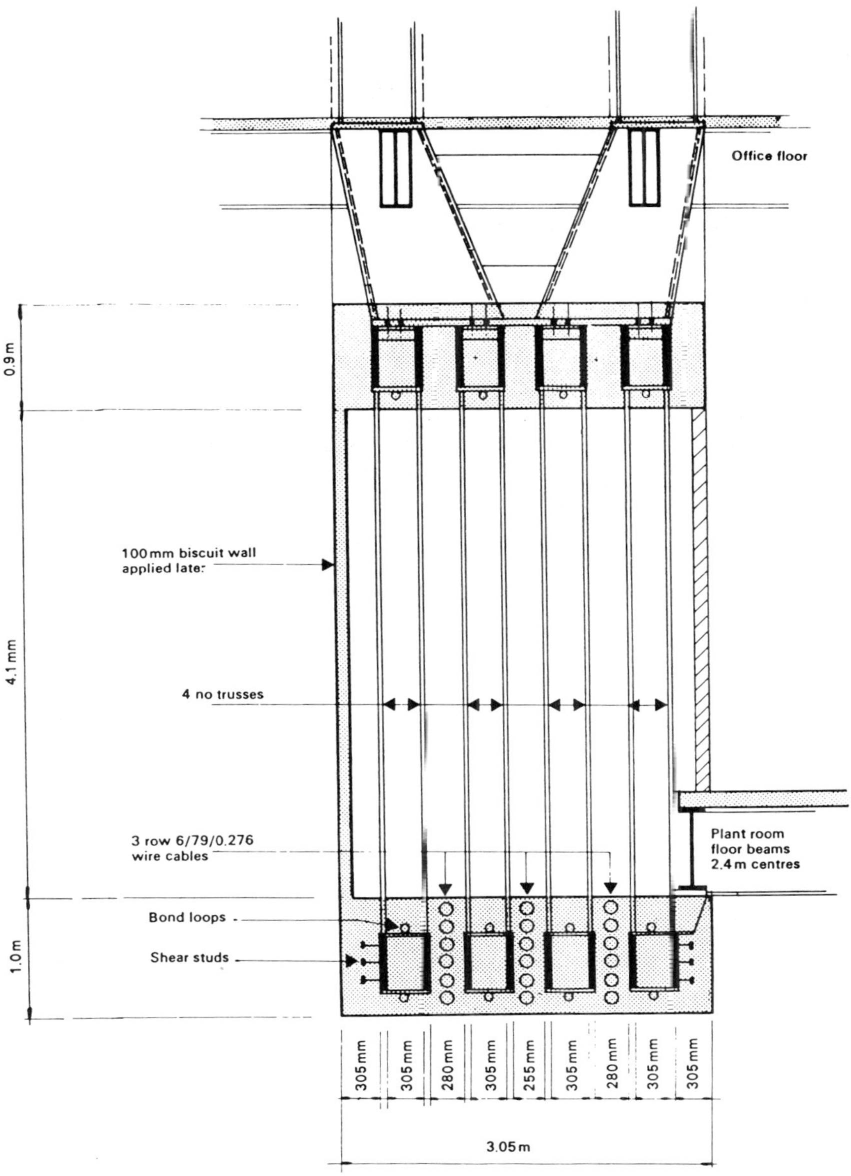

Fig. 7 Cross-section through transfer girders, OCBC Bank, Singapore

Fig. 8 ***Piloti*** **transfer structures, Grosvenor Place Development, Sydney**

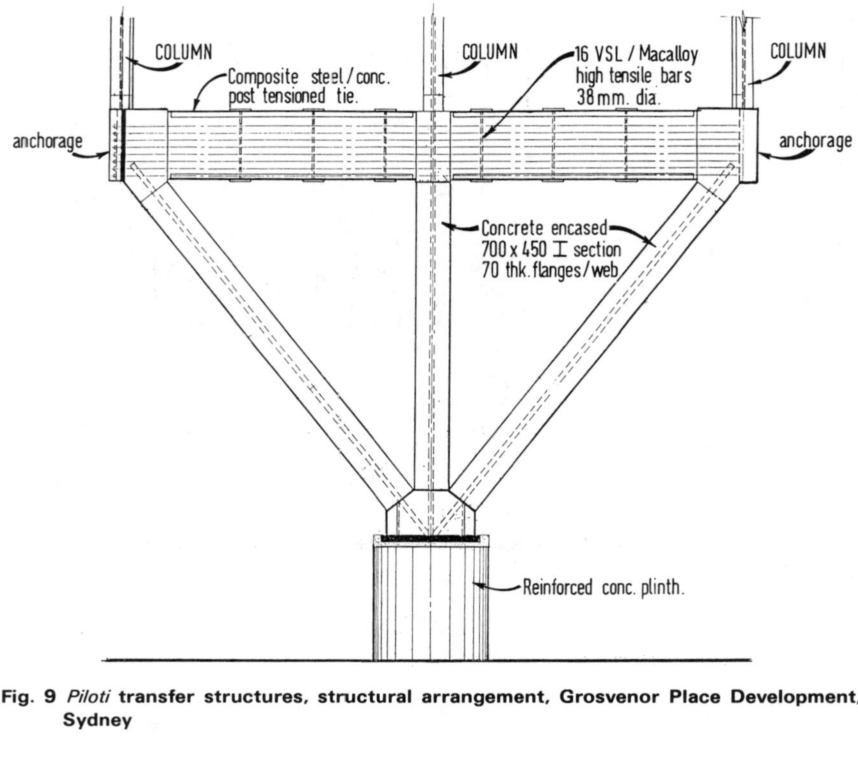

Fig. 9 ***Piloti*** **transfer structures, structural arrangement, Grosvenor Place Development, Sydney**

because of weather, industrial disputes, industry capacity, and construction problems. Nevertheless, early in the design process, a commitment to a particular structural system must be made.

Several construction methods have advantages and can be programmed into the building cycle. A structural steel frame with a steel pan deck floor system can be built very rapidly for large floor areas. Cycle times of 2 days for 2000 m^2 (21,500 ft^2) of floor have been achieved in high-rise construction. Slip forming, or self-climbing formwork systems for walls and service core elements are capable of short cycle times. Cranage and hoisting methods have improved, perhaps the most significant being the development of concrete pumps that can lift to heights of 200 m (650 ft).

Grosvenor Place Building, Sydney, utilizes a reinforced concrete core with composite steel perimeter columns. The typical floor area is 1900 m^2 (20,500 ft^2). The floor system spans 14.4 m (47.2 ft) and comprises a 125 mm (5 in.) concrete slab on metal pans acting compositely with 530 mm (21 in.) deep steel beams. The "optimum" cycle time was shown to be 4 days per floor and a 6-day program was planned and has been achieved. The cycle time for a traditional reinforced concrete structure was 12 days. A saving of 8 months in construction time was possible. The hoisting demands help explain this—134 tonnes (148 tons) of steelwork, pans, and reinforcement and 275 m^3 (9710 ft^3) of concrete as against 340 tonnes (375 tons) of formwork and reinforcement and 435 m^3 (15,360 ft^3) of concrete for a reinforced concrete system. On-site labor was reduced by 100 workers.

COST ASPECTS

Cost studies must provide a realistic comparison of the total cost including the effects of time saving. Rates will differ from locality to locality, country to country.

It is widely acknowledged that steel, or an expedited construction sequence, will have higher initial cost relying on speed of construction to achieve overall cost savings. The financial risk is also greater if the project is delayed. It is the author's view that the anticipated overall saving must be several times the added initial cost if the risk is to be acceptable.

Construction cost for a prestige high-rise building in Australia is \$1200 Aust/m^2 (\$800 US/m^2 or \$74 US/ft^2) of gross floor area. For Grosvenor Place, using composite construction, the additional direct construction cost was 1.2% and the forecast saving on interest and preliminaries was 3.3%, and on predicted inflation was an additional 4.4%.

FUTURE TRENDS

If, as economists predict, high inflation rates continue, there will be increased emphasis on speedy construction. Shorter documentation times will be available, leading to the adoption of "fast tracking"—the procedure where documentation and construction proceed concurrently. Documentation systems will become computerized. It will be possible to see computer software that will directly link the architect's plans, the engineer's analysis, the design team's documentation directly with a contractor's production of shop drawings, purchasing lists, cutting and manufacturing instructions for machinery, delivery schedules, and installation drawings. Management techniques probably will improve, as they must to achieve shorter construction times. Computer control of instructions and variations, drawing registers, programs, work schedules, and networks will be adopted. Improved materials will be developed. Steel and composite construction techniques are thus a trend gaining momentum. Expedited construction methods also will be examined. Such exercises will become more commonplace in the "Second Century of the Skyscraper!"

REFERENCES/BIBLIOGRAPHY

Allen, D. E. and Rainer, J. H., 1976
VIBRATION CRITERIA FOR LONG SPAN FLOORS, Canadian Journal of Civil Engineering, Vol. 3, No. 2, June.

Alves, M. and Berica, A., 1984
CINFAB—AN INTEGRATED STRUCTURAL STEEL FABRICATION SYSTEM, Seminar—Modern Structural Engineering, The Assn. Cons. Struct. Engrs. NSW, Sydney, Australia, August 15.

Clawson, W. C. and Darwin D., 1982
STRENGTH OF COMPOSITE BEAMS AT WEB OPENINGS, Journal of the Structural Division of ACSE, Vol. 108, No. ST3, Paper No. 16939, March, pp. 623-641.

Firkins, A., 1984
CITY BUILDINGS—THE STEEL SOLUTION, (Presented at International Conference on Steel Structures, Singapore, March 7-9), C1-Premier Ltd., Singapore.

Haryott, R. B. and Glover, M. J., 1984
DEVELOPMENTS IN MULTI STOREY BUILDINGS. TRENDS IN THE DESIGN OF STEEL CONSTRUCTION, Nat. Struct. Steel Conference "New Developments in Steel Construction Part 1," Br. Const. Steel Assn., London, December 11-12.

Mackenzie, I. M., 1980
THE CBA CENTRE, SYDNEY, Steel Construction Journal Australia, Steel Const., Sydney, Vol. 14, No. 1.

Mackenzie, I. M., 1981
THE CBA CENTRE, SYDNEY, The Arup Journal, London, Vol. 16, No. 4, December.

Nutt, J. G. and Haworth, D. P., 1976
CAPITAL TOWER, The Arup Journal, No. 3, Vol. 11, October, Ove Arup Partnerships, London.

Nutt, J. G., 1984
WIND AND THE DESIGN OF HIGH RISE BUILDINGS, Seminar—Modern Structural Engineering, The Association of Construction Engineers of NSW, Sydney, Australia, August 15.

Pascoe, G. W., 1984
STRUCTURAL STEEL FLOOR SYSTEMS, Seminar—Floor Systems in Building Construction, Institute of Engineers, Australia, Victoria Branch, Melbourne, September 14.

Taranath, B. S., 1982
COMPOSITE DESIGN OF FIRST CITY TOWER, HOUSTON, TEXAS, The Structural Engineer, London, Vol. 60A, No. 9, September.
Thompson, P. J. and Ainsworth, 1985
COMPOSITE BEAMS WITH WEB PENETRATIONS, GROSVENOR PLACE, SYDNEY (Third Conference on Steel Developments held in Melbourne, Australia, May 1985), Australian Institute of Steel Construction.
Thompson, P. J., O'Hea, R. S., Bergin, R. J., 1976
THE OCBC CENTRE SINGAPORE, The Arup Journal, Ove Arup Partnerships, London, No. 3, Vol. 11, October.

A Tall Building of Stacked Steel and Concrete Structures

R. Shankar Nair

Tall buildings of *mixed* or *hybrid* construction, in which the structure is partly concrete and partly structural steel, have been built for many years in many parts of the world. In general, this type of construction consists of either steel frames with concrete encasement or structures in which shearwalls, tube frames, or other major components are concrete while the remainder of the framing is steel. A new form of mixed construction is represented by a 70-story building in Chicago in which a 40-story concrete structure will be stacked on top of a 30-story steel structure.

The multiple-use development under construction at 900 North Michigan Avenue in Chicago will have a total of 186 thousand m^2 (2 million ft^2) of retail, office, hotel, apartment (condominium), and service space. The residential uses will be housed in a concrete structure, which will be seated on top of a steel structure that will house the other uses. This unusual configuration results in an efficient structure and provides optimum floor sizes and locations for each use. Figure 1 is a photograph of a model of the building.

Fig. 1 900 North Michigan Avenue *(Courtesy: Urban Investment and Development Co.)*

GENERAL LAYOUT OF BUILDING

The various uses, which occupy a total of two million square feet of space, will be accommodated in the 70-story building as follows (listed from top to bottom):

mechanical penthouses;
condominium apartments (18 floors);
mechanical room (2-story space);
hotel rooms and suites (18 floors);
mechanical room (2-story space);
office (20 floors);
retail, department store, hotel functions (8 floors);
parking and services (3 basement levels).

The layout of the building is shown schematically in Fig. 2. As indicated in Fig. 2, the size of floors is different for different uses. The floor configuration and the distribution of the various uses over the height of the building were chosen to provide the best floor size, shape, and location for each use, as follows:

> The basement and first eight above-ground floors (parking, service, department store, other retail, hotel functions) occupy the entire site of almost 9300 m^2 (100,000 ft^2). Large floors and proximity to street level are desirable for these uses.

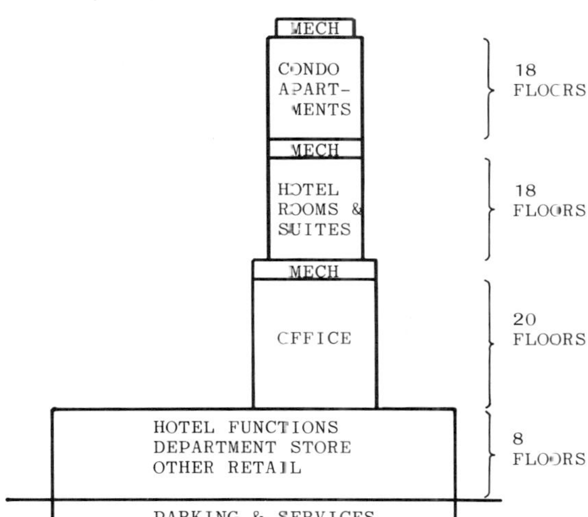

Fig. 2 Schematic layout of building

The 20 office floors, which are above the eight-story full-site "base" structure, have an area of 2500 m^2 (26,400 ft^2) each. A floor area of 2300 to 2600 m^2 (25,000 to 28,000 ft^2) is considered ideal in Chicago's office leasing market.

The residential floors (18 floors of hotel rooms and suites, 18 floors of condominium apartments) have an area of 1600 m^2 (17,200 ft^2) each and are H-shaped. This size and shape allows efficient space utilization without excessive depth of rooms and apartments from the exterior wall. The best location for residential areas is at the top of the building.

The bulk of the HVAC machinery is in a two-story space between the office and residential portions of the tower. This space is also available for components of a structural transfer system.

STRUCTURAL REQUIREMENTS AND CONSTRAINTS

The minimum spans and ceiling heights are different for the different uses in this building. The spacing of interior columns in the eight-story base and in the office portion of the tower was set at 9 m (30 ft) in each direction. Smaller column spacing is acceptable in the residential portions of the tower. If the 9 m × 9 m (30 ft × 30 ft) structural bays of the office floors were continued upward into the residential floors, a row of columns would be in the middle of the hotel and apartment corridors. To avoid this, a structural transfer system is required between the office and residential sections, regardless of the type of structural framing selected.

The clear ceiling clear requirement was 4.2 m (14 ft) in most areas of the eight-story base, 2.65 m (8 ft 8 in.) in typical office floors and 2.59 m (8 ft 6 in.) in typical residential areas. The HVAC and plumbing systems do not have horizontal runs under the typical residential floors. Therefore, suspended ceilings are not required in these areas (unless they are needed to conceal structural framing).

ALTERNATIVE STRUCTURAL SYSTEMS

Several alternative structural systems were evaluated. In all the designs, lateral load resistance was provided by frameworks of beams and closely-spaced columns on the perimeter of the tower. The alternatives studied in detail included two all-concrete designs and one hybrid steel/concrete structure, as follows:

Concrete structure. Spans 9 m × 9 m (30 ft × 30 ft) in base and office floors. Spans 9 m × 6 m (30 ft × 20 ft) in residential floors, with transfer system between office and residential sections. Concrete beam and slab framing for base and office floors. Either concrete beam and slab or concrete flat slab framing for residential floors.

Concrete structure. Spans 9 m × 9 m (30 ft × 30 ft) in base and office floors. Spans 4.5 m × 6 m (15 ft × 20 ft) in residential floors, with two-way transfer system between office and residential sections. Concrete beam and slab framing for base and office floors. Concrete flat plate construction for residential floors.

Mixed steel/concrete framing. Spans 9 m × 9 m (30 ft × 30 ft) in base and office floors. Spans 4.5 m × 6 m (15 ft × 20 ft) in residential floors, with two-way transfer system between office and residential sections. Steel framing (steel columns, steel beams, metal deck, and lightweight concrete fill) for base and office floors. Concrete construction (columns and flat plate) for residential floors. Steel transfer system.

EVALUATION OF ALTERNATIVES

Of the all-concrete alternatives, the design with the smaller spans in the residential floors was found to be distinctly more economical. The reduced floor cost in this design more than compensated for the additional transfer girders that it required. The short-span, flat-plate scheme also allowed smaller floor-to-floor heights, with resultant reductions in total building height and in those costs that are related to height.

Between the more economical all-concrete scheme and the mixed steel/concrete design, both of which utilized short-span concrete flat-plate construction for the residential floors, the difference in cost was found to be small. The cost analyses, which included consideration of differences in construction time and the resulting cost of money, generally showed the all-concrete design to be slightly more economical. Factors other than construction cost included the following:

The all-concrete design resulted in very large columns (about 1.5-m square [5-foot square]) in the department store and retail mall, even with 76,000 kPa (11,000 psi) concrete.

It was believed that certain potential tenant modifications could be achieved more conveniently with a steel-framed structure in the lower part of the building.

The project schedule called for the building to be opened in stages. The retail area, which was to be opened first, could be completed earlier with the mixed steel/concrete design. The all-concrete scheme could not satisfy the desired project schedule.

It was concluded that the mixed steel/concrete design offered the best overall solution to the particular economic, architectural, functional, and scheduling requirements of this project.

DETAILS OF DESIGN

The lateral load-resisting system is a perimeter moment frame or "tube." Since the building perimeter steps back to a smaller floor size between the office and residential sections, the "tube" is actually two separate tubes, one on top of the other. Load is transferred between the two tubes by a two-story high, three-dimensional transfer truss system in the mechanical space between the office and residential areas.

The total height of the building, including all penthouses, is 266 m (871 ft) above the ground floor. Floor-to-floor heights are 5.5 m (18 ft) in the eight-story base, 3.8 m (12 ft-4 in.) in the office section, and 2.8 m (9 ft-2 in.) in the residential part of the building. Typical column spacing in the perimeter frame is 4.5 m (15 ft). Diagonal bracing is added in the planes of the tube in the eight-story base.

Most of the columns and beams in the lower (steel-framed) tube are 1.0 m (36-in.) deep rolled sections. In the upper (concrete) tube, columns are 36 in. wide and beams vary in depth between 24 and 36 in. The steel/concrete interface at a typical exterior column is sketched in Fig. 3. At interior columns, an inverted base plate is welded to the top of the steel column; reinforcing bars for the concrete column are welded to the base plate.

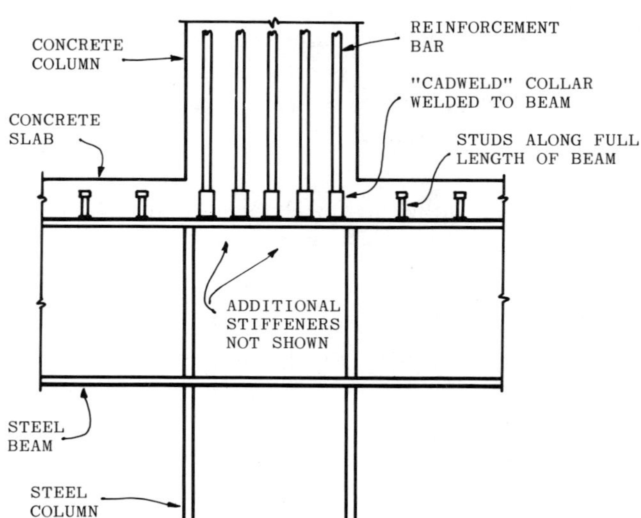

Fig. 3 Steel/concrete interface at exterior column

The foundations are drilled shafts, belled at the bottom, bearing on hardpan (hard silty clay) about 26 m (85 ft) below street level.

Structural analysis for vertical and lateral loads was performed using the techniques and computer program described in Nair (1975a). In some of the earliest analyses, lateral stability effects were included in a rigorous or "exact" manner by means of the procedure described in Nair (1975b). Subsequently, the simplified magnification-factor approach presented in Nair (1983) was found to provide results very close to the exact solutions for this building. The simple procedure was used for design.

The magnification factors for lateral stability effects were found to be in the range of 1.08 to 1.12 at service loads. (Greater magnifications—computed at appropriate higher load/resistance factors—were used in design.) These magnification factors are within the range that would be considered normal for all-steel and all-concrete buildings. Vibration characteristics also were found to be within the normal range for buildings of this height (871 ft). The fundamental periods are just under eight seconds for lateral oscillation in the principal directions.

Wind forces and the structure's response to wind loading were studied in the wind tunnel by high-frequency force-balance testing. The design criteria included an acceleration limit of 0.018g for a 10-year return period at the highest occupied floor. The wind study did not reveal any peculiarities in response that could be attributed to the building's hybrid steel/concrete design.

Vertical displacements and deformations in the building were analyzed in detail. All steel columns are to be fabricated longer than their nominal dimensions (by varying amounts) to compensate for anticipated column shortening under load. Floors in the concrete structure will be poured out-of-plane to elevations that have been calculated to compensate for the expected subsequent shortening of the structure below. It is estimated that if these special dimensional corrections were not made, some floors in the complete building would be out of plane by more than three inches.

CONCLUDING OBSERVATIONS

The building at 900 North Michigan Avenue is a mixed-use building in the truest sense of the term. Residential, office, and retail uses are equally important. No single use dominates the building. The structure must work well for all the uses.

There is a natural break in the structural continuity of this building between office and residential sections. The floor size (and building outline) changes here for sound, functional reasons. A system of transfer girders or trusses would be required at this location, regardless of the choice of structural system and material, because of necessary changes in interior column locations and the stepping back of the building's exterior wall to a smaller floor size.

Given the unavoidable break in the structural continuity of the building, it was appropriate to make a separate choice of structural system and material on each side of the break. Each of the two structures (above and below the discontinuity) is large enough that the premium due to a change in material is small compared with the advantages of choosing the optimum design for each structure. Computational techniques and expertise are available today for rational study of the potential problems involving secondary stresses and deformations that might arise as a result of a mixed steel/concrete design.

CREDITS

The developer, architect, and structural consultants for 900 North Michigan Avenue are as follows: *Developer:* Urban Investment and Development Co., Chicago; *Architects:* Perkins & Will, Chicago and Kohn Pedersen Fox, P. C., New York; *Structural Engineer:* Alfred Benesch & Company, Chicago; and *Structural Design Consultant:* R. Shankar Nair, Chicago.

REFERENCES/BIBLIOGRAPHY

Nair, R. S., 1975a
LINEAR STRUCTURAL ANALYSIS OF MULTISTORY BUILDINGS, Journal of the Structural Division, ASCE, March.

Nair, R. S., 1975b
OVERALL ELASTIC STABILITY OF MULTISTORY BUILDINGS, Journal of the Structural Division, ASCE, December

Nair, R. S., 1983
A SIMPLE METHOD OF OVERALL STABILITY ANALYSIS FOR MULTISTORY BUILDINGS, Developments of Tall Buildings 1983, Council of Tall Buildings and Urban Habitat, Hutchinson Ross Publishing Co.

A Tall Building Core-Hanger Interaction System

Marek W. Kwieciński
Adam Z. Pawłowski

A common feature of most European towns is that they were developed with respect paid to the existing buildings, which are generally low-rise and frequently of historical and architectural value. The nuclei of such towns had usually been castles enclosed by city walls around which some burghers' dwellings were gradually grouped. The castle towers and the church spires were usually two to four times as tall as the surrounding buildings, thus having been the skyscrapers of the past. This situation, together with the willingness to preserve a human scale of the towns' living tissue, has resulted in a tendency to keep tall buildings within a reasonable height.

Especially since the advent of the automobile and mass transportation, there has been a growing need to adapt the existing towns to contemporary demands while preserving, at the same time, their historical heritage and cultural values. The problem arose as to how to build tall in densely populated areas having an existing system of streets and allowing little or no demolition.

At least three conditions seem necessary to be fulfilled for the above outlined purposes: (1) the height must be kept within reason, (2) the height-to-width ratio must be appropriate, and (3) sufficient flexibility of the ground floor layout must be ensured.

The reasonable height of a European tall building which, being a kind of beacon, does not cease to interact pleasingly with its vicinity, seems to be somewhere between 60 and 150 m (200 and 500 ft). Some efforts to erect very high edifices with the prestige value have been observed to interfere with the spatial qualities of certain European cities.

The other skyline-shaping factor is the aspect ratio. Such ratios for some buildings erected in Warsaw and other Polish towns are shown in Fig. 1a compared with the building height. On average, Warsaw tall buildings are more slender than elsewhere in the country as shown by dashed and solid lines, respectively. The aspect ratios vs. height for some other European capitals are shown in Fig. 1b. It follows that the reviewed buildings are rather bulky and their aspect ratios seldom exceed $s = H/b_{min} = = 5$, partly as a result of energy conservation considerations and some building code requirements currently in force. Additional information can be found in Fig. 2: a shaded region is seen in which most of the points cluster, each representing a specific Polish building erected as a shear wall system or a steel frame system, some mixed structures also being accounted for.

One of the solutions to meet the third condition, ground floor layout flexibility, is to eliminate as many supports as possible on the ground floor. This may be particularly helpful when tall buildings are to be located in the city centers where only small sites are available and where harmony with short valuable houses and old streets is desirable. Typical situations at the street level are imagined in Fig. 3b. Some other advantages of such a ground level arrangement can also be pointed out: larger spans are often required on

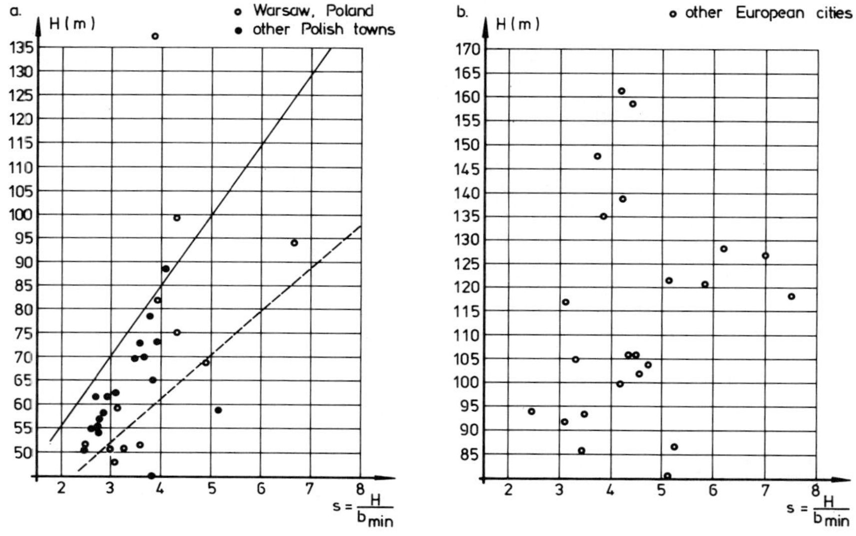

Fig. 1 Aspect ratios for Polish tall buildings

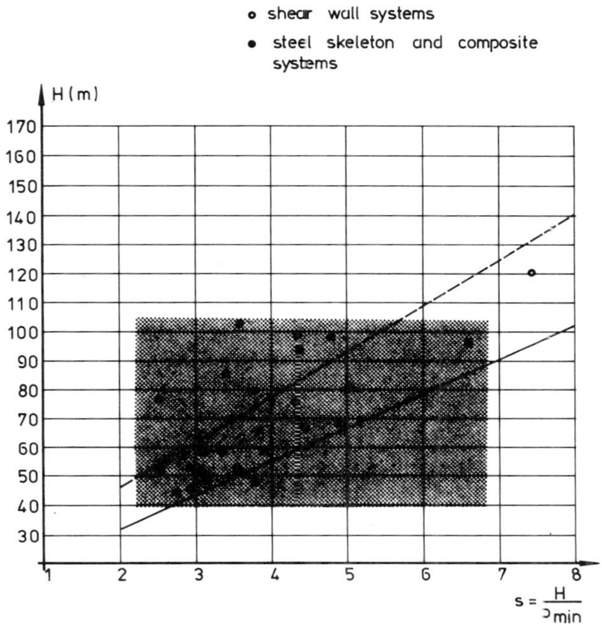

Fig. 2 Polish buildings erected as shear wall systems or steel or composite systems

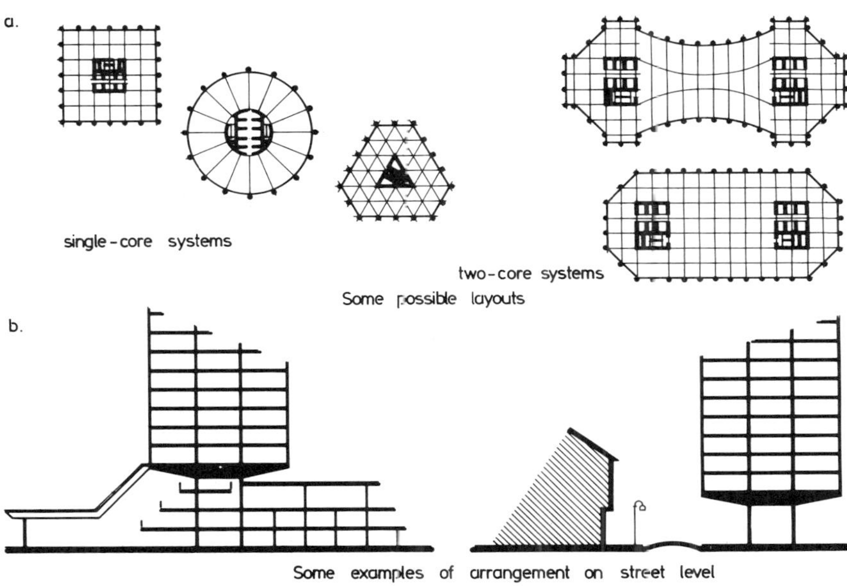

Fig. 3 Floor layouts and arrangements

the ground floor to accommodate shops, offices, exhibition areas, and such; growing transportation needs can be better satisfied by making an incision at the base of a tall building (such an undercut helps create what may be called a well-aired ecological corridor). On the other hand, cores can contain elevators, stairs, mechanical shafts, and so on.

What may be termed a European type of high-rise building can be envisaged as one that, having moderate height and limited aspect ratio, interacts agreeably with the historical surroundings at street level and, at the same time, consciously shapes a desired urban skyline.

The preceeding requirements are satisfied by the known suspended structural system. However, to make it more rigid laterally and better use the structural materials, a certain modification of the system is proposed. Consider, as an example, a single-core system in which the hangers are not only suspended from a heavy top hat (as shown in the inset of Fig. 4a) but are also attached to a similar, inverted hat placed directly above the ground floor, both hat levels serving as equipment stories (as shown in the inset of Fig. 4b). Thus, from the structural standpoint the hats, the exterior hangers, and the core interact in a combined way to resist the lateral loads. What is architecturally important is that no exterior tie-down columns penetrate to the ground level where only a confined space may be available to locate the core itself.

For a taller building, in which more than two mechanical floors are necessary, each of them can be designed as a massive structure and connected with hangers going down through the groups of floors from the top to the bottom hat. Each hat can be shaped as a grid of one-story height, a system of deep beams or trusses, a frame with oblique bars, or the like. The hangers hidden

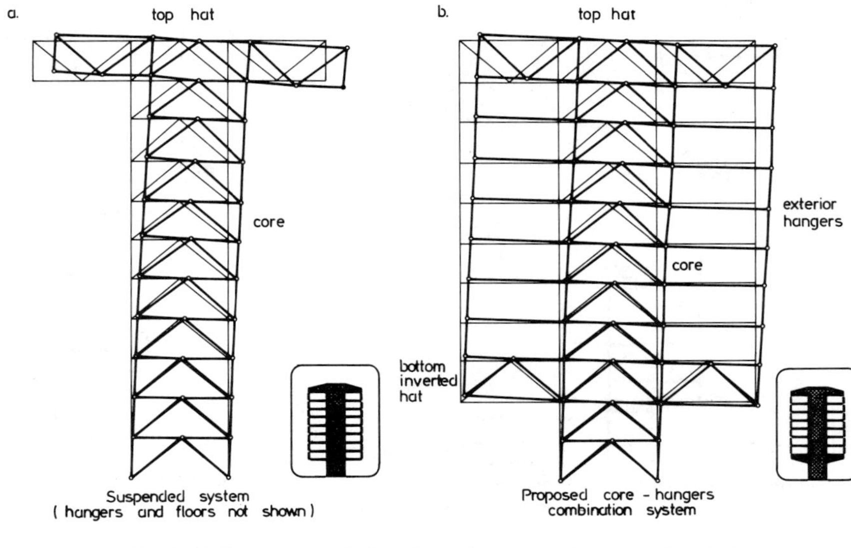

Fig. 4 Drift of suspended and core-hanger systems due to wind

in slender exterior columns interact with hats and reduce the drift of the building by creating a point of contraflexure to decrease the sway at the top.

To substantiate the structural concept of the proposed core-hanger interaction system, detailed finite element statical analyses of simple model structures with and without the bottom hats were made to show the expected effect. The numerical test structure was a single-core system with square layout (the first plan in Fig. 3a), 11 stories high, with the height-to-core width ratio $s_c = 41.25/9.00 = 4.58$ and the overall aspect ratio $s = 41.25/27.00 = 1.53$. One of the four plane core walls was analyzed together with the top hat (conventional suspended system, Fig. 4a) and with the top and bottom hats connected by hangers (proposed interaction system, Fig. 4b). Since the sole aim of the analyses was to compare drifts of the two systems due to wind pressure, in order to ease the calculations both model structures were discretized to become substitute elastic trusses as shown in Fig. 4.

The gravity and wind loads were duly determined for the two variants and, for definite strength characteristics of the trusses, the vertical and horizontal components of all the nodal displacements were computed and drawn to scale in Fig. 4a and 4b. The maximum sway of the proposed system turned out to be

$$(0.197/0.214) \times 100 = 92\% \qquad (1)$$

of that for the conventional suspended system. Next, the test structures 33 stories high were analyzed, having the height-to-core width ratio $s_c = 123.75/9.00 = 13.75$ and the overall aspect ratio $s = 123.75/27.00 = 4.58$. The maximum sway of the interaction system was found to be

$$(1.44/4.80) \times 100 = 30\% \qquad (2)$$

of that for the suspended system with no bottom hat.

Thus the interaction of both hats by means of hangers is found to decrease the drift of the building. Moreover, the effect is seen to increase considerably with the growing height of the structure. One can also contemplate prestressing the hangers to better control the lateral response of the building.

A natural development of the proposed system is to design two-core (Fig. 3a) and multicore combination systems which, preserving the main structural merits outlined above, allow for various spatial arrangements to please both the architect and the urban planner.

Transfer Structures

Jack Zunz
Chris Wise

Functional, esthetic, or planning needs predicate changes in column or wall layouts. In tall buildings these often create special and interesting engineering problems solved with some form of transfer structure. This paper considers transfer structures in multifunctional tall buildings and proposes some generic solutions, which are illustrated with examples. Possible future trends are briefly examined.

NEED FOR TRANSFER STRUCTURES

Obvious functional and economic reasons normally dictate the disposition of columns and walls in tall buildings in such a way that they are directly over their foundations. There are, however, often good reasons for changes of function with level that, in turn, lead to a modulation of the main vertical structure. This modulation usually results in some form of transfer structure. In tall structures the size and weight of the building to be supported means that these transfer systems constitute major elements of the building. Their effect on building cost and construction time can be considerable.

It is helpful to consider tall buildings under four broad functional headings.

Apartment Blocks. These generally have structures with regular arrangements of columns or load-bearing shear-walls between tenancies. They sometimes sit on a podium that makes provision for public and commercial facilities and often must accommodate the column grid of a car park in their lower levels. In congested urban environments they may also have to accommodate a mass transit facility.

Hotels. Hotels usually have standard repetitive floor configurations suitable for guest rooms, with a closely spaced column or wall grid for economy of floor construction. At lower levels longer spans are required for public amenities, shops, and banquet and conference facilities. Car parking provision also may be required.

Offices. Office buildings often have to make provision for column-free public plazas, commercial enterprises, banking halls, and conference centers. Again, a car parking facility may be required.

Multi-Use. In multi-use buildings, apartments, offices, hotel rooms, public amenities, and traffic facilities are stacked one on top of the other, each with its own structural sub-grid.

In such buildings, architects are faced with changing functional needs that demand equal responses from the engineer in the choice of appropriate structural systems for the transfer structure.

CHOICE OF TRANSFER STRUCTURE

Generic Types

Transfer structures may be rationalized into four generic forms.

Cantilever. Generally multitier or single-floor (Fig. 1)

Beam. Multitier or single-floor, sometimes low-level (Fig. 2)

Frame. Generally low-level (Fig. 3)

Plate or Grid. Generally low-level (Fig. 4)

Interaction with Lateral Stability System. It often makes sense to combine the lateral stability system with the transfer structure. In particular a perimeter stability system can be integrated with the transfer structure to form a vertical vierendeel or braced facade that can resist lateral as well as vertical loads (Fig. 5).

Interaction with Ground Conditions. Geotechnical considerations may influence the choice and configuration of transfer structures, which often imply a concentration of load at foundation level (Fig. 6).

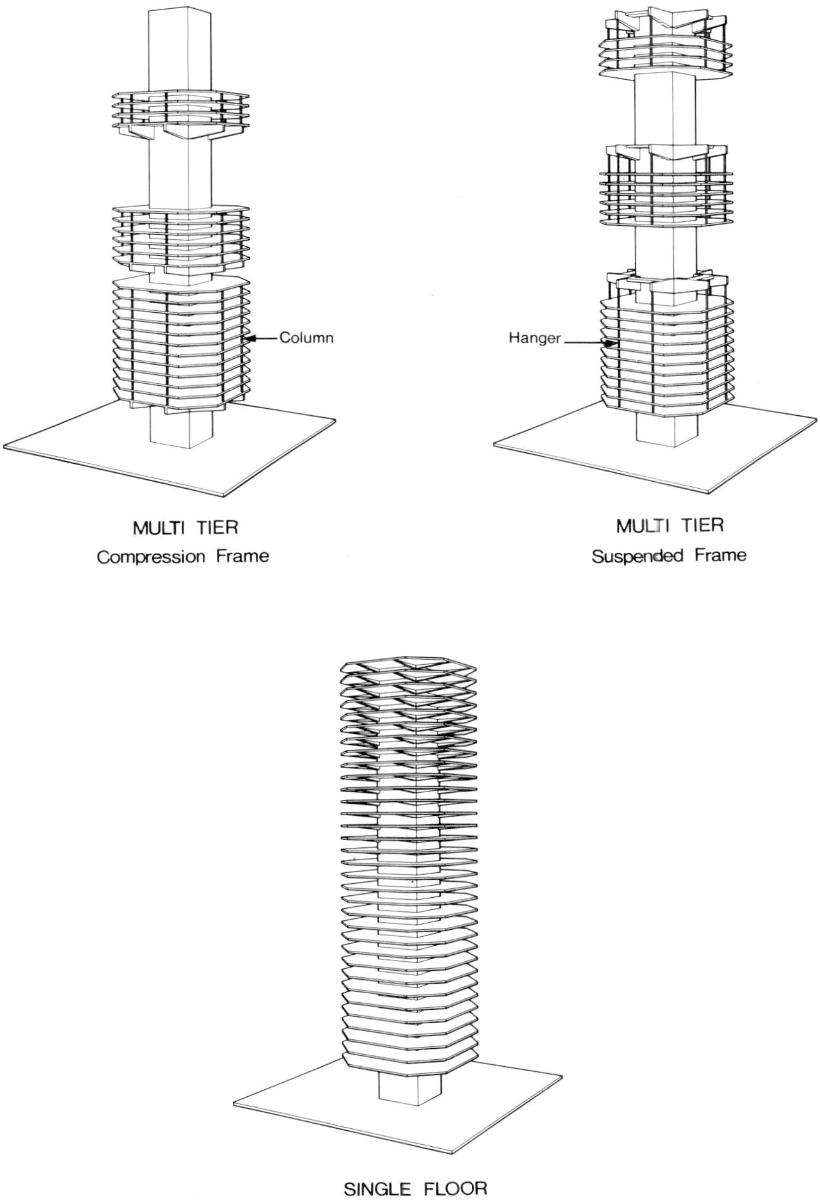

Fig. 1 Cantilever transfer structures

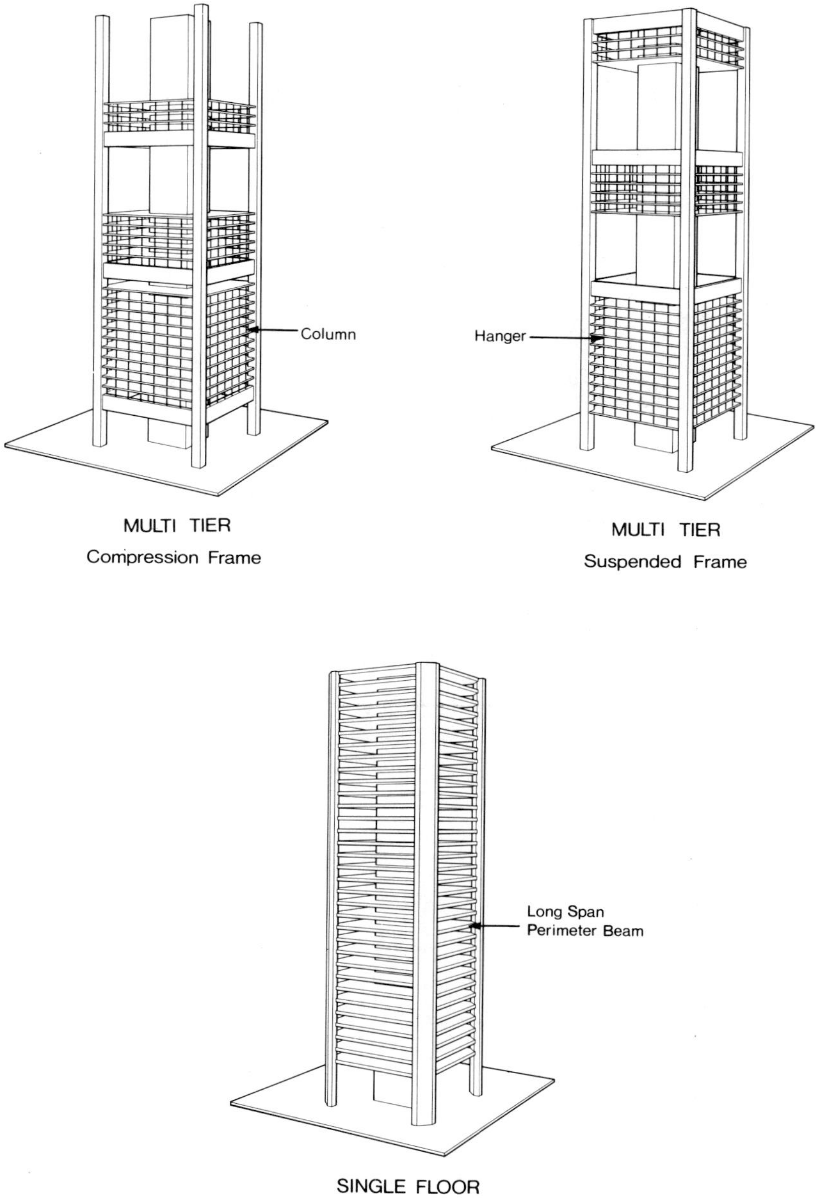

Fig. 2 Beam transfer structures

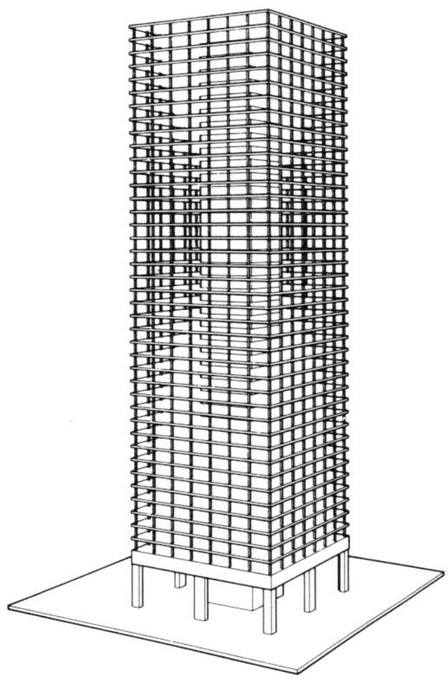

Fig. 3 Transfer frame

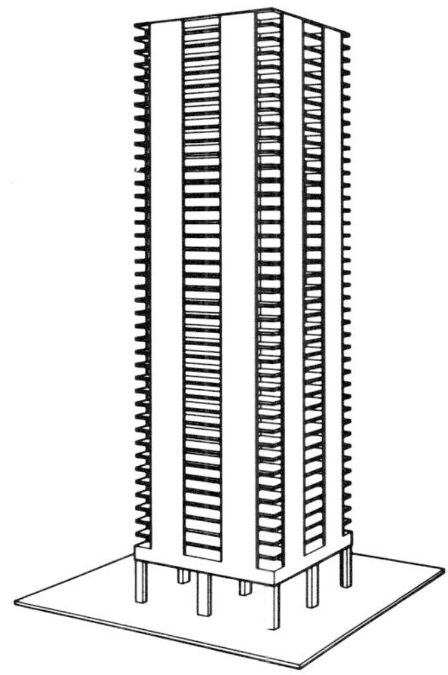

Fig. 4 Transfer plate

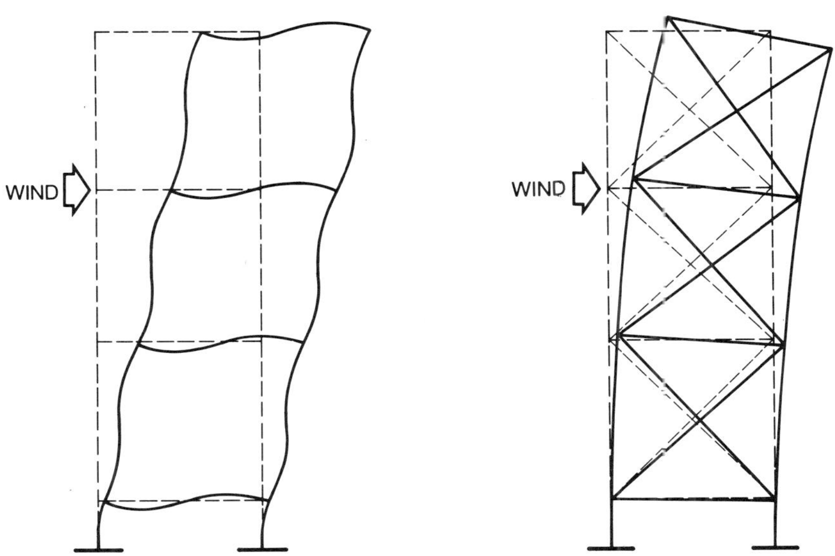

Fig. 5 Interaction with lateral stability system

Interaction with Services. Most modern tall buildings, particularly offices and hotels, have sophisticated mechanical and electrical installations. Appropriate service distribution is often achieved by dividing the building into a number of vertical zones, each served by its own mechanical plant. The service requirement for plant rooms is often integrated with a multitier transfer structure.

Construction Methods. Low-level transfer structures offer simplicity—they can be constructed using normal building techniques to carry a large number of super-structure floors.

Multitier systems require the construction of horizontal elements at high level. For reinforced or prestressed concrete structures, substantial temporary works are usually needed. The use of a tiered floor skeleton often permits the simultaneous construction of more than one floor. This can be significant for finishing trades, which can be committed to several floors at once, possibly saving construction time.

Single-floor systems, with their many similar horizontal members, lend themselves to the use of precast or prefabricated elements for quality control and speed of erection.

The Chosen Solution

While there cannot be hard and fast rules, most tall buildings require more than their central core to provide lateral stability when they reach the 40 to 50

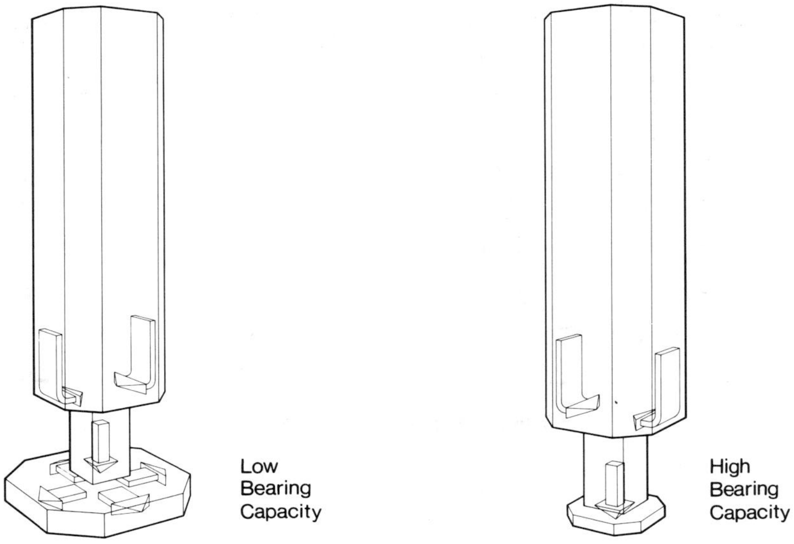

Fig. 6 Interaction with soil

story range. Core-supported cantilever options are consequently limited. Plate solutions, which can redirect loads in more than one direction, are particularly suitable for shear-wall buildings such as hotels or apartments where radical changes of grid are often called for. The choice of transfer structure may also be influenced by the shape of the building (Fig. 7).

Ultimately the chosen solution is the consequence of a compromise between economic, functional, engineering, and architectural requirements. The latter are often based on preconceptions, artistic license, and even fashion—all perfectly reasonable provided the end justifies the means—in other words, that the result is a good building.

CASE STUDIES

Cantilever Solution—Standard Bank, Johannesburg (Fig. 8)

Completed in 1970, the 35-story Standard Bank building uses a combination of multilevel cantilevers with a three tier suspended frame to create a distinctive elevation with the maximum clear space at plaza level. The three banks of precast lightweight concrete floors are hung from prestressed concrete cantilevers by prestressed concrete hangers which give fireproofing, finish, durability, and minimal extension while using the properties of high strength steel to the best advantage. The suspended frame was constructed downward from each cantilever. The cost of the heavily loaded core was offset by its improved performance as a prestressed cantilever under wind loads. The foundations to the core, originally designed as large raft, were eventually constructed as four 5m (16.4 ft) diameter concrete piers with 6m (19.7 ft) underreams when a heavily weathered contact zone between quartzite and diabase was discovered on site.

Beam Solution (1)—OCBC Centre, Singapore (Fig. 9)

The 51-story Oversea-Chinese Banking Corporation complex made use of multilevel transfer beams with a tiered compression frame to create an imposing Banking Hall at street level The tiered superstructure was chosen to take advantage of government incentives in the form of tax benefits for early completion in 1976. Each transfer girder was erected in stages as a composite of four 50-tonne (55-ton) steel trusses that were fabricated off site. They were winched up the slipformed cores, then encased in concrete and given a prestressed bottom chord. The entire superstructure floor load was transferred through the girders to the twin cores to prestress the stability system, enabling them to remain comparatively slender.

A. Circular Plan, Cantilever Solution

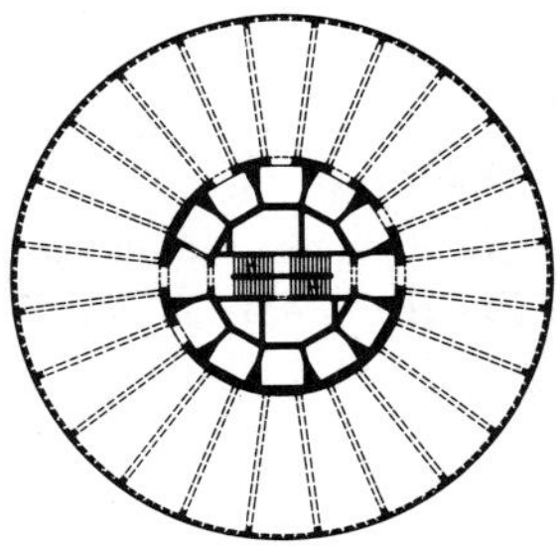

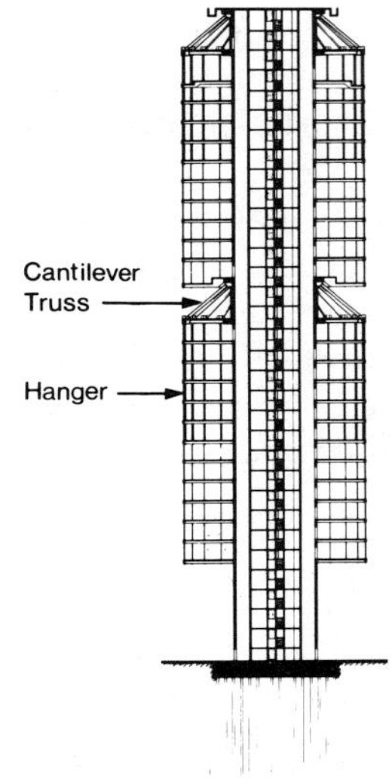

SABAH FOUNDATION HQ, EAST MALAYSIA

B. Square Plan, Beam Solution

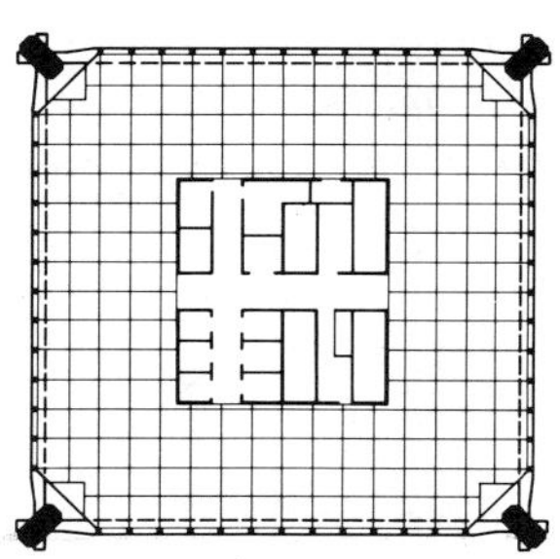

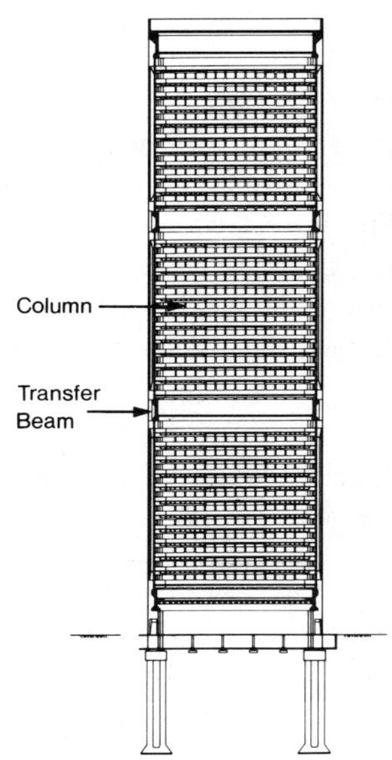

CAPITAL TOWER, MELBOURNE

Fig. 7 Influence of building shape

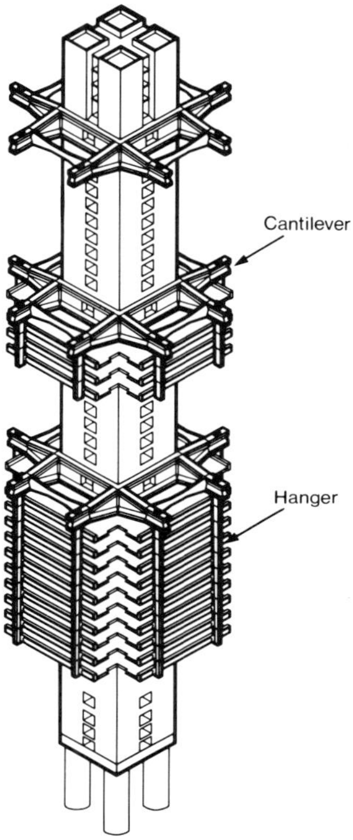

ISOMETRIC OF STRUCTURE

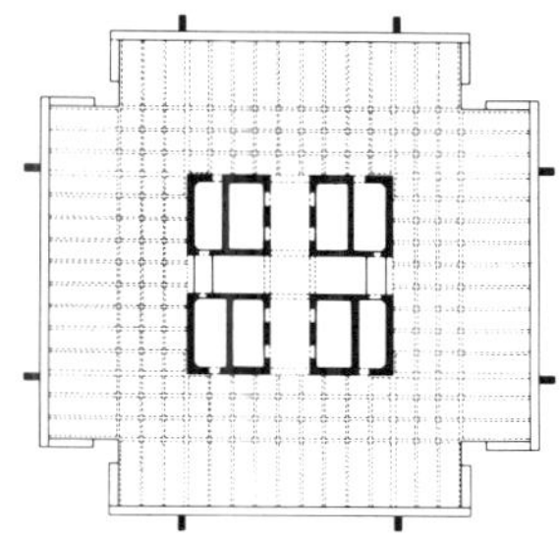

TYPICAL FLOOR

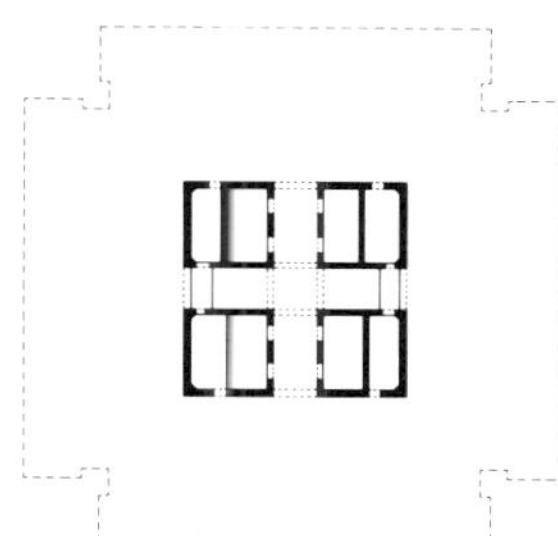

PLAZA

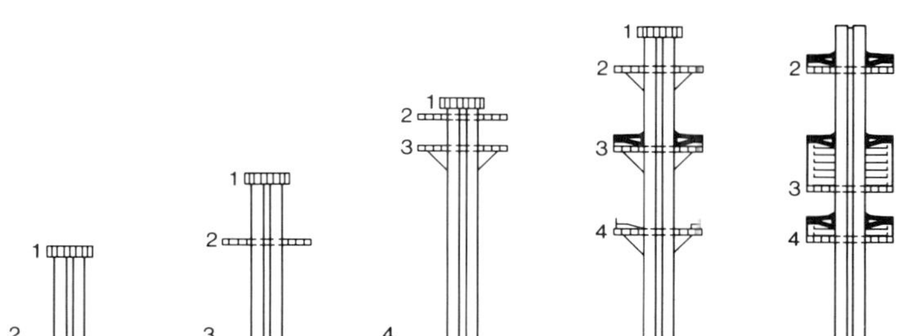

SEQUENCE OF CONSTRUCTION

Fig. 8 Standard Bank, Johannesburg

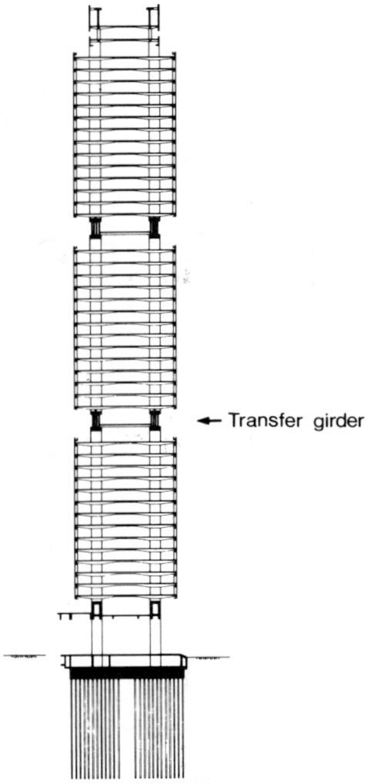

SECTION

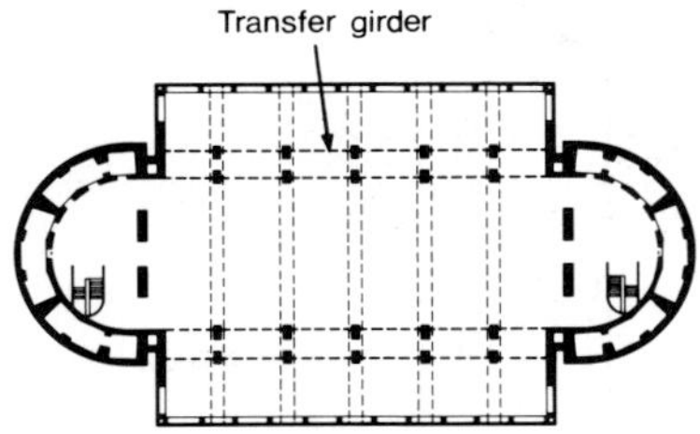

TYPICAL FLOOR

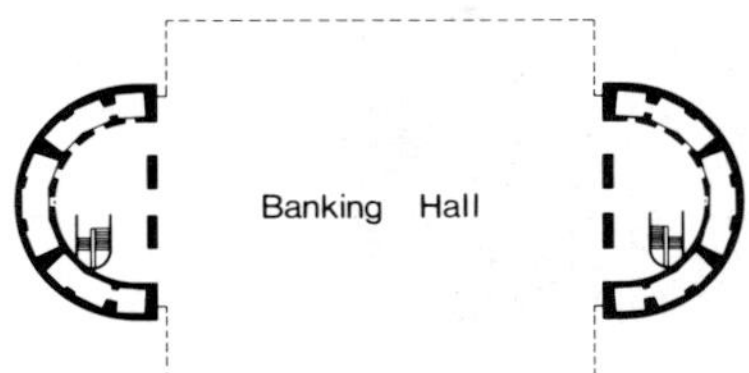

GROUND FLOOR

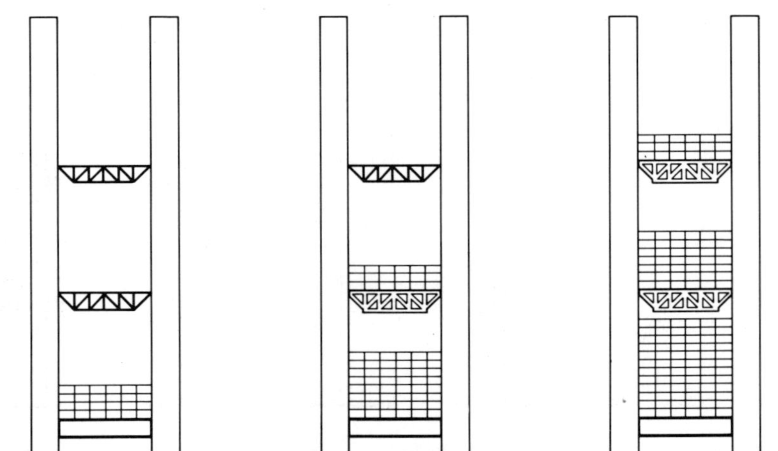

SEQUENCE OF CONSTRUCTION

Fig. 9 OCBC Centre, Singapore

Beam Solution (2)—Hongkong & Shanghai Bank, Hong Kong (Fig. 10)

The expressed steel structure of the 47-story Hongkong Bank is a considered response to the functional and architectural needs of the building. The five tier transfer trusses are integrated with the main masts to give the structure its lateral stability, with a deflection of approximately 300mm (1 ft) under the 100-year typhoon wind load. The suspended floor skeleton combines the minimum vertical structure with maximum flexibility of use, and gives a completely column-free ground floor space. The superstructure hangers were designed to act as columns during construction to allow the floor steelwork to follow closely behind the masts, without waiting for the completion of the suspension trusses. Exhaustive analytical and prototype studies were carried out to confirm the performance of the prefabricated steel components and to identify potential problems associated with the fabrication and erection process.

Frame Solution—Grosvenor Place, Sydney (Fig. 11)

The eight transfer frames at the foot of the 50-story Grosvenor Place project collect together the load from 24 perimeter columns and transfer it to eight primary columns at ground level. The curved facade and out-of-balance column loads contrive to create a complex system involving the interaction of the transfer frames with both first and ground floors and the core. Significant principal tensions are induced in the floor slabs, which were prestressed to ensure that their in-plane stiffness was not reduced by cracking. The composite steel and concrete transfer frames were designed to enable the construction of the superstructure floors to continue while the top boom of each frame was prestressed and link beams added.

Plate Solution—Luk Yeung Sun Cheun, Hong Kong (Fig. 12)

The 20,000 residents of The New Green Willow Village live in 17 towers above the 5.5-hectare (13.6-acre) Tsuen Wan Mass Transit Depot. Each 28 to 30 story shear-wall structure sits on a 2-m (6.6-ft) deep grillage of transfer beams which, in turn, are carried by primary columns passing through the depot. In analysing the interaction between the partially coupled shear-walls, the transfer structure, the depot, and its foundations, it was possible to examine the effect of the flexibility of the transfer structure on the stress distribution within the superstructure. As an example, wall stresses near the depot columns at transfer level increased from 4.7N/mm^2 (680 psi) in a

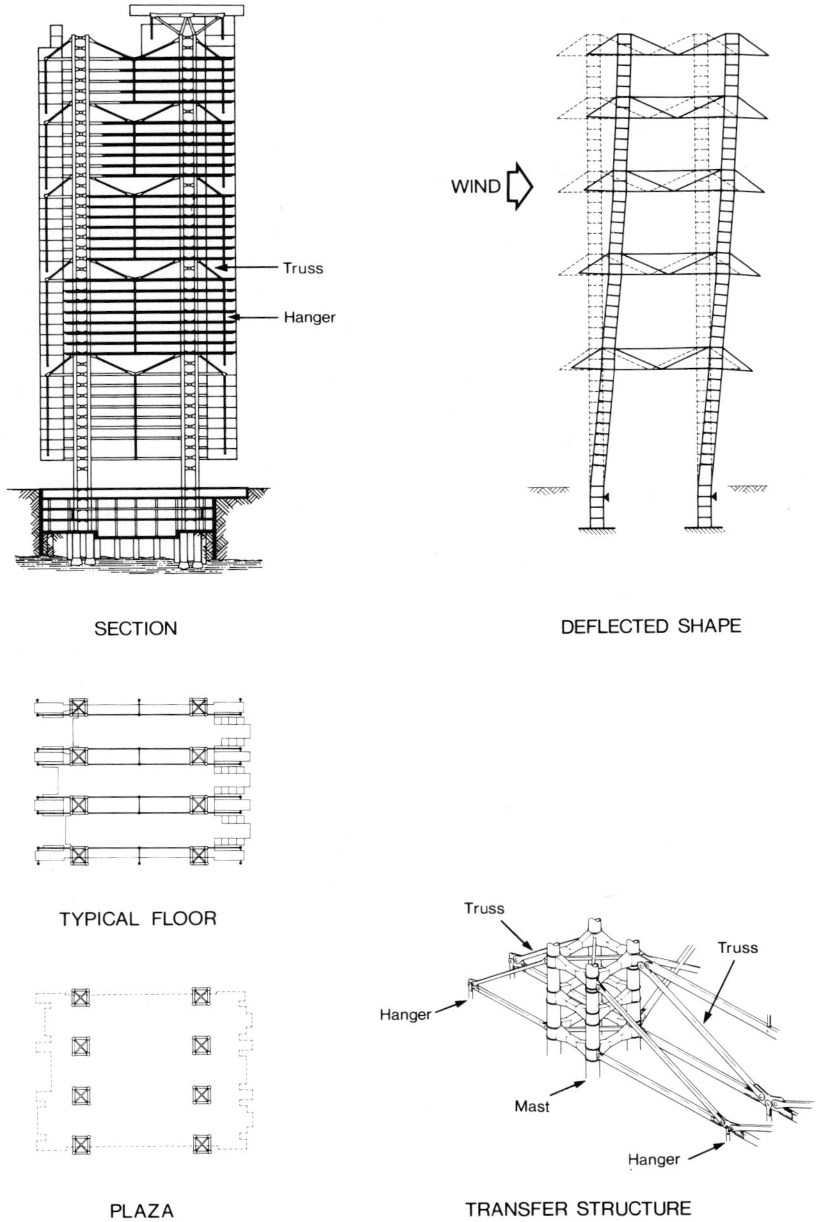

Fig. 10 Hong Kong & Shanghai Bank

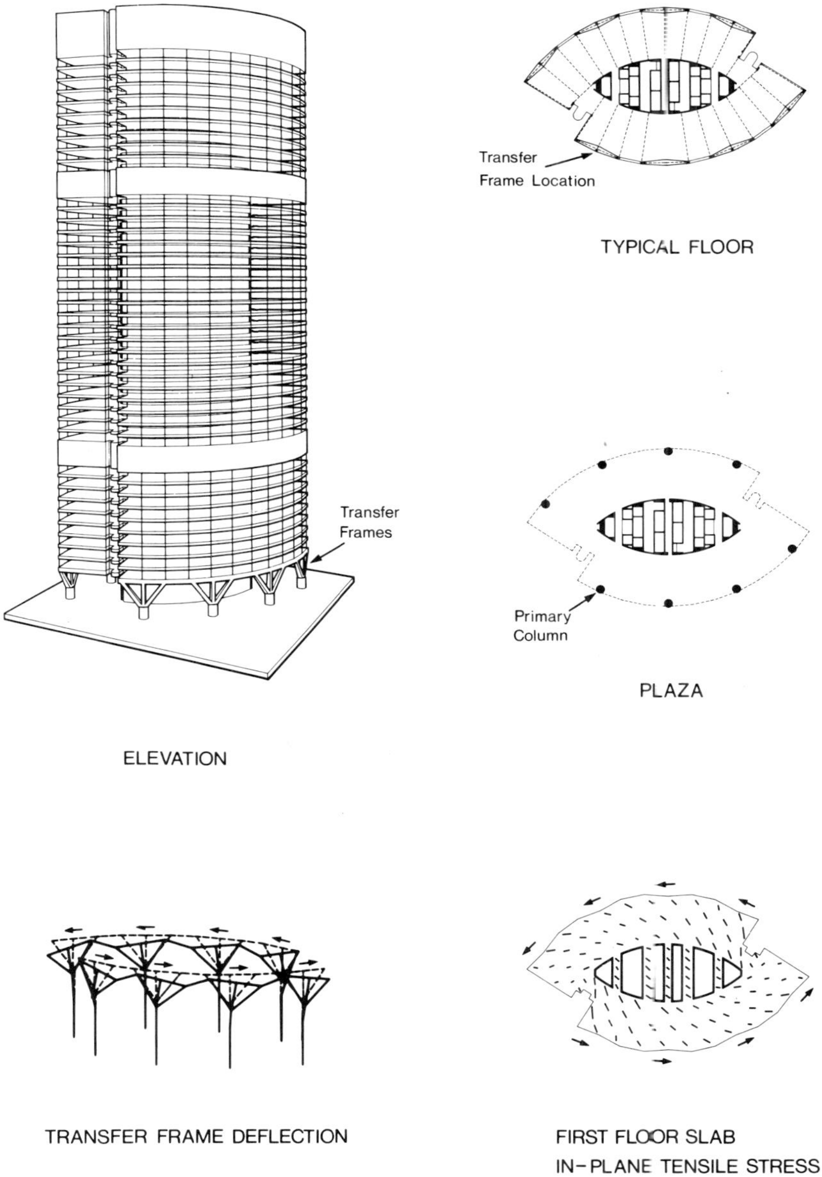

Fig. 11 Grosvenor Place, Sydney

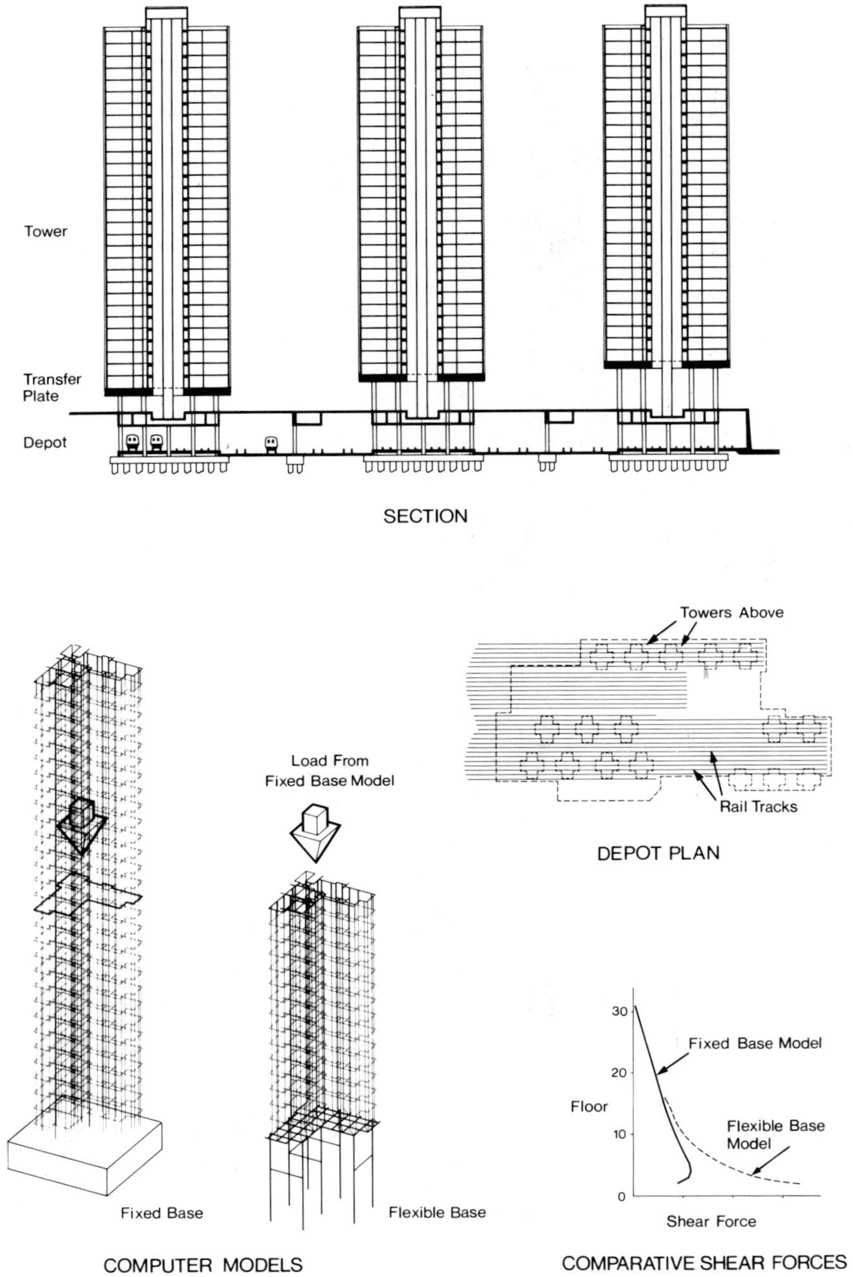

Fig. 12 Luk Yeung Sun Cheun, Hong Kong

30-story model with a fixed base, to 7N/mm^2 (1015 psi) in a 15-story model with a more representative flexible supporting structure. This was reflected in the shear forces and bending moments of the coupling beams nearby.

THE FUTURE

Someone said that you can never plan the future by the past. Even though predictions can be very wrong, there are some guidelines. Before looking at transfer structures in particular, what of their context in tall buildings generally? Post World War II euphoria and belief in the limitless potential of technology has given way to some cynicism, reaction, and much reappraisal. While this phase could be cyclical, the public for whom we build is more knowledgeable, articulate, and demanding. People are not sure whether they wish to be housed in high-rise buildings of questionable design. The professions associated with the building industry are, like all professions, under scrutiny and can no longer look forward to a privileged status in our society. They are being made to be totally accountable for all matters that may go wrong in the design and construction process.

Concurrently, this same public wants to participate and be consulted in what is to be built and is increasingly concerned about the environment, both natural and built. The property market is demanding more adaptable buildings. There are fashionable amenities like "full access floors" and "intelligent buildings," which are merely expressions describing an insatiable demand to accommodate accelerating if unpredictable change. These are only some of the contextual issues that put pressure on the designer to accommodate increasing adaptability in tall buildings. Transfer structures in their various forms are, therefore, likely to be an important tool in his armory.

Some of the generic forms of transfer structure have been described. Dramatic changes in construction materials are unlikely in the foreseeable future, although steel and its substitutes could well find applications in place of reinforced and prestressed concrete. Continuing development is needed. As for steel itself, connections are now more weldable and large members can be joined with relative ease. Quality assurance and control systems are becoming more sophisticated and reliable, a necessary development at a time of declining craft standards.

The combination of increased computing power and the development of modern analytical tools provide us with enough predictive capacity for the foreseeable future in tall buildings. This is especially relevant to transfer structures, particularly when they are combined with stability systems. However, we require more field data to ensure that our analytical tools are appropriate and that over-refinement in our calculations does not produce a false sense of security. We require feedback particularly on the dynamic behavior of clad buildings as opposed to the structural skeleton, and on comfort criteria for building inhabitants.

Most important of all, there is a perceptible tendency to question what is to be built, rather than how to build. In that context transfer structures are likely to play an increasingly important role, giving functional flexibility in tall buildings.

REFERENCES/BIBLIOGRAPHY

Henkel, D. J., Philips, A. B., and Gulager-Nielsen, E. 1984
NATIONAL BANK HOUSE, MELBOURNE—FOUNDATION DESIGN AND PERFORMANCE, Proceedings 4th Australian & New Zealand Conference on Geomechanics, Perth, Australia, May 14-18, pp. 300-304.

Michael, D. and Anderson, J. M. D., 1967
SUSPENDED FRAME BUILDINGS AND IN PARTICULAR THE STANDARD BANK CENTRE, Arup Journal, May, pp. 10-19.

Nutt, J. and Haworth, P., 1976
CAPITAL TOWER, Arup Journal, October, pp. 27-32.

Ove Arup & Partners, 1984
GROSVENOR PLACE, SYDNEY REPORT ON TRANSFER SYSTEM PART 1—INTERACTION WITH SURROUNDING STRUCTURE, October.

Ove Arup & Partners, 1984
GROSVENOR PLACE, SYDNEY REPORT ON TRANSFER SYSTEM PART 2—ASSESSMENT OF TRANSFER FRAMES, November.

Parkinson, B. and Hirst, J., 1981
TSUEN WAN RESIDENTIAL BLOCKS, Arup Journal, April, pp. 2-6.

Thompson, P., 1976
THE O.C.B.C. CENTRE, Arup Journal, October, pp. 3-11.

Zunz, G. J., Heydenrych, R. A., and Michael, D., 1971
THE STANDARD BANK CENTRE, JOHANNESBURG, Proc. I.C.E., February, Pamphlet 7346.

Zunz, G. J., Glover M. J., and Fitzpatrick, A. J., 1985
THE STRUCTURE OF THE NEW HEADQUARTERS FOR THE HONG KONG & SHANGHAI BANKING CORPORATION, HONG KONG, The Structural Engineer, September, Vol. 63A, No. 9, pp. 255-284.

Investigation and Prevention of Failures

Lev Zetlin

ENGINEERING INGENUITY

Exactly 20 years ago *Progressive Architecture* magazine devoted a special issue to the future of architecture, to which this author contributed an article on the Engineer's Third Millenium. Tall buildings of the future were featured prominently then. Even though some of the concepts for future tall buildings in that article appeared far fetched in 1966, it is encouraging to see that many of the creative far-fetched ideas have been realized in 1986. Prefabrication, modular construction, tube envelope, and dynamic stabilization of tall buildings are realities today.

Since prevention of failures is dependent on the knowledge of advanced technology of construction and on innovations that were widely discussed at the Second Century of the Skyscraper conference, it might be of interest to review some of the 1966 predictions.

Figures 1 a, b, and c. show tall structures stabilized by tension cables. Figures 1 d, e, and f show modular and tubulars envelopes tall structures, some of which are also stabilized by tension nets. Figures 2 a and b show stacked structures with integral transportation systems. Figures 2 c and d show stacked tall structures constructed over existing slums. The system is constructed as an arch with suspended platforms from the arch. The platforms provide space for promenades and shopping areas

Fig. 1 a–c. Tall structures stabilized by tension cables. d–f. Modular tubular envelopes for tall structures, some also stabilized by torsion nets.

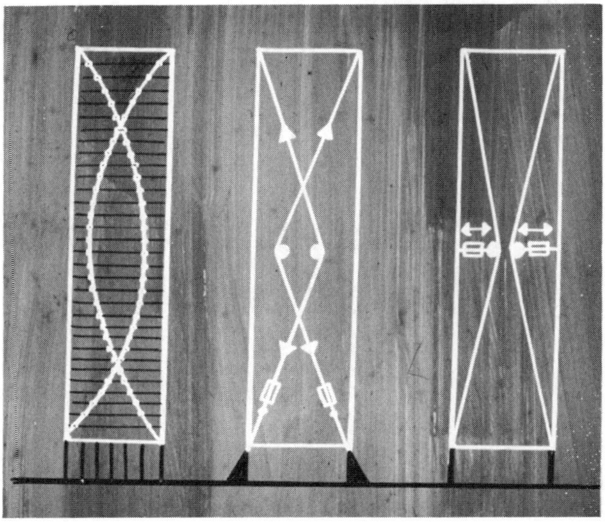

a

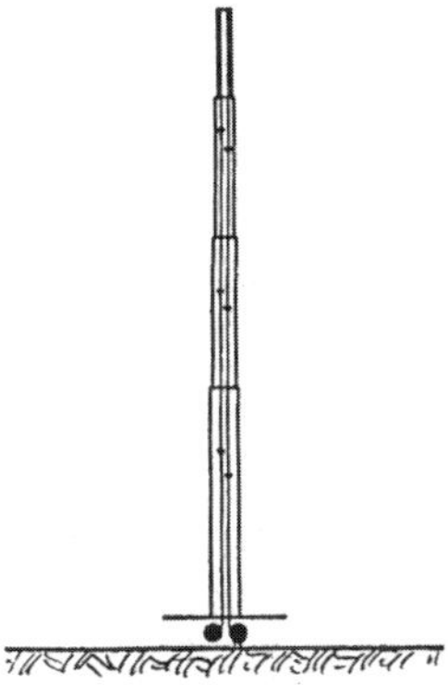

c

b

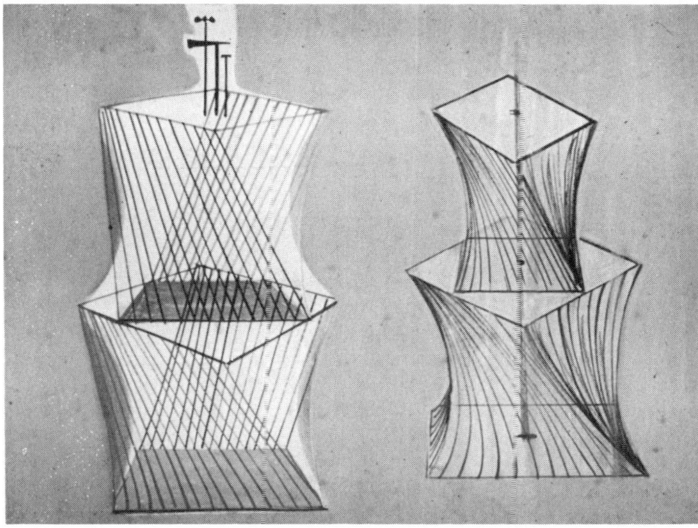

e

f

d

Fig. 2 a, b. Stacked structures with integral transportation systems. c, d. Stacked tall structures constructed over existing slums.

a

b

c

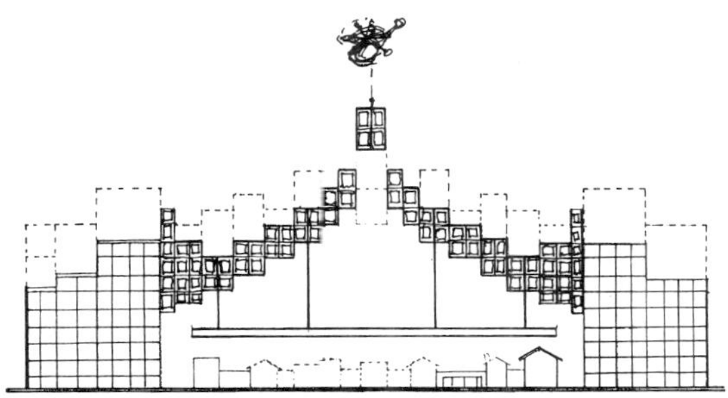

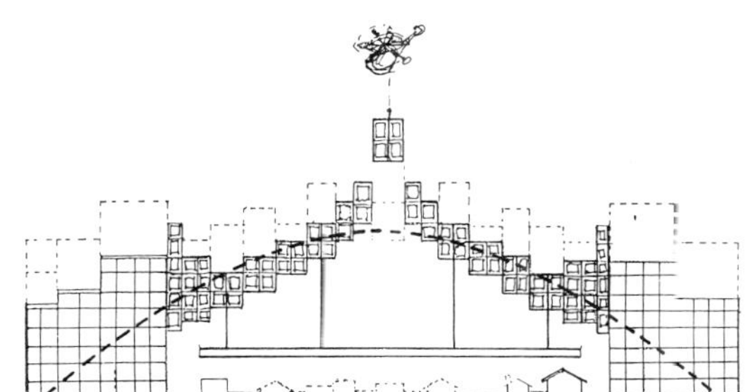

d

Tilted or inclined tall buildings offer many advantages for their occupants. However, the cost of a conventional building (rectangular in plan), if inclined as shown in Fig. 3a, is prohibitive. However, if the tall building is curved or has reentrant angles in plan, as in Figs. 3 b and c, the floors will provide stabilizing horizontal reactions against overturning moments created by the inclined structures, as shown in Fig. 3d.

Extrapolating the concept to the third millenium, it should be technologically possible to build a self-contained city, as in Fig. 3e, consisting of several levels of multistoried structures, a mile long and thousands of feet high, which form an ellipse in plans, and consisting of two interconnected inclined half-cities stabilized against overturning through the interconnection. Each thick strip in Fig. 3e is a 20-story building at one level, separated with a 300-ft space from another level, creating huge open spaces.

Figure 4 shows various stages of construction of a tall prestressed concrete building, completely prefabricated in a confined space on the ground, with all building components lifted up; this concept eliminates the need for scaffolds or cranes.

CHANGING TECHNOLOGY, INNOVATIONS, BUILDING FAILURES

Rapidly changing technologies of materials, methodologies of design and technique, and construction equipment, all of which are conducive to innovations, are a reality; all should be fostered and all should be taken advantage of. The increasing number of legal claims on building failures is also a reality. The two realities are interdependent.

Changes, improvements, and innovations in materials, construction techniques, and equipment have created a gap between the theories used in designing structures and the behavior of structures as actually constructed. The basic function of structural engineering is to predict the behavior of a structure (a building, an airplane, or the like) before it is constructed.

In the not too distant past, the accepted design theories of tall buildings, for example, were geared to structural steel of certain grades resulting in structural systems with a certain number of pounds of steel per square foot of building. In such buildings, vibrations and transmission of noise were not a problem and thus were not considered in the engineering design. Similarly, deformations and deflections of beams and of columns did not adversely affect window frames.

As improved steels have been introduced, however, it has become possible to reduce the weight of steel per square foot in a building. This reduction in the weight of steel offered many advantages: economic, esthetic, and functional (by providing more usable volume of space).

Unfortunately, buildings with light steel frames have been designed using the same theories as buildings with heavier steel frames. But in lightweight steel frames, vibrations and deformations of beams and columns do have a significant effect on the window frames and partitions. Problems occur, such as failure of window frames and glass, cracks in walls, and excessive sound transmission.

A structure with defects may have been well designed following manuals and building codes, but the manuals and building codes have not kept up with the advancing technology.

SETTING FOR CONSTRUCTION FAILURES

What most people in the construction industry are concerned with depends on what they are "structured" to be concerned with. A few decades ago, products were designed and made by the same person or company. Then came a separation between the designer of a product and the maker. In the last few years, with the advent of performance specifications, there is a new separation of the construction process. Four experts are involved: the designer of a total project who designates the function that the product is to have (the product could be a vital component of the entire structure); the product designer; the product maker; and the contractor, who is responsible for the quality of the product, its design, and its proper performance as part of a whole structure.

The author's experience investigating numerous structural failures indicates that the failures have not occurred because of errors in the traditional design. Most failures occurred because no attempt had been made to evaluate the impact of the gap between the traditional design methodology and the true behavior of a modern structure.

Fortunately, modern advanced design methodologies could serve as potent tools to "crystal ball" and prevent construction failures. These tools should be used in the initial phases of design.

EXAMPLES

Failures in tall buildings range through a wide spectrum: from leakage through exterior facade cladding to unusual geometric deformations of structural frames to catastrophic collapses.

Unusual Geometric Deformations

An L-shaped building is shown in Fig. 5a in elevation and by solid lines in Fig. 5c. Under wind every bent in a rectangular building (as in Fig. 5b) would bend equally to the adjacent bents. However, a building with truncated ends (as in Fig. 5c) would distort unequally, each bent bending differently from

a

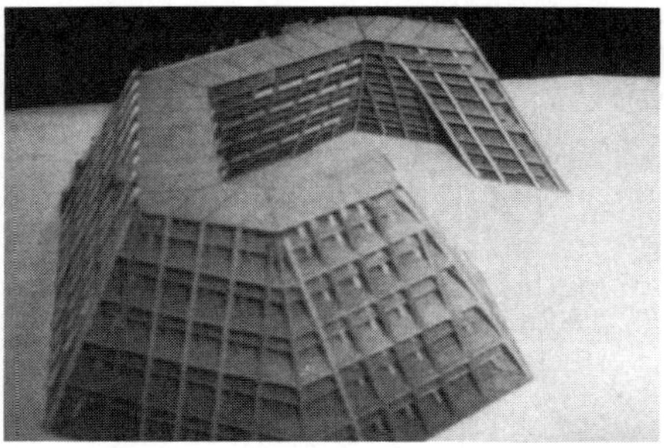

b

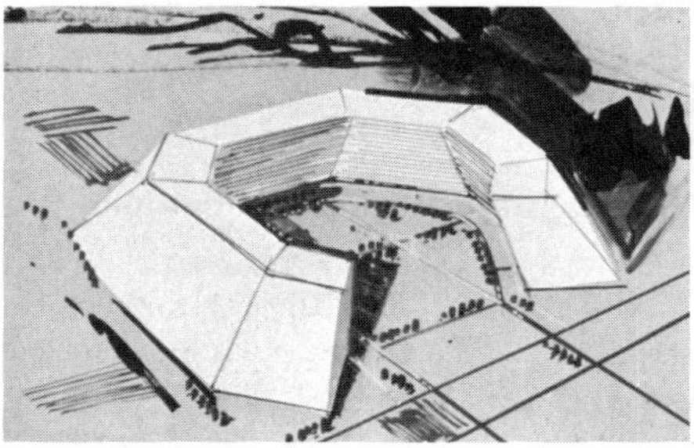

Fig. 3 Tilted or inclined structures. b

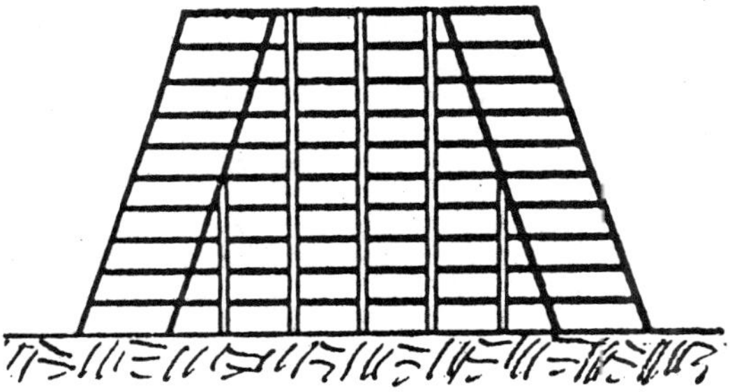

c

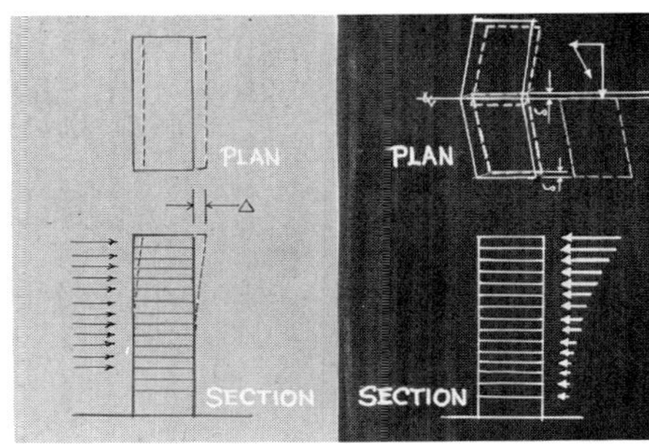

d

e

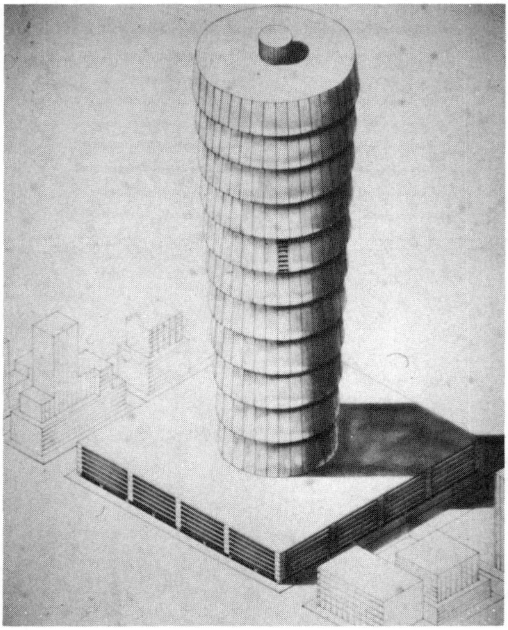

a

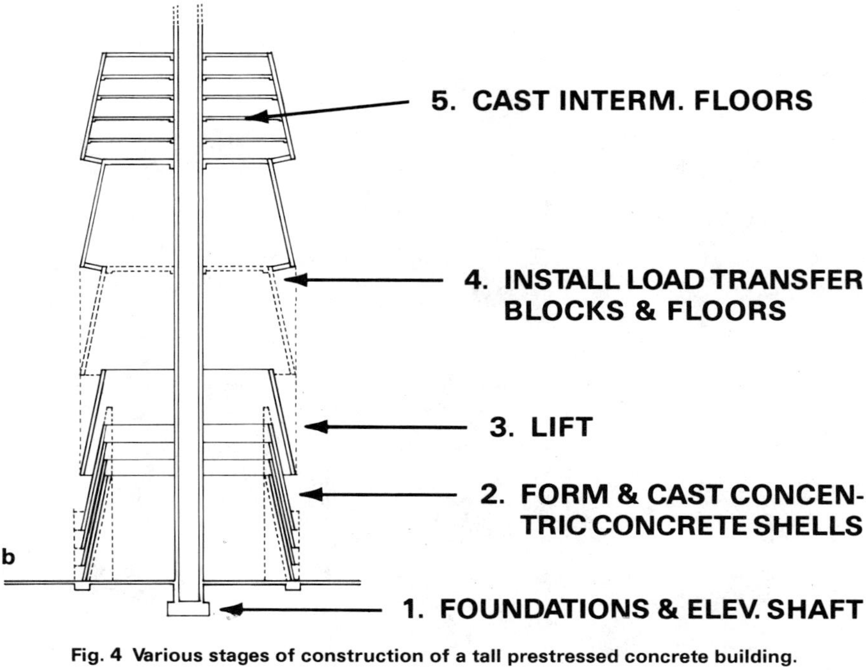

b

Fig. 4 Various stages of construction of a tall prestressed concrete building.

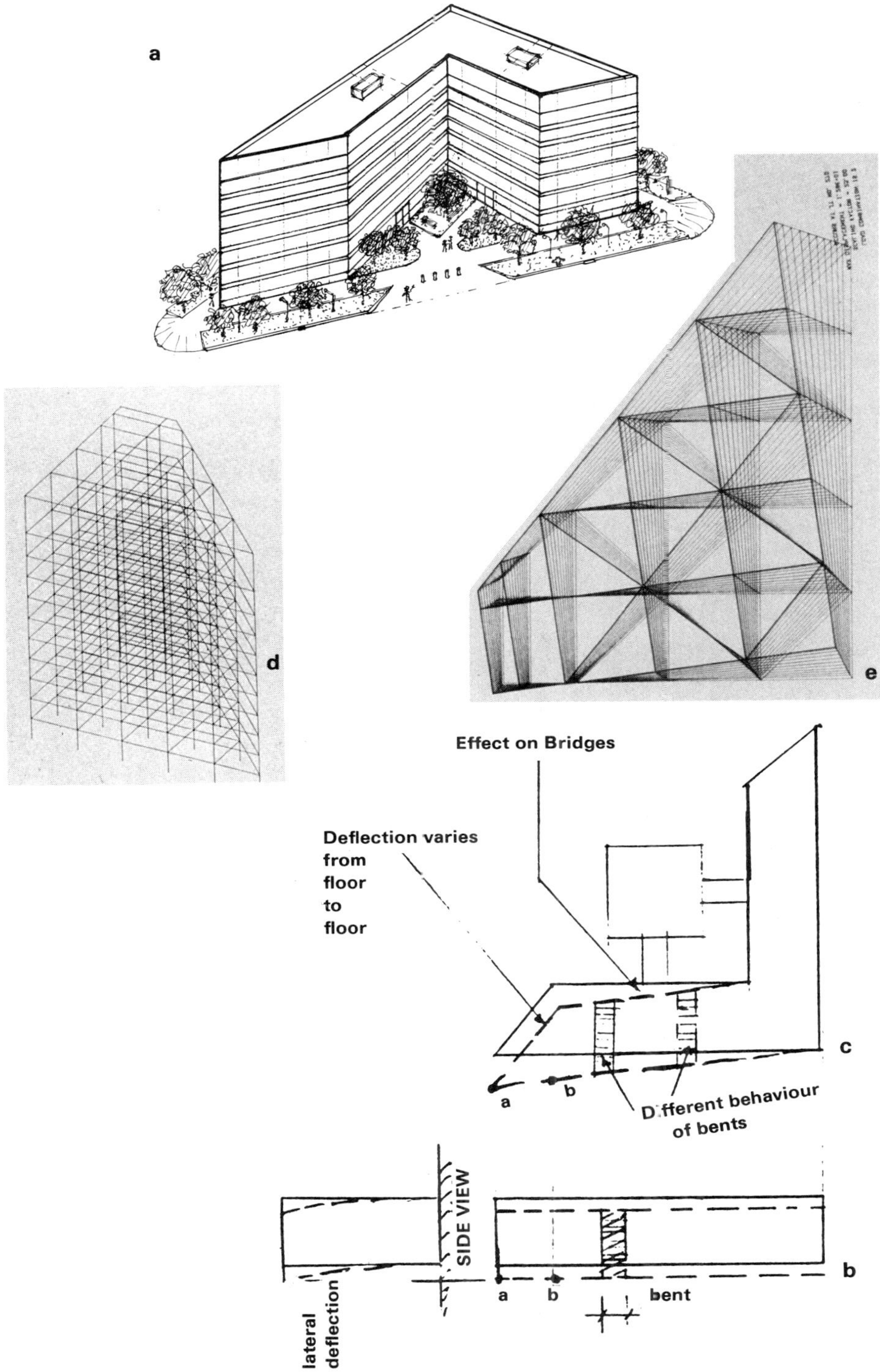

Fig. 5 Buildings with unusual geometric deformations.

the adjacent bents. If the unequal bending is not recognized during a design, damage to floors and to exterior walls would occur. The truncated ends of the building cause rotations of the ends of the floors in a fanning pattern, with rotation being largest at the top floor. Computer generated graphics of the rotations of the floors at the truncated ends of the building are shown in Fig. 5e. The frame of the truncated portion of the building used in the computer analysis is shown in Fig. 5d.

Exterior Brick Cladding

Numerous designs and extensive renovations occur in exterior brick facades because of rain water leakage and the misalignment of brick. Although commonly not recognized, the main problem is the unsymmetrical rotation

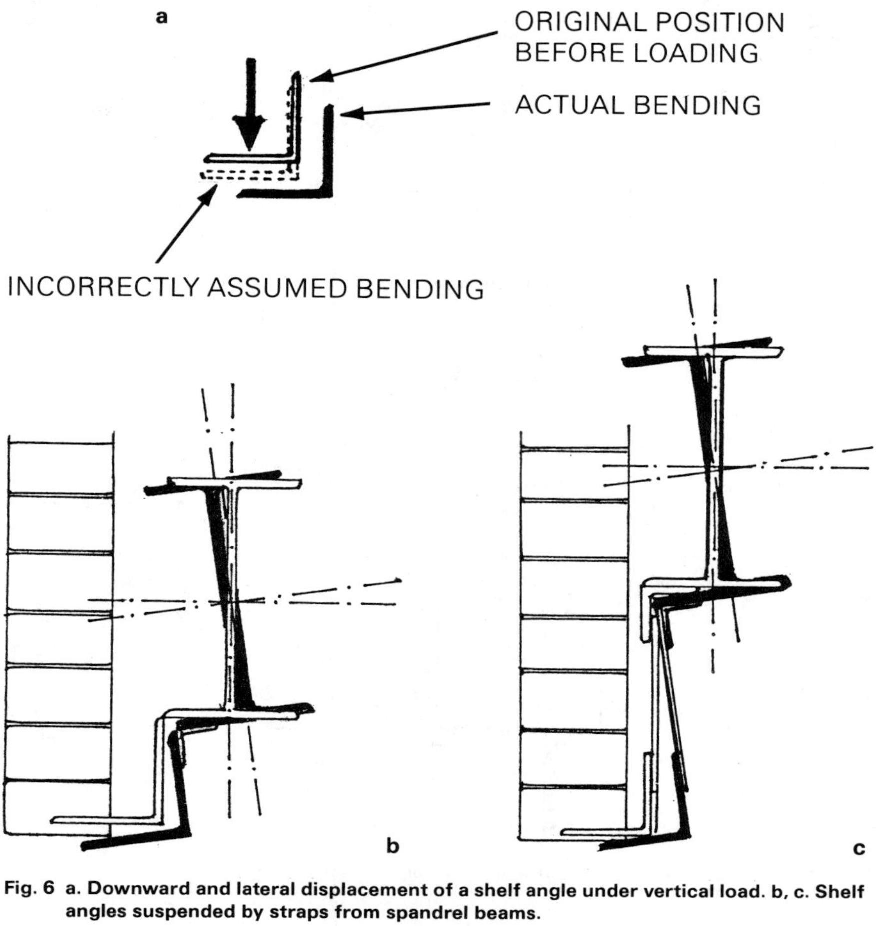

Fig. 6 a. Downward and lateral displacement of a shelf angle under vertical load. b, c. Shelf angles suspended by straps from spandrel beams.

of shelf angles whose principal axes are inclined with respect to the direction of the brick load. Downward and lateral displacement of a shelf angle under vertical load is shown in Fig. 6a.

In modern buildings the shelf angles very often are not attached directly to spandrel beams but are suspended by straps as shown in Figs. 6b and 6c. Such configuration increases the lateral movement of the shelf angles with the resulting loss of support for brick and creation of horizontal forces at the joints between the bricks, causing lateral misalignment of brick.

CONCLUSION

Because of the changing technologies and the new, previously unknown interactions between building components, the engineering profession should expect more building and structural failures and more claims in the future. A better understanding of the effects of changing technologies and their secondary effects would help pinpoint potential failures.

Design of Glass Against Breakage

Joseph E. Minor

Window glass breakage has become a sensitive subject for skyscraper designers, owners, and insurers. Once a minor nuisance handled by routine maintenance, now broken window glass may significantly damage building contents, instigate personal injury claims, and produce adverse publicity for building owners and designers. In addition, the once simple task of selecting a window from among a few clear glass products has become an arduous task involving structural mechanics, polymer chemistry, thermodynamics, ceramic science, coating technology, and hazards analysis. Finally, the modern skyscraper is within a city and houses a community within itself, making cladding performance in hurricanes, earthquakes, and other extreme events an important concern for public safety.

FACTORS TO BE CONSIDERED IN DESIGN

The designer of the building envelope designs against glass breakage. Since it cannot be assumed that failure will not occur, he should also design for the consequences of glass breakage. Considerations related to the former include factors listed in Table 1. The consequences of glass breakage listed in Table 2 also should be addressed.

Table 1 Factors considered in the design of glass against breakage

Wind effects	Human impact
• wind pressure	Falling and propelled objects
• windborne debris	Hail impact
Earthquake effects	Ice and snow
Thermal effects	Fire

Table 2 Consequences of glass breakage

Damage to building contents
Personal injury liability
Adverse publicity
Internal pressures
Damage to other buildings
Loss of structural integrity

These potential consequences are of such significance that they demand design attention. Petak and Hart (1980) report that in the year 2000 the cost of damage to building contents caused by windstorms will essentially equal the cost of damage to the building itself. In tall buildings, contents damage due to wind effects currently exceeds building damage. The increased use of overhead glazing in skylights and atriums has produced a concern for public safety in the areas below. Current standards for overhead glazing are written to virtually eliminate glass particle "fallout" in the event that overhead glazing is broken (GICC, 1984). This concern has heightened in recent years, and numerous personal injury claims have been filed. Surprisingly, similar concern has not been expressed regarding hazards to people on sidewalks and pedestrian walkways adjacent to tall building curtain walls, where the hazard presented by falling glass may be significant. While there were no injuries to people during the major glass cascading incident in Houston, Texas during Hurricane Alicia (Kareem, 1985), architects and building owners have stated that the adverse publicity drawn to their profession and to their community by this incident was unacceptable.

In addition to damage to building contents, possible personal injury, and adverse publicity, cladding failure can result in increased internal pressures, damage to other buildings, and loss of structural integrity. The Uniform Building Code (ICBO, 1985) is the only U.S. model building code that currently recognizes the effects of internal pressure in the event of glass breakage (see UBC Table 23-H). Several buildings in Houston were damaged by glass particles injected into the windstream from adjacent buildings (Kareem and Stevens, 1985). Finally, glass panels are being employed to provide wind bracing in some overhead glazing and skylight installations. Breakage of glass panels could result in loss on structural integrity in these structural systems.

The considerations outlined in Tables 1 and 2 make the selection of glass products an iterative process. A product that is well-suited for resisting breakage may not perform well if in fact it does break. For example, fully tempered glass resists wind effects (pressures and missile impact) to a degree that exceeds that of other types of glass, but may fall from the opening when it breaks (Fig. 1). Table 3 lists available glass products. Minor (1984) discusses the properties of various glass products in resisting breakage and in combat-

Fig. 1 Broken tempered glass fell from spandrels in Entex Building (background) while damaged annealed glass in vision areas remained in place (Hurricane Alicia, 1983)

Table 3 Glass products

Types	Units	Treatments	Installation
Annealed	Monolithic	Clear	Wet glazed
Heat strengthened (HS)	Insulating (IG)	Tinted	Dry glazed
Fully tempered (FT)	Laminated	Coated	Structurally glazed

Table 4 Glass product performance characteristics

	Design Requirement	
Design factor	Resist breakage	Resist consequences of breakage
Wind effects		
pressure	HS; FT; IG	Laminated
debris	FT	Laminated; film coated
Earthquake effects	(Unknown)*	Laminated; film coated
Thermal effects	HS; FT	Laminated; film coated; HS; annealed
Human impact	FT	FT; laminated
Falling, propelled objects (including hail)	FT	Laminated
Ice and snow	HS; FT	Laminated
Fire	HS; FT	Wired; laminated

*Research must be performed to assess relative abilities of glass types and products to resist in-plane shear forces imposed by earthquake effects.

ting the consequences of failure. Products are matched with their capabilities in this regard in Table 4.

The task of the designer becomes one of selecting products from those listed in Table 3 that can meet the requirements of the factors listed in Table 1 without breaking, while considering the consequences of glass breakage outlined in Table 2.

CONCERNS FOR FUTURE TALL BUILDING DESIGNERS

As the skyscraper enters its second century, two major new challenges are being presented to the designer of the building envelope. First, the design of the envelope has become a management, as well as a technical, problem. Second, extreme events (for example, hurricanes, earthquakes) must be considered in terms of their potential impacts on the communities within and adjacent to the skyscraper.

The Design and Construction Team

Design, construction, and installation of the building envelope has become a complex task involving many parties, including the following list of parties who may be involved:

Architect
Structural engineer
Curtain wall designer
Curtain wall consultant
Curtain wall fabricator
Glass manufacturer
Sealant manufacturer
Curtain wall installer
Contractor
Testing laboratory
Skylight manufacturer
Glazier

The party who integrates team activities to achieve an acceptable end product can be one of several of those listed above. Often there is no single party in responsible charge. If each party does its job in a professional and workmanlike manner, an acceptable end product is achieved. If a party or parties fail to act professionally, this arrangement becomes unraveled, and envelope failure may occur.

The failure of a connection in a skywalk at the Hyatt Regency Hotel in Kansas City in July, 1981 has implications for the design of the building envelope. In the Hyatt Regency skywalk failure, it appears that the structural engineers are going to be held solely responsible for the details of the connection design, notwithstanding the involvement of other parties in the design and construction (fabrication) process (*Engineering Times,* 1985). The American Society of Civil Engineers has, as a result of the Hyatt Regency skywalk failure, issued a policy statement on the responsibility for design of steel structures (*Civil Engineering,* 1985). While this policy statement addresses relationships between parties to the construction of the building frame, its implications for other components of the building are clear. A key observation offered in the statement is: "Contractual arrangements which fail to include clear assignment of responsibility and authority for design of structural elements and connections, which divide and fragment such responsibility and authority, or which tend to impose on one party the proper responsibilities of another party, are counterproductive and should be eliminated."

As we enter the second century of the skyscraper, more definitive responsibilities will be assigned to the various parties involved in the design, fabrication, and installation of the building envelope. Whether the responsible party becomes the architect, structural engineer, or curtain wall specialist, this party will now be addressing details of design and construction heretofore considered to be the sole responsibility of others.

Extreme Events

Future designers of skyscraper envelopes must consider the effects of extreme events on cladding performance. The failure of window glass in Houston, Texas skyscrapers during Hurricane Alicia produced significant damage and adverse publicity (Kareem, 1985) (Fig. 2). The special effects of

Fig. 2 Cladding damage to tall buildings in Houston, Texas (Hurricane Alicia, 1983)

Table 5 Turbulence characteristics of hurricanes

Recommended parameters for typhoons (hurricanes), coastal areas, 10 m height [a]			Parameters in current U.S. standard for coastal areas, 10 m height [b]	
Mean 10 min. wind speed (m/s)	Power law Exponent	Turbulence intensity (I)	Power law exponent (1/c)	Turbulence intensity (T_z)
20	0.33	0.200	0.10	0.13
30	0.29	0.176		
40	0.27	0.163		
50	0.26	0 153		
60	0.24	0 147		

[a] Ishizaki, 1983
[b] American National Standards Institute, 1972

earthquakes, hurricanes, and other extreme wind events must be considered during design if building envelope failures, property loss, personal injury, and adverse publicity are to be avoided.

Five properties of hurricane winds are distinctly different from properties of winds produced by other windstorms. Hurricane winds have a different probability of occurrence distribution, (Fig. 3), are more turbulent (Table 5), are more sustained (Fig. 4), change slowly in direction, and carry relatively large amounts of debris (Fig. 5). Taken together, these properties produce a wind environment that is more severe than the wind environment in other windstorms that produce the same peak wind speed. The effects of these

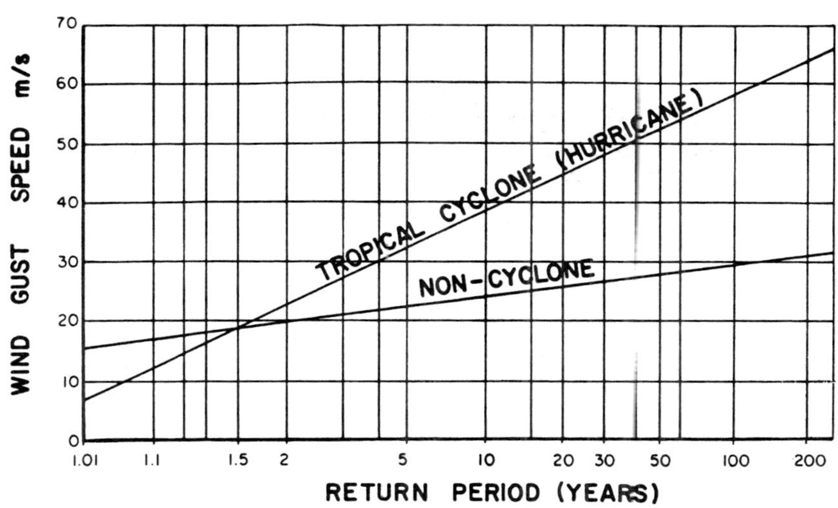

Fig. 3 Typical probability distributions for tropical cyclone winds (for Cairns, Queensland, Australia)

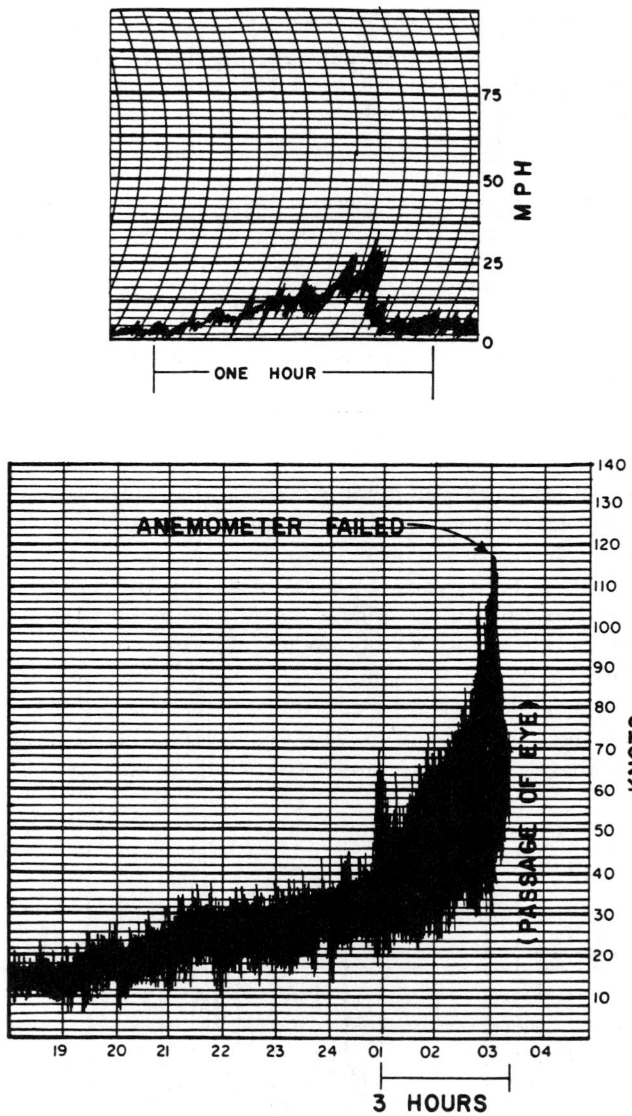

Fig. 4 Hurricane winds are sustained for longer periods of time than winds in other types of windstorms. a. Thunderstorm record (from Simiu and Scanlan) b. Record from cyclone Tracy, 24-25 Dec. 1974 (from Walker)

properties on window glass are discussed by Minor (1985) and were evident in Houston, Texas during Hurricane Alicia. Evidence from Hurricane Alicia and other hurricane occurrences clearly indicates that design criteria for buildings that may experience hurricanes must be established with appropriate recognition being given to the special character of hurricane effects. Sustained, turbulent winds that change slowly in direction may produce repetitions of relatively large pressure excursions (Fig. 6), as well as potentials for the occurrence of windborne debris, both of which are hazardous to window glass.

Earthquakes present special considerations for cladding design as well. Unlike the hurricane where advance warnings afford time to protect people from the hazards of falling glass, the earthquake strikes without warning. Spaces around glass edges are designed to accommodate dimensional changes that are produced by temperature effects and by interstory drift under wind loads. Under earthquake loadings, these spaces are not likely to be adequate for the interstory drifts that will occur; hence, glass plates will be loaded in-plane in shear, and may fail. If not designed to remain in the opening, broken glass may fall to the street injuring unwarned people below. Only limited research into the behavior of glass under these conditions has been conducted to date. This aspect of glass behavior is therefore a concern for the future.

Fig. 5 Debris produced by cyclone Tracy in Darwin, NT, Australia (1974)

CONCLUSION

The principal objective of the designer is to guard against window glass breakage. Consequences of failure must be considered, even though the premise is that glass failure under design loads will be "acceptably small". Responsibilities of parties within the building envelope's design and construction team will become more explicitly defined in the second century. More extensive consideration of the special effects of extreme events will be required of building envelope designers of the future.

ACKNOWLEDGEMENTS

The preparation of this paper was supported by the National Science Foundation (Grant No. ECE-8500620) and by Texas Tech University through the Glass Research and Testing Laboratory.

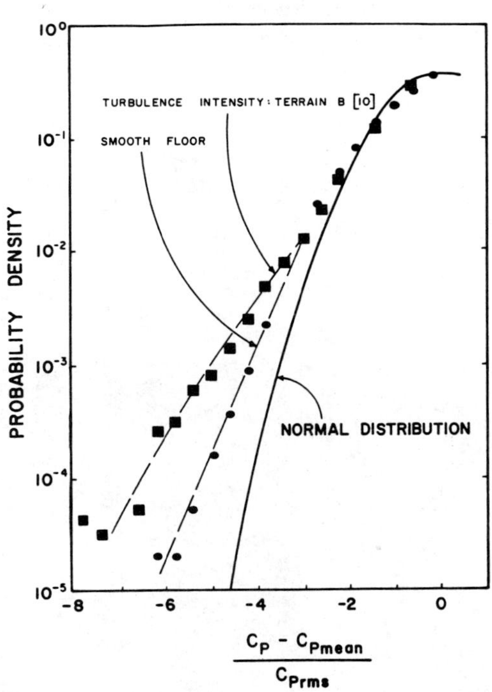

Fig. 6 Large pressure excursions are more likely to occur when winds are more turbulent (after Peterka and Cermak)

REFERENCES/BIBLIOGRAPHY

American National Standards Institute, 1982
MINIMUM DESIGN LOADS FOR BUILDINGS AND OTHER STRUCTURES, ANSI A58.1-1982, New York, NY.

Civil Engineering, 1985
ENGINEERS SHOULD TAKE RESPONSIBILITY FOR DETAILS, Civil Engineering, Vol. 55, No. 12, December, p. 12.

Engineering Times, 1985
ENGINEERS BLAMED IN SKYWALK DEATHS, Engineering Times, Vol. 12, No. 12, December, p. 1.

GICC, 1984
SLOPED GLAZING, proposed change to the Uniform Building Code, 1984 Code Change Proposals, International Conference of Building Officials, Whittier, CA.

ICBO, 1985
UNIFORM BUILDING CODE, International Conference of Building Officials Whittier, CA.

Ishizaki, H., 1983
WIND PROFILES, TURBULENCE INTENSITIES AND GUST FACTORS FOR DESIGN IN TYPHOON-PRONE REGIONS, Journal of Wind Engineering and Industrial Aerodynamics, Vol. 13, Nos. 1-3, December, pp. 55-66.

Kareem, A., 1985
HURRICANE ALICIA: ONE YEAR LATER, Proceedings of a Specialty Conference (Galveston, Texas, August 16-17, 1984), ASCE, New York.

Kareem, A. and Stevens, J. G., 1985
WINDOW GLASS PERFORMANCE AND ANALYSIS IN HURRICANE ALICIA, Proceedings, Specialty Conference, Hurricane Alicia: One Year Later (Galveston, Texas, August 16-17, 1984) ASCE, New York, pp. 178-186.

Minor, J. E., 1984
WINDOW GLASS DESIGN, Workshop Notes, Fourth Canadian Workshop on Wind Engineering (Toronto, November 19-20, 1984), Canadian Wind Engineering Association, National Research Council of Canada, Ottawa, pp. 195-212.

Minor, J. E., 1985
WINDOW GLASS PERFORMANCE AND HURRICANE EFFECTS, Proceedings Specialty Conference, Hurricane Alicia: One Year Later (Galveston, Texas, August 16-17, 1984), ASCE, New York, pp. 151-167.

Petak, W. and Hart, G. C., 1980
DAMAGE AND DECISION MAKING IN WIND ENGINEERING, Proceedings, Fifth International Conference on Wind Engineering (Ft. Collins, CO, July, 1979), Pergamon Press, Oxford, pp. 61-74.

Innovations in High-Rise Construction

John Norris

Innovation in construction is achieved by challenge. I've therefore structured this paper to describe Olympia and York's approach to innovation and flexibility in high-rise construction, then move on to a general assessment and conclude with what I think will happen in the future. But first I would like to give a brief picture of Olympia & York and how we approach high-rise construction.

The company started building low-rise industrial warehouses in suburban Toronto in the early 1960s, then shifted into high-rise office construction and has stayed in it ever since. Our flagship in Canada is First Canadian Place, a 72-story office tower in the heart of Toronto's financial district (Fig. 1). It allowed the opportunity to test construction methods in a number of ways because of its height and volume. This is the largest commercial development undertaken as one continuous project in Canada. Building in the clouds also created many problems of its own.

In Calgary we built Esso Plaza, which has unusually large and open public spaces for that city. In Boston, we built a 40-story tower and an atrium connected to the preserved facade of the old Stock Exchange Building. In Chicago, we built the 63-story Olympia Centre, a mixed-use development incorporating a four-level Neiman-Marcus Department Store, 18 floors of office space, and 292 luxury condominiums.

The company has, in various stages of construction: World Financial

Center (Fig. 2), a 743 thousand m^2 (8 million ft^2) development in New York; a 30-story office tower in Manhattan; a 25-acre development in downtown San Francisco; a 25-story building in Toronto and a 17-story mixed-use development in Orlando, Florida.

Olympia and York has built or bought, owns, and manages more than 150 projects in Canada and the United States with a total of more than 4.5 million m^2 (50 million ft^2). They are all different, with different construction and operations problems.

PREPLANNING

In examining how to construct a 72-story high-rise, we considered the following factors.

Fig. 1 First Canadian Place

Reduce time lost by inefficient handling of workers and material and achieve an early occupancy of the lower 22 floors and lower lobbies. To do so we had to make sure that all those hours of time lost by workers through material not being in the right place at the right time were reduced to the minimum.

We had to *eliminate haphazard storage and inefficient handling* of half a million tons of material from the base of the tower. To do this, we looked at the building as a vertical and horizontal factory assembly, requiring a highly

Fig. 2 World Financial Center site

disciplined control of workers and materials to cut down on labor waste and to increase productivity.

The other objective from a development perspective, was to prepare the lower 22 floors of the tower and lobbies for occupancy and normal use within 16 months of the steel going up and the completion of the rest of the tower six months after that. Naturally it called for a high degree of *preplanning and design modifications* to suit field requirements and greatly facilitate the construction process. If the first two objectives worked, the third would fall into place, and it did. A good factory owner studies the field problems before he commences manufacture.

MOVING WORKERS AND MATERIALS

After 18 months of planning and research, we developed an approach that we have been fine tuning ever since. It depends, quite simply, on ensuring that enough equipment, be it temporary elevators, cranes, conveyors, or hoists, are available to give quick and easy access for materials being handled from the street to the workplace. We also had to persuade the subcontractors that by organizing their transportation of materials and workers to conform to a monitored time frame of deliveries, they would be able to place their materials at the precise place of work, and on time. Obviously this does not happen without organizational and efficient controls. To achieve this, two things are essential: *one,* a traffic manager and staff acting as a fully autonomous service department to the project manager and having full control; *two,* in parallel and under the same traffic manager, a strict control and assignment of all means of hoisting, all of which calls for factory-like regimentation with a high degree of autonomy being implemented by our staff.

To accomplish those goals on the job we made sure we had a fully planned hoisting program for cranes utilized by all trades necessary for the frame that is, we hoisted by crane: steel deck reinforcing, electrical, mechanical, stairs, precast slabs for mechanical elevator, motor generator rooms, cladding, and prefabricated pipe clusters (Figs. 3 and 4). A horizontal and vertical rapid transit conveyor system was used for concrete, and an effective control system was utilized for temporary and permanent elevator systems until the project was complete.

It was essential to segregate deliveries of concrete, frame, and architectural materials from each other to ensure maximum flow of all materials up the building. For transportation we used nine temporary internal elevators in what eventually became permanent shafts. These were large enough to handle 52 men at a time as well as tenant materials (Fig. 5). Two of the elevators had a capacity of 3600 kg (4 tons). This elevator system distributed at peak hours nearly 1,500 men throughout the 72 floors with less than an

Fig. 3 Cranes were used to hoist materials in place directly from delivery trucks

average of half an hour a day of lost time per man. This was done by placing all elevators in peak up and down periods to hoist workers only. We also used 12-floor climbing elevators to feed the steel working level. These were a first in the construction industry. They were not a first in the mining industry.

SEGREGATING ARCHITECTURAL MATERIALS

Trucks bringing finishing materials were lowered in a 19-m (63-ft) long hydraulic elevator to a turntable 16 m (52 ft) in diameter below ground (Fig. 6) and lined up with one of the 11 basement loading docks that had been built

Fig. 4 Pipe assembly being hoisted atop First Canadian Place during construction

while the main tower was in the planning stage. The materials were then hoisted to a preplanned and assigned storage area for each trade on each floor. This last maneuver was vital to prevent double handling of materials. We guaranteed to the subtrades that if they arrived on site with their materials at a set time, they would have certain hoisting facilities available for the

Fig. 5 Internal construction elevators

Fig. 6 Truck on turntable is being aligned with basement loading dock

time requested. Every truck delivery and every lifting device was scheduled, tracked, and matched by computer in a manned control room. This traffic control operation was crucial to the success of our new construction systems. It is still being used by our operations staff in controlling the rapid delivery of tenant materials to both commercial and office floors, as well as controlling removal of rubbish. An added benefit was that trucks were not parked on city streets.

COST SAVINGS AND EARLY OCCUPANCY

What did we accomplish, after all the trucks and trades were gone?

1. We know our subcontractors saved far more money than we thought they would getting workers and materials to their workplaces, on time and in sequence.
2. By ensuring that only those materials for the first and second floor lobby areas were delivered and stored on those floors, we were also able to allow the first tenants to pass through finished lobbies to finished office space on the lower 22 floors only 16 months after the foundations 5 levels below the street were poured. The other 50 floors were completed on schedule as well.
3. A tremendous sense of accomplishment was experienced by both management and labor. Only those who live with major projects, day in and day out, and who know the potential for calamity will fully appreciate what that means.

The difference between success and failure in the field of high-rise construction is not unlike a military operation. The objectives are clearly spelled out by the developer, much like a general. The developer wants the best quality, at the right price, and on time. The planning is originated, coordinated, and executed by a corps of architects, engineers, and senior company management who, if they are wise, incorporate the knowledge of subcontractors as well as their own field and home office management personnel as needed, both in the design and construction phases.

Remember the workers need logistical support that puts them where they are needed, on time, and with the equipment they need, exactly where they need it. They also need to be motivated by an efficiently controlled transportation system to their workplace, which raises both productivity and morale. There are, of course, things we cannot control, such as the weather. And we cannot always predict delays of materials or union difficulties.

Efficiency brings its own rewards in improved safety (and therefore less cost) by having workers back on the job if their injuries can be treated satisfactorily on the site (Fig. 7). We do not accept the notion that speed is an excuse for poor quality and do not build faster and therefore cheaper, at the expense of quality. This, as a developer, would eventually become false economy when your aim is to lease high quality space.

Perhaps the best example of that quality principle is World Financial Center in lower Manhattan where the winter garden's domed roof is not exactly the ironworkers favorite place, given the pitch and the wind coming off the Hudson (Fig. 8). The dome and the tops of the four towers added millions of dollars in costs, but we are doing them because we will be judged by this project for years to come (Figs. 9 and 10).

PLANNING AND WORLD FINANCIAL CENTER

Needless to say a project of this scale in particular calls for intensive, extensive research. When you are bidding on a 743 thousand m² (8 million ft²)

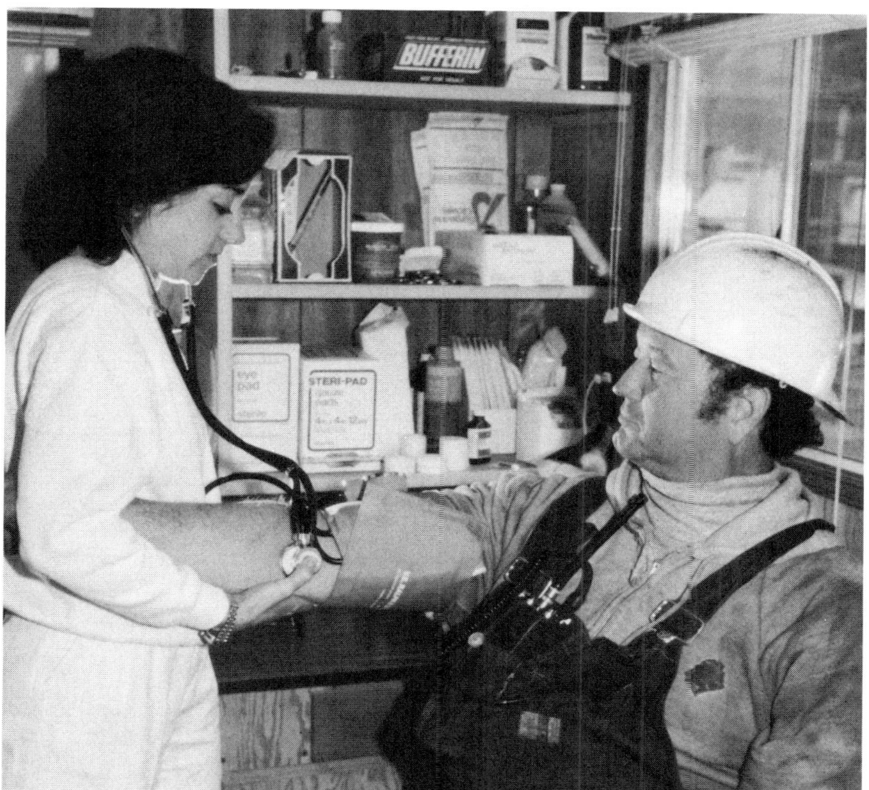

Fig. 7 An on-site medical facility enables workers to be treated immediately, saving time and cost

project, you strive to consider everything. After we were selected as the developer, our first priorities, after an overall review of the project, were access roads, security fences and entry gates, on-site access to unloading points for cranes, hoists and elevators, vertical movement of workers and material, and organizing construction services. The fences and gates depended on location, accessibility, city traffic regulations, volume of anticipated deliveries and numbers of trucks, and sequence of construction, which must all be reviewed. It is also vital to a developer to provide the least irritating separation of construction and tenant access if the building is to be occupied before it is completed.

Access roads and unloading points on the site (Fig. 11) are influenced by several factors: one, the phasing of exterior work; two, location and the use of climbing and ground cranes; three, location of inside and outside hoists; four, permanent or temporary loading and unloading docks, the numbers and positions of which are among the most complex decisions and that should be made at the earliest possible time; five, a review of what a tenant is going to need and its effect on design and construction services, including trailers for workers and materials, temporary or permanent facilities for power, telephone, water and sewer, first aid, parking, cafeterias, and so on. There is a lot to think about. Planning is only as good as the implementation, which is only as good as the cooperation and motivation of management and labor working as a team.

Fig. 8 World Financial Center domed roof

Fig. 9 Access roads, outside elevators, and cranes are all an important part of the construction process.

Fig. 10. Construction of domed winter garden

Fig. 11 Deliveries to the site

SCHEDULING IN THE COMPUTER AGE

A high-rise construction project needs a competent team and technology as well. We have standardized the use of a computerized construction scheduling system for planning and logistical control of the flow of materials and labor. This system records all past, present, and future deliveries. It also provides print outs of those deliveries for those people implementing the control on-site and accounts receivable people off the site. It is a 3,000-function system that took three years to perfect and is still being fine-tuned. The same equipment can produce an up-to-date progress position, when it is needed. It also reconciles workers' time sheets with the payroll and resolves logistical errors.

Before the last of the four World Financial Center towers is completed, the system will be adapted to orchestrate deliveries of office supplies and equipment to tenants and handed over to our building operations management. By effective control through this computerized system, we prevent traffic congestion, allowing the trades to plan more effectively. It also offers a quick and precise checking of overlapping workers, eliminates costly mistakes and delivers a level of control, speed, and accuracy that a manual system can't match.

THE CHALLENGE OF STEEL AGAINST CONCRETE IN THE HIGH TECH ERA

In my opinion, steel is more flexible than concrete in making changes to accommodate the new wave of high technology, which in turn makes for flexibility in tenant-inspired changes. Time is important in construction, too, and a steel frame, especially if it is designed with bolted connections, is faster to erect for a high tech building than concrete (Fig. 12).

Fig. 12 Construction of steel-framed tall building

That does not mean to suggest that concrete is not appropriate in certain circumstances. The choice should be based on a detailed analysis at the preplanning stage of the contract.

THE HIGH TECH ERA: NOW

A century ago, engineer William Lebaron Jenney had the ingenious idea of hanging masonry walls on a steel framework. He did it for the 10-story Home Insurance Building in Chicago, and the skyscraper was born. In many parts of the world a 10-story building is still considered to be a skyscraper.

Today we are on the threshhold of another turning point in the evolution of skyscrapers. The construction of a 200-story skyscraper is feasible. Whether or not it is financially advisable or even desirable is another matter. Frank Lloyd Wright's Mile High Building is one example. In New York Donald Trump plans to build a tapered 150-story tower on Manhattan's west side. The sky is literally the limit.

The turning point I refer to is the incorporation of high technology communication installations in high-rise office towers. The so-called "smart" office building is one of the most significant construction-related developments of the last 10 years. New office tower construction must consider data processing, telecommunications, electronic mail, teleconferencing, centrex switches, extra power requirements, and all the other high tech features that major, and even minor, tenants expect. That applies to old buildings as well as new. These factors must all be considered at the predesign stage.

The differences between "smartness" and "dumbness" in a building is not usually apparent. The intelligence is built into the framework and the systems inside the conduits. It is out of sight behind walls, under floors, and between beams and ceilings. Some buildings are incorporating sleeves through which fiber optic cables would be threaded, against the day when the cost of fiber optics becomes more attractive. Most buildings are having to consider extra height between floors and spaces between ceilings to allow for the tremendous increase in wiring telecommunications rooms, additional power requirements, and other such features.

Architect Robert Reich has been quoted as saying that for the next few years 90 to 95% of all intelligent buildings will be old buildings retrofitted to the new technology. The critical issue, he says, is not new power but the redistribution and expansion of existing power facilities.

Apart from the leasing considerations, preplanning of power requirements has enormous financial implications. IBM is promoting a cabling system which, it says, will make buildings more flexible for computer users. IBM also points out that the labor required to prewire a building is one third the cost of adding cable after the building is finished.

The cost of adding provisions for built-in intelligence more than repays the investment, a theory shared by Joseph Newman, president of Tishman

Research Company. Other considerations are the cost of double bank elevators against single bank elevators and the size and capacity of these elevators, permanent and temporary, bearing in mind the need to take high tech equipment to floors occupied by incoming tenants. This has to be evaluated at an early stage.

THE HIGH TECH ERA AND THE FUTURE

The cost and return of going higher than 70 or 80 stories, in terms of elevators, ducts, life safety, services, and urban neighborhood considerations will probably rule superskyscrapers out for some time in the future, which means that to achieve maximum cost effectiveness of more traditional height, building planning will take on a more pivotal role.

High tech materials and those affecting their installation will have to be ordered well in advance because of their scarcity or their complexity. The supply has to be assured.

Another recent development is one hundred percent, preset metal deck cellular floors, which allow for maximum flexibility for wiring and the installation of power and communication systems. The preset outlets are integral to the metal deck and provide access to power sources on a convenient and predictable basis. They also accommodate rapid changes to keep tenants up-to-date.

Prewiring of office towers for high tech equipment will become as basic as elevators. Heights between floors will be greater to allow for additional services required for a high tech age. Adequate riser space to give maximum flexibility for tenants will also be provided. Space for high tech infrastructure will become a vital planning and construction factor.

Developers, owners, and managers will have to plan as far ahead as possible for power and communication requirements of tenants, and exert greater control over the consumption of this power. Unexpected and unusually high energy consumption can be financially burdensome, if not catastrophic.

The location, routing, and continuous upgrading of life safety, security, and environmental systems controls will have to be considered more closely. The installation and location of a high tech central system control console will have to be determined at the earliest stage to avoid expensive alterations.

There will be a growing use of prefabrication, perhaps not on a European scale, but it is coming to North America, because of ever increasing labor costs and the need to be competitive. In addition, we try to pre-cast as much as possible our elevator machine and mechanical room floors, to ensure minimum delay in the installation of equipment, and to get the systems on line at an earlier date.

Greater use should be made of helicopters, especially at roof levels (Fig. 13).

Engineers will have to find ways to deal with increased structural loads in buildings, both planned and under construction, to support high tech communications and security installations. For example, computer floors require a 450 N (100 lb.) live load, which means an expensive retrofit in an old building designed for a 225 N (50 lb.) live load. The areas allocated for vertical power and communications services will have to be increased at the design stage and power requirements should be reviewed accordingly.

More use will possibly be made of raised floors to allow for greater flexibility in changes to communications systems without affecting the tenant below. These changes we find run with our tenants at 30% a year. Because of the volume of cabling required for high tech requirements, the height of raised floors has been increased from 230 mm to 460 mm (9 in. to 18 in.). This means that the floor to floor height should be evaluated at the design stage.

Mixed-use developments will increase, with office space, shops, theaters, hotels, condominiums and restaurants, mixed in varying proportions. And with increasing frequency, which will make construction more interesting but more difficult, each has its own special requirements.

Fig. 13 Helicopter hoisting materials to top of First Canadian Place.

My message is that innovation in high-rise construction is a question of saving time and money and ensuring increased efficiency without losing quality. Part of the equation of time equals money is that the flow of materials and labor is a logistical supply issue that becomes critical for developers anxious to provide early tenant occupancy of their high-rise buildings. Finally, I suggest that all of us in the high-rise construction industry will have to plan and design for greater flexibility to keep up with the explosive rate of change in high technology as we head into the second century of the skyscraper.

REFERENCES/BIBLIOGRAPHY

Anderson, S. D. and Woodhead, R. W., 1980
PROJECT MANPOWER MANAGEMENT: MANAGEMENT PROCESSES IN CONSTRUCTION PRACTICE, John Wiley and Sons, Inc., New York, NY.

Barrie, D. S., ed., 1981
DIRECTIONS IN MANAGING CONSTRUCTION: A CRITICAL LOOK AT PRESENT AND FUTURE INDUSTRIAL PRACTICES, PROBLEMS AND POLICIES, John Wiley and Sons, Inc., New York, NY.

Frein, J. P., ed., 1980
HANDBOOK OF CONSTRUCTION MANAGEMENT AND ORGANIZATION, second edition, Van Nostrand Reinhold Company, New York, NY.

Godfrey, R. S., 1982
BUILDING SYSTEMS COST GUIDE, second edition, Robert S. Means Company, Inc., Kingston, MA.

Goldberger, P., 1985
THE PROSPECT OF BIGGER TOWERS CAST A SHADOW, New York Times, December 29.

Good, K. O. and Carolina, J., 1980
CONSTRUCTION FOR PROFIT, Reston Publishing Company, Reston, Virginia.

Halpin, D. W. and Woodhead, R. W., 1980
CONSTRUCTION MANAGEMENT, John Wiley and Sons, Inc., New York, NY.

McGraw-Hill Information Systems Company, 1986
DODGE BUILDING COST CALCULATOR AND VALUATION GUIDE, Quarterly, McGraw-Hill Information Systems Company, New York, NY.

Newell, J. E., 1979
BUILDER'S GUIDE TO CONSTRUCTION FINANCING, Craftsman Book Company, Solana Beach, CA.

R. S. Means Company, Inc., 1986
BUILDING CONSTRUCTION COST DATA (Annual), R. S. Means Company, Inc., Kingston, MA.

Royer, K., 1981
THE CONSTRUCTION MANAGER IN THE 80s, Prentice-Hall, Inc., Englewood Cliffs, NJ.

Tucker, J. B., 1985
SKYSCRAPERS: AIMING FOR 200 STORIES, High Technology, January.

Times Square Office Center The Story of A High-Rise Complex

L. V. Shute
H. R. Corry

In December, 1983 Turner Construction Company was retained by Park Tower Realty to prepare an in-depth preconstruction study of the proposed four building, 372 thousand gross m^2 (4 million gross ft^2), high-rise office development complex at Times Square in New York City (Fig. 1). As of this writing our studies continue, as the drawings and specifications continue to be finalized and tenants brought on board, while the true critical path remains acquisition of the property through condemnation proceedings.

The architect is John Burgee with Philip Johnson, the structural engineer is the office of Irwin G. Cantor, the mechanical/electrical engineer is Cosentini Associates, and Jenkin & Huntington are the elevator consultants.

The purpose of this paper is twofold: first, to go back to basics to list all of the variables, and second, to apply these variables as they have affected the final or semifinal selection of the design at Times Square. There are few, if any, more complicated case studies encompassing all of the many considerations going into the design and construction of a high-rise complex in a congested metropolitan area.

There is really nothing new here, except perhaps the evolution of the "smart buildings" concept. However, there have been and will continue to be fine tuning of such things as the exterior granite support system and flashing details, the solution to the window washing problem, the building management system, and all of the various means of communication available in today's market, from the basic telephone to rooftop dishes.

We started with certain givens:

1. The four footprints;
2. Number of stories;
3. Permissible gross area;

Fig. 1 42nd Street Development, New York, New York (John Burgee Architects with Philip Johnson)

4. Preferred facade treatment;
5. Existing site conditions;
6. Schedule constraints.

The Times Square Redevelopment Corporation had earlier taken development proposals for several discrete elements of the total development along 42nd Street and in their planning had effectively transferred much of the "air rights" of the full land mass to the eastern edge of the site to allow for a lower profile of buildings—both old and new—on the north and south sides of 42nd Street between 7th and 8th Avenues. Park Tower Realtors were selected in competition to be the developers of the Times Square site.

This fixing of the geometry set the stage and locked the project into a massing and logistics scenario that allowed for no elbow room. The single most important condition affecting planning of this project was the existing subway system surrounding these four buildings and in fact "trespassing" on one of the blocks, which was to be expanded and upgraded during construction of the towers. In the order of construction:

1. Foundations
   Rock is 7.6 to 15.0 m (25 to 50 ft) below street. Existing subway bottom elevations vary 6 m to 18 m (20 ft to 58 ft) below street.
   a. How many basements?
   b. How handle off-street deliveries in the finished buildings?
   c. Parking facilities?
   d. If cogeneration, where put the plant?
   e. If ice or chilled water storage, where?
   f. If central refrigeration, where?

2. Structural frame
   a. Preferred column spacing?
   b. Story heights
   c. Transfers, if any?
   d. Live load?
   e. How take wind load? Exterior? Interior?
   f. Structural steel versus reinforced concrete
   g. Decking—formed, metal, or concrete plank?
   h. Electrification, if any
   i. Fireproofing?

3. Exterior Facade and Roof
   a. Alternate designs
   b. How support stone?
   c. Double glazing?
   d. Fixed versus operable windows
      Window washing
   e. Weather proofing

4. Interior space layout
    a. Public lobbies
    b. Core
    c. Tenant
    d. Flexibility for special tenancies
    e. Selection of Materials

5. Vertical Transportation
    a. Conventional versus sky lobby, versus double deck
    b. Banking/intervals
    c. Car capacity
    d. Service cars
    e. Controls
    f. Moving stairs

6. Mechanical Systems
    a. Cogeneration
    b. Ice or chilled water storage
    c. Central plant 1 versus 4, 4 versus local
    d. Fan Rooms-central versus individual floor, central cooling versus self contained units
    e. Exterior heating and cooling
    f. Controls
    g. Sprinklers
    h. Plumbing

7. Electrical Systems
    a. Distribution voltage
    b. Equipment location
    c. Underfloor electrification
    d. Lighting
    e. Special

Although the above list may be somewhat elementary, it is well on occasion to step back and review the big picture starting at square one.

Traffic congestion and the one-way traffic pattern severely limit the choices in planning site logistics and sequencing of construction operations (Fig. 2). In addition, existing subway construction surrounding the site and planned subway renovations within the site create a jigsaw puzzle adding further complications. As can be seen on the plan of Building A (Fig. 3) planned locations for hoist facilities of both workers and materials are virtually dictated by site conditions.

The area of site logistics is by far the most critical feature of this project.

Purchasing Considerations
Market studies
Foreign versus domestic

Timing and sequence
Keeping options open
Controlling escalation

Construction Scheduling
Big picture—sequence of starts
Horizontal schedule
Vertical schedule
Manpower requirements
Cash flow/manpower flow

In summary, the future of high-rise construction will likely lead to situations where project sites are even more dense and more complicated than

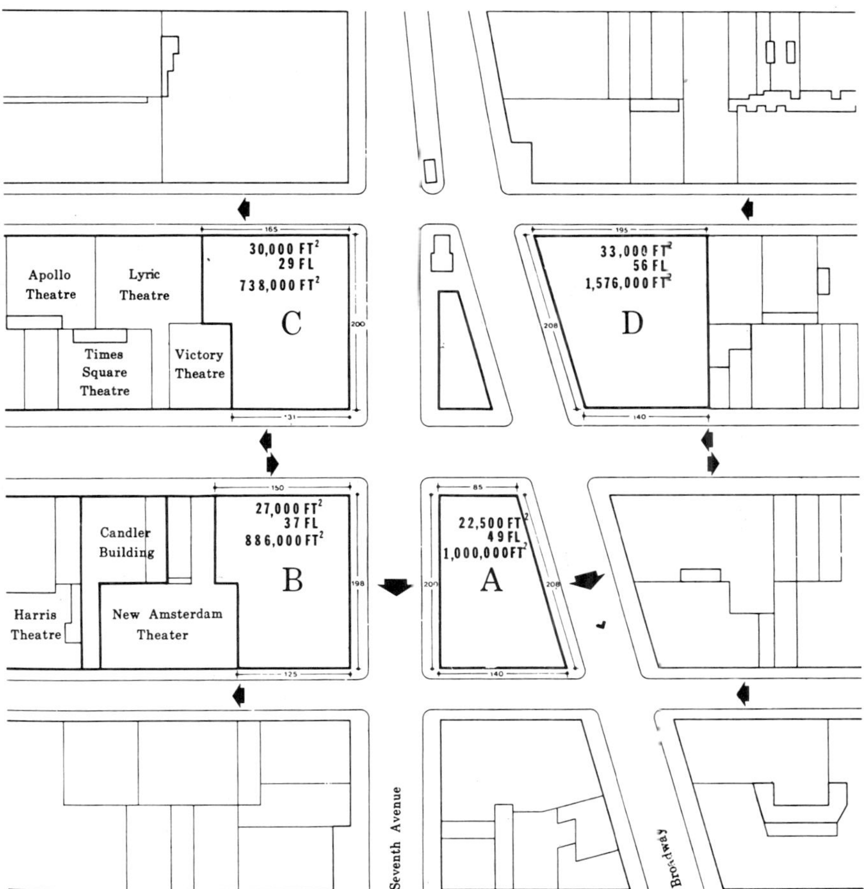

Fig. 2 Site traffic pattern—the Office Center at Times Square

they are today. As land values increase and technology permits us to design even higher structures with even more sophisticated communications systems, we are likely to see more "Times Squares." Ingenuity and flexibility in approach to both design and construction planning will be in greater demand than ever.

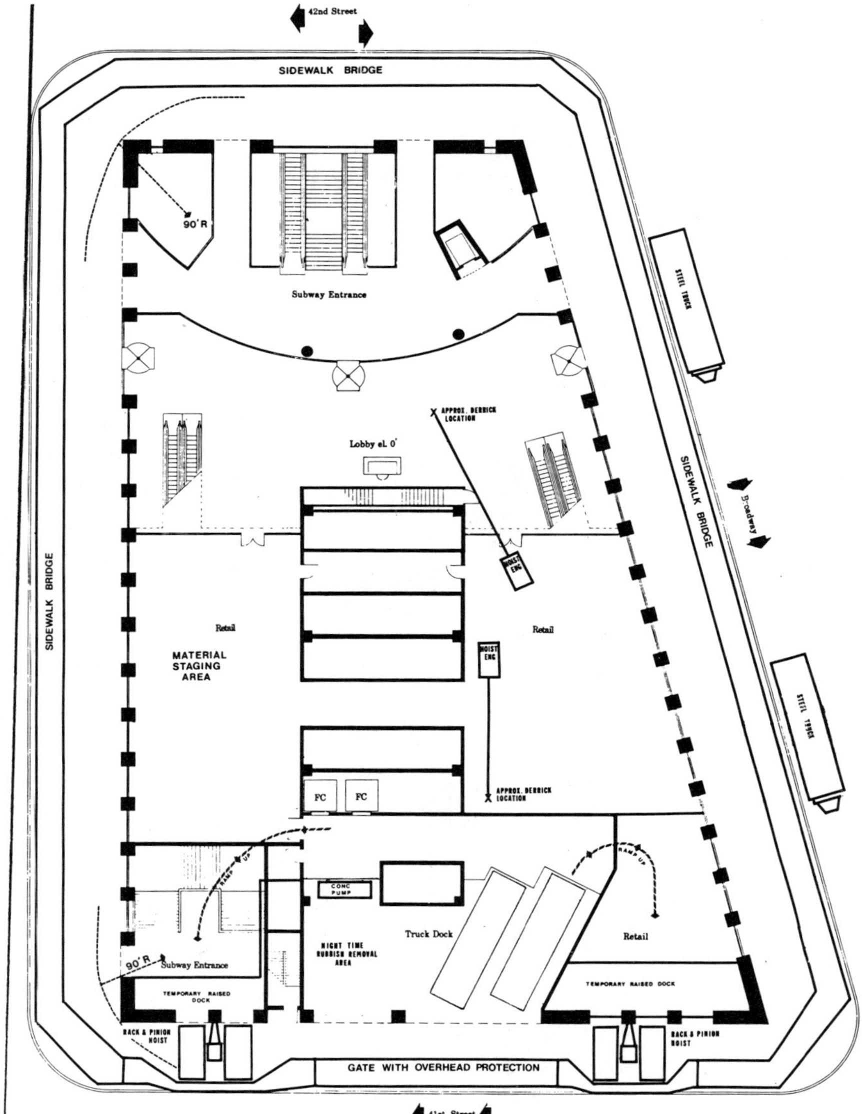

Fig. 3 Office tower A plant layout—the Office Center at Times Square

Recent Improvements in Foundations

Edmund M. Burke

There have been many changes in design, techniques, and methods of constructing building foundations in recent years, so much is available to discuss. This paper attempts to highlight significant recent advances and improved construction methods, and to provide field observations on foundation construction.

The whole scene, of course, is made up of the many individual bits and pieces of improvements that have brought us to our present level of foundation capabilities. Some of the more significant aspects will be discussed briefly.

One important development in foundation construction has been the increased use of the slurry or diaphragm wall technique. After a period of development in Europe, the first major foundation use in the United States was for the World Trade Center in New York in 1967 and 1968 where it formed the permanent basement wall. After the tiebacks and excavation were completed, reinforced concrete floor slabs were constructed into keyways formed into the perimeter slurry walls. This technique has had continued use. Figure 1 depicts typical slurry wall construction for a new building. The economic justification here involved the avoidance of underpinning the adjacent building where the soil strength permits use of this procedure. Notable, too, is the drainage chase formed by the interior block wall construction. Any leakage through the slurry wall is directed down into the drainage course through

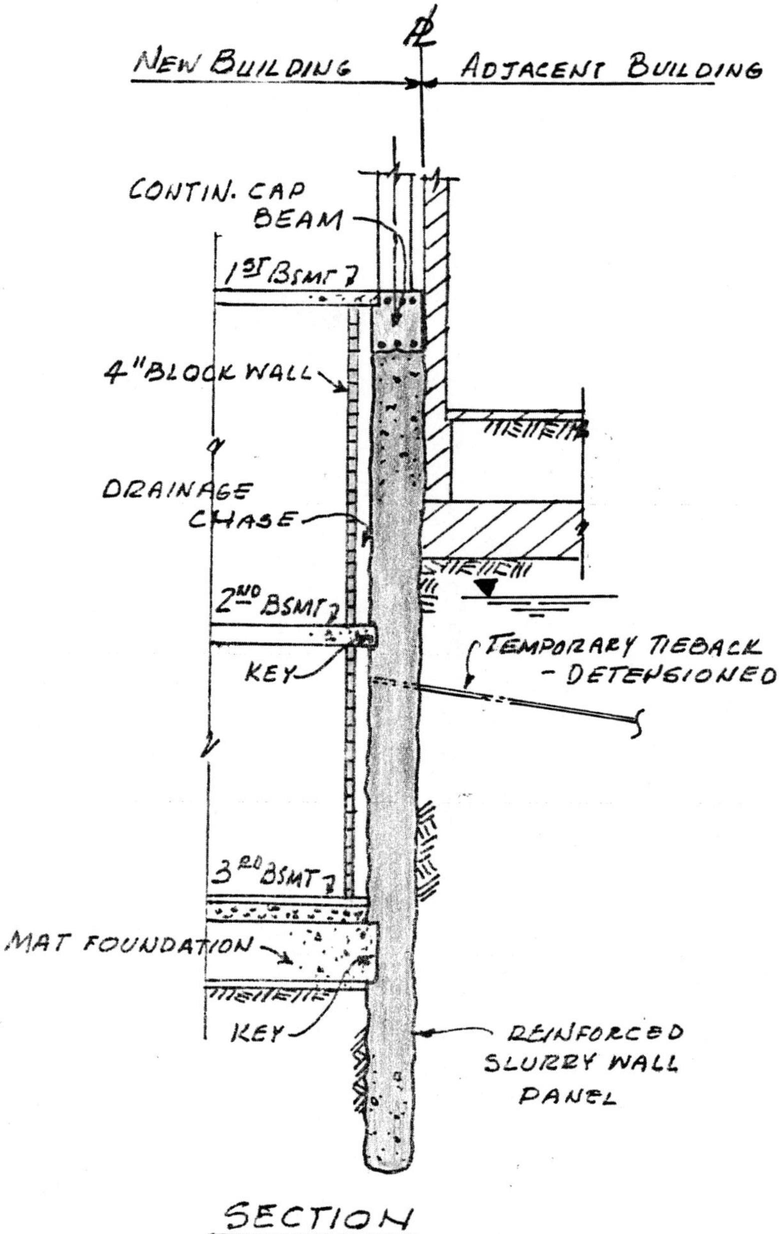

Fig. 1 Typical slurry wall construction for a new building

slots in the floor slabs. There also have been recent use of the walls, constructed in segments in a slurry trench, as the basement walls and to support perimeter building columns. Examples of this approach are the perimeter basement walls for a hotel and for two of the buildings that form the World Financial Center in New York City. Figure 2 shows an isometric view of such use. The design for the permanent condition includes final lateral loading conditions in the rebar design, in addition to the temporary condition, as well as provision for bearing capacity. To conserve bearing capacity, the tiebacks are detensioned when the adjacent floor slabs are capable of carrying the earth pressures. As shown in Fig. 2, perimeter columns can be located directly over the slurry wall. Therefore, a continuous heavily reinforced concrete cap is cast over the completed slurry wall to distribute the concentrated column loads and to tie the wall panels together at the top.

Naturally, the high construction costs involved in the use of slurry walls must be justified to those who pay the bills. In some cases, it can offset the cost of an alternative expensive cofferdam through rubble and other obstructions in soft river sites, while in other cases it can reduce or eliminate the costs for underpinning adjacent higher buildings, such as was shown in Fig. 1.

Permanent building tiedowns are another significant and recent tool for designers. The tiedowns resist hydrostatic uplift pressures on a building with multiple basements and are economically justified as the alternative to a thick structural mat or a beam system to span to widely-spaced columns. Figure 3 shows the foundation conditions representative of two recent hotel construction projects in Washington, D.C. The mat foundation handles the load bearing requirements of the building, but it is only capable of resisting the 79 kPa (1,660 psf) uplift pressure due to ground water, at the typical column bays. As shown in Fig. 4, at the 27m (90 ft) wide ballroom the tiedown scheme was used and permitted slimming the mat locally to 0.75 m (2.5 ft) thickness. The tiedowns are designed similar to tiebacks for temporary cofferdam support or to augercast piles that carry tensile loads. The submerged weight of soil engaged by the tiedowns exceeds the design uplift with a safety factor.

The space program, the computer explosion and other technological advances have spawned a series of monitoring devices that are becoming important tools for foundation engineers. These instruments are extremely sensitive and usually dependable. The list includes:

1. The dynamic pile driving analyzer (PDA), which is attached to a pile and measures strain wave behavior imposed on the pile during driving. The tensile and compressive stresses produced in the pile during driving are imputed parameters that are monitored to protect a concrete pile from the effects of overdriving or from cracking under easy driving. The energy level developed in the pile butt is also useful to evaluate hammer efficiency. Bearing capacity is estimated by assumed pile and pile/soil interaction

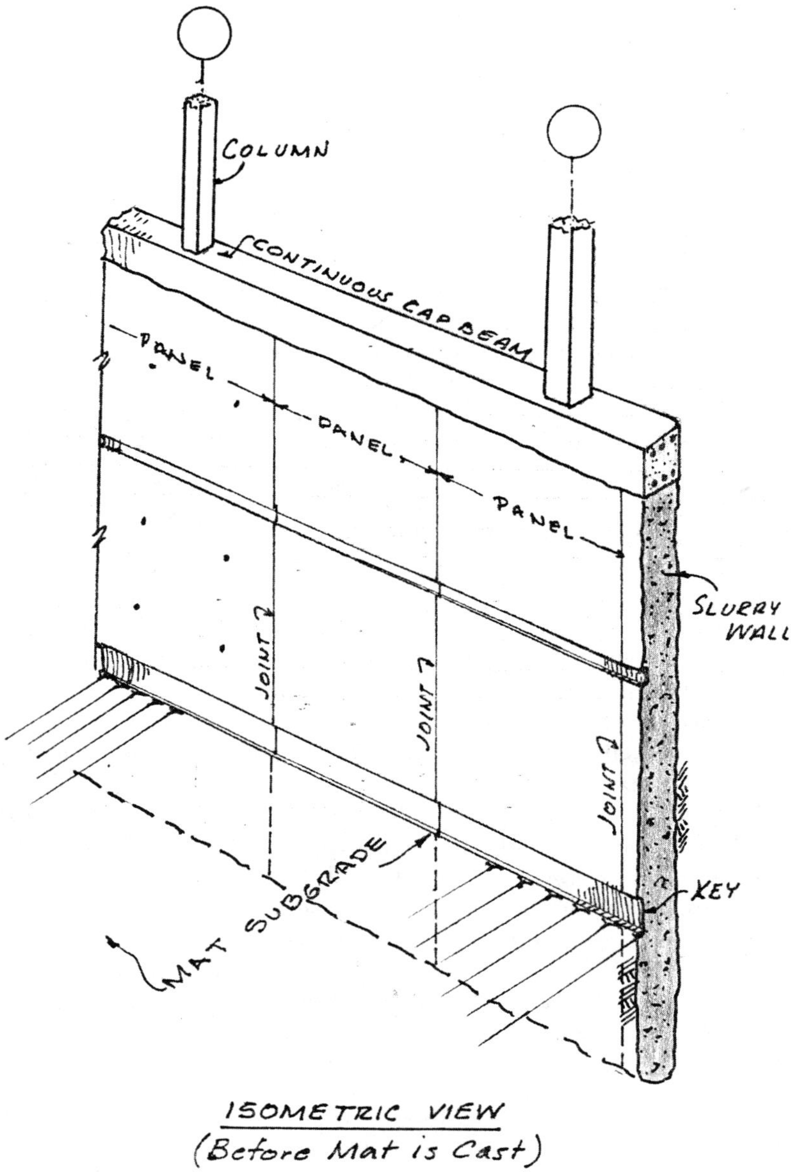

Fig. 2 Isometric view of use of walls constructed in segments in a slurry trench

properties. These values can be reanalyzed by computer later using additional measured parameters for a more accurate bearing capacity determination. The PDA is quite important for inspecting prestressed concrete piles, especially those with high bearing capacities. However, for higher capacity piles, the predicted bearing capacity by PDA does not always mesh with static load test results. Another drawback is the tremendous volume of data generated since routinely five parameters are measured directly for each hammer blow. This is contrasted with the single group of measurements generated by a static load test. The PDA is a useful tool, however.

2. A family of devices, basically sensitive extensometers, has emerged. The linear variable differential transducer (LVDT) is a device that measures the movement of the extensometer plunger by electrical field changes. It can be hooked onto a deep benchmark to read building settlements or to the end of a fixed invar tape to measure horizontal movements. Since the LVDT has an electrical output it can be connected readily to a small computer to monitor a series of such devices in a single control room, which was done successfully for the recent complicated underpinning of two nuclear plant buildings. Tape extensometers are useful to monitor diametric divergence/convergence in tunnels, lateral movement of piers during underpinning, and other applications. Slope inclinometers are appearing more frequently

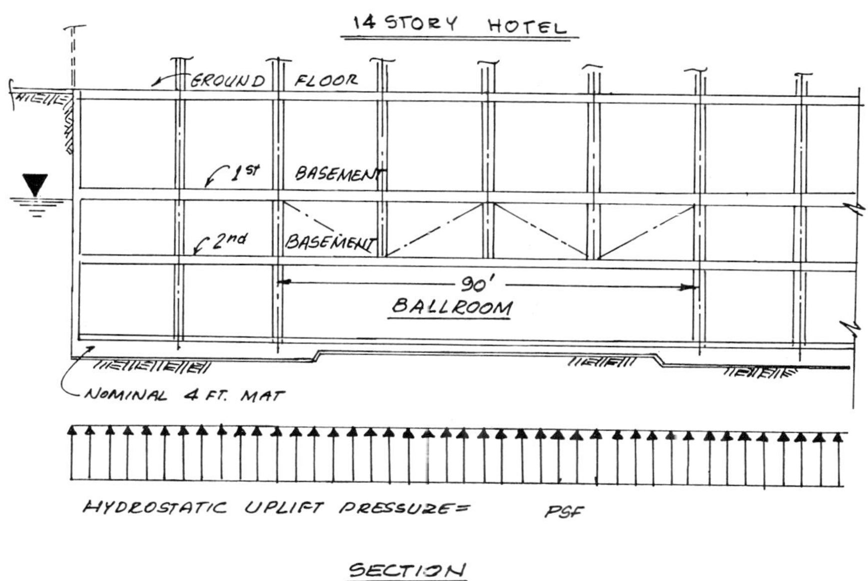

Fig. 3 Foundation conditions of recent hotel construction project

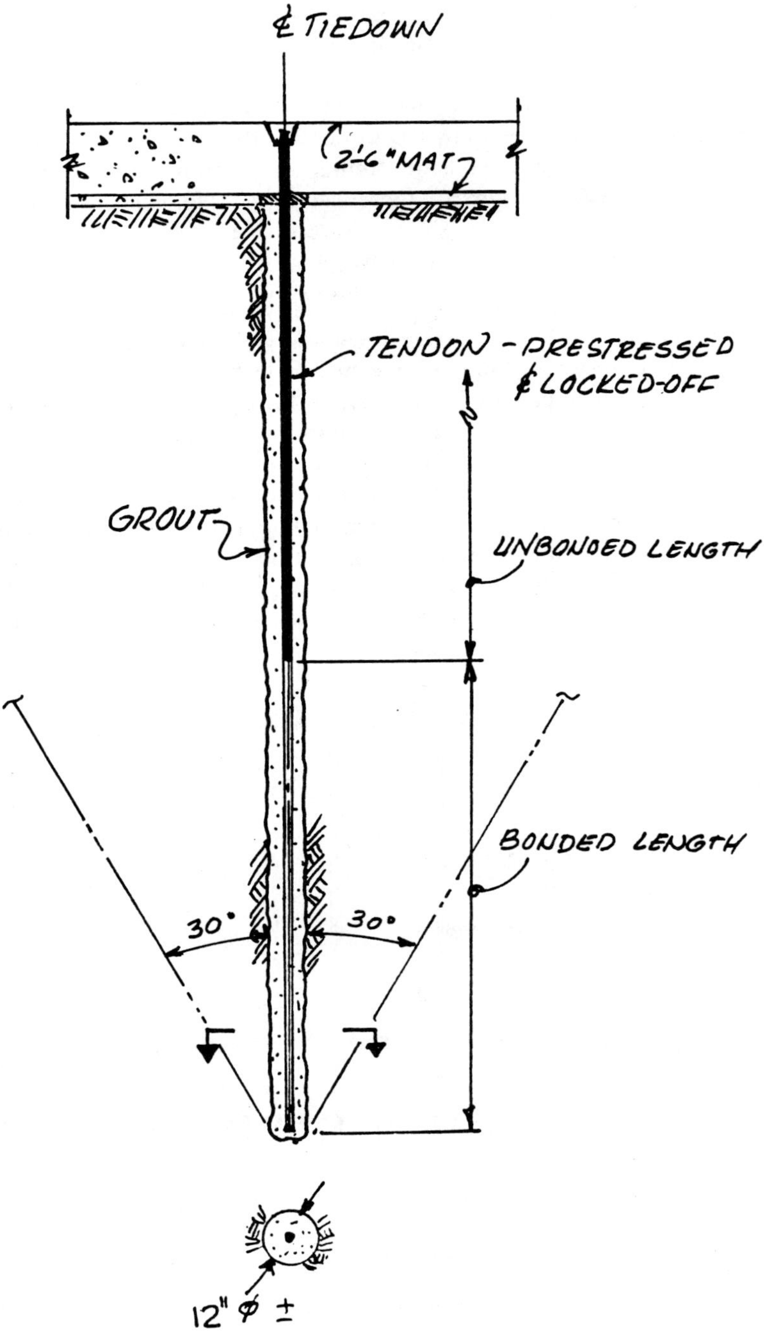

Fig. 4 Tiedown scheme

at the deeper foundation cofferdams to detect and measure soil movements and patterns in the retained soil behind a cofferdam.

A recent and significant advance in rock coring techniques is the nonmagnetic oriented core. Warren George, Inc. has patented a method whereby a modified drill rig holds the core barrel in a fixed position while markings are scribed into the rock core as it moves up into the core barrel. One scribe line is made on one side and three on the opposite side. The orientation of the scribes is recorded in the field by the inspector or driller with respect to north or some other base. The original in-situ orientation of the rock core is thus known. The strike and dip of the bedding of the rock and any jointing encountered can be described. This ingenious and simple device is useful in designing measures to protect adjacent structures against the effects of rock slippage during rock excavations due to both vertical and horizontal surcharge loadings, as well as in analyzing the effects of new buildings on adjacent deeper structures, such as tunnels.

Basement waterproofing is another area where recent developments have given the foundation engineer new tools. As shown in Fig. 5, a drilled and

Fig. 5 Drilled and blasted rock face

blasted rock face is quite rough. When a foundation is cast against such a face, frequent shrinkage cracks occur due to the keying of the green concrete into the rigid rock face. The in-plane drainage system shown in Fig. 6 is very useful in rock cuts for new buildings. The panels of drainage material placed against the rock face act as a bond-breaker and materially lessen shrinkage

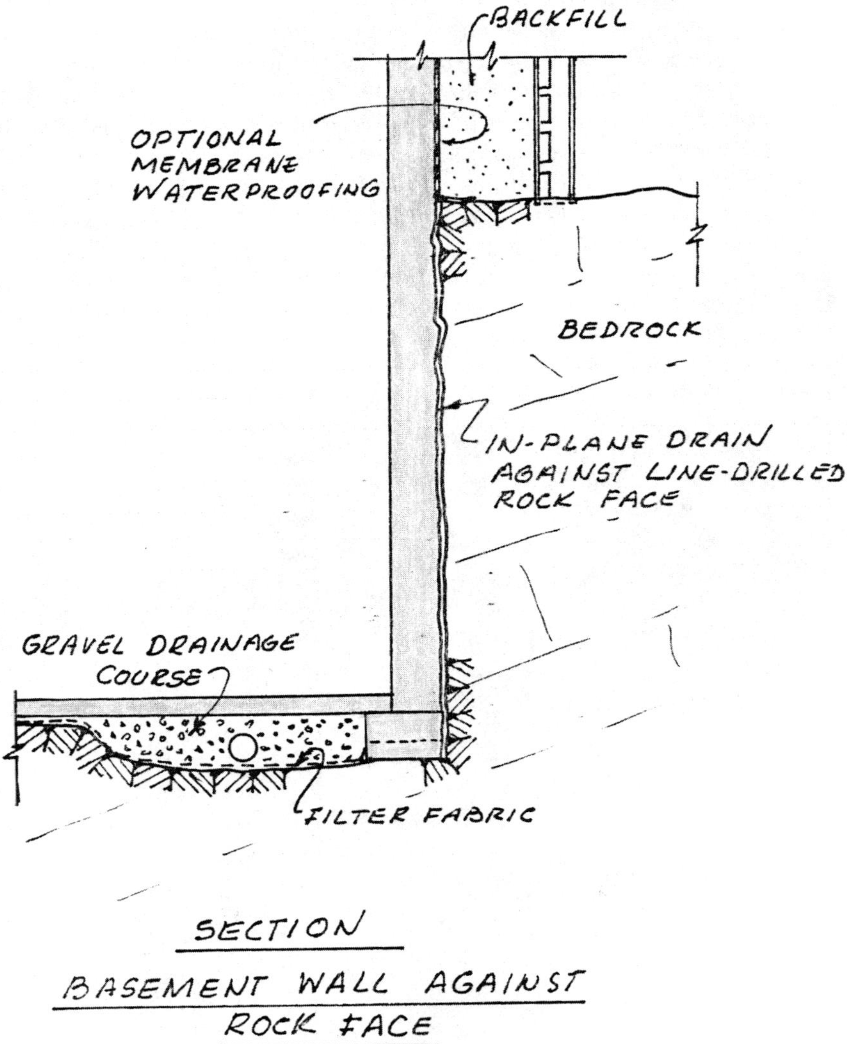

Fig. 6 In-plane drainage system

cracking in the wall concrete. The material also collects seepage from the rock face and carries it to the bottom of the wall. Figure 7 is a detail at the bottom of wall and shows how the drainage core is formed when the wall concrete is cast. The filter fabric against the rock keeps the core free of soil sizes that would tend to clog the bottom, where the drain pipes through the footing are located. Figure 8 is a photograph of "Miradrain" being applied against rock at a job in New York. The filter fabric is white in the photograph.

Ingenious engineers and constructors will continue to initiate techniques and designs that will improve quality and, of course, lower construction costs. The most easily foreseeable advances will be in monitoring, measuring, and surveying—those areas where heavily-endowed research efforts for the space program and the offshore petroleum industry will filter down, after modification and simplification, to devices that will be useful at the foundation sites of the world. These monitoring devices are important because they offer the means to use what Ralph Peck has named the observational method,

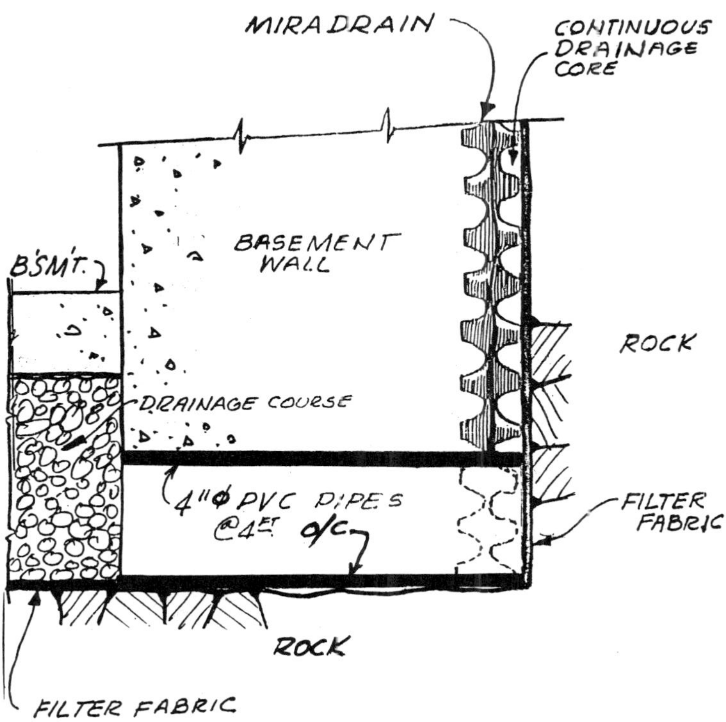

Fig. 7 Detail at bottom of wall, showing how drainage core is formed when the wall concrete is cast

in other words, learning during actual construction on how and where improvements can be made. Equally important is the use of sensitive and reliable monitoring devices as a tool to permit more daring and innovative designs while retaining control of the situation in the field. However, the author cautions that daring engineers should plan in advance the contingency measures that they might need on short notice. This precautionary measure might even extend to having the materials and equipment on-site so that there is no delay in putting the contingency plan into operation.

Monitoring devices have permitted higher and higher levels of load bearing in piles, for example, by establishing how and where the soil or rock is supporting the pile both during pile load testing and PDA monitoring.

Fig. 8 "Miradrain" applied against rock

High-Rise Public Housing Development in Singapore

Khee Poh Lam

HISTORICAL BACKGROUND

Singapore's public housing dates back to the formation of the Singapore Improvement Trust (SIT) in 1927 under the British Colonial Government. However, as far as housing development was concerned, there was no significant progress, as "SIT was much less a housing authority than a municipal body, its work was confined largely to constructing and widening roads and creation of open spaces" (Housing and Development Board, 1970).

In the postwar years of the 1950s the living conditions in Singapore were described as follows: "Acute overcrowding in dilapidated slums, appalling conditions of squatter settlement; bedding on wooden banks or in rented cubicles, high rate of urbanization, grossly inadequate housing delivery system, rapid deterioration in available space standards. . . ."(Goh, 1956).

In 1959, Singapore was granted self-government. Realizing the urgency of the situation, the Government dissolved SIT and formally established the Housing and Development Board (HDB) on February 1960 as a statutory board under the Ministry of National Development. The board's primary function then was to tackle the problem of the acute shortage of accommodations and to formulate and launch an urgently needed comprehensive housing policy providing low cost decent housing for the public.

Since then, five successive Five-Year Housing Programs have been implemented, each out-performing its predecessor in the number of flats completed. At the end of the last Program in 1985, over half a million housing units had been completed. (Fig. 1).

HOUSING AND DEVELOPMENT BOARD

The success of public housing development in Singapore is primarily attributed to the fact that the HDB has not limited itself merely to providing

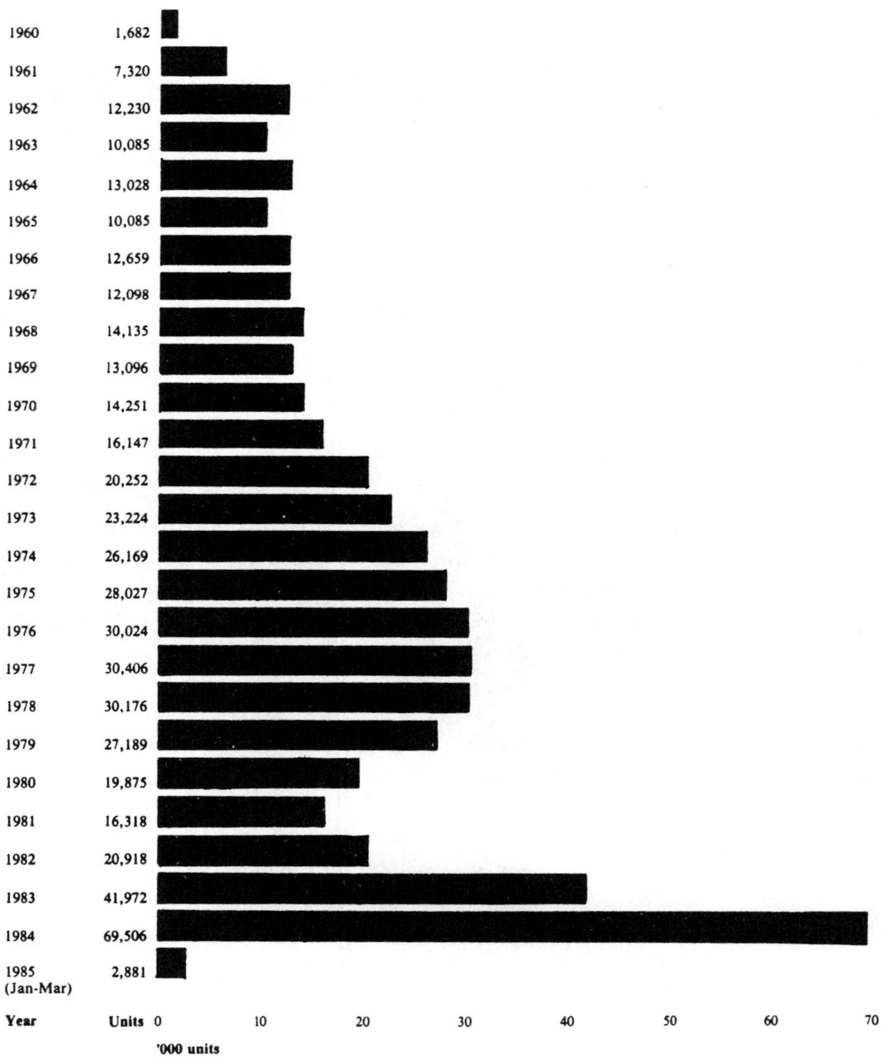

Fig. 1 HDB Building Statistics *(Housing and Development Board. 1984/85).*

acceptable living accommodations but has responded to the challenge of building a total environment suitable for community development and improvement in the quality of life. An extensive and well-coordinated organization is required to administer the entire development.

The HDB is now the sole national public housing authority in Singapore and its current scope of responsibility extends to the planning of new towns and town centers, the design of offices, shops, industrial buildings, and sports facilities, as well as providing parks, playgrounds, and community facilities. Land reclamation and resettlement of squatters are also undertaken by the board. Two other significant secondary functions include the active promotion of the advancement of construction technology and playing a supportive role in community development in the public housing estates. All these operations are implemented through three divisions and two departments headed by a Chief Executive Officer (Fig. 2).

The Building and Development Division is primarily responsible for the physical planning and implementation of all public housing and its related facilities. Six departments work in tandem with each other in undertaking the formidable task (Fig. 3).

The main issues affecting public housing development in Singapore involve land use, finance, and the performance of the construction industry.

LAND USE

In a small island nation such as Singapore with a total area of only 614 km^2 (246 mi^2), land is always a premium in any development. The initial decision for high-rise high-density development met with criticism, some of which was justifiable. Notwithstanding the ongoing debate, Singaporeans in general have accepted the high-rise high-density living as a way of life. Perhaps the assets that make this life style acceptable are that "residents are by and large adaptable to change. . . . The co-operation and discipline of our people also help keep human conflict and vandalism to a very low level. . . . The tropical weather in fact makes high-rise living desirable and the open-to-sky courtyard usable"(Liu, 1979).

The necessity to acquire land at affordable cost was accepted as an absolute prerequisite for effective long-term development. Hence, tough legislation was introduced initially to acquire land and properties for public purposes. In the process, huge areas of slums and squatters were cleared causing considerable hardship in the early days. However, the benefits were soon realized through the continued success of delivering large numbers of housing units to the public at affordable prices.

The HDB must review its long-term land use plan for public housing periodically to ensure effective use of the limited supply of land. To overcome the constraint and to meet the targetted number of dwellings, buildings are spaced closer together but generally kept to heights below 13 stories to provide a more intimate environment. The net residential density in public housing is around 195 units per hectare with plot ratio ranging from 1.7 to 2.3.

FINANCE AND ECONOMICS

The HDB development programs are entirely financed from government loans. Loans for properties developed for sale are repayable over 10 years

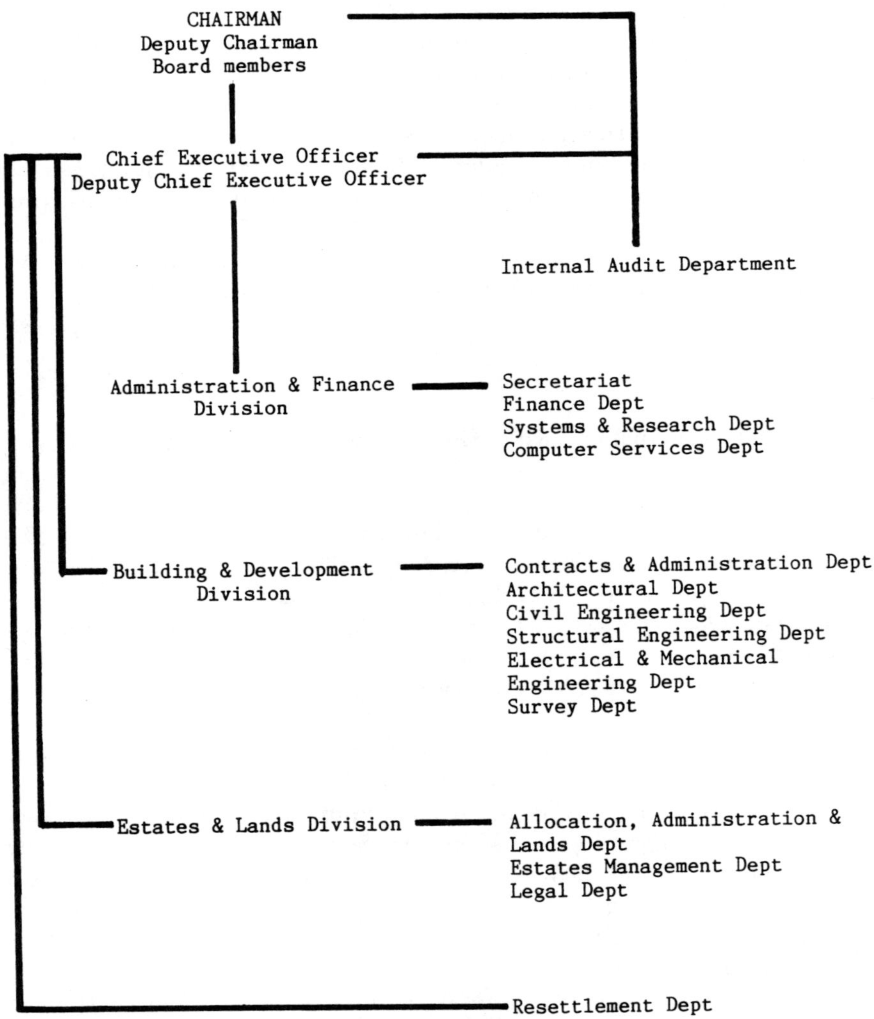

Fig. 2 HDB Organizational Structure

with 6% interest p.a. and loans for properties for rent are repayable over 60 years at 7¾% p.a. The financial highlights for the last Five-year Housing Program are given in Table 1. The resulting deficit of revenue income less revenue expenditure is fully subsidized by the Government. This figure has been kept relatively modest, varying between 0.5% to 4% of the budgeted

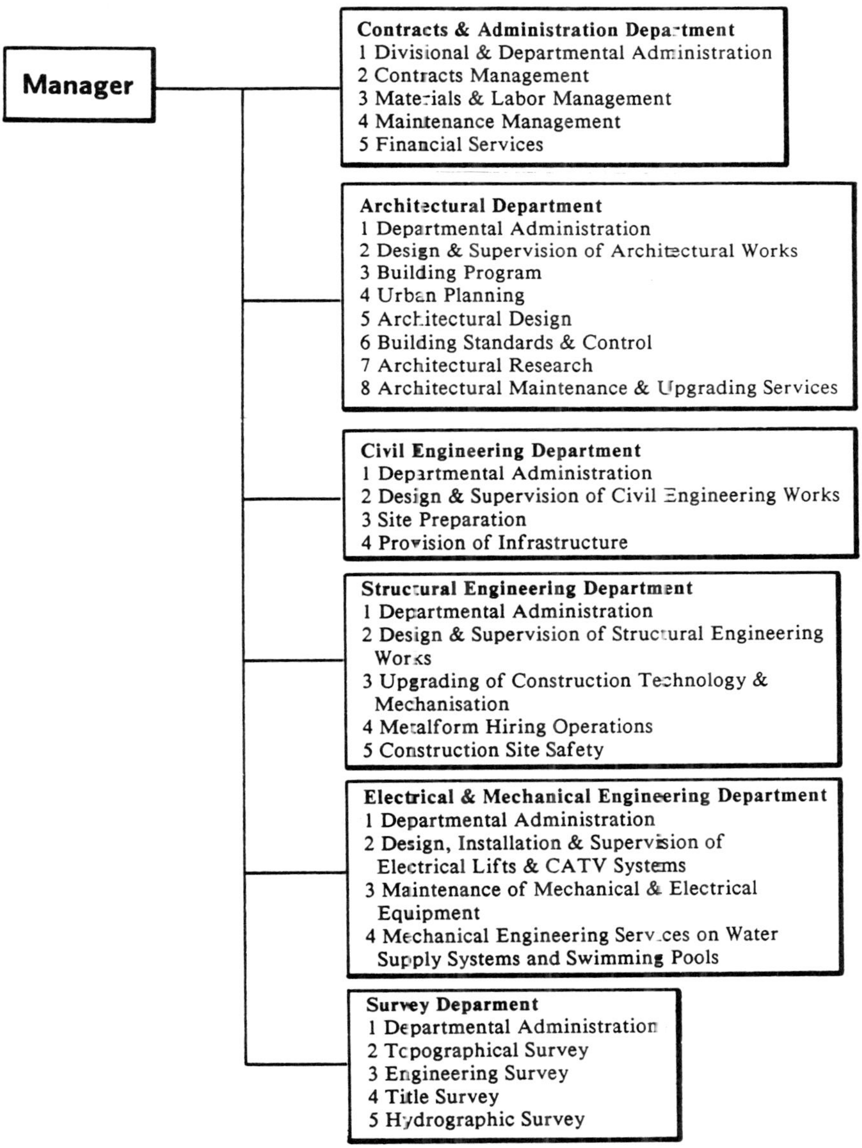

Fig. 3 Functions of the Building and Development Division

capital expenditure over the last five years. To date, the cumulative amount of subsidy received from the government has amounted to S$684 million.

Despite the current downturn in the construction industry, the development loans to HDB approximate 7% to 8% of the gross domestic product and more than 40% of government estimates. This is possible as Singapore has a very high savings rate especially through the Central Provident Fund (CPF), and the government has healthy reserves. Furthermore, the required percentage of contribution to the CPF has increased substantially over the years to a current 35% of the monthly salary, with 25% from employees and 10% from employers up to a per month maximum.

THE CONSTRUCTION INDUSTRY

The construction industry is ultimately responsible for the physical realization of the Housing Program and therefore it has a major influence on its development.

Between 1979 and 1982, Singapore had an unprecedented boom in the construction industry. There was an acute shortage of building workers and the response from contractors to the HDB's tenders was poor. Construction costs escalated substantially (48% in the case of HDB flats), which forced the Board to increase the selling price of dwelling units by an average of 38%. Furthermore, the HDB had to increase the target number of dwelling units in 1981 because of a sudden surge of applicants on the waiting list. To meet this accelerated program and to ameliorate the poor response from contractors,

Table 1 Financial Highlights—Five-Year Summary (Housing and Development Board, 1984/85)

	1984/85	1983/84	1982/83 (S$ Million)	1981/82	1980/81
Budgeted capital expenditure	3,855	3,036	2,254	1,505	1,133
Percentage of government's development estimates	43%	39%	33%	30%	31%
Loan drawings from government	2,980	3,514	2,253	1,464	1,071
Capital expenditure	3,530	4,045	2,651	1,581	1,061
Revenue expenditure	1,144	954	759	548	507
Revenue income	1,112	833	667	542	467
Gross deficit for the year	32	121	92	6	40
Sale of flats: value of flats sold during the year	4,719	1,363	797	537	581
Accumulated value of flats sold since 1964	11,068	6,349	4,986	4,189	3,651
Mortgage loans granted during the year	2,781	906	539	371	338
Mortgage loans outstanding to lessees	5,460	2,811	2,135	1,647	1,458

the Board initiated the awarding of contracts through negotiations in June 1981. At the same time, incentives were offered to contractors to encourage them to take on more projects and to complete contracts on schedule.

Two particular innovations were made: (1) An interest-free financing scheme was implemented in June 1981 to help local contractors acquire approved labor-saving and productive plant equipment. A loan of S$1 million was granted for every S$12 million worth of contract secured. (2) A core-contractors scheme was introduced in January 1982 to select a group of established HDB building contractors and to commit them to undertake the Board's building projects. These contractors would be assured of continuity of work at attractive negotiated prices and a minimum workload of 1,000 units per contractor per year.

Materials Management

The Board's ultimate objective is not only to provide an adequate and continuous supply of building materials to meet its own need but also to achieve self-sufficiency in certain essential items for the country. To meet these goals, the Board has embarked on a long-term program of resource planning. To centralize the production and supply of essential building materials under one body, the Resources Development Corporation (Pte) Ltd, a government-owned company, was appointed as managing agent for the brickworks, granite, and sand quarries as of October, 1981.

To ensure continuous availability of building materials, half-yearly projections of requirements, bulk purchase, stockpiling, and constant communication with manufacturers/suppliers are being carried out

Labor Management

The construction workforce in Singapore is comprised of a large proportion of foreign workers. As of March 1985, the HDB (1985) employed a total of 32,415 workers on their sites. Of these, only 26.3% are local workers.

The long-term objective of the Board is to build up a permanent pool of local construction workers. In August 1981, the government launched a National Service Scheme known as the Construction Brigade to build up civil defense capabilities and to provide vocational skills. The intention was to encourage enlistees to remain in the construction industry after completion of their national service.

In 1984, the Government established the Construction Industry Development Board to upgrade the industry. Soon after, the CIDB set up a Construction Industry Training Center to train local workers to meet the industry's needs.

Mechanization Program

Another major development in counteracting the labor shortage problem (especially skilled labor) was the implementation of a comprehensive mechanization program. The aim was to increase worker productivity and efficiency of construction operations, resulting in reduced labor and time and improved work quality.

The use of metal formwork for conventional construction, appropriate

Table 2 Summary of Prefab Contracts *(CIB 84 Singapore, 1984)*

Contract	Contractor	Number of Apartments	System of Construction	Contract Sum	Contract Period
Prefab Contract 1	M/s White Industries Ltd (Australia)	15,000	Precast, frame and slab	$600 M	Jul. 1981 to Jul. 1988
Prefab Contract 2	M/s GTM Coignet Joint Venture (France)	15,000	Large panel prefabrication	$600 M	Nov. 1981 to Nov. 1987
Prefab Contract 3	M/s SGE-Construction (France)	7,000	Semi-precast (insitu wall and slab with precast facades, toilets, balcony slab, and staircases	$220 M	Jun. 1982 to Dec. 1985
Prefab Contract 4	M/s Marubeni-Shimizu Consortium (Japan)	15,000	Large panel, prefabrication	$550 M	Dec. 1982 to Dec. 1987
SUMMARY OF SYSTEM CONTRACTS					
System Contract 1	M/s Lee Kim Tah (Pte) Ltd (Singapore)	5,000	Progressive strength (in situ beam and slab with precast lightweight concrete partition and facade walls	$100 M	June 1982 to Oct. 1985
System Contract 2	M/s Daewoo Corporation (Korea)	8,000	Semi-precast (in situ wall and slab with precast refuse chute, staircase, and parapet)	$310 M	May 1983 to Dec. 1984

mechanized systems, and industrialized methods of construction are included in the program. Table 2 summarizes the industrialized building contracts. The feedback so far indicates an increase of 30% average labor productivity over conventional construction techniques. Furthermore, there is ongoing technology transfer from the foreign contractors to the local industry that has long-term benefits.

HOUSING TO MEET THE BASIC NEEDS

Meeting the basic need (shelters for the needy) may be measured in terms of eligibility, affordability, waiting time, and floor space standards (Liu, 1984).

Eligibility

Only Singaporeans are eligible for public housing. Table 3 shows the distribution of the population within specific gross monthly household income brackets and Table 4 shows their corresponding eligibility for HDB flat types.

Flats are allocated on a first-come-first-serve basis. Available flats are

Table 3 Household income groups (estimated)

Monthly income	Percentage of population	
$500 and below*	28%	Eligible for HDB's Low Cost Housing
$501–$1,500*	54%	
$1,501–$4,000	14%	
$4,001–$6,000	2%	Eligible for HDB's Middle Income Housing
$6,000 and above	2%	
	100%	

*Source: Singapore 1985 (Information Division, Ministry of Communications and Information)

Table 4 Income eligibility ceiling for home ownership scheme

Monthly household income of applicants	Type of flat
Up to $1,500	Any type of flat (3-, 4-, 5-rooms or executive flat)
$1,501–$4,000	4-rooms, 5-room or executive flat
$4,001–$6,000	Middle income housing

balloted to ensure there is no favoritism or queue-jumping. However, priority is given to families affected by resettlement schemes as well as applicants in uniformed services such as the Singapore Armed Forces, the Police Force, and the Fire Brigade.

Affordability

Since the launching of the "Home Ownership for the People" scheme in February 1964, the number of applications to purchase flats continues to increase. Initially the response was poor since it was still beyond the financial resources of most HDB tenants. However, since 1968, CPF contributions can be used for the initial downpayment and subsequent installment payments for the purchase of flats. The compulsory savings of the purchasers have therefore been effectively used to provide financial stability for their families through this long-term property investment.

Table 5 shows the trend in the demand for flats during the period 1980–1985. The increase in the total number of applications is attributed to a number of factors including the accelerated resettlement program and the relaxation of the HDB's eligibility conditions.

Over the last three years, the prices of HDB dwellings have been increased only between a modest 2.5% to 5% on the average. The selling prices of HDB units as of 31 March, 1985 according to the zones are shown in Table 6.

Waiting Time

The HDB tries to maintain the waiting time at between 1½-2½ years. Table 7 compares the percentages of applicants with the waiting period. In the year 1984-1985, most applicants who had been on the waiting list for more than two years were offered flats. The resulting percentage of applicants still on the list

Table 5 Demand for Flats 1980/1981 — 1984/1985

Year	To Buy	%	To Rent	%	Total	%
1980/81	37,924	77	11,095	23	49,019	100
1981/82	38,232	76	12,370	24	50,602	100
1982/83	32,869	79	8,791	21	41,660	100
1983/84	39,037	85	6,632	15	45,669	100
1984/85	49,459	90	5,570	10	55,029	100

for waiting periods of 1 year or more is due primarily to fact that applicants are being more selective of the location and the design of the flats.

Floor Space Standards

Floor space standards vary from 8 to 10 m^2 (86 to 107 ft^2) per person for 3-, 4- and 5-room flats (Table 8). These standards compare favorably with other Asian countries as shown in Table 9.

Table 6 Prices of HDB flats (as of 31 March 1985)

Zone		Improved	Simplified	New Generation	Model 'A'	
New Town	3-room	22,700	26,800	30,400	37,300	—
	4-room	—	42,700	46,400	57,300	—
	5-room	70,500	—	—	82,400	—
	Executive	—	—	—	—	103,700
Outer Suburban	3-room	25,000	29,500	33,400	40,800	—
	4-room	—	46,900	51,000	63,100	—
	5-room	77,700	—	—	90,700	—
	Executive	—	—	—	—	114,000
Inner Suburban	3-room	27,900	33,100	37,400	45,800	—
	4-room	—	52,300	57,100	70,500	—
	5-room	87,000	—	—	101,600	—
	Executive	—	—	—	—	127,700
Outer Urban	3-room	31,800	37,700	42,600	52,200	—
	4-room	—	59,600	65,100	80,500	—
	5-room	99,200	—	—	115,700	—
	Executive	—	—	—	—	145,600
Inner Urban	3-room	36,200	42,800	48,500	59,500	—
	4-room	—	67,700	74,200	91,700	—
	5-room	113,000	—	—	132,000	—
	Executive	—	—	—	—	166,000

Table 7 Waiting period and percentage of applicants

Period of Waiting	1983/1984	1984/1985
Less than 1 year	32%	49%
Between 1 and 3 years	51%	33%
3 years and more	17%	18%

HOUSING IN CONTEXT

In opting for high-rise housing, it is absolutely essential to consider the wider environmental context within which its development occurs. The HDB has structured its development on a hierarchy of communal spaces, which is described as a "flow from the courtyard-in-the-sky" (a term used for access balconies) to precinct square, to neighborhood center and lastly, to the town center (Lim, 1983).

The entire town is divided into neighborhoods of about 6,000 dwelling units. Each neighborhood is divided into precincts of four to eight residential blocks with each block housing 100 to 120 families. Facilities such as children's playgrounds and small shops are found at the precinct level. Supporting

Table 8 Floor space standards

Type of flat	Average flat size (m^2)	Average household size (persons)	Average floor space per person (m^2/person)
1 room	30– 33	3.8*	7.9– 8.7
2 room	40– 45	4.7*	8.5– 9.6
3 room	60– 75	5.1*	11.8–14.7
4 room	80–105	5.4*	14.8–19.4
5 room	123–135	4.7*	26.2–28.7
Executive	145	4.5**	32.2
HUDC	155	4.0**	38.7

*HDB 1981 Sample Household Survey Report
**Estimate

Table 9 Floor space standard in Asian countries

Countries	Floor area (m^2/person)
Hong Kong* (1980)	2.2– 5.7
Indonesia* (1975)	9.0–20.0
Japan** (1973)	11.0–16.0
Malaysia* (1975)	4.6–13.3
Philippines* (1975)	5.0–10.0
Sri Lanka* (1980)	3.4– 7.1
Singapore (1983)	7.7–38.7
Thailand* (1975)	9.0–20.0

Sources:
*Housing Asia's Million (1979)
**Supplementary textbook for the Group Training Course in Housing '80

essential services such as banks, medical clinics, administrative area offices, and neighborhood parks are in the neighborhood center. Large emporiums, cinemas, and supermarkets are found in the town center. In addition to these facilities, schools, community centers, and sports complexes are scattered throughout the town at selected locations. The coherent organization of these facilities enables the new town to sustain a defined level of socio-economic self-sufficiency.

Early versions of HDB development may be justifiably criticized as regimental, dull, and unimaginative. However, it is only fair to evaluate them in the light of the constraints in terms of available resources and pressure for urgent demand at that time. Greater attention is now being given to urban and architectural design quality of new towns to enhance their imageability. Nature is introduced wherever possible to soften the impact of the intensive urban environment while civic squares and urban vistas are created to provide a sense of *place*. The streetscape is imbued with a greater variety of urban texture offered by buildings with varying height, length, and configuration. To restore the sense of human scale, about 5% to 10% of the buildings are 4-story, interspersed among the normal 10 to 13-story blocks and the occasional 20 to 25-story point blocks. Attempts are also being made to use traditional building forms such as pitch roofs, deep overhanging eaves, and tall windows typical of a tropical building to promote a regional identity.

ROLE OF RESIDENTS

The success or failure of a public housing program may ultimately be gauged by the quality of social life that exists within the housing estates. It is acknowledged that "being physically close together does not automatically generate community feelings. . . . Architectural design and planning layout can obstruct but cannot in themselves, produce social interaction" (Chua, 1981). The input of residents will determine the extent to which the desired social interaction can be realized within the housing estates.

Active measures have been taken to encourage public participation in community development. In 1978, the government initiated the formation of Resident's Committees (RC) in public housing estates throughout the country to promote a community spirit among residents and to provide a better channel of communication between the residents themselves and between the residents and the government. In addition to organizing social and recreational facilities aimed at fostering neighborliness and social cohesion, the RCs also help to generate greater public awareness of national issues and policies. General good housekeeping is also encouraged through implementation of Residents Education Programs. At the end of May 1985, there were some 267 Residents' Committees formed in 76 out of 79 constituencies.

FUTURE TRENDS

The population in Singapore is estimated to increase by another million to 3.5 million by the year 2030 before it stabilizes. In view of this, the sixth Five-Year Building Program (1986–1990) has been formulated to provide another 160,000 dwelling units.

It is increasingly apparent that residents of Singapore's public housing estates are no longer simply concerned with basic shelter and essential services but with better design, quality of workmanship, and estate services. To meet the growing expectations, the HDB will have to continue to be innovative in all aspects pertaining to the implementation and administration of future public housing developments.

In the next decade, it is expected that Singapore's economy will mature and the frantic pace of development will slow down. The demand for land for new development will also be moderated. It is then important and necessary to continuously study land use allocation on an island-wide basis to assess the possibility of allocating more land for public housing and the desirability of low-rise high-density housing.

With an increasing building stock in a high-rise high-density environment, the demands on the level of services to all mechanical equipment as well as general maintenance operations will become more intense. In anticipation of an escalating repairs and redecoration program each year (projected 100,000 units per year from 1988 onward), manpower and other contracting resources must increase to manage this workload.

In the area of social administration, consultation services to the HDB area offices will be systematized in the near future. Much needed and relevant social service resource building in the community will also be encouraged.

CONCLUSION

The history of the Housing and Development Board (or for that matter the nation of Singapore) has covered a relatively short time span of 25 years. Today, some 2.09 million of 81% of the total population are housed in HDB flats, of which 76% are living in flats that they own. This achievement is undeniably laudable despite criticisms at various levels. Ongoing attempts by the HDB to enlarge its scope to include community development through a creative environmental identity is indeed encouraging. The task is not easy, but this venture will certainly be worth pursuing.

REFERENCES/BIBLIOGRAPHY

Chua, B. H., 1981
HDB'S INTEREST IN SOCIAL MANAGEMENT, a talk given at the Social Awareness Seminar, JTC Auditorium, October 3 & 4, p. 3.

CIB 84 Singapore, 1984
HIGHRISE CONSTRUCTION TECHNIQUES AND MANAGEMENT FOR THE 1990's, conference, 23-25, February.

Goh, K. S., 1956
"HOUSING" URBAN INCOMES AND HOUSING; A Report on the Social Survey of Singapore, Singapore Government Printers, pp. 61-68.

Housing and Development Board, 1970
FIRST DECADE IN PUBLIC HOUSING 1960-1969, Housing and Development Board, Singapore.

Housing and Development Board, 1984/85
ANNUAL REPORT, Housing and Development Board, Singapore.

Housing and Development Board, 1985
DESIGNED FOR LIVING—PUBLIC HOUSING ARCHITECTURE IN SINGAPORE, Housing and Development Board, Singapore.

Lim, W. S. W., 1983
PUBLIC HOUSING AND COMMUNITY DEVELOPMENT: THE SINGAPORE EXPERIENCE, MIMAR 7, Architecture in Development.

Liu, T. K., 1979
PUBLIC HOUSING—THE SINGAPORE EXPERIENCE, paper prepared for the 17th IFAWPCA Convention, Singapore, 23-28 September.

Liu, T. K., 1984
HOUSING POLICIES AND LIFE STYLE, Paper presented at the 1983 Singapore Professional Convention "High-Rise, High-Density Living", Singapore Professional Centre, Selected Papers, August, pp. 10-24.

Plaza at Tao Payoh Town Center. Disposition of low-rise amenity buildings (cinema, branch library, department store) and a residential 'point' block around an ornamental pool provides a distinct identity

A market/hawkers' center in one of the Tao Payoh Neighborhood Centers strategically located within walking distance for all residents within the neighborhood

Tao Payoh Town Center Mall. The rhythm and articulation of the buildings and the use of the shuttered windows are reminiscent of old shophouses in Chinatown

Bukit Merah Town Center. Introduction of natural landscape provides shade and softens the impact of the intense urban development

Jurong East Neighborhood Center—design with a 'Chinese village' flavor in response to the context of the adjacent Chinese Garden

Precinct at Kim Tian Place—an 'outdoor room' for play and recreation

Delta Avenue. Variations on a theme provide visual relief to avoid the sense of monotony in a high-rise high-density development

Jurong West. Layout of articulated housing blocks along tree-lined avenues generates interesting visual rhythms and adds a new dimension to street architecture

Bukit Merah Estate. Low-rise housing blocks flank the road to humanize the scale of the street architecture

Demand for High-Rise Housing in the United States

Richard Kateley

Overall demand for housing is declining in the United States. There is a simple demographic reason for this: Household formation, the key component of housing demand, will drop during the remainder of this decade and fall further in the 1990's. As a result, most economists and housing analysts believe that total new residential construction will be at relatively low levels for the next decade (Real Estate Research Corporation, 1985; Apgar, 1985).

However, other countervailing forces are at work in the economy and in social and lifestyle preferences that play a role in the choices of where people live. Many of these factors point to a continued and, in some cases, expanding appeal of high-rise apartments.

Market-rate rental and ownership housing are dealt with in this paper. Public housing is excluded. Many market-rate projects, however, are subsidized through federal, state, and local programs.

CHARACTERISTICS OF HIGH-RISE RESIDENTS

To assess future demand for high-rise residential units, we need to know something about existing tenants. Unfortunately, accurate national data on the characteristics of high-rise dwellers do not exist. Although they are as diverse as the total population, and their units range from upscale penthouses to moderate-income housing, some generalizations can be made:

A large majority are renters, as opposed to condominium or cooperative owners.

Renters are more likely to be single or divorced than married. Among those who are married, less than 20% have children.

On an average, renter households are headed by younger people, though the aging of the baby boom generation will change the statistics somewhat over time.

Incomes of renter households are lower than those of owners. However, some households—especially single adults and seniors—simply choose to rent and avoid the management burdens of home ownership.

High-rise housing ownership is found only in the largest cities, and even in those areas represents a declining share of the market. New condo construction is limited to high-end units, and the demand for such luxury space is very limited. Even in midtown Manhattan, with its concentration of affluent U.S. and foreign purchasers, condo projects are moving very slowly. (The condominium craze of the 1970s was fueled in large measure by investor speculation rather than owner-occupant demand. Low appreciation has led investors to lose interest in condos.)

Real Estate Research Corporation (RERC), in the conduct of market feasibility studies for developers and investors, has surveyed residents and workers in a large number of cities to determine the "profile" of prospective downtwon residents. High land costs mean that downtown housing will in almost all cases be high-rise, so these data approximate the characteristics of typical residents of today's new high-rise developments.

A large majority (89%) are renters rather than owners.

A majority are singles (never married or divorced), living alone or with a roommate, who are just starting a career.

Two-person households (married or unmarried) are also common, but there are few households with children.

Yuppies are visible high-rise dwellers, but middle-aged single working women and empty nesters are almost equally important segments.

High-rise residents, as expected, also have relatively large incomes, well above the national average and higher than residents of suburban garden apartments.

Onterie Center, a mixed-use high-rise in Chicago

Given these general trends, two questions arise: (1) Will there be enough small, young, relatively well-off households to support high-rise residential projects? and (2) Are there other household types that may emerge to support demand for tall buildings? The answer to both is a qualified "Yes."

DIMENSIONS OF FUTURE DEMAND

The conclusion that there will be modest demand for the next decade rests on the likelihood that selected market segments within the broad housing spectrum will desire high-rise living. Four sets of factors must be considered: economic, demographic, lifestyle, and financial.

Economic Factors

Over the last 10 years the United States has been in transition from a manufacturing-based to a service-based economy, which has meant that the country's net new employment is concentrated in office rather than factory settings. Consequently, downtown and suburban office construction is booming and the white-collar labor force is mushrooming. Living close to an office is much more attractive than living near a factory, and many time-conscious office workers are interested in eliminating or drastically reducing their automobile, bus, or train commutes. Also, although many service jobs are low paying, a significant proportion of the new office workers earn the salaries required to cover the high costs of high-rise buildings.

Demographic Factors

A second major trend is the aging of the American population and the shift from predominantly child-oriented households to a growing majority of adult households. As shown in Table 1, 58% of U.S. households did not contain children under 18 years of age in 1984. Single persons now represent one out of every four households in the country. Half of all the new households formed between 1980 and 1984 were nonfamily units, either singles or people living with unrelated individuals. In the four-year period, we added 4.6 million households in the United States, but only 850,000 of them contained children. Because high-rises appeal to small, childless, adult households, this trend reinforces continuing demand.

It is not surprising to find that the growth in adult households has been accompanied by an increase in multifamily housing construction, both rental apartments and condominiums. In 1975, less than one-quarter (23.1%) of the housing units built were multifamily; by 1984, the proportion was 34.8%. (Kateley and Lachman, 1985).

Lifestyle Factors

Even in a zero-growth environment, people's housing requirements change, and demand results from new consumer preferences. Two lifestyle or consumer/preference trends will affect high-rise residential construction. The first is the renewed interest in downtown living. The second is the emergence of the over-65 segment as a fast growing part of our population.

Downtown residential development—like any real estate venture—requires the right mix of ingredients to succeed. In monitoring trends and working in

Table 1 Composition of U.S. households

	1980		1984	
Total Households	100%		100%	
Non-child-oriented	56%		58%	
Married couples		30%		30%
Single persons		22%		24%
Other		4%		4%
Child-oriented	44%		42%	
Married couples		31%		28%
Male-headed		2%		2%
Female-headed		11%		12%

Source: U.S. Bureau of the Census, March 1984

High-rise housing complex in Florida

downtowns across the United States, RERC has discerned four characteristics that indicate an environment conducive to development of downtown housing: (1) Large white-collar and professional work forces—as evidenced by substantial office construction; (2) A recent history of successful renewal and new development—especially retail projects; (3) A concentration of cultural institutions and entertainment activities; and (4) Geographic compactness and/or an efficient public transit system.

All of these characteristics help to create the "liveliness" that is so important to residential downtowns. Downtown housing has succeeded, not just in cities that, like New York, San Francisco, Chicago, Boston, and Washington, D.C., have long traditions of affluent in-town living. Projects have now been marketed in Charlotte, Milwaukee, Des Moines, Raleigh, St. Louis, Hartford, Baltimore, Norfolk, San Diego, and a host of other cities.

The elderly represent America's fastest growing population group; this group will double over the next 30 years. In fact, the number of people over 65 is now larger than the total number of teenagers. Also, seniors are more affluent than in the past. About 31% of the over-65 households have incomes of more than $20,000; seniors have a total net home equity in excess of $700 billion; and a small but significant segment of older people can afford very fancy housing. As a result, retirement housing represents the single most important residential market.

Nearly nine out of ten (87%) of the elderly live in their own housing units. Seniors are not very mobile. Less than 5% move to another state on retirement, and most older people (75%) stay in the same county if they move. However, many are looking for options that provide some services as well as a place to live. Providing good housing alternatives for this group will require enormous creativity. Developers will need to understand health care, social services, and transportation as well as housing.

High-rise construction, especially in downtown settings, may work well for younger seniors, 55 to 75 years old, who are independent and do not need health care. This group is already well represented in existing high-rise buildings, both rental and ownership.

Financial Factors

While owning a single-family home in the suburbs may still be the American dream, it is beyond the financial reach of many households. RERC estimates that, of current renters, fewer than one-third can afford a $50,000 home. Less than 10% could purchase an $80,000 home even at today's low interest rates. This means that many people must continue to rent. While a larger proportion of renters cannot afford high-rise apartment rents, a substantial number of them can. The affordability problem is expected to last through the mid-1990s therefore assuring a pool of renter-oriented households.

Presidential Towers, a mixed-use complex featuring four 49-story apartment buildings in Chicago

Of course, others who can afford to buy, choose not to. People who expect to move within five years may find it not financially advantageous to buy in today's low-appreciation environment. Many of the young professionals who will occupy high-rise apartments fall into this category.

CONCLUSIONS

The future success of high-rise housing is by no means assured. Population growth will not automatically create demand. Developers and design professionals will have to carefully monitor economic, demographic, lifestyle, and financial trends and target their product to carefully defined groups. Selective opportunities do exist because of changes in the composition of our population and, importantly, because of the spreading acceptance of downtown living.

REFERENCES/BIBLIOGRAPHY

Apgar, W., 1985
HOUSING FUTURES FORECAST MODEL, Joint Center for Housing Studies of MIT and Harvard University, Cambridge.

Kateley, R. and Lachman, M., 1985
INVESTMENT OPPORTUNITIES IN APARTMENTS, John Hancock Realty, Boston.

Lachman, M. and Miller, R., 1985
DOWNTOWN HOUSING—WHERE THE ACTION IS, Journal of Real Estate Development, Vol. 1, #1, pp. 15-27.

Real Estate Research Corporation, 1985
EMERGING TRENDS IN REAL ESTATE, Equitable Real Estate, New York.

A Comparison of High-Rise Housing in India with Industrialized Countries

Jashwant B. Mehta

More than 40 cities in the Third World are expected to have a population of more than 15 million by 2000 A.D. It is estimated that by 2000 A.D. India will contain 300 to 400 million in urban population and will rank as the country with the largest urban population in the world. The massive population increase has put great pressure on land in big cities such as Bombay, Delhi, Calcutta, and others, calling for its optimum use. Tall multistoried buildings would be the inevitable answer. This paper gives a comparison of certain essential aspects of urbanization and high-rise living between India, a leading developing country, and industrialized countries. While the migration in the industrialized countries of the West took place for better opportunities, it has been different in many countries of the Third World including India. It has often been a question of survival for the much of the rural population who leave their native villages seeking job opportunities in big cities such as Calcutta and Bombay. It is a combination of *pull* and *push*. This tremendous population explosion has created acute housing shortages, severe transportation problems, and extreme crowding of people both on the streets and in dwellings. Hopefully this comparison will lead to better understanding and awareness for professionals in both developing and industrialized countries.

EFFECT OF ACUTE HOUSING SHORTAGE

Indian Conditions

Large cities. Obtaining housing is a big problem. A large percentage (40%–60%) of the population in metropolitan areas such as Bombay, Calcutta or Madras, live in slums. More than 80% of the population live in one-room tenements in Bombay. Under the circumstances, for a large majority of the population, the effects, ill or otherwise, of high-rise living is not a major issue. Because there is a seller's market, tall buildings are often ill-planned and badly designed.

Small cities. In smaller cities where land values are not very high and housing is relatively cheap, the trend is to own a one-or-two story house with independent land ownership even though the area of land owned may be small (as small as 150 to 200 m^2 [1600 to 2200 ft^2]).

Industrially Advanced Countries

In metropolitan cities of developed countries, such as New York, London, and Chicago, the housing situation is much better than their counterparts in developing countries. The consumer has more choice. The taller buildings are better designed and have better amenities.

IMPACT OF TECHNOLOGY AND LOCAL CONDITIONS

Indian Conditions

Technology is less advanced. The quality of services is less efficient. The security arrangements and, to a large extent, the operation of elevators and other equipment is manual. Building construction technology is largely manually oriented. Workmanship is still below par in most cases. Problems of water seepage through the outside walls, windows, and such are very common. Utensils and clothing are manually washed on bathroom floors, leading to more wear and tear of the flooring. Bathroom and toilet flooring becomes a major source of leakage within a few years of occupancy, reducing the overall life of the building. The quality of plumbing fittings, such as pressure-reducing valves, is not high. Maintenance problems are acute. There is also more wear and tear of fittings and the like due to the higher occupancy rate. More research is necessary to find a solution adaptable to local conditions.

Industrial Advanced Countries

Technology is more advanced and the operation of all essential services including security is automated. The quality of building materials, especially cladding, window frames, and such, is superior. There are very few leakage problems in bathrooms, laundries, and kitchens because utensils are washing in the sink or dishwasher and clothes are washed in washing machines. The shower is taken in the bathtub. The use of toilet paper leaves the floor dry most of the time. All these mean that the problem of water seepage is practically nonexistent.

TRANSPORTATION PROBLEMS AND THEIR EFFECTS ON DENSITY, LAND PRICES, AND PERMITTED FAR

Indian Conditions

Public transportation is very much below standards. The rapid transit facilities, such as those available in cities like New York, London, Tokyo, and Paris, are either nonexistent or extremely inadequate. As a result there is tremendous pressure on land near the central business districts (CBDs). While it is possible to commute 20 or 30 km (13 or 19 mi) comfortably and in a reasonable time in cities such as New York or London that have excellent rapid transit facilities, it is difficult and very strenuous in metropolitan cities of the developing world such as Bombay, Delhi, Calcutta, Dacca, or Jakarta. Consequently the population is extremely dense in downtown areas. The reader will get some idea from the fact that in spite of a low floor area ratio (FAR) of 3 to 4, a particular segment around central business district in Bombay has as many as 3 lac (0.3 million) people cramped in an area of a mere 1.75 km^2 (0.7 mi^2). Delhi has as many as half a million people packed into the old walled city in an area of only 5.2 km^2 (2 mi^2).

The land values are also relatively high in or around areas near the CBD. Thus, against a per capita income of $600 per annum, the land rate per m^2 in south Bombay and central Bombay ranges from $500 to $1000 per m^2 (for a permitted FAR of 1.33). This is probably the highest in the world if weight is given to the per capita income.

Even those who can afford the high cost of automobiles and fuel ($3.00 for 5 l (1.3 gal) of gasoline), find it difficult to stay far away from their work place because of crowded and inefficient road systems. Although the number of automobiles is far less than that in other developed countries (there is only one automobile for every 10 families in Bombay), crowded roads and lack of freeways cause the average speed in peak-hour traffic to be 10 to 15 km per hour (6 to 10 mph). Thus even the affluent prefer to stay in high-rise apartments near their work place.

Because of the ever increasing urban population in large metropolitan areas, providing the bare minimum necessities, in other words water supply and sewerage, uses most of the municipal funds. Of the total funds allocated for the Greater Bombay's development plan (the period 1964–1978), 65% was utilized merely to augment water supply schemes. Still the per-capita water availability is a mere 100 l (25 gal) per head per day. Obviously, other amenities (e.g. transportation, housing, and recreation) are low priority and the situation is worsening. A peculiar problem of transportation is the intermingling of slow-moving and fast-moving vehicles on the same roads. Hand-carts, bullock-carts, and bicycles travel the same roads with trucks, automobiles, and giant buses, resulting in much higher accident rates both for vehicles and pedestrians.

Industrially Advanced Countries

Well-developed rapid transit facilities and road systems (along with the cheaper cost of automobile transport), make commuting easier and a lot more comfortable to downtown and midtown areas even for those living in far away suburban areas. Those who can afford it prefer to stay in houses in quiet suburban areas rather than in apartments, especially if they have children. Land values are higher but considering per capita income and higher permitted FAR, they are much lower when compared to metropolitan cities of developing countries.

BUILDING CODES, OPEN-SPACE REGULATIONS, AND PLINTH AREA

Indian Conditions

Because of the dependency on natural light and ventilation, building codes require more open spaces around the building, resulting in much lesser plinth areas per floor. High-rise buildings of 10 to 20 floors are quite common even with a permitted FAR of 1.33 to 2.00. Automobile parking is generally preferred at ground level or in basements. In view of the smaller plot sizes (downtown areas of most of the metropolitan areas were already developed in the late nineteenth century or earlier part of twentieth century when city planning was practically nonexistent), a slim tall building sometimes becomes a necessity to allow even the low permitted FAR to fulfill the requirements of leaving open spaces all around for light and ventilation and to provide for parking and other amenties (Fig. 1). To reduce congestion, there is a tendency to restrict FAR. In Bombay, for example, the maximum permitted FAR is a mere 1.33* against 2.45 permitted earlier (Fig. 2). Even the lift and staircase areas also must be included in this low FAR in places like Bombay. A vicious cycle has been created wherein the old buildings (because of their higher FAR) are being maintained rather than being demolished, throttling the redevelopment of downtown and midtown areas. Rather than reducing congestion, it has reduced the available built-up area per person, which has gone down to as low as 2.9 m^2 (31 ft^2) per person is south Bombay. More problems are thus created than solved. The upgrading of building codes is still not done for tall buildings. Thus, for example, the design for proper lift systems is not required in the building codes of most cities. Similarly, fire-fighting regulations are not strictly enforced.

*My friends in the United States are surprised when I tell them that the maximum permissible F.S.I. in Bombay as per current building regulations is only 1.33 and the maximum in large metropolitan cities, even in commercial districts, is up to 3.00. On the other hand my colleagues in Bombay get a shock when I tell them Chicago permits FAR up to 40.00 and even the Sears Tower has been able to consume only up to 34.00. For many of them it is news to learn that even well-developed Asian cities like Singapore, Hong Kong, and Tokyo have permitted FAR up to 10.00 on net plot.

Fig. 1 High-rise buildings in suburban areas in a city like Bombay with a low FAR of 1.00 create space for semiprivate bungalows or row-houses.

Industrially Advanced Countries

The permitted FAR is generally much higher. The dependency on natural light and ventilation is much less. As a result, the open space regulations are less stringent. The average area of the tall building is greater. In fact the very purpose for which the tall building technology has been developed, creating more built-up areas as the city grows, is served to a large extent.

THE ROLE OF TALL BUILDINGS FOR LOW INCOME AND MIDDLE INCOME HOUSING

Indian Conditions

The role of tall buildings—to fill the need for creating more spaces at ground level as the city grows—is yet to be recognized. Because of relatively

Fig. 2 **The current building codes of Bombay permit a FAR of 1.33. A situation is created whereby the owners or developers with the help and cooperation of the occupants, maneuver and manage to maintain the old FAR under the guise of repairs-cum-renovation. The illustration shows a reconstructed 'old' building. The old higher floor height of 4.3 m (14 ft) or so is also used to advantage by subdividing it into two floors, providing the developer a FAR of 7 or 8 against the permitted 1.33 and old existing FAR of 4 or so. Permission for taller buildings of the same or slightly higher FAR than existing built up area, leaving adequate open spaces, would be a desirable and practical proposition.**

higher unit costs for maintenance and construction, it is generally felt that tall buildings will not be a solution for the housing needs of the masses. Because of the high costs and the large number of ill-planned, ill-designed tall buildings (Figs. 3–5), sociologists, journalists, and some planners have a prejudiced and biased attitude against high-rise buildings. In fact, time and again there is a hue and cry to restrict the height of buildings to 24 m (80 ft) or so in India. They have already succeeded in doing so in cities like Bangalore and Pune.

Industrially Advanced Countries

There is more maturity and probably a less biased attitude. Singapore and Hong Kong are two cities that have passed through the transition phase from a 'developing' city to an 'industrially advanced' city in a short period of 25 to 30 years. They have successfully provided accommodations for low and middle income groups in multistoried buildings. Their success might provide examples to other metropolitan cities in developing countries.

Fig. 3 A hospital building, a poor example of conversion of an old low-rise building into a high-rise building. No care had been taken in giving the upper floors even a semblance of integration with lower floor treatment.

Fig. 4 The beautiful Taj-Mahal Hotel (a Gothic style development) along with the new Taj-Continental building constructed a few years ago. The architect has tried matching at ground level but otherwise the overall development does not create a good match.

Fig. 5 A poor example of a high-rise development. The nearby high-rise building on the same plot detracts attention and reduces the esthetic value of dome-shaped Legislative Assembly building of Maharashtra state.

IMPACT OF SOCIAL AND CULTURAL FACTORS

Indian Conditions

Situation within the apartments. Due to higher occupancy and close family ties, there is more interaction in an average Indian household. In most of the households the family consists of husband, wife, children, and some elderly person or persons. There is thus less feeling of isolation especially among the children when the parents are out.

Situation within the building. There is also more mixing and social interaction among the residents in a high-rise building (Fig. 6). Most of the families staying in an apartment house know each other more intimately. This phenomenon could also be attributed partly to the fact that there are more religious, social, and cultural festivities due to the heterogeneous culture of Indian society.

Small children are better off. Due to easy availability of domestic labor it is possible for small children to leave the apartments more often (accompanied by the maid-servants) and play in the open compound at ground level or outside the building. The parents are less worried and the need for surveil-

Fig. 6 The Panorama building in Bombay is an interesting conversion from low-rise to high-rise. On an existing 4 story tall building (constructed in 1935-1937), 5 more floors were added in 1966-1967. (The building is named Panorama because it commands a panoramic view of Bombay city, being located at a high point.)

lance is less. The incidence of crime as (witnessed in interior public spaces including lobbies, corridors and open spaces around the building in cities like New York) is practically nonexistent.

Industrially Advanced Countries (of West)

By and large, living in high-rise buildings is not preferred by upper middle class families having children. Public housing experiments in high-rise buildings have not been very successful in cities like London or Chicago. Either the very affluent or couples without children and old people have found it convenient to live in high-rise buildings.

USE OF OPEN SPACES AROUND THE BUILDING

Indian Conditions

In tropical climates, the open spaces in the building compound are more popular and used more intensively both by the children as a playing area and by the occupants in general.

Industrially Advanced Countries

In more extreme climates, especially during the long winter (October–April) in Europe, North America, Japan, and Russia, the open areas in the compound are not very popular as a play area for children or for other common use by the occupants.

Prefabricated Concrete Systems

Bohdan Lewicki

The beginning of prefabricated concrete systems goes back to the 1930s, but their real development actually began after the Second World War with the growing use of cranes at building sites. The technical progress offered by the new building methods, operating with large size precast elements, was enormous and so, too, was the interest of European countries faced with a postwar housing shortage. The architectural ideas expressed in the Charter of Athens also contributed to the fascination.

Thirty years of experience allow us to discuss the highlights of prefabricated concrete systems. Wherever there has been the combination of low labor costs, a demand for multistory buildings, and funds available to cover the demand, prefabricated construction has proved highly competitive. It is, however, an industrialized building method, requiring the requisite discipline in design, performance, and site organization.

Thirty years of experience have also clarified the essential technical problems. Some of these will be briefly discussed here.

JOINTS IN EXTERNAL WALLS

Watertightness of joints in external, large-panel walls, cracks in the external leaves of sandwich panels and thermal bridges in this kind of external wall are among the problems that have caused most trouble. The essence of the problem is the substantial difference between the thermal behavior of a solid masonry wall and that of a sandwich (multilayer) wall.

The entire thermal deformation of the external leaf of a sandwich wall is concentrated in the joints (Fig. 1). Consequently, the joints should be designed to remain tight during such deformation, which can vary up to 2 mm (0.08 in.). The external leaf should be free to expand and contract with the fluctuating temperature.

The thermal insulation should be continuous throughout the joint, in other words, without any interruption that may become manifest at the surface of the internal wall at low temperatures, finally producing wet spots and dark mold. A classic example of such a joint is called the "open joint" because the tightness of the joint is secured only by its geometry (Fig. 2). Rainwater may enter the joint but will not penetrate all the way through and has an open way out (Lewicki, 1966).

The open joint concept, first developed in Denmark, is the most economical and reliable solution. The prerequisite is, however, very dense concrete and exact execution of both the sandwich panels and the site work. Sealing the joints with mastic is the next best solution—perfect as long as the mastic retains its required properties.

The principles for joints in large panels to fulfill the user's essential requirements are rather simple and now well known to all engineers specializing in the design of system buildings. This was not the case, however, in all countries during the early years, when intensive use was made of this technology. Several buildings from that period are now leaking and showing mold at the corners and are poor publicity for buildings with large panels.

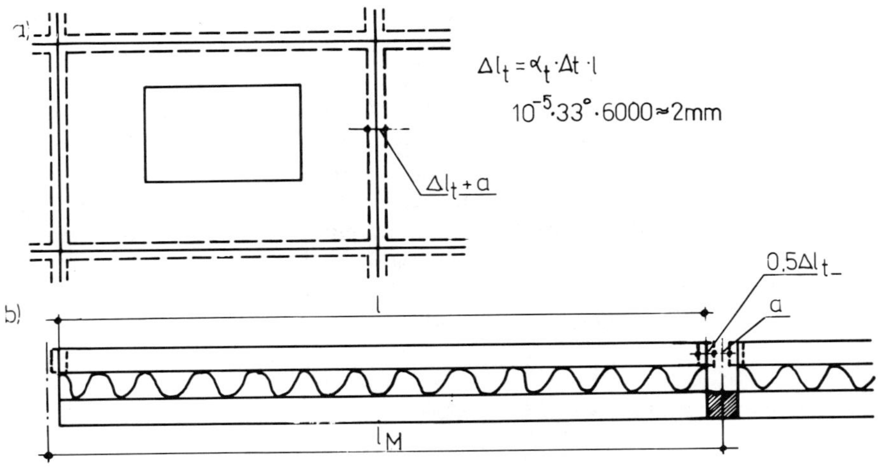

Fig. 1 **Thermal deformation of joints in an external sandwich panel wall; a. deformation, b. concept of unrestricted expansion and contraction**

LIMITING THE PROBABILITY OF PROGRESSIVE COLLAPSE

During the lifetime of a building some abnormal loadings may occur, such as gas explosions and the like, which the building should resist, although sometimes not without local failure. The widely known gas explosion at Ronan Point, London, resulting in progressive collapse of one entire corner of the building, focused attention on the problem and led to new code provisions (Ministry of Housing and Local Government, 1968). Properly designed and executed, large-panel buildings resist abnormal loads just as well as cast-in-situ buildings. The basic idea is to provide the structure with the ability to develop an alternative stability system after a local failure

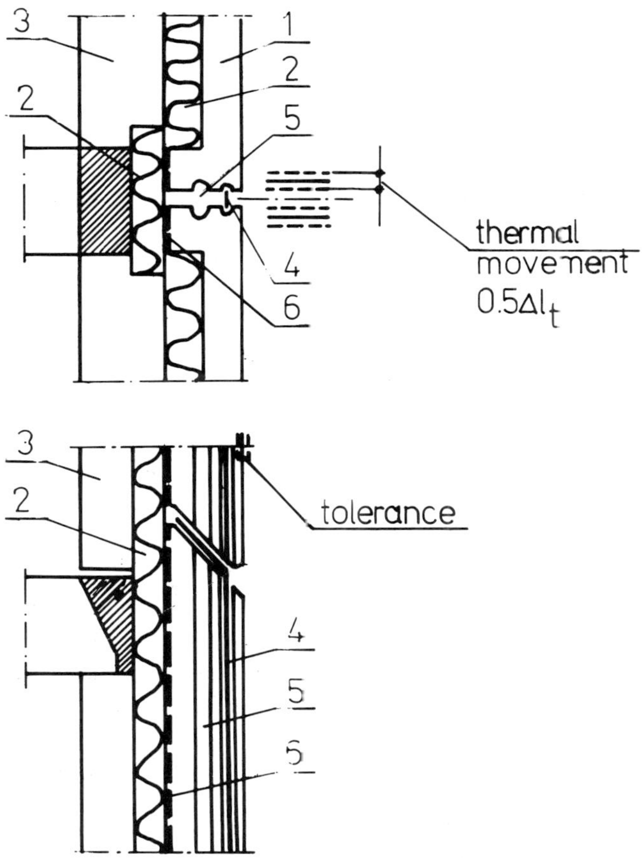

Fig. 2 **"Open joint" in large-panel sandwich wall 1. external concrete leaf; 2. thermal insulation; 3. internal structural leaf; 4. neopren strip to impede direct rainwater access; 5. decompression channel; 6. air tightness screen**

because of blast or similar loading, thereby preventing extension of the primary failure. Two examples of such secondary stability systems are shown in Fig. 3.

Tie beams placed at floor level play a particularly important part in the development of the secondary stability system. The floor slabs must be well anchored in the tie beams, both to prevent them from falling down after the support is lost and for structural interaction with the tie beams.

The requirements for development of alternative stability systems result in a somewhat greater quantity of tie-beam reinforcement. However, a simple statical diagram (Lewicki et al., 1982) demonstrates that in the case shown in Fig. 4, the shear force T_3 acting in the floor diaphragm may considerably reduce the force N_1 acting in the tie beams, particularly after the yield point is reached in the tie-beam reinforcement, and this contributes considerably to the economy of the tie-beam reinforcement required. Full-scale laboratory tests have verified the theoretical model and proved that even with relatively weak wall-to-floor connections, savings on tie-beam reinforcement exceed 50% (Lewicki, Cholewicki, and Makulski, 1983).

BUILDING IN SEISMIC REGIONS

Large-panel buildings are wall structures and as such are relatively stiff. The horizontal seismic force F_h, which is the structural response to earthquake motion, is therefore larger than for framed structures without bracing walls. The interstory drift is smaller, however, and the damage to the adjacent nonstructural elements (partitions, external walls) is consequently smaller. It

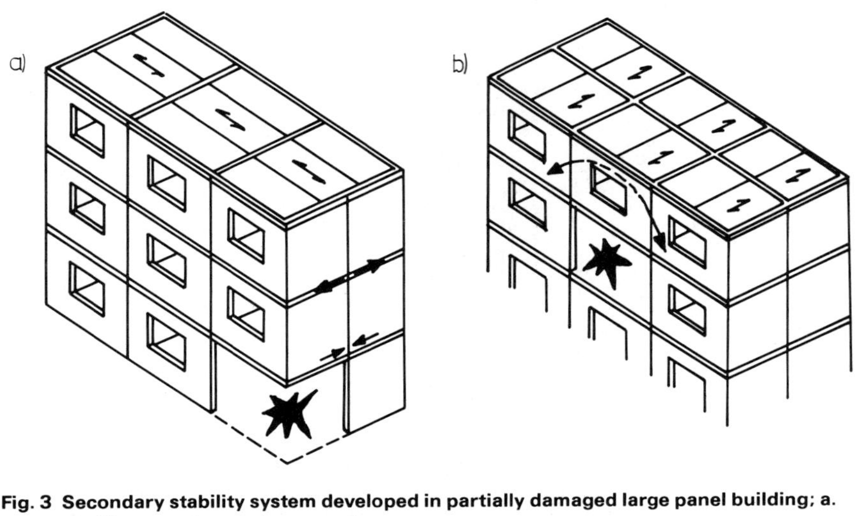

Fig. 3 Secondary stability system developed in partially damaged large panel building; a. cantilever system; b. beam system

follows that the repair cost after an earthquake is much smaller for a well-designed large-panel building than for a frame building.

The difference in the behavior of the two classes of buildings and in the extent of repair became particularly evident after the 1973 earthquake in Bucharest. Following that event, the Rumanian Ministry of Construction decided to increase the construction of multistory buildings with large panels from 10% before the earthquake to 35% in 1983.

According to Japanese experience, the cost of panel systems in earthquake regions is about 3% less than that of conventional cast-in-situ structures (Council on Tall Buildings, 1978).

The behavior of a large-panel structure subjected to earthquake motions does not differ significantly from the behavior of an identical cast-in-situ structure (Lewicki et al., 1982). Therefore, the magnitude of the horizontal force F_h acting on a large-panel structure can be calculated in the same way as for a monolithic structure. The difference between the two forms of structure is the existence of joints in the large-panel structure. The fact that the joint stiffness is lower than in a monolithic structure is not very important, although by reducing the overall rigidity of the structure, the presence of joints also reduces the magnitude of F_h. It is the detailing of the joints to secure the integrity of the structure under earthquake loads and the possible reduction of the carrying capacity of the joints that are critical for the design of the structure.

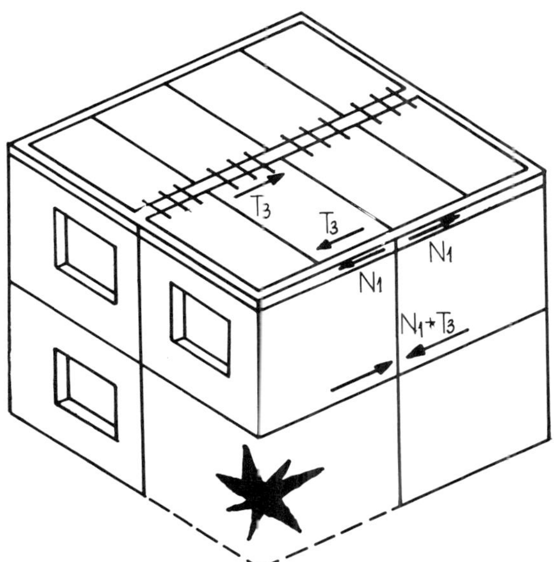

Fig. 4 Floor-tie beam interaction in developing the secondary stability system

The general requirement for large-panel structures—to connect the precast elements into rigid horizontal and vertical diaphragms—is of particular importance for buildings in seismic regions. The wall and floor diaphragms should be firmly jointed together with adequate reinforcement in the joints and with tie beams.

The substantial role played by the tie beams in the integrity and strength of the large-panel structure has been mentioned in the discussion about limiting progressive collapse. This role must also be stressed here, from the point of view of satisfactory behavior of large-panel structures in earthquake conditions (Lewicki et al., 1982). If tensile forces are likely to occur in the structural walls, vertical ties should be provided as well. The structural joints should be provided with sufficient reinforcement to behave in a ductile fashion.

In principle, the extent of the aseismic provisions depends on the expected intensity of the earthquake motion, but brittle failure of any structural member should be avoided in all cases. The uncertainty of prediction of both earthquake intensity and the behavior of the structure under earthquake loading is relatively large, and for that reason, a conservative design approach as well as design for an additional mechanism of energy dissipation is very necessary.

Vertical joints in structural walls should be designed as keyed joints. The reinforcement in the joints, which must not be less than that in the wall

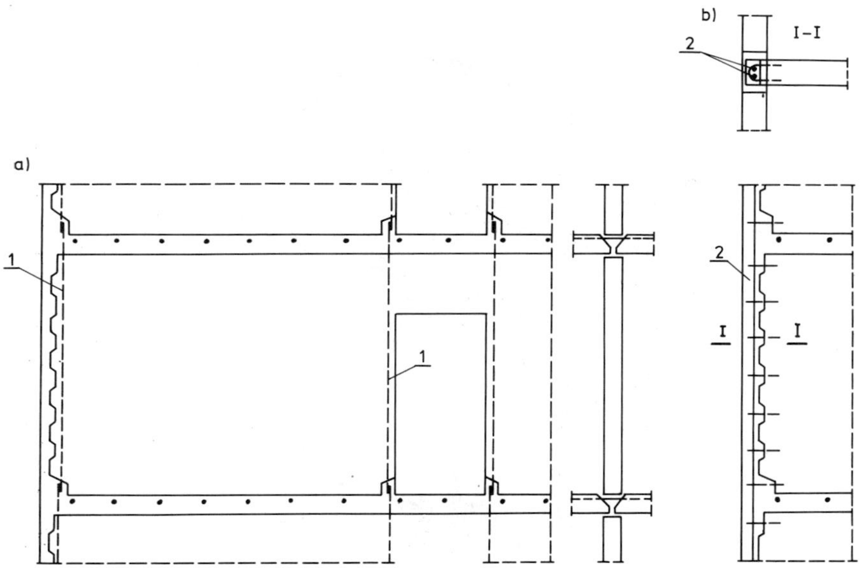

Fig. 5 Joints in walls entirely under compression, a. with reinforcement [1] cast into the wall panels, and b. placed in cast-in-situ concrete [2]

panels, may be either distributed or concentrated. The American Code (NBS, 1978) and, following it, the CEB Model code (Comite Euro-International du Beton, 1985a) require a reinforcement ratio of ≥2.5% in both directions. The Soviet regulations limit this requirement to walls subjected to tensile stresses (Council on Tall Buildings, 1978). Where this is not the case, ≥1% is considered sufficient. Because of better interaction with concrete, distributed reinforcement is generally recommended. The joints are mostly designed with keys the width of the wall, which allows a large crosssectional area through the keys and easier connection of the reinforcing bars projecting from the wall panels into the joint. This method also allows better control of the joint during concreting.

Horizontal joints that are expected to be entirely under compression may be designed as plain joints, as in buildings in nonseismic areas. However, in building in zones of moderate and high seismicity, it is recommended that these joints be provided with transverse reinforcement/vertical ties, placed close to the edges of the walls. The vertical tie reinforcement may be cast into the wall panels, as shown in Fig. 5a, or placed in situ, as shown in Fig. 5b. In the second case, it is necessary to ensure the interaction of the reinforcement in the cast-in-situ concrete with the wall panel by loops (stirrups) projecting from the wall panel.

For horizontal joints that are expected to be partially in tension and hence in unfavorable conditions, it is recommended that keys be provided, as shown in Fig. 6, to counteract sliding of the wall in the horizontal direction. Properly arranged transverse reinforcement in horizontal joints/vertical ties should also be provided. Concentrated keys, projecting from the lower wall panel (Fig. 6b) or cast-in-situ (Fig. 6c) are normally reinforced in order to increase their resistance to shear. In this case, the resistance to local pressure of the joined wall should be checked. Monolithic keys developed by cutting the corners of the wall panels (Fig. 6c) may serve as keys in both vertical and horizontal joints.

The resistance of joints subjected to earthquake motions depends on the geometry of the wall (Comite Euro-International du Beton, 1985b). In high-rise buildings with shear walls consisting of relatively narrow (compared with the height of the building) wall strips connected with coupling beams (mostly lintels over doors and windows), the joints are in some respects protected by the ductility of the coupling beams and do not enter the elastoplastic stage during the earthquake motion. The design resistance of such joints may be assumed to be about 0.8 of that adopted in nonseismic areas. However, in wide walls without rows of openings, the design resistance of joints should be assumed to be much lower (Comite Euro-International du Beton, 1985a; Tassios and Tsoukoutas, 1983).

Although experience of the behavior of prefabricated structures under earthquake conditions is somewhat limited in some respects, enough is still known to build safely with large panels in seismic regions.

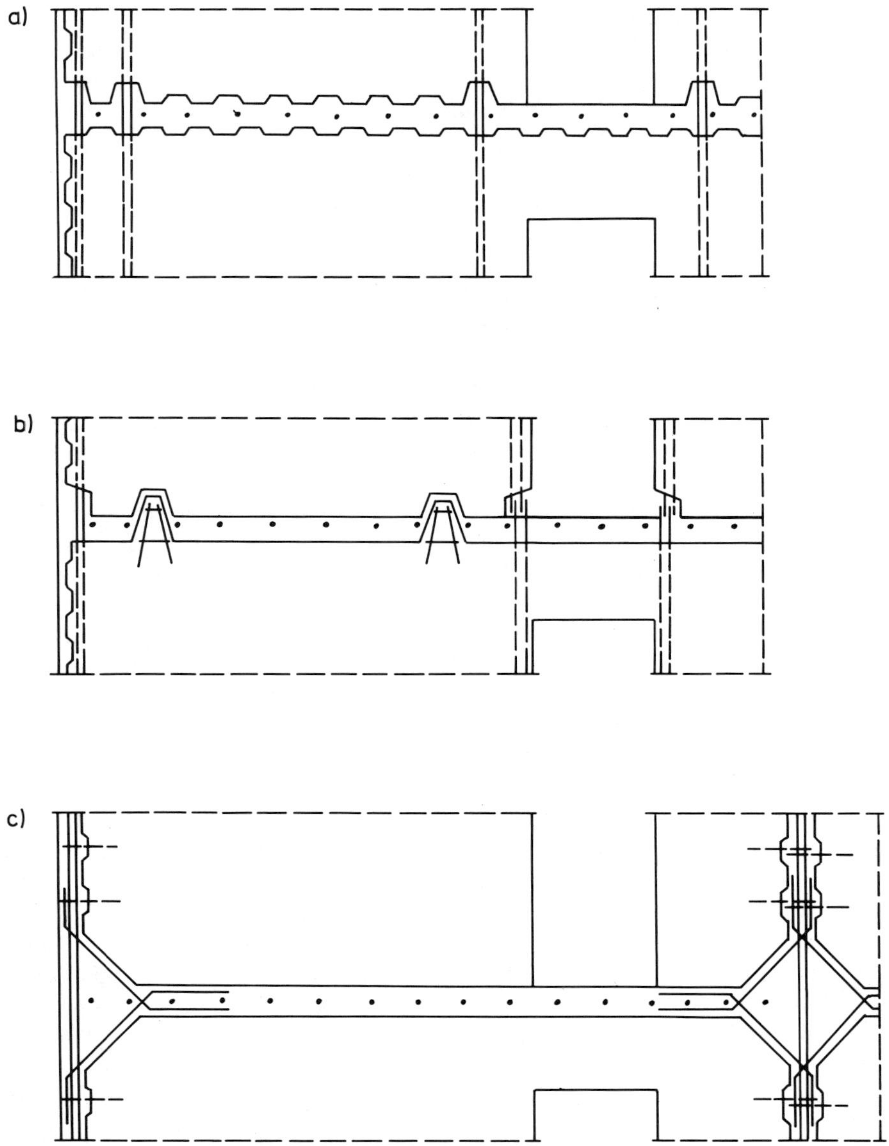

Fig. 6 **Joints in walls partially under tension, a. keyed horizontal joint; b. with reinforced individual keys, projected from the lower panel; c. with monolithic keys, produced by cutting the corners of the wall panels**

REFERENCES/BIBLIOGRAPHY

Comite Euro-International du Beton, 1985a
CEB MODEL CODE FOR SEISMIC DESIGN OF CONCRETE STRUCTURES, CEB Bulletin No. 165.

Comite Euro-International du Beton, 1985b
CEB DRAFT GUIDE FOR THE DESIGN OF PRECAST WALL CONNECTIONS, CEB Bulletin No. 169.

Council on Tall Buildings and Urban Habitat, 1978
STRUCTURAL DESIGN OF TALL CONCRETE AND MASONRY BUILDINGS, Monograph on Planning and Design of Tall Buildings, Volume CB, ASCE, New York.

Lewicki, B., 1966
BUILDING WITH LARGE PREFABRICATES, Elsevier Publishing Company, Amsterdam.

Lewicki, B. and coll., 1982
PROGETTAZIONE DI EDIFICI MULTIPIANO INDUSTRIALIZZATI, ITEC editrice, Milano.

Lewicki, B., Cholewicki, A., and Makulski, W., 1983
LARGE-PANEL BUILDING: BEHAVIOUR IN PARTIAL DAMAGE, Building Research & Practice, July/August.

Ministry of Housing and Local Government, 1968
REPORT OF THE INQUIRY INTO THE COLLAPSE AT RONAN POINT, London.

NBS, 1978
TENTATIVE PROVISIONS FOR THE DEVELOPMENT OF SEISMIC REGULATIONS FOR BUILDING, NBS Special Publication 510, June, National Bureau of Standards.

Tassios, T. P. and Tsoukoutas, S. G., 1983
BEHAVIOR OF LARGE PANEL CONNECTIONS, Building Research & Practice, July/August.

WSN 321-75, 1976
RECOMMENDATIONS FOR DESIGN OF LARGE PANEL BUILDINGS—IN RUSSIA (Instrukciya po proyektirovaniyu panielych zdanly), Moscow.

INTEGRO: Open-Element Prefabricated Construction System

Pavel Cízek

INTEGRO is an open-element load-bearing prefabricated construction system for public and industrial buildings. Its principle of openness is based on a strict unification of the load-bearing element crosssections and joints, providing the production of a variety of precast members, differing in length, type of reinforcement, and grade of concrete which can be produced, requiring minimum adjustment of equipment or formwork modifications (Fig. 1).

All these conditions ensure precast members a high versatility of application. The structure consists of two types of columns: H-section and rectangular; two types of beams: twin-beams and cladding beams, and two types of floor panels: ribbed and flat slabs (Fig. 2).

The length and/or height of elements vary in multiples and combinations of the basic module of 300 mm or 150 mm (12 in. or 6 in.). It enables the choice of different spans from 6 m to 12 m (20 ft to 40 ft) in the direction of beams and from 6 m to 18 m (20 ft to 60 ft) in the direction of floor slabs (Fig. 3).

Construction heights of floors are flexible dependent upon structural composition, load, and types of columns employed. The principle of openness of the load-bearing system INTEGRO enables the structural design to meet easily the user's requirements regarding composition, scale of elements,

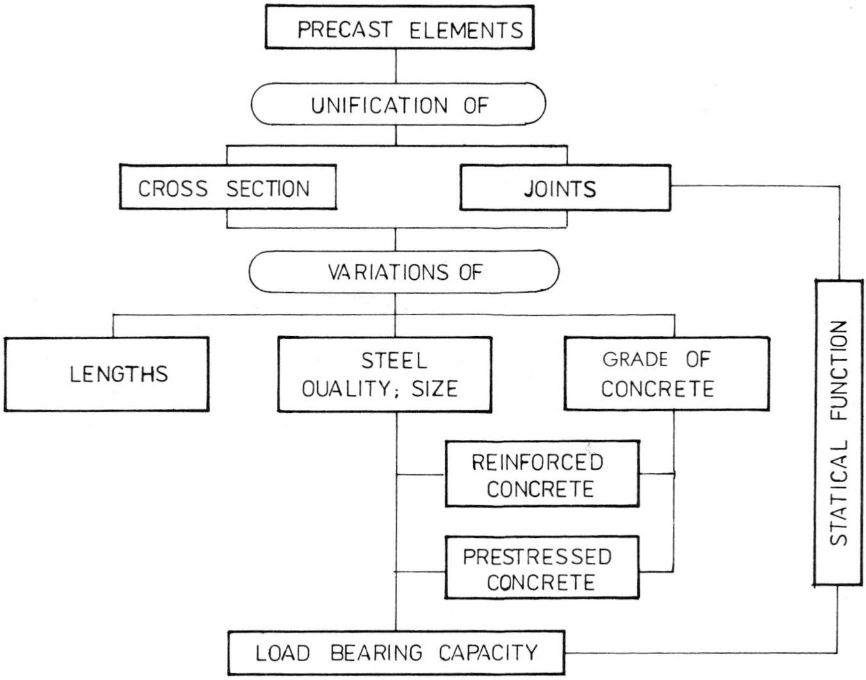

Fig. 1 Scheme of open-elements—prefab system INTEGRO

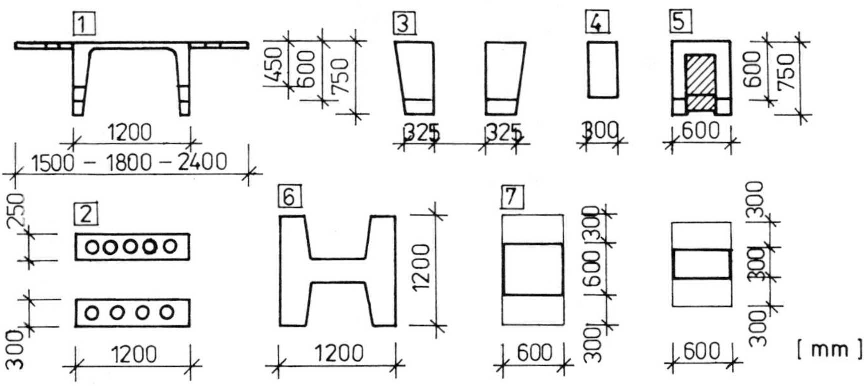

Fig. 2 Basic cross section of the precast elements (1) ribbed slabs; (2) flat slabs; (3) and (4) twin beams; (5) contour beams; (6) H-section column; (7) columns of rectangular cross section

loading capacity, accommodation of wiring and piping ducts, and fire resistance, etc. (Figs. 4 and 5).

COLUMNS

There are two types of columns. One is the H-section column inscribing a 1.2 m (4 ft) modular square and consisting of blocks of modular heights 1.2-1.5-1.8-2.1 m (4-5-6-8 ft) including vertical continuous cavities intended for connecting reinforcement to be grouted. The grouting mix is pumped into the cavities from the bottom upward, usually to fill in the cavity of one or more story height. These columns provide space for hidden ducts of vertical piping and tubing (Fig. 6).

The other column type is one of constant crosssection 0.6/0.6 m (2.0/2.0 ft) or 0.6/0.3 m (2.0/1.0 ft) with cantilevering projections intended for supporting the floor beams. It is manufactured, transported, and erected as a whole in a height over several stories.

BEAMS

Twin beams are a characteristic of the INTEGRO system. Depending on the type of connection with the columns these beams may behave as frame cross-bars, or as simple or continuous beams. The sharing of the load by two beams of the same size and shape allows the cross-section depth to be either

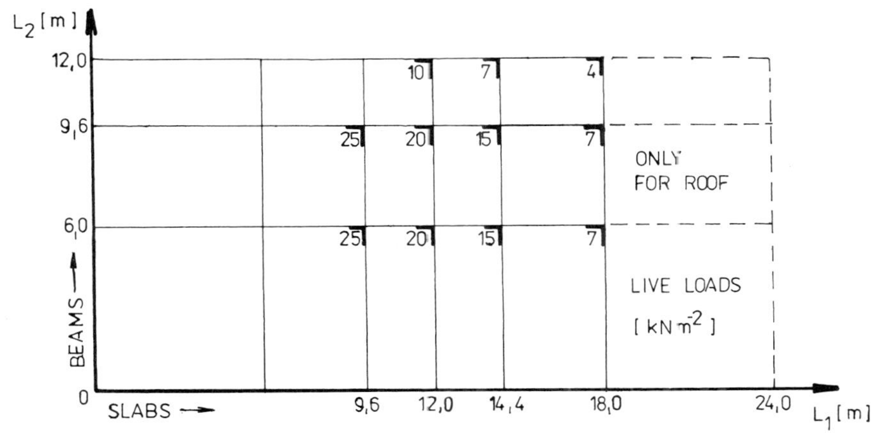

Fig. 3 Plan dimensions of structures using INTEGRO system

0.6 m (2.0 ft) or 0.75 m (2.50 ft) as required. The space between the twin beams is sufficient for accommodating hidden horizontal ducts for wiring and piping or for horizontal shafts.

FLOOR SLABS

Two types of prestressed concrete floor slabs are in use. They are ribbed slabs of modular depths 0.45 m (1.5 ft), 0.6 m (2 ft) and 0.75 m (2.5 ft) and flat slabs SPIROLL or SPAN DECK. The ribbed slabs have modular widths of 1.2-1.5-1.8-2.4 m (4-5-6-8 ft).

The choice of the slab type depends on the span, the required load-

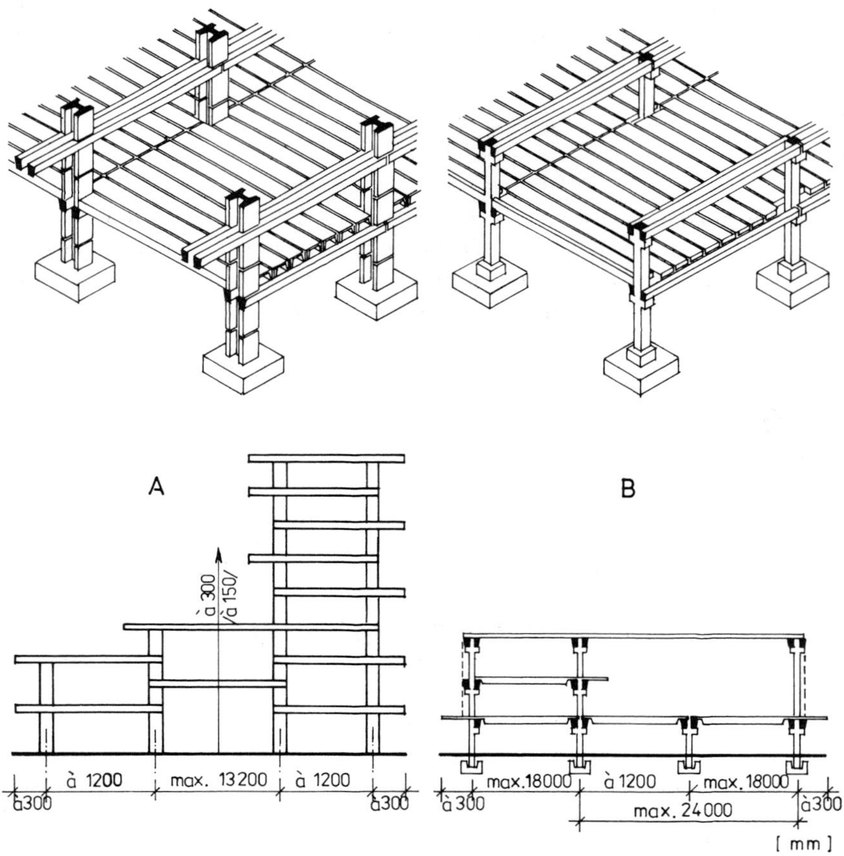

Fig. 4 Use of structure with (A) columns of H cross section; (B) continuous columns

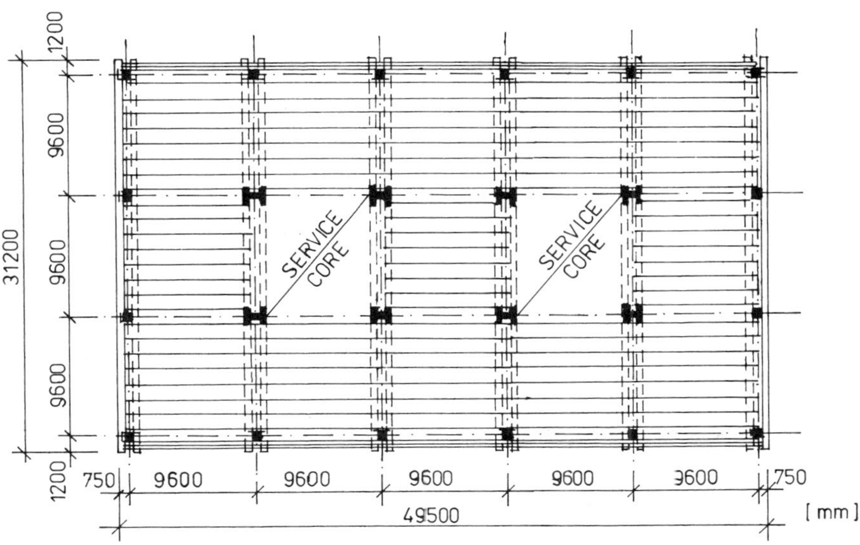

Fig. 5 Examples of both column types used in plan for high-rise office building

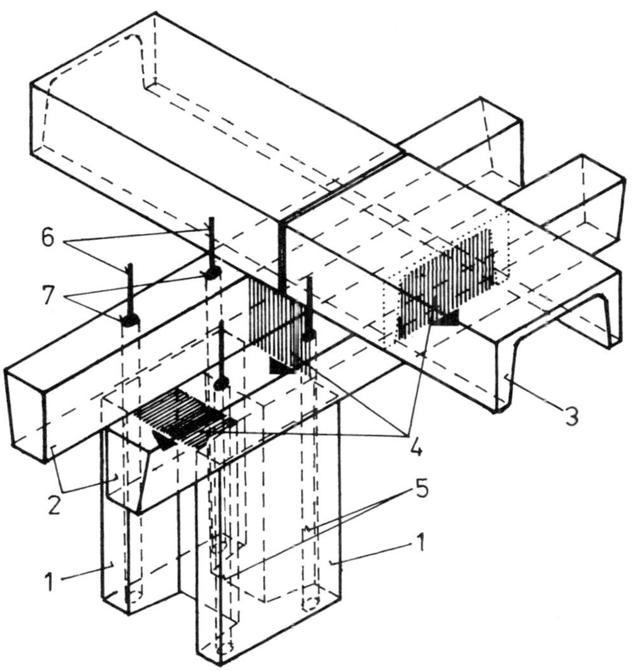

Fig. 6 Main detail of skeleton—isometric view. (1) column with H cross section; (2) twin beams; (3) ribbed slab; (4) possible hidden piping and wiring; (5) vertical cavities; (6) connecting reinforcement bars; and (7) grouted mix Vusokret

carrying capacity, modular composition, mass, and relation to horizontal ducts. The floor slabs are put on the upper surface of beams and from the statical point of view their behavior is that of simply supported beams. At the point of bearing, the ribbed slabs may have their full depth, or they may be embedded in the support by means of a 300 mm (12 in.) high spur (Fig. 7).

This load-bearing system is appropriate in the construction of single-purpose department stores and warehouses, cultural centers, office buildings, schools and other public buildings, as well as for multistory industrial production halls. The inherent structural characteristics and parameters of the INTEGRO system make it suitable also for the construction of multiuse buildings designed for reconstruction or completion in central urban areas as well as for some buildings in new housing estates with vertical functional zoning: apartments placed on the upper floors and community facilities, shops, catering, and parking areas placed in the bottom section of the building.

The use of the load-bearing system INTEGRO provides two entirely different designs of multiuse buildings. The first design makes use of ordinary production units for the upper floors of apartments, that is, transverse load-bearing wall systems which may be precast, cast-in-place, or combined. The bottom section is built with the INTEGRO system, providing the layout of columns according to the degree of required interior flexibility in this section of building (Fig. 8). According to the existing situation, the transverse or the longitudinal load-bearing system, columns in pairs may be arranged under every wall, every other wall, or every third wall. It is ensured by the transition support structure (Fig. 9) with beams enabling the formation of

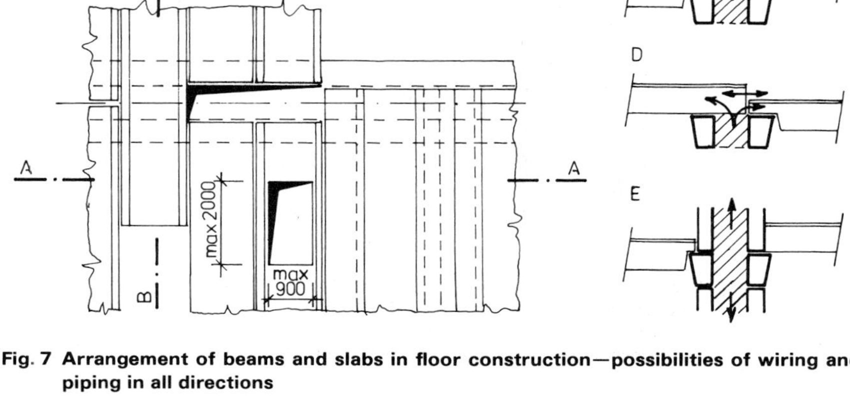

Fig. 7 Arrangement of beams and slabs in floor construction—possibilities of wiring and piping in all directions

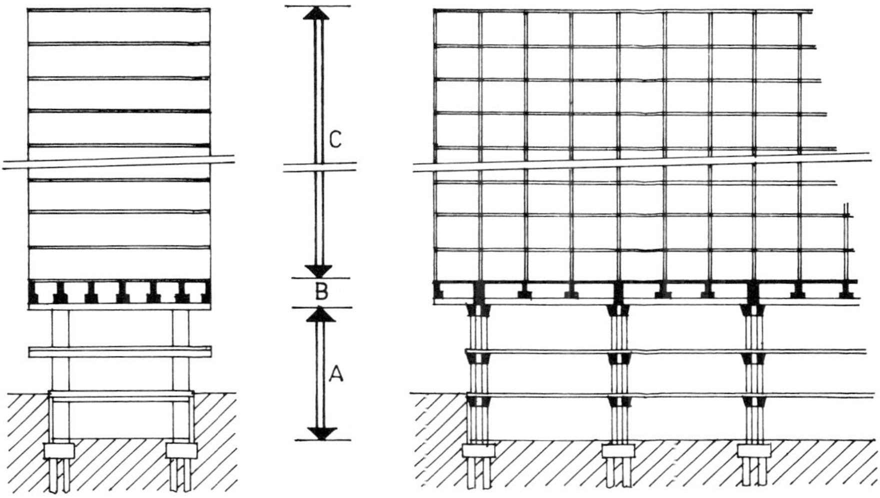

Fig. 8 Structure for multiuse buildings with change of loadbearing system in vertical arrangement (A) INTEGRO system/community facilities; (B) transition support structure; and (C) transverse load-bearing wall system/residential part

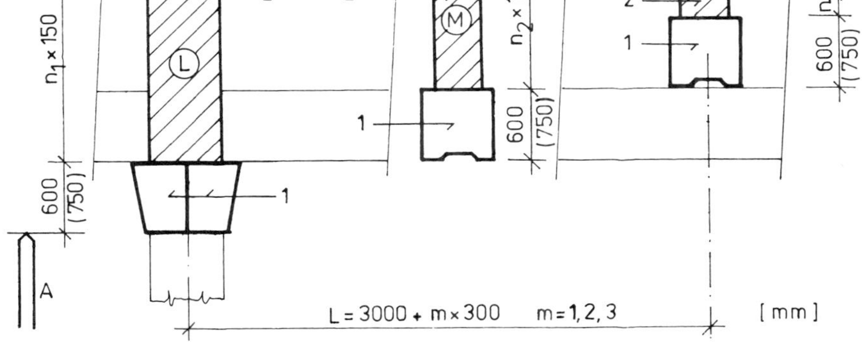

Fig. 9 Orthogonal load-bearing system of the transition support structure (L) main beam; (M) secondary beam; (N) supplementary beam; (1) precast portion of cross section; (2) cast-in-place portion of cross section

orthogonal load-bearing systems. The beams consist of precast load-bearing INTEGRO elements with composite cast-in-place portions, their height, width, reinforcement, and concrete mix being chosen in accordance with the statical structural design.

Alternatively, the INTEGRO load-bearing system can be used as a primary structure with variable height of floors, and at the same time serve as a permanent structural skeleton into which a building-block system can be placed. The building blocks afford great flexibility to the interior arrangement of the building. They can be removed, rebuilt, or changed depending on the temporal changes of function. A detailed analysis showed that ground plan modular units 10.8 × 10.8 m (36 × 36 ft) or 10.8 × 14.4 m (36 × 48 ft) are optimal for the primary structure. The principle of this design concept is illustrated in Fig. 10.

REFERENCES/BIBLIOGRAPHY

Cizek, P., 1985
OPEN PREFAB LOADBEARING SYSTEM, Stavebnicka rocenka, ALFA, Bratislava.

INTEGRO, 1986a
CATALOGUE OF PROJECTS AND BUILDINGS, ZIPP, Bratislava.

INTEGRO, 1986b
Proceedings of Conference in Prague.

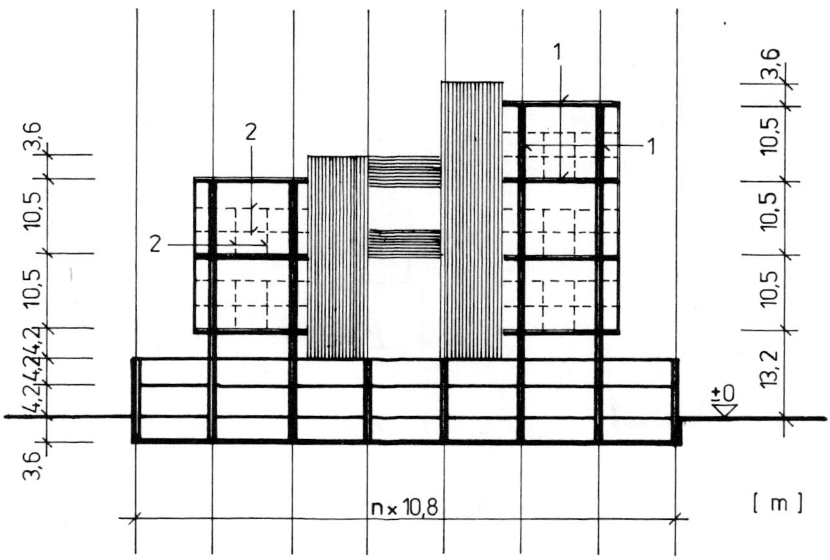

Fig. 10 Example of possible arrangement of permanent structure (1) with flexible interior structure (2) of residential block—cross-section

Multiuse department store ZOD Ruzinov in Bratislava

Multiuse department store ZOD Ruzinov in Bratislava under construction

Multiuse cultural centre Ruzinov in Bratislava

Structure for multiuse buildings

Skeletal system of multiuse commercial center Lamac in Bratislava

Cladding of Lamac commercial center, Bratislava

Printing house under construction in Prague

Multistory industrial building TESLA in Vrable—under construction

Construction Robotics in Japan

Seishi Suzuki
Tetsuji Yoshida
Takatoshi Ueno

Three years ago a robot was applied to construction for the first time in Japan (Yoshida et al., 1984). Since then, several experimental construction robots have been developed (Paulson, 1984). Research projects on construction robots have included technical subjects (such as robot mechanisms and sensor signal processing) and social subjects (such as possible unemployment caused by robots changing the labor-intensive character of the construction industry). The first participants in these projects in Japan were mainly from general contractor companies, but now subcontractors, related professional societies, and governmental organizations participate in construction robotics.

There have not been any fundamentally innovative improvements in construction engineering for a long time, so construction robots have attracted the attention of construction engineers. It is widely recognized that there is a large gap in production engineering between the construction industry and other industries, for example, the electronics industry. The application of construction robots is motivated by the need to introduce this recently developed technology and to innovate construction engineering.

The future of construction robots has not yet been clarified. This paper reports on the current situation concerning construction robots, including their background, and some examples in Japan, with the aim of helping to predict future trends.

BACKGROUND OF CURRENT RESEARCH AND DEVELOPMENT

Construction robots have mainly been developed by general contractors. The most common motivations for the development of construction robots are to improve working conditions and to reduce accidents, to overcome the shortage of skilled workers, and to improve construction productivity.

Improvement of Working Conditions and Reduction of Accidents

Generally speaking, construction is an outdoor activity done under dusty or dirty conditions. Furthermore, the materials handled are heavy and vary widely. However, some recent construction machines have been remarkably improved. Working and safety conditions are also being improved, and productivity is increasing.

Still, it is impossible to change a construction site into an automatic factory and to completely eliminate accidents or injuries caused by prevailing working conditions on the site. About three hundred thousand workers were killed or injured by accidents in the last 10 years in Japan. One-third of these are related to construction (Figs. 1 and 2). The rate of workers killed in the construction industry is higher than in other industries, and 40% of the fatalities are construction workers. Accidents in the construction industry are mainly caused by falls, construction machines, and automobiles (Fig. 3). Accidents occurred where people work with machines.

The incidence of occupational diseases has been decreasing year by year, and in 1983 there were about 15,000 cases. This includes about 3,000 construction workers. Records show the number of sick workers per thousand workers in the construction industry is two times that for all industries. These data clearly show that working conditions on construction sites definitely need to be improved.

The introduction of robot technology seems to be a most practical method to decrease accidents on construction sites. Improvement of working conditions on construction sites by ordinary means is nearly impossible.

Shortage of Skilled Workers

Economic progress has brought general changes to Japanese society, including a change in the value of work by employees. The work attitude of young workers, especially, has changed dramatically in the last 10 years. Young people normally prefer easy, high-technology work. Employment in service and high technology industries, such as the electronics industry, is showing an upward trend while employment in the construction industry is decreasing.

In the construction industry, the number of engineers and managers trained in high technology is not sufficient to meet the demand. Skilled workers who are trained for manual work on site are also in short supply. Furthermore, the average level of skills is declining. The fact is that there are few newcomers to the construction industry, and the average age of skilled workers is going up. The general shortage of construction workers is illustrated in Figs. 4 and 5. There has been a shortage for the last 10 years, a shortfall of between 10% to 40%.

A close relationship exists between the shortage of construction workers and wages: a shortage of workers usually causes wage increases. However, wage increases are limited by the added value of the work. Figure 6 shows the relation between working hours and wages in the manufacturing and construction industries. Workers in the construction industry obviously work more hours and are paid lower wages. This may be difficult to change in the near future.

Productivity in the manufacturing industries has been increasing year by year as a result of increased high-level automation. However, it is difficult to find effective means for automation in the construction industry. Wages to be paid to construction workers are similar to those of other workers. In other words, they are paid well in spite of their low productivity. In order to keep the construction costs stable and within a socially acceptable level, standards and specifications have to be down-graded. This gives construction the image of being too expensive, and so-called low-cost, poor-quality buildings are becoming more popular, causing a decrease in investment for construction.

To deal with this problem, effective means are required to cut labor costs and maximize use of worker skills. Construction robots are considered one of the best solutions.

Improving Labor Productivity

General contractors and subcontractors are always looking for improvements in labor productivity. Changes in labor productivity in Japan for manufacturing and construction are shown in Fig. 7. The gap in labor productivity has steadily widened since 1971. Despite the fact that the total value of construction investment has increased (Fig. 8), labor productivity has remained constant. This is one of the most serious problems in the construction industry in Japan.

It is difficult to compare labor productivity in construction in Japan with that in foreign countries, because the categories of statistical data available are generally different. However, there do not seem to be big differences between Japan and other advanced countries. The Japanese government has been cooperating with private companies with the aim of improving labor productivity in the construction industry. All agree that there should be a concrete, nationwide campaign to accomplish this.

The consensus is that automation must proceed quickly. Making use of robots, as well as mechanical and electronical technology normally applied to robots, could prove effective. Even though rapid productivity improvement as observed in manufacturing is unlikely, such productivity improvement as has been achieved by other industries using robots is very desirable.

Dramatic changes of workers, machines, labor, wages, and skills will occur in the construction industry while this improvement of productivity is under way. It is possible that traditional facilities could become useless one day, as the manufacturing industries have experienced. It is necessary to be prepared for such an event and thus guide the construction industry onto the right path.

SUCCESSFUL APPLICATION OF CONSTRUCTION ROBOTS

There is no standard definition for construction robots in Japan. In general, machines that can replace workers are called construction robots. Using information obtained from a survey conducted recently, a list of so-called

Table 1 Construction robots in Japan (1985)

No.	Name	Task	Features
1	SSR-2 (Shimizu Site Robot-2)	Fireproofing for steel structures	1. Self-positioning 2. Self-moving 3. Playback robot 4. The thickness of sprayed material is equal to that applied by a skilled worker 5. Continuous spraying even under poor working conditions

construction robots in Japan has been prepared, excluding nonworkable types (Table 1). Specifications for robots are also listed in the appendix.

CURRENT RESEARCH AND DEVELOPMENT PROJECTS

The construction industry is basic and very important to Japan. The government has made the research and development of construction methods and materials a major priority, regarding construction as necessary for all social activities, maintaining civil life and economic activity. The latest research activities in government and universities have mostly concerned the introduction of robot technology to construction. Typical examples are as follows:

WASCOR Project

The WASCOR (WASeda COnstruction Robot) Project (Hasegawa and Tamki, 1985) was a multiclient research project, in which Waseda University played the leading role. The application of robotics to the construction industry has been studied in Waseda's System Science Institute since 1977.

(Text continues on page 516.)

Mechanical Structure	Control system	Size (mm) L∗W∗H	Weight (kg)	Developer
1. Arm: 6 degrees of freedom 2. Omni-directional vehicle 3. Light weight	1. Electro-hydraulic servo 2. CP/PTP control 3. Detecting position and correcting itself by position sensor 4. Memory capacity CP: 128min(MAX) PTP: 3800 points	1750 ∗ 1350 ∗ 2500 | 3000	805	Shimizu[1] Kobe[2]

(continued)

Table 1 Construction robots in Japan (1985) (continued)

No.	Name	Task	Features
2	Mighty Jack	Steel beam positioning for erection	1. Fixed on column head by two grippers 2. Eliminates dangerous work
3	Concrete Slab-Finishing Robot	Slab concrete finish	1. Programmed travelling 2. Travel on wet concrete 3. Area of finishing is equal to skilled worker's
4	Drilling Robot	Drill for Tunnelling	1. Memorizes drilling patterns 2. Drills accurately
5_{-1} 5_{-2}	Automatic Concrete Sprayer	Concrete spraying for NATM[5]	1. Replace concrete worker under poor working conditions
6	Abrasive Jet Cutting Robot	Concrete cutting by abrasive jet	1. Remote control by touching sensor 2. Can trace any line
7	Automatic Heavy Reinforcing Bar Arranging Robot	Heavy reinforcing bar arranging	1. Reduces labor for iron bar handling 2. Handles staggered or radial arrangements
8	Stud Welding Robot	Stud dowel welding	1. Reduces labor for welding 2. Speeds up welding work

Mechanical Structure	Control system	Size (mm) L*W*H	Weight (kgf)	Developer
1. Crane attachment 2. Two column gripper 3. Beam positioning winch	1. Column gripping is sequence controlled 2. Radio control beam positioning	6700–7800 * 1000 * 1400	1800	Shimizu
1. Arms: 2 degrees of freedom with trowel 2. Travelling device with four rollers	1. Controlled by gyrocompass and measuring roller	1260 * 1270 * 1750	560	Kajima[3]
1. Five-arm jumbo excavator	1. Teaching playback control 2. Hydraulic on-off control			Kajima
1. Tractor attachment or truck mounted	1. Hydraulic on-off control 2. Program control			Kajima Taisei[4] Kobe
1. Arm-multiple degrees of freedom				Kajima
1. Arm: multiple degrees of freedom 2. Carrying or positioning steel bars	1. Manual operation 2. Remote control 3. Programmed control	4000 * 2200 * 2600	5500	TEPCO[6] and Kajima
1. Versatran type				Kajima

(continued)

Table 1 Construction robots in Japan (1985) (continued)

No.	Name	Task	Features
9	Exfoliated Wall Tile Detector	Detecting tile exfoliation	1. Saves labor and cost of tile inspection 2. Raises the detection accuracy
10	Auto Clamp	Clamp work for structural steel column erection	1. Eliminates dangerous work 2. Speeds up work
11	Clean Room Testing Robot	Leak testing of clean room filter	1. Measures automatically 2. Eliminates contamination by workers
12_{-1}	Concrete Distributing Robot	Fresh concrete distribution	1. Saves labor 2. Speeds up work
12_{-2}	Concrete Placing Crane	Fresh concrete distribution	1. Saves labor 2. Speeds up work
13	Multi-purpose Travelling Vehicle	Cleaning, grinding, etc. of concrete slab surface	1. Saves labor 2. Eliminates repetitive work

[1]Shimizu Construction Co., Ltd.
[2]Kobe Steel Ltd.
[3]Kajima Corporation
[4]Taisei Corporation

Mechanical Structure	Control system	Size (mm) L*W*H	Weight (kgf)	Developer
1. Suspended travelling			20	Kajima Takenaka[7]
	1. Radio control		270	Ohbayashi[8]
1. Traveler 2. Telescopic shaft on X-Y table	1. Probe positioning by touch sensor 2. 27ch FM remote control	750 * 1000 * 1620		Ohbayashi
1. Arm: 4 degrees of freedom horizontal type	1. On board operation		4513	Takenaka
1. Arm: 5 degrees of freedom vertical type	1. Remote operation			Ohbayashi
1. Travelling device 2. Cleaning and grinding module	1. Controlled by gyrocompass and measuring 2. Self-scheduling of travel path	1400 * 700 * 900	235 (cleaning) 270 (grinding)	Shimizu

[5]New Austrian Tunneling Method
[6]The Tokyo Electric Power Co., Inc.
[7]Takenaka Komuten Co., Ltd.
[8]Ohbayashi-Gumi Ltd.

Professor Yukio Hasegawa was the leader of WASCOR, started in 1982. Researchers were from 11 companies, including general contractors and construction machine manufacturers. The objective of the project was basic research and investigation for robotization in construction to improve productivity. The main results of the project were:

– Methods of work analysis, system design and robot modularization were developed;

– Requirements for robotization in construction work and problems caused by robotization were clarified;

– Two construction methods for robotization were designed.

Some of the results were reported at the ISIR (International Symposium on Industrial Robots) held in Tokyo in September, 1985.

"Ministry of Construction" Research Project

The Ministry of Construction started a 5-year project in 1983 for the "development of an advanced system for construction technologies with proper use of electronics" (Kobayashi, 1985). This is the biggest project in Japan concerning research of new construction systems (which include robots). The Ministry of Construction has the lead role in this project. The project is proceeding in cooperation with the Building Contractors Society, related professional groups, universities, and general contractors.

The main theme of the project is the advancement of construction technology, covering many related subjects including characteristics of the construction industry, requirements for utilizing electronics, development of electronics technology for construction, and proposals for new construction systems using electronics. The results of the study are expected to have great impact, eventually calling for a complete review of conventional construction systems. The budget allocated for this project is about 500 million yen, with development of actual construction robots performed in a separate project.

Research and Development Project of Robots for Critical Environment

The Research and Development Project of Robots for Critical Environment (Yamamoto, 1985) is one of the biggest projects run by MITI (Ministry of International Trade and Industry). Its goal is to develop robots and robotic technologies for critical environments. Therefore, robotization of construction is not the main objective; however, the results will be used in construction.

The national project was initiated in 1983 with a 20 billion yen budget, partly supported by the participants' own voluntary research. The objectives, master plan, and schedule of this project are shown in Tables 2, 3 and 4.

Development of a Decommissioning System for Nuclear Facilities

Robotic technologies for treating radioactive materials or working in radioactive environments have been developed by the government and some companies. Decommissioning of radioactive concrete shield walls in nuclear power plants is one of the main objectives in robotization.

Table 2 Research and development objectives for major technologies (MITI)

Item	Objectives
Robots for work in nuclear power plant	Develop robots which can inspect facilities and equipment and conduct sophisticated work such as repair by moving within the working environment with remote support given by an operator from a distance at nuclear related facilities such as nuclear power generation plants.
Robots for undersea operations	Develop robots which can conduct sophisticated operations such as maintenance, inspection and repair by moving 3-dimensionally in the ocean, while keeping the work position and posture with remote support given by an operator from a distance at facilities related to offshore oil field development.
Robots for fire fighting and rescue operations	Develop robots that can conduct sophisticated operations such as maintenance, disaster control and rescue by moving within the stricken area with remote support given by an operator from a distance at industrial and social facilities in the case of disasters.

SOURCE: Agency of Industrial Science and Technology

Table 3 Research and development objectives for major technologies (MITI)

Item	Technology
Robots for work in nuclear power plants	Locomotion Manipulation Perception Radiation tolerance
Robots for undersea operations	Technology related to position keeping Manipulation Perception
Robots for fire and rescue operations	Perception Tolerance

Table 4 Research and development schedule (MITI)

Research item	FY	1983 1984 1985 1986		1987 1988 1989 1990	
Robots for work in nuclear power plants	Unique technology	Conceptual design; Trial production and experiment	Interim assessment		Total assessment
	Total system			Conceptual design; Trial production and experiment; Detailed design	
Robots for undersea operations	Unique technology	Conceptual and basic design; Trial production and experiment		Experiment	
	Total system			Conceptual design; Trial production and experiment; Detailed design	
Robots for fire fighting and rescue operations	Unique technology	Conceptual and basic design		Trial production and experiment	
	Total system			Conceptual design; Trial production and experiment; Detailed design	
Fundamental technologies	1. Structural technology	Development of basic structural technology		Establishment of structural technology	
	2. Control technology	Development of basic control technology		Establishment of control technology	
	3. Support technology	Development of basic support technology		Establishment of support technology	

General contractors are developing methods and facilities for this purpose. Japan Research Reactor No. 3 (JRR-3) and the Japan Power Demonstration Reactor (JPDR) run by the Japan Atomic Energy Research Institute at Tokai, will be the first research plants for actual field tests of these machines in Japan. The JRR-3 reactor will be removed in one piece, like the Shippingport Reactor removal in the United States, requiring no dismantling work.

The concrete shield wall of the JPDR will be dismantled by using a cutting robot which has been under research since 1981. The decommissioning of the JPDR is scheduled for 1988–1989. The project is being led by the Agency of Industrial Science and Technology. The main robot manufacturers, civil general contractors, and nuclear engineering firms are participating in these projects.

CONCLUSION

The two main approaches to the introduction of robotic technology for construction work in Japan are introducing robots into traditional construction sites and into new construction fields.

Examples of the former are shown in this paper, but there are only a few examples that can strictly be called robots. The others are automatic equipment rather than robots, which have flexible and high-level functions. The decommissioning of a nuclear power plant is the only example of application to new fields. However, the technology achieved is expected to overcome severe conditions, and advance the field of construction robotics for difficult environments in addition to general construction work. A level of technology must be used which general contractors normally do not have. However, they will be able to develop their capabilities as well as overall project management. They are expected to change from general contractors to engineering service firms using new construction technology and systems.

The development of the construction robot in Japan has just begun and diligent research should continue in the future. Developers of construction robots expect big changes within 10 years. There will be many difficulties in bringing about a new era, but many opportunities as well

ACKNOWLEDGMENT

The authors would like to thank Professor Irving J. Oppenheim of the Civil Engineering Department, Carnegie-Mellon University, for his suggestions regarding a survey of robotic development in Japan. We would also like to express our gratitude for the kind cooperation given by Japanese companies concerning current projects.

APPENDIX

SSR-2 (Shimizu Site Robot-2)

Spraying fireproof material, rock wool, and cement slurry is dirty and uncomfortable work because small particles of rock wool fill the surrounding area. This robot system is designed to maintain enough distance between a worker and the spray nozzle to improve conditions. It is also designed to perform work at a higher speed equal in quality to that of a skilled expert. The SSR-2 Robot (Fig. 9) characteristics are as follows:

- With a position sensing system that adjusts the posture of the robot in relation to steel beams to be sprayed, the robot can stand at the optimal position to repeat a spray action;
- It is an omni-directional programmable vehicle;
- Quality of the spray work (thickness of the sprayed material) is equal to that of a skilled worker.

Specifications are given in Table 5. Figures 10–13 give other details of the SSR-2 robot system.

Mighty Jack

This manipulator lifts two or three steel beams and sets them in the right position by sequential control (Figs. 14 and 15). When lifting beams, the manipulator itself is hung from a tower crane. While setting beams, the manipulator grasps the tops of columns and does not need to be lifted by a tower crane. The tower crane can be used for other jobs while the manipulator is working. The positioning and assembly work with the manipulator is as follows:

Puts cables on steel beams
Hangs up the manipulator with beams by a tower crane
Puts the manipulator on the top of 2 columns
Releases tower crane cables from the manipulator (manual)
Adjusts the distance between 2 columns
Sets beams in the right position one by one (Figs. 16 and 17)
Connects beams to columns (manual)
Lifts the manipulator to perform the next cycle.

Characteristics are safer steel beam assembly, rapid assembly work, and noise from tapping the jig to adjust bolting holes is eliminated. Specifications are given in Table 6.

Table 5 Specifications for the SSR-2 (Shimizu)

		Item	Specification
Travelling device		Degrees of freedom	3 (travelling revolution)
Travelling device	Travelling function	Power Travelling speed Positioning precision	hydraulic actuator min. 1.2 m/min max. 6 m/min ±5 mm
Travelling device	Revolution function	Power Revolution angle Revolution speed Revolution precision	hydraulic actuator ±102.5° (manual) 1±90° (automatic) min. 0.4 rpm max. 2 rpm ±0.5°
Travelling device	Outrigger function	Power Elevating speed Elevating precision	hydraulic actuator 200 mm/min ±1 mm
Travelling device		Height	max. 500 mm
Travelling device		Control sequence	automatic travelling method programmed by travelling distance and revolution angle
Travelling device	Correcting function	Confirming precision	±5 mm
Travelling device	Correcting function	Method of correcting position	Detecting the position and correcting itself by position sensor at the end of the manipulator arm.
Manipulator		Degrees of freedom	6 (Right-left turning, up-down traverse, In-out, Right-left swing, Up-down swing, Revolution)
Manipulator		Position precision	±5 mm
Manipulator		Sequential mode	Electro-hydraulic servo CP/PTP control
Manipulator		Memory capacity	CP: 4 ~ 128 min PTP: 3800 points
Others		Weight	805 kg (Manipulator 335 kgf Tractor 470 kgf)
Others		Measurement	Length: 1750 mm Width: 1350 mm Height: 2500 ~ 3000 mm
Others		Safety function	• Collision protecting device with tape switch • Optical device for detecting obstacles • Alarm device with rotary light

Table 6 Specifications of Mighty Jack Robot

Degree of freedom	1
Column gripper	2
Adjusting stroke of column grippers	1000 mm
Control system	Fixed sequence control and wireless teleoperation
Power source	Hydraulic
Dimensions	L 6700–7800, W 1000, H 1400 mm
Weight	1800 kgf
Hanging load capacity	1500 kgf

Concrete Slab-Finishing Robot

The concrete floor floater (Figs. 18 and 19) is for finishing concrete floor surfaces by trowel after pouring concrete. In the past, this work was always done by skilled workers. Because of the crouching posture required, the work is exhausting. Particularly in the winter when it takes a long time for concrete to harden, late night work is often unavoidable. This robot reduces the number of night shift workers by half and ensures quality equal to that provided by skilled workers. It features the following:

Trackless travelling robot with arm and trowel

Equipped with a gyrocompass and a travel distance sensor, the robot has a self-navigation device that enables it to determine its position, automatically adjusting its track if it runs off the right track

It can travel freely on concrete that has not yet hardened

Table 7 Specifications for Concrete Slab-Finishing Robot

Robot	
Dimensions Travel device	L 1270, W 1260, H 1750 mm (when stored: L 1450, W 1050, H 1570 mm)
Weight	480 kgf
Travel speed	0–200 mm/sec
Work speed	200–300 m^2/hour
Trowel	
Dimensions	Dia. 1020, H 690 mm
Weight	80 kgf
Type	4-blade revolving type
Blade revolutions	0–80 rpm
Blade angle	0–30°
Diameter of blade revolution	900 mm
Dimensions of blade	L 350, W 200 mm
Arm	
Type	Horizontal articulate type
Revolving angle	
First arm	−75–+75 deg.
Second arm	−105–+105 deg.
Safety devices	
Touch sensor	ON/OFF switch
Obstruction sensor	Ultrasonic wave senor
Floor opening sensor	Optical switch
Warning device	Patrol light (orange when travelling) Flash light (red, emergency)

Entirely controlled by computer, the travel and movement of the robot's arms can be operated using simple instructions

High quality finish (Fig. 20).

Table 7 gives the specifications for the Concrete Slab-Finishing Robot.

Drilling Robot

This robot can store the excavating hole patterns in its memory and operates automatically with a fully automatic 5-arm jumbo excavator (Fig. 21). The New Austrian Tunneling Method (NATM) has been widely used in Japan for tunnel construction. With additional mechanization, automation, and robotization, NATM tunnel construction has been increased. The automatic excavation machine (5-arm jumbo excavator) was developed for the purpose of increasing the excavation speed regardless of worker skill and maintaining a flat, smooth surface without over-excavation. The machine is positioned on the surface and then repeats the excavation procedure in accordance with the pattern in its memory.

Automatic Concrete Sprayer (Kajima)

Concrete spraying is one of the most important jobs of the NATM. This work accounts for about 20% of the hours required for tunnel construction. The operation of the nozzle normally requires a skilled nozzleman. There are also problems such as dust and the collapse of sprayed concrete. The automation of the nozzle work was, therefore, in urgent demand.

This sprayer controls the quantity of concrete sprayed according to the consistency of concrete (density and concentration) (Fig. 22). The quantity discharged is determined by the amount of air and air pressure and the quantity of accelerator added. These factors are controlled by computer, so the concrete spraying work can be carried out without a technician.

Automatic Concrete Sprayer (Kobe Steel)

The automatic concrete sprayer is one of the playback robots for concrete spray work for the NATM for tunnel construction (Figs. 23 and 24). Many manual manipulators have already been applied to this work, helping to free workers from hazardous working environments. This robot can control its arm to set the spray nozzle at the optimal position. The quality of sprayed concrete is thus better than that done by other manipulators. The set-up procedure is as follows:

Set the robot near the center of a tunnel,
Enter the three parameters of the tunnel section (right, left and top),
Input the conditions for spray work (range of arm rotation, rotating velocity).

Specifications are given in Table 8.

Abrasive Jet Cutting Robot

The abrasive jet method for cutting concrete structures by ultra high pressurized water jet is being used in constructing the tunnel between Aomori and Hakodate in Japan (Fig. 25). The manipulator, an essential part of industrial robots, has been used to work as an arm for moving the nozzle, enabling it to cut concrete structures in any direction. As the device is equipped with a "touch sensor" that maintains constant distance between the nozzle and the surface of the concrete, it provides very accurate cutting.

Table 8 Specifications of Automatic Concrete Sprayer (Kobe Steel)

Manipulator	
Arm rotation	260°
Arm transverse	1500 mm
Arm telescopic	1400 mm
Nozzle roll	90°
Nozzle pitch	180°
Load capacity	60 kgf
Repetitive precision	+50–−50 mm
Working area	1860 mm
Nozzle waving pattern	3 patterns
Vehicle	
Travelling velocity (MAX)	3 km/hour
Turning velocity (MAX)	3 rpm
Climbing angle (MAX)	57.5% (30°)
Turning radius (MIN)	400 mm
Controller	
Control system	Electro-hydraulic servo
Input system	Point to point
Power source	AC200/220V, 50/60Hz, 3 phase 17KVA or DC24V, 40A
Weight	8000 Kgf

Automatic Heavy Reinforcing Bar Arranging Robot

The foundation of a nuclear power plant has a huge number of long large-diameter reinforcing bars placed in layers. The weight of these (38 mm (1.5 in.) in diameter and 8 m (26 ft) long) is over 980 N (220 lbf). This work is executed by five to seven workers on the fixed bars. It is very difficult, must be performed with utmost care, and takes a long time. This robot can carry 20 reinforcing bars and travel on the laid bars, placing them automatically at prearranged intervals (Fig. 26). The robot moves the bars to the right and left, shifting the position of joints to form a zigzag or radial arrangement. With a change of arms, the robot can also perform vertical and horizontal placement of reinforcing bars for walls. This robot can reduce labor costs by 40% to 50%. As the cycle time for arranging reinforcing bars is about 1.5 minutes per bar, the total time for arranging reinforcing bars is cut by 10%.

Stud Welding Robot

In building nuclear power plants, large quantities of devices, reinforcing bars, ducts, and the like are embedded in concrete. In order to place these devices, stud dowels must be welded to reinforcing bars. The welding of several million stud dowels to reinforcing bars takes a long time and a great deal of labor. The welding robot (Fig. 27) executes this work effectively. The number of workers is reduced from three to two, and their working posture is improved. The ratio of welding defects has also been reduced from 1.9% to 0.7% with this robot, and the speed of the work increased by 10%.

Exfoliated Wall Tile Detector (Kajima)

The examination of exterior wall tiles for adhesion has conventionally been done by tapping tiles lightly with a hammer to detect loose tiles by sound. In 1974, Kajima Corp. developed a device that taps each tile mechanically and began using it for inspections (Fig. 28). So far, this inspection device has been applied to 40 construction projects. The newly developed detector has overcome most of the defects of previous machines and raised detection accuracy. This device has been awarded a certificate of high technical excellence by the Ministry of Construction.

The detector consists of four components: a travelling device set on a roof, a tapping device that travels along walls, a control device, and an exfoliation measuring device placed on the ground. The measuring device consists of a wave analyzer and a microcomputer, applying the principle of sound recognition. Since two parameters, maximum amplitude and frequency, are used, reliability is high. When the device finds loose tiles, the microcomputer signals "exfoliation," and at the same time automatically records the place where the exfoliation has been found. The detector adopts the nondestructive inspection method, with the following features:

Using this device is safer than performing manual inspections

As the tapping device is designed to climb walls by itself, no scaffolding is needed. It can be used for buildings up to 50 m (165 ft) high

The result of the measurements and the place of exfoliation are recorded automatically

The type of exfoliation, either behind the tile or between the concrete and bed mortar, can be estimated.

A system that can automatically draw a chart of the location and depth of exfoliation is under development.

Exfoliated Wall Tile Detector (Takenaka Komuten)

The exfoliated wall tile detector is operated remotely to detect weak bonds between tiles and external walls of the buildings (Figs. 29–31). This machine records where these weak bonds are on the wall. Its characteristics include low cost of tile inspection, safety, rapid inspection, and high reliability. Specifications are given in Table 9.

Auto Clamp

The function of this crane attachment (Fig. 32) is to release the cable from a steel column in high places using a wireless remote device (Fig. 33). This

Table 9 Specifications of Exfoliated Wall Tile Detector (Takenaka Komuten)

Dimensions	L 864, W 513, H 230 mm	
Weight	20 kgf	
Load capacity	6 kgf (MAX)	
Velocity	Upward	5–7 m/min.
	Downward	10 m/min.
Power	100v AC, 20A	
Mobility	2 caterpillars	
Clinging	2 blowers	

attachment, hung from the hook of a tower crane, has been used at two sites for steel column assembly work. First, a pair of magnetic devices with a shear pin are attached to the top of a steel column. After the column is set in place, the attachment is magnetically and mechanically released from the column by a PM teleoperation device. This machine is composed of a pair of magnetic and mechanical clamps, a control system, and a wireless communication system. Its characteristics are the elimination of manual releasing of the cable, and time and labor savings in assembly. Specifications are given in Table 10.

Clean Room Testing Robot

This robot checks for leaks in filters on the clean room ceiling (Fig. 34) and eliminates the additional spread of dust caused by manual inspection. Under the control of software programming, it moves from point to point sequentially. This robot is composed of a vehicle, telescopic measuring mast, and an X-Y table on which the telescopic measuring mast is mounted. Its characteristics are that it saves labor and time in inspections, and it is highly applicable to other inspection work by rearranging measuring devices and reprogramming. Specifications are given in Table 11.

Table 10 Specifications of Auto Clamp Robot

Hanging load (MAX)	1.5 tons
Weight of the clamps	33 kgf each
Weight (total)	270 kgf
Control	manual (when putting wire) & wireless teleoperation (when releasing wire)
Electric linear cylinder	
Load (MAX)	25 kgf
Stroke (MAX)	70 mm
Wireless operating system	multiplex transmission
Power unit (battery)	12 V, 65 A

Table 11 Specifications of Clean Room Testing Robot

Dimensions	L 1000, W 750, H 1620 mm
Composition	Vehicle, X-Y table (stroke of X-axis: 700 mm, Y-axis: 1000 mm) and mast (telescopic stroke: 300 mm)
Control	CPU Z80, ROM 4K, FM remote controller (27 ch.)
Sensing	5 touch switches (at the top of the probe) 4 ultrasonic sensors and touch switch (vehicle)
Power	4 dry battery 12 V, 36AH

Concrete Distributing Robot

The concrete distributing robot rapidly distributes fresh concrete (Fig. 35). In actual use the number of workers required is reduced from 15 to 10. This robot has been applied to distribute 100,000 m^3 (130,800 yd^3) of concrete. It frees workers from heavy work, improves the quality of reinforcement of concrete slabs, and speeds distribution. Specifications are given in Table 12.

Concrete Placing Crane

The concrete placing crane machine has been developed to automate concrete placement but has additional functions as a crane (Fig. 36). This machine went into operation in November 1984 and has already placed 10,000 m^3 (13,000 yd^3) of concrete. It is composed of 4 booms driven and controlled by a hydraulic servo. The control system has a microcomputer and is programmable for automatic operation. Its operation is very easy, even in the manual mode. The concrete placing crane rapidly pours concrete to wall forms, eliminates dirty heavy work, prevents misplacement of reinforcement bars by workers, reduces concrete placing labor costs, and is highly efficient and adaptable as construction equipment. Specifications are given in Table 13.

Multipurpose Travelling Vehicle

This vehicle can execute many kinds of finishing work on concrete slab surfaces, travelling automatically and avoiding obstacles such as columns

Table 12 Specifications of Concrete Distributing Robot

Working area	880 m^2
Reach (MAX)	18 m
Number of joint	4
Concrete pipe	5 inches dia., Steel
Operator	1
Power	AC200v, 7.5kw, Hydraulic
Weight	4,500 kgf

and walls. The work modules of this vehicle perform cleaning, grinding, and other related functions (Figs. 37 and 38). The work module and vehicle can be separated, making it easy to attach the new work module like a measuring device, with a handling arm and a tool for work on ceilings. Easy to operate, the vehicle recognizes its own path using the gyrocontrol device and the software program mounted in it. Thus no one is needed to supervise the vehicle's work. It is small and light enough to be carried by two men. Specifications are given in Table 14.

Table 13 Specifications of Concrete Placing Crane

Reach (MAX)	30 m
Height of the boom end (MAX)	58 m (when the mast height is 25 m)
Elevation	0–90° (0–85° as a crane)
Rotation	400°
Rotation of the distributor arm	140°
Operation	Automatic or manual teleoperation
Number of joints	5
Concrete pipe	5 inch dia., steel
Load capacity (as a crane)	2.7 tons × 10 m, 1.0 ton × 21 m
Lift stroke	50 m

Table 14 Specifications of Multipurpose Travelling Vehicle

Travelling vehicle	
Dimensions	L 600, W 700, H 900 mm
Weight	180 kgf
Travel speed	30–500 mm/sec
Power source	DC 24 V (battery)
Cleaning module	
Dimensions	L 700, W 700, H 900 mm
Weight	55 kgf
Work speed	8 m^2/min
Grinding module	
Dimensions	L 700, W 700, H 900 mm
Weight	90 kgf
Work speed	2 m^2/min

REFERENCES/BIBLIOGRAPHY

Engineering News Record, 1983
JAPAN TAKES EARLY LEAD IN ROBOTICS, Engineering News Record, July 21, USA.

Fenves, S. J. and Rehak, D. R., 1984
ROLE OF EXPERT SYSTEM IN CONSTRUCTION ROBOTICS, Proceedings of Workshop Conference on Robotics in Construction, Carnegie-Mellon University, Pittsburgh, PA.

Gatton, T. M., 1984
ROBOTIC ASSEMBLY FOR MOBILIZATION CONSTRUCTION, Proceedings of Workshop Conference on Robotics in Construction, Carnegie-Mellon University, Pittsburgh, PA.

Goto, K., Fukuda, T., Tanba, T., and Otsubo, Y., 1983
SELF CLIMBING INSPECTION MACHINE FOR EXTERNAL WALL, Robot No. 38, Japan Industrial Robot Association (JIRA), Tokyo.

Hasegawa, Y., 1981
ROBOTIZATION OF REINFORCED CONCRETE BUILDING CONSTRUCTION, Proceedings of 11th International Symposium for Industrial Robots (ISIR), JIRA, Tokyo.

Hasegawa, Y., 1983
ROBOTIZATION OF CONSTRUCTION WORK, Robot No. 38, JIRA, Tokyo.

Hasegawa, Y., 1984
ROBOTIZATION OF CONSTRUCTION WORK, Proceedings of Workshop Conference on Robotics in Construction, Carnegie-Mellon University, Pittsburgh, PA.

Hasegawa, Y. and Sugimoto, N., 1982
INDUSTRIAL SAFETY AND ROBOTS, Proceedings of 12th ISIR, Paris, ISF Publications, England.

Hasegawa, Y. and Tamaki, K., 1985
MODULES FOR CONSTRUCTION ROBOTICS SYSTEMS, Proceedings of 2nd Workshop Conference on Robotics in Construction, Carnegie-Mellon University, Pittsburgh, PA.

Hasegawa, Y. and Tamaki, K., 1985
ROBOT MODULE APPLICATION FOR COMPLICATED CONSTRUCTION SYSTEM, Proceedings of 15th ISIR, Tokyo.

Kangari, R., 1985
ROBOTICS FEASIBILITY IN THE CONSTRUCTION INDUSTRY, Proceedings of 2nd Workshop Conference on Robotics in Construction, Carnegie-Mellon University, Pittsburgh, PA.

Kano, N., 1985
CONSTRUCTION PLANNING BY A ROBOT, Proceedings of 2nd Workshop Conference on Robotics in Construction, Carnegie-Mellon University, Pittsburgh, PA.

Kobayashi, K., 1985
DEVELOPMENT OF ADVANCED SYSTEM FOR CONSTRUCTION TECHNOLOGIES WITH PROPER USE OF ELECTRONICS, Proceedings of 15th ISIR, Tokyo.

Matsubara, S., 1983
FLOOR CLEANING ROBOT, Robot No. 38, JIRA, Tokyo.

Mori, M., Nakamura, T., Matsushita, H., and Tase, Y., 1983
SOME EXAMPLES OF ROBOTIZATION IN OUR COMPANY, Robot No. 38, JIRA, Tokyo.

Murai, T., Aogyagi, H., and Kawamura, T., 1983
CONCRETE DISTRIBUTING ROBOT, Robot No. 38, JIRA, Tokyo.

Nof, S. Y., 1985
HANDBOOK OF INDUSTRIAL ROBOTICS, John Wiley & Sons, New York.

Paulson, Jr., B. C., 1984
THE POTENTIAL FOR ROBOTICS IN CONSTRUCTION, Technical paper of Manufacturing Engineers, Robots West Conference, Anaheim, CA.

Rader, M., 1985
POTENTIAL FOR ROBOTIZATION AND AUTOMATION IN GERMAN CIVIL ENGINEERING AND

CONSTRUCTION INDUSTRY, Proceedings of 2nd Workshop Conference on Robotics in Construction, Carnegie-Mellon University, Pittsburgh, PA.

Saito, M., Tanaka, N., Arai, K., and Banno, K., 1985
THE DEVELOPMENT OF A MOBILE ROBOT FOR CONCRETE SLAB FINISHING, Proceedings of 15th ISIR, Vol. 1, Tokyo.

Sangrey, D. A., 1985
QUALITY AND RELIABILITY AS MOTIVATIONS FOR CONSTRUCTION ROBOTICS, Proceedings of 2nd Workshop Conference on Robotics in Construction, Carnegie-Mellon University, Pittsburgh, PA.

Sangrey, D. A. and Warszawski, A., 1984
CONSTRAINTS ON THE DEVELOPMENT OF ROBOTS FOR CONSTRUCTION, Proceedings of Workshop Conference on Robotics in Construction, Carnegie-Mellon University, Pittsburgh, PA.

Skibniewski, M. and Hendrickson, C. T., 1985
EVALUATION METHOD FOR ROBOTICS IMPLEMENTATION: APPLICATION TO CONCRETE FORM CLEANING, Proceedings of 2nd Workshop Conference on Robotics in Construction, Carnegie-Mellon University, Pittsburgh, PA.

Tanaka, H., 1985
HUMAN IMPLICATIONS OF ROBOTIZATION IN THE WORKSITE: THE JAPANESE EXPERIENCE, Robotics, Vol. 1, No. 3, Elsevier Science Publishers B.V., Netherlands.

Terai, T., 1986
THE CONSTRUCTION INDUSTRY IN THE ADVANCED INFORMATION SOCIETY, Science & Technology in Japan, Vol. 5, No. 17, Tokyo.

Ueno, T. and Yoshida, T., 1983
ROBOTIZATION OF SPRAY WORK FOR FIREPROOFING STEEL STRUCTURE, Robot No. 38, JIRA, Tokyo.

Warszawski, A., 1984
APPLICATION OF ROBOTICS TO BUILDING CONSTRUCTION, Proceedings of Workshop Conference on Robotics in Construction, Carnegie-Mellon University, Pittsburgh, PA.

Warszawski, A., 1984
ECONOMIC EVALUATION OF ROBOTICS IN BUILDING, Proceedings of Workshop Conference on Robotics in Construction, Carnegie-Mellon University, Pittsburgh, PA.

Warszawski, A., 1985
ECONOMIC IMPLICATIONS OF ROBOTICS IN BUILDING, Building and Environment, Vol. 20, No. 2, Great Britain.

Yamamoto, K., 1985
CURRENT CONDITIONS AND PROSPECTS OF RESEARCH AND DEVELOPMENT ON THE MOST ADVANCED ROBOTICS TECHNOLOGY IN JAPAN, Proceedings of 1985 ICAR, Tokyo.

Yoshida, T., Ueno, T., Nonaka, M., and Yamazaki, S., 1984
DEVELOPMENT OF SPRAY ROBOT FOR FIREPROOF COVER WORK, Proceedings of Workshop Conference on Robotics in Construction, Carnegie-Mellon University, Pittsburgh, PA.

Yoshida, T. and Ueno, T., 1985
DEVELOPMENT OF A SPRAY ROBOT FOR FIREPROOF TREATMENT, Shimizu Technical Bulletin, No. 4, Shimizu Construction Co., Tokyo.

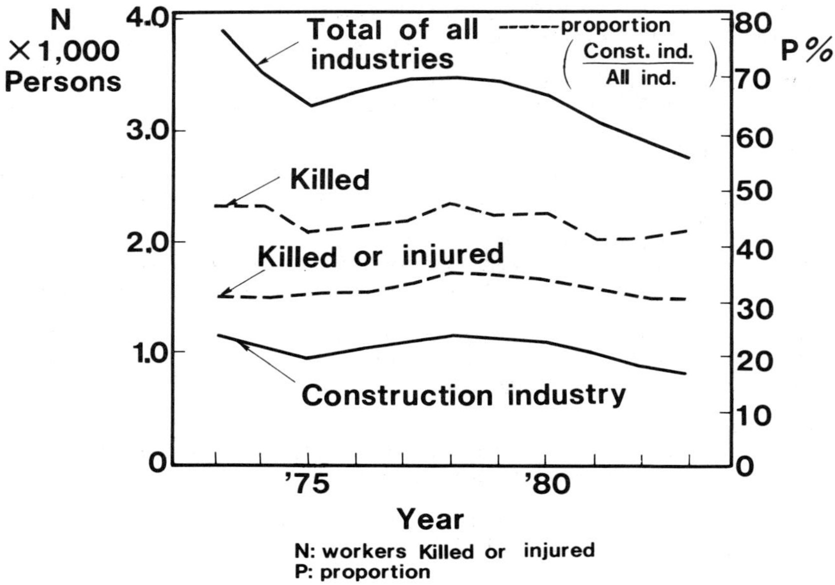

Fig. 1 Construction employees killed or injured in accidents

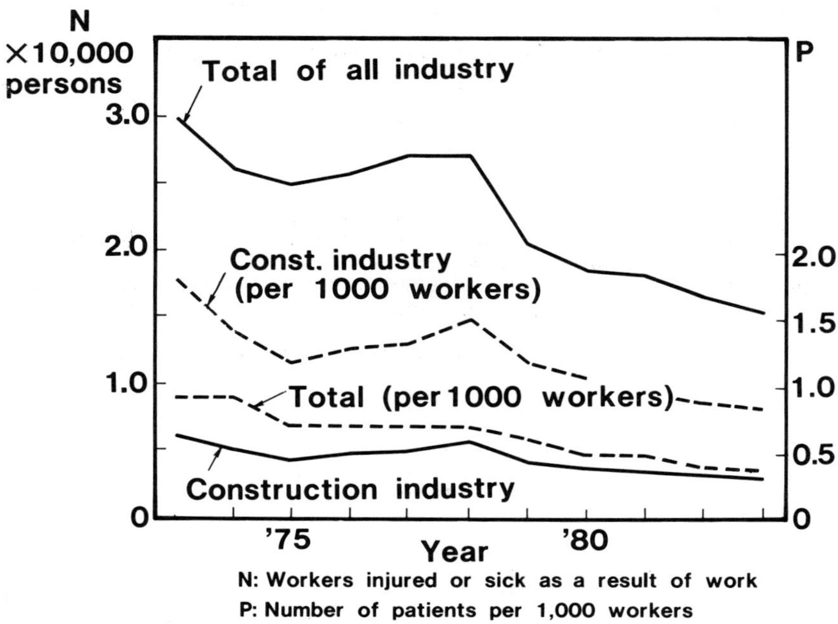

Fig. 2 Trends in occupational injuries or sickness

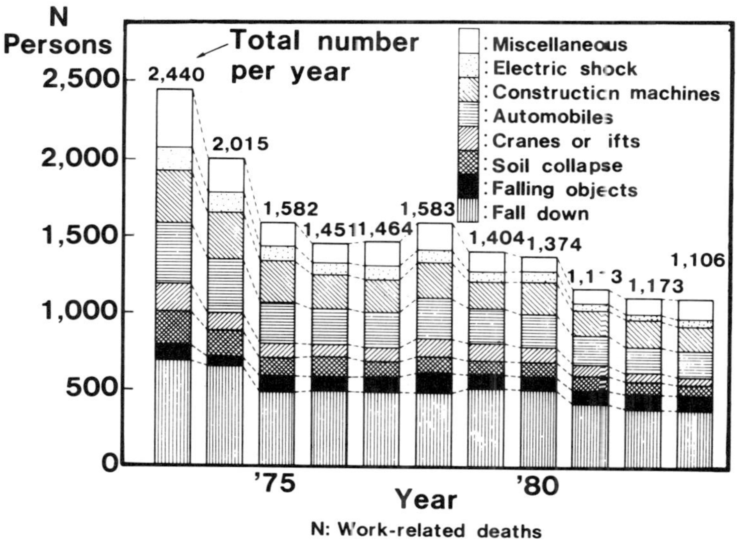

Fig. 3 Deaths on construction sites

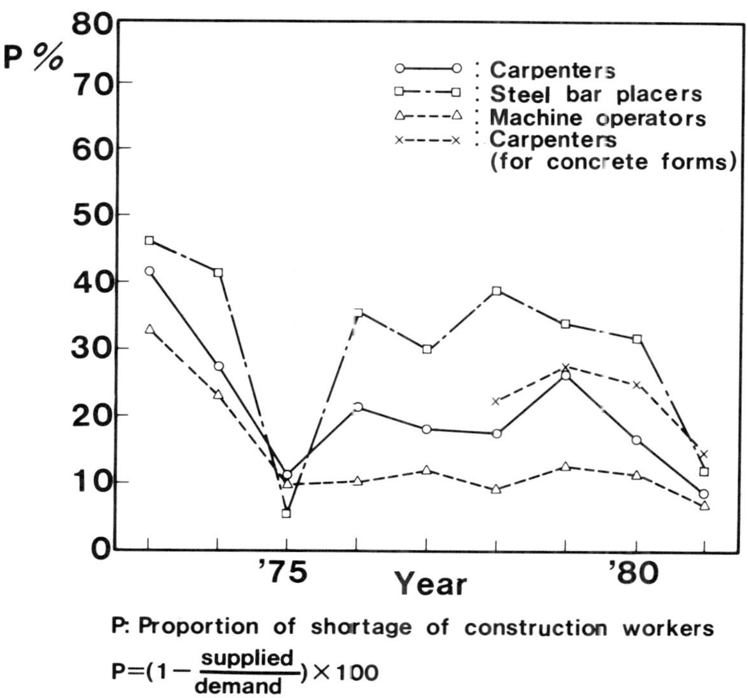

Fig. 4 Shortage of construction workers *(Yoshida et al., 1984)*

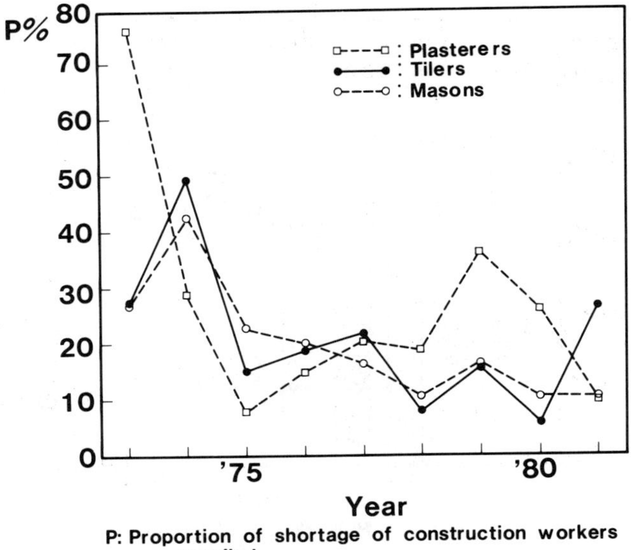

$$P = (1 - \frac{\text{supplied}}{\text{demand}}) \times 100$$

Fig. 5 Shortage of construction workers *(Paulson, 1984)*

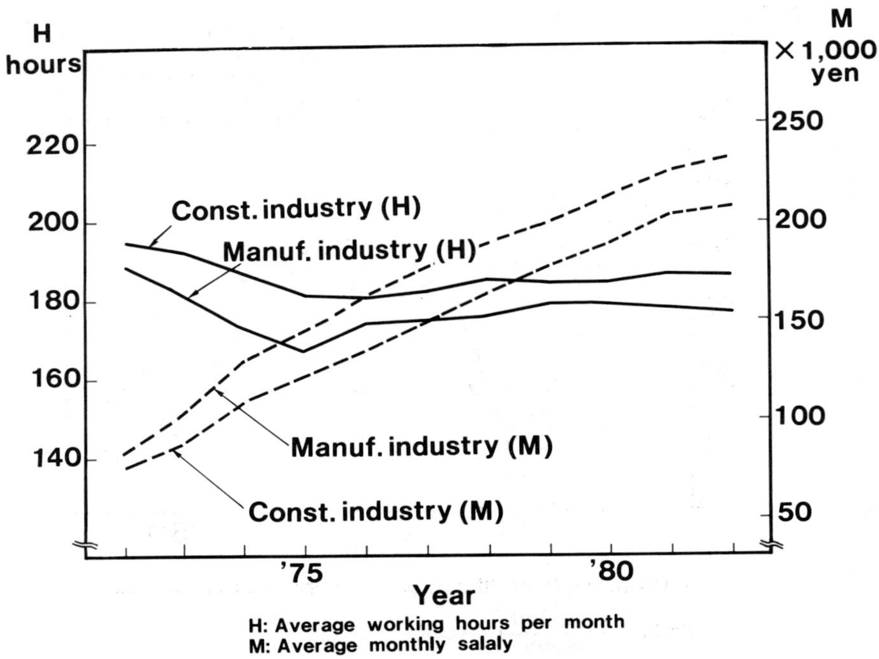

Fig. 6 Working hours and earnings

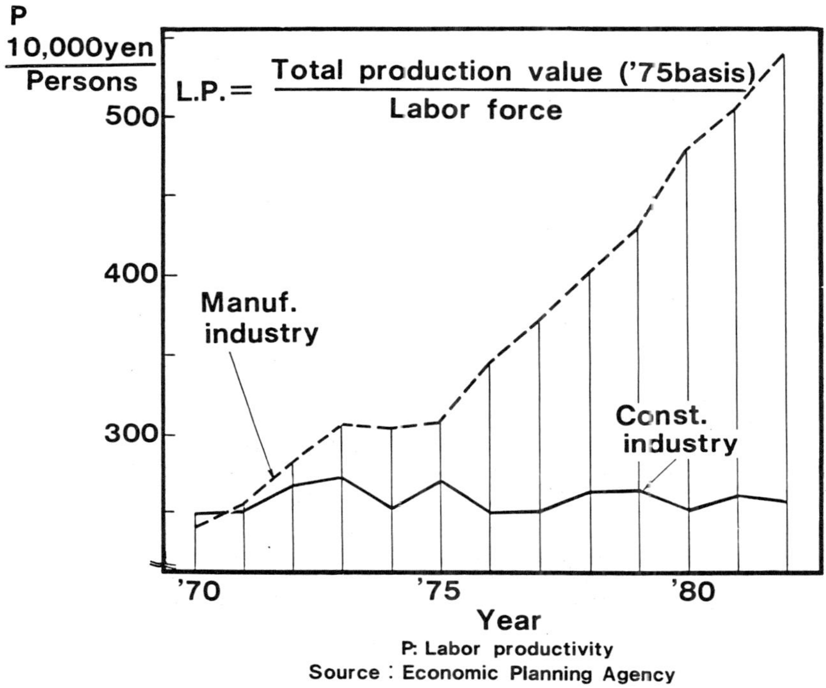

Fig. 7 Trends in labor productivity in Japan

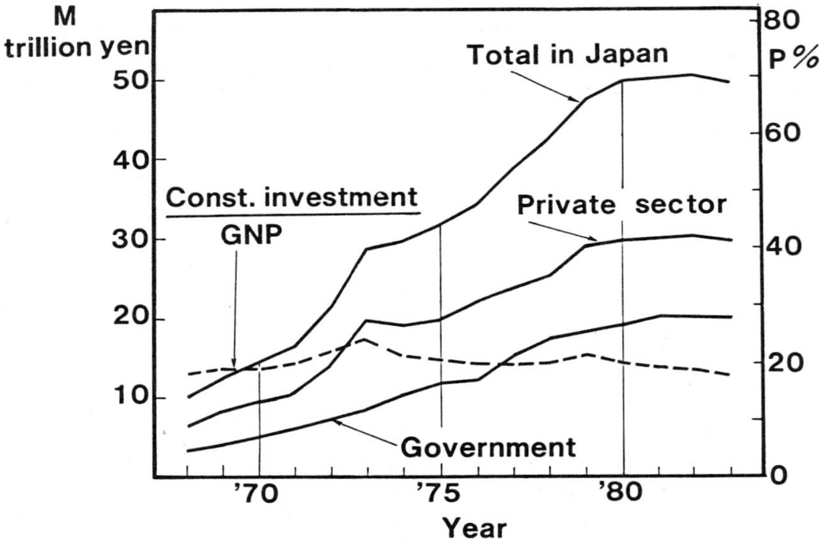

M: Total value of construction investment
P: Proportion of construction investment to GNP
Source: Science and Technology Agency

Fig. 8 Trends in construction investment

Fig. 9 SSR-2 (Shimisu Site Robot -2)

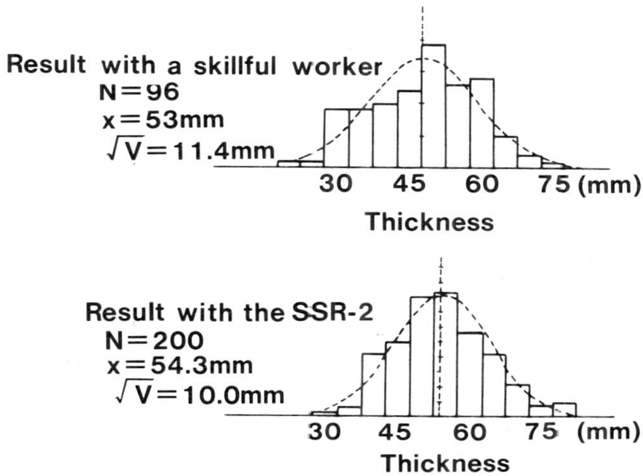

Fig. 10 Distribution of fireproofing material thickness *(Shimizu)*

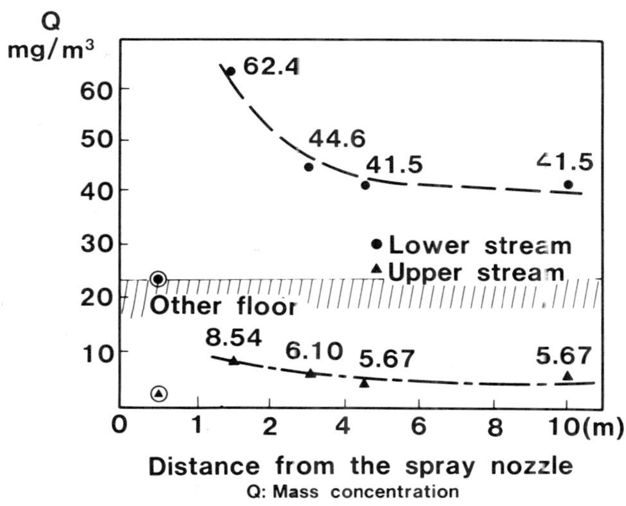

Fig. 11 Dust concentration *(Shimizu)*

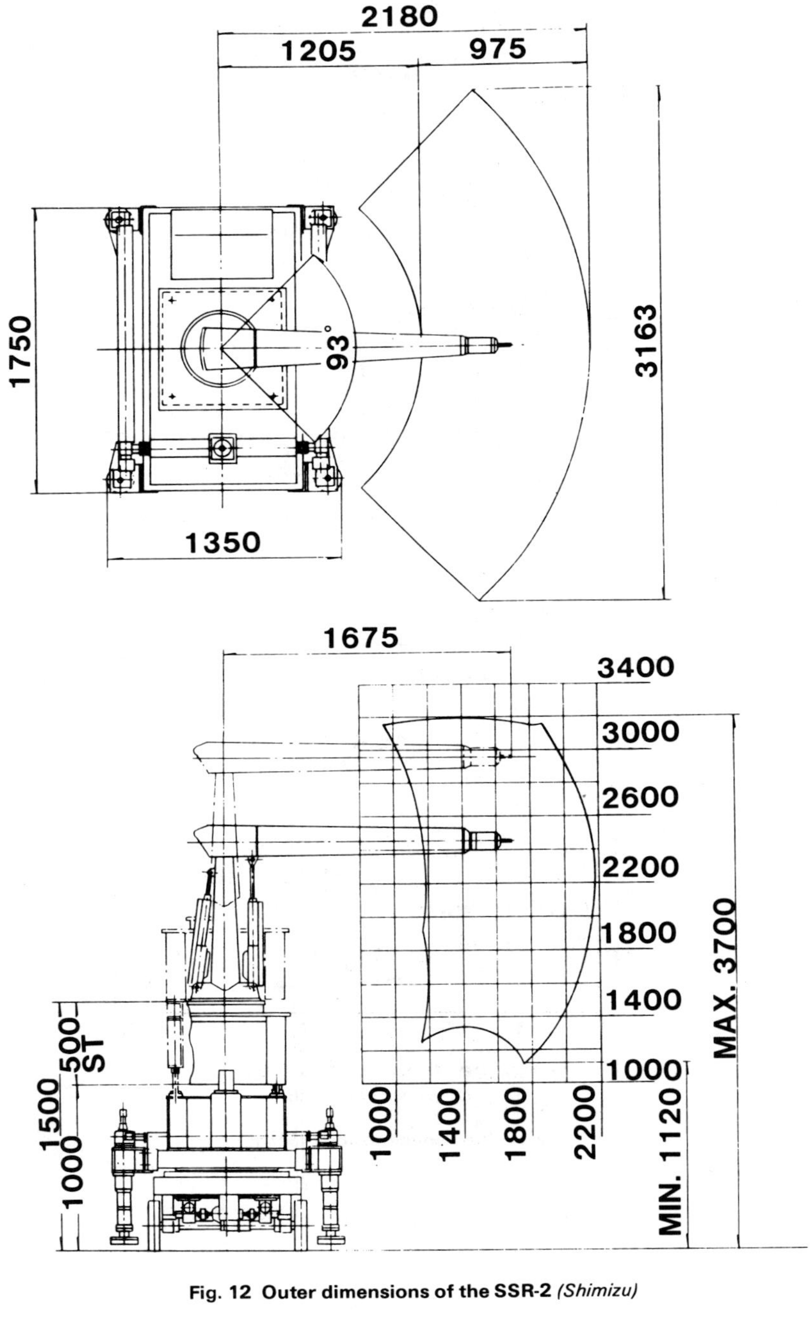

Fig. 12 Outer dimensions of the SSR-2 *(Shimizu)*

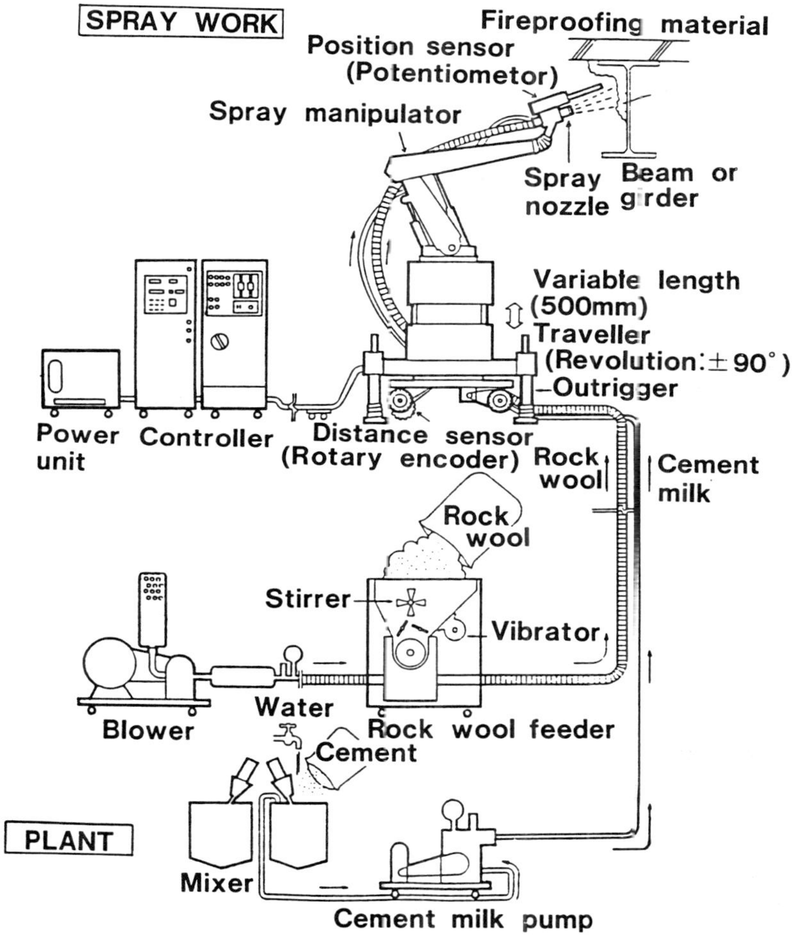

Fig. 13 Rock wool spray system of the SSR-2 *(Shimizu)*

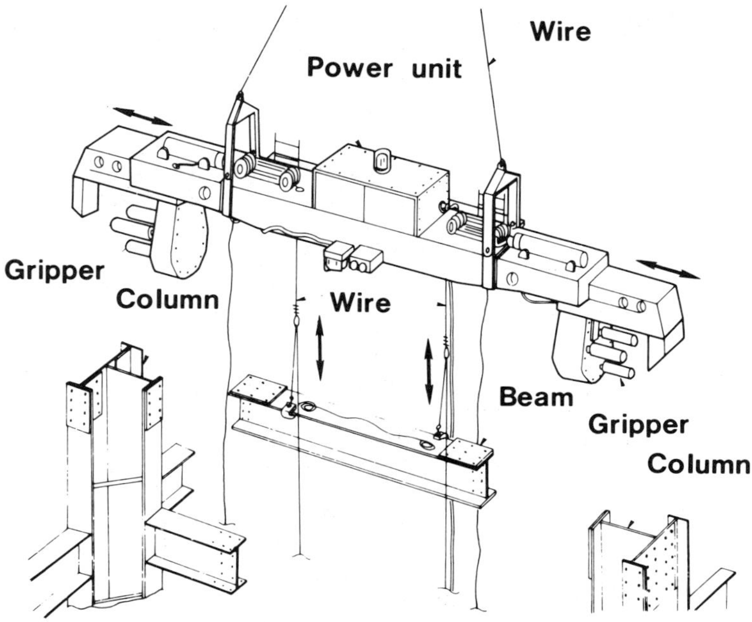

Fig. 14 Mighty Jack *(Shimizu)*

Fig. 15 Mighty Jack *(Shimizu)*

Fig. 16 Fixing upper beam

Fig. 17 Fixing lower beam *(Shimizu)*

Fig. 18 Concrete slab-finishing robot *(Kajima)*

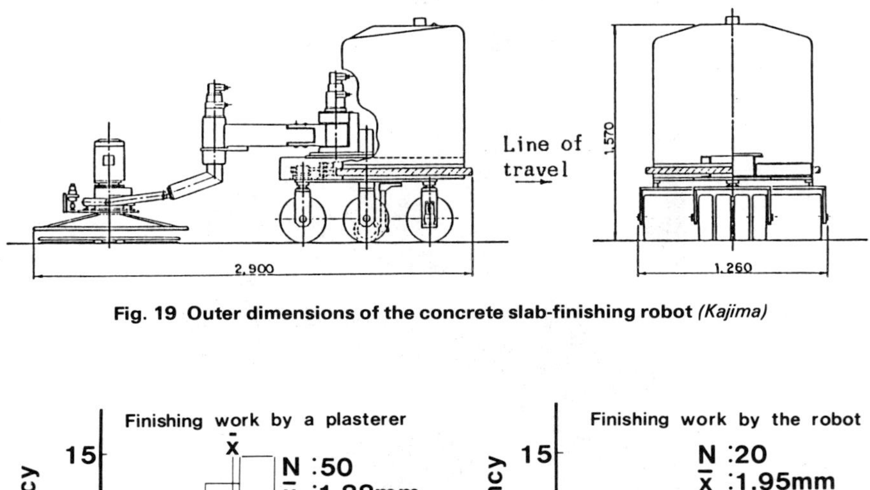

Fig. 19 Outer dimensions of the concrete slab-finishing robot *(Kajima)*

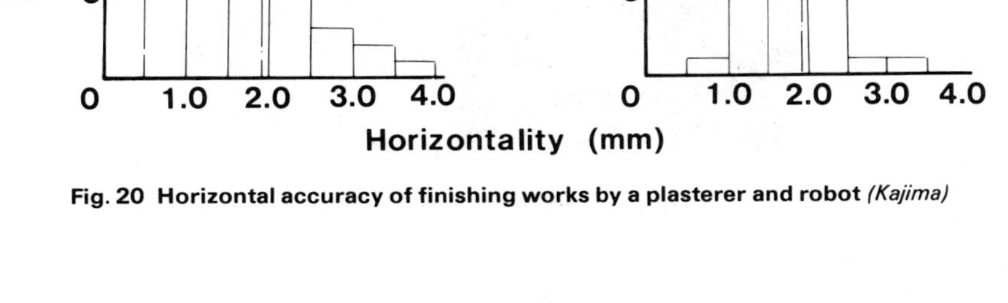

Fig. 20 Horizontal accuracy of finishing works by a plasterer and robot *(Kajima)*

Fig. 21 Drilling robot *(Kajima)*

Fig. 22 Automatic concrete sprayer *(Kajima)*

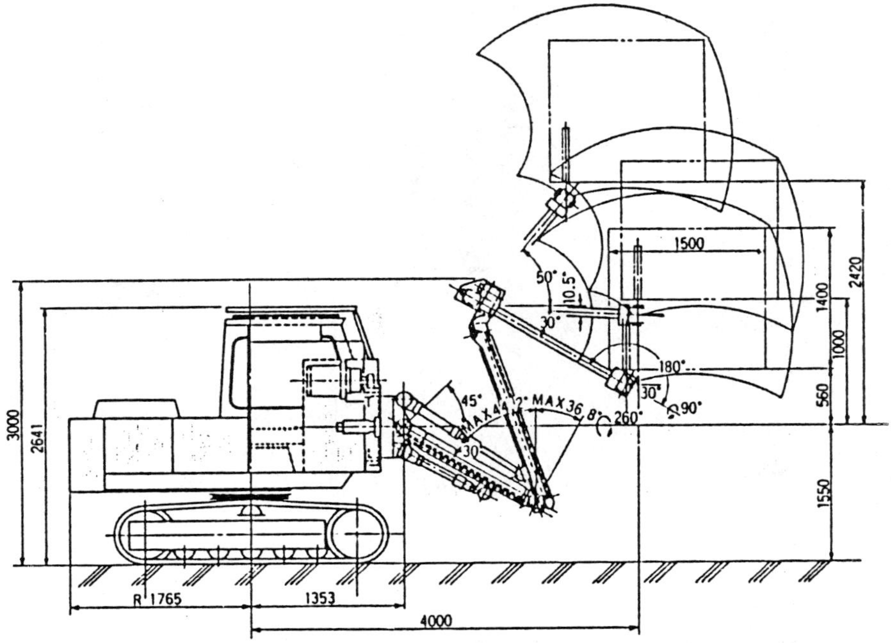

Fig. 23 Outer dimensions of the automatic concrete sprayer *(Kobe Steel)*

Fig. 24 Automatic concrete sprayer *(Kobe Steel)*

Fig. 25 Abrasive jet-cutting robot *(Kajima)*

Fig. 26 Automatic heavy-reinforcing-bar-arranging robot *(Kajima)*

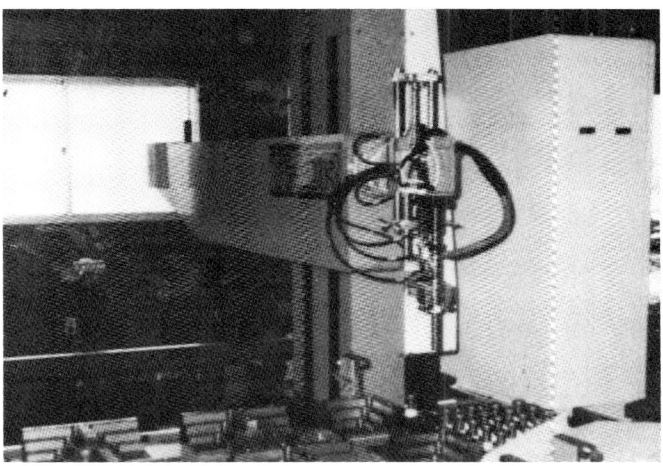

Fig. 27 Stud-welding robot *(Kajima)*

Fig. 28 Exfoliated wall tile detector *(Kajima)*

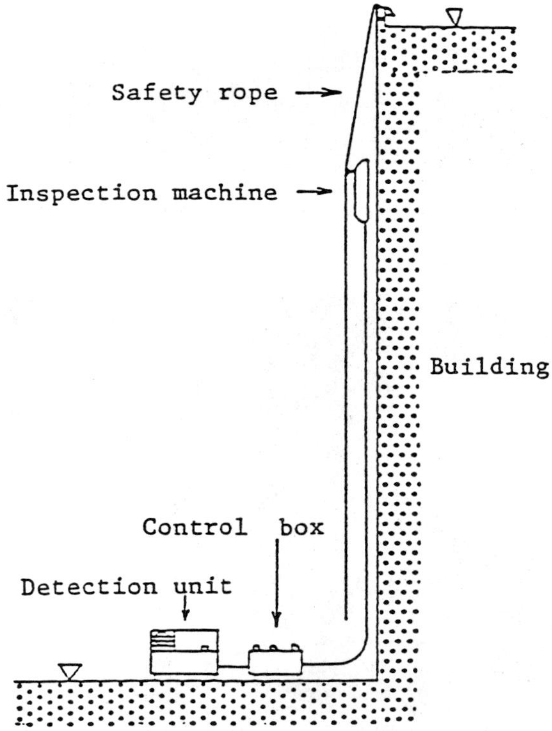

Fig. 29 General features of external wall inspection device *(Takenaka Komuten)*

Fig. 30 **Exfoliated wall tile detector** *(Takenaka Komuten)*

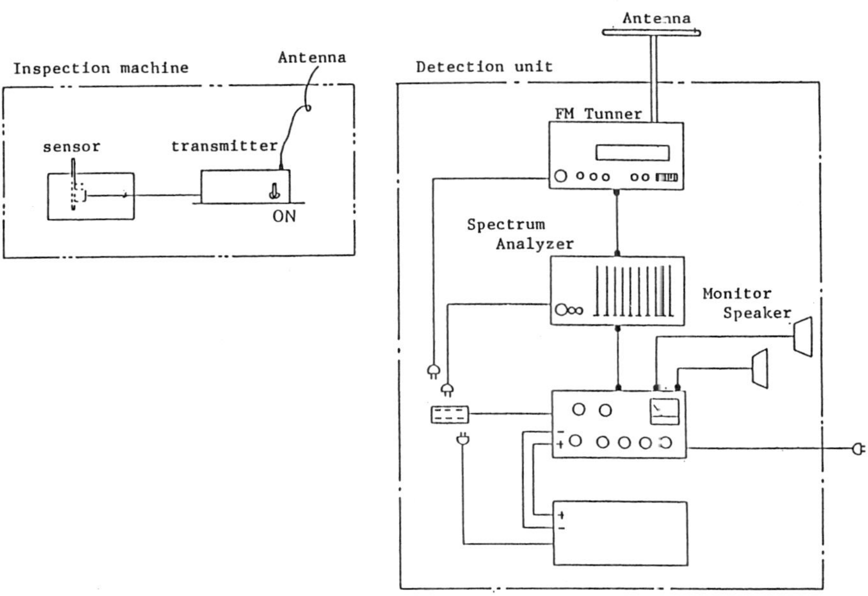

Fig. 31 **Block diagram of the detecting circuit** *(Takenaka Komuten)*

Fig. 32 Auto clamp

Fig. 33 Auto clamp placing a steel column *(Ohbayashi Gumi)*

Fig. 34 Clean room testing robot *(Ohbayashi Gumi)*

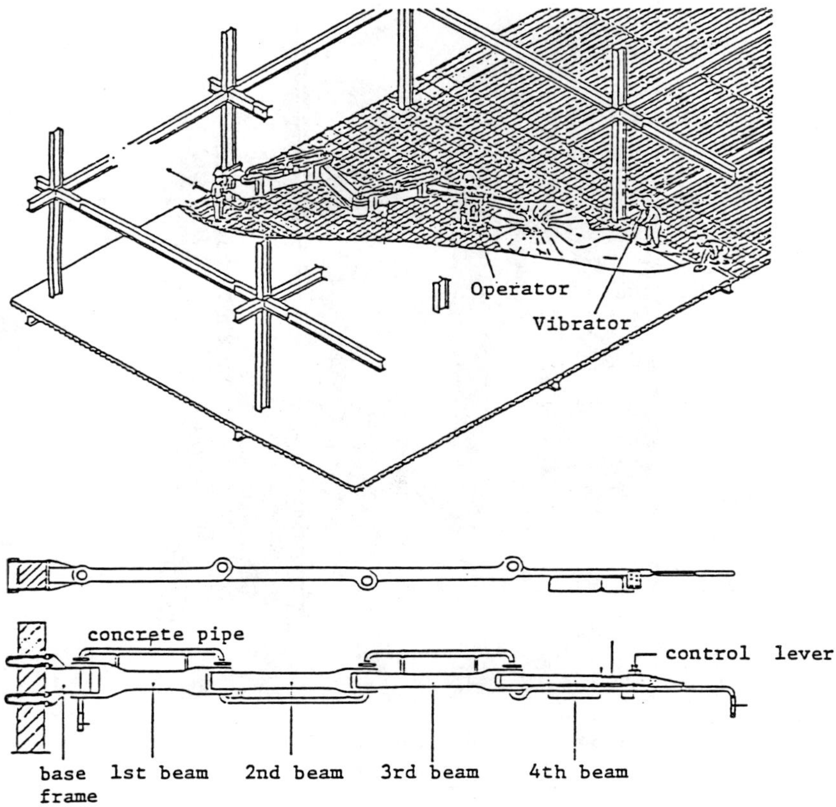

Fig. 35 Concrete distributing robot *(Takenaka Komuten)*

Fig. 36 Concrete placing crane *(Ohbayashi Gumi)*

Fig. 37 Multipurpose travelling vehicle—slab surface cleaning *(Shimizu)*

Fig. 38 Multipurpose travelling vehicle—slab surface grinding module *(Shimizu)*

Application of Systems Methodology

Computer Integrated Facilities Management

Jeffrey E. Harkness

Facilities management, according to the International Facilities Management Association (IFMA), is concerned with the design, construction, maintenance, and management of the physical environment as it relates to people and work processes. This definition is very general but includes a number of distinct and interrelated areas. Add to this definition the large amount of space that is involved in the management of tall buildings and it becomes obvious that to manage these large amounts of space effectively, a computerized system that integrates all areas is required.

Some of the areas that are important to facilities management are listed in Table 1. In this table, the areas are divided into two levels: those directly under the jurisdiction of the facilities management department such as facilities planning and space management, and those areas that are not directly under the jurisdiction of the facilities manager, such as human resources, accounts payable/receivable, fixed asset management, and purchasing. Table 1 illustrates how these functions interrelate with each other. For example, facilities planning needs information from human resources with respect to the number of personnel currently assigned to a given department; the facilities planner also needs information from space management to identify how much space currently is being used by a given department and from lease management to determine the current lease rate. In order for facilities planning to meet the needs of an organization, the above information is necessary for the facilities planner to design or modify space to meet the projected needs of an organization. Another example is space management. In order to manage a space effectively, information from the construction drawings such as electrical, HVAC and building layout is needed. Also, information from facilities maintenance and communications management is needed to meet the needs of the people who occupy the space.

FACILITIES PLANNING

The purpose of facilities planning is to develop a facility specification that meets an organization's strategic plan and maintains a high level of productivity. This specification consists of where the space will be, how much space is required, and how long the space will meet the organization's needs. The specification will also include the physical placement of departments with respect to each other (stacking and blocking) and the design of spaces to insure a high level of productivity.

SPACE MANAGEMENT

The purpose of space management is to maintain an organization's strategic plan and productivity goals as they relate to the physical environment at a minimum cost. Space management is generally concerned with facilities maintenance, energy management, security management, communications management and lease management. In general, the facilities maintenance system will have a major impact on the operating cost of a facility as will

Table 1 Areas important to facilities management

UNDER JURISDICTION OF FACILITY MANAGER
Facilities Planning
Space Management
Facilities Maintenance
Lease Management
Communications Management
Security Management
Energy Management
Construction Management
Project Management
NOT UNDER JURISDICTION OF FACILITY MANAGER
Human Resources
Fixed Assets
Accounts Payable/Receivable
Purchasing

energy management. Effective management of communications enables the facilities manager to better meet the voice and data needs of an organization at minimum cost. Security management is responsible for the safety and security of its personnel as well as protecting its assets.

The effective management of large facilities will result in reduced operating costs, increased productivity, and substantial improvement in meeting the needs of those occupying the facility. To obtain these benefits, a computer system is necessary since a large amount of space means a large amount of information and this information needs to be accessible, manipulated, and reported. Because of the nature of facilities management, this information consists of both traditional alpha-numeric as well as graphic information. Because of the inherent interrelationship between the various facility management tasks, computerization alone will not give the maximum benefits. In order to maximize these benefits, an integrated information base is required. With an integrated data base, the facilities management staff has all the necessary information available. The information being maintained by the space management group is available to the facilities planners to aid them in their projections and analyses. Without this integration, having accurate, up-to-date information available for the decision making process is impossible and leads to ineffective, inefficient, and costly management of the facility.

Until recently, the computerization of facilities management has proceeded from two fronts. One, the graphics front in which the graphics vendors were trying to incorporate nongraphic information into their graphic database. This, in general, has been very unsuccessful since the graphics vendors have expertise in graphic systems but very little expertise in database management of traditional information and facilities management. The other direction has been strictly nongraphic database management addressing those areas such as facilities planning and space management, however, neglecting the important graphics requirements of the information base. Both these approaches have been unsuccessful in meeting the needs of the facilities management department.

Recently, a new approach to the computerization of facilities management has emerged. In this approach both the graphic system and the database management systems are viewed as tools for the development of a computer integrated facilities management system. This approach allows for maximum flexibility and a system that is able to incorporate new technologies in graphics and database management as these technologies become available to the users.

Reduce Operating Expenses by Computerizing Facilities Management

Grant S. Farquhar

The operating income of the facilities manager today is coming under increasing pressure from two major fronts. First, the marketplace is demanding increased levels of service to support existing and prospective tenants, and second, the cost of operating facilities is increasing each year without interruption.

As reported in the 1985 *Experience Exchange Report* prepared by the Building Owners and Managers Association (BOMA), following after increases in property taxes/insurance and energy, repair and maintenance expenses showed the largest increase of all operating expenses in U.S. private sector office buildings. Unfortunately, managers have little control over increases in taxes and insurance. Substantial capital expenditures are typically required to reduce energy expense. Therefore, repair and maintenance costs, which now average \$13.67/m^2 (\$1.27/ft^2), present the facilities manager with the best opportunity to further control and reduce operating expenses. In order to realize this opportunity, management must rethink many of the policies and practices used to control and deploy resources. Other segments of American industry have made great strides in this area and the facility manager can do the same with the proper perspective. Realistically, an automated system is required to provide the information needed to exercise the added control.

Nearly half of all operating expenses can be impacted by automating the facilities management functions and simultaneously reexamining the internal practices used to deploy and control resources.

An automated facilities management tool should encompass all of the major functions of the facilities manager, as presented in Fig. 1, in a user friendly fashion and with an assurance of accuracy. Fundamental to exercising additional control, the facilities manager must be able to readily access many different types of information from the systems data bases. Some examples of the information required include locations and drawings of space and equipment; contracts in place; available and required personnel/materials; vendors and contractors; predictable work requirements; budgeted and actual expenditures; and occupant changes. Availability and responsiveness are necessary characteristics of the system if it is to play a key role in supporting the daily activities of the facility manager. The information must be available at the time situations come up where decisions are made.

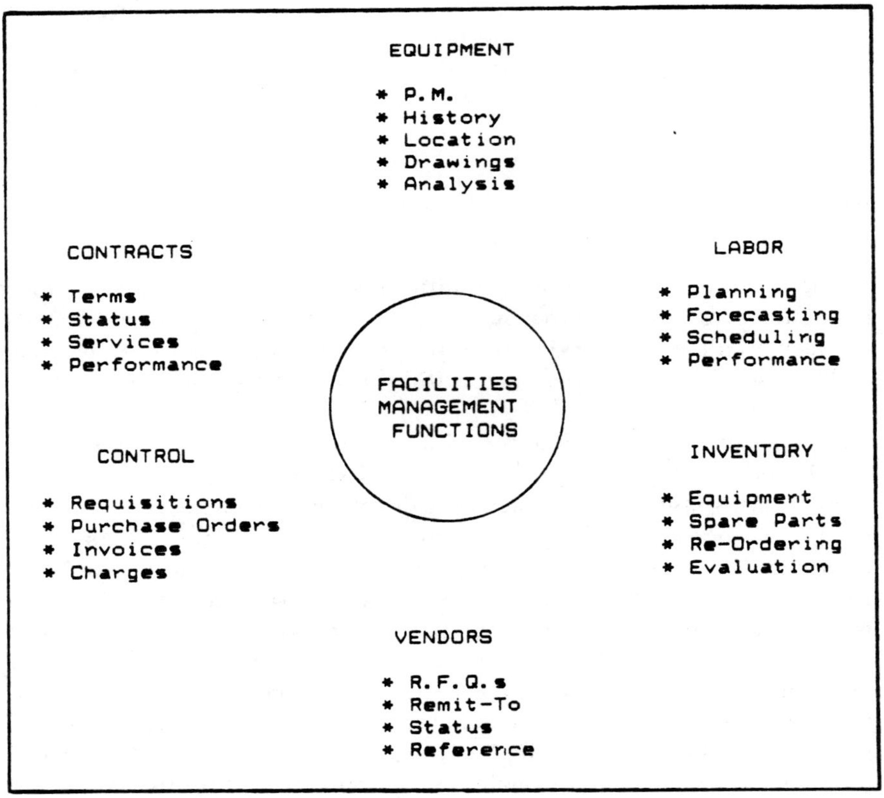

Fig. 1 Major functions of facilities manager

Batch systems and smaller systems with limited capacity have their place, but are not able to offer the user anything more than a reporting of activities and performance of the fact. While beneficial, this will not provide the support needed to markedly change the operating environment.

Areas offering the primary opportunity for cost reductions are contracts administration, labor productivity, procurement practices, and equipment performance. Other additional expense areas that can be influenced by automation are utilities, administration and payroll, janitorial services, and roads/grounds/security. Design, construction, and rearrangement subsets can reduce the cost of renovation and reduce occupancy delays.

The effective property manager of tomorrow must be aggressive with these operating costs to ensure economic viability of the building and to attract tenants by offering operating costs projected at below industry standards. This might involve challenging practices and standards of the past and necessitate the implementation of new disciplines and approaches geared to delivering better service at a reduced cost.

FUNCTIONS OF A COMPUTERIZED FACILITIES MANAGEMENT SYSTEM

The functions that the facility manager should include in the system are those of many disciplines, from engineering, to space management, to design, to accounting. The scope of these functions are as follows:

Equipment: Equipment specifications, preventive maintenance and inspection requirements, drawings, location, vendor, repair and performance history, warranties, and replacement cost analysis.

Labor: Workload quantification by skill and frequency, assignment to specific crews or contracts, sequencing and scheduling of tasks, forecasting and performance indicators.

Vendors and Contractors: Reference look up by services or materials provided, invoicing specifications, history, contracts, and current status reports.

Contracts: Terms and conditions, request for quotes and competitive bids, scope of work, insurance, invoicing and payment status, and performance reports.

Inventory: Furnishings and spares tied to vendor and contracts files, order status by job, and performance of select materials or equipment.

Control: Budget reports, requisitions, and authorization for expenditures, departmental charges, invoicing and accounts payable, direct and indirect cost accounting, blanket purchase order release, and contract payments.

These functions can be integrated in a logical form with numerous other systems: accounting, security and lock control, design, construction and rearrangement, energy performance monitoring systems, payroll, cabling, and much more. A sophisticated computerized system will allow on-line, real time data access when and how the managers need it.

BENEFITS OF A COMPUTERIZED FACILITIES MANAGEMENT SYSTEM

The cost effective facilities manager must compensate for ever increasing labor rates, energy, inflation, and other externally controlled variables while providing the tenants with the service they expect. A computerized system can allow the manager to challenge these costs by providing on-line inquiry, comparative and exception reports, and timely, detailed information from these diverse sources. With a comprehensive system and a disciplined manager, significant, lasting cost savings (as presented in Fig. 2) can result.

Many building managers have only obsolete, inaccurate equipment specifications and history. Preventive maintenance and inspection requirements are often lax, outdated, and poorly adhered to. Consequently, labor costs are inflated to respond to failure while the equipment performance suffers.

COMPUTERIZED FACILITIES MANAGEMENT

BENEFITS

EQUIPMENT

* Reliability	* Cost Analysis
* Energy	* Performance
* Tenant Satisfaction	* Uptime

LABOR

* Productivity	* Efficiency
* Quality	* Cost Effectiveness
* Projections	* Workload Analysis

VENDORS AND CONTRACTORS

* Performance	* Contracts Administration
* Warranties	* Terms and Conditions
* Reference File	* Auditing

INVENTORY AND CONTROL

* Budget Control	* Reduced Delays
* Procurement Practices	* Performance
* Better Contracts	* Auditing

Fig. 2 Computerized facilities management benefits

Tenants become dissatisfied with the time required to locate and correct the cause of their complaints. Samples of reports that can aid the facility manager to more effectively manage are shown in Fig. 3.

The productivity of the labor force can be increased by:

1. clearly defining the work requirement;
2. accurately diagnosing the causes of malfunctions;
3. identifying skills and materials required;
4. disciplined adhering to inspection requirements;
5. analyzing equipment performance.

Aggressive attention to these issues can result in labor productivity increases from 10%–30%.

Material costs and administrative time can be reduced by:

Replacing outdated manual files with comprehensive cross referenced data bases. Information can be readily accessed by type of material, specific equipment required, vendor performance data, lead times, and terms.

PM DELINQUENT REPORT - DAYS DELINQUENT 1

EQUIPMENT ID	EQUIPMENT NAME	PM ID	PM JOB DESCRIPTION	FREQ	TYPE	LAST PM DATE	OPEN WO	DATE WO ENTERED
00400PH431	SLURRY PUMP MOTOR	000102	ROUTINE LUBE	15	I	12/23/85	000134	1/1/86
00400PH432	GOULDS SLURRY PUMP	000104	PUMP PM	30	I	12/6/85	000135	1/1/86
00400MS433	H.G. AIR CHILLER	000103	INSPECT CHILLER UNIT	15	S	12/23/85	000137	1/1/86
00400MR432	N. PUMP SLURRY MOTOR	000104	PUMP PM	30	S	12/6/85	000162	1/1/86

COSTED ISSUE BY EQUIPMENT-ID INQUIRY

PART#	DESCRIPTION	WO	QTY	AVG. PRICE	AMOUNT	TRANS DESC
12256	GOULDS PUMP 16" SHAFT	000318	1	426.0000	426.0000	ISSUE
11058	PUMP COUPLING JOINT	000319	1	20.0000	20.0000	ISSUE
12363	GOULDS PUMP REAR SEAL (DURAMET)	000320	2	321.0000	642.0000	ISSUE
12416	GOULDS OUTER SEAL GASKET	000320	1	134.5000	134.5000	ISSUE
				TOTAL:	1212.50	

EQUIPMENT COST HISTORY

EQUIPMENT ID: 00400MR431 EQUIPMENT NAME: SLURRY PUMP MOTOR
LAST WO: 000111 LAST WO DESC: REPLACE WORN CLAMPS

	MTD EXP PM	MTD EXP RP	MTD EXP TOT	YTD EXP PM	YTD EXP RP	YTD EXP TOT	LYR EXP PM	LYR EXP RP	LYR EXP TOT
LABOR	520	480	1000	520	480	1000	5700	3000	8700
LABOR OVERHEAD	52	48	100	52	48	100	570	300	870
STORES	100	50	150	200	100	300	150	150	300
STORES OVERHEAD	10	5	15	20	10	30	15	15	30
INVOICES	100	100	200	100	100	200	600	550	1150
CONTRACTOR	400	300	700	400	300	700	3500	4200	7700
TOTAL COST	1182	983	2165	1292	1038	1330	10135	8215	18750
NUMBER WO'S	23	17	40	23	17	40	132	56	158

DOWNTIME/ON-STREAM SUMMARY

	MTD	YTD	LYR
DOWNTIME FREQUENCY:	2	1	13
DOWNTIME HOURS:	4	4	40

Fig. 3 Sample management reports

Providing information to support the procurement process. More favorable contracts can be negotiated with the aid of comparative reports, histories, and other information.

SUMMARY

The decade of the 80s has been characterized by increasing competition, not only domestically, but on a world-wide scale. Methods of doing business and practices once considered acceptable are being challenged today, in industries' attempt to improve their competitive position. Management must be willing to change and accept change as positive and in the best interest of the company and its customers.

In the area of facilities management, costs can be reduced substantially if the facilities manager is willing to change. This involves reexamining current practices and procedures with an objective perspective and then identifying the system needs to support the changes. By looking at practices, procedures, and the system to support the change together as necessary ingredients of the process, the facilities manager can be successful in making substantial progress in reducing operating costs.

Perceptions of the Computer in Design

Peter Hoyt
Bob Stockdale

Myths about the usefulness of computers have discouraged some designers from taking advantage of the power of a computer-aided design and drafting (CADD) system. Some think the computer limits the creative process by restricting the designer to geometric abstractions, while others look upon a computer simply as a drafting tool. Neither myth is correct

The computer is to the designer what the typewriter is to the writer: a tool to explore and develop creative ideas. A CADD system allows the designer to generate alternate design models quickly, to explore ideas in great detail early on, and to give the client more information. Time is less of a factor than with manual techniques because changes can be made so quickly.

We use the CADD system for quantitative testing, such as handling cut and fill analysis for site design, and calculating areas and volumes for an entire building or a specified part of a building. We also have sunview, solar access, and shadow study programs in the CADD software These programs visually inform the viewer of the amount, the direction, and the angle of sunlight or shadow in a given environment. The speed with which the computer performs these calculations allows us to do more testing early, and to work with precise figures throughout the project.

Esthetic testing is another advantage computers offer us. Graphic data about the central business districts of major urban areas is stored in the computer. We enter design data for a current project and manipulate the views, studying a building in its urban context. We can look at the design from any angle, and look out the windows of the building to the surrounding cityscape. We also use the computer to store topographic data and generate contours for site planning and landscape design.

One of the major advantages of the computer is the increased communication between the architect and clients, and among the members of the project team. The clients get more detailed three-dimensional drawings at every stage of the design process, giving them a fuller understanding of the building design. With animation capability a client can be "flown" around a city or building, taken up to the 23rd floor and shown the view out their window, on the computer. The clients can thus provide feedback early on, and one is able to respond to their suggestions for changes rapidly.

Within the design team the CADD system makes it possible for all disciplines to communicate and collaborate on complex design issues. Software can be integrated, so everyone on the project team has quick reference to identical data. When architecture makes a change, engineering has it immediately. Project management is more aware of the work being done by all the disciplines, so quality control is improved. The end result is more accurate drawings with detailed information for the client and the builder. Quantity take-offs and bills of materials are generated directly from the database, so budget projections are accurate.

Traditionally, preparing presentation-quality drawings has taken several weeks. All design work would be frozen and preparations of the final drawings would begin. CADD has dramatically cut down the time allowance prior to deadlines for production of the final drawings. It provides a complete set of presentation quality drawings at all times. This increases the amount of time available to think creatively about a project and make changes.

The computer has touched architecture with far-reaching effects. The clients see what they are getting before it is built. The implications of every line are understood in two and three dimensions. The esthetic and technical accuracy of CADD drawings help architects devise better solutions to design problems and to communicate ideas more effectively to clients.

The ultimate role of the computer is to serve as the sketch pad of the designer. CADD allows us to explore a variety of ideas quickly, making intelligent decisions based on quantitative and esthetic testing, and visualizing and communicating design ideas.

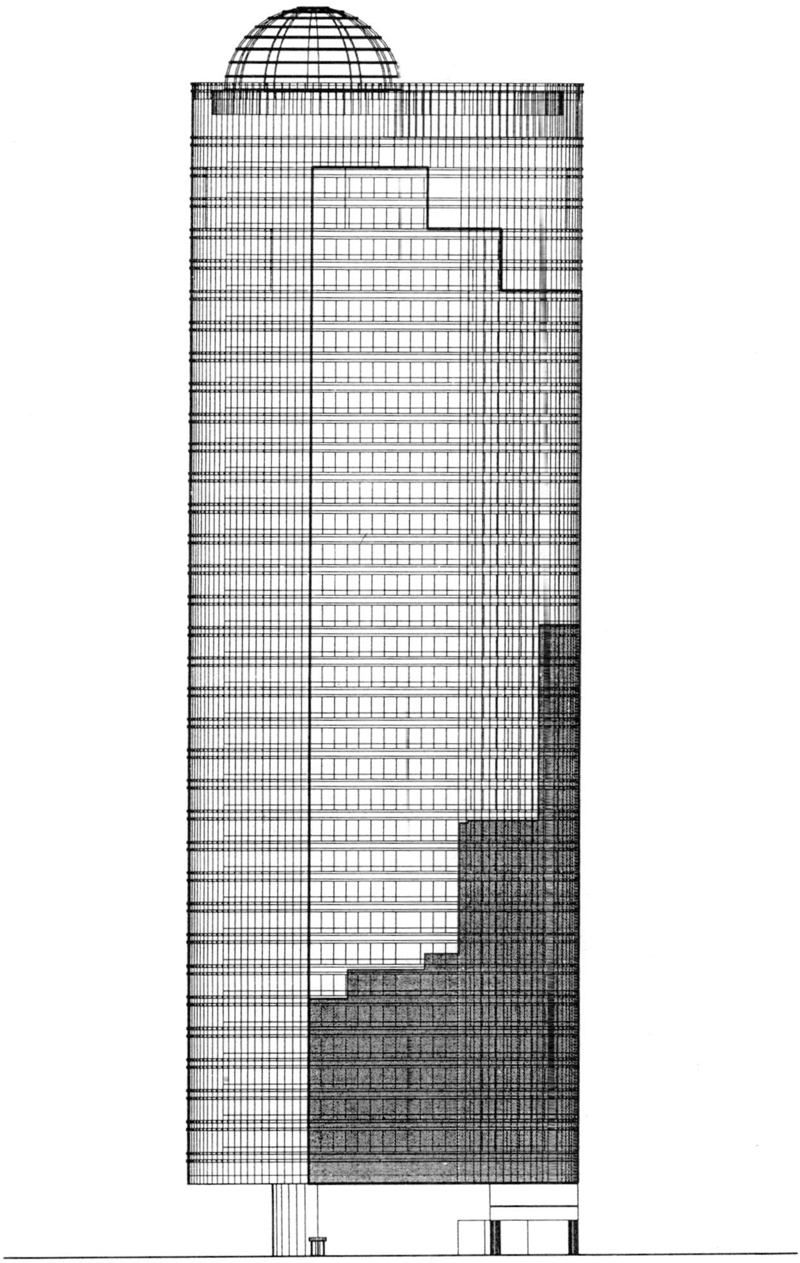

Shadow study of east facade, 9:00 a.m., December 21

Perspective and skyline studies

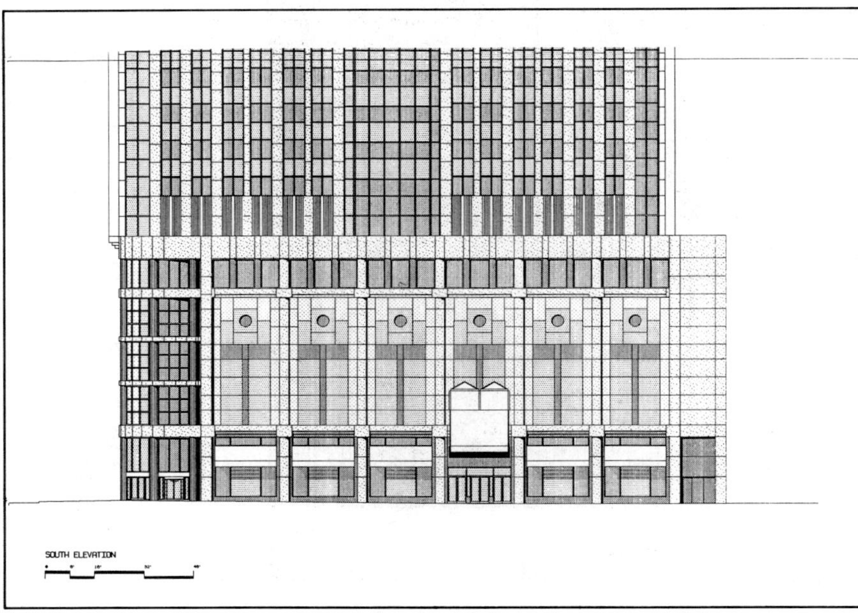

Street level detail

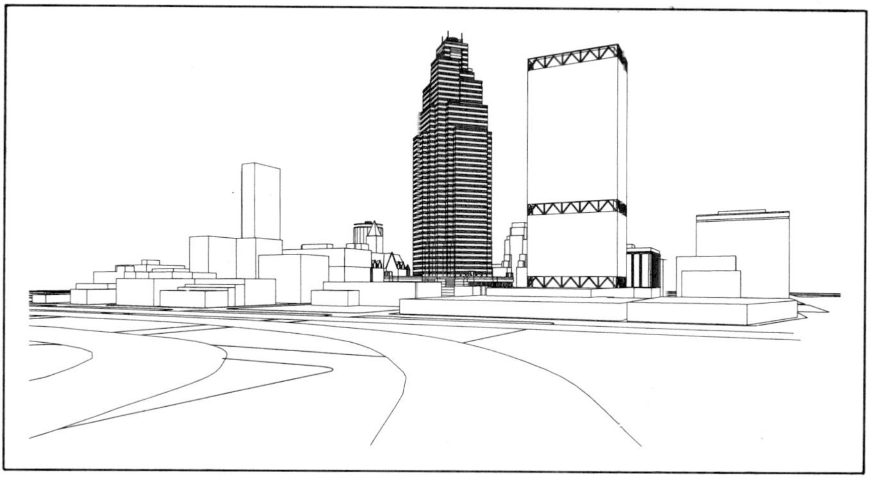

Visual impact study

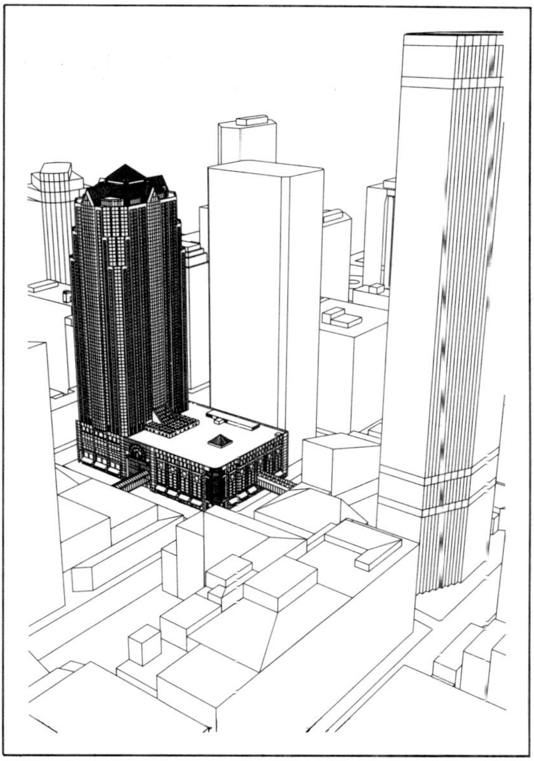

Aerial perspective of tall building

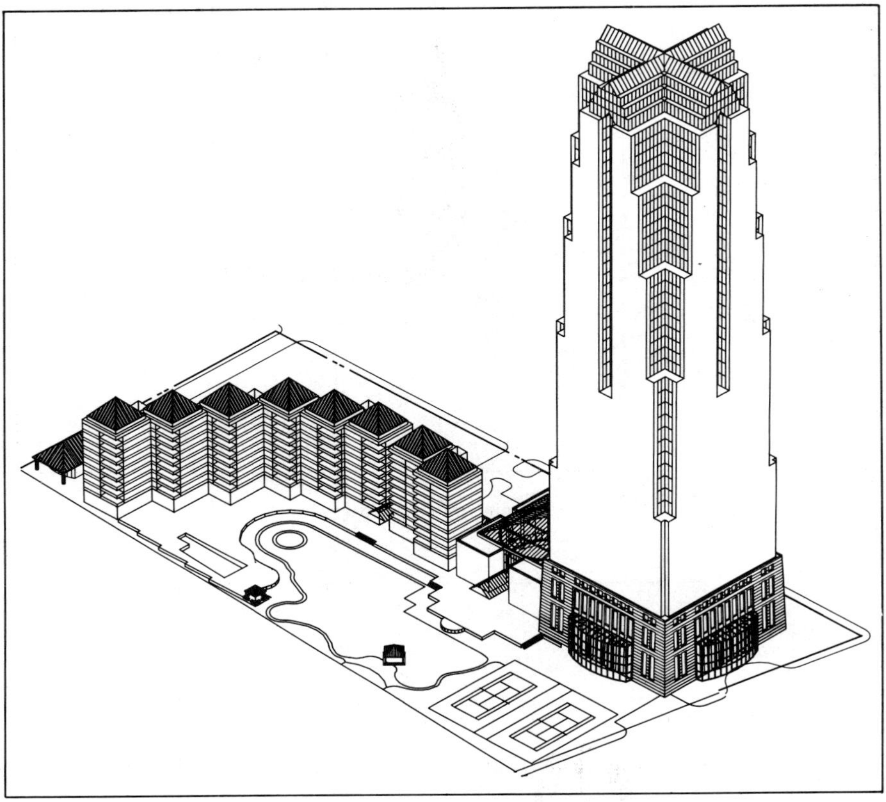

View of a tall building complex

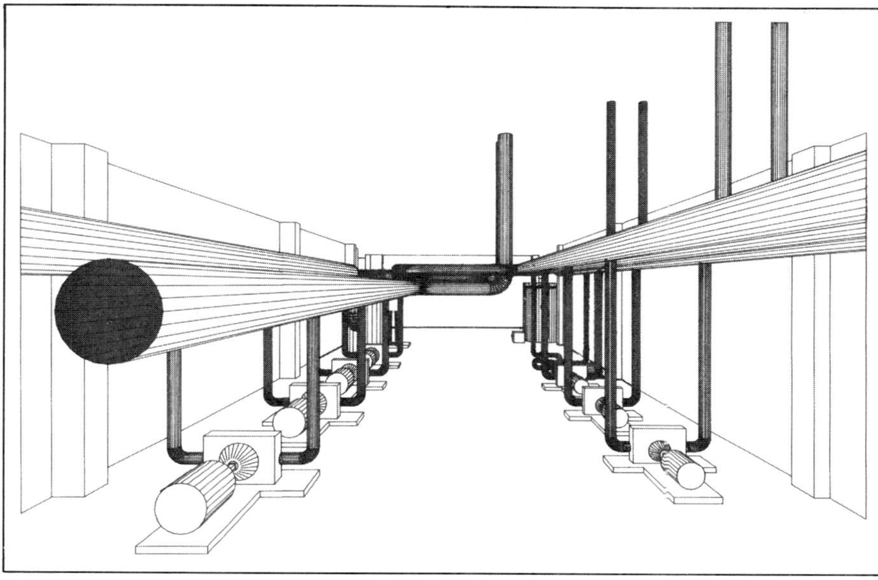

Piping network produced through CADD technology

Application of Systems Methodology

Role of Expert Systems in High-Rise Building Design

M. L. Maher
S. J. Fenves

INTRODUCTION

The emergence of expert systems, interactive computer programs that incorporate the knowledge and judgment of experts in appropriate domains, promises to extend significantly the use of the computer in the design of high-rise buildings. In current design practice, the use of computers is limited to areas in which algorithmic solutions are available, such as analysis of structural systems and proportioning of components. Expert systems provide a means for using the computer in the solution of problems that have resisted being formalized as algorithms. This paper examines the use of expert systems in the preliminary design of high rise buildings through a discussion of the prototype expert system HI-RISE. The nature of expert systems is described, followed by a discussion of HI-RISE. Using HI-RISE as a starting point, potential expert systems in building design are described and the impact of such systems on the building design process are discussed.

NATURE OF EXPERT SYSTEMS

Expert systems have developed into practical problem-solving tools that can reach a level of performance comparable to that of a human expert in some specific problem domain. Expert system concepts are surveyed in Duda and Gasching, 1981; Duda and Shortliffe, 1983; Nau, 1983; and Hayes-Roth et al., 1983. In this paper, expert systems are described in terms of their components and their range of applications.

Components of Expert Systems

An expert system can contain from three to six of the components illustrated in Fig. 1. All expert systems contain the following three basic components:

The *knowledge base* contains the knowledge specific to the domain of the problem to be solved. The knowledge in an expert system can be classified according to a spectrum ranging from *deep* to *surface* knowledge. Deep or causal knowledge is knowledge of basic principles, such as Newton's laws or static equilibrium. Surface or heuristic knowledge is knowledge developed from basic principles through experience. Analysis procedures lie close to the deep knowledge end of the spectrum, while knowledge about combin-

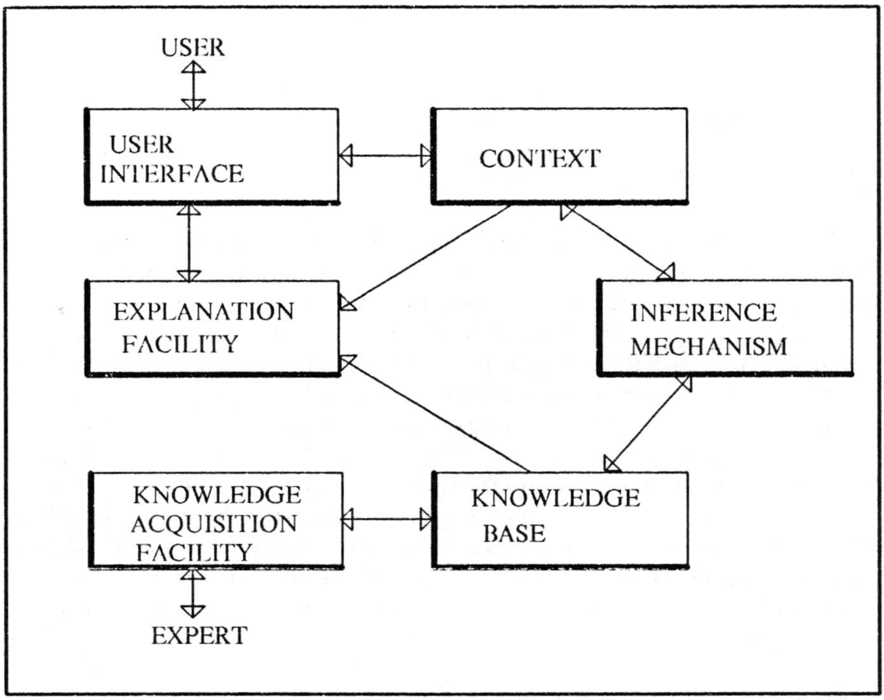

Fig. 1 Schematic view of an expert system *(Sriram et al., 1985)*

ing and placing structural systems in a given building is closer to surface knowledge.

The *context* contains facts that reflect the current state of the problem. The organization of the context depends on the nature of the problem domain. The context builds up dynamically as a particular problem is being considered and is used by the inference mechanism to guide the decision making process.

The *inference mechanism* manipulates the context using the knowledge base. Typically, the inference mechanism applies the knowledge base to the context using an approach suitable for a class of problems The inference mechanism can embody a number of problem solving strategies, such as forward chaining, where the system reasons about the initial state of known facts until a goal state or conclusion is determined to be true or appropriate. The problem solving strategy serves as a formalization of the process used to solve a problem; it defines the focus of attention at any point in the solution process. More detailed descriptions of problem solving strategies can be found in Nillson, 1980; Rich, 1983; Stefik and Martin, 1977; Maher and Fenves, 1984.

There are three other components that are not necessarily part of every expert system but are desirable in a final product.

The *knowledge acquisition module* serves as an interface between the expert(s) and the expert system. It provides a means for entering knowledge into the knowledge base and revising this knowledge when necessary.

The *explanation module* provides explanations of the inferences used by the expert system. This explanation can be a priori—why a certain fact is requested—or a posteriori—how a conclusion was reached.

The *user interface module* provides an interface between the user and the expert system, usually as a command language for directing execution. The interface is responsible for translating the input as specified by the user to the form used by the expert system and for handling the interaction between the user and the expert system during the problem solving process.

Range of Applications

The range of potential expert system applications covers a spectrum bounded by *derivation* problems and *formation* problems at the ends (Amarel, 1978). In derivation problems, the solution to a problem is derived as one of the many possible solutions contained in the knowledge base. The input to derivation problems is used to determine the most appropriate solution. In formation problems, elements are combined to formulate a solution. The input to a formation problem consists of the conditions that the solution must satisfy. Most problems fall between these two extreme categories. Classes of problems normally encountered at the derivation end of the spectrum are described below.

Interpretation. The problem is to analyze given data to determine their meaning. Examples are the Dipmeter Advisor (Davis et al., 1981), for interpreting dipmeter data and extracting information about geological patterns and trends, and PROSPECTOR (Duda et al., 1979) for identifying ore-bearing geological formations.

Diagnosis. The problem is to find the state of a system based on the interpretation of data (which may be imprecise and incomplete). In this category are medical diagnosis programs, such as MYCIN (Shortliffe, 1976) for infectious diseases, and CATS for diagnosing diesel-electric locomotive malfunctions (Bonissone, 1982).

Monitoring. The problem is to interpret signals continuously and make required changes depending on the state of the system. An example is the Ventilation Manager (Fagan et al., 1979) for monitoring patients' ventilation therapy.

Formation problems are usually examples of the *generate-and-test* paradigm: a possible candidate solution is generated by one part of the system and is then tested for suitability by another part. Formation problems fall into two subclasses: *constraint satisfaction* and *optimization* problems. In constraint satisfaction, it is only required that the solution satisfy a set of constraints, while in optimization an attempt is made to find the optimal solution. Classes of problems usually encountered at this end of the spectrum are as follows.

Planning. The problem is to set up a program of actions to achieve certain goals. An outstanding example in this category of MOLGEN (Stefik, 1981) for planning experiments in molecular genetics.

Design. The problem is to generate a description of a physical system to perform a certain function. A very successful example, in a limited definition of design, is R1 (McDermott, 1980) that designs VAX computer configurations, including selecting the needed components and determining their physical layout and interconnections.

HI-RISE

Design Process

The structural design process starts with a definition of a need to transmit loads in space to a support or foundation, subject to constraints on cost, geometry, and other criteria. The final product of the design process is the detailed specification of a structural configuration capable of transmitting these loads with the appropriate levels of safety and serviceability.

The design process may be viewed as a sequence of three stages:

1. *Preliminary design* (conceptual design) involves the synthesis of a potential structural configuration satisfying the key constraints.
2. *Analysis* is the process of modelling the selected structural configuration and determining its response to external effects.
3. *Detailed design* is the process of selecting and proportioning the structural components such that all applicable constraints are satisfied.

There may be significant deviations between the properties of components assumed at the analysis stage and those determined at the detailed stage, which would necessitate a reanalysis. Other major and minor cycles of redesign may also occur.

The structural design of high-rise buildings is based on the designer's experience as well as his understanding of the behavior of structural systems. The process of configuring a structural system for a given building uses a combination of structural system knowledge, experience, and creativity. In practice, only a single or a few alternative configurations may be considered for a given building because of the time required to determine structural feasibility and efficiency of each alternative. The use of an expert system consultant during the preliminary stages of high-rise building design can serve two purposes: (1) Several alternative configurations can be considered, modified, and optimized; and (2) the experience of senior structural designers can be made available to more junior designers.

HI-RISE is an expert system that configures and evaluates several alternative structural systems for a given three dimensional grid. The expertise in HI-RISE is derived primarily from a recent book on preliminary structural design (Lin and Stotesbury, 1981) containing approximate analysis techniques and applicable design heuristics. HI-RISE addresses the preliminary structural design stage using hierarchical planning to represent the preliminary design process, so that the design proceeds from the abstract to the detailed.

HI-RISE has been restricted to a relatively small class of buildings in order to study the formalisms needed for the development of a prototype design system. The class of buildings HI-RISE can design consists of rectangular commercial or residential buildings. Buildings rectangular in plan and elevation were chosen because the geometric representation of such buildings is easier to manipulate than buildings with setbacks, skewed column lines, curved edges, and the like. There is no lower limit on the number of stories; the term, high-rise, was chosen because the design of high-rise buildings can be divided into preordered tasks, with the design of the lateral load resisting system typically governing due to the vertical cantilever effect. Also, there is no upper limit on the number of stories, however, the current version of HI-RISE only contains knowledge for the design of buildings less than 50 stories. HI-RISE has been limited to residential or commercial buildings, since the design considerations for these types of buildings are similar.

Scope

The scope of HI-RISE is best clarified by describing the input to the system and the output the user of the system can expect. The input to HI-RISE assumes that space planning has been completed. This allows HI-RISE to be concerned only with the structural aspects of the design. The output, or solution, presented to the user by HI-RISE is a description of the feasible structural systems, detailed to the level necessary for selection among alternatives and for further modeling and analysis of a selected alternative using formal analysis techniques. The input and output are handled by a graphics based user interface (Barnes, 1984).

Input. The input to HI-RISE is a three dimensional grid. The representation of the input grid is illustrated in Fig. 2. The input grid specifies to HI-RISE the spatial constraints the building must satisfy. The topology of the grid is defined by the number of stories and the number of bays in each direction, where the directions are referenced by *narrow* and *wide*. The narrow direction is parallel to the face of the grid with the smaller total dimension, and the wide direction is parallel to the face of the grid with the larger total dimension. The geometry is defined by the dimensions of the bays and the minimum required clearance for a typical story. Other spatial constraints, such as the location of vertical service shafts or internal spaces, are specified in terms of their location on the input grid. Currently HI-RISE provides for the specification of vertical service shafts and mechanical equipment floors. The location of vertical service shafts is significant because they can typically be enclosed by walls with few or no openings. This implies that structural subsystems such as solid shear walls or braced frames may be used on the perimeter of a service shaft without detrimentally affecting the architectural space plan. The location of mechanical equipment floors is significant because such floors do not need windows implying that story-high trusses may be used to tie the lateral load resisting subsystems together. (The consideration of the mechanical floors in the structural design is not yet implemented in HI-RISE.) Other input information required by HI-RISE is the intended occupancy of the building, and the wind and live load.

Output. Once the input has been specified, the interaction between the user and HI-RISE is graphical. The context representing the feasible structural alternatives is displayed in tree form as it is generated by HI-RISE providing a mechanism by which the user may monitor the activities of HI-RISE. Currently, the user can not gracefully interrupt HI-RISE, as all interaction is initiated by HI-RISE. HI-RISE evaluates each feasible alternative according to its relative cost and other features. The user is presented with HI-RISE'S calculation of relative cost and evaluation of all feasible alternatives.

The user participates in the selection of a structural alternative from the set of feasible alternatives generated by HI-RISE. Each feasible alternative can be presented to the user graphically, using the original grid and indicating the type and location of the feasible system under consideration. Information about the structural components of any feasible alternative may be requested. This information includes the preliminary dimensions and designations of the components that make up the structural systems. The user then has the option of choosing a feasible system for further consideration or letting HI-RISE choose on the basis of its own evaluation.

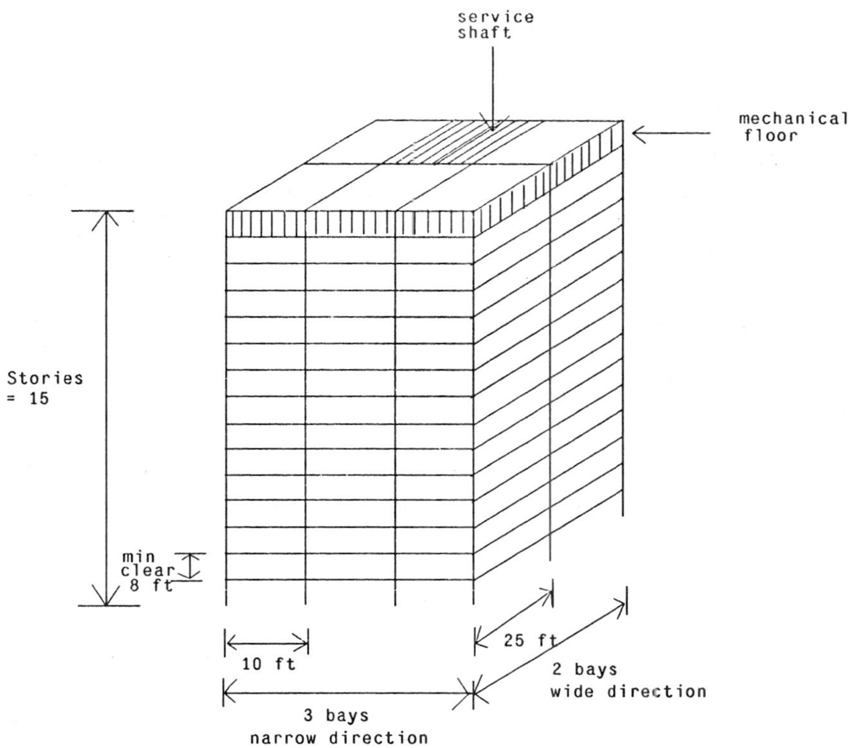

Fig. 2 Graphical representation of input

Process Decomposition

Within HI-RISE, the preliminary design process is divided into two major tasks; each task addresses the design of a functional system. The functional systems are designed in a fixed order: first the lateral load resisting system is designed, followed by the design of the gravity load resisting system. The design of a new functional system is not started until the previous functional system design is completed. This convention has the advantage of confident access to information generated in a previous task and the disadvantage of needing information from a task not yet started. In the above task order, results from the design of the gravity system, namely, the type, depth, and weight of the floor system, are needed for the design of the lateral system. This information is estimated within the lateral system design task with heuristics using the input information of occupancy and bay span. The task order enumerated above was chosen because, as a first approximation, the design of high-rise buildings is governed by the design of the lateral load resisting system.

Each of the two major tasks is decomposed into a set of similar subtasks. The subtasks have the same goals for each functional system, however, the details of reaching these goals differ. The remainder of this section describes the general goals of the subtasks.

Synthesis. The first subtask is to synthesize a set of alternatives for the functional system under consideration. The synthesis is performed as a depth-first search through the appropriate generic subsystems stored in the knowledge base. The search space is pruned using heuristic elimination rules.

Preliminary Analysis. The purpose of the analysis subtask is to evaluate the feasibility of an alternative and to define its component groups. Feasibility is evaluated by the formulation and evaluation of one or more feasibility constraints. Analysis provides one set of ingredients of the constraints, namely, the required capacity of the system components. The analysis is implemented as a procedural representation of approximate, heuristic preliminary analysis techniques. Component groups are defined so that preliminary sizing or proportioning is performed only for one component in a group.

Parameter Selection. The parameters of the structural components are initially selected using heuristics. These initial parameters are used to evaluate the feasibility constraints. If a constraint is violated, some heuristic recovery rules are applied to revise the parameters. Once satisfactory parameters are selected, in other words all applicable constraints are satisfied, the alternative is considered feasible.

Evaluation. Evaluation of a structural design may be based on many diverse features of the design. Evaluation is usually done by designers in an abstract form. Some of the features that may be considered are esthetics, economics, efficiency, and structural integrity. HI-RISE evaluates alternatives with a linear evaluation function. There is a distinct evaluation function for the two functional tasks. The variables in the function are features of the system that may be quantified or, at least, ranked.

System Selection. HI-RISE presents all structurally feasible systems to the user indicating which system has been determined to be the "best," selected as the system with minimum value of the evaluation function. The user may either accept the recommended design or override the decision of HI-RISE and choose one of the other structurally feasible systems.

Representation of Design Information

The development of HI-RISE required the selection of a representation for the design information. As the design of a building progresses, the amount of information generated increases rapidly. The efficient representation of this information is necessary for ease of storage and access. The representation used by HI-RISE went through many revisions until it became clear that the design information fell into three major levels: the global level, the functional level, and the physical level. The physical level is further subdivided into the subsystem level and the component level. The organization of these levels is illustrated in Fig. 3.

The *global level* contains information global to the entire building design. In HI-RISE, this includes information such as the intended occupancy of the building, the topology and geometry of the three dimensional grid, and the location of vertical service shafts. This information is divided into two sublevels: building information and grid information.

The *functional level* separates the design information according to function. In HI-RISE, the design information is grouped into two functions, the lateral load resisting system and the gravity load resisting system. The lateral load resisting system defines the structural systems, subsystems, and components that are responsible for resisting lateral loads such as wind or seismic forces. In HI-RISE, only wind loads are considered. The gravity load system defines the floor system and components responsible for transmitting the gravity loads to the foundation.

The *physical level* is the hierarchical representation of the structural system alternatives and is composed of the subsystem and component levels.

The subsystem level represents the structural subsystems used by the synthesis task to configure an alternative functional system. The details of these levels differ for the two functional systems.

A lateral load resisting system is configured from four hierarchical levels of information, as shown in Fig. 4. The first level is the representation of alternative three dimensional subsystems, such as a core. The second level is

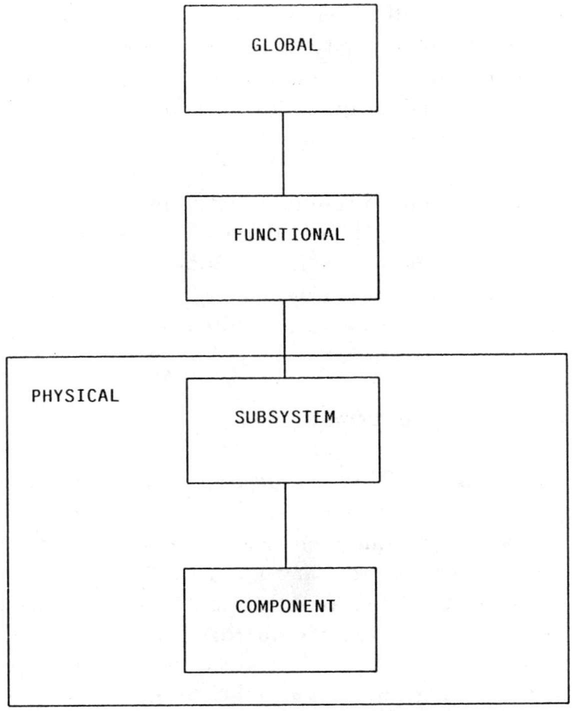

Fig. 3 Overall organization of design information

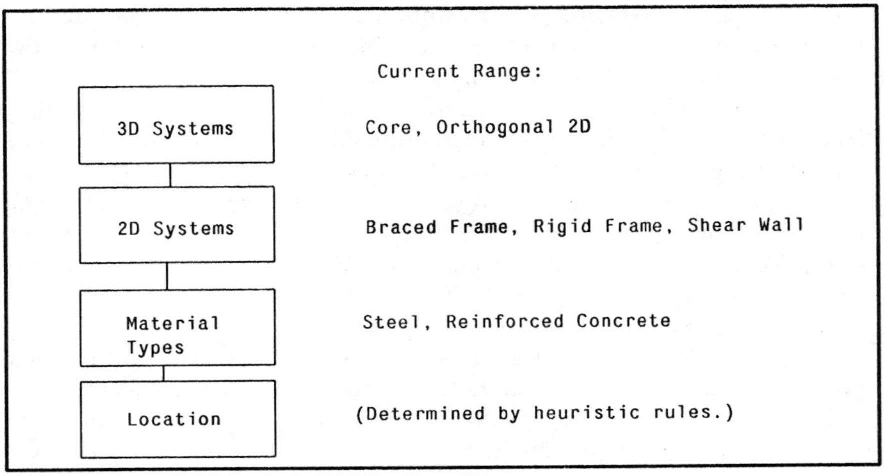

Fig. 4 Levels of synthesis for lateral system

the representation of alternative two dimensional vertical subsystems, such as a braced frame. The third level is the alternative material, such as steel. The fourth level is the alternative location of the lateral load resisting system.

The gravity load system is configured by selecting a two dimensional horizontal subsystem, such as a reinforced concrete slab or a concrete-topped steel deck. This selection combines the second and third levels of the lateral system; it also includes the determination of how the slab will be supported and whether intermediate beams will be needed. The levels of synthesis for gravity systems is shown in Fig. 5.

The component level is the representation of the information associated with the components in each subsystem. Components include beams, columns, diagonals, walls, and slabs. Beams and columns are divided into groups containing components with similar design requirements and are represented in two levels: the first level contains information about all the beams or columns in a subsystem, the second about a typical beam or column in a group.

The overall organization of the context tree is illustrated in Fig. 6. An example of a partial context tree of feasible structural configurations generated by HI-RISE is shown in Fig. 7. There are three different relationships between the nodes in the tree: *is-alt, part-of, uses.* The is-alt relationship indicates the descendents of the node are alternatives in the solution. The part-of relationship indicates that the descendents of a node are all part of a solution. The uses relationship is a lateral link used, for example, to connect a constraint and the subsystem or component it constrains.

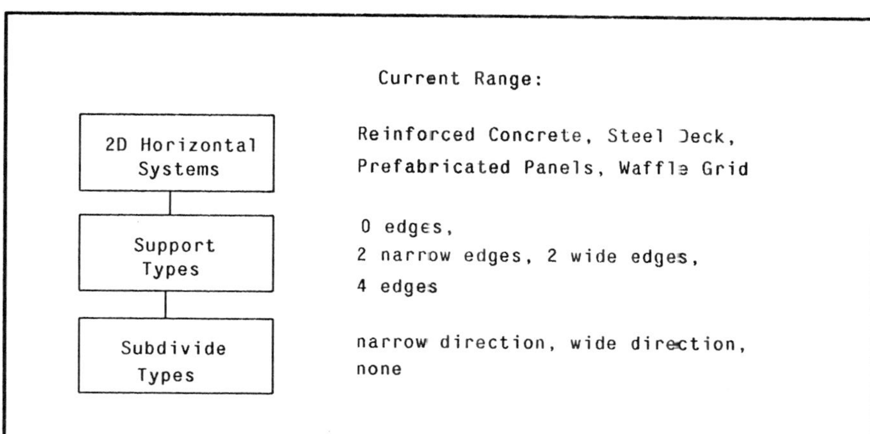

Fig. 5 Levels for synthesis of gravity system

Evaluation of HI-RISE

Although HI-RISE was not intended to mimic the structural designer's thought processes, we believe that the overall structure of HI-RISE roughly parallels that of a good designer, in that

HI-RISE systematically generates potential structural configurations by proceeding from the most abstract (3D scheme) to the more detailed levels;

at each of these levels, it "knows" about a number of possible alternatives to consider; and

it efficiently prunes the search space by rejecting potential alternatives as soon as a constraint is violated.

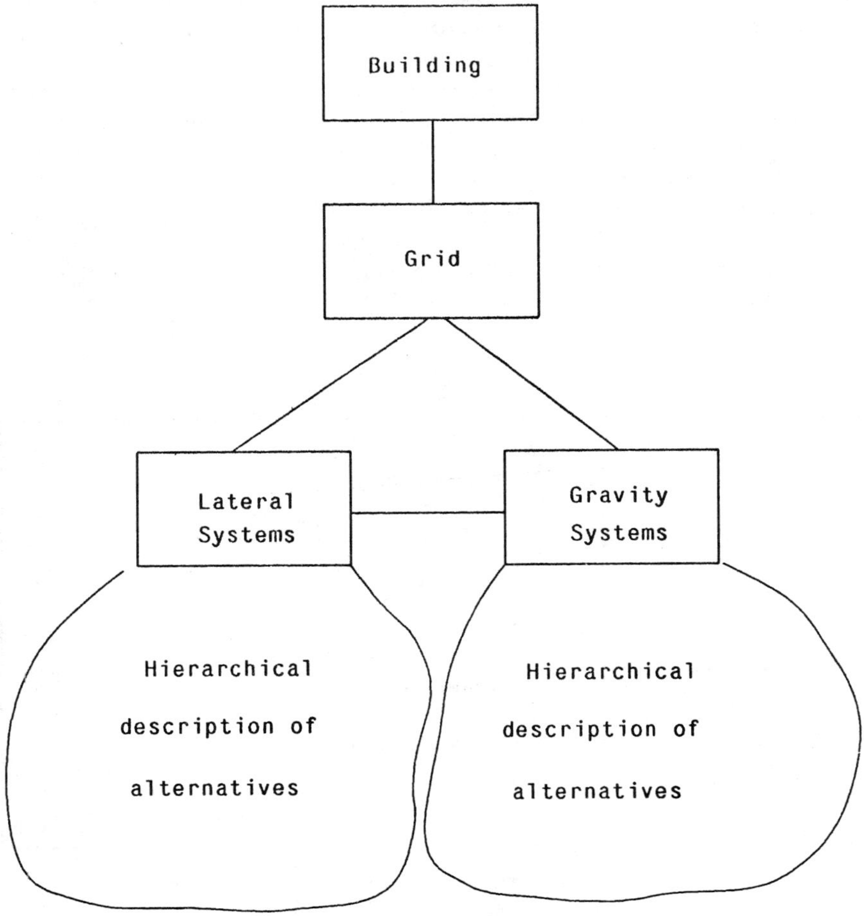

Fig. 6 Organization of context tree

It should be clear from the description of HI-RISE presented that HI-RISE does not produce "innovative" designs, incorporating novel structural concepts or subsystems. On the other hand, it can produce more than "routine" designs by exploring a wider range of alternatives at multiple levels of abstraction than a designer can normally accomplish; and "customizing" generic subsystems to the specific building at hand. Furthermore, the hierarchical structure of HI-RISE's knowledge base provides for the rapid incorporation of new alternatives. For example, in order to add "semi-rigid frame" or "coupled shear wall" to the present 2-D systems, one would have to add one approximate analysis procedure and the appropriate constraints such a 2-D subsystem imposes on the remaining levels of synthesis.

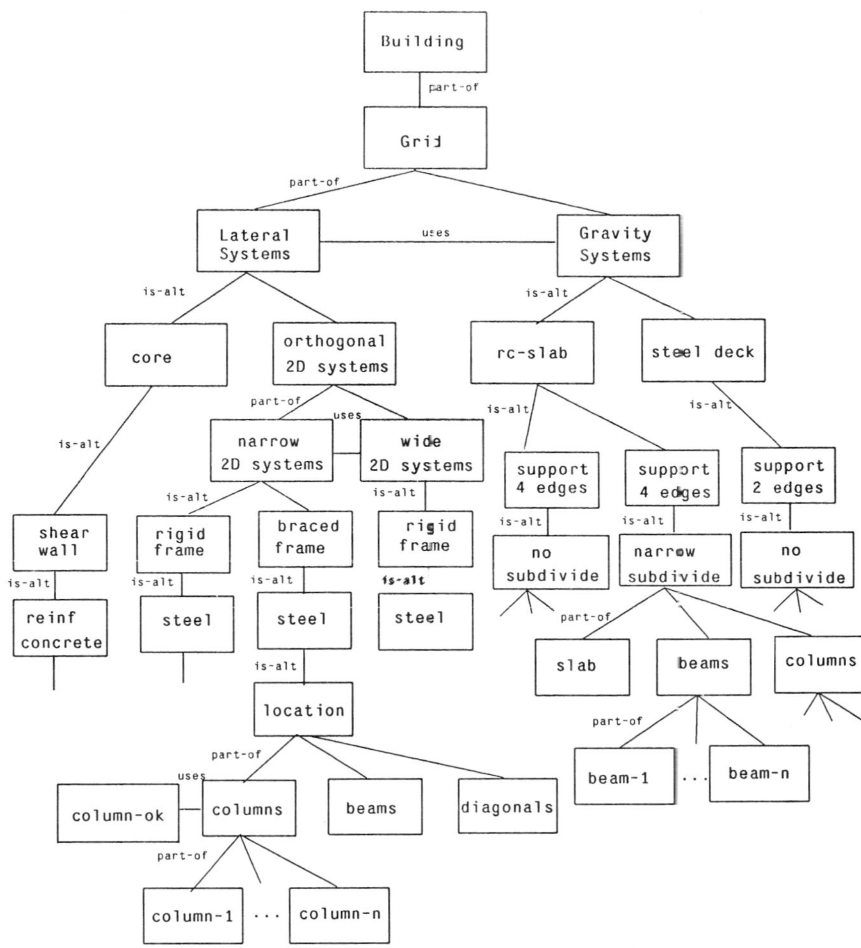

Fig. 7 Example of a context tree

Since HI-RISE is the first attempt to formalize a very complex process, its current implementation has several shortcomings, as follows:

As stated before, HI-RISE always designs the lateral load resisting system first, and then selects compatible gravity systems. To be realistic under a wide range of conditions, the system should be able to proceed either way, possibly guided by a rule set which determines the best order.

HI-RISE currently fills the entire grid; it will not accommodate setbacks or interior openings, nor structural schemes that use only a subset of the gridlines.

The number of alternatives at each level is limited, and needs expansion; furthermore, it is assumed that a given scheme expands to the full height of the building.

Improvements to overcome the above limitations are possible. We are currently working on a new synthesis strategy that can start with either functional system and post constraints on the other system, to be subsequently resolved (Sriram, 1983).

A large improvement could be realized in the configuration of alternatives if HI-RISE contained more realistic placement rules, including the option of only using a subset of the grid lines for structural subsystems.

Currently, the placement of the gravity load resisting system is implicitly based on the design of a typical bay. This placement assumption should be revised to allow for a more global view of the horizontal plane of the building.

The placement of the lateral load resisting system is based on very simple rules that assume a single selection from each level of synthesis. The current set of placement rules should be revised and expanded into two classes: horizontal and vertical placement. Horizontal placement rules are similar to the rules currently used by HI-RISE, where a single selection from each level of synthesis constitutes a lateral system placed on the horizontal plane of the input grid. Vertical placement rules would stack different lateral systems within the vertical plane of the input grid, allowing a subset of the lateral load resisting systems to extend only to a partial height of the building.

POTENTIAL APPLICATIONS OF EXPERT SYSTEMS TO HIGH-RISE BUILDING DESIGN

The design of high-rise buildings can greatly benefit from the incorporation of expert systems in the design process. Many of the considerations during the total system design process are based on previous experience. The scope of such considerations encompasses the entire building as an integration of many different systems. Expert systems could be applied to the design of a single system, such as the structural system, and to the compatibility between systems, such as the interaction of the structural system with the architectural space plan. Some of the design considerations amenable to an expert system approach are described below, proceeding from specific to more general issues.

Structural Engineering Applications

Preliminary or conceptual design is only one aspect of the structural design process. Other areas of structural design amenable to expert system approaches include the following:

Modelling assistants. In order to analyze a structure, the design engineer has to

1. select the analysis method appropriate to the task at hand;
2. convert the representation of the physical structure into the mathematical model needed by the method chosen;
3. perform the analysis of the mathematical model; and
4. interpret the model response in terms of the physical situation.

Many programs are available for the third step, which is purely algorithmic. The other three steps are largely heuristic, and depend on the analyst's experience and expertise. Expert systems could provide substantial assistance in these areas. An early expert system acting as an automated consultant to advise on the selection of analysis options was SACON (Bennett and Engelmore, 1979), but it has been put into production use. In building design, such expert systems would also have to assist the engineer in developing a progression of appropriate analytical models as the design progresses.

Detailing/proportioning assistants. The design of structural components, whether for preliminary sizing or detailed proportioning, is not performed solely based on the governing design standards. Component and connection design also involves knowledge about structural behavior, material properties, design strategies and style, and knowledge about which standard provisions are applicable and *why,* suitable approximations and shortcuts for specific standard provisions, and boundaries of applicability of standard provisions, and so forth. Currently used detailing and proportioning programs implement all of this heuristic knowledge as algorithms. These programs are notoriously difficult to modify (say, when design standards are updated) and provide little or no explanation. Detailing programs implemented in an expert system framework can eliminate these shortcomings. We are currently working on a prototype of such an expert system for structural component sizing, proportioning and standard conformance checking (Fenves and Garrett, 1985).

Constructability critics. The "ultimate" evaluation of a proposed structural design is in determining whether it can be economically and conveniently constructed. Unfortunately, more design engineers don't have much field experience, and thus don't know what can or cannot be conveniently constructed. It would be desirable to incorporate constructability criteria into the earliest design stages, for example as heuristic elimination rules in HI-RISE's synthesis subtask. This is not practical today, primarily because the knowledge base of such criteria in not formalized. However, it does seem feasible to develop expert system constructability critics that would evaluate proposed designs with respect to a variety of criteria, providing suggestions to the designers for modifications.

Applications in other disciplines

At the risk of trespassing into areas outside our expertise, it is clear that the design of electrical, mechanical, HVAC and other systems in high-rise buildings has many characteristics similar to structural design: possible alternatives have to be synthesized, evaluated, analyzed, and detailed. While the specific domain knowledge is different, it appears to us that the concepts presented in the context of structural design are directly applicable to these design disciplines as well.

Substructure design is definitely one of the most promising application areas of expert systems. Not only are the design processes comparable to those for structural design; there are also several other processes that intimately depend on accumulated geotechnical knowledge, expertise, and heuristics. Chief among these are subsurface exploration and interpretation of field and laboratory tests, and monitoring of excavation and substructure construction processes. We have developed a prototype expert system for interpreting cone penetrometer field date that even in its present version can reproduce experts' interpretation 85% of the time (Mullarkey, 1983).

The role of expert systems in architectural design and space planning is not yet clear to us. Aspects of spatial layout are being approached with expert systems techniques (Flemming et al., 1986).

Integrative Applications

Integrative applications of expert systems in building design could aid in the coordination of the design of the architectural, structural, and mechanical systems within a building, thereby improving the efficiency of the building as a whole. Much too often, the design of each system is performed in isolation from the design of the other systems, and the integration of the architectural, structural, and mechanical systems is performed through individual interpretations of drawings and design documents. The use of expert systems to integrate the design of these systems could incorporate a global view of the performance of the whole building in the design of its parts.

There are two possible approaches for the development of expert systems in this area. One approach is to use the expert system as an active design aid, considering the compatibility issues during the generation of design alternatives. In order to be effective, the expert system would require access to design information from each discipline, so that advice relevant to the current state of the total design could be formulated. A second approach is to use the expert system as a passive check on the design alternatives to indicate how well each alternative satisfies the compatibility requirements. In this approach, the expert system could provide advice in the form of recommended revisions to a system design so that it better accommodates the designs of other disciplines.

IMPACT OF EXPERT SYSTEMS ON BUILDING DESIGN

Expert systems could have an impact in two major areas: improving the entire building design and training inexperienced designers. The improvement of building designs could be realized through the interaction between the designers and a source of accumulated expertise. The expert system can provide an unfailing memory of design alternatives, coupled with experience in combining and fitting design solutions for each unique situation.

Expert systems are excellent environments for training inexperienced designers due to their ability to trace or explain their reasoning process. An expert system can provide advice or suggest alternative solutions that a novice designer currently obtains from experienced designers. Assuming that the expert system would be more available than an experienced designer, the inexperienced designer would have the opportunity to use and learn the heuristics applicable to designing a building in a shorter period of time.

The use of expert systems can also encourage cooperation between design groups. The use of expert systems for integrating the design considerations among the different disciplines could provide a means of communication that is not easily accommodated among individual human experts. The use of a machine to provide advice or design revisions could eliminate the antagonism that may be present when that advice comes from a person.

SUMMARY AND FUTURE DIRECTIONS

Expert systems are a new class of computer programs, providing a viable mechanism for dealing with the subjective, heuristic, or "ill-structured" aspects of building design which cannot be handled by algorithmic programs. Expert systems, as all computer programs, are expensive to develop. Several of the expert systems cited earlier required 5 to 10 man-years to develop. Development of expert system may be even more expensive than for conventional programs, because the knowledge base has to be elucidated from experts, formalized, and compiled.

Furthermore, expert systems have to be highly idiosyncratic, in the sense that they must embody the specific expertise and "design style" of the individual organizations that intend to use them. Unlike conventional algorithmic programs, expert systems cannot be "general purpose".

There appear to be two possible options for the future. One is to have each organization attempt to develop its own set of expert systems. At this early stage of expert system technology such wide-scale experimentation is highly desirable, but in the long run this approach promises to lead to the same proliferation as has occurred with algorithmic programs. The second option, intellectually and professionally more challenging, is to attempt to develop cooperatively a "core" expert system for a discipline, and couple it with excellent knowledge acquisition facilities, so that each organization can enhance, expand, and "customize" this core to reflect its own set of heuristics.

REFERENCES/BIBLIOGRAPHY

Amarel, S., 1978
BASIC THEMES AND PROBLEMS IN CURRENT AL RESEARCH, in Ceilsielske, V. B. (editor), Proceedings of the Fourth Annual AIM Workshop, held at Rutgers University, June, New Holland Publishers, New York.

Barnes, Sandra, 1984
DICE DESIGN INTERFACE FOR CIVIL ENGINEERING, Master's thesis, Carnegie-Mellon University, September.

Bennett, J. S. and Engelmore, R. S., 1979
SACON: A KNOWLEDGE-BASED CONSULTANT FOR STRUCTURAL ANALYSIS. In Proceedings of Sixth International Joint Conference on Artificial Intelligence (IJCAI), August 20-23 William Kaufmann Inc., Los Altos, pages 47-49.

Bonissone, P. P., 1982
OUTLINE OF THE DESIGN AND IMPLEMENTATION OF A DIESEL ELECTRICAL ENGINE TROUBLE-SHOOTING AID. Technical Conference of the BCS SGES, U.K.

Davis, R. et al, 1981
THE DIPMETER ADVISOR: INTERPRETATION OF GEOLOGIC SIGNALS. In Proceedings Seventh IJCAI, pages 846-849, William Kaufmann Inc., Los Altos, CA.

Duda, R. O., et al., 1979
A COMPUTER-BASED CONSULTANT FOR MINERAL EXPLORATION. Final Report SRI Project 6415 SRI International edition.

Duda, R. and Gasching, J. G., 1981
KNOWLEDGE-BASED EXPERT SYSTEMS COME OF AGE. BYTE 6(9):238-279, September.

Duda, R. O. and Shortliffe, E. H., 1983
EXPERT SYSTEMS RESEARCH. Science 220:261-268, April.

Fagan, L. M., Kunz, J. C., Feigenbaum, E. A., and Osborn, J. J., 1979
REPRESENTATION OF DYNAMIC CLINICAL KNOWLEDGE: MEASUREMENT INTERPRETATION IN THE INTENSIVE CARE UNIT, In Proceedings Fifth IJCAI, pages 1014-10 9, William Kauffmann, Los Altos.

Fenves, S. J. and Garrett, J. H., Jr., 1985
STANDARDS REPRESENTATION AND PROCESSING. In IABSE-ECCS Symposium on Steel in Buildings, pages 107-114, ETH, Hongger, Berg, CH, Zurich.

Flemming, U., Coyne, R., Glavin, T., Rychener, M., 1986
A GENERATIVE EXPERT SYSTEM FOR THE DESIGN OF BUILDING LAYOUTS, Proceedings of the First International Conference on Applications of AI in Engineering April, Computational Mechanics Publications, Southampton.

Hayes-Roth, F., Waterman, D., and Lenat, D., 1983
BUILDING EXPERT SYSTEMS. Addison-Wesley, Reading, Mass.

Lin, T. Y. and Stotesbury, S. D., 1981
STRUCTURAL CONCEPTS AND SYSTEMS FOR ARCHITECTS AND ENGINEERS. John Wiley and Sons, New York.

Maher, M. L. and Fenves, S. J., 1984
HI-RISE: AN EXPERT SYSTEM FOR THE PRELIMINARY STRUCTURAL DESIGN OF HIGH RISE BUILDINGS, in Knowledge Engineering in Computer-Aided Design, North-Holland, Netherlands.

McDermott, J., 1980
R1: A RULE-BASED CONFIGURER OF COMPUTER SYSTEMS. Technical Report CMU-CS-80-119, Carnegie-Mellon University, Pittsburgh, Pa.

Mullarkey, P. M., 1983
DEVELOPMENT OF AN EXPERT SYSTEM FOR INTERPRETATION OF GEOTECHNICAL CHARACTERIZATION DATA FROM CONE PENETROMETERS. Ph.D thesis proposal, Department of Civil Engineering, Carnegie-Mellon University, November.

Nau, D. S., 1983
EXPERT COMPUTER SYSTEMS, Computer 16:63-85, February.

Nillson, N. J., 1980
PRINCIPLES OF ARTIFICIAL INTELLIGENCE, Tioga Publishing Company, Palo Alto, California.

Rich, E., 1983
ARTIFICIAL INTELLIGENCE, McGraw-Hill, New York.

Shortliffe, E. H., 1976
COMPUTER-BASED MEDICAL CONSULTATIONS:MYCIN, American Elsevier, New York.

Sriram, D., 1983
KNOWLEDGE-BASED APPROACH TO INTEGRATED STRUCTURAL DESIGN. Unpublished Ph.D Thesis Proposal, Department of Civil Engineering, Carnegie-Mellon University.

Sriram, D., Maher, M. L., and Fenves, S. J., 1985
KNOWLEDGE-BASED EXPERT SYSTEMS IN STRUCTURAL DESIGN, Computers and Structures, Vol. 20, No. 1-3, pp. 1-9.

Stefik, M. and Martin, N., 1977
A REVIEW OF KNOWLEDGE-BASED PROBLEM SOLVING AS A BASIS FOR A GENETICS EXPERIMENT DESIGNING SYSTEM. Technical Report STAN-CS-77-596, Computer Science Department, Stanford University, March.

Stefik, M., 1981
PLANNING WITH CONSTRAINTS (MOLGEN 1), Artificial Intelligence 16:111-140.

Knowledge-Based Systems for Tall Buildings

V. Tuncer Akiner

The life span of a tall building is made up of five stages: planning, design, construction, operation and finally demolition. The nature and magnitude of the entire project is such that it requires the involvement of many different professionals such as developers, economists, town planners, architects, interior designers, engineers, landscape architects and so on. The height, size, density, and the various impacts of the tall building within the urban context increase each day. The technical advances in many disciplines provide new tools for handling problems in the tall building process.

Knowledge about tall buildings is extremely abundant and exists in many disciplines. It is dynamic knowledge. Therefore it is becoming impractical to keep it in books, because books do not provide a dynamic enough medium for encapsulating, manipulating, and processing this type of knowledge. It is suggested then that the knowledge source should be shifted from books to disks. Also the existing knowledge of the experts should be captured in the computer medium.

In the above-mentioned five stages in the tall building life span, many of the issues that the professional handles are ill-defined, nonstable, nonhomogeneous, nonlinear, and subjective. Moreover, in many cases the professional must cope with missing and fuzzy information. The conventional algorithmic programming methodology is inadequate to process such information. A conceptual framework is proposed for a comprehensive knowledge-based system for tall buildings. It is a system that will capture and manipulate information via knowledge engineering techniques. In this system a declarative approach to computation will be adopted, knowledge will be represented explicitly, and necessary inferences will be made. It will allow the user to have an intelligent assistant in decision making within the various tasks associated with tall buildings.

This paper commences with an explanation of knowledge engineering. Expert systems and their advantages over the conventional data processing systems are discussed. For the development of an integrated knowledge-based system for tall buildings, the sources of facts and knowledge are identified. Some examples of representation of facts and knowledge from the existing sources are presented. Then a conceptual framework for a knowledge-based system is proposed. Finally this paper concludes that such a system would provide a powerful tool for those professionals involved with tall buildings.

KNOWLEDGE ENGINEERING

Knowledge engineering is a sub-field of artificial intelligence that is concerned with the art of creating computer programs by employing appropriate techniques for knowledge representation, knowledge deployment (processing), and knowledge acquisition (Akiner, 1986).

Knowledge Representation

In knowledge engineering, a *representation of knowledge* is a combination of data structures and interpretive procedures that, if used in the right way in a program, will lead to "knowledgeable" behavior. Work on knowledge representation has involved the design of several classes of data structures for storing information in computer programs, as well as the development of procedures that allow "intelligent" manipulation of these data structures to make inferences. In this context a data structure is not knowledge, any more than an encyclopedia is knowledge. Metaphorically, it could be said that a book is a source of knowledge, but without a reader, the book is just ink on paper (Barr et al., 1981).

Researchers in knowledge engineering have developed various methods for representing knowledge about the world in their programs. There are five basic methods of representing knowledge in the computer medium: Production Systems, Semantic Networks, Procedures, Frame and Script Systems, and Logic Systems (First Order Logic/Calculus) (Lansdown, 1982; Winograd, 1971). None of these is entirely satisfactory for all problems, but all have provided frameworks for representing knowledge. Generally, knowledge-based systems are built on a combination of these methods.

Deployment of Knowledge

The choice of a representation of knowledge does not necessarily help in deciding how the knowledge is going to interact with a problem. For example, in cases where knowledge is encoded as production rules, even the straightforward production system paradigm is strongly influenced by factors that have not been mentioned. In the system design process, the sequence of putting the rules into action and the question of whether the rules are part of the database (and thus can be modified by other rules) must be investigated. Depending on the purpose and the way in which the knowledge-based system is to be used, appropriate software architectures are implemented.

Knowledge Acquisition

This is the most time-consuming, expensive, and critical task in the development of a knowledge-based system, because it includes the entire process of extracting knowledge from knowledge sources, ensuring that all cases are covered, integrity is maintained, and thus the knowledge is encapsulated (Quinlan, 1980). In the construction of large knowledge-based systems, hundreds of rules and thousands of facts required by these systems are generally obtained from books. An additional source is obtained by interviewing experts knowledgeable in the field; and it is from this important resource that the term *expert systems* got its name.

EXPERT SYSTEMS

Expert systems is a sub-field of knowledge engineering. Feigenbaum (1981) defines an expert system as "An expert system is an intelligent computer program that uses knowledge and inference procedures to solve problems that are difficult enough to require significant human expertise for their solution." Knowledge in any field is usually of two types: public and private. Public knowledge includes published definitions, facts and theories associated with the field or domain in question. Private knowledge is the type of knowledge that may not be publicly available through books, but rather acquired over years of experience and is thus in the hands of human experts. These are mainly rules of thumb that have come to be called heuristics (Hayes-Roth et al., 1983). Heuristics enable the human expert to make educated guesses when necessary, to recognize promising approaches to both well-defined and ill-defined problems, and to deal effectively with erroneous or incomplete data.

EXPERT SYSTEMS VS. CONVENTIONAL DATA PROCESSING SYSTEMS

Expert systems differ in important ways from conventional data processing systems. In contrast to the traditional data processing systems, expert systems generally possess several distinguishing features such as symbolic representation, symbolic inference, and heuristic search. The main disadvantage of conventional programs is that the knowledge of the subject area that they embody is bound to the program in such a way as to be inaccessible and virtually unalterable without a major overhaul of the entire program. They lack generality and flexibility, and they require the presence of a programmer in the process of posing questions to the system. Moreover, they cannot handle incomplete data. In expert systems, knowledge is explicitly represented and thus it is separate from control.

SOURCES OF FACTS AND RULES (KNOWLEDGE) ABOUT TALL BUILDINGS

Turning attention now to tall buildings, there are several sources for the facts and rules that will be transferred into a knowledge-based system for tall buildings:

The five-volume Monograph published by the Council on Tall Buildings and Urban Habitat (CTBUH) (1978–81):
- *Planning and Environmental Criteria for Tall Buildings* (Vol. PC)
- *Tall Building Systems and Concepts* (Vol. SC)
- *Tall Building Criteria and Loading* (Vol. CL)
- *Structural Design of Tall Steel Buildings* (Vol. SB)
- *Structural Design of Tall Concrete and Masonry Buildings* (Vol. CB)

"Update" volumes (3)
Proceedings of the numerous conferences on tall buildings
The forthcoming topical volumes
The surveys conducted by the Council
Knowledge resulting from other committee work of the Council
The results of new research
Knowledge obtained from human experts through interviews

REPRESENTATION OF FACTS AND RULES ABOUT TALL BUILDINGS

Some examples of the representation of facts and rules extracted from the above-mentioned five Monograph volumes are presented below. First, a copy of the fact or rule as it exists in the Monograph is given. Second, that fact or rule is expressed in pseudo logic programming form.

Ventilation

Table 1 is a part of the "Gertis' Rule" for the design of tall buildings without air conditioning in a moderate climate zone (Gertis, 1975):

1. IF the climate is moderate, and the building is to be without air conditioning THEN the width of the building should not exceed 15 m to 20 m (50 to 65 ft).
   In pseudo logic programming form:
   width(T_Building, W) if
      without_air_conditioning(T_Building) and
      climate(T_Building, moderate) and
      W =< 20.

2. IF the climate is moderate and the building is to be without air conditioning THEN natural window ventilation is required.
   ventilation(T_Building, natural_ventilation) if
      without_air_conditioning(T_Building) and
      climate(T_Building, moderate).

Structural Systems

Khan's research findings (1974) about the most frequently used structural systems for steel and concrete structural frames based on their tallness criteria are given in Fig. 1. Some of these findings are as follows:

1. The structural system of a building should be a rigid frame IF the structure is steel AND the number of stories of the building ≤ 30.
   structural_system (T_Building, rigid_frame) if
      material(T_Building, steel) and
      number_of_stores(T_Building, NS) and
      NS =< 30.

Table 1 Gertis' Rule (Monograph Vol. PC, p.435.)

1. Width of tall building not to exceed 15 m to 20 m (50 ft to 65 ft).
2. Natural window ventilation.
3. Windows in corner rooms facing toward one direction only.

2. The structural system of a building should be a frame-shear truss IF the structure is steel AND the number of stories of the building is $\leq$ 40 AND $>$ 30.

```
structural_system(T_Building, frame_shear_truss) if
        material(T_Building, steel) and
        number_of_stories(T_Building, NS) and
        NS =< 40 and
        NS > 30.
```

3. The structural system of a building should be a bundled tube IF the structure is steel AND the number of stories of the building is $\leq$ 110 and $>$ 100.

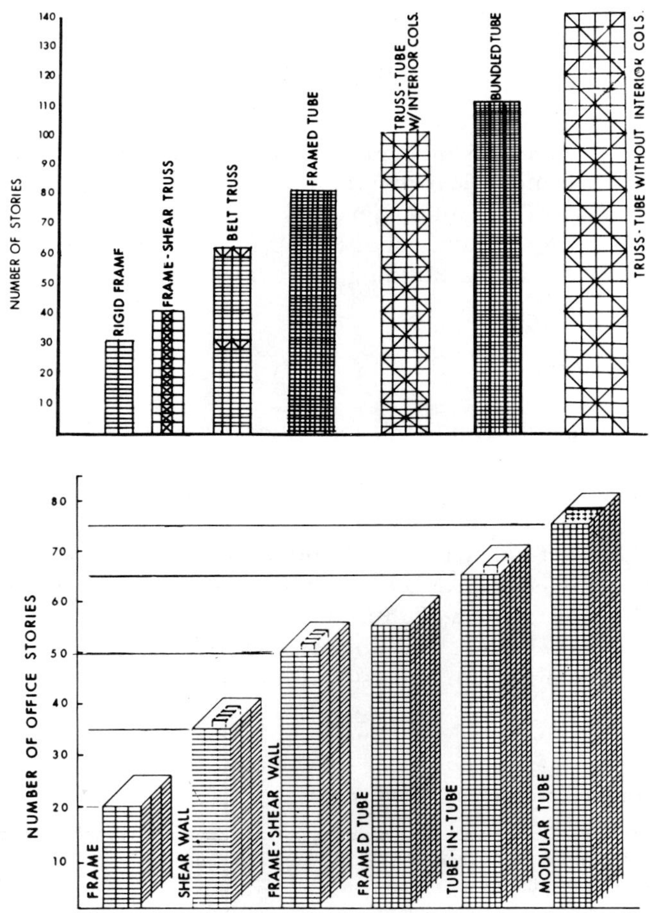

Fig. 1 Types of steel and concrete structures (Monograph Vol. SC, p. 5)

```
structural_system(T_Building, bundled_tube) if
        material(T_Building, steel) and
        number_of_stories(T_Building, NS) and
        NS =< 110 and
        NS > 100.
```

Tallness

Table 2 contains facts about some of the tallest buildings in the world. The first two rows of facts about the tallest buildings in the world can be expressed in pseudo logic programming form as:

```
tall_building(sears_tower, chicago, 1974, 110, 443, 1454, steel, office).
tall_building(world_trade_center_south, new_york, 1973, 110, 412,
   1350, steel, office).
```

Earthquakes

Figure 2 shows the causes of damage which are taken from the section on Evaluation of Earthquake Damage in Monograph Vol. CL (Council on Tall Buildings, 1978–1981). The facts from Fig. 2 may be encoded as follows:

1. cause_of_damage(mexico_city_earthquake, 1957, relative_rigidity_of_elements_not_considered).
2. cause_of_damage(mexico_city_earthquake, 1957, reinforcement_in_reinforced_concrete_beams_inadequate_in_center_portion_of_span).
3. cause_of_damage(mexico_city_earthquake, 1957, pounding_between_adjacent_elements).

Table 2 Tallest Buildings in Major Cities (Monograph Vol. SC, p.423.)

Building	City	Year completed	Number of stories	Height meters	Height feet	Material	Use
Sears Tower	Chicago	1974	110	443	1454	steel	office
World Trade Center South	New York	1973	110	412	1350	steel	office
World Trade Center North	New York	1972	110	412	1350	steel	office
Empire State	New York	1931	102	381	1250	steel	office
Standard Oil (Indiana)	Chicago	1973	80	346	1136	steel	office

Table 3 illustrates the categories of damage caused by earthquakes. This knowledge may be written in English and in pseudo logic programming form as follows:

1. Level of damage is 0 IF there is no damage AND ratio to replacement cost is 0.
   level_of_damage(T_Building, 0) if
       no_damage(T_Building) and
       ratio_to_replacement_cost(T_Building, 0).
2. Level of damage is 1 IF there is minor structural damage AND ratio to replacement cost is 0.001.
   level_of_damage(T_Building, 1) if
       minor_non_structural_damage(T_Building) and
       ratio_to_replacement_cost(T_Building, 0.001).

   The definition of "minor non structural damage" could further be encoded as:
   minor_non_structural_damage(T_Building) if
       cracked(walls, T_Building) and/or
       cracked(partitions, T_Building) and/or
       damage(mechanical, T_Building) and/or
       damage(electrical, T_Building).

Wind

Table 4 illustrates the response to wind storm of four buildings studied by Robertson (1973). The facts expressed in the first row of the table may be encoded as follows:

wind_storm(world_trade_center, new_york, 116, 438, 1435, 9.2, 11.0, 0.091).

Mexico City Earthquake of 1957. The earthquake of July 28, 1957 in Mexico City affected tall building construction under modern earthquake design principles (Zeevaert, 1957; Rosenblueth, 1960). The main conclusion that was reached from this earthquake was the relatively large amount of damage to tall buildings in comparison to that found in short, rigid buildings. Some of the specific items noticed in relation to damage are as follows:

1. Some damage resulted because the relative rigidity of resisting elements was not considered in the design.
2. Reinforcement in reinforced concrete beams was often inadequate in the center portion of the span.
3. Pounding between adjacent elements was a cause of damage.

Fig. 2 Mexico City earthquake of 1957 (Monograph Vol. CL, p. 111)

Concrete Structural System

Table 5, extracted from Section 3.3, p. 54 of Vol. CB, illustrates built-in rules about selection of concrete structural systems. The rule within the first row of the table may be expressed as:

```
structural_system(T_Building, flat_plate) if
        span_of(T_Building, S) and
        S >= 7.3 and
        S =< 7.9 and
        loading(T_Building, 30.2) and
        type_of_occupancy(T_Building, OT) and
        member(OT, [motel, hotel, hospital, dormitory, apartments])
        and
        under_floor_duct(requires_much_coordination).
```

Table 3 Damage States (Monograph Vol. CL, p.123.)

Level of damage (1)	Description of level of damage (2)	Ratio to replacement cost[a] (3)
0	No damage	0
1	Minor nonstructural damage—a few walls and partitions cracked, incidental mechanical and electrical damage	0.001
2	Localized nonstructural damage—more extensive cracking (but still not widespread); possibly damage to elevators and/or other mechanical/electrical components	0.005
3	Widespread nonstructural damage—possibly a few beams and columns cracked, although not noticeable	0.02

Table 4 Robertson's Schedule of Building Properties (Monograph Vol. SB, p.389)

Building and location (1)	Levels (2)	Height, in meters (feet) (3)	Density (4)	Period, in seconds (5)	Frequency, in hertz (6)
World Trade Center, New York	116	438 (1435)	9.2	11.0	0.091
Seattle First National Bank, Seattle	50	183 (600)	9.6	6.7	0.149
United States Steel Building, Pittsburgh	78	275 (900)	8.9	8.5	0.118
Theme Towers, Los Angeles	50	183 (600)	9.3	4.1	0.244

The need is first to provide a knowledge-rich, complete, dynamic and state-of-the-art medium for encapsulating all that is "known" about tall buildings by human experts and all that is being kept in any form of storage—books, monographs, magazines, computer programs and so on. This may be viewed as a comprehensive, new source of knowledge.

The second objective is to offer to the professionals involved with tall buildings a large number of knowledge-based systems that act as intelligent assistants, advisors, evaluators and troubleshooters that will be built upon the above-mentioned sources.

CONCEPTUAL FRAMEWORK FOR KNOWLEDGE-BASED SYSTEM FOR TALL BUILDINGS

The eventual goal is to provide tall building professionals with a tool kit of fully integrated, cross checked, reliable knowledge about tall buildings. Since the amount of knowledge is extensive, a similar structure is proposed for the creation of the system as was successfully implemented in the preparation of the Monograph published by the Council on Tall Buildings and Urban Habitat. That is, for each committee, a knowledge-based system module would be developed. A conceptual framework for such a system is illustrated in Fig. 3. It is to be built by knowledge engineers who will encapsulate the knowledge of the topical committee members, other experts in the area of tall buildings, published material, and computer programs into the system. The knowledge-based system consists of four major components:

1. Natural Language Interface: It allows the user to pose queries on the system and receive answers to them.

Table 5 Selection of Concrete Structural System (Monograph Vol. CB, p.54.)

Description (1)	Span range, in meters (feet) (2)	Loading, in kilograms per square meter (pounds per square foot) (3)	Type of occupancy (4)	Under floor duct (5)
Flat plate	7.3 to 7.9 max. (24 to 26) up to 10.66 (35) pre-stressed	light to 30.2 (60)	motel, hotel, hospital, dormitory and apartments	requires much coordination
Waffle flat plate	up to 15.2 (50)	moderate to 50.4 (100)	monumental, libraries	very difficult

2. Inference Mechanism. A mechanism that manipulates the rules base and the facts base and ensures that the correct rule is applied to the facts in the correct instance.

3. Facts Base: Contains a set of facts about tall buildings.

4. Rules Base: Contains a set of rules that may be domain dependent or domain independent. The domain dependent knowledge is about tall buildings, and the domain independent knowledge is physical, logical, and mathematical knowledge.

The process of operating the system is as follows: The user poses a query to the system about a particular problem associated with tall buildings through the natural language interface facility. The inference mechanism identifies and uses the appropriate facts and/or rules and makes the necessary inferences.

For example in *Planning and Environmental Criteria* (Vol. PC) there are sixteen subject or topical areas. (For each one there is a committee.) Therefore 16 knowledge-based system modules will be developed for Vol. PC, one of which might be, perhaps, landscape architecture. Similarly for each topical volume, knowledge-based systems related to each committee will be developed. Every system will have its own four components as mentioned previously. Thus, they will be developed independently, in the early development stage of the entire system. However, eventually all the "sets of knowledge-based systems," (represented now by the topical monograph volumes of the Council on Tall Buildings and Urban Habitat) which are com-

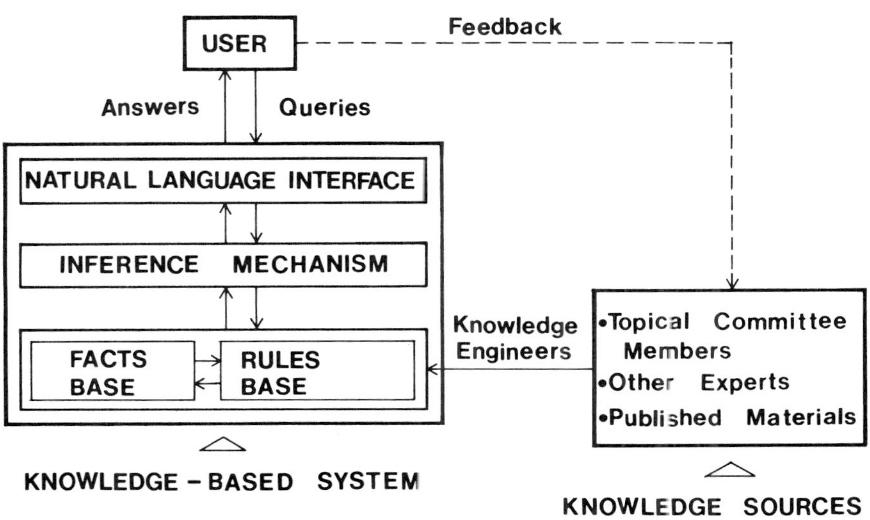

Fig. 3 Structure of a knowledge-based system for tall buildings

posed of individual knowledge-based systems, will be linked to a general inference mechanism as shown in Fig. 4. In this case, the user interacts with the system by natural language interface facility of the system. The general inference mechanism also allows each "knowledge-based system set" to be linked to other knowledge-based system sets. This provides the knowledge-based systems within the integrated knowledge-based system to "communicate" with one another in the process of inference making. Thus a truly, comprehensive, knowledge-rich environment is proposed within such a framework.

CONCLUSION

The speed of development of technology must be reflected in the construction industry. Particularly in the area of tall buildings such developments must be implemented rapidly. The vast quantity of knowledge that exists in books becomes obsolete very quickly. Updating such printed materials is costly and slow. It is an appropriate time to start building a state-of-the-art knowledge-based system to encapsulate and manipulate facts and rules about tall buildings. Also the rules of thumb, the experiences of the experts in this field must be captured.

The development of an integrated, knowledge-based system that will act both as a source of facts and rules, and also as an integrated set of knowledge-

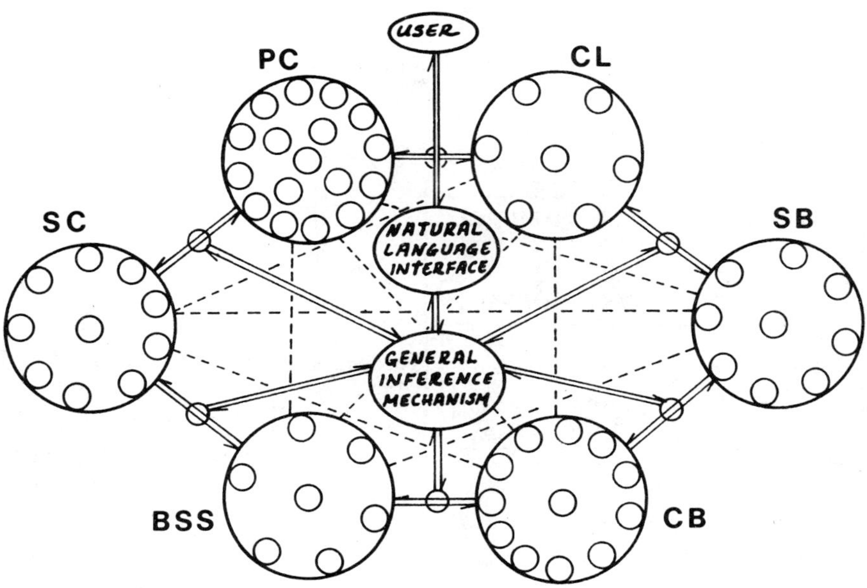

Fig. 4 An integrated knowledge-based system for tall buildings

based systems that will act as an intelligent assistant, advisor, and cross-checking facility, will increase the creativity and productivity of those professionals involved with tall buildings.

ACKNOWLEDGEMENTS

The author is indebted to Lynn S. Beedle, the director of the Institute for the Study of the High Rise Habitat of Lehigh University, for his valuable critiques and encouragement and to all the committee members of the Council on Tall Buildings and Urban Habitat who are providing the wealth of facts and rules which the new technology will make available for use in more timely fashion.

REFERENCES/BIBLIOGRAPHY

Akiner, V. T., 1986
KNOWLEDGE ENGINEERING IN OBJECT AND SPACE MODELING, in Expert Systems in Civil Engineering (Kostem, C. N. and Maher, M. L., eds.), ASCE, New York, pp. 204-218.

Barr, A., Cohen, P., and Feigenbaum, E. A., 1981
THE HANDBOOK OF ARTIFICIAL INTELLIGENCE, William Kaufmann, Palo Alto, Vol. 1, pp. 143-216.

Council on Tall Buildings, Beedle, L. S., 1978-81
PLANNING AND DESIGN OF TALL BUILDINGS, A Monograph in 5 volumes, ASCE, New York.

Feigenbaum, E. A., 1981
EXPERT SYSTEMS IN THE 1980s, in Bond, A. (ed.), Infotec State of the Art Report, Pergamon-infotec.

Gertis, K., 1975
TALL BUILDINGS AND THERMAL INSULATION, (Hochhäuser und Sinnvoller Warmeschutz), Proceedings of German Conference on Tall Buildings, Mainz, October 2-4, Deutsche Gruppe der Internationalen Vereinigung fur Brückenbau and Hochbau (IVBH), Wiesbaden/Köln, German Federal Republic, pp. 257-271.

Gertis, K., 1980
NON-STEADY HEAT AND MOISTURE TRANSFER PROBLEMS IN BUILDING PHYSICS, 11th Congress—Vienna, Introductory Report, Theme VIC, (Proceedings of Congress held in Vienna, Austria, August 31-September 5), IABSE, Zurich, Switzerland, pp. 127-132.

Hayes-Roth, F., Waterman, D. A., and Lenat, D. B., 1983
AN OVERVIEW OF EXPERT SYSTEMS in Hayes-Roth, F., Waterman, D. A., and Lenat, D. B. (eds.), Building Expert Systems, Addison-Wesley, Reading, Massachusetts.

Khan, F., 1974
A CRISIS IN DESIGN—THE NEW ROLE OF THE STRUCTURAL ENGINEER, Proceedings of the Conferences on Tall Buildings (Kuala Lumpur, Malaysia, December), Institution of Civil Engineers, Kuala Lumpur, Malaysia.

Lansdown, J., 1982
EXPERT SYSTEMS, THEIR IMPACT ON THE CONSTRUCTION INDUSTRY, Report in RIBA Conference Fund, London.

Quinlan, J. R., 1980
FUNDAMENTALS OF THE KNOWLEDGE ENGINEERING PROBLEMS, Technical Report No. 149, Basser Department of Computer Science, The University of Sydney.

Robertson, L. E., 1973
LIMITATIONS ON SWAYING MOTION OF TALL BUILDINGS IMPOSED BY HUMAN RESPONSE FACTORS, Proceedings of the Australian and New Zealand Conference on the Planning and Design of Tall Buildlings, Sydney, Australia, August, pp. 171-180.

Winograd, T., 1971
PROCEDURES AS REPRESENTATION FOR DATA IN A COMPUTER PROGRAM FOR UNDERSTANDING NATURAL LANGUAGE, Technical Report AI TR-17, MIT, Cambridge, MA.

Building Service Systems

Introductory Review

Gordon Rigg

The papers that make up this and other related chapters address two key issues: How well have we served the building owner in the past, and, what does the future hold? The second issue captures our imagination, but answers to the first provide the foundation for our forward thinking.

Much good work has been performed in the field of HVAC systems over many years. But some of the benefit has been negated by inadequate attention to ensuring that the end product measures up to expectations. So we find that, on completion, too many problems exist. For example, commissioning is not carried out to finality and records of the as-installed conditions are not carefully prepared. As a result, subsequent operations are impeded.

There is a critical need to ensure that equipment and systems can be maintained effectively by reasonably trained personnel. Expectations must recognize the limitation of skills available in the industry.

Essentially we are talking about quality control. The ability to control quality begins with the designers. Systems and equipment must be laid out such that maintenance personnel can work with adequate clearances and safety; provisions must be made in the systems for the installation of devices that allow commissioning to be performed quickly and accurately.

Past shortcomings have also developed through inadequate understanding of the behavior of buildings and their external envelope under extreme environmental conditions. For example, the effects of wind and temperature difference can cause serious degradation of HVAC system performance. This again is a design issue that requires much more than calculation by rote—even to the extent of scale modelling.

This, then, is the perceived state of the art. Enquiry will reveal that it is not a phenomenon limited to North America nor any other region; it is a worldwide issue. HVAC designers who respond to the challenges presented by these inadequacies will indeed be putting in place the foundations for the future.

The *Second Century of the Skyscraper* embodies the concept of extreme tallness. As such, it captures the imagination of all people and, in particular, building designers. For building services the issues of tallness are compounded by those of intelligence—that is, the requirement of the so-called "smart building".

Much has been said about the "smart building", and while evolving technologies allow us to understand in the short term what we are dealing with, there is, beyond this, a question of how we will deal with the unknown in the long term.

What will communication systems be like in 10, 20 or 30 years? And what environmental systems will be needed to support them? These systems, whatever they are, will need to be installed in today's tall building, many of which, in view of their size, will not be demolished for many years.

So building system designers must respond to the short term needs as well as the long term, largely unknown future. In essence, it behooves us to design for flexibility to create spaces by thinking about the organization of building elements, each with the other, such that facilities can be altered and added around the essential structural skeleton or framework. The concept of modularity, even for building systems, is likely to intensify as a fundamental approach as we move forward.

To achieve success in the future even greater emphasis will need to be placed upon equal participation between all project team members. Architects, engineers, and constructors will work together to achieve the end result. We have seen buildings evolve architecturally from an initial structural concept. In the future, demands of building services and construction requirements will add their weight to the initial concept and thus to the ultimate architectural solutions so that the truly integrated building can be attained.

High-Rise Office Buildings: Changes Anticipated by the Year 2000

I. A. Naman

What is wrong with mechanical and electrical installations in large buildings today and why should we expect them to change in the next fifteen or twenty years? The answer is a great deal is wrong, certainly as they are perceived by the occupants of the buildings for whom they are created. Present air conditioning systems are complex and difficult to operate properly, are often inefficient, and frequently cannot provide satisfactory levels of comfort. Electrical power is not used efficiently, either for lighting or general purposes; plumbing systems waste large volumes of water; and there is no general agreement on design of fire suppression and smoke control systems. Designs must be changed to overcome the deficiencies, and creative new methods and techniques will be sought to meet present requirements as well as new ones, thereby rendering some present designs obsolete. The economics of the market place will compel the changes; an attractive incentive, return on investment, always moves us toward improvement. The big changes will be to do better what we have already been doing and to accommodate the new requirements of a computer and communication age. This discussion looks first at what the needs are and then to where these may take us.

AIR CONDITIONING

System Design

An ideal air conditioning system would

be under complete control of an individual tenant,
be energy efficient,
require no ductwork,
require no central cooling or heating plant,
require no operating personnel,
have complete automation and energy control,
require little maintenance, and
be economically advantageous to install.

Many of the systems designed today do not use energy efficiently even with present day capabilities, although this is not generally recognized. System selection has come to reflect a current "style", frequently not thoroughly researched, analyzed, expertly considered, nor even reviewed for application on the particular project. It is almost as if narrow neckties are in, wide ones are out, so everyone wears narrow neckties.

Variable Air Volume Systems (VAV). An example of current "style" is the popular variable air volume (VAV) systems that are being used almost to the exclusion of all else (Fig. 1). VAV is considered by many design engineers to be the ultimate in creative design, offering the best of everything, with lowest possible energy use. Whereas such a solution may be the best for a specific application, it is not the general rule. The VAV solution has many disadvantages, one of the most significant of which is occupant discomfort during periods of low air circulation. High efficiency is anticipated with VAV, based on the assumption that fan horsepower decreases as the air volume required for cooling decreases. However this ignores any other factors. Since pure VAV systems (not with fan-powered terminals) normally provide only cooling, energy may be wasted during the heating season, and this waste may exceed any possible saving that could be realized in fan horsepower. With separate perimeter heat, an increase in heating energy results from the need to have some air continuously supplied to a space for ventilation purposes. With a VAV system this will always be cold air, so that zones exposed to the exterior surfaces of a building will have simultaneous cooling and heating in winter. VAV with reheat (Fig. 2) gives the same result. Further, for either of these systems, to minimize simultaneous cooling and heating, it is necessary to have a properly coordinated cooling and heating control system that reduces cooling to the minimum needed for ventilation, prior to bringing on the heat. Many designers overlook this fact, but even when considered, the difficulty of keeping temperature controls carefully adjusted and accurate

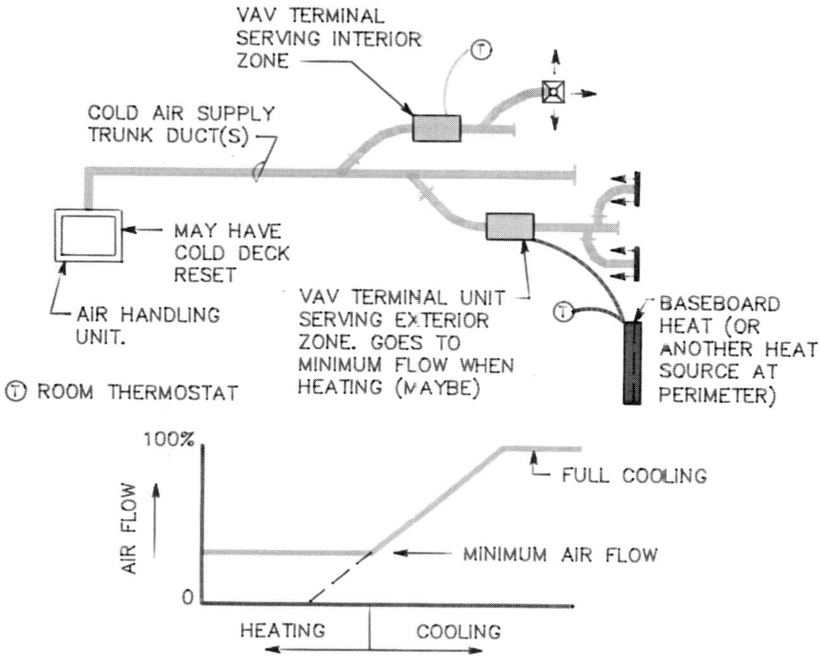

Fig. 1 VAV with separate perimeter heat

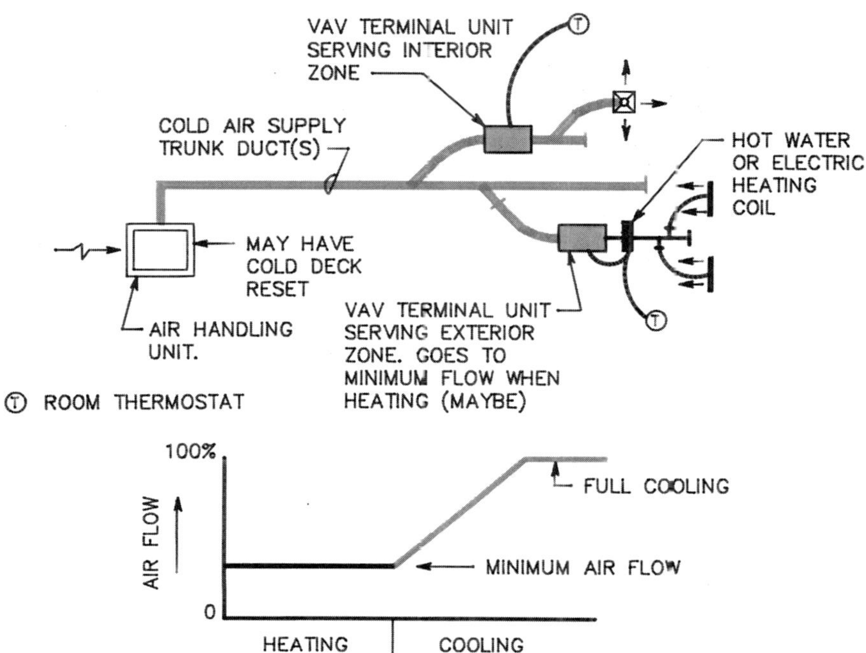

Fig. 2 VAV with reheat

over a period of years is difficult. It is simply not likely to be achieved, so that system efficiency tends to reduce with the passage of time.

VAV does provide an ability to install an unlimited number of zones of control, and in such cases its application may be valid notwithstanding other factors. The fan-powered terminal system (Fig. 3), a variation of the pure VAV, also provides only cooling from the central system but delivers the cool air to a combined fan-powered box and heating coil which, in turn, distributes warm or cool air to the zone served. The added electrical input for the additional, inefficient motors required to continuously operate the terminal boxes increases the total fan energy for each floor by some 25%, assuming properly designed, low-energy air handling systems are installed. The intermittent fan-powered box (Fig. 4) uses somewhat less energy, but causes a change in air volume when it cycles due to thermostat control, and occupants become aware of it.

Economizer Cycle. The so-called "economizer cycle," which uses outside air for cooling in winter (Fig. 5), is another example of a design solution for which the economics can be questioned. This solution is widely used since, intuitively, it seems obvious that avoiding operation of refrigeration equipment during cold weather must result in energy savings. It is worthwhile to question this premise. In many cases a return air fan is added to the system. This fan runs whenever the system is in operation. A review of the kilowatt hours consumed by adding return air fans is interesting. In a typical 1,000,000 ft^2 (92,900 m^2) building which operates for 55 hr/wk, an estimated 1,500,000 kW/hr will be used in a year just to operate the return air fans. This is equivalent to running a 300 kW (1,000 ton) chiller, including auxiliaries, for 1,650 hr/yr, hence energy savings begin to occur only after the "economizer cycle" has operated to save 1,000 tons of load for 30 weeks in a year. In most climates this would never occur. If the "economizer cycle" is designed properly, without a return air fan and with some means of controlling volume changes due to wind pressure, it may be made to operate satisfactorily and efficiently. Systems that require a return air fan for normal operation are too inefficient and should never be considered except in some special and unusual circumstances.

When we consider that the majority of installations in the USA utilize one or another of these systems, and that they have the inefficiencies just described, it becomes obvious that more thought and consideration must be given to system design. There are more efficient systems available which function well, cost less, and perhaps should be used more.

Installation

The installing contractor has some responsibility for many of the operating difficulties experienced. Even many "responsible" mechanical contractors install their work carelessly, so that properly designed systems cannot be made to function satisfactorily.

Operation

Comfort conditions are not being maintained in many buildings because installation is poor, operational practices by building personnel are poor, design solutions have been flawed, and erection of the buildings themselves have not been satisfactory. These matters are discussed in the following.

In most cases building operating personnel do not have the basic engineering skills and understanding to provide the central operation that maintains a reasonable comfort level while achieving desired economies. A well prepared and on-going operations training program is needed, with appropriate consideration for the particular project design. Yet few sources are available to provide a competent training program.

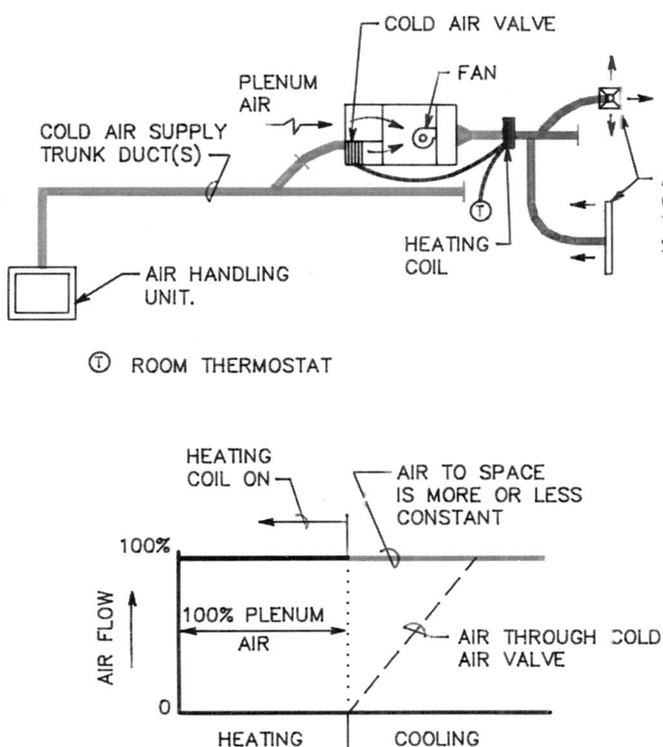

Fig. 3 Continuous fan-powered box

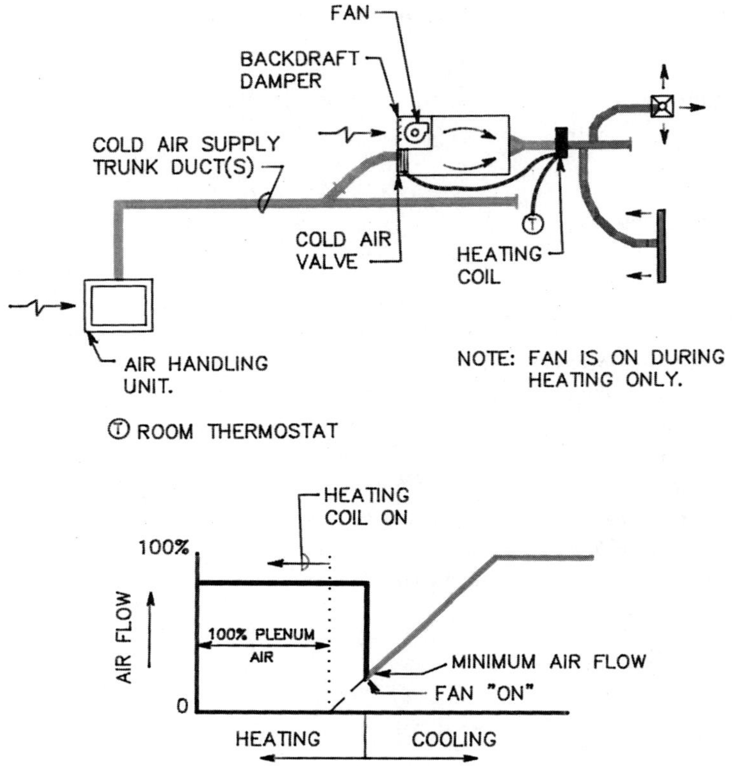

Fig. 4 Intermittent fan-powered box

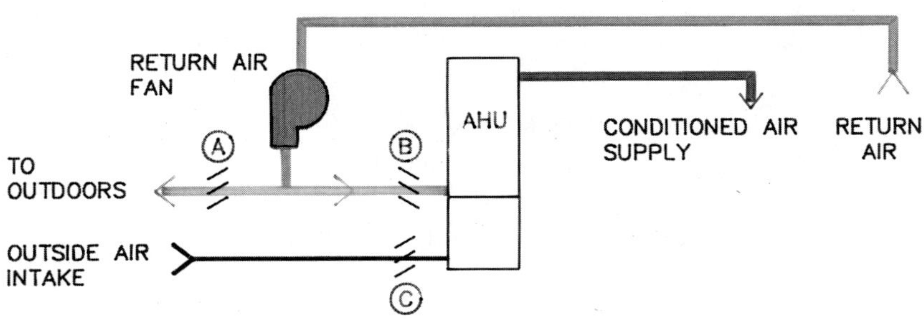

	DAMPER Ⓐ	DAMPER Ⓑ	DAMPER Ⓒ
NORMAL OPERATION –	PARTIALLY OPEN	PARTIALLY OPEN	PARTIALLY OPEN
FULL ECONOMIZER CYCLE – 100% OUTSIDE AIR	OPEN	CLOSED	OPEN

Fig. 5 Schematic of typical economizer cycle

New Trends—Ice Storage Systems

A new trend seems to be developing toward refrigeration systems that utilize ice storage for the purpose of reducing utility cost. This has the advantage of reducing electrical demand during the normal daily periods of high energy use, thereby helping to decrease investment in utility generating stations, and permitting utility rates to be reduced. In addition, the chilled water temperature may be reduced about 5°F, permitting a reduction of perhaps 20% in air handling unit horsepower. However, this energy saving is more than offset by the increased energy used in producing ice, so that the actual overall cost effect will depend entirely on the utility company rate structure and the increased investment required. Lower charges for off-peak use may permit a cost saving, even though actual energy consumed may be increased and more investment will be required.

Task for the Future

If all these matters were unusual, it would not be so important. However the industry has lost much of the skill and dedication that has been exhibited in the past. Creative new ways must be found to overcome the deficiencies that are becoming increasingly obvious.

We expect that more expert engineering services and design can overcome the inherent deficiencies now observed in energy use and functional capabilities, reduce the need for skilled installation crafts, and provide the rugged simplicity required for operating personnel to keep installations functioning properly and efficiently. Additionally, creative design solutions and new concepts in equipment may be expected, with computers providing complete automation to reduce labor requirements and to improve efficiency and comfort.

RESEARCH AND DEVELOPMENT

Electro-Thermal Cooling

Future development of a method to provide local cooling efficiently by direct use of electrical energy may become possible, but proper materials to make it practical are not yet available. Nonetheless this direct, electro-thermal cooling method is in use today for cooling small refrigerators and for exotic requirements in the NASA space program. Direct electric cooling and heating would eliminate much of the air duct distribution system for most installations and would take away the central cooling and heating plant, along with the need for most operating personnel. If electrical cooling were used for all environmental conditioning, the plenum space above tenant space ceilings would take on a new character, perhaps becoming a large, horizontal ventilation space with outside air being used for removing the heat.

Accommodating the Computer Age

The introduction of computers into the office work place is just beginning and will expand rapidly. A new environmental factor is introduced by these computers, though perhaps not the very small personal types; the small ones appear to dissipate such a small amount of heat, about 200 watts, that a few scattered in a large office area do not seem to be causing significant cooling loads or zoning problems, though they could become a significant factor if a large number of units were involved.

Other types of computers dissipate more heat and are being used in work stations where they may add about 5.5 W/ft^2 (54 W/m^2) to the cooling load. This is of particular interest for several reasons. They must be accommodated electrically; for a typical office floor of 25,000 ft^2 (2,300 m^2), it is possible they could add 125 kW of load to the electrical system per floor. This load must be cooled with 35 tons of additional refrigeration capacity along with increased air handling capability and, additionally, special zoning if the computers are not placed and used uniformly throughout an area. If one office has such a computer heat source and another identical office does not, or it is not in operation, the environmental needs differ and separate room temperature control must be provided for each of the spaces.

At the present time an air supply system is being tried on an experimental basis utilizing a floor supply system for air conducted directly into furniture where computer heat loads exist. Though this method attempts to provide a solution for the obvious and immediate need, and may become the best solution available, the real, long-term solution should be directed to a major reduction in heat generated by the equipment. An alternative solution would be a local cooling system incorporated directly in the computer as an inherent part of the computer design utilizing direct electrical energy cooling. This electro-thermal method would still need to discharge the heat to a point outside the occupied area, perhaps to the plenum space above or below the computer.

THE BUILDING

In addition to changes in the building to accommodate new office technology, other improvements in construction may be anticipated, the greatest being to permit present systems and designs to function as intended. Environmental systems function as integrated and interrelated parts of the whole building. The interrelationship affects comfort, energy use, space utilization, and cost of installation, so that it is appropriate to mention certain of the building components and conditions that are of particular interest:

- Infiltration through outside enclosing walls
- Condensation on exposed surfaces
- Mullion design and fenestration
- Insulation of the building envelope
- Security
- Fire safety.

Infiltration

Infiltration is a good example of the need for careful analysis as changes and improvements are made in building construction. Many modern buildings have such high infiltration rates that the energy use and comfort conditions are seriously affected.

We have all observed that most buildings have a very large amount of air movement through the structure in cold weather, as evidenced by the inrush of air when a street level door is opened. If the enclosing walls were really tight, this could not occur.

Proper design and careful erection of all parts of the building, with special emphasis being given to the enclosing walls, as well as ventilating systems, can reduce the infiltration to a very low figure, and the greatly improved results are worthy of the effort.

Condensation

By reducing infiltration to a very low level, condensation on interior surfaces may become of concern; condensation control is achieved by a combination of design for warm interior surface temperatures and low dewpoint temperature (the surface temperature at which condensation occurs). Infiltration acts to lower the dewpoint.

Moisture is introduced into the building air by the presence of people and by other sources, such as restaurant kitchens and humidifiers The amount of moisture concentration in the air determines the temperature at which condensation will occur on cold surfaces. More moisture causes condensation at higher surface temperature, so the design must seek to achieve surface temperatures as high as reasonable, and to reduce the moisture concentration to a level low enough to avoid condensation.

Moisture removal cannot be achieved in cold weather by utilizing the air conditioning cooling coils that function so well in summer. It is necessary to limit the introduction of moisture from internal sources, and to limit the concentration, in general, by the controlled use to outside air. In cold weather, outside air is always dryer than inside air and is thus capable of serving this dehumidification purpose. Thus, buildings with high levels of infiltration have a reduced problem with condensation.

An example of the calculated outside air requirements related to condensation will illustrate this point, assuming a cold climate. To limit the moisture concentration on an office floor having a population density of about 300 ft^2 (28 m^2), so that condensation will not occur on surfaces at 1°C (34°F), it will require through the introduction of outside air at a rate of about 0.05 cfm/ft^2 (15.24 l/min./m^2) of floor area. However, many inside metal surfaces may be well below freezing point, as observed when frost is formed, and either a higher volume of outside air will be needed, or a building design change will be required to raise the temperature of cold metal surfaces and windows.

In some geographic areas, for example along the southern coast of the United States, there is a different and at times more serious condensation problem in tight buildings with low rates of infiltration. In winter, cold weather fronts move rapidly through the area, coming out of the northwest, but are immediately preceded by low barometric pressure that brings warm, moist sea breezes from the south. During the warm period, the building materials and furnishings absorb a surprising amount of moisture, which is released quickly as outside temperatures fall rapidly and building surface temperatures fall below the dewpoint. The consequence of this weather pattern has been observed where sufficient moisture condensed on mullions to cause extensive damage to drapes and furnishings. Only relatively large amounts of outside air introduced during the initial cold period will limit the condensation. It is obvious that mullions designed for higher surface temperatures have advantages in cold weather.

Mullion Design

The metal mullions that provide structural stability and strength to support large glass areas are often overlooked as significant factors in the energy use of the building. In fact, the winter heat loss through mullions alone may equal the loss through the entire glass area that they support.

For a basic mullion having no thermal break, factors affecting inside surface temperature and heat transfer include the geometry of cross-section, such as relative areas on the inside and outside portions, and surface treatment. Coatings or other treatment of surfaces can help to reduce heat loss. Figure 6 shows the same mullion in two positions. With the larger area 94.5 in.^{2} (0.06 m^2) in the room, (toward the source of heat) surface temperature is approximately 2°C (35°F) and heat transfer is 38.3 Btu/ft (12 kJ/m) of length. With the smaller area 43.5 in.^{2} (0.03 m^2) in the room, surface temperature is minus 7°C (19°F), but heat transfer is only 24.8 Btu/ft (8 kJ/m) of length. In the summer, with the source of heat outside, more heat will be conducted with the larger area outside. In winter, more heat will be conducted with the larger area inside. Since the design preference for one direction of heat flow is reversed when summer-winter temperatures reverse, and what is good for cold weather is not good for warm weather, an economic analysis must be made to indicate which to favor, along with consideration of surface temperatures. There is much being overlooked in mullion design.

Insulation

Insulation of the building walls has increased in importance as energy costs have multiplied but anticipated results are not being achieved in the field. Changes in building construction methods and design require more thoughtful consideration to achieve proper insulation values. This is demon-

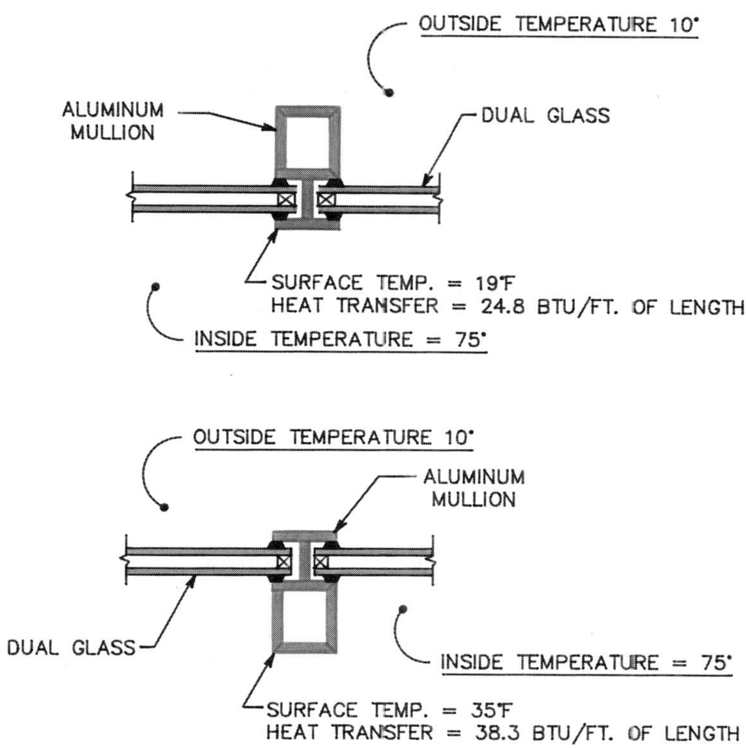

Fig. 6 Effect of mullion orientation on surface temperature and heat transfer

strated by exterior wall designs in use today that do not permit effective insulation. As architects become more aware of the need and the technical aspects of energy flow, they will make the necessary changes. The desire for a comfortable environment and the urgent need to reduce costs, both for original construction and operation, bring into focus the cooperative effort required of architects and engineers, material suppliers, and construction contractors.

SECURITY

Security against terrorist attacks, assault and intrusion are becoming of increasing concern. New systems will provide control of access and movement within buildings. Immediate detection of flammable liquids or explosives and methods for handling such emergencies will become part of normal automated procedures.

FIRE SAFETY

Fire safety is of growing concern and many changes can be anticipated. It takes fuel to feed a large blaze and to produce volumes of smoke, so we can be assured that construction materials and furnishings will be eventually eliminated as a fuel source, though further research will be needed to produce satisfactory substitutes for some present materials. Combustible materials needed in a local area, such as a library or paper supply closet, will be locally protected or housed in special enclosures.

In fires, smoke and fumes usually pose the greatest danger to life and these can be controlled by limitations on the materials permitted. In spite of the improvements that can be anticipated, however, arsonists or unique accidents are always a threat, hence methods of fire suppression and smoke control will still be needed for the foreseeable future. Smoke exhaust systems and air supplies controlled by computer will help limit the spread of fumes and smoke to relatively small areas.

Stair pressurization now appears to be the best protection for escape routes during fire emergencies, and many building codes have requirements for such pressurization. However, there is wide variation in these code requirements as to the test conditions under which pressurization systems must function, such as the number of stair doors that must be open while maintaining desired pressure. Standards will be developed that address this matter on a more orderly and factual basis, and calculation methods will be provided for pressurization systems so as to achieve a reliable design procedure to meet the criteria. Though the design calculations may not appear to be particularly complex, actually they are difficult to make. Variables such as wind force, infiltration rate, interior partitioning, stack effect in a tall stairwell, stair configuration and tightness of the stair enclosure are just some of the factors affecting stair pressurization system performance. A good pressurization design should utilize a standard computer program that will provide a system having an inherent capability of functioning reasonably satisfactorily in spite of the variables. The basic capability to accomplish this is already available.

PLUMBING

Water is a limited resource and needs to be conserved. In typical large office buildings, most of the water consumed is used for sanitary purposes in flushing fixtures and for makeup to air conditioning cooling towers. The water for the entire building may be reduced to approximately one-half by utilizing a sanitary system that uses no water; it recirculates a suitable, non-flammable, non-toxic liquid and reclaims that liquid after each cycle. Similar systems are in operation now in remote, water-short areas and will surely become more needed as water supplies are depleted in populated areas.

ELECTRICAL AND LIGHTING

Overload protection may be provided more accurately with electronic controls, permitting improved protection. Desired electrical frequencies will be produced electronically to permit more efficient lighting without ballasts, and to give ease of motor speed control. Lighting sources will be improved to provide a more desirable color spectrum, more eye comfort, and greater efficiency.

BUILDING AUTOMATION

Computers can be expected to take over the complete operation of building systems, performing all functions now done manually, as well as some not now possible.

During fire emergencies, the computers can be expected to activate and control all emergency systems in order to limit spread of fire and smoke and to permit safe exit. Elevators are involved in the emergency procedures and can be expected to have computer control to bring all passengers to a safe landing, sensing which elevators are occupied and giving those priority.

Lighting is being wasted when there is no occupant and this will be corrected by sensors to turn the lights off when no one is present. Lighting use during building cleaning normally requires a quarter to a third of the total lighting energy consumed. With lighting energy using about 50% of the total building energy consumed, 12–14% of the total building energy consumption goes to lighting during the cleaning period, and this can be reduced substantially.

Computer control of the operation of all energy-consuming equipment will be developed and will provide related improvement in both comfort conditions and energy use. Major changes can be expected in temperature control systems. Room temperature controls will have an internal computer for each instrument to coordinate and control sequencing of cooling and heating. This may even extend to controlling solar heat gain through glass (increasing gain as needed when available, reducing when desirable). The control could be self-compensating for changes in cooling/heating loads to maintain desired room temperatures as outdoor conditions change (otherwise, thermostats would normally control at a lower room temperature in winter, warmer in summer, even though the set point is unchanged).

COMMUNICATIONS

The new technology in communications is rapidly developing. Future buildings will accommodate and provide for the new systems that tenants will require. At present this is limited to being mainly a matter of distribution of cables to an increased number of locations. As accommodation and profit center, certain communications and computer services may become part of the package offered to tenants by the building owner.

CONCLUSIONS

The design of buildings and their systems will change to suit the new requirements of tenants. In particular these will be brought about by the developing computer and communication technology. Automation of building systems operation will reduce labor costs and improve efficiency of energy use. Basic design improvements and better construction methods will reduce the heat gains or losses from the building, and will contribute to a more comfortable environment, along with reduced operating costs. Improved mechanical, electrical and electronic components and system design will provide more simple solutions with greater convenience, improved operating efficiency, and a more comfortable environment. Fire suppression, smoke removal, and safety designs will be improved and standardized for maximum protection benefits.

Energy Consumption and Power Requirements of Elevators

Joris Schroeder

THE TYPICAL TRIP

The energy consumption of elevator systems is difficult to define since many parameters are involved. In a study performed in 1984, Hans Bosshardt of Schindler Management AG showed that there is a "typical trip" representing the average daytime energy consumption of an elevator system. The typical trip time in seconds must be multiplied by the motor rating in Kw and the number of trips performed to arrive at the energy consumption:

$$E(\text{Kwh/h}) = \text{Motor(Kw)} \times \text{Starts(1/h)} \times \text{TP(s)}/3600\text{(s/h)} \tag{1}$$

The typical trip time (TP) is different for different drive systems. Also, it is affected by the number of floors served, motor size, roping, and the like. The mean value, however, is a good indication of the quality of the drive system in terms of energy consumption (Table 1).

Table 1 Typical trip time mean value

Drive	Floors served	TP(s) Range***	TP(s) Mean
Gearless, DC Thyr.	12–18	3–5	4
Gearless, MG	12–18	4–6(8)*	5(6)*
Geared, ACVV low-mass	6–12	5–8	6.5
Geared, ACVV high-mass	6–12	7–10	8.5
Geared, AC 2-speed	4–8	9–12	10.5
Hydraulic, w/o cwt.	3–4	6–8**	7**

* = for low traffic intensity (<1000 St/day)
** = motor Kw = 3 × AC 2-speed, at equal load × speed
*** = lower end for 1:1 roping, large motor, few stops
DC Thyristor = DC drive machine with thyristor (SCR) converter
MG = Motor-generator converter, AC to DC (Ward-Leonard System)
ACVV = AC induction motor with variable voltage control
AC 2-speed = 2-speed AC induction motor
w/o cwt = Without counterweight

TYPICAL CONSUMPTION

Typical consumption data for equipment serving buildings are usually quoted in Kwh/m^2 × year. To determine typical elevator consumptions, three steps were required: (1) elevatoring calculations; (2) total power (Kw) installed calculations; and (3) energy consumption calculations. The elevatoring calculations required certain assumptions, given in Table 2.

Table 2 Assumptions for calculating elevator consumption

Handling capacity/5 minutes, in % of population	15%
Passengers/floor (floor size)	100
Office space/passenger (population density)	15m^2
Office/gross space	.85
Drive: DC Gearless, SCR converter	
Sky lobbies at mid-point of building	
Traffic intensities were assumed to be high: Starts/day	1500

The assumptions in Table 2 resulted in the following equations:

$$\text{Rated Kw} = \text{Rated Speed} \times \text{Load}/160 \tag{2}$$

$$\text{Kwh/m}^2 \times \text{yr.} = 19.2 \times \text{Rated Kw/Population} \tag{3}$$

The equations were used to produce the data for Fig. 1.

The typical consumptions shown in Fig. 1 are based on near-optimal elevatoring solutions. The curves are typical values, in other words they can be applied to buildings of any floor size.

It is interesting to note that double-deck elevator systems save some energy, but relatively small amounts. Shuttle elevators serving a sky lobby at the mid-point of the building and dividing the building into two identical sections are the more energy-efficient solution. They can, in some cases, save 20% to 50% of energy, depending on building height. The reasons become quite clear when the characteristics of shuttle versus local and express groups are examined:

> Local and express elevators operate at intervals of 20 to 30 seconds. For the average six-car group this results in roundtrip times of 120 to 180 seconds, or 2 to 3 minutes. On a 10-hr, or 600-min day, this produces 600/2 to 3 = 300 to 200 roundtrips/day. Each roundtrip requires 5 to 7.5 stops. Consequently, the starts/day will be 300 × 5 or 200 × 7.5 = 1500 (starts/day), local or express.
>
> Shuttles travel at shorter intervals, say at 15 seconds. Also, groups are larger, say eight cars. This results in an average roundtrip time of 8 × 15 = 120 sec, or two minutes. On a 10-hr, or 600-min day this produces 600/2 = 300 trips. Each roundtrip requires two starts and stops only. Consequently, the starts/day will be 300 × 2 = 600 (starts/day), shuttle.

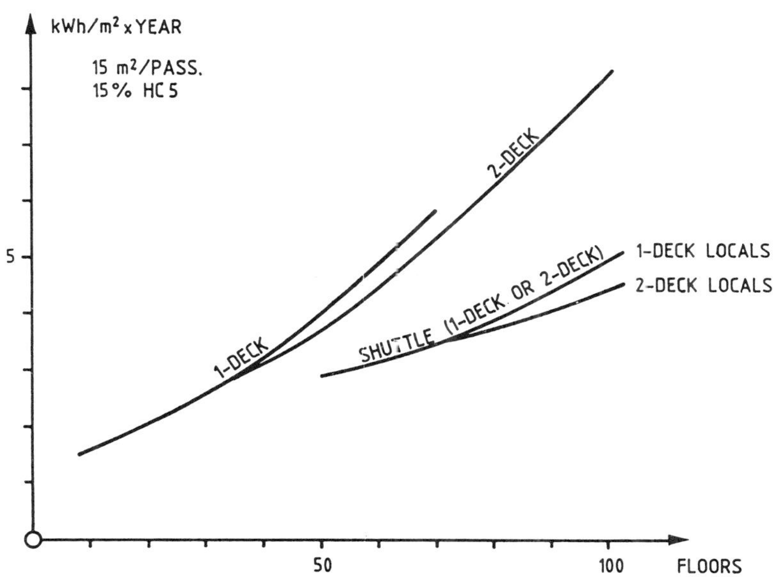

Fig. 1 Typical energy consumption based on one or two decks

Since shuttle elevators travel long distances, their starts/day are only 40% of the starts of conventional local or express elevators. With energy consumption a linear function of starts, shuttles are clearly better energy converters than conventional elevators. Elevators operate more efficiently while running greater distances at high speed than when continually starting and stopping.

When using the typical consumptions shown in Fig. 1, the following rules must be observed:

1. For MG-based Ward Leonard drives, 50% should be added (see Table 1).
2. For medium traffic intensity (1200 starts/day), 20% can be subtracted.
3. Handling capacity of 5-minute is 15% of population. For different capacities, energy consumptions change proportionately (See Fig. 2).
4. Floor populations do not affect typical consumptions are long as the office space/passenger remains at $15m^2$ ($160\ ft^2$) (See Fig. 3).

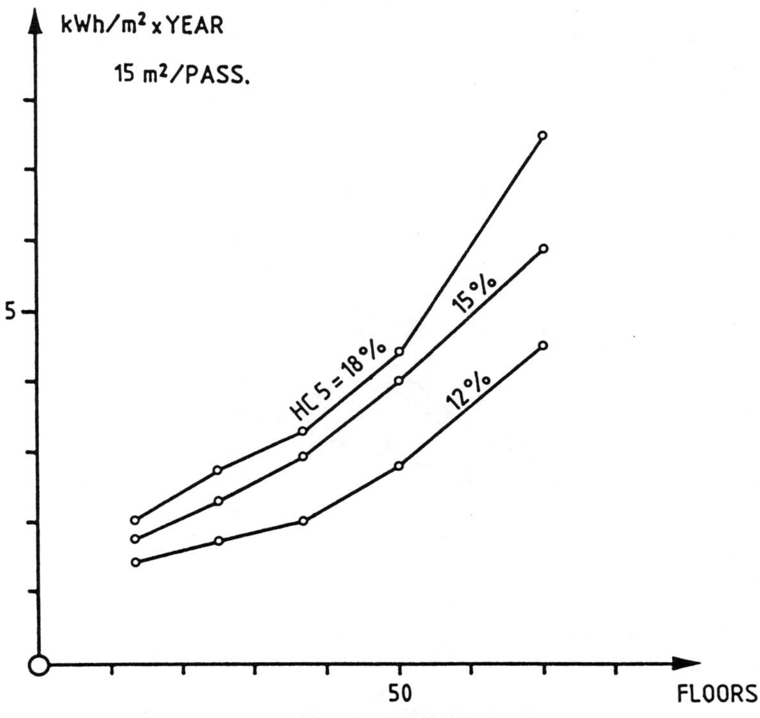

Fig. 2 Typical energy consumption based on handling capacity

TYPICAL POWER REQUIREMENTS

In addition to consumption data, building planners are usually interested in the power requirements of the elevator system. Fig. 4 shows the typical power requirements that were derived as a by-product of the energy consumption calculations. It should be noted that the typical power requirements of double-deck configurations are, again, lower than those for conventional single-deck arrangements. Shuttle-based systems require less power than all other configurations, again, but the saving is not quite as impressive as the energy saving.

MACHINE ROOM HEAT RELEASE

Like any machine, elevator drives release their losses in the form of heat, which must be removed from the machine room by adequate venting, or even cooling.

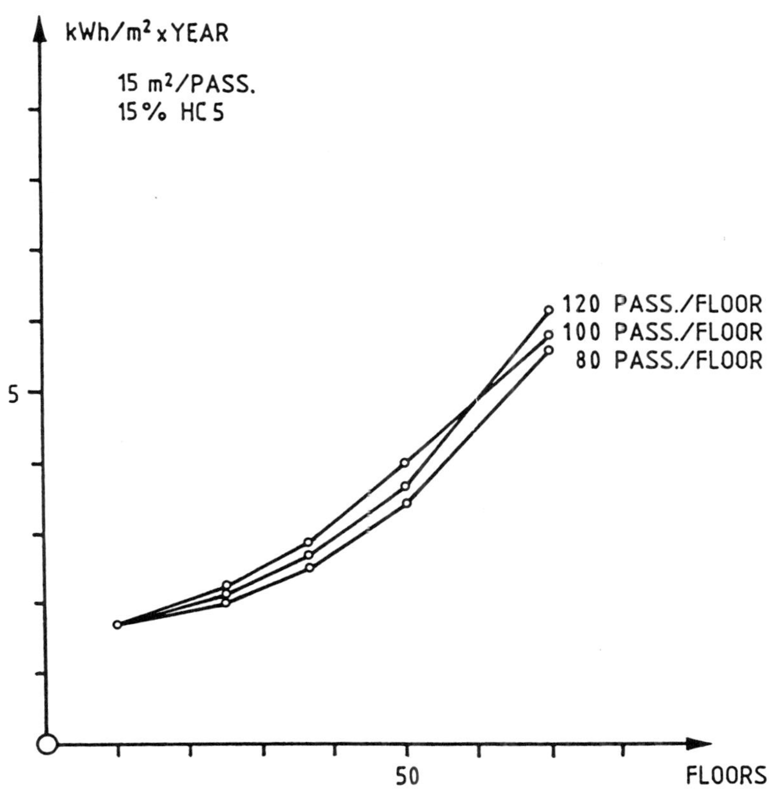

Fig. 3 Typical energy consumption based on passengers/floor

The task of an elevator is to carry passengers into the building and out again. The net work done each day is zero. Consequently, energy consumed will purely cover losses E, of which $c \times E$ are dissipated in the machine room, $(1 - c) \times E$ in the hoistway and into the floors.

For DC gearless machines with SCR converter, the machine room heat release factor c becomes .75 and the machine room heat release E_{MR} will be:

$$E_{MR} = .75 \times Kw \times 1500/10 \times 3.25/3600 = .1 \times Kw\ (Kwh/h) \qquad (4)$$

In applying the equation to other drives, different c factors must be used to reflect poorer efficiency drives, dissipating larger shares of the total losses into the machine room. Machine room heat dissipation of the most widely-used drive systems thus becomes:

$E_{MR} = .1 \times Kw\ (c = .75)$, DC Gearless, SCR
$E_{MR} = .16 \times Kw\ (c = .80)$, DC Gearless, MG
$E_{MR} = .21 \times Kw\ (c = .80)$, ACVV, low-mass
$E_{MR} = .28 \times Kw\ (c = .85)$, ACVV, high-mass
$E_{MR} = .38 \times Kw\ (c = .90)$, AC 2-speed
$E_{MR} = .22 \times Kw\ (c = .90)$, Hydraulic w/o cwt.

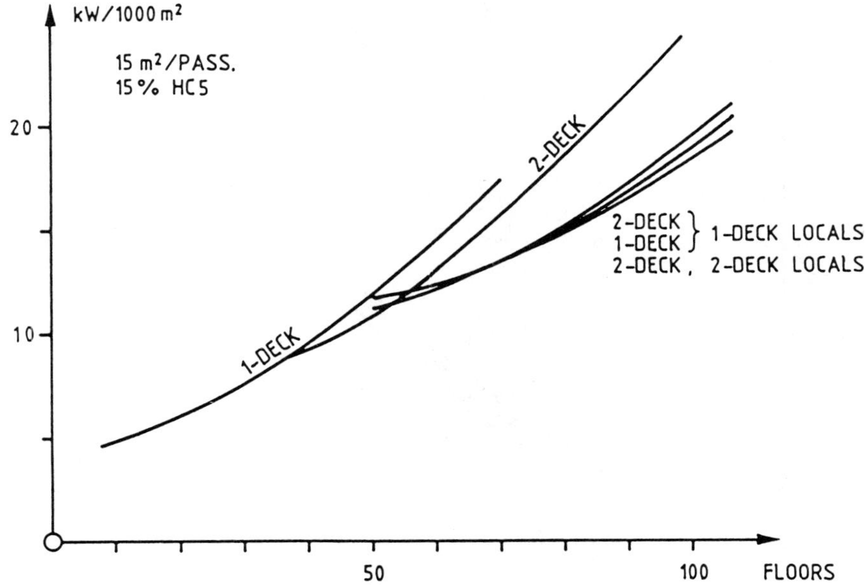

NOTES: 1) FOR MG BASED DRIVE ADD 50%
2) ALLOW +15% FOR INEFFICIENCY OF ELEVATORING SOLUTION

Fig. 4 Typical power requirements, derived from energy consumption calculations

Note: Hydraulic motor Kw = 3 × AC 2-speed

The extremely high amount of heat released by the hydraulic elevator is a result of the heat release during each down trip, during which the total potential energy of load + car weight must be dissipated into the oil tank. Oil heating is the greatest single contribution to the heat release of the hydraulic elevator.

CONCLUSION

Numbers are meaningless unless they are put into perspective: typical energy consumption of 4 $Kwh/m^2 \times$ year means annual cost of $\$0.40/m^2 \times$ year. This should be compared to typical rents of $\$100/m^2 \times$ year. Elevator energy cost amounts to less than ½% of the rent, or approximately $0.03/day for each 15 m^2 office. It is evident that elevators are highly energy-efficient, low-cost transportation.

Vertical and Horizontal Transportation

Space-Saving Elevators for the Very Tall Building

Joris Schroeder

In a very tall building, the space required for the elevator core becomes a significant part of the total building space. By selecting space-saving elevator configurations, the space requirements can sometimes be reduced by almost 50% compared to conventional solutions.

In order to put space requirements into perspective, a total of 14 buildings of up to 214 floors have been analyzed. For each building, up to five elevator configurations were examined. The study covered a total of 36 elevatoring "case studies."

To permit a fair comparison, the following assumptions were made in all cases: population was assumed to be 100 passengers/floor; handling capacity was set at 15% of population/5 min; elevator core area = hoistway + (hoistway width × car depth); office space was assumed to be 15 m^2/passenger (160 ft^2/passenger). Based on these assumptions, the ratio of core to office area was then calculated and plotted as a percentage over the height of the buildings, expressed in floors. The plot is shown as Fig. 1. The elevatoring configurations investigated were single-deck, no sky lobbies; single-deck, one 1-level sky lobby; double-deck, no sky lobbies; double-deck, one 2-level sky lobby; quadruple-deck, one 4-level sky lobby. In addition, two solutions based on the use of two sky lobbies were investigated: single-deck elevators, two 1-level sky lobbies; double-deck elevators, two 2-level sky lobbies. A mixed configuration of double-deck shuttles and single-deck local and express elevators was also examined. The space requirements were slightly higher than for the double-deck, no sky lobbies configuration.

The plot projects a rather consistent picture of the situation. The elevator core requires an increasing amount of space, and the ratio elevator/office space increases as buildings grow taller. Double-decking can save a considerable amount of space. The use of sky lobbies served by shuttle elevators will also save space, but not quite as effectively as double-decking. Using two sky lobbies does not improve space requirements, but increases core space over the one sky lobby arrangement. The use of quadruple-deck shuttles permits further savings (the arrangement will be described later).

Figure 1 illustrates the overall situation, showing the space required by the core in comparison with the office space served. Obviously, the core space ratio is much worse at the floors of the low-rise zone of the lower section of the building. All groups serving the lower section of the building and all upper section shuttles must be accommodated. Table 1 lists the low-zone space ratios for three very tall buildings.

Table 1 indicates clearly that the core space requirements of the 214-floor building become rather monumental, at 34% to 43% building average and

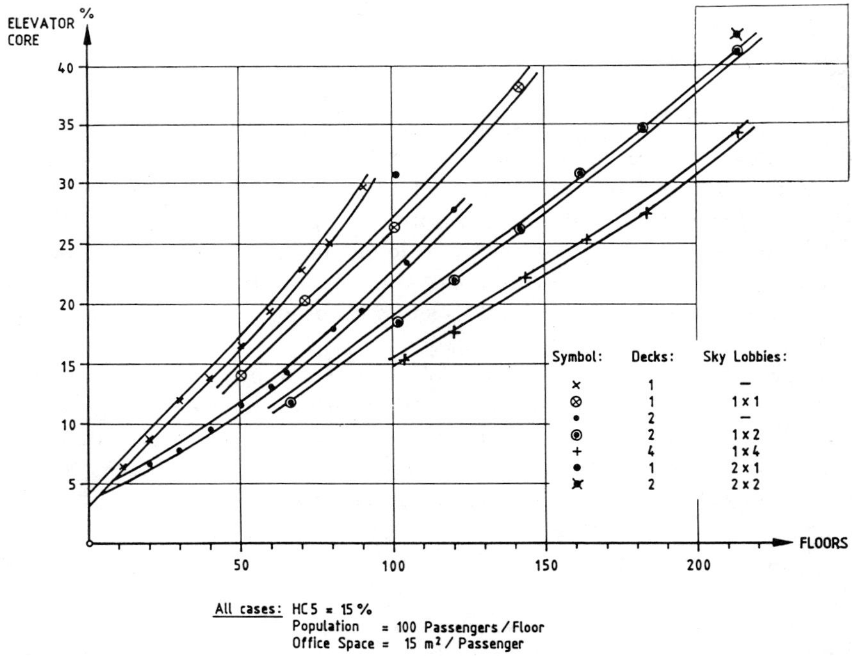

Fig. 1 Elevator core space, in % of office space served

58% to 73% in the low zone. This definitely raises the question of whether buildings of 200 or more floors can still be considered economically feasible.

The space ratio study included a quadruple-deck configuration. Since this arrangement has not been used in the past, it will be examined in more detail. Figure 2 illustrates the proposed configuration while Fig. 3 shows the conventional double-deck equivalent using double-deck elevators throughout. As will be noted, the quadruple-deck solution is actually a double double-deck arrangement which will be as easily understood by its passengers as a double-deck arrangement. To avoid passenger confusion in the 4-level lower terminal boarding area, the upper boarding levels L1 and L2 should be assigned to groups serving upper zones in their respective sections. Accordingly, lower boarding levels B1 and B2 would then be assigned for access to lower zone groups. This division of access levels would thus equally apply to the two sections of the building as well as to the quadruple-deck shuttle group. The arrangement with four boarding levels offers the additional advantage of distributing incoming and outgoing traffic evenly over four floors, thus reducing crowding and the risk of congestion. Obviously, the four boarding levels must be connected by escalators to facilitate traffic into and out of the building, as well as transfer traffic.

Quadruple-deck elevators may become a necessity for very tall buildings of 140 floors, or more, but they definitely will require new designs:

New, compact machines capable of high horsepower, but still fitting the limited space above relatively small hoistways

Heavier safeties and oil buffers.

Table 1 Core space ratios (in % of office space served)

Floors	Decks	Sky lobbies	Core %: Building	Core %: Low zone
102	1	2 × 1	31	50
	1	1 × 1	27	44
	2	1 × 2	19	29
	4	1 × 4	15	22
142	1	1 × 1	38	65
	2	1 × 2	36	44
	4	1 × 4	22	34
214	2	2 × 2	49	73
	2	1 × 2	41	72
	4	1 × 4	34	58

The 140-floor building seems to be well within the range of existing elevator technology; even 200 floors may be feasible today, from an elevator technology standpoint, but will definitely require new, heavier components. The mile-high building of 400+ floors seems to be beyond the range of feasibility, both for reasons of elevator space economy and for other reasons related to elevator technology.

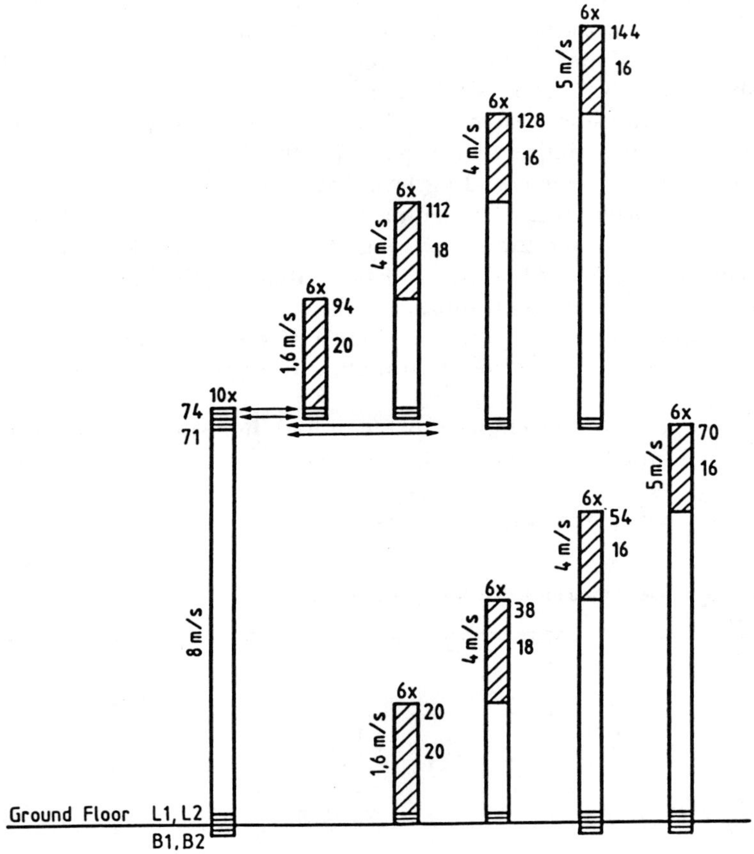

All decks 1 600 kg
58 elevators (20 shuttles)
Core 47 822 m² (22 % of office space served)

Fig. 2 144-floor building, 4-deck shuttles

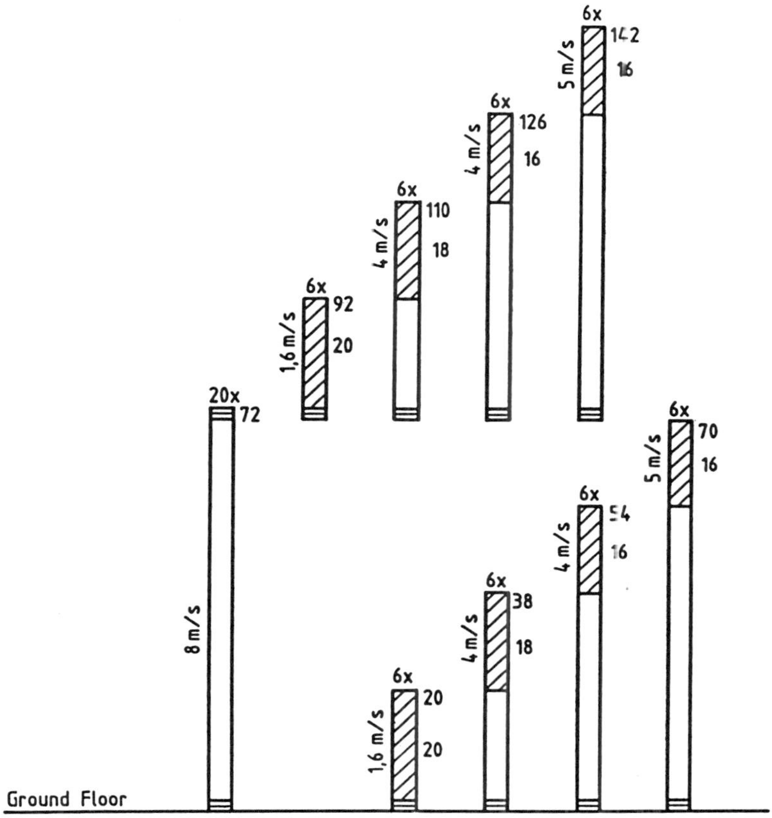

All decks 1600 kg
68 elevators (20 shuttles)
Core 56 102 m² (26 % of office space served)

Fig. 3 142-floor building, 2-deck shuttles

Current and Future Trends: The New Age of Buildings

Joseph H. Newman

In the New Age of Buildings, we will have better places in which to work, live, and interact. Buildings will run more efficiently, be more responsive to the needs of their occupants, and will more effectively accommodate the new high-tech tools of business. Materials and products will be more durable, better designed, easier to maintain, and of higher quality. More attention will be paid to increased construction productivity and an expeditious building cycle. Not only will buildings be more "intelligent," providing more effective life safety, security, energy conservation, communication capability, and environmental control, but those who design, build, manage, and own them will be smarter.

In the New Age of Buildings, more of the imperatives of the previous ages such as energy conservation and life safety will not fade away. They will intensify and be coupled with new imperatives. The owner will become increasingly sophisticated and re-emerge as a major decision-maker on the building team, setting higher standards for the built environment and demanding more value for the investment. The owner will have more influence and leverage than ever before.

In the New Age of Buildings, competition will be as keen as ever and owners will continue to buy at the best possible price, but that price will be higher than ever before because owners will be buying products, systems, and services with more added value than ever before. While this paints a rosy picture of the New Age of Buildings, there are some clouds. In this new era the regulatory flood will not abate. It will move from the federal level to the local and state level and become more of a burden than ever. The risk factor due to a litigious society will likely remain.

Considering all of these factors—both positive and negative—the question is whether the building community is prepared for the momentous and complex changes—for the tough competition—for the pitfalls as well as the new opportunities. To provoke your thinking, this paper reflects a little on the past, on what is happening today, and the implications and challenges springing from the changes now or soon to be underway, citing some of the factors that are likely to influence decision-making and suggesting some things worth doing.

THE PAST

After World War II, most of us in the United States were preoccupied with rebuilding America, increasing the manufacturing and labor capacity to accomplish this. Then we entered the era of increased comfort and improved performance when manufacturers introduced modern air conditioning systems, better lighting, new appliances, and a host of more functional and attractive interior and exterior products. Following that, the Age of Construction Management helped improve the process of building. Then the Age of Life Safety came along, when seismic, fire safety, and environmental regulations proliferated. Following the oil embargo of 1973, we entered the Age of Energy Conservation.

THE NEW AGE

Now we are entering a new age, with the emphasis on improving the productivity of activities that take place within buildings—on improving the quality of the built environment itself. We are entering a new age with greater attention to alternatives and style. This is the era of the microprocessor, telecommunication, and building automation. This is the electronic/computer/information age. This is the era of better materials and improved design. One could refer to it as the *era of higher expectations.* However, none of the issues and concerns of the past have gone away. They add to the complexity and urgency of dealing with the new expectations. Juggling priorities has become increasingly difficult. Because of these factors, today's executives and managers must consider more variables, jump more hurdles, and look deeper into their crystal balls than ever before.

Fortunately, to help decision-makers, more information than ever before is readily available by voice, video, in print, and by computer, enabling the decision-maker to be more knowledgeable about what is available and what is possible. This, coupled with greater first-hand exposure to the rich variety of available alternatives (there is more to see worldwide than ever before), feeds the expectation level of the decision-maker.

The new generation of decision-makers has surrounded or soon will surround itself with executives and specialists who can digest the plethora of available information and evaluate available alternatives. Marketing, sales, and manufacturing executives will have to be as sophisticated as the new breed customer or specifier. They will have to provide information on electronic format as well as in hard format, including formats to permit direct transfer of information to drawings, specifications, and cost estimates. This is the wave of the future.

Another factor to consider is global competition, which has resulted in a significant focus on improving productivity, which in turn has led to a re-examination of the workplace and the factory in the United States. Eric Bloch, the director of the National Science Foundation, in a talk on manufacturing technologies, stated that the factory of the future must be flexible and efficient and able to deal with great complexity. Computer-integrated manufacturing offers the potential to reach this goal.

He pointed out that historically, as manufacturing improved, flexibility was lost. Machines and the assembly line vastly improved speed and quality but at the expense of much reduced flexibility. Economies of the past were economies of scale and it just was not possible to make a few units of a specialized design at a reasonable cost. The challenge, according to Bloch, is to restore the flexibility of the individual craftsman while maintaining the efficiency and quality of mass production.

This flexibility will have dramatic benefits:

> By relaxing the requirement to standardize in order to produce efficiently, it will encourage customization of the product to meet the specific needs of the client.
>
> The ability to respond rapidly to changing market demands will encourage frequent and rapid product innovations
>
> By allowing efficient production of small lots, it will make possible production on demand in place of large inventories.

Why is this important? Because it seems the trend in the future is toward specialty custom high-performance and upscale products. The decision-maker is willing to trade bottom line dollar—particularly where it shows and can improve his business—for individuality and freedom of choice. To a degree, this desire for individuality is fueled by younger decision-makers—the so-called baby boomers. One pollster said that the baby boomers are suspicious of both big government and big business and are pro choice on everything. To a degree, this phenomenon is fueled by an ever increasing user affluency. Some commentators have said that it has become popular since the Carter era to show one's affluence. The decision-makers of today and tomorrow listen to a different drummer than yesterday's decision-makers.

The developer and builder competes today by being different than the competition and by being able to adjust quality to meet marketplace changes. The building product producer who is able to respond to these changes will be well served. The issue is becoming clear: lower volume at higher profit margins or greater volume at lower profit margins. Specialty custom high performance and upscale products can give rise to the former.

Of course, mass-produced lower cost products will always be needed, but the ratio of specialty to mass produced product is changing and the building producer who wishes to succeed must change too. It is the author's opinion that about 20% of the buys are now or soon will be going in this direction and that by the end of the decade it will approach 35%.

Signs of this happening can be seen in buildings. More money is being spent on natural stone exteriors than ever before. The amount of money being spent on interior space, in some cases, equals that being spent for the remainder of the building. So-called luxury is no longer limited to the executive suite. It has moved down a notch or two in the pecking order. Look around you at the ever growing percentage of retail space that has been upscaled. It is even happening in the percentage of hotel and institutional space that has been upscaled. In the New Age of Buildings this trend will continue and accelerate.

THE ENERGY FACTOR

The cost of energy continues to rise and will be a major problem again in the not-too-distant future. All the energy savings that have resulted from energy conservation and improved design practice will be of little comfort. None of this has any real bearing on the future price of electricity because utility companies must comply with more and more environmental regulations; must fix aging generating and distribution facilities; must add new generating facilities to meet growth; and in some cases must discontinue nuclear generating plants, all of which is very costly.

Keep in mind that the computer revolution, while bringing improved productivity to the workplace has all but wiped out the energy savings achieved during the last dozen years of conservation because computers and ancillary equipment and the additional air conditioning required use a great deal of energy. We have increased productivity at the price of increased energy utilization.

Also, as we become more of a service economy, the associated tasks require more lighting. Likewise, as we get older, we need more light to do our tasks. A typical 40-year-old needs more light than the 20-year-old; the 60-year-old needs more than both. The fact is that lighting levels and lighting quality have been increasing. This does not mean that we will return to the levels of the 1960s, but we have reversed the trend of going lower.

What can be done? We have improved the efficiency of HVAC equipment and appliances and buildings are better insulated. Much of the unnecessary energy fat has been removed. It is unlikely that there are quantum leap opportunities in this area. The last major opportunity is to control energy use when no one is present. Shutting lights off automatically; increasing or decreasing temperature automatically depending upon the season; automatically reducing use of other equipment when not needed. This is where the focus must be. This approach is not only common-sense, it is based on the philosophy that one should have adequate lighting, and a comfortable environment when working—to maximize productivity. When no one is present is the best time to change the workplace environment to reduce energy utilization.

The advent of occupancy sensors is making all of this happen. This technology could not be given away several years ago. Today, companies such as Tishman Research have sold over 100,000 sensors under the trade name of Infracon[1]; and the revised ASHRAE Standard, now out for public comment, recognizes occupancy sensors by allowing increased lighting wattage if sensors are utilized in a space. In the New Age occupancy sensors will also control temperature. An important side message is that an owner/builder faced by a need becomes a building product producer. In the New Age of Buildings, owners will find more ways to get what they want if the building product or service community is not responsive.

THE SO-CALLED "INTELLIGENT BUILDING"

In response to this new phenomenon, we are seeing more companies offering packages of integrated systems, the brain of which is both a communications and control system that continuously monitors and performs building functions. Voice and data are being transmitted simultaneously within a building over fiber optic "highways," eliminating redundant wiring. Central computing power is starting to be available to all building tenants at wholesale prices. The responsibility for security, fire safety, telephone, data and work processing, energy management, environmental control, elevator control, and other functions is being consolidated with the objective of making managers and operating personnel more efficient and effective.

Every building owner and user has different needs shaping his or her definition of an intelligent building and what it should do. Generally speaking, most developers would be likely to agree that a building is intelligent if it is

[1]Tishman Research sold its Infracon business to JWP Inc. in June, 1987.

fully rented, provides an optimal environment for productivity, operates without headaches at optimum cost, and uses the most advanced technology available. Significantly, however, the majority would probably add that building intelligence involves the integration of all building systems. While applicable in many cases, intelligence is not always simply a function of integration. Any number of permutations and combinations are possible, depending on an owner's needs. A building with just one or two integrated systems is not automatically less intelligent than a building with more. The important message is that in the New Age we will be seeing more and more automatic control of the building environment and better ways of facilitating business functions. In the New Age there will be a better balance between localized and central control of functions. An occupant will be able to adjust his or her environment within limits as desired.

In the New Age the personnel to service all the electrical, electronic, computer, heating, cooling, and telephone equipment, among others, will soon be based in each building or complex nearby. These service people will have advanced diagnostic tools that will pinpoint problems before they actually occur. Preventive maintenance will become a reality.

Services will be sold or charged according to use, reducing costs. Billing for services will be automated. Specialized service companies will take over employee education programs designed to instill greater productivity.

Not only will these companies teach employees, they will help select the right balance of equipment and support services and help corporations cope with changing needs. Work simplification, equipment optimization, and access to information of all kinds will be among the normal services provided. The critical mass needed to make these services and related equipment affordable is large enough to require a pool of users and a wholesaler—giving rise to new tenant business arrangements—sometimes referred to as shared service deals.

In the New Age of Buildings, some manufacturers will maintain partial ownership of its products in return for a prorata share of equity. This will help sell technological benefits, address service/maintenance considerations more effectively, and foster creative financing.

REDUCING TIME TO BUILD

Many opportunities are springing from the New Age emphasis on increasing productivity and improving quality. One of these is reducing the time to build. Extended construction duration is due primarily to the decision making process being too long, too inefficient, and too complex—not due to inefficient labor output or to actions of the building product producer.

Some of the major barriers to trimming cycle time, are:

Inadequate information on and understanding of the ripple effect of delays
Unnecessary or improper regulations
Too many required approvals and permits and/or too long a time to gain them
Inadequate tools to make timely decisions and to be able to explore more alternatives than ordinarily possible

To reduce these barriers, consolidation of responsibility and authority is inevitable. It will clearly help shorten cycle time. A greater decision-making role can be predicted for the professional designer and the professional construction manager that will come about as government delegates more and more of its prerogatives.

In the New Age we will have better diagnostic tools. The field of diagnostics is not limited to analytical tools. It embraces devices that signal deterioration. For example, a device that emits a noise when an air filter is clogged sufficiently to be ineffective alerts the user that it is time for replacement.

In the New Age of Buildings owners will want to know the economic and practical implications of using products or systems with different degrees of obsolescence, with different maintenance or service options, and with different replacement options. They will want know levels of quality available, price ranges, and expected performance. They will want to know what they are paying for that may be unnecessary but that is mandated by regulation. The owner's new desire to see a broader picture and to consider new alternatives provides new opportunities and this too will come to pass in the New Age.

CREATIVE NEW BREED

Not too long ago, decision-making was driven by the construction trades. The decision-makers listened to the carpenters, electricians, plumbers, and other tradespersons for their "how to do" or "what to do." Rarely were many alternatives examined. Today's decision-making is driven by a new breed who are more creative, demand more information, are less inclined to take one's word, write longer and tougher contracts, and examine more options. Their educational and experience credentials are exceptional. They are entrepreneurs in every sense of the word. They are the advance guard of the New Age professional.

It appears we can take a cue from the growing use of more durable products such as ceramic tile, stone and marble, synthetic floor covering, high pressure laminates, decorative metals, glass, and certain other abuse-resistant finishes. The style and esthetics of these products have been improved substantially. This trend suggests that continuing efforts to upgrade and to create more durable attractive surfaces and products will pay off. When a decision-maker does not like the look or esthetics of a product he will not inquire about the performance or the price.

In the New Age of Buildings owners will be using more durable, better designed, and high quality products—and will pay a premium for real added value while seeking the best price among the improved products. Also, in the New Age we will see major developments in ceramic surfaces reminiscent of the development of organic coatings of the post World War II era.

We will see other things as well. More thought will go into the design of vertical and horizontal plenums and shafts to improve space efficiency, get multiple-duty benefits and provide greater flexibility for change.

There will be more discretionary time for leisure, volunteerism, and creative endeavors during the work week, helping to accelerate a current trend towards multi-use buildings. Much of the discretionary time will be spent in multi-use buildings. There will be better pedestrian transportation among and between buildings by under or above ground links.

However, some things will not change. Flexibility to accommodate changes in work force size and function will always be needed. It has been said that building interiors are rebuilt completely every 20 to 25 years. The author predicts this will be reduced to every 10 to 15 years.

CLOUDS ON THE HORIZON

A cloud hangs over our heads, resulting, in large part, from regulators who took the approach that restrictions can solve problems instead of the engineering and the scientific approach of analysis and synthesis. The result has been overlaps, conflicts, and lack of uniformity in building regulations that pose almost insurmountable barriers to those in the building community who seek to operate beyond the borders of a given political jurisdiction. Regulatory constraints have dampened the zeal for innovation and have frustrated the efforts to make productivity gains.

The time has come to support requirements that governmental departments and agencies must show cause why a regulatory action is needed before rule-making procedures are implemented—rather than allowing departments and agencies first to initiate the rule-making process and then ask the public to show why a rule *should not* be promulgated. Though the safety and best interests of the public must be safe-guarded, the burden of proof must be reversed. The process for establishing regulations must include adequate documentation on cost benefits and feasibility presented in a uniform way and in accordance with ground rules that members of the building community must agree upon. This will come to pass in the New Age.

In the New Age, when regulations threaten the use of designs, materials, or procedures that could be used to improve efficiency and productivity, more members of the building community than in the past will get involved to determine if the proposed limitations are really justified. If not, they will fight them, and, if justified, accelerate development activities to overcome the shortcomings. Take the toxicity problem. If New York state prevails in requiring every producer to test every organic or organic-containing product with a test that is flawed and has no bearing on reality, every major state will follow. Chaos and costs will be mind boggling. The fact that the building community stopped the toxicity monster at the federal door will be to no avail. The New York state building community and others are fighting this while seeking technological solutions.

Product liability, litigious consumers, high deductible insurance, indemnification, and limited warranties are just a few of the clouds that hover over the New Age. Collectively, they may be referred to as the *risk factor.*

Several things can be done to help reduce this factor. We must develop a new generation of products that are less subject to misuse. In many product liability cases, it is not the basic product, but the way it is installed or used that is the problem. In most cases, a small increase in strength or durability, a change in chemical properties, development of an improved assembly, or installation procedure, or self-inspection, may be all that is needed to overcome potential problems. Additionally, there is a need to support legislation that will put a cap on liability awards—high enough to deter irresponsibility and low enough to lower the *risk factor.* Ed David (1985), writing on technology and risk, argues that human judgment rather than rote application of scientific principle must decide where technology lies on the learning curve. He feels that many of today's successes do not come from companies who have advanced their technologies but cannot get off dead center, but from competitors with less developed technologies who understand the market. Hopefully the culture to exploit new technologies among building product manufacturers and designers will change. Fear of risk is debilitating. Fear of the scale of economies mentioned earlier is debilitating. Fear of lawsuits is like a cancer eating away at the building community's future. The New Age professionals will probably not be as fearful as their predecessors and they may seize the opportunity to reduce the *risk factor* through improved products and systems and through a hard fight to change the ground rules.

Hopefully you have gained some insight of future change and opportunities—from one viewpoint. Others may see things differently, but there is a general consensus among professionals, owners, users, and businessmen in the building community that if they try hard enough and cooperate; and if they think more positively, more broadly, and develop new skills, they can shape change.

The historical adversarial approach to problem solving is generally counterproductive. No longer can we live with an approach that pits special interest against special interest and government against industry. We must come to the table in a cooperative spirit—as equal partners—to remove the barriers that impede progress. The skyscraper building team has always been at the forefront of initiating constructive action. As we enter the second century of the skyscraper, they will rise tall to the occasion.

REFERENCES/BIBLIOGRAPHY

David, E., 1985
ARTICLE FROM *THE BRIDGE,* National Academy of Engineers, Vol. 15, No. 2, Summer, p. 4.

Criteria and Loading

Introductory Review

Alan G. Davenport

As coordinator for the criteria and loading group I was frequently reminded during our deliberations at the Third International Conference in January, 1986 of the important benchmark, provided by the Monograph, published a few years ago. When one returns to the document, it establishes our state of knowledge; one can see what progress has been made since and what problems are still outstanding.

That conference provided us with opportunities to look at various specific problems. To begin with, we dealt with some questions relating to structural safety and quality assurance—overriding considerations both of philosophy and practicality. During the last few years, safety methods, based on a probabilistic concept, have been emerging. They provide a framework in which we can rationally express the inevitable uncertainties affecting almost every single physical factor with which we have to deal. These have led to the successful emergence of the LRFD and limit states design codes; a step forward has been achieved.

Outstanding problems remain nevertheless. The rules for applying these methods—such as the second moment reliability concepts—need to be fleshed out more fully, and extremely important questions of safety remain. The safety of an individual structure versus the safety of large numbers of structures built to similar specifications is an emerging question of insurability because of the enormous consequences if a defect is found in a large number of structures. Two cases were mentioned. One was Ronan Point; the other, a case in which large estates of apartment buildings were built under the same contract using an unwashed sand. All were suffering the same deterioration.

Graham Armer made the important distinction between the safety of personnel and of property and contents of buildings. With contents of increasing value, structural elements, of apparently superficial cost and importance, can emerge as being highly significant elements. This is certainly true of the cladding of structures.

The question of quality assurance was treated by Carl Turkstra who presented an illuminating paper about the need and the difficulties of assuring quality. He cited a number of examples in which failures in tall buildings involved not only the designer but also the supplier, the owner, and the contractor whose joint responsibilities are sometimes extremely poorly defined. A rational structure can be attached to the joint administration of a final technical product; certain practices we take for granted, such as always giving the job to the lowest bidder, may be questioned.

Fire safety was ably dealt with by Margaret Law, who raised a number of problems. Perhaps one of the most topical was that of the fire protection required in large open atria. The apparent lack of adequate technical knowledge to define the complex flow situations with both hot and cold smoke could lead to excessive caution by fire departments.

The problems of earthquake loading exemplified by the status report on the Mexico City earthquake of 1985 are ably addressed in the earthquake chapter. Many questions of great importance are highlighted: the effects of focussing and amplification of the seismic waves by the land form; the selective frequency due to the resonance of the geological formation that severely affected buildings of a certain height; and the influence of pounding.

The super tall building was a topic that spilled over into many sessions, and the description of several people's ideas of what the super tall structure looked like was perhaps one of the most exciting aspects.

In the limited context of wind loading, this author's paper suggests that the supertall building—from the Home Insurance Building to the Sears Building—has proved to be a trailblazer in defining the wind loads, the benefits of which have spilled over into other buildings, less tall than the "super" tall. Structures such as the Empire State Building have provided information that corroborates today's wind tunnel tests. Other full-scale measurements are now providing us with confidence in the diminutive wind tunnel models of extremely large buildings. The emergence of wind tunnels is hitting its stride.

Several authors refer to the dominant influence of dynamic wind response. These dynamic influences increase as height increases. Economic pressure towards lightness increases the susceptability to dynamic loading. The principle solutions that the structural engineer uses to deal with the problems of wind are an increase in mass, an increase in stiffness, and an increase in damping. If mass is lightened, then increases in stiffness and damping must compensate. The highly efficient structural systems described in this volume are a major contribution to the solution of these future problems.

Damping in structures is highly uncertain and was given special treatment in two papers. Alan Jeary, who measured damping in a number of structures, shows how it increased with amplitude. He gives some tentative equations that quantitatively relate damping to properties of the structure, a base level of damping. In the discussion of wind loading, this author referred to the possibility of damping of aerodynamic origin being negative and offsetting some positive damping from the structure, which could lead eventually to large amplitudes and instability. To avoid this, one either modifies the shape of the structure to change the characteristics of negative aerodynamic damping, or one adds damping to the structure. The latter is the subject of a paper by Ken Wiesner, who has experience with certain types of damping; he describes well the potential for additional damping systems of both the passive and active kind; he also discusses the possibility of techniques such as compensating tendons and also aerodynamic baffles operated through a feedback control system.

A consequence of dynamic motion is the effect of occupants, the subject of a paper by Andrew Irwin. He discusses the influences affecting perception and draws from experience in other fields such as offshore rigs.

Bill Melbourne deals with the description of the wind pressures; post-modernism in architecture very clearly makes it more difficult for any kind of systematic description of pressures such as one would expect to find in the code and points to the usefulness of the wind tunnel testing. He also refers to the peculiar characteristics required in loading glass, and the question of static fatigue of glass, the phenomenon related to the way in which glass breaks gradually under repeated dynamic loads.

Part of the loading on the glass is due to internal aerodynamic pressures to which Alan Dalgliesh refers. There is a link between the loading question and questions that are perhaps still unasked by mechanical, heating, and ventilating engineers. The internal pressure is determined by the infiltration in the building; in turn the infiltration is determined by the external pressures and the equilibrium pressure inside; the latter is, in fact, determined by external pressures. The question of infiltration affects the heating and ventilating and could well emerge as important in conserving energy in the future.

One of the interesting topics reported on by Alan Dalgliesh is a full-scale study of how well rain screen systems manage to balance pressures on the outside and the inside of the exterior cladding panels.

Finally, Joe Minor illustrates some of the great advances that have taken place in the last very few years in studying the strength of glass. He refers to the action of debris, to the effect of having a succession of glass failures, and to the pressure regime when one glass breaks and unloads and changes the internal pressure. A question is how far should one go in designing for the consequences of the initial failure of a glass panel and the precipitive failure that may happen to successions of panels. He also raises the question of glass failure in hurricane winds when there is a warning of the imminent hurricane. He raises the question of the requirements for the design of glass panels for earthquake loading in the plane of the glass and the curtain wall. Several countries now have a rational basis for glass design, which perhaps takes into account some of the unusual features of static fatigue and the apparent significant reduction of strength after a few years. These matters need addressing and we are hopeful that new codes will fill the breach at just about the right moment.

Behavior of Structures Retrofitted with Diagonal Steel Bracings During the 1985 Mexico City Earthquake

Enrique Del Valle Calderón

A strong earthquake affects many structures, possibly leaving them in a weakened condition from the stability point of view. The first decision to be made is whether a building can be repaired or must be demolished; this decision depends on the type of damage, on the feasibility and cost of the repair work and on the confidence of the owner that the building will be safe once it is repaired.

Several methods of retrofitting have been used, including increases in the size and reinforcement of columns and beams or the addition of shear walls or bracings.

Del Valle C. (1980) presented the reparation criteria for two buildings using steel diagonal braces. Their behavior during the September 19, 1985 earthquake in Mexico City is described here including some additional details of the work performed.

DESCRIPTION OF THE BUILDINGS

As may be seen in the reference cited above, one building which will be called "A," is a 9-story flat slab condominium apartment building. Extreme flexibility in the transverse direction and a very small separation from adjacent buildings caused Building A to pound against its neighbors. It had severe damage in the longitudinal walls in the stories where it impacted those buildings. Damage in the transverse partition walls caused by excessive deformations occurred as well as fracture in a column in the fourth story. Neither the longitudinal nor the transverse partition walls were considered structural in the original analysis of the building; however, there was no gap between the walls and the structure to allow free movement during earthquakes, so these walls became structural and were damaged.

Figure 1 shows a typical plan and a cross section of the building, including the bracing system adopted to increase lateral stiffness in the transverse direction, which were installed in the central bay of three of the "equivalent" frames formed by the columns and the waffle slab. As the bay was relatively narrow, lateral forces would produce large vertical forces in the columns of this bay; therefore, additional reinforcement for them was provided using four angles at the corners joined with plates, as shown in Fig. 2. It was not possible to give continuity to these angles along the height of the building, and the forces acting on them during earthquakes were transmitted through the solid slab portion around the columns by means of bolts, using a sort of

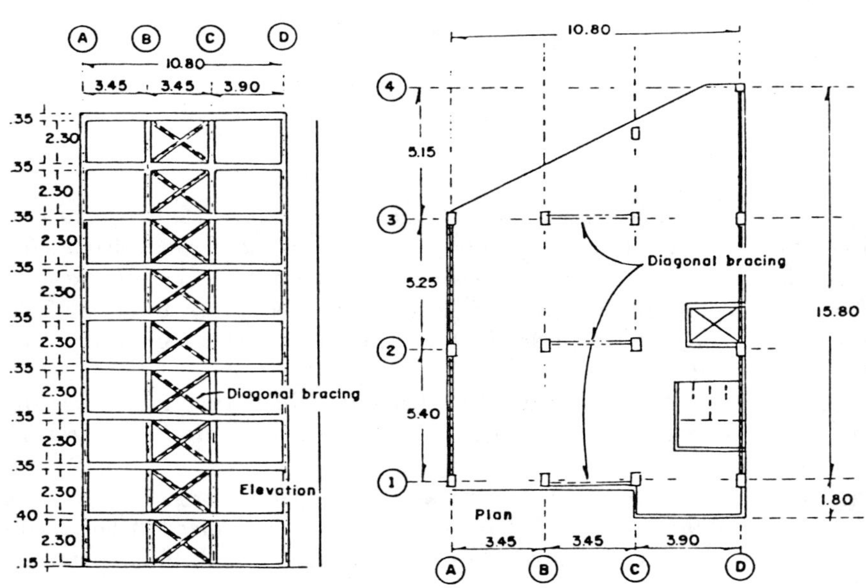

Fig. 1 Typical plan and cross section of building A

footing made of steel plates at each end. The space between the slab and the plates was grouted to guarantee good contact and transmission of these forces. It was also verified that the longitudinal reinforcement bars in the columns could help resist the possible tension developed in a strong earthquake, and that the foundation could withstand the new distribution of forces. In all stories the longitudinal walls along both property lines were strengthened with a concrete cover 50 mm (2.0 in.) thick, reinforced with wire mesh (see Fig. 1).

Vibration periods of the building were measured with a portable seismometer before and after retrofitting. In the transverse direction, first mode period diminished from 1.05 sec to 0.8 sec representing a significant increase (more than 70%) in the stiffness, which would reduce lateral displacements of the structure in future shocks. It eliminates most of the pounding problems with neighbors, as the calculated displacements were approximately one-third of those that would occur without the braces.

Fig. 2 Reinforced columns, using four angles at corners joined with plates

Building "B" is a 12-story medical condominium described by Del Valle C. (1980). It suffered greater damage than building "A" in 1979, as several columns had diagonal and vertical cracks and beams were also cracked due to shear and flexural effects (Figs. 3 and 4).

After studying the problem it was concluded that the damage could be attributed to a defective analysis that underestimated the stiffness of the front and rear frames having deep spandrel beams because the model that was used assumed that the moments of inertia of the beams and columns had constant values along the theoretical center-to-center axis, neglecting joint effects that, in this case, increase column stiffnesses considerably. This effect was difficult to take into account when the building was designed, because no computers were available at that time.

As the actual stiffness of these two frames was larger than the one assumed

Fig. 3 Cracks due to shear and flexural effects in building B caused by the 1979 earthquake

in the analysis, they attracted forces much larger than those that had been used in the design and therefore failed.

Because the building was fairly regular and damaged elements were mainly on the extreme frames, it was decided to reinforce both frames using new facades that had diagonal steel elements along the whole width of the building, as shown in Fig. 5. This solution provided enough stiffness in the new elements to reduce the participation of the damaged original structure to a minimum in future earthquakes. The slabs had to be reinforced to increase their strength to transmit seismic forces to the new facades. Also some additional piling was necessary for these elements. Cracks in spandrel beams were repaired using epoxy resins. Heavily damaged columns in the lower levels of the building were reinforced with steel plates. Connections between the original structure and the new facades were made using steel plates and bolts.

Fig. 4 Beam cracks in medical condominium building caused by the 1979 earthquake

Periods of Building B were also measured before and after retrofitting; they changed from 1.87 to 1.15 sec; therefore the increase in stiffness was approximately 164%.

Retrofitting of both buildings was based on forces estimated from the 1976 Mexico City code.

BEHAVIOR OF THE BUILDINGS DURING THE SEPTEMBER 19, 1985 EARTHQUAKE

Both buildings (A & B) are located in the soft soil zone of Mexico City where damage was severe.

Building "A" had some problems again, as it struck its neighbors, although to a lesser extent than in 1979. Partition walls in the short direction were also cracked at some levels, and bracings at the levels of pounding loosened. This was expected because the separation between adjacent buildings was very small, and though stiffness was increased more than 70%, it was not enough to

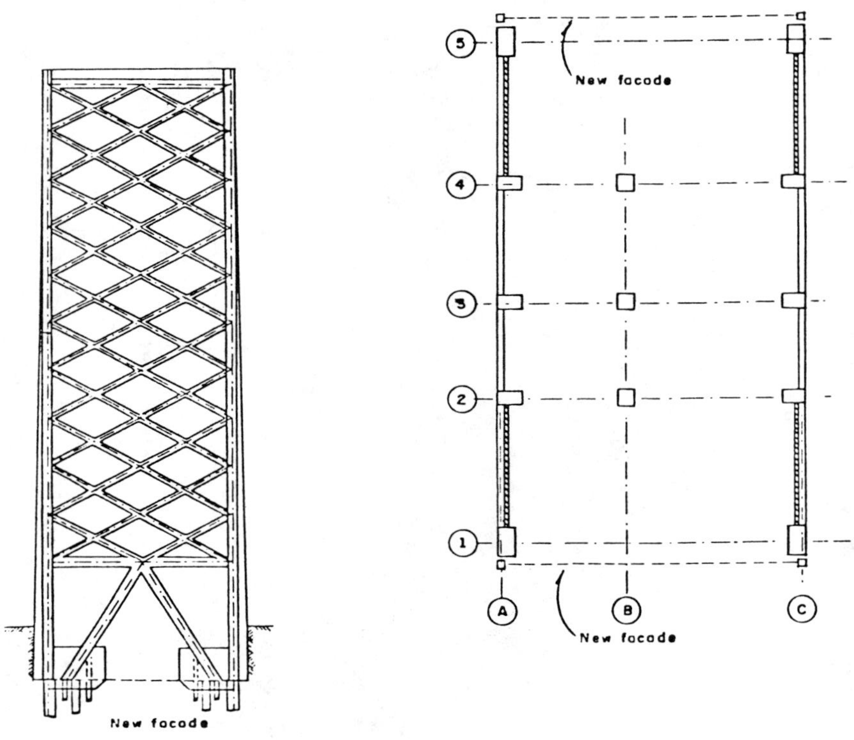

Fig. 5 Typical plan and new facades used to reinforce frames, with diagonal steel elements along whole width of building B

avoid pounding, especially considering that retrofitting was calculated on the basis of the 1976 Mexico City code, and design accelerations recommended there were exceeded by more than three times. Therefore, the behavior of this building was good. It has been already repaired and this time people did not have to be evacuated.

Building "B" had excellent behavior as it suffered no damage. This was really a surprise, taking into account the severity of motion around it that caused several buildings to collapse less than 300 ft (100 m) away and, as was mentioned before, that retrofitting was based on a code whose recommendations were exceeded.

One of the probable reasons for the good behavior of building "B" is that, in the way the new structure was connected to the original one, additional damping was included, which could probably be determined by a forced vibration test. Another fact that may explain the good behavior is that the reinforcement was continuous through the height, because it was placed on the outside of the original building.

CONCLUSIONS

From the behavior described above, it can be concluded that diagonal steel bracing may be a good solution for the reinforcement of structures damaged by earthquakes. One advantage of this solution is that the building might have been damaged due to a relative coincidence of its dynamic properties with those of the soil, and the added stiffness changes the situation. Another advantage is that very small additional seismic forces would result. When retrofitting is made by adding concrete shear walls, the increased weight reduces the efficiency of the solution, because the seismic forces would increase proportionally.

It is always necessary to make a complete analysis of the original structure plus the reinforcement. It has to be verified that the seismic forces have the appropriate path to reach the reinforcement. Slabs have to be reinforced at times or special collectors of force should be added.

REFERENCES/BIBLIOGRAPHY

Del Valle C., E., 1980
SOME LESSONS FROM THE MARCH 14, 1979 EARTHQUAKE IN MEXICO CITY, Proceedings of the Seventh World Conference on Earthquake Engineering, Volume 4, Istanbul, Turkey.

Popov, E., Takahashi, K., and Roeder, C. W., 1976
STRUCTURAL STEEL BRACING SYSTEMS. BEHAVIOR UNDER CYCLIC LOADING, Earthquake Engineering Research Center, Report UCB/EERC-76/17, University of California, Berkeley.

Rea, D., Shah, H. C., and Bouwkamp, J., 1971
DYNAMIC BEHAVIOR OF A HIGH RISE DIAGONALLY BRACED STEEL BUILDING, Earthquake Engineering Research Center, Report UCB/EERC-71/5, University of California, Berkeley.

Damage Statistics of the September 19, 1985 Earthquake in Mexico City

Gilberto Borja Navarrete
Manuel Díaz Canales
Alejandro Vázquez Vera
Arturo A. Bello Maldonado
Luis B. Rodríguez González
Sergio Escamilla Aguilar
Gerardo Muñoz González
Enrique Del Valle Calderón

Several studies have been made to assess the type and magnitude of the damage caused by the earthquake that shook Mexico City on the morning of September 19, 1985. Grupo ICA (Civil Engineers Associated Group) has participated in all stages of analysis of the effects of this quake, and has prepared a complete study that includes seismicity aspects, characteristics of the earthquake, geological aspects, existing instrumentation, evolution of the codes, definition and behavior of structures, and damage evaluation backed by photogrametric studies and direct inspection of buildings. This paper summarizes part of the study and describes the types of structures that existed in the city and how they were affected by the quake. An important feature of the study is that an inventory of all the existing buildings in the zone of maximum damage was obtained, including such information as type of structure, construction material, and number of levels, in order to compare damaged buildings against existing buildings.

GEOLOGICAL ASPECTS AND CHARACTERISTICS OF THE MOTION

Stratigraphically, Mexico City is founded on three different types of soils: very soft and wet clays typical of the ancient lake area, compact sands and silts characteristic of the hill zone, and transition soils consisting of soft clay with less water content, interbedded with lenses of sands and gravels.

The zone of maximum damage was located entirely in the ancient lake bed zone (Fig. 1). Acceleration value in the different types of soil ranged from about 20% gravity in the lake zone, considerably exceeding design accelerations specified by the code, to 4% gravity in the zones of firm soil.

The most severe destruction occurred in the lake zone and was mainly caused by filtering and amplification of the seismic waves coming from the distant epicenter, resulting in an almost harmonic, long duration motion with a dominant period of 2 sec. The amplitudes of the motion were very

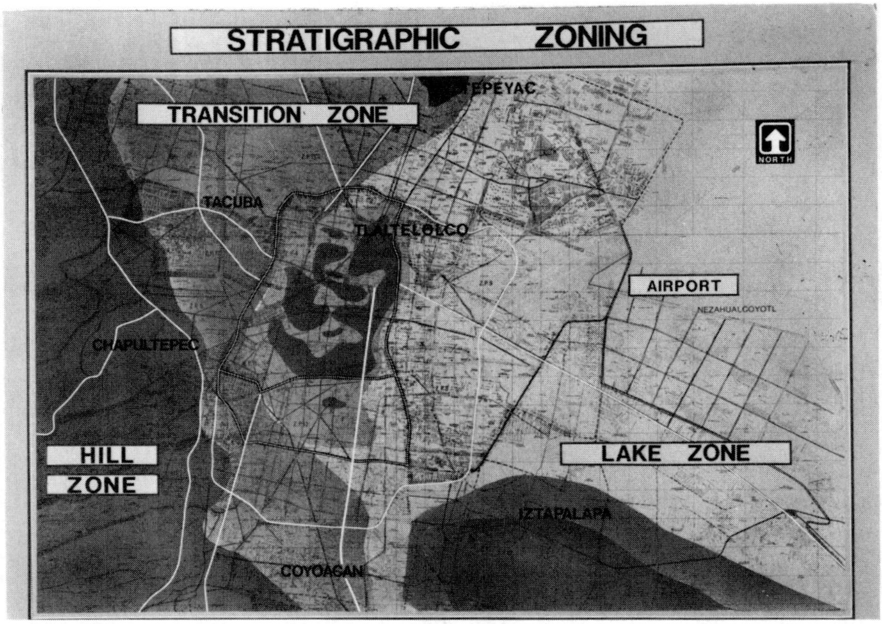

Fig. 1 Stratigraphic zoning

large (Fig. 2). The intense phase of motion in that zone lasted for about 45 sec and affected mainly those buildings that had oscillation periods in the vicinity of 2 sec, which were close to resonance, as shown in Fig. 3. In this figure the design spectrum for the soft soil zone of the city is compared with the response spectrum obtained for that zone from the E–W component of the accelerograph records shown in Fig. 2, for a damping value of 5% of critical.

The earthquake caused undulations and permanent buckling of the surface as well as bending of rails, something never before observed in the city. Underground pipes cracked and the sudden settlement of some buildings and overturning of some others was caused by soil or pile failure.

STRUCTURAL BEHAVIOR AND DAMAGE EVALUATION

Types of existing structures

The six basic types of structures in the city are shown in Fig. 4. The first corresponds to old buildings and consists of very thick masonry bearing walls and floor systems with wood or steel beams, and brick or wood floors or stone vaults and arches. The second is a modern version of the other, with thinner

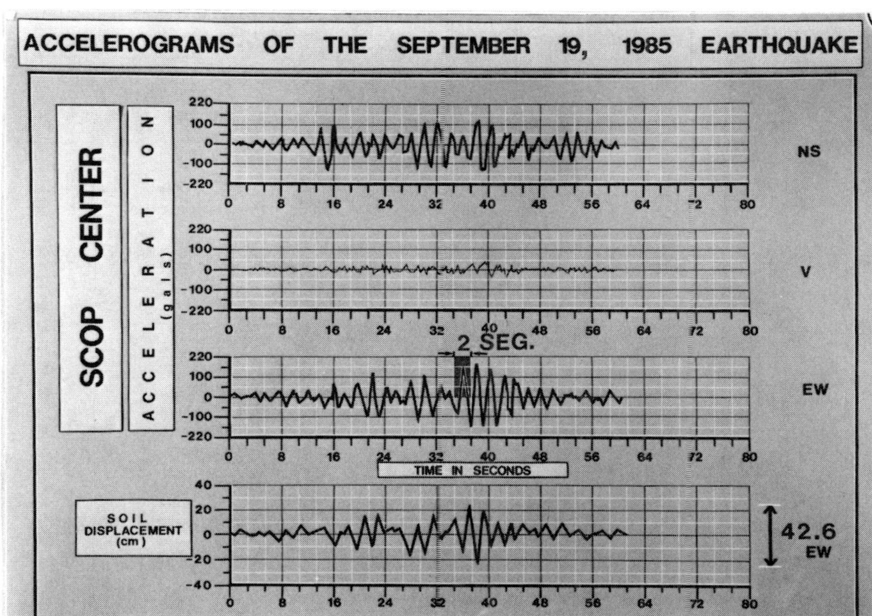

Fig. 2 Accelerograms of the September 19, 1985 earthquake

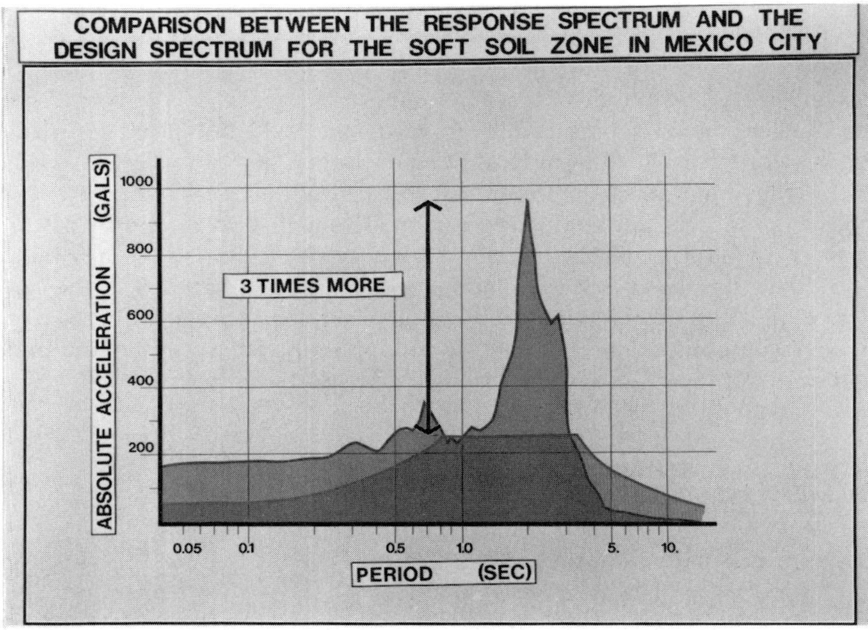

Fig. 3 Comparison between the response spectrum and the design spectrum for the soft soil zone in Mexico City

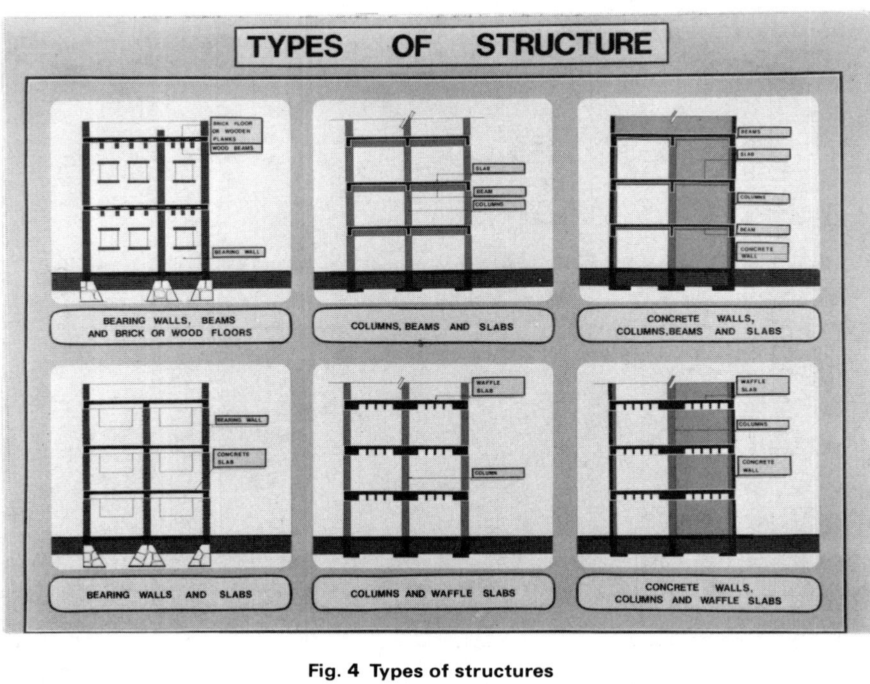

Fig. 4 Types of structures

masonry bearing walls reinforced with vertical and horizontal concrete elements and with concrete floor systems. These two types are mainly used for housing with a maximum of 6 to 8 stories. As they have many walls in both directions, they are rather stiff and their periods of vibration are relatively small, usually less than 0.5 sec. Therefore, their dynamic response in the soft soil zone was small and they suffered little damage, in general, as will be seen later.

The third and fourth types of structure are framed systems, with deep beams and slabs for the former and waffle slabs for latter, connected to columns to conform rigid frames. The partition or facade walls have no theoretical structural function. They are mainly made of reinforced concrete, but in some cases, for the taller buildings, steel is used. The number of stories varies from 2 or 3 to more than 40, as in the well-known Latino-Americana Tower, and are used for housing or office buildings. These two types were the most affected by the earthquake, as their periods of vibration, in many cases, were close to those of the motion and their displacements were large.

The final two types of structures have additional stiffening elements such as concrete or masonry shear walls or diagonal bracings, either of concrete or steel, placed in some bays to reduce displacements and improve the general behavior of the structure. Few buildings with these systems were damaged.

Types of damage commonly found

Concrete was found to be the predominant material in the seriously affected structures. Few steel buildings were damaged, mainly because this material is used only in taller buildings whose dynamic response was lower because their periods of vibration are longer than those of the soil. Only one 21-story steel building collapsed; its measured period before the earthquake was 2 sec, therefore it was certainly a case of resonance.

The main types of failure found in concrete buildings were diagonal cracking in beams, columns and walls due to shear; cracking and loss of concrete in beams and columns due to compression caused by flexure or a combination of axial forces and flexure, with buckling of the reinforcing bars; and shear failures in waffle slabs around columns, with several cases of punching.

A larger incidence of failures occurred in buildings from 6 to 15 stories high, mainly with structural systems of the types 3 and 4 described before. Many failures in upper or intermediate levels were caused by pounding with adjacent buildings, changes in stiffness and/or mass, or excessive loads. A large number of corner buildings suffered heavy damage due to torsional oscillations produced by collaboration of asymmetrical nonstructural walls, as the deformations caused by the quake exceeded the gap left between the structure and nonstructural brick walls, forcing them to take seismic forces.

Buildings with waffle slabs and columns, forming "equivalent" frames, had twice as many failures as buildings with deep beams and columns developing rigid frame action. "Soft" first story, caused by the existence of nonstructural walls in upper levels of apartment buildings with parking garages without walls at the base was also the reason for some partial or total collapses.

STATISTICS OF DAMAGED BUILDINGS

To make a rapid evaluation of damage just after the earthquake, aerophotogrammetric surveys, to back up direct observation of damaged buildings, were made. The most affected zone has an approximate surface of 17 mi^2 (43 km^2) and is limited by the Circuito Interior to the north and west, by the avenues Eugenia and Division del Norte to the south and by Calzada de la Viga to the east (Fig. 5). This zone was divided into 10 sectors. The whole metropolitan area covers 440 mi^2 (1100 km^2).

For a preliminary evaluation, three types of damage were considered: total collapse, partial collapse, and severe structural damage. Figure 6 shows that in the 10 sectors mentioned above there are 53,358 buildings, from which only 757 had those types of damage, representing only 1.4% of the existing buildings. Of the 757, 133 totally collapsed, 353 partially collapsed, and 271 sustained heavy structural damage. Sectors 1 through 4 contained 86% of the total.

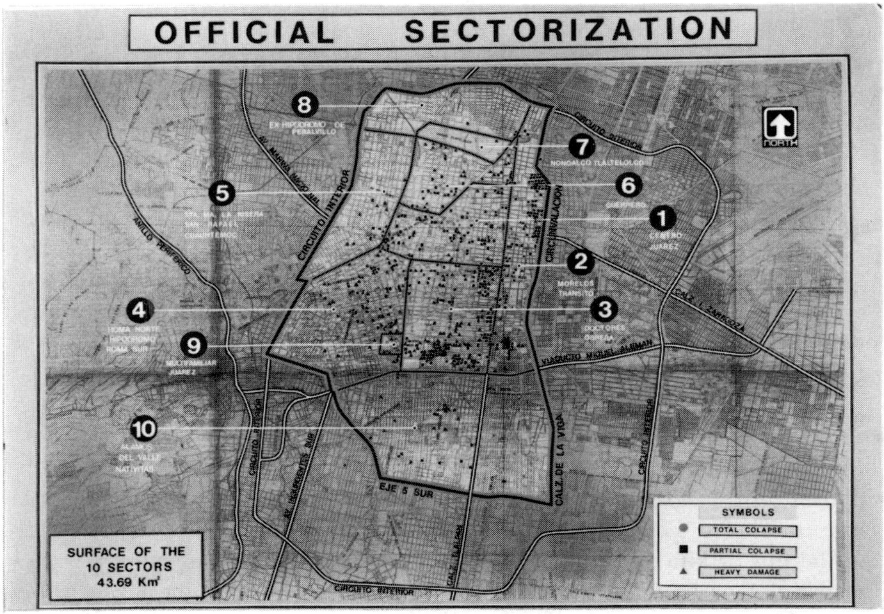

Fig. 5 Official sectorization

A very interesting statistic was obtained in relation to the number of levels of the affected buildings, as shown in Fig. 7. The observed failures, with respect to the total number of each type of building that existed in the 10 sectors ranged as follows: 1% of the structures of 1 to 2 stories, 1.3% of those with 3 to 5 stories, 8.4% of those with 6 to 8 stories, 13.5% of the ones with 9 to 12 stories, and 10% of the buildings with more than 12 stories. The number of buildings with more than 15 stories with severe damage is very small, confirming that the most affected types of structure were those with vibration periods close to the predominant periods of the soil motion. It is surprising that these averages for the 10 sectors change very much from sector to sector; for example the percentage of damaged buildings against those existing in the range from 9 to 12 stories in sector 2 is as high as 67%, probably due to soil conditions. This will be investigated in future work.

Although the number of 1- to 2-story buildings affected is relatively large, (46% of the total number of cases) it was found that most of them were seriously deteriorated by lack of maintenance before the earthquake.

In regard to building use, Fig. 8 shows that 55% of the damaged buildings were for housing, 23% private offices, 9% public office buildings, and 13% of different uses: factories, schools, hospitals, hotels, recreational, commercial, banks, or religious buildings. Certainly many other buildings were less seriously affected. Official records mention more than 3,300 affected buildings,

TYPE OF DAMAGE

	SECTORS												RELATIVE COMPARISON
	1	2	3	4	5	6	7	8	9	10	SUM	%	%
TOTAL COLAPSE	24	24	20	48	3	5	1	-	1	7	133	18	0.252
PARTIAL COLAPSE	59	102	106	41	6	28	-	4	2	5	353	47	0.658
HEAVY DAMAGE	31	39	99	57	3	21	1	-	2	18	271	35	0.490
SUM	114	165	225	146	12	54	2	4	5	30	757	100	1.4%

Sectors 1–4: 650 86%

EXISTING BUILDINGS: 53 358

Fig. 6 Type of damage

HEIGHT OF THE BUILDINGS

NUMBER OF LEVELS	SECTORS 1	2	3	4	5	6	7	8	9	10	SUM	%	EXISTING BUILDINGS SUM	RELATIVE COMPARISON %
HOUSING UP TO 2	13	111	128	35	1	44		2		12	346	46	37 484	1.0 %
FROM 3 TO 5	30	24	61	44	5	8		1		6	179	24	13 498	1.3 %
FROM 6 TO 8	38	18	18	47	3	2		1	1	8	136	18	1 616	8.4 %
FROM 9 TO 12	26	6	12	18	3				3	4	72	9	531	13.5 %
MORE THAN 12	7	6	6	2			2		1		24	3	229	10.4 %
SUM	114	165	225	146	12	54	2	4	5	30	757	100	53 358	1.4 %

Fig. 7 Height of the buildings

USE OF BUILDINGS

	SECTORS 1	2	3	4	5	6	7	8	9	10	SUM	%	
SINGLE HOUSING	4	37	30	26		14				8	119	16	55%
MULTIPLE HOUSING	23	72	93	64	5	21	2	1	5	9	295	39	
PUBLIC ADMINISTRATION	16	12	20	8		3				6	65	9	45%
PRIVATE ADMINISTRATION	49	17	48	35	5	10		3		6	173	23	
HOSPITALS			11								11	1.5	
SCHOOLS	2	1	9	5		1					18	2	
BANKS	7		3								10	1	
RELIGIOUS		1		1		1					3		
HOTELS	5		3	1	2						11	1.5	
FACTORIES	1	25	6	3		3				1	39	5	
THEATERS	1		1	3		1					6	1	
GATHERING CENTERS	6		1								7	1	
SUM	114	165	225	146	12	54	2	4	5	30	757		

Fig. 8 Use of buildings

but more than half of them had only nonstructural damage and may be repaired easily.

CONCLUSIONS

The earthquake that occurred on September 19, 1985, with a Richter magnitude of 8.1 and maximum intensities on the modified Mercalli Scale of IX and X in Mexico City, caused the most severe historical recorded destruction. Considering the severity of the motion in the zone of soft soil of the city and the peak of the response spectrum for structures whose vibration periods is near or equal to two sec, it is surprising that the number of severely damaged buildings was not larger, as design code accelerations were exceeded considerably, as shown by Fig. 3. This suggests that actual ductility and damping in many buildings was larger than the currently assumed values or the actual strength is larger than that predicted by models. This should be investigated.

The marked influence of the soil upon which a building is constructed will modify previous concepts regarding microzonation of the city. Also, the large amplifications of the soft soil will be of concern, as there are other earthquake-prone cities in the world with soft soils.

Future buildings should include stiffening elements in some bays, in order to reduce lateral deformations and obtain better performance under strong earthquakes. It is recommended to increase the number of accelerographs installed in the city, to measure the effects of future earthquakes in the different types of soil, and to improve microzonation.

Partially affected buildings should be carefully studied by highly qualified specialists who should design the necessary retrofit and rehabilitation that guarantees their safety.

ACKNOWLEDGEMENTS

The study carried out by Grupo ICA was made possible with the collaboration of a large number of engineers, architects, and technicians, who work in different engineering firms of the group; mainly ISTME, GEOSISTEMAS and AEROFOTO. They performed the field surveys, inventoried existing buildings, organized the information to be processed by computer, made the drawings to illustrate the study, and many other jobs necessary to make this study a reality.

REFERENCES/BIBLIOGRAPHY

Del Valle C., E., 1984
EARTHQUAKE DAMAGE TO NON-STRUCTURAL ELEMENTS, Proceedings of the Eighth World Conference on Earthquake Engineering, Vol. V, San Francisco, California

International Association for Earthquake Engineering, 1984
EARTHQUAKE RESISTANT REGULATIONS. A WORLD LIST, Tokyo, Japan.

Prince, J., et al, 1985
INFORMES IPS-10A, IPS-10B, IPS-10C, IPS-10D, IPS-10E, GAA-1A, GAA-1B and GAA-IC, Instituto de Ingenieria, UNAM, September-October.

Rosenblueth, E., Meli, R., and Resendiz, D., 1985
EL TEMBLOR DEL 19 DE SEPTEIEMBRE DE 1985 Y SUS EFECTOS EN LAS CONSTRUCCIONES DE LA CIUDAD DE MEXICO, Informe preliminar del Instituto de Ingenieria de la Universidad Nacional Autonoma de Mexico, September.

Effects of the September 19, 1985 Earthquake on the Buildings of Mexico City

Roberto Meli

As a first step in evaluating the effects of the earthquake, the gathering of evidence was a matter of high priority, as it could disappear in a few days because of the demolition and removal of collapsed or severely damaged buildings.

In the three days immediately following the quake, a first survey was made of those buildings that showed evident signs of major damage when viewed from the street. Therefore damage to a significant number of buildings was not detected in this first survey. Moreover, little information could be collected, first because in collapsed buildings only few general structural characteristics could be identified, and second because many severely damaged buildings could not be inspected from the inside.

The evaluation was limited to buildings, as damage to other civil engineering works was minor. Practically no industrial facilities existed in the affected zones. Underground structures and bridges behaved remarkably well and only water pipelines buried at shallow depths suffered severe damage.

DAMAGE CLASSIFICATION

The types of damages were classified as (1) total or partial collapse, (2) major structural damage, (3) significant structural damage, and (4) minor damage.

Buildings in the first category included those that failed by structural collapse, foundation failure, or induced failure by an adjacent building. The second category included buildings damaged to a degree that assumed their repair was not economically feasible. Buildings in the third category require major structural strengthening, whereas the last category refers to local damages that presumably do not affect the overall safety of the structure.

DAMAGE ZONIFICATION

Figure 1 shows the location of the collapsed or severely damaged buildings (damages of types 1 and 2). Figure 2 identifies a zone of very high density of damaged structures and a second one where this density is lower but still significant. The boundaries of this second zone are not well defined. North and south of the zones shown in the figure are spots where few buildings were severely damaged. The approximate area of high density of damages is 9.2 mi^2 (23 km^2) and that of the zone with significant damages is roughly 26 mi^2 (65 km^2). Both zones are located in the part of the city with the highest density of medium to high story buildings. The total inhabited area of the Mexico City valley is estimated at 1000 km^2 (400 mi^2).

A clear correlation exists between the geographical distribution of damage and type of subsoil. Figure 3 shows the most commonly used zonification of the subsoil in Mexico City.

Three main zones are defined: the hill zone of firm soil, the lake zone of deep deposits of a very soft clay, and, in between, a transition zone where the clay deposits are shallow. Fig. 4 shows soil profiles in north-south and east-west lines crossing the most affected area of the lake zone. The following layers are identified: a shallow superficial layer of dessicated clay or fill material, then a first soft clay layer followed by a first hard layer of compacted silty clay; then a second layer of soft clay and finally the deep firm soil deposit. Curves of equal depth of the second hard layer are shown in Fig. 5.

Areas of significant damage are located in the lake zone; outside this zone only minor damage and mainly some nonstructural damage in tall buildings is reported. The area of highest damage is located at the west of the lake zone where the depth of the first firm layer is between 26 and 32 m (85 and 105 ft) and that of the firm deposit is between 26 and 46 m (85 and 151 ft).

The lack of damage in the zone of firm soil and in the transition zone is explained by the low intensity of the ground movement that was not amplified by the layers of soft soil. The low damage in the areas of thicker soft soil layers is partially caused by the fact that the natural period of vibration of these layers is significantly higher than the dominant periods of vibration of the movements that arrived from the subjacent firm deposits, thus giving rise

Fig. 1 Location of collapsed (●) or severly damaged (x) construction

to a less severe amplification; it is also caused by the lower density of tall buildings in these areas.

Fig. 2 compares the zone where some structural damages occurred in two previous earthquakes that affected Mexico City (1957 and 1979) with those corresponding to this earthquake. The area of damage is much larger in this last earthquake and there is a certain coinicidence between the zone of maximum intensity of damage in 1985 and the total affected zones in the previous quakes.

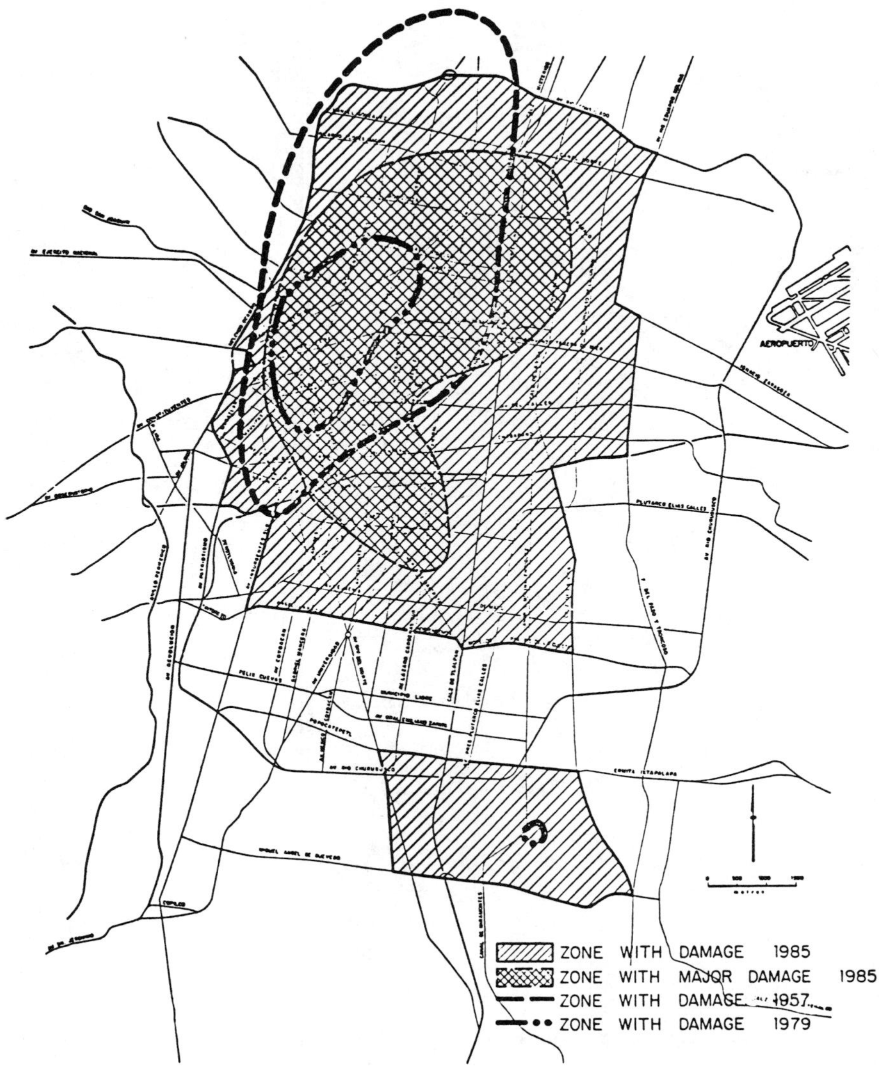

Fig. 2 Location of damage zones

CHARACTERISTICS OF THE DAMAGED BUILDINGS

Some characteristics of the damaged buildings that are relevant from a structural point of view and that could be determined even in collapsed buildings were identified: the number of stories, the structural system, and the approximate date of construction. The statistics are summarized in Table 1.

The highest incidence of damage occurred in buildings of 6 to 15 stories. The number of damaged buildings lower than five stories was rather small,

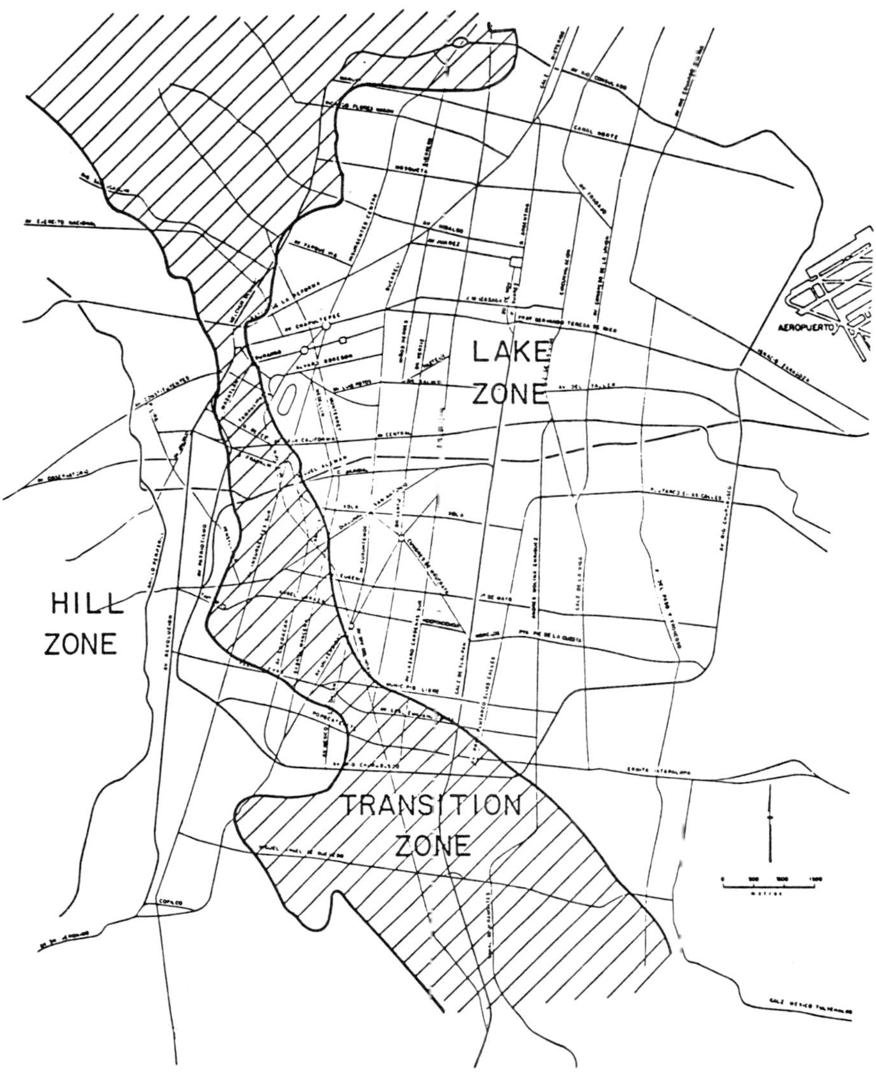

Fig. 3 Subsoil zonation of Mexico City

considering that the great majority of existing buildings in the area are in that story range. Very few buildings of more than 15 stories were damaged.

An approximate census of all existing buildings in the most affected area was made. The data were used to estimate the percentage of damaged buildings with different heights. In this case damages of the first three types defined earlier were included. The percentages are as follows:

Buildings up to 2 floors	2%
Buildings between 3 and 5 floors	3%
Buildings between 6 and 8 floors	16%
Buildings between 9 and 12 floors	23%
Buildings higher than 12 floors	22%
Total damaged buildings	3%

The variation in percentage of damage with the number of floors is due to the fact that in the area of maximum damage the vibration of the ground had dominant periods in the range of 1.5 to 2.5 sec. Buildings of low height, mainly with load bearing masonry walls, have short natural periods that correspond to low ordinates of the acceleration response spectrum. Buildings of medium height fall into a zone of the spectrum where the ordinates

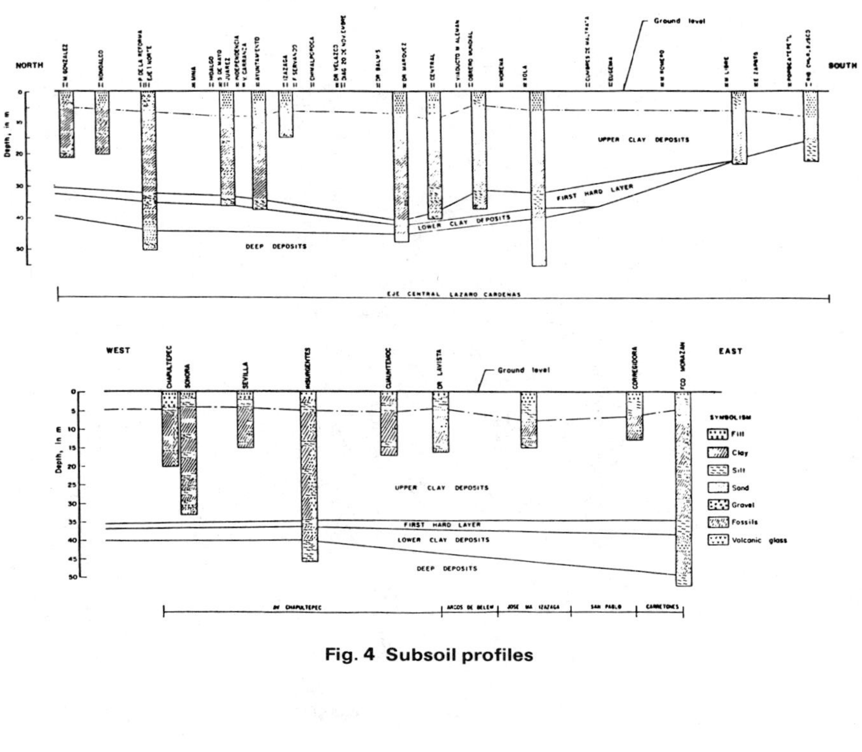

Fig. 4 Subsoil profiles

were large and, if damaged, their reduced stiffness produces a shift toward zones of even larger ordinates. On the other hand, very tall and flexible buildings have natural periods that exceed those where the spectral ordinates were large.

Regarding the age of the buildings, the three ranges used in Table 1 correspond to periods in which different buildings codes were enforced.

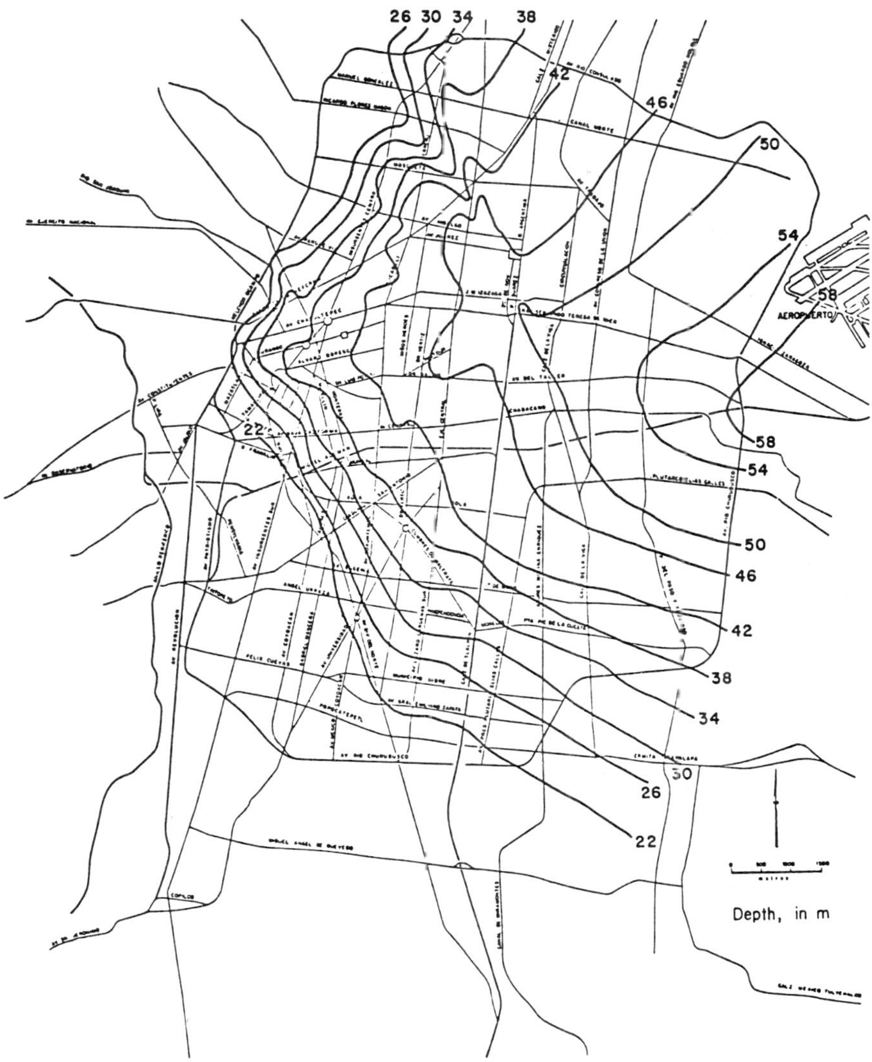

Fig. 5 Equal depth curves of the deep deposits

Before 1957 seismic regulations were very simplistic and unconservative. At the end of 1957 a rather comprehensive set of seismic design recommendations was enforced and from 1976 the present code governs the structural design. Due to the lack of information about the number of buildings constructed in each period, data of Table 1 are not conclusive. It is possible that the lower number of damaged buildings of the more recent period could be attributed to an improvement of design and construction practices after the 1976 code.

The classification regarding structural systems include concrete frame, waffle slab and concrete column, and steel frame. In the three cases the main structural system was commonly stiffened by masonry walls and only infrequently by concrete shear walls. The fourth case was that of load bearing masonry walls, which was most commonly used for low-rise construction. The small number of damaged buildings of this last case reflects the more favorable response of low, stiff buildings. Regarding the other three types of building the statistics are not conclusive: very few steel buildings existed in the affected area, and the waffle slab system was the most popular for medium height buildings after the sixties.

Table 1 Statistics of damaged buildings

		Year of construction			Numbers of floors				Total
Type of structure	Type of damage	←1957	76–76	1976→	≤5	6–10	11–15	>15	Collapsed or very severe
Concrete frames	Collapse	35	59	13	36	62	9	0	107
	very severe	9	19	7	8	23	4	1	36
Steel frames	Collapse	5	4	0	4	2	1	2	9
	very severe	1	0	0	0	0	1	0	1
Flat slab	Collapse	3	35	12	23	23	4	0	50
	very severe	5	20	11	9	18	8	0	35
Masonry	Collapse	7	4	1	10	2	0	0	12
	very severe	2	3	0	4	1	0	0	5
Others	Collapse	0	1	1	1	1	0	0	2
	very severe	2	4	2	6	2	0	0	8
Total	Collapsed and very severe	69	149	47	101	134	27	3	265

TYPES OF STRUCTURAL FAILURE

The main reason for the large number of failures is the extraordinary severity of ground motion attained in an area of the city where the characteristics of the subsoil were such that the movements transmitted by the subjacent firm soil were greatly amplified. The structures were subjected to many cycles of ground movement of large amplitudes and with an approximately constant vibration period. The effects of this ground movement were significantly larger than those contemplated by the present code, especially for medium height flexible buildings.

Besides this main reason, in many cases certain characteristics of the buildings contributed to intensify the effects of the ground movement, giving rise to some prevailing modes of failure.

Brittle Behavior Due to Column Failure

In most cases the collapse of concrete frame buildings was originated by failure of the columns by bending and axial compression, by shear, or by a combination of these factors. The inspection of beams and waffle slabs after collapse indicated very little evidence of yielding of the steel, therefore no ductile behavior could be developed. The more common mode of failure was a progressive loss of capacity of the columns to withstand vertical loads due to the deterioration of concrete enclosed by the reinforcement preceeded by the buckling of longitudinal reinforcement. The amount and distribution of stirrups and longitudinal bars was not sufficient to provide the confinement that could avoid the progressive deterioration of the concrete subjected to many cycles of very large stress.

Effect of Infill Walls

Masonry infill walls saved many buildings from failure. This occurred when these walls were numerous, symmetrically distributed, and well tied to the main structure. In this case they absorbed a great portion of the lateral loads and protected the structure, and even if diagonally cracked they still contributed to the strength and helped to dissipate the energy introduced by the quake. Nevertheless, in several cases these same masonry walls contributed to the failure, as in the following situations.

Non-Symmetric Array of Walls. A remarkable number of corner buildings failed, generally having brick walls on two sides and open facades on the others. This is an extreme case of torsion introduced by non-symmetric distribution of walls that greatly increased the forces in some column lines and contributed to their failure.

Soft Ground Floor. Many buildings had numerous masonry walls in the upper floors and an open ground floor to be used for parking, halls, or stores. This situation gives rise to a great demand for energy dissipation concentrated in the first level and to large forces and lateral displacement in the columns, thus contributing to their failure.

Lack of In-plan Symmetry Caused by Wall Destruction. In a large number of cases masonry infill walls that were poorly reinforced or badly tied to the main structure were destroyed by flexural forces normally in their plane or by shear; this situation reduced the resistance of the structure to lateral loads and frequently gave rise to lack of symmetry that increased the forces introduced in some column lines.

Damage Due to Previous Earthquakes

A certain number of buildings that failed had been damaged by previous earthquakes, and in several cases, this damage had not been repaired correctly.

Short Columns

Some columns had their lateral displacements restricted by masonry walls or by concrete parapets up to a certain height. This situation made them stiffer than those of other lines in such a way that a major part of the lateral force is transmitted to them, frequently producing a brittle failure because shear prevails over bending due to the small height-to-depth ratio.

Pounding of Adjacent Buildings

Generally the separation between adjacent tall buildings was very small, thus pounding was rather frequent. The impact was responsible for several cases of local damage but it is also believed to be responsible for a certain number of failures in the upper floors of buildings where the columns were affected by the pounding of the floors of an adjacent building. In other cases the failure of upper floors is attributed to a sharp reduction of strength and stiffness at this level.

Overload

In several buildings the upper floors were used as a warehouse or as filing areas, thus originating larger live loads than those considered in the design and contributing to the production of large inertia forces and the failure of the buildings.

P-Δ Effect

The excessive lateral flexibility of some structures and the rotation of the foundation gave rise in several cases to a significant increase of the bending moments in the columns of the lower floors. This effect was noticeable in buildings that collapsed with their floor slab displaced to one side.

Punching of Flab Slabs

Most flat slab buildings collapsed because of the failure of the columns with no signs of previous damage in the slab (commonly a waffle slab). Some cases of shear cracking caused by the combined effect of vertical and lateral load were found. In the few cases of complete punching failure, it seems that the solid zone around the column was very poorly reinforced.

FOUNDATION FAILURE

Only infrequently could the collapse of the building be totally attributed to a foundation failure; the few collapses of this kind are related to very slender buildings where large overturning moments were produced on raft foundations or friction piles.

It is assumed that in several cases the settlements and rotations at the building bases due to incipient foundation failure significantly contributed to the collapse of the building. The P-Δ effect due to the movement of the foundation could be an important factor in the failure of the columns at the ground level especially in buildings with an open ground floor without infill or shear walls.

The possibility that the collapse of some buildings was occasioned by a reduction of structural capacity due to prior differential settlements cannot be excluded. Damage due to differential settlements can be divided into two groups: excessive settlements due to the building's own weight and regional settlements of the subsoil of the valley.

The second case could have occurred especially in the parts of the city where the soil profile is rather irregular or in the case of adjacent buildings with different types of foundations (piles vs. surface foundation, overcompensated foundation vs. net overweight).

In other cases classified as severe damage, the incipient foundation failure was the main cause of damage, as in tilting of slender buildings or as in almost uniform and severe settlement of squat heavy constructions.

It is a matter of further investigation to determine the marginal load capacity that should be left to the soil in order to avoid incipient failure under earthquake loading. The study of how much the load capacity of friction piles is reduced under cyclic loading and the structural capacity of piles at their connection with the foundation basis and in sections of sudden changes in the soil profile is also considered of great relevance.

Lessons of the Mexico City Earthquake

Ignacio Martin

We should dispel any feeling of complacency that the tragedy of the September, 1985 Mexico City earthquake could not occur in an American city in a seismic zone. It can. This tragedy happened in a city with one of the most modern seismic codes in existence. We have a lot to learn. It is very fortunate that Mexican engineers have provided us with vast information on this earthquake. This earthquake is probably one of the best documented in history, so we will long continue to study its effects.

We desperately need ductility in earthquake zones, but we should not abuse ductility in the loading for design. At this time we are reviewing our seismic codes, and we are talking about ductility reduction factors of eight. In view of the lesson of Mexico City, a second look should be taken at those ductility reduction factors.

The importance of soil conditions cannot be dismissed by saying that Mexico City was built on a lake. It may be a worst case scenario, but other cities of the world have poor soils and are in seismic zones. We tend to think that during an earthquake, as the structure degrades by cracking and drifting, the period of the building increases. Under normal conditions that will tend to reduce the equivalent seismic loads. But in very poor soil conditions, increasing the period increases the equivalent seismic loads. This problem must be explored, and we have to review what are we going to build on poor soils. Perhaps there is also a lesson here for planners: should zoning reflect that condition? Should we be making studies of the periods of soils in order to determine what types of buildings we should be building in those areas?

There are lessons in analysis too. It is imperative to study the need to have three-dimensional analysis for seismic considerations. Mathematical models must be reviewed. We should not be satisfied by the output of a computer. What are we inputting into that program? What is happening with the changes of stiffness? What is happening with the change of the neutral axis of a shear wall during an earthquake?

When we assess the damage of this earthquake, we are entering into a second generation in the study of the performance of buildings. Instead of looking at buildings that have collapsed (probably an experienced engineer would know immediately why that building collapsed), we should look for the buildings that perform well to discover why those buildings performed well, and what we should be doing that will make a building withstand a strong earthquake.

Fifteen percent of the buildings were damaged in Mexico City by hammering. Should we have joints in our buildings? How should we design those joints?

Most of the tall buildings of the past that have been examined have lasted for a hundred years. We have to think of the changes that occurred in those hundred years. What happens when the loading of buildings changes because the use of the building changed? Can the building withstand the change?

Sometimes after an earthquake buildings are repaired and the profession is satisfied. We even publish the way we repaired the buildings. Are we doing a good job? What systems will do the trick?

Finally, the most important aspect of seismic design to be considered is the selection of the structural system. No code can prevent the failure of a poorly selected structural system. This has been evident in Mexico City.

Dampers show promise, but we must be very cautious as to the long-time performance of dampers because of the possibility of failure. In the life of the building, it must work. So we have to admire Mexican engineering and the scientific way they are studying this tragedy. There is a wealth of information that has already been gathered in this earthquake. We have much to do, and quite evidently more research is needed.

Definition of Wind Pressure on Tall Buildings

W. H. Melbourne

Design pressures for cladding elements on tall buildings, in particular glass, have undergone a minor revolution in the past decade. Pressures of 1 kPa (20 psf) thought to be sufficient in many parts of the world up to 1960 have given way progressively to design pressures typically between 2 and 3 kPa on tall buildings. This has been brought about by the combination of the knowledge of existence of higher pressures from wind tunnel model tests in scaled turbulent boundary layer flows and from the cladding failures that have occurred on a number of tall buildings.

The future in respect to the definition of wind pressures on tall buildings will not require gazing into the crystal ball so much as disseminating what is already known.

PRESSURE REGIONS

The pressures on a tall building can be generalized into three distinct regions as illustrated in Fig. 1.

1. The upstream face where pressures are positive and reflect the maximum dynamic pressures in the incident wind flow.

2. The rear faces (well downstream of a primary reattachment), in the wake region where pressures are negative, but only moderate in magnitude.

3. The streamwise faces near an upstream corner under a reattaching shear layer, where pressures are highly negative and intermittent.

The maximum positive pressures on an upstream face and the negative pressures on a downstream face are about half the magnitude of the maximum negative pressures that can occur on a streamwise face under a reattaching shear layer. Hence, in terms of defining design wind pressures for cladding on tall buildings, it is necessary to concentrate the majority of our efforts on those regions near leading edges and on the freestream flow and leading edge characteristics that control the magnitude of the pressures generated under the reattaching shear layer.

Some understanding of the mechanism of the generation of the very high negative pressures under reattaching shear layers has been given by Melbourne (1979) and there is continuing research in a number of research establishments to gain both further fundamental understanding and a means of reducing the magnitude of these pressures. Avoidance of edge discontinuities, use of curved shapes and slotted (vented) leading edges go a long way towards reducing the pressure magnitude, but it is unlikely that such knowledge would have much influence on an architect who is developing a building shape to meet esthetic factors. Given that certain critical conditions can be achieved, it is up to the designer to recognize the potential for high wind-generated cladding loads and to be prepared to pay the inherent cost penalty.

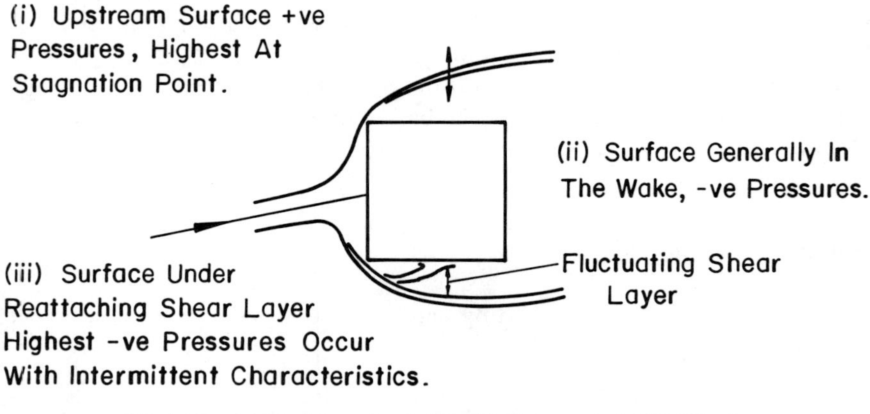

Fig. 1 Pressure regions associated with flow around a bluff body

CODIFICATION OF WIND PRESSURES

Modern wind loading codes are based either on a design mean wind speed or mean maximum gust wind speed. The exception is the U.S.A. Code based on a fastest mile wind speed, which adds a further complication. The use of a design gust wind speed implies a quasi-steady approach to the design process, but this is not true in general because the coefficients used in such codes have been back-worked from real peak coefficients obtained from model tests based on a mean wind speed. The definition of the duration of such a peak is of considerable importance to the definition of a design load, particularly for glass.

There is a simple connection between peak pressure coefficients measured in model tests, based on mean wind speeds (usually defined as being the freestream value at the height of the top of the building), and the equivalent pressure coefficients in the quasi-steady codes. It is, in terms of a design pressure,

$$\hat{p} = C_{\hat{p}_{\text{model}}} \tfrac{1}{2}\rho \bar{V}_h^2 = C_{\hat{p}_{\text{code}}} \tfrac{1}{2}\rho \hat{V}_h^2 \tag{1}$$

where

p = a peak pressure

ρ = air density (= 1.2 kg m^{-3})

$C_{\hat{p}_{\text{model}}}$ = a peak pressure coefficient determined from model measurements referenced to mean velocity at building height h

$C_{\hat{p}_{\text{code}}}$ = the peak pressure coefficient from the code (obtained by using a pressure coefficient and local pressure factor)

$\bar{V}_h$ = the design mean wind speed at building height h

$\hat{V}_h$ = the design maximum gust wind speed at building height h.

From which it can be seen that

$$\frac{C_{\hat{p}_{\text{model}}}}{C_{\hat{p}_{\text{code}}}} = (\hat{V}_h/\bar{V}_h)^2 \tag{2}$$

This ratio is also the Gust Factor implied by the quasi-steady code.

Using a wind model, the relationship between hourly mean and maximum gust wind speeds can be derived. The relationship between peak pressure coefficients, based on the respective wind speeds, can also be derived. An example is given in Table 1. Also given in Table 1 is the ratio $(\hat{V}_h/\bar{V}_h)^2$ which is the ratio, Eq. 2, of $C_{\hat{p}_{\text{model}}}/C_{\hat{p}_{\text{code}}}$. Obviously it is not possible to generalize about the ratio between the mean velocity based peak pressure coefficients from the model measurements and the maximum gust wind speed based peak pressure coefficients in the code, but an example can be given as follows:

For a 100 m (328 ft) high building in suburban terrain

$$(\hat{V}_{100}/\bar{V}_{100})^2 = 2.60$$

$C_{\hat{p}_{code}}$		$C_{\hat{p}_{model}}$
−1.0	is equivalent to	−2.6
−1.5	is equivalent to	−3.9
−2.0	is equivalent to	−5.2

DURATION AND ATTENUATION OF PRESSURES

The definition of wind pressures must contain recognition that the effective pressure on a cladding element varies as a function of spacial extent of the element and of time and therefore the definition of a design load (applied statically) must take into account the area and response time of the cladding element. The real dynamic process is very complex and quite varied, as can be illustrated by the following examples:

1. Massive masonry curtain wall elements have a relatively long response time with respect to the development of stresses in reinforcing or in attachment details. In addition, most of these elements are vented such that the mean component of the external pressure is effectively cancelled and the peak pressure developed across the panel is attenuated with increasing wall porosity.
2. Lightweight wall panelling, by contrast, has a relatively short response time, with natural frequencies of the order of 10 Hz, and hence (with

Table 1 Wind speed ratios referenced to a unit 3 second mean maximum wind speed at 10 m in open country terrain, derived from the Deaves and Harris wind model for a gradient mean wind speed of 50 ms^{-1} for suburban terrain roughness.

Height z (m)	$\hat{V}_z/\hat{V}_{10\,m}$	$\bar{V}_z/\hat{V}_{10\,m}$	$(\hat{V}_h/\bar{V}_h)^2$ (where $h = z$)
5	0.73	0.36	4.11
10	0.83	0.44	3.56
20	0.94	0.52	3.27
50	1.07	0.63	2.88
100	1.16	0.72	2.60
200	1.24	0.82	2.29
500	1.35	0.99	1.86

respect to the development of stresses) a load duration of 1/10 sec is significant. This type of panelling is also commonly vented, again cancelling the mean pressure contribution and attenuating the peak pressure.

3. The loading and consequent stressing of glass elements to failure is even more complex because, in addition to the response time involved, glass as a material has a complex cumulative stress duration dependence and a panel deformation dependence. Further, there is frequently a long-term reduction in the strength of glass due to weathering. The definition of load duration for glass not only has to take into account the area and response time of the panel, but also the duration of major storm events and the deformation mode of the panel.

Four particular aspects of the duration of loading of cladding elements can be identified for definition, two of which can be dealt with simply and two, while complex, could, in the future, be accommodated without introducing too much complexity into the actual design process.

The dependence of design load on elemental area is to a large extent already covered by using various area averaging techniques in the wind tunnel modelling process. However, for cladding elements on tall buildings the relative areas concerned are rarely large enough to warrant any reduction of the point pressure data.

The load duration in respect to response time of the element is calculable on a probabilistic basis. Given a probability distribution of the pressure (in terms of time) and an upcrossing distribution (for a given cycling rate) then the former divided by the latter for a given probability level gives the average duration of pressure above that level for each upcrossing. It is rare that such a process is considered in the design process. The current general procedure is to use peak pressure coefficients measured on a wind tunnel model with a system response that implies that in full scale terms the peak pressure so determined is about a 1 sec mean maximum pressure. Such a determination cannot be regarded as conservative for many of the smaller and less massive cladding elements. For glass panels the cumulative loading, as will be discussed, tends to take over. However we cannot ignore this aspect for much longer, particularly as it is a relatively simple matter to be explicit about the effective duration of peak pressures from model measurements or as given in wind loading codes, and then to provide a factor to convert to a different duration that would be specified by the designer to suit the response time of the cladding element under consideration.

The reduction in design load due to the effects of panel venting cannot, as yet, be simply accommodated in the design process. However work at Monash University for the Australian wind loading code is aimed at providing a very simple design load reduction factor as a function of wall panel porosity. It is hoped that a solution of the type shown in Fig. 2 will be achievable.

Finally the accommodation of the effects of cumulative loading on glass panels, while very complex, has, with the aid of modern computers, a potentially simple solution for specific applications such as on major tall building projects. Brown (1974) derived an equation from previous studies which can be reduced to a form which represents a cumulative damage criterion, in other words

$$\int_{o}^{T}[\sigma(t)]^{n}dt = \text{constant} \tag{3}$$

where T = the time over which the glass is stressed; $\sigma(t)$ = the time varying stress; n = a high power (nominally 16), and which has been discussed recently by a number of authors (Dalgliesh, 1979; Beason, 1980; Walker, 1984; Holmes, 1984; and Calderone, 1985). Unfortunately, the cumulative damage criterion is not easy to apply to panels of window glass due to their nonlinear behavior; however Calderone, working at Monash University, has shown that it may be possible to provide a single, approximate, log-log relationship between pressure and stress to make effective evaluation of the cumulative damage criterion possible. Given this possibility it is then a simple matter, with the aid of computers, for wind tunnel pressure data to be integrated at the time of collection and then presented in a form suitable for direct application to the glass manufacturers charts. Calderone has given examples of this process using model pressure data from Cheung and Melbourne (1984) to give equivalent constant pressures for 3 and 60 sec for various storm durations and pressure-stress relationships, and has also shown, by further

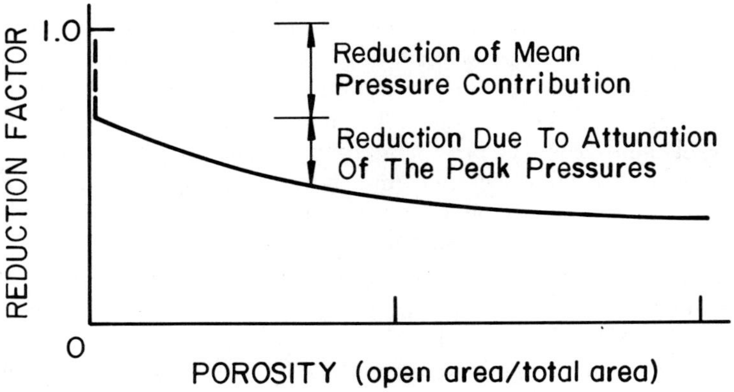

Fig. 2 Factor to be applied to a design peak cladding pressure to account for the effect of wall panel porosity (venting)

example, that a 60 sec ramp load is virtually equivalent to a constant load of 3 sec duration.

THE HIGHEST DESIGN PRESSURES

As noted earlier, the maximum design pressures on tall buildings are the negative pressures that occur on a streamwise face under a reattaching shear layer and depend critically on leading edge configuration and the turbulence intensity of the incident flow. Cheung and Melbourne (1984) have studied a range of typical tall building configurations to provide basic data for codification purposes. This study was brought about because of the move by the Australian Wind Loading Code to increase the maximum local cladding design pressure coefficients (quasi-steady approach based on gust wind speed) from −1.5 to −2.0 to accommodate the effect of potential edge discontinuities. They showed that for typical rectangular tall buildings the highest negative pressure coefficients (based on mean wind speed) did not exceed −4.0 but that with low level edge discontinuities (in other words, in the region of highest freestream turbulence) values of about −5.0 were shown to occur. In terms of equivalent, quasi-steady code, pressure coefficients based on a 3 second mean maximum gust wind speed, these values were close to −1.5 and −2.0, respectively. As such this justifies the use of the higher design pressure coefficient when the potential edge condition is not known (due to possible neighboring development) but puts a significant penalty on a rectangular building standing alone and free of edge discontinuities, and an even high penalty on a building with curved surfaces.

REFERENCES/BIBLIOGRAPHY

Beason, W. L., 1980
A FAILURE PREDUCTION MODEL FOR WINDOW GLASS, Inst. for Disaster Research, Texas Tech. University, Lubbock.

Brown, W. G., 1974
A PRACTICABLE FORMULATION FOR THE STRENGTH OF GLASS AND ITS SPECIAL APPLICATION TO LARGE PLATES, Nat. Res. Council of Canada, Publ. No. NRC 14372, Ottawa, November.

Calderone, I., 1985
DIRECT INTEGRATION OF TIME LOAD HISTORY FOR GLASS DESIGN, thesis submitted for M. Eng. Sc. degree, Monash University, November.

Cheung, C. K. and Melbourne, W. H., 1984
CLADDING PRESSURES ON RECTANGULAR HIGH RISE BUILDINGS, workshop on Wind Engineering and Industrial Aerodynamics, C.S.I.R.O. Division of Building Research, Australia, pp. 108-114.

Dalgliesh, W. A., 1979
ASSESSMENT OF WIND LOADS FOR GLAZING DESIGN IAHR/IUTAM, Symposium on Practical Experience with Flow Induced Vibrations, Springer-Verlag, Berlin, pp. 696-708.

Holmes, J. D., 1984
WIND ACTION ON GLASS AND BROWN'S INTEGRAL, workshop on Wind Engineering and Industrial Aerodynamics, C.S.I.R.O. Division of Building Research, Australia, pp. 35-38.

Melbourne, W. H., 1979
TURBULENCE EFFECTS ON MAXIMUM SURFACE PRESSURES—A MECHANISM AND POSSIBILITY OF REDUCTION, Proc. 5th Int. Conf. on Wind Engineering, Fort Collins, Perganon Press, pp. 541-552.

Walker, G. R., 1984
INTERACTION OF WINDOW GLASS AND WIND, workshop on Wind Engineering and Industrial Aerodynamics, C.S.I.R.O. Division of Building Research, Australia, pp. 32-34.

Air Infiltration and Internal Pressures in Tall Buildings

W. Alan Dalgliesh

Wind forces are through walls via leakage paths and larger openings, and the resulting air movement alters pressures inside buildings. Leakage paths can occur as narrow cracks and other small defects in the seals joining windows and wall panels. Whatever the situation with regard to larger openings, all buildings have some degree of uniformly distributed leakage through which wind can influence internal pressures.

This paper explores how wind loads are affected by air flow through the walls. The wind load on a wall is the external pressure minus the internal pressure, and even for uniformly distributed leakage, the internal pressure contribution can be substantial. Moreover, one window-size opening may lead to a transfer of most of the wind load from the windward wall to the leeward and side walls or vice versa, depending on the location of the opening. How much load is transferred depends on the flow resistance of the opening relative to the flow resistance of the distributed leakage; the tighter the building envelope, the greater the wind load transfer for a given opening.

The flow resistances of distributed leakage through exterior walls have been measured and related to the pressure differences causing flow in several tall buildings in Canada (Shaw et al., 1973; Shaw, 1979). With this information, one can deal quantitatively with the effects of large openings on cladding pressures for any given building. Cladding pressures across each wall of a typical tall building were calculated using a computer program originally developed to assess air infiltration (Sander, 1974). Results of ten cases are examined, five for each of two wind directions. Two cases are for distributed leakage alone while the other eight include one opening, small or large.

FLOW RESISTANCE OF DISTRIBUTED LEAKAGE

Distributed leakage is evaluated in terms of a flow coefficient C and a flow exponent n. Air flow Q is m^3/s for a pressure difference P in Pa and "tributary" area A in m^2 is

$$Q = C\,A\,P^n \tag{1}$$

The exponent n varies from 0.5 for turbulent flow through a sharp-edged orifice to 1.0 for laminar flow through very narrow passages, and is found to averge about 0.65 for exterior walls.

Experiments on eight buildings ranging from 9 to 25 floors in height are consistent with the following value of C for average exterior wall construction: 0.9×10^{-4} $m^3/(s \cdot m^2 \cdot Pa^{0.65})$. Typical values for "tight" and "loose" construction are one third and double this value, respectively.

COMPUTER INPUT: MODEL OF A 56-STORY BUILDING

The example building, 30.5 m × 61 m in plan, has a story height of 3.8 m and an overall height of 221 m. For purposes of calculation, it is considered sufficiently accurate to divide the building into compartments of six floors except for the 45th, where an opening was assumed to occur for eight of the ten cases studied. The opening is represented by a shaft with an external vent. Floor-to-floor leakage, with flow exponent 0.5, occurs through stairwells and

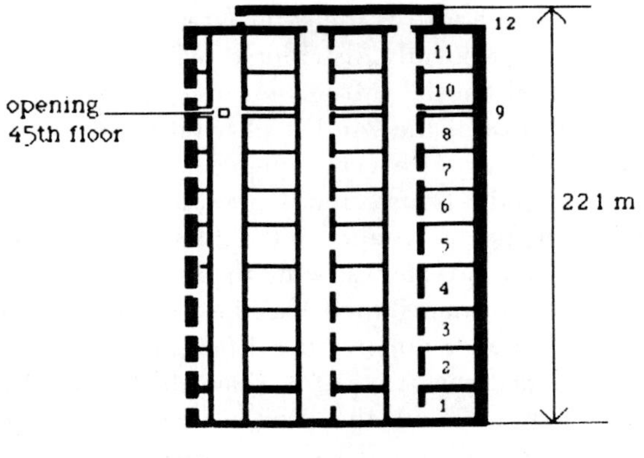

Fig. 1 Schematic of 56-story building showing the 12 compartments and the 3 vertical shafts used in the computer analysis

service shafts, represented by a second shaft, and through elevator doors, represented by a third shaft (Fig. 1). Distributed leakage for average exterior wall construction is used for all cases. Four cases have an opening of 1.5 m^2 added, and four more have an opening of 3.0 m^2. Other details of the computer input are similar to the example given by Sander (1974).

COMPUTER OUTPUT: NET PRESSURES ACROSS EXTERIOR WALLS

The net pressures across each wall for each of the ten leakage conditions are listed in Table 1. Figures 2 and 3 show that each of the ten cases results in an internal pressure different from the rest. Increasing the internal pressure transfers some of the cladding load from the windward wall to the other three walls. Similarly, decreasing the internal pressure transfers load from the three walls (under negative external pressure) to the windward wall. Note that the differences in net pressures between wind on the long wall (Fig. 2) and wind on the short wall (Fig. 3) tend to disappear for cases with the larger opening.

INTERNAL PRESSURE COEFFICIENTS

Internal pressure coefficients are derived by dividing the net pressure for any wall (windward, side, or leeward) in Table 1 by the reference wind pressure (1.33 kPa), and subtracting the resulting net pressure coefficient from the external one (windward, 0.78; side, −0.65; leeward, −0.26).

Simple approximations to internal pressure have been proposed. Davenport and Surry (1983) suggest a linear average of external pressure coefficients, C_{pe}, for all surface areas, weighted by the corresponding "equivalent orifice areas," described below. Newberry and Eaton (1974) used the orifice equation in which flow is proportional to the square root of pressure difference, requiring an iterative solution to balance inflow and outflow.

Table 1 Net pressures across walls at the 45th floor

Wind Normal to Wall		Long Wall Windward kPa	Long Wall Sides kPa	Long Wall Leeward kPa	Short Wall Windward kPa	Short Wall Sides kPa	Short Wall Leeward kPa
Opening Area (m^2)	Opening Location						
3.0	Windward	0.16	−1.74	−1.22	0.21	−1.68	−1.17
1.5	Windward	0.43	−1.46	−0.95	0.58	−1.32	−0.80
No opening		1.15	−0.74	−0.23	1.46	−0.44	0.08
1.5	One Side	1.62	−0.28	0.24	1.75	−0.14	0.37
3.0	One Side	1.80	−0.10	0.42	1.84	−0.05	0.46

The equivalent orifice area *(A_{eo})* for the distributed leakage of a wall is

$$A_{eo} = \frac{A\ C\ (r/2)^{0.5}\ P^{0.65-0.5}}{C_d} \tag{2}$$

Air density $r = 1.3\ kg/m^3$ and discharge coefficient $C_d = 0.60$ were used to make the linear and square root approximations for comparison with results from the more detailed computer model (see Table 2). The three estimates of

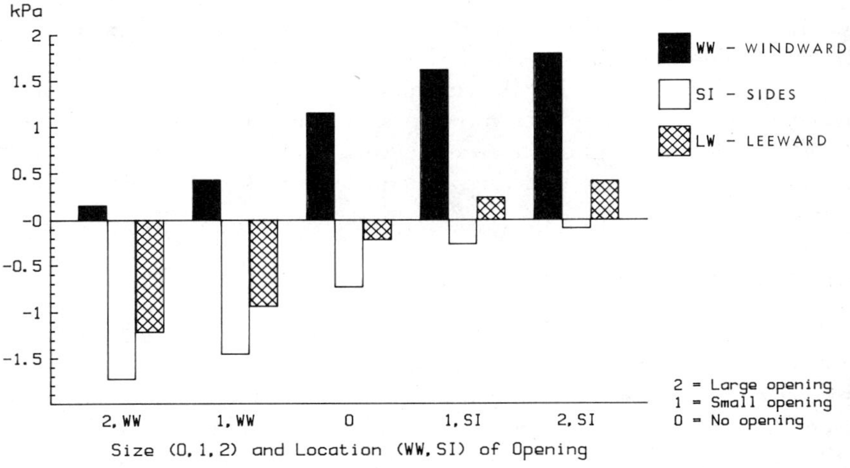

Fig. 2 Effect of an opening on wall pressures; wind normal to long face of building

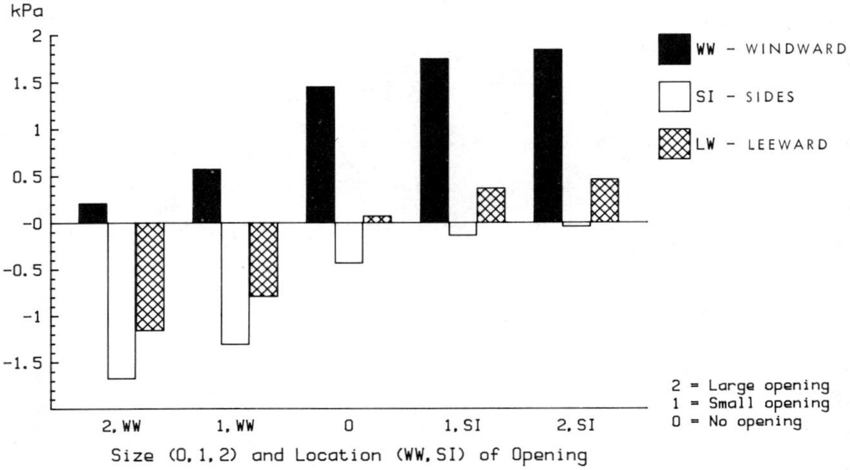

Fig. 3 Effect of an opening on wall pressures; wind normal to short face of building

internal pressure coefficient, C_{pi}, are reasonably similar for the cases of no opening and large opening. However, both simple approximations overestimate the effects of the small opening, a feature emphasized by the line chart of Fig. 4 in which each case is labelled with the internal pressure coefficient of the computer model result. The simpler linear approximation is closer to the computer model result in all but one case.

EFFECT OF OPENING AREA

The area of the opening ranged from 1.3% to 5.2% of the area of the 45th floor wall in which it was located. However, the effectiveness of the opening

Table 2 Approximate and Detailed Estimates of C_{pi}

Wind Normal to Wall		Long Wall			Short Wall		
Opening		Sq. Root	Linear	Detailed	Sq. Root	Linear	Detailed
Area (m^2)	Location						
3.0	Windward	0.78	0.72	0.66	0.77	0.69	0.62
1.5	Windward	0.77	0.66	0.46	0.76	0.62	0.34
No opening		−0.13	−0.01	−0.09	−0.47	−0.31	−0.32
1.5	One Side	−0.64	−0.56	−0.44	−0.65	−0.61	−0.54
3.0	One Side	−0.65	−0.60	−0.57	−0.65	−0.63	−0.61

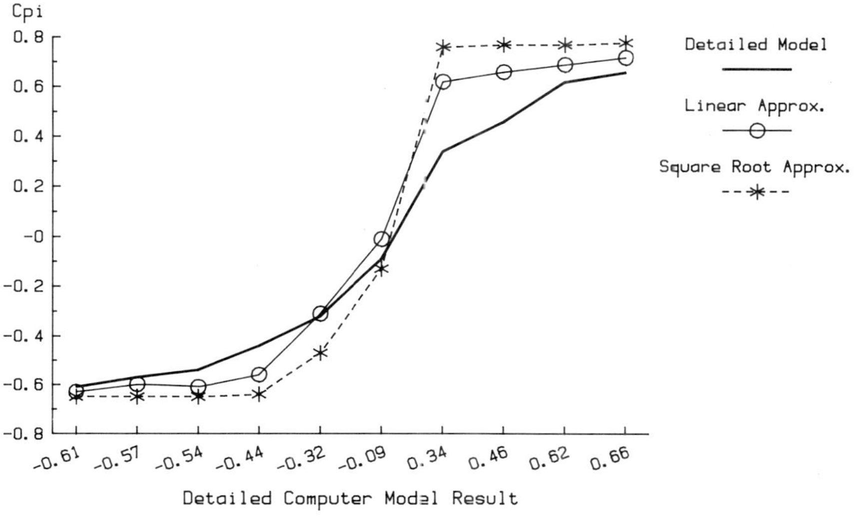

Fig. 4 Mean internal pressure coefficient calculated for 45th floor of 56-story building

depends more directly on its area relative to the equivalent orifice area of the distributed leakage. The sum of the A_{eo} for all four walls of the 45th floor is only about 0.25 m^2.

The effect of relative area for wind normal to the long wall can be seen in the third column of Table 2. The small opening is six times the total A_{eo} and changed the internal pressure coefficient C_{pi} from −0.09 to −0.44 when the external pressure coefficient C_{pe} outside the opening was −0.65, and to 0.46 when C_{pe} was 0.78. The large opening is 12 times the total A_{eo} and brought C_{pi} from −0.09 to −0.57 when C_{pe} was −0.65, and to 0.66 when C_{pe} was 0.78.

Similar trends can be followed for the five cases of wind normal to the short wall (see last column, Table 2). A surprisingly small opening is sufficient to cause significant shifts of wind load between the windward wall and the other three walls.

CONCLUSION

A single opening the size of a large window may well have the effect of increasing the maximum cladding pressure by 60% over the case in which uniformly distributed leakage is assumed. The transfer of cladding load from the windward wall to the other three walls (or vice versa) is mainly the result of the wind itself altering the internal pressure, acting through the overall leakage of the building.

A computer program of the sort used in this study could be used in design to alert both the mechanical and the structural consultants to the consequences of deviating from a uniform distribution of leakage. The same computer model will predict air infiltration effects under calm or moderate wind conditions and cladding pressures for design wind speeds.

Pooling of information available to mechanical and structural consultants could benefit both contributions to design by making possible a more realistic assessment of internal pressure. For example, use could be made of the more detailed patterns of external pressure coefficients available to the structural designer. The program would have to be modified to allow compartments in both the horizontal and the vertical directions because the flow around a building produces large horizontal gradients in external pressure. The model used in this study gives only mean pressures, and should be extended to deal with fluctuating external pressures.

REFERENCES/BIBLIOGRAPHY

Davenport, A. G. and Surry, D., 1983
THE ESTIMATION OF INTERNAL PRESSURES DUE TO WIND WITH APPLICATION TO CLADDING PRESSURES AND INFILTRATION Presented at the ASCE Structural Engineering Conference, Houston, Texas.

Newberry, C. W. and Eaton, K., 1974
WIND LOADING HANDBOOK, Department of the Environment, Building Research Establishment, London, H.M.S.O.

Sander, D. M., 1974
FORTRAN IV PROGRAM TO CALCULATE AIR INFILTRATION IN BUILDINGS, National Research Council of Canada, Division of Building Research, Computer Program No. 37, Ottawa.

Shaw, C. Y., Sander, D. M. and Tamura, G. T., 1973
AIR LEAKAGE MEASUREMENTS OF THE EXTERIOR WALLS OF TALL BUILDINGS, ASHRAE Transactions, Vol. 79, Part 2, pp. 40-48. Also reprint of National Research Council of Canada, Division of Building Research, NRCC 13951.

Shaw, C. Y., 1979
A METHOD FOR PREDICTING AIR INFILTRATION RATES FOR A TALL BUILDING SURROUNDED BY LOWER STRUCTURES OF UNIFORM HEIGHT, ASHRAE Transactions, Vol. 85, Part 1, pp. 72-84. Also reprint of National Research Council of Canada, Division of Building Research, NRCC 18029.

Effects of Higher Modes on Response of Wind-Excited Tall Buildings

B. Samali

Wind-excited vibrations of a high-rise building can cause failure of structural members, unsightly cracks on the interior claddings, and more. However, under strong wind conditions most buildings do not have safety problems. On the other hand, the acceleration response of buildings causing discomfort of tenants is of considerable concern. Such undesirable effects due to excessive accelerations may reduce the serviceability of the building drastically.

Traditionally, the response of a building is expressed in terms of normal modes. In practical applications usually the first vibrational mode is taken into account for approximation. However, unlike the displacement response, the force and especially acceleration type responses involve more vibrational modes and a higher accuracy is required by considering more vibrational modes. The determination of normal modes can be difficult or extremely time consuming; therefore, the true advantage of a normal mode formulation depends on whether or not it is adequate to represent the structural response by only the first very few modes.

The purpose of this paper is to demonstrate the significance of higher modes in calculating the force and acceleration type responses of wind-excited buildings using transfer matrix formulation. This alternate formulation dispenses with the intermediate step of computing the normal modes and deals directly with the desired structural responses. The approach is especially suitable for tall buildings having many identical stories in construction.

In this study the wind turbulence is modeled as a stochastic process with a mean wind velocity that varies with the building height plus a random fluctuation with a cross-power spectral density. A random vibration analysis is carried out to determine the stochastic response of tall buildings. In particular, the statistics of the floor acceleration response is emphasized. A numerical example using a 40-story building is worked out to demonstrate the effects of higher modes on structural response of wind-excited tall buildings and the proposed analytical approach.

FORMULATION

Consider the structural model of an N-story building, shown in Fig. 1, with the mass of the building lumped at the floor levels. To simplify the analysis, it is assumed that each story unit is identically constructed—in other words, the mass as well as the stiffness is identical for each story unit. Linear elasticity is assumed to be provided by massless columns or shear walls. Wind excitations are applied at discrete floor levels. The response of the structure to external excitations can be described by a displacement variable and a force variable in each construction unit. Referring to Fig. 1(c), let X_j and Y_{j-1} be, respectively, the displacement at the floor and the shear force in the columns or shear walls immediately below the jth unit. By taking Fourier

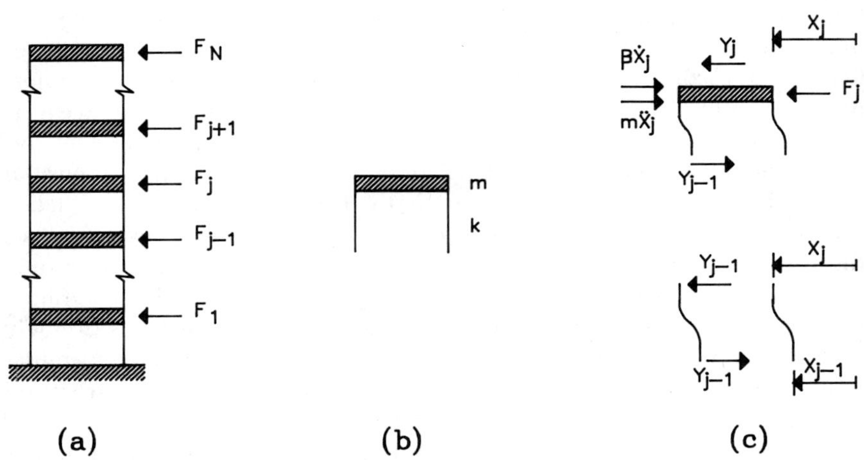

Fig. 1 Structural model
(a) N-story building under wind loads
(b) Constuction unit
(c) External and internal forces on *j*th unit

transforms of the equation of motion and the force-displacement relation of the jth story, the resulting equations may be cast in a matrix form as follows (Yang and Lin, 1981):

$$\begin{Bmatrix} \bar{X}_j \\ \bar{Y}_j \end{Bmatrix} = \begin{bmatrix} 1 & K^{-1} \\ (-m\omega^2 + i\beta\omega) & K^{-1}(-m\omega^2 + i\beta\omega) + 1 \end{bmatrix} \begin{Bmatrix} \bar{X}_{j-1} \\ \bar{Y}_{j-1} \end{Bmatrix} - \begin{Bmatrix} 0 \\ \bar{F}_j \end{Bmatrix} \quad (1)$$

or more simply:

$$\{Z\}_j = [T]\{Z\}_{j-1} - \begin{Bmatrix} 0 \\ \bar{F}_J \end{Bmatrix} \quad (2)$$

In Eq. 1, m = the mass of the jth floor; β = damping coefficient; K = the stiffness; ω = frequency in rad/sec and $i = (-1)^{1/2}$ and in Eq. 2, $\{Z\}_j = \{\bar{X}_j, \bar{Y}_j\}$, is a state vector; an over-bar denotes the Fourier transform, prime indicates the transpose of a matrix or vector; and $[T]$ is known as a transfer matrix. Equation 2 can be applied repeatedly to obtain a relation between $\{Z\}_0$ and $\{Z\}_N$ namely:

$$\{Z\}_N = [T]^N\{Z\}_0 - \sum_{j=1}^{N} [T]^{N-j} \begin{Bmatrix} 0 \\ \bar{F}_j \end{Bmatrix} \quad (3)$$

The boundary conditions at the base and at the top floor of the building dictate that

$$\{Z\}_N = \begin{Bmatrix} \bar{X}_N \\ 0 \end{Bmatrix}, \{Z\}_0 = \begin{Bmatrix} 0 \\ \bar{Y}_0 \end{Bmatrix} \quad (4)$$

By substituting Eq. 4 into Eq. 3 and considering the 2nd equation of the resulting equations, one obtains the following:

$$Y_0 = \sum_{j=1}^{N} \frac{\tau_{22}(N-j)\bar{F}_j}{\tau_{22}(N)} \quad (5)$$

in which $\tau_{22}(N)$ = the (2,2) element of $[T]^N$; and $\tau_{22}(N-j)$ = the (2,2) element of $[T]^{N-j}$. An analytical approach to compute $[T]^n$ where n is any integer number is given in Yang and Lin, 1981. Once $\bar{Y}_0$ and therefore $\{Z\}_0$ is computed in terms of the random excitations $\bar{F}_j$ and the elements of transfer matrix $[T]$, the response vector $\{Z\}_m$ at the mth floor, in other words, Fourier transforms of the displacement at the mth floor and the shear force immediately above the mth floor, $\bar{X}_m$ and $\bar{Y}_m$, can be calculated from Eq. 3 with N being replaced by m;

$$\{Z\}_m = [T]^m \{Z\}_0 - \sum_{j=1}^{m} [T]^{m-j} \begin{Bmatrix} 0 \\ \bar{F}_j \end{Bmatrix} \tag{6}$$

Spectral Relationship and Wind Load Spectra

Wind loads on a building are often modeled as stochastic processes that are stationary in time and nonhomogeneous in space. Thus, Eqs. 5 and 6 can be used to construct the relationships between the cross-spectral densities of the inputs, F_j ($j = 1,2, \ldots N$), and those of the outputs, X_j and Y_j ($j = 1,2, \ldots$ N), using the standard definitions, such as the following for Y_j:

$$\phi_{Y_iY_j}(\omega) = \lim_{T \to \infty} \frac{E[\bar{Y}_i \bar{Y}_j^*]}{2\pi T} \tag{7}$$

in which the $E[\quad] =$ the ensemble average, and an asterisk denotes the complex conjugate. The cross-spectral density of base shear force, Y_0, can be obtained by substituting Eq. 5 into Eq. 7:

$$\phi_{Y_0Y_0}(\omega) = \sum_{i=1}^{N} \sum_{j=1}^{N} Q_1(N-i)\, Q_1^*(N-j)\, \phi_{F_iF_j}(\omega) \tag{8}$$

in which

$$Q_1(N-n) = \frac{\tau_{22}(N-n)}{\tau_{22}(\mathrm{N})}, \; n = i, j \tag{9}$$

Similar expressions can be obtained for $\phi_{X_mX_m}(\omega)$ and $\phi_{Y_mY_m}(\omega)$ by expanding Eq. 6 and utilizing similar definitions (Yang and Samali, 1983). The spectral densities of structural responses given by Eq. 8 and alike require the cross-spectral densities of the wind loads, $\phi_{F_iF_j}(\omega)$. For illustrative purposes, the spectral density proposed by Davenport (1961 and 1966) is used in which the one-sided cross-spectral density is converted into a two-sided spectrum (Yang and Lin, 1981). The standard deviation of base shear force is given by

$$\sigma(Y_0) = \left[\int_{-\infty}^{\infty} \phi_{Y_0Y_0}(\omega)\, d\omega \right]^{1/2} \tag{10}$$

Standard deviations of other response quantities, like X_m and Y_m $(m = 1, \ldots N)$, may be computed similarly. The standard deviation of top floor displacement is obtained from the following relation.

$$\sigma(X_N) = \left[\int_{-\infty}^{\infty} \phi_{X_N X_N}(\omega)\, d\omega \right]^{1/2} \tag{11}$$

The power spectral density of the acceleration response is equal to the power spectral density of the displacement response multiplied by ω^4, that is,

$$\phi_{\ddot{X}_m \ddot{X}_m}(\omega) = \omega^4 \phi_{X_m X_m} \tag{12}$$

The standard deviation, $\sigma(\ddot{X}_N)$, of the top floor acceleration is given by

$$\sigma(\ddot{X}_N) = \left[\int_{-\infty}^{\infty} \phi_{\ddot{X}_N \ddot{X}_N}(\omega)\, d\omega \right]^{1/2} \tag{13}$$

NUMERICAL EXAMPLE

A 40-story building is chosen for illustrative purposes. The structural and aerodynamic data used in the computation are as follows: floor mass $= 1.29 \times 10^6$kg; floor stiffness $= 10^6$ kN/m, damping coefficient $= 21550$ kg/s; area exposed to wind $= 192$ m^2, gradient height $= 300$ m; mean wind speed at 10 m above ground $= 11.46$ m/s; drag coefficient $= 1.2$. The spectral densities of variance of the top floor displacement, base shear force and top floor acceleration are plotted in Figs. 2–4, respectively. It is observed from these figures

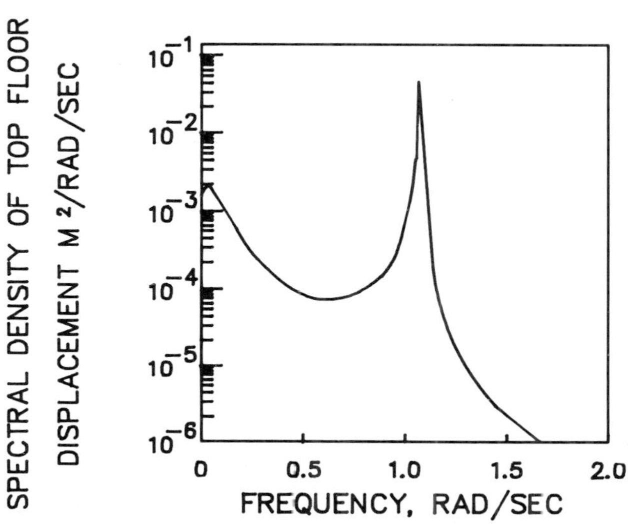

Fig. 2 Spectral density of top floor displacement

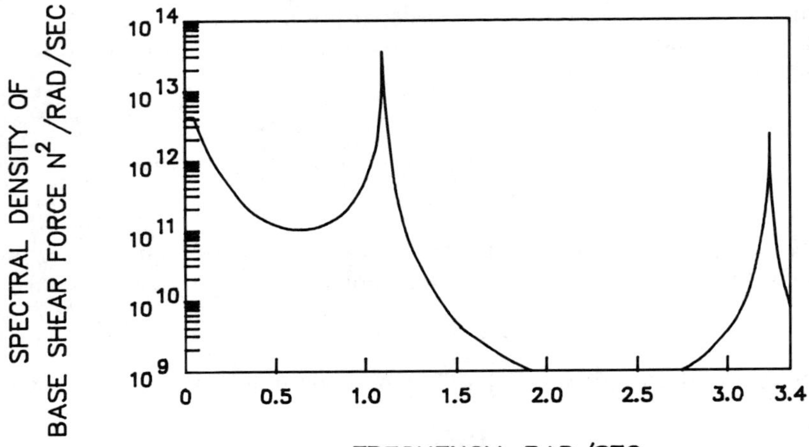

Fig. 3 Spectral density of base shear force

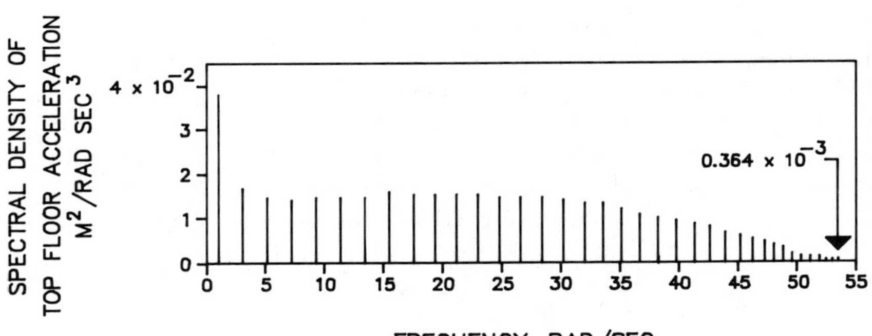

Fig. 4 Spectral density of top floor acceleration

Table 1 Standard deviations of response quantities

$\sigma(X_{40})$, mm	$\sigma(Y_0)$, kN	$\sigma(\ddot{X}_{40})$, mm/sec^2	$\sigma_1(\ddot{X}_{40})$, mm/sec^2
46.8	1917	190	46.5

Table 2 Mean values of response quantities

X_{40}, mm	Y_0, kN	$\ddot{X}_{40}$, mm/sec^2
101.3	3890	0

that the displacement response is dominated by the first mode only. For base shear, the contribution of the second mode is also significant, and for acceleration response, the contribution of higher modes is quite significant. The first peak at $\omega = 0.04$ rad/sec. in Figs. 2 and 3 is due to the maximum of the wind spectrum, whereas other peaks coincide with the natural frequencies of the building. The standard deviations $\sigma(X_{40})$, $\sigma(Y_0)$, and $\sigma(\ddot{X}_{40})$ were computed by integrating numerically those spectra presented in Figs. 2–4 and taking the square roots. The results are presented in Table 1. Also presented in Table 1 is the standard deviation, $\sigma_1(\ddot{X}_{40})$, of the top floor acceleration calculated by taking only the first mode into consideration. The mean values of response quantities are given in Table 2.

CONCLUDING REMARKS

A methodology to analyze the stochastic response of tall buildings excited by strong wind gusts has been presented and the significant contribution of higher modes especially for acceleration type responses demonstrated. The acceleration response contributed by higher vibrational modes requires further experimental investigations, since the mechanism of wind-structure interaction in higher mode responses is observed to be different from that in the first mode response.

ACKNOWLEDGMENT

The author gratefully acknowledges the contributions made by Prof. J. N. Yang of the George Washington University, Washington, D.C.

REFERENCES/BIBLIOGRAPHY

Davenport, A. G., 1961
THE SPECTRUM OF HORIZONTAL GUSTINESS NEAR THE GROUND IN HIGH WINDS, Quarterly Journal, Royal Meteorological Society, Vol. 87, April, pp. 194-211.

Davenport, A. G., 1966
THE TREATMENT OF WIND LOADING ON TALL BUILDINGS, Proceedings of Symposium of Tall Buildings, University of Southampton, Pergamon Press, Inc., New York.

Yang, J. N. and Lin, Y. K., 1981
ALONG-WIND MOTION OF A MULTI-STORY BUILDING, Journal of Engineering Mechanics Division, ASCE, Vol. 107, No. EM2, April, pp. 295-307.

Yang, J. N. and Samali, B., 1983
CONTROL OF TALL BUILDINGS IN ALONG WIND MOTION, Journal of Structural Engineering, ASCE, Vol. 109, No. 1, January, pp. 50-68.

The Response of Supertall Buildings to Wind

Alan G. Davenport

Tall buildings have had a strong influence on the development of wind engineering and vice versa. As structures have grown taller the need has increased for more precise descriptions of the wind forces on them and of their responses.

This paper describes the responses of structures that have historically been the tallest of their time—"the supertall." In this select group (Fig. 1) are the 100-year-old Eiffel Tower, the Empire State Building, the World Trade Center Towers, the Sears Building, and the CN Tower.* The last three of these have been the subject of wind tunnel and meteorological studies at the Boundary Layer Wind Tunnel Laboratory, The University of Western Ontario. The response to wind of the others have been closely studied and are, in retrospect, topics of more than passing interest.

The final section reflects on some of the difficulties that lie ahead for even taller buildings and the direction of their solution.

THE EIFFEL TOWER

Nearly 100 years ago the Eiffel Tower was completed to mark the occasion of the Paris Exhibition of 1889. With a height of 300 m (985 ft) this became the

*The Ostankino television tower in Moscow has been omitted from this group.

world's tallest structure—almost double the height of the previous tallest, the 168 m (500 ft) Washington Monument.

It is interesting to compare the wind load assumptions Eiffel used with the 278 kg/m^2 (56 lb/ft^2) used in the Forth Bridge two years earlier. In this account presented to the Societé des Ingenieurs in 1885 Eiffel states:

> With regard to the intensity (of the wind pressure) we have made two assumptions: one which supposes that the wind has a constant force of 300 kg/m^2 (61.5 lb/ft^2); the other that this pressure increases from the base where it is 200 kg/m^2 (41 lb/ft^2) to the top where it attains 400 kg/m^2 (82 lb/ft^2).
>
> With regard to the exposed surfaces, we have not hesitated in assuming, in spite of the apparent severity of the assumptions, that on the upper half of

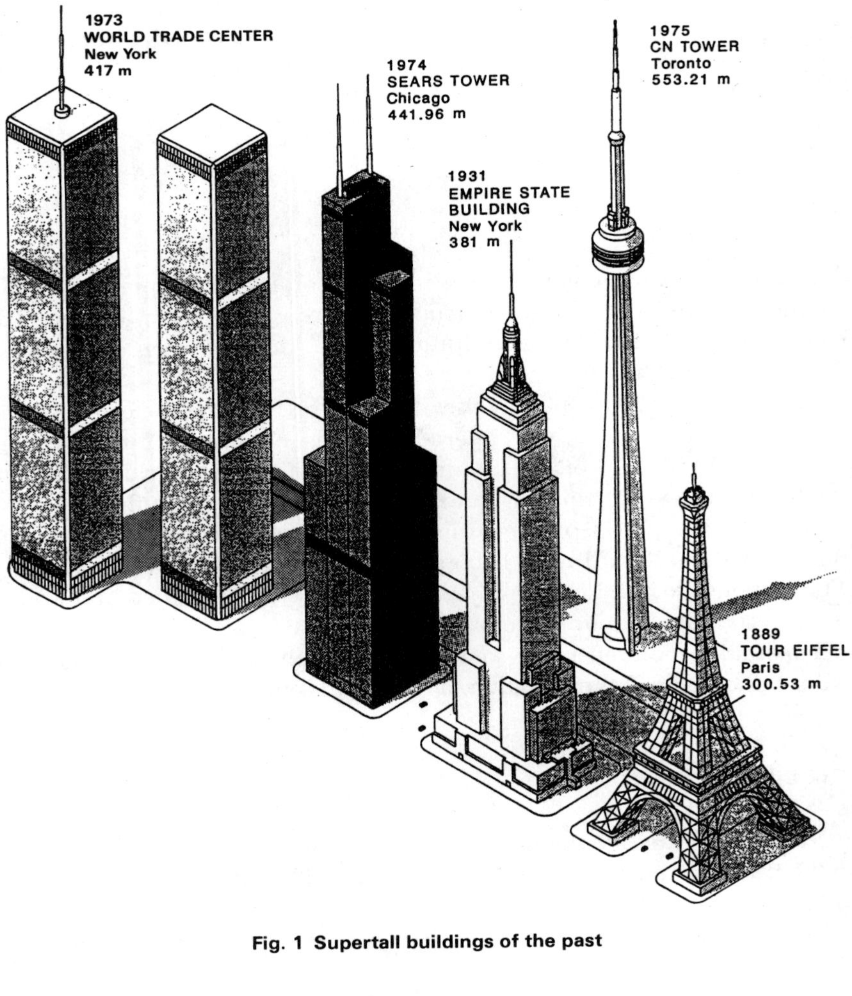

Fig. 1 Supertall buildings of the past

> the tower all the lattice work is replaced by solid surfaces; that in the intermediate section, where the openings become more important, the frontal area is taken as four times the actual area of iron; below this (the first stage gallery and the upper part of the arcs of the legs) we assume the frontal area is solid; finally at the base of the tower we count the legs as solid and struck twice by the wind (i.e. each leg separately exposed to the full force of the wind).

The deflection of the tower in a 24 m/sec (54 mph) wind in which wind pressure was taken as 78 kg/m^2 (16 lb/ft^2) was calculated to be about 200 mm (8 in.). Eiffel also assumed that the vibration would be very slow so that effects of movement on sightseers would be imperceptible. The comfort of sightseers in a 120 mph (54 m/sec) full design wind was regarded as purely academic.

Soon after completion of the tower, Eiffel began experiments from his laboratory at the top of the tower on many topics of medical, physical, meteorological, and engineering interest. His study of wind at this height was a pioneering achievement. Measurements, such as those in Fig. 2, indicated mean wind speeds two or three times as great as those at the Bureau of Meteorology taken at a height of 20 m (65 ft). Also of immediate interest were his observations of the tower movement using a vertical telescope aimed at a target on the top level of the tower. Traces of the movement of the tower are shown in Fig. 2. The center of the target represents approximately the equilibrium conditions under uniform temperature. The elliptical movement of the tower is represented by the envelope curves shown. The mean wind speed at the summit is shown in the record for 12 November 1894. The wind in both cases is more or less diagonal to the tower. We can first deduce from the design figures quoted and the measurements, that the ratio of the deflections/(velocity)^2 assumed in design is roughly 3.5 times that measured. Second, the gust factors (ratio maximum to mean deflection) are of the order of 1.4 to 1.7. This compares with a gust factor for the CN Tower in Toronto of the order of 1.7.

SKYSCRAPERS

The 17-story Monadnock Building built in 1891 in Chicago was the subject of much concern during a violent wind storm in 1894. Observations were made using both a transit and a plumb bob suspended down the central stairwell of the building. Both instruments indicated movements of 12 to 6 mm (1/2 to 1/4 in.) double amplitude in an 80 mph wind.

Suspended from a height of 45 m (150 ft) or so, the plumb bob behaved as a pendulum with a natural period of vibration of approximately 14 sec. This period would be too long for the plumb bob to follow the quicker, harmonic movements of the building resonant with its own natural frequency. The transit, on the other hand, would reveal resonant movements clearly. The rough agreement between the two instruments suggests that resonant movements were in fact small. This fits with current expectations of how this broad, heavy masonry building should behave. It is not characteristic of modern buildings. These observations would reflect slowly varying (or "background") movements but not the static effects caused by the sustained wind.

Much later, during the skyscraper boom in the 1930s there was renewed interest in the full scale behavior of tall buildings, particularly in the United

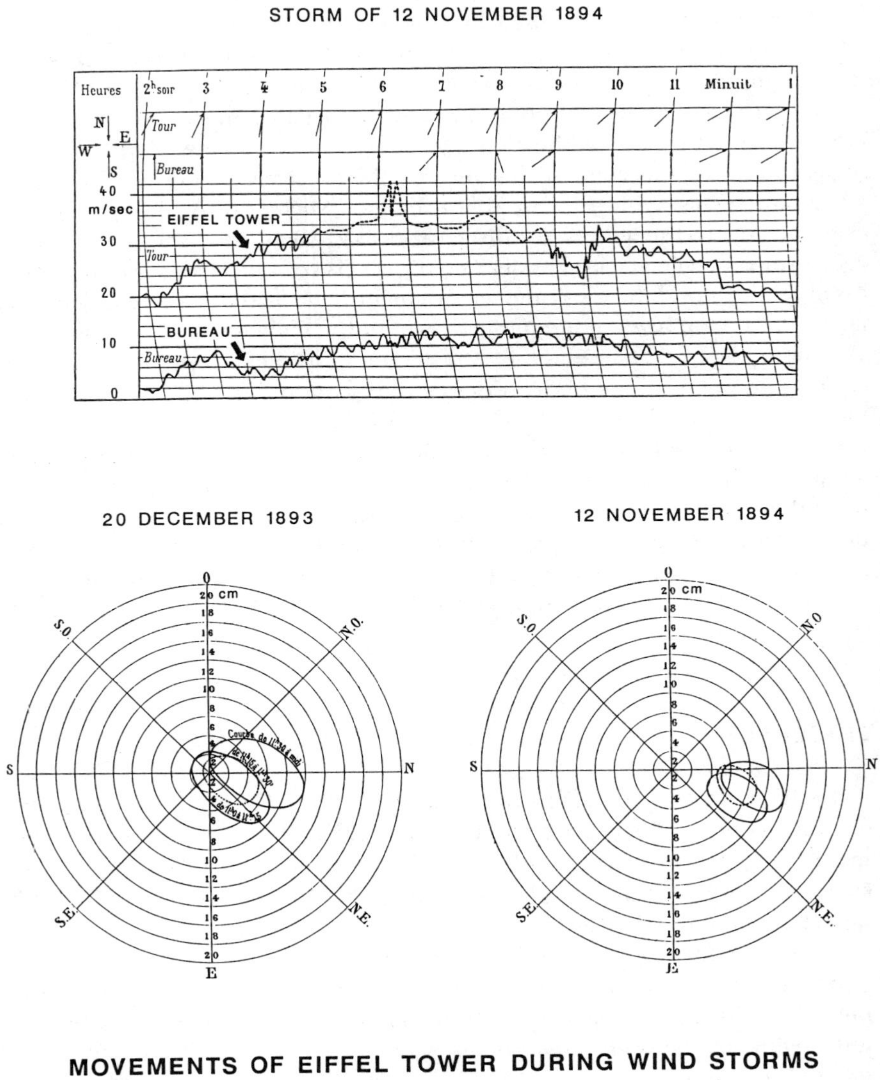

Fig. 2 Movements of Eiffel Tower during wind storms

States. There were a number of ad hoc, full scale experiments undertaken. Cushman Coyle (1931) for example used an ingenious portable horizontal pendulum to record the motion of New York skyscrapers. Typical records are shown in Fig. 3. These clearly reveal the oscillatory character of the response. From these and other such experiences, general impressions were formed relating the merits of stiffness to control dynamic amplitude. Coyle for example observed in 1929 in "The American Architect":

> As a rule the bracing of tall buildings has been regarded as a problem in safety. As less and less bracing is used, and they do not blow over, the feeling grows that the heavy construction of the past was unnecessary. Almost anything, it seems, will stand up.
>
> But the older towers, which were designed to be safe according to a theory which was perhaps ultra-conservative, achieved incidentally a result which was of more immediate importance. Their vibration in time of storm is slight, ranging generally within a quarter of an inch total travel. They are not only safe, but they are on the safe side of the motion which most people can feel which for the owner is the point where the trouble begins Some of the more recent towers, on the other hand, have two or three times as much swing and are therefore, by that much, closer to the "threshold of sensation" of the average man or woman.
>
> It may as well be taken for granted that all the high towers will produce a sensation in certain people, and that their appeal must be in any case only to

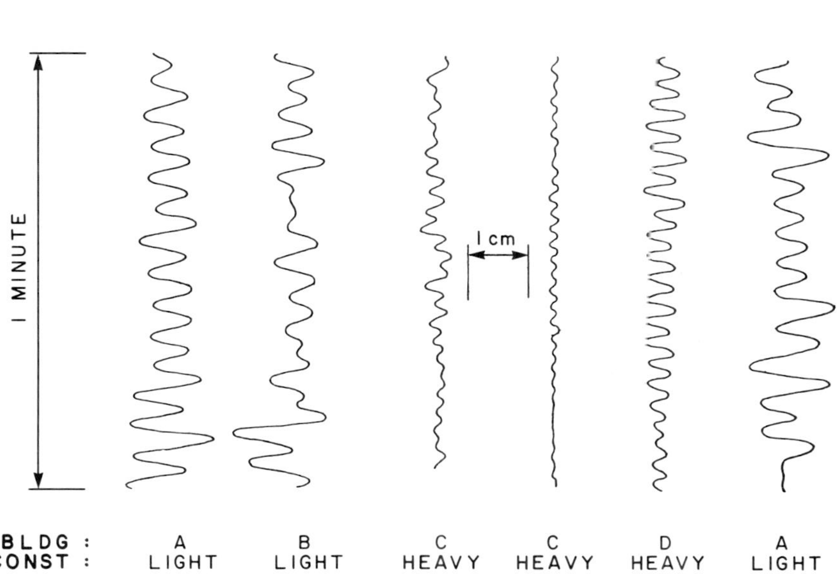

Fig. 3 Building sway in gale winds *(after Coyle, 1929)*

the usual individual whose nerves are not abnormally sensitive. It is obvious, however, that the people who can feel a three-quarter inch swing must be about ten times as numerous as those who can feel a quarter inch movement of a building, and therefore ten times as likely to cause trouble to the renting agent. A great deal can be done, of course, to avoid difficulties by having no fixtures that will sway and stimulate unnecessary feelings in the tenants; but if all extraneous factors are eliminated, the fact remains that it is easier to feel a large motion than a small one.

The conclusion was:

In the case of high buildings the frame must be designed to resist wind pressure with sufficient stiffness to keep the vibration caused by the wind within limits that inspire the occupants with confidence in the strength of the structure.

The literature of the time and the extensive discussion in *Engineering News Record* and elsewhere revealed many interesting observational facts. Around these, rules of thumb developed limiting deflections under wind to certain fractions of the height, and it was on the strength of these that many impressive structures were built. In his influential treatise on "Wind Bracing," Spurr (1930) stated somewhat pessimistically "the whole question of vibration in buildings from the effects of variable wind pressures is complicated by the indeterminate nature of the pressures themselves as well as by the great variation in size, shape, weight, height and location of buildings."

EMPIRE STATE BUILDING

Rathbun (1940) undertook the most significant full scale study in this era. His study on the Empire State Building followed the wind tunnel testing by Dryden and Hill (1933) and was for its time, a monumental piece of full-scale experimentation. The experiments used an anemometer, 30 manometers, 28 cameras with remote operating mechanisms, 22 extensometers, 1 collimator and target, and 1 plumb bob. Valuable data were collected and published in the paper, but there was evident difficulty in extracting definite conclusions. Perhaps Spurr's pessimism was contagious!

In reviewing the paper today with the benefits of hindsight, a number of aspects buried in the paper appear noteworthy. The following are a few.

The records of wind speeds at various weather stations in New York and at the Empire State Building (Fig. 4) reveal generally the significant increase in wind velocity with height in a heavily built up urban area. This increase extends, with an approximate "power law" exponent of 0.45, up to the full height of the Empire State Building. The boundary layer thus measured differed significantly from the assumption made by Dryden and Hill (1933) in their wind tunnel study, that above 60–90 m (200–300 ft) the flow would be uniform, thus justifying the use of an aeronautical type of wind tunnel. This 0.45 power law wind profile implies that the mean wind pressure (the square of the mean wind speed) increases with height roughly with a power law exponent of 0.9—or almost linearly.

Pressures were measured at 10 stations on each of three floors using manometer boards and flash cameras. An analysis of the individual pressure readings suggests they contain anomalous values. One has the suspicion that water leakage into the manometer tubes may have been a consistent problem.

What is perhaps noteworthy is that the magnitudes of the (mean) pressure coefficients are modest and, generally speaking, lie between ± 0.5. These are quite consistent with pressure coefficients measured today in boundary layer wind tunnels but are significantly less than those, for example, measured by Dryden and Hill in uniform flow wind tunnel experiments. Pressure coefficients measured in these experiments have been integrated to yield aerodynamic force coefficients and the latter are plotted in Fig. 5 and are referred to later. The wind tunnel experiment also indicated that the anemometer on the Empire State Building, mounted 15 feet above the stubby mast on top, read approximately 23% too high.

Rathbun also observed the movement of the building using a plumb bob extending the entire height of the fire shaft from the 86th to the 6th floor (Fig. 4). The plumb bob was damped in an oil bath and its position read relative to a grid system. By observing the somewhat erratic nature of the 'neutral' position of the building Rathbun concluded that, for small deflections, the building was not completely elastic and appeared to be restrained by frictional effects. In addition, wind currents in the fire shaft were suspected to

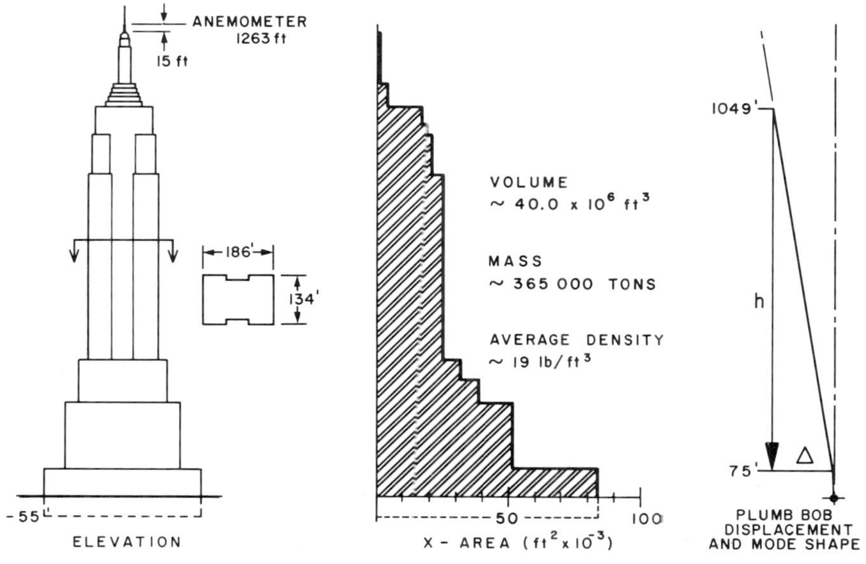

Fig. 4 The Empire State Building: elevation, mass and displacement

have caused some deviation of the plumb bob string. (In fact these offsets might also be affected by differential solar heating of opposite sides of the building.)

In spite of this, a reanalysis of the plumb bob data plotted in Fig. 6 contains some interesting features. The building symmetry was used and the mean deflection was computed from the mean position under light wind conditions. The points are plotted on a velocity squared scaling and fitted with straight lines. Admittedly the scatter is large, particularly at lower velocities, nevertheless the trends are apparent.

In Fig. 7a, the slopes of the lines (Δ/V_a^2) in Fig. 6 are plotted on the right hand scale against the wind azimuth. These data provide direct evidence of the full scale aerodynamic overturning moments, represented nondimensionally by the aerodynamic moment coefficient, C_M. For overturning in the north-south, or y direction, this is defined as

$$C_{M_y} = \frac{M_y}{\frac{1}{2}\rho_{air} V_H^2 H^2 B} \tag{1}$$

In the key diagram in Fig. 7, M_y is the overturning moment at the base, ρ_{air} is the density of air, V_H is the velocity at the reference height H, and B is the reference width. The overturning moment can be related to the deflection of the top of the building and a stiffness factor K, by

$$M = K\,(\Delta/h) \tag{2}$$

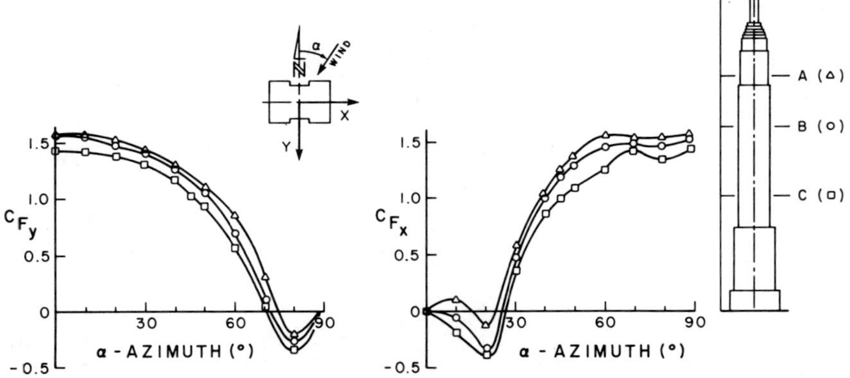

Fig. 5 The Empire State Building: wind tunnel measurements in uniform flow by Dryden and Hill

Here the load distribution corresponds to that of the mean wind load and Δ/h is the "deflection index". Substituting Eq. 2 into Eq. 1 gives

$$C_{M_y} = \frac{K_y}{\frac{1}{2}\rho_{air} H^2 Bh} \frac{\Delta_y}{V_H^2} \tag{3}$$

A similar expression for C_{M_x} can be written with D replacing B.

The problem now becomes that of defining the stiffness K. As Rathbun

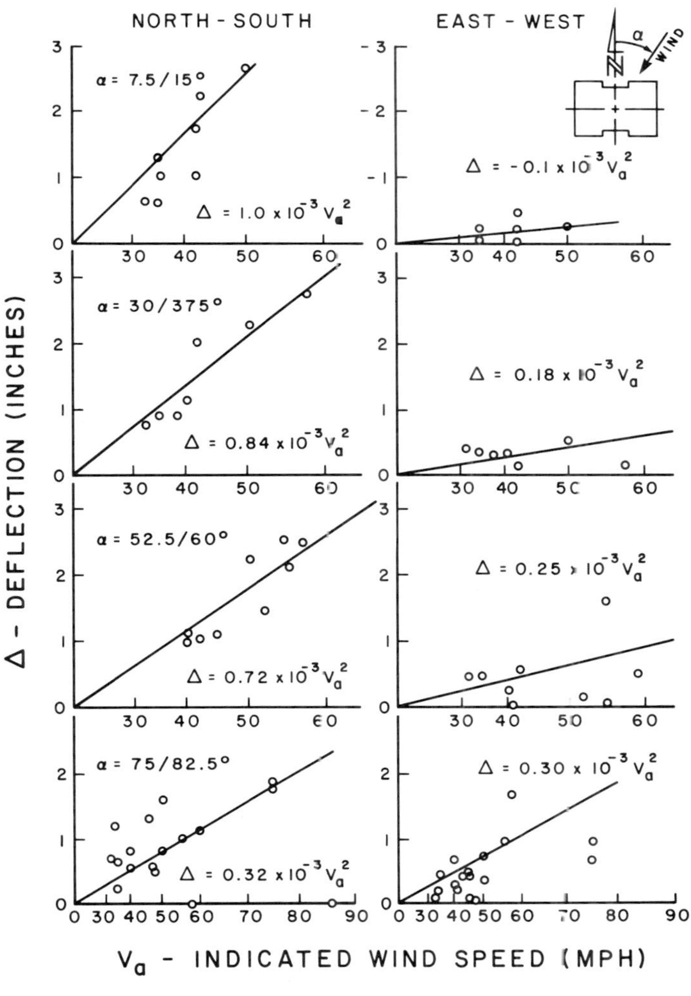

Fig. 6 Observations of plumb bob deflection

points out it would be difficult to calculate this from structural drawings because of the significant but uncertain contribution to stiffness made by the masonry infill. One clue however is provided by the relative deflections in the normal wind directions, which from Fig. 6 are seen to be $(1.00 \times 10^{-3})V_a^2$ and $(0.30 \times 10^{-3})V_a^2$ in the north-south and east-west directions respectively. If it is assumed that the moment coefficients C_{M_x} and C_{M_y} are roughly equal, and we adopt the mid-height plan dimensions of 186 ft × 134 ft it follows from Eq. 3 that

$$K_x/K_y = (0.3/1.0)(186/134)$$
$$\approx 0.42$$

A more direct estimate of the stiffness can be found from the sway of the building in its fundamental modes of vibration. Rathbun observed these vibrations both by means of a vertical collimator from the 6th floor to a target on the 86th, and by extensometer readings of strain on a corner column.

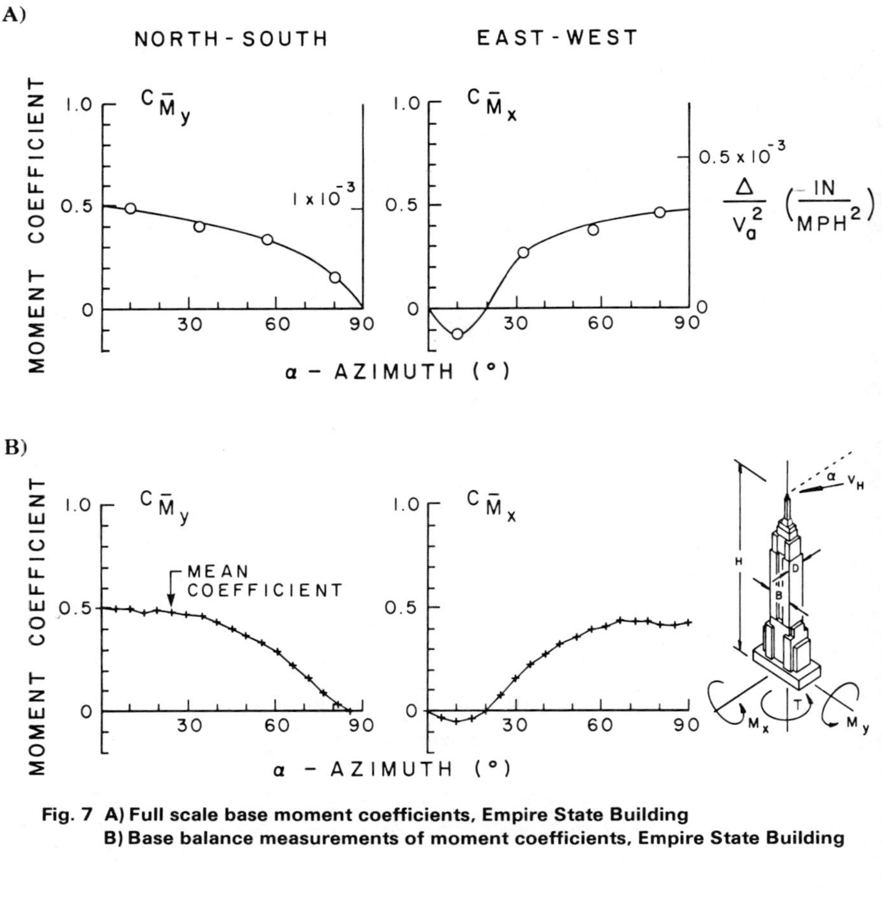

Fig. 7 A) Full scale base moment coefficients, Empire State Building
B) Base balance measurements of moment coefficients, Empire State Building

Samples of these observations are shown in Fig. 8. Both types of observations indicated that the north/south period of vibration was approximately 8.2 sec. Assuming that the ratio of the north/south stiffness to the east/west stiffness is roughly 0.42, the period of vibration in the east/west direction should be shorter by a factor of $\sqrt{0.42} = 0.65$, that is a period of 5.3 sec. There is clear evidence of such a period of vibration—5.5 seconds—in the lower record of the strain of the corner column that would "see" both N-S and E-W movements.

To use this information to estimate the stiffness we note in tall slender buildings the fundamental mode shape is invariably close to a straight line. Thus the mean wind load, discussed earlier, and the inertial load have

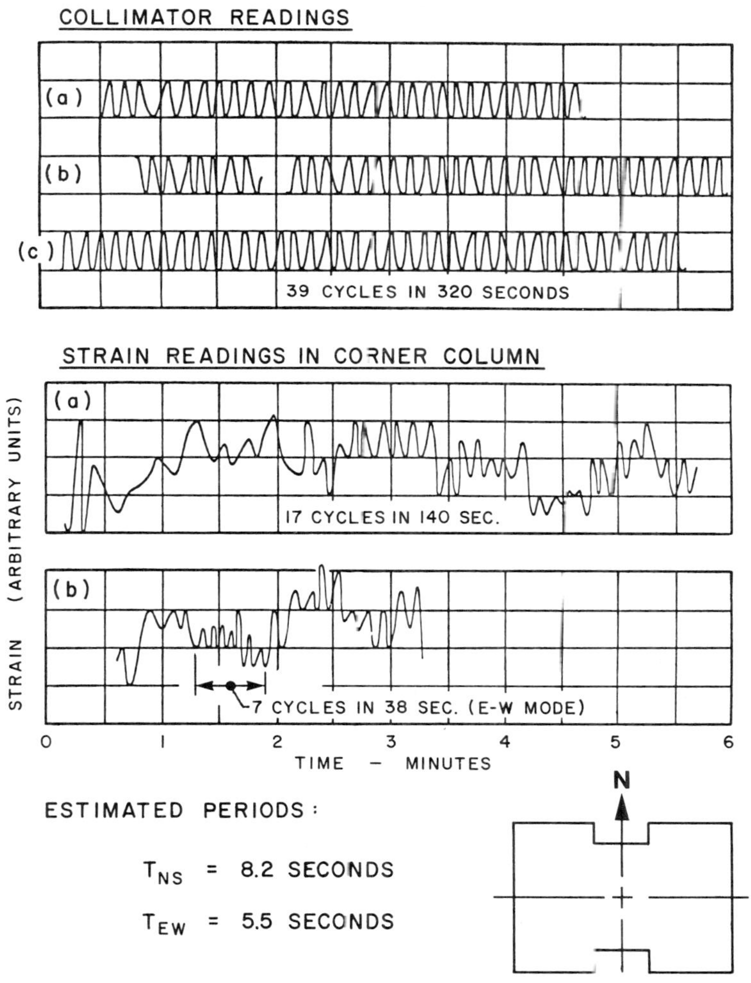

Fig. 8 Sway of the Empire State Building

similar straight line distributions and so do the deflections. This fortuitous result leads to a simple expression for the period of vibration, $T = 2\pi\sqrt{I_o/K}$ in which, I_o is the moment of inertia about the base, T the period of vibration and K the stiffness we are seeking. The stiffness of the building for either inertia or the mean wind loading is then approximately

$$K = 4\pi^2 I_o / T^2 \tag{4}$$

Referring to Fig. 4. the average density of the building, ρbld, is approximately 19 lb/ft^3 and the mass moment of inertia, I_o, is approximately equivalent to that of a prism of dimensions $(B \times D \times h)$ of 186 × 134 × 1050 ft. The stiffness can then be expressed,

$$K = 4\pi^2 \left(\tfrac{1}{3} \rho_{bld}\, h^3 BD \right) / T^2 \tag{5}$$

Substitution in Eq. 3 leads to the final expression for the moment coefficient (y direction)

$$C_{M_y} = \frac{8\,\pi^2}{3} \; \frac{\rho_{bld}}{\rho_{air}} \; \frac{h}{H}^2 \; \frac{D}{T^2} \; \frac{V_a}{V_H}^2 \; \frac{\Delta}{V_a^2} \tag{6}$$

The term (V_a/V_H) is the position correction factor for the anemometer, which, as stated earlier, is approximately 1.23. In the X direction B replaces D.

From the data it follows that with (Δ/Va^2) expressed in inches/(mph)^2 the moment coefficients are

North-South: $(T = 8.2\,\text{sec};\, D = 134\,\text{ft})$ $C_{M_y} = 0.50 \times 10^3\,(\Delta/V_a^2)$

East-West: $(T = 5.5\,\text{sec};\, B = 186\,\text{ft})$ $C_{M_x} = 1.55 \times 10^3\,(\Delta/V_a^2)$

The resulting moment coefficients are then as shown in Fig. 7.

Deflection in the east/west direction appears to reverse in sign for small angles of attack but not the north/south direction. Dryden and Hill's wind tunnel tests (Fig. 5) also indicate this reversal in the force, but for both north/south and east/west directions. This reversal is in fact indicative of a potential galloping instability. Fortunately, because of the mass and damping the critical speed turns out to be very high.

These results provide a useful comparison with more recent boundary layer wind tunnel tests on the Empire State Building using the base balance technique. A model of the building, machined accurately from a stiff foamed plastic, is mounted on an ultra-sensitive balance capable of measuring the base shears, moments, and torques. These force components can be expressed through coefficients as in Eq. 3. The mean and root mean square (RMS) base moment and torque coefficients for various wind directions are shown in Fig. 9. From these the moment coefficients have been replotted in Fig. 7. The comparison with the full scale results derived above is striking and reveals not only the same magnitudes for C_M but also similar details in the reversal of side force for small angles of attack. This provides very important full scale confirmation of the modelling.

Full scale readings of the dynamic building movements were also obtained by observing a target at the 86th floor through a vertical collimator set up at basement level. Their relationship with windspeed is more difficult to interpret and they show no obvious variation with direction. They have been replotted in Fig. 10 and, as may be seen, the dynamic deflections appear to be somewhat less than the peak static deflections and lie between two lines $\Delta = .00025\ V^2$ and $\Delta = .00055\ V^2$. This suggests that the maximum dynamic excursions are of the order of 55% of the largest mean excursions ($\Delta = .001\ V^2$). One must of course allow for the fact that this is likely to underestimate matters because of the difficulty and patience that would be required to really observe the actual maximum. Nevertheless it indicates that the gust factor (the ratio of the absolute maximum to the mean) would be about 1.55 or perhaps higher. This value is low compared with recent structures not having the masonry infill and heavy stone facade of the Empire State Building. It is also suggestive of somewhat higher damping.

The model results also indicate the dynamic behavior. The root mean square moment coefficients in Fig. 9 show only slight variation with direction,

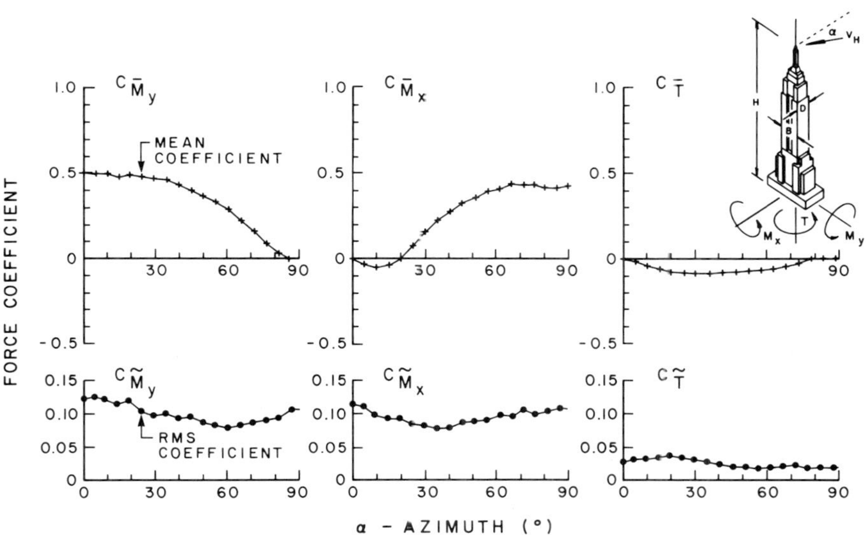

Fig. 9 Base balance measurements of moment coefficients, Empire State Building

a not uncommon behavior. In Fig. 11, the spectra of the dynamic force are shown as a function of the reduced wind speed, $V_H T/\sqrt{BD}$. When substitutions are made this reduced wind speed for north-south motion is equal to 0.091 V_a where V_a is the indicated wind speed at the top in mph.

If we assume that the observed full scale dynamic response was mostly the resonant component, and that the reduced velocities are in the range 4–7 (50–80 mph) we can estimate the ratio of the peak dynamic response to the mean response from the relation

$$\frac{\textit{Peak dynamic response}}{\textit{Mean dynamic response}} = \frac{\hat{\Delta}}{\bar{\Delta}} = g\frac{C_{\tilde{M}}}{C_{\bar{M}}} \quad \frac{\pi}{4\zeta}\frac{FS(F)}{VAR}$$

Assuming the values $g = 3.0$, $C_{\tilde{M}} = 0.10$, $C_{\bar{M}} = 0.50$, $FS(F)/VAR = 0.02$, $\zeta = 0.02$, we find $\hat{\Delta}/\bar{\Delta} = 0.53$—almost identical to the observed value of 0.55. In this case the uncertainties are larger and could amount to a factor of two.

We can also consider the accelerations with which these dynamic displacements are associated. In recent years suggestions have been made (in the National Building Code of Canada, for example) that for comfort, accelerations should not exceed 10–20 milli-gs every 10 years. Meteorological studies in New York have suggested that the 10-year mean wind speed at the height of the Empire State Building is about 80 mph—or 98 mph as would be indicated on the Empire State anemometer. Extrapolating the lines in Fig. 10, the deflections at this wind speed are between 2.5 to 5.5 inches. The corresponding

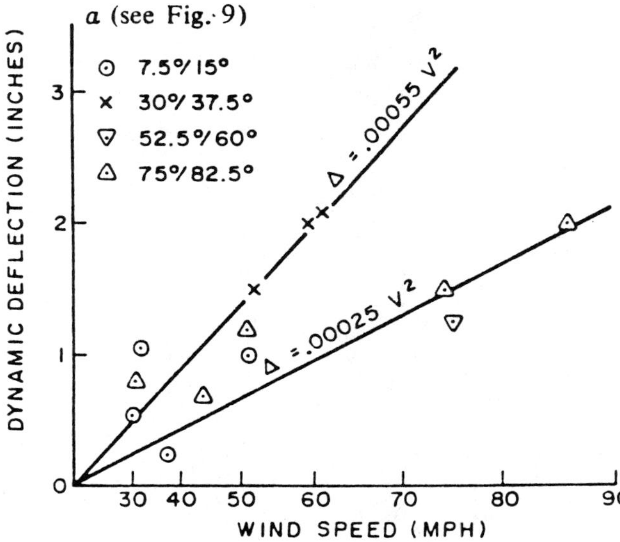

Fig. 10 Dynamic sway measurements of the Empire State Building. V = uncorrected windspeed mph. Δ = deflection inches

accelerations, {given by $(4\pi^2/T^2) \times$ displacement}, are then found to be between 4 and 8.5 milli-gs. This is near the 85th floor; at the top of the building the upper figures may be nearer 10 mg. The apparently satisfactory performance of this structure adds some credibility to the criterion mentioned earlier, at least as a lower bound.

These values of acceleration are less than those encountered in modern steel buildings of similar size in which accelerations may be twice as great.

THE WORLD TRADE CENTER

When the twin towers of The World Center in New York were built nearly two decades ago, an unusual degree of attention was given to their behavior in wind. At the time, wind tunnel studies were rarely carried out for buildings; if they were, they involved static pressure models tested in the normal uniform steady flow of aeronautical wind tunnels. Several aspects of the wind studies for the World Trade Center represented radical departures from previous practice; one was the use of aeroelastic models (hitherto used only for highly flexible structures such as bridges and chimneys). Another was the modeling of the turbulent boundary layer flow characteristics of the natural wind, achieved partly through modeling the building topography of Manhattan.

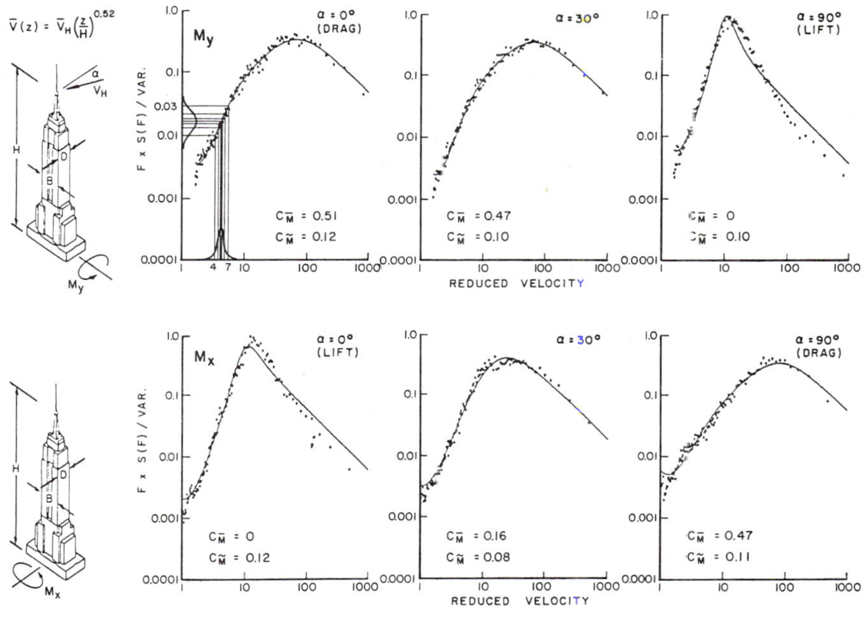

Fig. 11 Empire State Building: measurements of power spectral density of base overturning moments

The experimental study for the project, which the writer directed for the consulting engineers, employed three wind tunnels: the new micrometeorological wind tunnel at Colorado State University, designed by Cermak and Plate; the National Physical Laboratory by Scruton; and the Boundary Layer Wind Tunnel Laboratory, The University of Western Ontario. The results affirmed the important role of dynamics and turbulence in tall building response to wind.

Experiments were carried out on a single square tower of similar proportions to the World Trade Center, the results of which are shown in Fig. 12. In these experiments the ratio of the height to width was 6.5, the mean density ratio of the building model to the air was approximately 182, the moderately rough boundary was representative of open country and the rough boundary represented a built-up urban area. The mean velocity and turbulence profiles are shown in Fig. 13. The responses are given in terms of the ratio of the rms and mean responses for the X and Y direction (σ_x, σ_y, $\overline{X}$, $\overline{Y}$, respectively) to the building height H. The mean wind speed, V, at the top of the building is expressed through the velocity profile in Fig. 13 and the reduced velocity, V/fD, where f and D are the natural frequency and diameter of the model. Reduced velocities encountered in practice would seldom exceed 10 and are

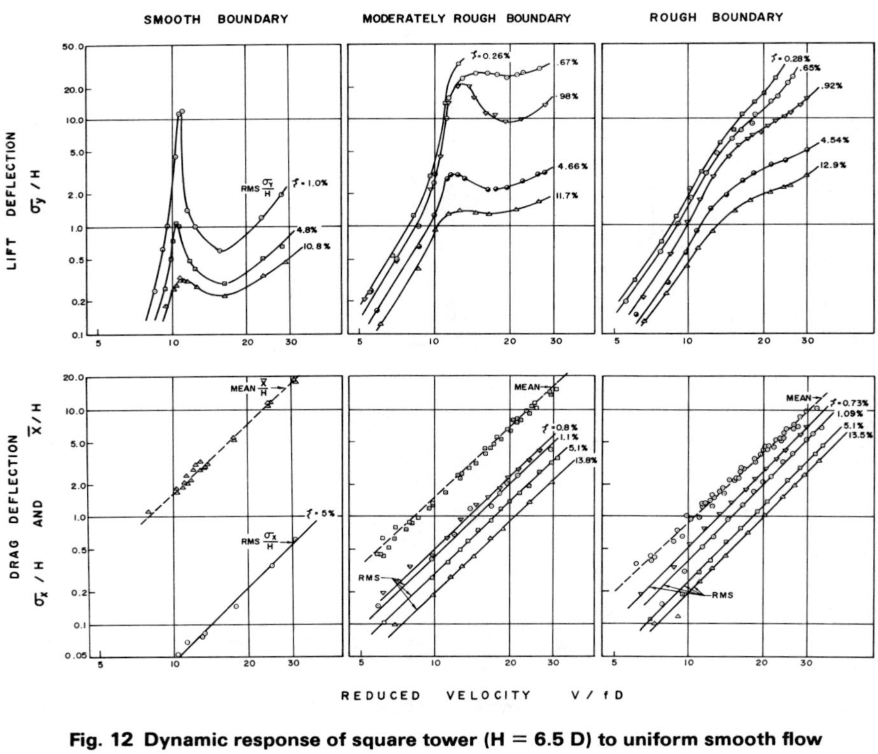

Fig. 12 Dynamic response of square tower (H = 6.5 D) to uniform smooth flow

generally much below. When the experimental results are examined in detail the following points are noteworthy.

In the drag direction the slope of the mean response indicates the usual proportionality to the velocity squared. However, the dynamic rms drag response increases with a power slightly in excess of a square law; increases systematically with the roughness of the boundary; and gradually decreases with damping.

In the lift direction the dynamic response is considerably more complex. There is, of course, no mean response due to symmetry. With the smooth boundary (negligible turbulence) a marked peak occurs at a reduced wind speed of approximately 11, the amplitude of which collapses rapidly with increased damping. In the turbulent boundary layer flows the peak becomes blurred; the dynamic response of the off-peak wind speed increases, and the damping reduces the responses more gradually.

It is obvious from these results that the peak dynamic response, which may reach three times σ_X or σ_Y, are highly significant compared to the mean and the influence of the turbulent boundary layer is dramatic. These facts have changed the direction of wind tunnel testing and it would now be quite inappropriate to consider wind tunnel testing of buildings without modeling the boundary layer.

This experience was a turning point in the wind tunnel testing of structures and tall buildings in particular. It has, in parallel, led to the acceptance of analytical approaches to allow for the dynamic effects—the gust response factors.

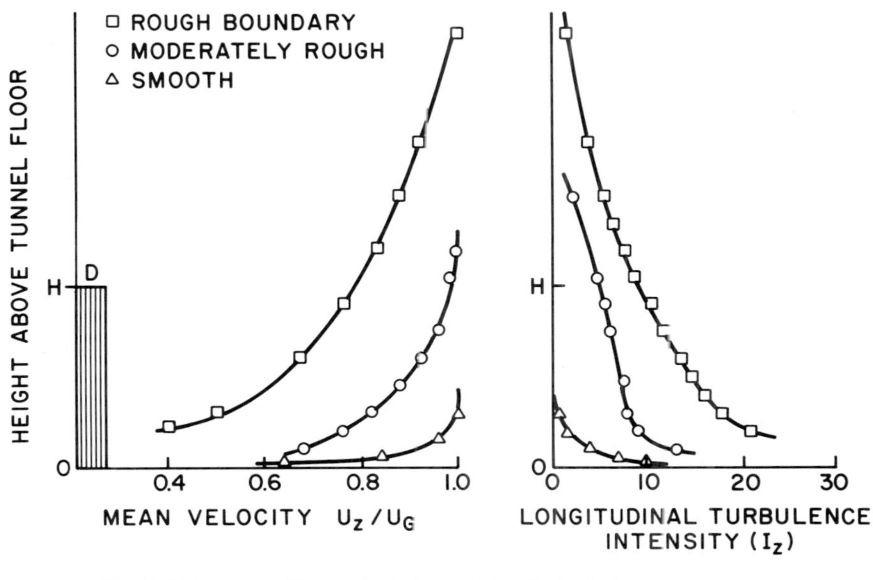

Fig. 13 Velocity profile and turbulence intensity in tests on square tower

THE CN TOWER

The approximately 555-m (1,820-ft) high CN Communications Tower in Toronto has now been operational for nearly a decade. The action of wind on this unique tower, which remains the world's tallest free-standing structure, was extensively studied at the Boundary Layer Wind Tunnel Laboratory during the design of the tower. Including investigations on early preliminary designs, this study continued over a period of some 6 years and included the following main parts:

- Aeroelastic and section model wind tunnel tests were carried out to evaluate an earlier 3-legged version of the tower. It was subsequently replaced by the present configuration, which was found to be aerodynamically and economically more effective.
- A 1:450 scale aeroelastic model of the tower was tested in turbulent boundary layer flow representative of wind at the project site. This defined the wind loads on the concrete shaft and the steel antenna and provided information on the overall response of the tower, including the displacements and rotations of the antenna mast and the accelerations of the tower at the restaurant and other levels.
- A 1:450 scale static pressure model was tested to provide information on the local peak pressures and suctions on the elevator shaft glazing, at the restaurant and upper observation levels and at the lower accommodation levels, including the pool lobby and other lower buildings at the tower base. It is noteworthy that some of the highest local exterior suctions found in the entire study occurred on the pool lobby.
- Pedestrian level winds at the base of the tower were studied using the pressure model of the tower and lower accommodation levels. These measurements indicated relatively high wind speeds, particularly near the three tapered legs of the shaft.
- A 1:60 scale partial model of the upper accommodation levels and adjacent parts of the shaft was studied to examine the action of wind on the air-supported radome enclosing the microwave transmission equipment just below restaurant level. The radome was modelled aeroelastically and information was provided on the internal pressurization required to limit deformations and assure its performance at high wind speeds.
- Analytical estimates were made to evaluate the effectiveness of two tuned mass dampers attached to the antenna mast. These two auxiliary mass dampers were designed to ensure a minimum level of damping in the second, fourth, and fifth modes of vibration of the

tower. While the aeroelastic model study did not indicate excessive movements of the antenna mast, these dynamic absorbers were added to increase the reliability of performance.

- Some initial observations of the full-scale response were made during the construction of the tower. These measurements provided important validations of assumed tower properties. The most significant of these were made after the completion of the concrete shaft. These confirmed the structural damping used in the aeroelastic model study and the anticipated mass and stiffness properties.
- Meteorological data for the Toronto area were analyzed to arrive at a statistical model of the probability distribution of gradient wind speed and wind direction for the area. This model provided an estimate of the probability of exceeding particular levels of wind speed from different azimuth directions and was used in the synthesis of the wind tunnel model findings to provide predictions of various wind induced full scale effects.

Following completion of the tower, full scale measurements of the wind and the response of the tower were made. From the results, there appear to be no surprises in the observed response of the tower. Its dynamic properties tend to be consistent with its design values and indications suggest that the wind tunnel model study provided representative estimates of the wind-induced response of the tower. Somewhat more unexpected are the properties of the wind. The boundary layer appears to be deeper and the turbulence intensity higher than conventionally assumed.

FURTHER OUTLOOK

In recent years taller buildings have been discussed. One of these proposals was for a 150 story building using a "mega frame," to increase the efficiency of the structural system and hence the stiffness. Although the structure itself was never built, some of the ideas were later used in a proposal by Skidmore, Owings & Merrill for the Bank of the South West in Houston. As well as the highly efficient structural system, the cross-section was slightly tapered and openings were punched through the upper portions. The purpose of maintaining maximum stiffness is to raise the natural frequency and raise the critical speed for the onset of vortex shedding to higher wind speeds with lower levels of probability. The widening of the base (or taper) of the structure, spreads the vortex shedding over a broader range of frequencies and reduces the amplitude of the excitation. The venting of the upper portions of the building inhibits the perturbations in the wake near the tip associated with vortex shedding and reduces lateral the excitation. Experiments showed reductions of 20% of the dynamic amplitude because of these features.

Naturally any modification of the structure must complement the architecture. The tapered shape conforms to the reduced elevator shaft requirements and the possible use of upper levels for apartments. The openings might lend themselves to a feature—a "sky garden" perhaps.

The problems that confront the construction of taller buildings are, in summary,

- The prevalence of higher winds and the uncertainty of their prediction;
- The greater variability of the turbulence structure; and
- The relatively higher energy level in the turbulence at the lower frequencies.

The challenges in the structural engineering are to maintain the mass, damping, and stiffness, to minimize the influence of P-Δ effects, and to develop a geometry which is not too slender and is aerodynamically stable.

ACKNOWLEDGEMENT

The writer acknowledges with sincere thanks the structural engineers and architects of the "supertall," with several of whom the writer has had the great privilege to work; his colleagues at the Boundary Layer Wind Tunnel Laboratory; and Dr. Tony Tschanz former graduate student, who designed and built a high resolution base balance that has had an important impact on the conduct of building aerodynamics.

REFERENCES/BIBLIOGRAPHY

Coyle, D. C., 1929
MUSHROOM SKYSCRAPERS, American Architect, pp. 829-832.

Coyle, D. C., 1931
MEASURING THE BEHAVIOUR OF TALL BUILDINGS, Engineering News Record, pp. 310-313.

Davenport, A. G., 1967
GUST LOADING FACTORS, Journal of the Structural Division, ASCE, Vol. 93 (ST3), pp. 11-34.

Davenport, A. G., Isyumov, N., and Jandali, T., 1971
A STUDY OF WIND EFFECTS FOR THE SEARS PROJECT, The University of Western Ontario, Engineering Science Research Report, BLWT-5-1971.

Davenport, A. G., 1977
WIND ENGINEERING—ANCIENT AND MODERN—THE RELATIONSHIP OF WIND ENGINEERING RESEARCH TO DESIGN, Proceedings of the Sixth Canadian Congress of Applied Mechanics, May 29-June 3, Vancouver, B.C.

Dryden, H. L., and Hill, G. C., 1933
WIND PRESSURE ON A MODEL OF THE EMPIRE STATE BUILDING, Bureau of Standards Journal of Research, Vol. II, pp. 493-523.

Eiffel, G., 1885
PROJECT D'UNE TOUR EUFER DE 300 M DE NAUTEUR. MEMOIRES DE LA SOCIETE DES INGENIEURS CIVILS (Project for a Tower in Iron of 300 m Height. Memiors of the Civil Engineering Society), Paris, pp. 345-370.

Eiffel, G., 1900
TRAVAUX SCIENTIFIQUES EXECUTES A LA TOUR DE TROIS CENTS METRES DE 1889 A 1900 (Scientific Experiments Conducted on the 300 m Tower from 1889 to 1900), L. Maretheux, Paris, France.

Isyumov, N., Davenport, A. G., and Monbaliu, J., 1984
CN TOWER, TORONTO: MODEL AND FULL SCALE RESPONSE TO WIND, International Association for Bridge and Structural Engineering (IABSE) 12th Congress, Session VI, September 3-7, Vancouver, B.C.

Rathbun, J. C., 1940
WIND FORCES ON A TALL BUILDING, Transactions ASCE Vol. 105, pp. 1-41.

Robertson, L. E., 1973
DESIGN CRITERIA FOR VERY TALL BUILDINGS, Proceedings of the Australian and New Zealand Conference on the Planning and Design of Tall Buildings, August 14-17.

Tschanz, T. and Davenport, A. G., 1983
THE BASE BALANCE TECHNIQUE FOR THE DETERMINATION OF DYNAMIC WIND LOADS, Sixth International Conference on Wind Engineering, Gold Coast, Australia, March 21-25.

Spurr, D., 1930
WIND BRACING—THE IMPORTANCE OF RIGIDITY IN HIGH TOWERS, McGraw Hill Publishing Co., New York.

Fire Safety Design For Tall Buildings: Recent Developments

Margaret Law

One of the most significant changes in approach to fire safety design is the growing recognition of the importance of fire engineering by design. The traditional prescriptive approach cannot be extended much further; more money is being spent on fire safety measures, yet most fire engineers would hesitate to say that it is being spent in the most effective way or that it is achieving the desired results. Traditionally also, the desired result has been defined in vague terms—safety of life or property—but only in the field of property safety has quantification of costs and benefits had an explicit effect on the measures required. Action is now required at the national level to establish criteria for acceptable standards of public safety. An international organization can initiate efforts to achieve this by showing how the considerable amount of available technical and statistical data might be used.

STRUCTURAL FIRE PROTECTION

The structural stability of modern tall buildings exposed to fire has been shown to be satisfactory. For most uses—office, hotel, residential, hospital—the fire loads are relatively low, and any areas with higher fire loads such as storage are treated selectively. In terms of fire resistance periods, standards vary significantly from country to country, probably as the result of historical accident. Now that calculation methods are established and the benefits of good detailing can be recognized explicitly, an opportunity to reassess the design of structural stability in fire exists. In addition, a reliability-based approach has been proposed (CIB, 1983). Initially, this approach could be tried out in a given country using the existing regulations for calibration.

CONTROL OF FIRE SPREAD

Structural stability does not necessarily prevent the passage of fire and smoke through many stories of a tall building. Traditional compartmentation into fire-tight cells is difficult to achieve in a building that relies on many service systems for the normal operation of the building. Other measures are therefore needed to reduce the chance of fire—if it occurs—becoming large.

Detection

The sooner a fire is detected, the easier it is to control. A review in the United Kingdom (Home Office, 1980) of the effectiveness of various fire protection measures concluded that little benefit would be derived from reducing attendance times of the local authority fire brigades (which are low) or increasing the initial weight of attack; the factor most likely to lead to the development of fires into large-loss fires is delay in discovery of the fire. Some provisional estimates for 1978 indicated that for occupied buildings other than dwellings, when fires were discovered within 5 minutes of ignition, only about 5% spread beyond the room of origin; whereas 17% did so when discovery was later. These statistics take into account detectors designed and installed before 1978. Recent developments in design would be expected to give improved performance and greater benefits.

Automatic Sprinklers

The benefit to be obtained from automatic sprinklers depends on the use of the building. In industrial buildings they can reduce the chance of a large-loss fire by a factor of about 6 (Dunn and Fry, 1966) but in shops Rogers (1977) gives the factor as 2.

United Kingdom fire brigade reports include an assessment of the floor area damaged by fire. An analysis (Law, 1985a) of statistics for shops (Morgan and Chandler, 1981) indicates that for large fires, (fires of 10 m^2 (108 ft^2) area or more) the proportion of fires P(%) exceeding an area A_f (m^2) is given by

$$P = 62\ A_f^{-0.95} \text{ for sprinklered shops} \tag{1}$$

$$P = 107\ A_f^{-0.77} \text{ for nonsprinklered shops} \tag{2}$$

Similar statistics (Morgan and Hansell, 1985) for large fires in offices can be arranged to give

$$P = 60\ A_f^{-0.63} \text{ for sprinklered offices} \tag{3}$$

$$P = 180\ A_f^{-0.78} \text{ for nonsprinklered offices} \tag{4}$$

However the nonsprinklered data for "daytime" fires in offices are significantly different from those for "nighttime."

$$P = 130\ A_f^{-0.78} \text{ day} \tag{5}$$

$$P = 340\ A_f^{-0.78} \text{ night} \tag{6}$$

This illustrates the value of people being present. If they are, then sprinklers in offices would reduce the chance of a large fire by a factor of about 1.5. If they are not, then by a factor of about 4.

CONTROL OF SMOKE SPREAD

It has been estimated that one person in 10 will refuse to move through smoke if the visibility is reduced to 8 m (26 ft). The combustion products from burning only 1 kg (2 lb) of wood, when dispersed in a volume of 1000 m^3 (35,000 ft^3), can reduce the visibility to this level (Hinkley, 1975). Near the fire source, the smoke will be hot and tend to form a layer under the ceiling, above head level. As the smoke moves away and cools, it can fill corridors down to floor level and may enter rooms remote from the fire. This smoke movement is caused initially by expansion and buoyancy of the heated gases from the fire; subsequently it can be significantly affected not only by the ventilating and air conditioning systems within the building but also by pressure differences caused by stack effect and wind.

The traditional methods of smoke management—physical barriers and natural ventilation openings in protected routes—have been found inadequate for those tall buildings that are subject to these air flows and pressure effects. A knowledge of the building physics is desirable and alternative smoke management methods should be employed.

Smoke Control

The term smoke control is normally used to describe the method that establishes favorable pressure differences across barriers using mechanical pressurization. The spaces selected to be pressurized are usually the protected escape routes—the staircases, the lobbies and occasionally a corridor. The system is designed by identifying the size and location of the leakage paths and calculating the air flow that will be needed to maintain the required pressure differential. Data and calculation methods are available (Klote and Fothergill, 1983; BSI, 1978).

The advantages of this type of active control are that escape stairs and lobbies need not be on external walls; leakage paths in barriers can be allowed for; the stack, buoyancy, and wind effects can be taken into account; exit doors propped open do not let in as much smoke as those in a traditional building; the design is not sensitive to the size of fire that is postulated.

Smoke Extract

An early example of smoke extract design is the stack over the stage of a theater. The roof vent of the stack is sized so that all air flow is from the auditorium to the stage, the neutral plane being at the level of the top of the proscenium (Lie, 1972). Another example is the design of automatic roof vents for industrial buildings that provide a clear layer of air at ground level so that the fire brigade may tackle the fire (Thomas et al., 1963). These are examples of natural venting, which relies entirely on buoyancy and can be designed to evacuate large volumes of smoke. Mechanical extract is also possible, but is normally only practical for evacuating smoke from small fires. Mechanical extract is most practical when the size of the fire can be controlled by automatic sprinklers and the smoke plume kept as compact as possible to reduce entrainment.

The extract system can be designed to protect escape routes, in which case it is necessary to define the time for onset of hazardous conditions and the time available for escape; to protect adjacent accommodation, in which case it is necessary to postulate a likely hazardous condition; to release smoke after a fire, in which case a rate of six air changes an hour may be adequate. There are no very clear guidelines on the performance required in these different circumstances.

ATRIUMS

The modern atrium, being roofed, can increase the risk of smoke and heat spread through the building. If the floors overlooking the atrium are glazed, then the risk of fire spread from floor to floor via the atrium is not likely to be greater than via the external facade, and with automatic sprinklers installed, the chance of vertical fire spread is low (either within the atrium or along the facade.) However, even with sprinklers there is a chance of smoke entering the atrium, thus putting the other floors at risk, unless positive smoke management measures are adopted. The two approaches favored are either to extract smoke from the fire before it enters the atrium or to allow the smoke to enter and then vent it safely at roof level (Saxon, 1983; NFPA, 1981). As mentioned above, clearer guidelines on the performance requirements are needed, particularly where production of escape is the primary objective.

ESCAPE

Escape codes differ in detail from country to country. However they do share some common approaches: a person confronted by a fire should be able to turn away and find an exit to a protected route in the other direction, which means there should be at least two exits from any floor; there should be a limit on travel distance to the exit; once having entered the protected route (usually a staircase), the person must be protected until a place of safety is reached. The place of safety is ultimately the open air at ground level, but in a tall building it may initially be a designated refuge zone. A concept has been developed for estimating the time from ignition before hazardous conditions are attained in a room, which is compared with the time needed for people to escape from that room. A developing fire can be postulated and the time of operation of detectors can be calculated (Cooper, 1983). A beyond-the-room model would obviously be relevant to tall buildings.

Most studies of escape movement have concentrated on entry into a staircase and the subsequent flow. Measurements have been made of crowd movement along corridors and on staircases, notably in Japan, Britain, and Canada (Melinek and Booth, 1975; Fauls, 1977). Formulas for evacuation time usually assume fixed values of flow density or velocity. A more flexible model, taking into account the pedestrian movement in a particular building, has also been developed (Kendik, 1985).

RESEARCH

The UK Home Office review mentioned earlier confirms that the United Kingdom fire grading requirement of the building regulations has been broadly successful in relation to life safety and structural stability (most deaths and injuries being caused by the contents). This is generally true for the other developed countries. Whether the requirements are rational and economic is a topic that is not really addressed in the report, although there are some interesting assessments of the economic benefits to the community of installing active fire protection measures in various sectors. With the background of buildings being relatively successful in surviving fires, research has been devoted more to the behavior of the contents. In relation to life safety, the areas of research have been domestic dwellings and institutions. In relation to property safety, the major research interest has been in the contents of industrial and storage buildings. However in practical building design the major constraints and the increased expenditure on fire safety needed to comply with regulations are incurred in quite different areas. For life safety, the buildings that contain large numbers of people are the major concern, while for property, the standards of protection, both active and passive, are being increased in buildings that are not particularly valuable, either in structure or in contents (Law, 1985b).

One reason this has happened is that the authorities perceive that the larger the number of people at risk, the higher the standard of public safety should be. However there is no agreed definition of the standard and no assessment of whether the measures required are effective and make the best use of resources. The authorities also perceive that the public can become alarmed by a large property loss, even when there are no deaths or injuries. What is the critical size of loss and how does it vary with sector? How much should public regulations mandate the safety of private property? Mandatory automatic sprinkler protection is increasing at the same time that the standards of structural fire protection are rising.

The improvement in structural fire protection is being achieved not by changing the regulations, but by rewriting the rules for the design of elements to achieve a specified grade of fire resistance. Ought not the grading be reduced to maintain the status quo? If the beneficial effect of sprinklers can be recognized how is it quantified?

CONCLUDING REMARKS

The technical basis for fire engineering design of tall buildings has increased very significantly in recent decades. If this information is to be used effectively, then safety criteria must be established and the performance of various fire safety measures must be quantified accordingly.

REFERENCES/BIBLIOGRAPHY

BSI, 1978
SMOKE CONTROL IN PROTECTED ESCAPE ROUTES USING PRESSURIZATION, BS 5588, Part 4, British Standards Institution, London.

CIB, 1983
A CONCEPTIONAL APPROACH TOWARDS A PROBABILITY BASED DESIGN GUIDE ON STRUCTURAL FIRE SAFETY, CIB/W14, Fire Safety Journal, Vol. 6, No. 1.

Cooper, L., 1983
A CONCEPT FOR ESTIMATING SAFE AVAILABLE EGRESS TIME IN FIRES, Fire Safety Journal, Vol. 5.

Dunn, J. and Fry, J., 1966
FIRE FOUGHT WITH FIVE OR MORE JETS, Fire Research Technical Paper No. 16, London, HMSO.

Hinkley, P., 1975
WORK BY THE FIRE RESEARCH STATION ON THE CONTROL OF SMOKE IN SHOPPING CENTRES, Building Research Establishment CP33/75, Borehamwood.
Home Office, 1980
FUTURE FIRE POLICY: A CONSULTATIVE DOCUMENT, Home Office, Scottish Home and Health Department, London, HMSO.

Kendik, E., 1985
ASSESSMENT OF ESCAPE ROUTES IN BUILDINGS AND A DESIGN METHOD FOR CALCULATING PEDESTRIAN MOVEMENT, Paper at SFPE 35th Anniversary Engineering Seminar, Chicago, May.
Klote, J. and Fothergill, J., 1983
DESIGN OF SMOKE CONTROL SYSTEMS FOR BUILDINGS, ASHRAE, September.

Law, M., 1985a
FIRE PROTECTION IN TERMINAL BUILDINGS, Paper in Symposium on Building Services for Airports, Gatwick, CIBSE, November 6-7.
Law, M., 1985b
TRANSLATION OF RESEARCH INTO PRACTICE: BUILDING DESIGN, Paper in First International Symposium on Fire Safety Science, Gaithersburg, October 9-11.
Lie, T. T., 1972
FIRE AND BUILDINGS, Applied Science Publishers, London.

Melinek, S. and Booth, S., 1975
AN ANALYSIS OF EVACUATION TIMES AND THE MOVEMENT OF CROWDS IN BUILDINGS, Building Research Establishment CP96/75, Borehamwood.
Morgan, H. and Chandler, S., 1981
FIRE SIZES AND SPRINKLER EFFECTIVENESS IN SHOPPING COMPLEXES AND RETAIL PREMISES, Fire Surveyor, October.
Morgan, H. and Hansell, G., 1985
FIRE SIZES AND SPRINKLER EFFECTIVENESS IN OFFICES—IMPLICATIONS FOR SMOKE CONTROL DESIGN, Fire Safety Journal, Vol. 8, No. 3.

National Fire Protection Association, 1981
LIFE SAFETY CODE, National Fire Protection Association.

Paul, J., 1977
MOVEMENT OF PEOPLE IN BUILDING EVACUATIONS, Human Response to Tall Buildings, Chapter 21, P. J. Conway, ed., Dowen, Hutchinson and Ross, Stroudsburg, Pennsylvania, Community Development Series, Vol. 34.

Rogers, F., 1977
FIRE LOSSES AND THE EFFECT OF SPRINKLER PROTECTION OF BUILDINGS IN A VARIETY OF INDUSTRIES AND TRADES, Building Research Establishment CP9/77 Borehamwood.
Rutstein, R., 1979
THE ESTIMATION OF THE FIRE HAZARD IN DIFFERENT OCCUPANCIES, Fire Surveyor, April.

Saxon, R., 1983
ATRIUM BUILDINGS: DEVELOPMENT AND DESIGN, Architectural Press, London.

Thomas, P., et al., 1963
INVESTIGATIONS INTO THE FLOW OF HOT GASES IN ROOF VENTING, Fire Research Technical Paper No. 7, London, HMSO.

Safety, Quality Assurance, and Performance

Carl J. Turkstra

At the Third International Conference on Tall Buildings, where the second century of the skyscraper was celebrated, we could also have begun to celebrate another anniversary. Nineteen eight-seven marks the fortieth anniversary of the beginning of structural safety analysis in the United States which began with Freudenthal's classic paper (Freudenthal, 1947).

At first there was great reluctance to abandon the traditional concept of absolute safety. Gradually, however, probabilistic models have been accepted to describe loads and provide a basis for rational decision.

At the time of the 1972 Tall Building Conference, safety analysis was in an exceptionally fertile period (Joint Committee on Tall Buildings, 1973). Second-moment analysis theories had been devised and a way had been found to develop practical design codes. The 1972 papers on structural safety were important from an intellectual point of view but were rather academic in scope. Following the conference, a number of major studies were completed including the work of Galambos and Ravindra (1978) for LRFD in steel and the work of Ellingwood et al. (1980) supporting the new ANSI load factors.

Now, 40 years after Freudenthal's paper, structural reliability analysis is well established even though we have unsolved problems, such as system reliability, and poorly solved problems, such as seismic analysis. We have also new problems such as serviceability and the treatment of existing buildings, but it seems clear that reliability analysis is set on a path of orderly evolution.

Given such a success story, it is unfortunate that we have to add a caveat. Although we have formulated a number of sophisticated models that can be turned over to designers or code groups for "rational decision," we have investigated only a small part of the problem.

We have formulated the random game against nature with physical variables, but this is only a small portion of the total picture. The major position is the human issue, where the variables are human nature and organizations comprised of humans.

To understand the total picture we must consider human behavior, which we now know is a primary variable in structural response. We must begin to make realistic assumptions about human beings, who are not rational optimizers but rather survivors looking for solutions that will do the job—trying to succeed, trying to survive.

One of the most interesting studies of human error was made by Melchers and Harrington (1984) in Australia, who sent out a very simple design survey to 325 licensed engineers. About one in ten replied.

The first question involved a single story gable rigid frame building (two columns; two rafters; fixed joints) with specified design codes, location, opening patterns, materials, and dimensions. Design moments at the nodes were requested.

A critical design case was the combination of dead + live + wind loads in the roof. Only nine responses seemed to conform to the code without gross errors. For these nine, the estimated design moment at the ridge had a coefficient of variation of 52%. Another nine responses tried to comply with the codes but had gross errors, usually linked to internal pressures. If one combines the nine "correct" solutions with the nine "gross error" solutions, the design bending moment at the ridge varied from −107 to +70 with a coefficient of variation of 1,182%. The average was of the wrong sign. Only 10% bothered to use the 0.75 load reduction factor in the code for the load combination.

With results like these, we must ask: Why do so many buildings stand up?

To evaluate structural safety realistically we must go beyond the statistics of human error and consider the entire design, construction, and utilization process. The scope of the problem can be illustrated through an example related to the Port Authority of New York and New Jersey (1984), which recently released a remarkably candid report on the Journal Square failure.

The failure involved the collapse of a suspended ceiling over a public thoroughfare after a number of years of service. The ceiling was supported by wire hangers attached to tabs anchored in the floor slab. Upon investigation it was found that the drawings indicated a maximum hanger load of 27 kg (60 lbs.). No one seems to have designed the hanger locations and the ceiling contractor seems to have used a pattern for lightweight suspended ceilings. The ceiling was heavy plaster.

Over the years, some hangers broke and maintenance or service people tied them back up—sometimes 2 wires to a single tab under the floor slab. During a maintenance operation, part of the ceiling was observed to drop. Eventually, an engineering team took up the problem. The failure occurred during an inspection.

A simple naive explanation for this failure is that the ceiling support was inadequate. A stronger support system should have been provided.

In reality there was a flow of failures:

1. An inappropriate suspension system was provided for a heavy plaster ceiling which was required to support maintenance crews on the ceiling.

2. Responsibility for the design fell into the cracks between organizations.

3. Maintenance people did not report important incidents.

4. The organizational response to a significant partial failure was confused.

5. The engineering response did not recognize the seriousness and potential consequences of the failure.

Such a failure can be called a *system* failure. Consideration of such failures leads us to the broad question of quality assurance.

When we begin to consider the complete failure system, we introduce many new variables: *people* (How are they motivated? What training should be provided?); *organization* (Who are the people involved? How do we delegate responsibilities? How do we ensure communication?); and *risk* (How should it be assigned?). We begin to question our classical concepts of blame. The common situation where, in the face of construction problems all those involved assume defensive positions, has been widely questioned by, for example, a committee of the Corps of Engineers. The concept of the "lowest bidder wins" can lead to appalling results.

When the concept of safety is broadened, many allies are found. Industrial psychologists in management circles, for example, can explain a great deal about motivation.

Not long ago Charles Perrow, a sociologist at Yale University, published a book called *Normal Accidents* (1984), which arose out of his involvement with the investigation of nuclear accidents. Perrow's thesis is that one must assume that "accidents will happen" and that certain systems are inherently prone to catastrophe. Error proneness is measured by two system properties. The first critical property concerns system interactions. Do elements interact in a simple linear (in other words, predictable) way or are there many complex interactions that are unplanned and not considered explicitly? Are there many secondary effects?

A second critical question concerns the coupling within the system. In tightly coupled systems, a change in one element directly affects another (like

Figure 1 Quality assurance in construction (a preliminary research agenda)

1. Project formulation

   the role of players
   utilization plans
   design criteria
   communications

2. Quality Assurance in Design

   trends in codes (LSD)
   choice of systems/error proneness
   policies regarding checking
   policies regarding independent review
   division of responsibilities
   documentation

3. Quality Assurance in Construction

   policies regarding material testing
   inspection strategies
   how to inspect/key elements
   communications, feedback

4. Human Errors

   concentration, haste effects
   motivation, morale
   incentives
   auto-control, ethics
   education

5. Miscellaneous

   alternative organization
   - USA vs. Europe
   - private vs. government
   - military vs. civil

   case studies, forensic engineering
   legal constraints

the position of the steering wheel in a car and the direction of the front wheels). In a loosely coupled system, fuzzy relationships exist between responses.

According to Perrow, a system is error prone if it is tightly coupled and has many complex interactions. His analysis was prophetic with respect to chemical plants, which he correctly diagnosed as highly prone to catastrophe even before the 1985 Bhopal, India toxic gas leakage accident, which killed thousands of people and injured thousands more.

His analysis also suggests why structures generally succeed in spite of naive analysis and many errors. There is a very loose coupling between design calculations and real building response Many complex effects are not considered but, in general, secondary effects, such as the contribution of partitions to stiffness, work in our favor.

This is where the field of structural safety analysis seems to be headed at the moment. The questions asked are being broadened to include all the physical, organizational, legal, and practical considerations that affect the quality of a structure during its lifetime. As a preliminary agenda for related studies in the field of tall buildings, one possible list of areas for consideration is shown in Fig. 1.

An attempt to develop a useful approach to quality assurance in tall building design and construction will involve many new questions. At present for example, the art of design checking is very poorly understood. The selective advantages and disadvantages of alternative project organizations have never been well established.

It is evident that the results of a broad approach to quality assurance can only lead to guidelines for action and perhaps a series of case studies that can be used for educational purposes. Such qualitative analysis is used routinely in medical, legal, and management education and should be an essential feature in engineering education as well.

REFERENCES/BIBLIOGRAPHY

Council on Tall Buildings and Urban Habitat, 1980
CRITERIA AND LOADING, Vol. CL, Monograph on the Planning and Design of Tall Buildings, ASCE, New York.

Ellingwood, B., Galambos, T. V., MacGregor, J. G. and Cornell, C. A., 1980
DEVELOPMENT OF A PROBABILITY BASED LOAD CRITERION FOR AMERICAN NATIONAL STANDARD A58, NBS Special Publication 57, June.

Freudenthal, A., 1947
THE SAFETY OF STRUCTURES, Trans., ASCE, Vol. 112.

Galambos, T. V. and Ravindra, M. K., 1978
LOAD AND RESISTANCE FACTOR DESIGN, Journal of the Structural Division, ASCE, No. ST9, September.

Joint Committee on Tall Buildings, 1973
PLANNING AND DESIGN OF TALL BUILDINGS, (Proceedings of ASCE-IABSE International Conference held at Lehigh University, August, 1972), ASCE, New York (5 volumes).

Melchers, R. E. and Harrington, M. V., 1984
HUMAN ERROR IN STRUCTURAL RELIABILITY—I, INVESTIGATION OF TYPICAL DESIGN TASKS, Report No. 2, Civil Engineering, Monash University.

Port Authority of New York and New Jersey, 1984
JOURNAL SQUARE TRANSPORTATION CENTER CONCOURSE CEILING COLLAPSE, Special Investigation Report, June.

Perrow, C., 1984
NORMAL ACCIDENTS, Basic Books, Inc., New York

Structural Safety and Quality Assurance

Structural Safety: Some Problems of Achievement and Control

G.S.T. Armer

The continuing development of codes and regulations for structural design in the United Kingdom and elsewhere has highlighted several areas of design where the knowledge and understanding of the process necessary to develop these documents constructively could be improved. One such area is structural safety. This paper contains a simple analysis of the concept of safety within the context of building construction and also a discussion of the problems associated with its use as an objective in the building design and control system.

OBJECTIVES OF STRUCTURAL SAFETY CONTROL

Broadly speaking, the objective of structural safety is to prevent loss of life and limb and to prevent significant economic costs such as compensation, rebuilding, or some consequential activity. This multiple objective causes difficulties arising from the problem of precedence; in other words, which objective should control the safety level? Undoubtedly there will be situations where limitation of economic losses due to structural failure will not be compatible with the protection of life.

A resolution to this problem may be found by recourse to the agencies that establish the objectives. These agencies can be categorized in the following way: (1) public authority (central government, local authority); (2) financier (mortgager, insurer); and (3) owner (corporate, individual).

Each of these categories represents an independent function that exerts control on safety levels in structures. The fact that government may also own and finance a particular project will not affect this discussion since it must still reflect the three separate roles. Likewise, there may be mixing of responsibilities between the other categories.

First, a public authority may have a universal responsibility for structural safety of all construction. This responsibility is primarily for public health and safety in economically advanced countries. In the United Kingdom this principle is established in several public health Acts of Parliament and the responsibility for its implementation is devolved upon local authorities.

Second, the financier can have two quite differently founded interests in the structural safety of any building. The first is that of the mortgager who will wish to protect the loan of money using the property as security. In this case, any property used for such a purpose must be secure in itself. The second is that of an insurer of the occupants and the fabric of a building.

The relatively low damages awarded for loss of life or injury in British courts compared with those in the United States (as reported in the British press), suggest that for individual lives, the insurers' risk is not great and is unlikely to have any significant effect on design. However, in places where many people congregate, such as tall buildings, the insurer's liability will be much greater and an insurer might be justified in looking at the design and current condition of any public assembly building. Insurers may thus influence the performance of buildings used for this purpose in an indirect way.

Third, the building owner will have many reasons for being concerned with the safety of the occupants and the security of the structure itself, since he will be servicing any mortgage and paying the insurance premiums and will be using the building for business or domestic purposes. It is unlikely that the owner will be able to affect structural safety in a building as reflected in its design, unless he employs his own engineer to undertake that design.

If the government establishes a publicly acceptable risk to life caused by structural failure, it is then questionable whether variations to individual risk from one class of building to another are acceptable. The person as an individual has the right to be equally safe from structural failure in his place of work, in his place of entertainment, or indeed, in any other public place. It is difficult to see any economic consideration that can justify any degrading of this risk for particular cases. Thus a universal requirement for public safety establishes a lower bound for the structural safety of all construction.

The reality of the dichotomy in the concept of structural safety that follows from the foregoing discussion must be reflected in the formulation of building controls and codes. It is essential to use consistent terminology to identify the difference between *safety of people* and the associated characteristic for structural elements and structures, *safety of structures,* which together comprise *structural safety.* There is no general but explicit relationship between human safety and structural security, vide section on probabilities of failure.

ECONOMIC CONSEQUENCES OF STRUCTURAL FAILURE

Each structure is both a single entity and part of a population of similar structures. The failure of a structure as a single entity generates direct replacement costs and consequential costs due to loss of business, provision of alternative accommodation, and so on.

A structure has many characteristics, each of which might entitle it to membership of a population of structures. For example, these populations might comprise the following:

1. All structures owned by a single individual or corporation such as a local authority or public utility.
2. All structures made of a common material.
3. All structures made by a common construction method, for example large panel construction.
4. All structures with a common structural form, for example long span roofs.
5. All structures performing a common function, for example offices.
6. All structures designed to a single code or standard.
7. All structures designed by a single consultant.

Clearly, any particular structure may be an element of several different populations. The size and nature of the populations to which a failed structure belongs is of major economic importance. Consider, for example, the situation where the failure in a particular structure is attributable to a characteristic that relates that structure to a large group of other structures. Provided it is relatively easy to identify the whole group, then the integrity of all structures within the group must be questioned as happened, for example, in relation to the failure of Ronan Point, shown in Fig. 1 (HMSO, 1965). At best, a review of the design might be required and at the worst, demolition of the population, both of which could prove expensive for large populations.

This particular problem illustrates the paradoxical effect of standardization whether through a prescribed design process such as a code of practice, or through a construction method such as prefabrication or any other unifying device.

The merits of standardization are understood to be the easy maintenance of uniform safety levels and optimization of costs. However, the rarely mentioned concomitant demerits are that any deficiencies in the standardized construction or unforeseen changes in the operating environment put the whole standardized population at risk.

The economic significance of a failure and hence the level of safety of a structure required by the agencies listed above will therefore depend on (a) their direct interest in the failure itself, or (b) their direct interest in the population of similar structures.

For example, often the owner-occupier of a single property, be it domestic or business, will not be overly concerned with failures of buildings similar to his own, usually because he will have none of the necessary information with which to generate such concern. However, failure of his own property could be catastrophic for him. Equally, the owner of a large population of different buildings may not be very worried by the loss of one building, if this only

Fig. 1 The failure of a Large Panel System structure, Ronan Point, following a gas explosion 5 stories from the top of the building

represents a small percentage of his capital investment. If, however, it is implicit in the failure of one structure that the integrity of many of his other buildings may be at risk, then a single failure could have considerable economic significance for his investment. Thus the level of safety of the structure for any particular building dictated by economic constraints will depend on its economic significance to its owner, its relationship to other structures whether under the same ownership or not, and upon its financing.

For some structures, requirements other than structural or economical (in the sense used above) may control the strength of a structural element. For example, the size of a column section at the top of a structure may be the same as at the bottom purely to satisfy practical construction requirements, or fire resistance requirements may lead to the use of extra structural material and so on. In these circumstances, the safety of a particular structure or structural element may be controlled by a facet of the design process that superficially has little to do with structural performance.

Summarizing, the political requirement for public health and safety establishes a minimum level of safety for all buildings and any modifications to raise this standard will depend on the economic status of the particular structure, or on nonstructural constraints.

PROBABILITIES OF FAILURE

The modern generation of structural codes and standards attempts to invoke an extended concept of safety using statistical philosophy. Thus the terms *probability and reliability, characteristic values, upper and lower fractiles* and so on, have received a widespread application in Europe in the previously assumed purely deterministic process of structural design. It is therefore necessary to discuss the elements of the previous section in terms of probability.

When considering the relationship between the probability of structural failure and the probability of death or injury due to structural failure, three cases must be examined:

1. *When structural failure is due to the presence of people, as in Ronan Point* (an occupant lighting the match to ignite the gas). In the extreme, it is possible to argue that the two probabilities are the same for this case.

2. *When the environmental change that causes structural failure makes people either enter or leave a building.* For example, fire will make people leave, whereas a thunderstorm would normally make them take shelter. Since in this case a strong correlation exists between the presence of people in a structure and its failure (but not always in the same sense), it is not possible to make a general statement about the joint probability since there is an arbitrary relationship between them.

3. *When no connection exists between the cause of structural failure and the death or injury of the occupants of the structure due to that failure.* In this case there is a subtle but nevertheless arguable proposition that the two probabilities are independent in a statistical sense and therefore that the joint probability is simply their product.

When considering the relationship between the probability of failure of a single structure and the probabilities of failure of the populations to which that structure belongs, again three cases must be considered:

1. *Where the nature of failure in a structure pertains to its uniqueness and there is therefore no risk to any population.* An example is failure due to a lightning strike or failure of the first structure to be made from a new material caused by failure of that material. In this case, there is no population and therefore no associated probability of population failure due to that particular case.
2. *Where the cause of a structural failure physically affects a population of structures,* for example a hurricane affecting a large urban area. In this case, there is a strong argument for identical probabilities of failure for both the individual structure and the population.
3. *Where the cause of a structural failure has a nonphysical connection with other structures.* An example is a particular material used in many buildings. The importance of the connection here depends very much on communication of information, public opinion, political, and economic pressures. In this situation it is difficult to establish any formal relationship between the probabilities of failure of the structure and the corresponding population.

This discussion has demonstrated that there is no unique relationship between the loss of life and limb due to structural failure and the probability of structural failure and also that there is no unique relationship between the probability of failure of a single structure and of a population of similar structures. The previous section established the differing safety goals required for a particular structure by the agencies responsible for it. Consequently, the term *safety level* or *level of probability,* often referred to in relation to the implementation of new codes or standards, is indeterminate.

FAILURE PATHS WITHIN THE STRUCTURE

So far, the safety and probability of structural failure has been discussed from the point of view of the owner, the financier, and the public health controller. The analysis now continues from the designer's viewpoint with discussion of the relationship between structural failure, element failure, and material failure and the multifarious input data assumed in design and what exists in reality.

Assuming for the moment that it is possible to establish the probability of failure of a structural element, it is then possible to establish qualitatively the possible associations between this probability and that for failure of the whole structure. They are:

1. When a single element failure causes the physical collapse of the structure. In this case, the two probabilities are identical.
2. When a single failed element is identical or very similar in type and function to many others in a structure, so that the structure is deemed to have failed in spite of the other elements still being intact, as in Stepney School (Bate, 1975) (Fig. 2).
3. In all other circumstances, the relationship between the probabilities of element and structural failure is either an inequality or it is indeterminate.

Fig. 2 The failure of a precast roof over a swimming pool at Stepney School, London

Moving further into the system, the possible relationships between the probabilities of occurrence of certain values or combinations of values of the basic variables for design and the probability of element failure can be considered. Provided that the probabilities of occurrence of the design values can be established and either their mutual correlations or independence (other than through the element itself) and that an accurate (perfect) model of the mechanical behavior of the element can be established for the particular load case, then it should be possible to deduce a probability of failure for the particular load case under consideration and likewise for other load cases. Inevitably, however, these load cases will generate different levels of probability that cannot, because of the mechanics of the problem, be controlled to a defined value for all load cases or mechanisms of load transfer. Therefore, even with this ideal case there is no such thing as the probability of failure (or alternatively reliability index) of an element; it is by the nature of things, multivalued.

It is recognized in codes and standards that use the partial factor of safety method for structural design that many critical features of design, such as workmanship, detailing, modelling, and so forth, cannot be quantitatively assessed. As a direct consequence, any probabilistic target values for factors corresponding to these features cannot be prescribed or deduced. It follows, therefore, that any attempt to arrive at a rational probability of failure for a given element in a structure cannot succeed.

DESIGN FOR SAFETY AND SECURITY

The foregoing apparently negative assessment of the prospects of relating the clients' requirements for the safety of their structure, government requirements for human safety and the engineers' ability to quantify much of the basic data for structural design must not be construed as being detrimental to the cause of structural safety. The analysis reflects only the world as it is. The hierarchical nature of the design problem involving quasi-indeterminate connections between the levels or stages in the design process allows the independent treatment of safety at every appropriate level. Were this not possible, of course, then the safety of structures would indeed have always been a matter of chance.

A demarcation in the passage from nondeterministic concept to deterministic construction must exist somewhere in the building process. Fortunately, it is possible to identify a practical demarcation where that part of the design process involving random variable input (loads, material strengths, etc.) that can be dealt with using statistical techniques is distinct from the deterministic, explicit design/construction parts.

Safety of People

The protection of life and limb within and around structures comes within the province of the building regulations or building codes. It is not easy to define exactly what should be contained in such control documents, since on one hand, the principles they embody have been established already by a legislative act and on the other, they should not comprise complete design instructions, since design is concerned with other matters in addition to safety.

It seems reasonable to accept that a set of functional requirements or performance specifications should form part of the regulations. As with all control systems there is a need for enforcement and also a need for both the designer and the control officer to understand the intent and be able to derive a legitimate interpretation of the requirements. These practical needs have led to the use of schedules, deemed-to-satisfy clauses and the introduction of approved documents (such as codes) containing prescriptive solutions. A difficulty with the use of these devices is that the regulatory systems then usurp the designer's role since they increasingly become de facto prescriptions for design. Furthermore, only parts of these prescriptions are actually needed by the authorities to implement with the law.

Only part of an engineer's objective when preparing the design of a structure is to meet regulatory requirements, therefore, it seems sufficient to limit regulations for structural safety to the following controls: (1) a specification of the loads to be carried; (2) a statement that the foundation should be adequate to carry the loads; and (3) a statement that materials should be durable and appropriate for their application.

Since stability must be ensured following local damage, it may be reasonable to include a control calling for stability that would call up the design strategies outlined elsewhere (Armer, 1983). Arguments that say that only the relevant parts of any code need to be called up by any legislation, do not avoid the fact that code drafting committees put a great deal of effort into creating an integrated methodology for the designer. In consequence, the selection of only a small part of a code in the hope that it will be implemented in a fashion that is independent of the remainder can be nothing but impossible.

Safety and the Evaluation of Basic Variables

Our knowledge of the behavior of structures and our ability to predict the environment in which they will have to exist is limited. It is therefore necessary to accept that any design process will only indicate a minimum level of performance during the life of a structure and that for some or even most environmental states, it may actually be far stronger than is necessary for adequate structural performance. This apparent lack of economy is not only a desirable feature in design for reasons of system stability, it is impossible to avoid, since only some of the several parameters of design can be equally controlled.

Modern research has enabled the designer to develop his methodology on two fronts. The first is concerned with structural behavior. Our state of knowledge in this field is such that it is not unreasonable to consider the performance of a structure in several different conditions. These have been called limit states and can be broadly split into two categories: serviceability and ultimate. The second front is concerned with data on basic variables, such as loads (or actions) and material properties. With these new data has come the possibility of using powerful statistical techniques appropriate for their analyses. A further consequence of these developments has been a reconsideration of the function and magnitude of numerical factors in design. In particular, it has encouraged the explicit partitioning of the safety element of the design process into parts related to the various constituents of that process.

The development of limit state design and partial factor of safety formats has been the inevitable result of much research and experience. However, the pressure to introduce these ideas into practice has brought about some confusion of understanding of what can and cannot be achieved with the new methods. Some difficulties and misconceptions are discussed below.

Safety in Design

The concept of safety in the context of current design is complex. The danger or risks inherent in structural design arise from the designer's limited knowledge of the exact environment in which the structure may have to operate and also of the material properties and performance of the structure itself. In the context of building design and construction, safety is sought by using the best design and construction techniques available, the best materials and construction expertise and the best environmental data. With the design calculations, margins are ensured more specifically by the use of appropriate numerical factors. For the purpose of this discussion, it is accepted that all pratical controls on material properties, workmanship, details, and so forth, are exactly as specified by the designer. Such an assumption is necessary to establish any analytical (or more properly, synthetic) process of design in which formal controls over safety are incorporated.

Consider the limit state or the Load and Resistance Factor Design (LRFD) approach to design. These methods appear to place equal importance on meeting serviceability requirements as on meeting those for ultimate limit states. It is arguable, however, that the principal objective of design is to produce serviceable structures and further, that completed structures complying with serviceability requirements are ipso-facto safe. The question arises therefore; what function does the establishment of ultimate limit states perform vis-a-vis the security of structures? It can be said that the requirement of stability under extreme load states provides a specific security margin over the serviceability condition for the explicitly considered load cases. The evaluation of the margin does not, within the conventional meaning of such a factor, result in a partial safety factor. Indeed, this part of the safety control mechanism is more akin to design requirements in that it affects primarily the nature of structural behavior.

The explicit use of numerical factors in the design process is to circumvent the differences between the design model and its assumptions and the real structure and its environment. Factors in this class reflect ignorance of the real world. Consider the ideal "perfectly safe" building. Such a building would be one for which the material and structural properties were known exactly and could be maintained throughout its life and also for which environmental, loading, and support conditions were known and controlled within the prescribed levels. No factors would be required in the design of this building.

Traditionally, designers have incorporated factors into their analyses by assuming pessimistic values of basic variables, by the use of conservative analytical models and quasi-global factors at appropriate stages of a design. In such an approach to design, it is only possible to decide qualitatively whether the method is successful or not by a study of the nature and rate of failure in the population of buildings constructed according to its principles. There is no practical way of evaluating the actual factor for any given load case.

The introduction of a partial safety factor system into the design process requires the possession of a significant body of data related to the basic variables to be considered in a design. This data should allow the assessment of ignorance of each variable independently. Each factor can only reflect the state of knowledge of one specified parameter. The many other factors known to affect design, such as modeling errors, cannot be evaluated individually; allowance for them has to be made by some lumped or global design factor. This class of factor differs from those associated directly with particular variables in that determination of magnitude involves evaluation by some form of calibration in relation to existing design practice. Thus the introduction of partial factors of safety requires two following conditions to be satisfied: a large body of well-chosen data related to each basic variable it is required to factor, and clear and unequivocal rules to quantify individual partial factors of safety.

For pragmatic reasons, where data for a particular variable is not available, realistic values of basic variables must be used whether chosen intuitively or by some analytical process.

The partial factor of safety systems incorporated into some modern codes or standards do not comply with the second condition. The use, for example, of pessimistic load values defeats the object of a partial factor of safety approach.

The justification for changing the format of a code must be that it becomes more self consistent and conforms more properly with the design process and that new data and methods can be introduced in the future without requiring modification of the factors; also, ideally it should be easier to use.

CHOICE OF BASIC VARIABLES

General

The choice of values for basic variables to be used in a combined limit state/partial factor of safety design method is conceptually quite different from that appropriate for a permissible stress or load factor method. Great care must be exercised to ensure that pessimistic values ("to be on the safe side") are rejected, since the employment of such values defeats the object of using partial factors associated with the variables. The values chosen must depend on their function in the design process and consequently on the methods used to compile the data base. Critical assessments should be made of each variable data bank in order to evaluate its potential role as a guide to the establishment of design standards. Consider, for example, snow loading in the United Kingdom. Data in this field is limited, but it is probable the critical extreme values associated with the heaviest snow falls have not been measured because the manual techniques of depth and density assessment require the on-site attendance of a technician. This, more often than not, is prevented by the snow itself.

A study of the potential causes of extreme values of basic variables show that very often they are generated by mechanisms that are quite different from those that generate normal distributions. Whether or not these extreme values are important in design terms must depend upon the particular variable under consideration. However, two important points follow from this discussion: (1) Measurement techniques designed to gather data in the normal range will often not be appropriate to record extreme values of basic variables that may be generated by quite different mechanisms; (2) Where extreme values are generated by abnormal mechanisms, the probability of their occurrence cannot be deduced by extrapolation from distributions representing normal variations.

It is necessary to choose the "best" figure to represent the variable, in other words, one that does not implicitly include a safety margin, and also, in the case of loads, to choose a figure from the appropriate distribution.

Material Properties

Data on material properties are collected from a variety of sources and over considerable periods of time. These data represent the variability that has existed in the material properties manufactured and supplied by the industry.

For use to be made of this information, it must be accepted first, that material specified for a new construction would be produced in the same general environment and consequently be legitimately included in the same data distribution, and second, that the designer makes a statement along these lines:

> a value has been chosen for material strength that, it is possible, might occur with current practice and that only a small percentage has been weaker than this. Therefore should this building be manufactured from the material specified, it will function properly under the prescribed loads.

It is irrelevant that should the actual material used be nearer to the average strength met in practice, the completed structure may be stronger than the designer aimed for. A difference between assumed and actual strength so generated cannot be treated as a safety factor for design purposes since it is unknown at the time of design and, until the construction is completed, indeterminate. This strength difference has, of course, great importance in the reassessment of the performance of existing structures and is the source of many problems of rationalizing the apparent high strength of structures as deduced from load tests, which cannot be simply correlated with the lower values generated by the substitution of low measured material strength in current design methods.

Loads

The choice of loading values is constrained by an extra consideration not included in the general analysis given above. This constraint arises because each principal limit state can be defined basically by a loading condition. The choice of any particular load value must reflect its function in a limit state design. It was pointed out above that the main objective of design is to satisfy serviceability conditions, meaning that the normal loading states likely to pertain at some time during the life of the structure should be carried satisfactorily. Now, available loading data relates almost exclusively to normal loads and has been collected from a wide range of sites. It is therefore proper for the designer to choose an upper fractile and accompany his design with the implicit statement that:

> Loading data collected from many similar structures have shown that this high load value is possible. This structure has been designed to remain serviceable should it be subjected to this load.

In practice, many structures will never be subjected to their design loads, but as with the material values, the indeterminate difference between peak actual and design values cannot be used as a safety factor at the design stage.

Very careful inspection of the nature of loading data must be undertaken before it is assumed that simple extrapolation of data gathered in serviceability conditions is appropriate to ultimate limit state conditions. Extreme floor loads may, for example, be generated by fires, floods, office parties and so on, and it is proper to choose ultimate load values from the nonnormal distribution so generated.

A further problem to be resolved before choosing a design value from the data bank of a variable load pertains to the relationship between a building life as reflected by that data bank and the prospective life of the building being designed. Considering first the floor loading data available, such data have been collected over a relatively short period (a few years) when compared with the life of a permanent building. It has, however, been collected simultaneously from many similar buildings, and the argument is propounded that the variation in loads indicated by the spatial distribution of measurements is very closely analogous to the temporal variation that might occur during the life of a single building. Such an assumption is insupportable except by resort to heuristics. Fortunately, it is not necessary to follow such a difficult path. Consider the question: What significance has a standardized design load for a class of floors in the performance of a particular floor? Examine the following answers to this question:

If the design value is conservative or a high characteristic value and is used in a permissible stress design method or a load factor design method, then it can only be said to indicate a conservative lower bound for the performance of a floor.

If this design value is used in a properly constituted partial factor of safety design method, its effect is indeterminate.

These somewhat unsatisfactory responses can be resolved. The distribution of values deduced from the data bank reflect the nature of loads on a whole population of floors and that any specific value chosen to characterize this distribution and subsequently to act as a design value will serve to unify the performance of any new population of floors. Put simply, standard design values for variables control the performance of the population of structures in which they are used.

The paucity of information relating to the temporal variation in floor loads during the life of a building means that there is no logical basis to justify any variation in design load to suit an expected building life. It seems however reasonable to accept that to all intents and design purposes, floor loads are invariant with time for most structures.

Turning second to wind loading on structures, much effort has been directed towards the collection of both temporal and spatial (geographical) variations in wind speed data and to the conversion of this data into equivalent pressures on buildings. Because of the nature of this load, it has proved convenient to express characteristic values for the wind data in relation to "return periods" and to discuss 50- or 100-year winds when establishing design values for this variable. In the BRE Digest on Wind Loading (HMSO,

1984) by implication and by assumption in the BS Code for the structural design of farm buildings (British Standards Institution, 1978–1981), these return periods have been related to an expected building life or, more explicitly, to design life. No simple connection exists between the two concepts even though they are expressed in an identical form (Moore, 1982). Consider the design of a one-off special building to be constructed in a known locality with an expected life of say 50 years; prudence might suggest a 95% characteristics value for the design wind load, which in turn would imply perhaps a 150–200 year return period.

The design of more run-of-the-mill construction leads to a different problem of data interpretation. As before, the choice of a standard design value is directly related to the performance of the population of structures designed using that value. Therefore, in the case of data such as that for the wind, the period to be considered is that for the population as a whole, for example, while buildings within that population are extant. For example, if a particular form of short life structure (say two years) such as a specialized farm building is required for a particular production system, which itself might be expected to have an economic life of say 15 years, then the characteristic value should be chosen in relation to the 15 years. If this were a 95% value, then the resultant return period would be very much greater than the two-year life expected for a particular building.

DURABILITY

A special problem occurs when the strength of a structure or structural element is reduced by some form of degradation. In this situation the previously-defined serviceability loads become de facto ultimate loads if the reduction in strength is sufficient to cause failure. The options open to a designer faced with the problem of durability are quite limited. In those cases where a particular element is vulnerable, this element can be designed for a given life from a material with a known rate of deterioration. Such an approach obviously requires close supervision of the structure during use to ensure replacement of the element at the end of its designed life. Another option is to assume, say, a blanket 30% reduction in element/joint strength and to complete the design on the basis of the reduced strength. This latter solution could, of course, prove costly. It is arguable that durability problems could be avoided by good detailing, the use of stable materials, and the like.

CONCLUSIONS

This survey of structural safety and of some of the devices available to the designer engineer to ensure the security of his structure and the safety of those who are in or around it has produced a number of conclusions.

1. The standard of construction required by a government to satisfy its responsibility to ensure the safety of the individual establishes a minimum that should be inviolate.

2. Raising the minimum standard for economic reasons should be based on the particular structure under consideration with special attention being paid to the size and nature of the population of existing structures it will join.

3. It is not possible to produce a logical argument connecting the statistical interpretation of the environment in which a structure will have to operate during its lifetime and the elements and materials of which it is made, with the reliability of performance of that structure.

4. The principal objective of the limit state design is to produce serviceable structures.

5. Partial factors of safety can only be introduced validly into a design process if the following conditions are satisfied: A large body of well chosen data related to the appropriate basic variable is available; and clear and unequivocal rules exist to quantify the factors.

6. In general, extreme values of loads should be based on appropriate surveys since they cannot necessarily be deduced by simple extrapolation from existing knowledge of normal or serviceability loads.

7. The design process aims only to ensure that a lower bound for structural performance is achieved; it is not a model for the actual behavior of a real structure.

8. Data banks for design variables should contain information related to temporal dependence if an expected building life is to be established.

9. Standard values for basic variables specifically control the performance of the population of structures for which they are used and only indirectly affect the performance of any particular building.

These conclusions apply to all structures, but they naturally play a particularly important role in designing tall buildings.

ACKNOWLEDGEMENTS

The ideas propounded in this paper have been developed with the aid of several of the author's colleagues at the Building Research Station, in particular, Mr. J. R. Mayne has contributed significantly. The paper is published by permission of the Director of the Building Research Establishment.

REFERENCES/BIBLIOGRAPHY

Armer, G. S. T., 1983
THE STABILITY OF STRUCTURES, Building Research and Practice, July/August, pp. 216-221.

Bate, S. C. C., 1975
REPORT ON THE FAILURE OF ROOF BEAMS AT SIR JOHN CASS'S FOUNDATION AND RED COAT CHURCH OF ENGLAND SECONDARY SCHOOL STEPNEY, Building Research Establishment Current Paper CP58/74, June.

British Standards Institution, 1978-1981
CODE OF PRACTICE FOR THE DESIGN OF BUILDINGS AND STRUCTURES FOR AGRICULTURE, British Standards Institution, BS5502, London.

HMSO, 1965
COLLAPSE OF FLATS AT RONAN POINT, CANNING TOWN, Report of the Inquiry, HMSO, London.

HMSO, 1984
THE ASSESSMENT OF WIND LOADS, BRE Digest 119, Garston, UK.

Moore, J. F. A., 1982
DISCUSSION ON PAPER BY ROSE AND BURBAGE, The Structural Engineer, V60A, No. 8, August.

Motion in Tall Buildings

Andy W. Irwin

For many years engineers, architects, and planners have struggled with structural, esthetic, amenity, social, and service problems concerning tall buildings. Considerable advances have been made in understanding and solving whole areas of difficulty but often these solutions reveal latent problems that were previously masked by other factors or that are introduced by the changes in practice. The constant endeavor by designers and researchers to provide economic solutions to problems involving tall building structures and long span floors has led to the introduction of a selection of slender and lightweight structural forms, which often incorporate newly introduced high strength materials. These structures, although not normally prone to any excessive deflections, stresses, fatigue or other forms of damage, are generally more dynamically responsive than their predecessors such that perceptible motion of overall buildings can be caused by wind storms and sometimes by everyday wind forces. Perceptible vibration of individual floor spans can also be a regular occurrence and may even be continuous in some cases.

The increasing awareness of the problems of human perception of motion in structures, and in tall buildings in particular, has aroused interest in the subject in an attempt to determine perception thresholds of vibration and magnitudes below which motion in tall buildings is found to be satisfactory. Attention has also been drawn to aspects of structural design other than those of deflection limits, crack-width criteria, fatigue, ultimate strength and stiffness considerations, by the adoption of serviceability criteria in limit state design.

Vibration in fixed structures may be predominantly in one direction but more often motion occurs simultaneously in several axes, and rotational components of vibration may be present. Many other factors may influence the response of the occupants of a building to its overall performance. For example, noise intrusion and mechanical vibration of humans often act together and, although at upper levels damage to hearing from exposure to noise and physical injury from the action of mechanical vibration may be treated separately, each contribute inextricably to the subjective response of people when both noise and vibration are present.

In this paper, a number of factors that may influence the response of human beings to motion in tall buildings are discussed. Methods are put forward for the prediction or assessment of the probable response of occupants of tall buildings to single and multiaxis linear and rotational vibration at various magnitudes, and resulting from several types of occurrence, both in the presence and absence of associated noise.

INFLUENCES ON HUMAN PERCEPTION AND RESPONSE TO VIBRATION

A whole science has developed around the subject of human perception of vibration. Many theories have been put forward as to which body organs or senses are sensitive to or are disturbed by motion (see References). Experiments have shown that identical symptoms can be induced in humans by stimulation of the body's central nervous system by any of a large variety of effects. For example, motion sickness can be induced by purely visual effects (Steele, 1961) as shown by research in recent years and as exploited in the "Gay '90s" amusement park attraction last century (Gibson, 1984). The exact body organs or receptors that initiate perception of motion are of considerable interest to medical researchers and the definitive reasons for vibratory effects on performance or disturbance of humans are a matter of curiosity to psychologists. However, for practical purposes, and considering the present state of medical and psychological research on the subject, it is sufficient to acknowledge that perception of and effect on mental performance and the general response of humans to lower magnitudes of vibration depends largely upon the degree of stimulation of the body's central nervous system by whatever medium. In other situations involving delicate work or the use of sensitive instruments, the criteria for upper magnitudes of vibration that can be accepted may simply be interference with the performance of manual dexterity tasks or with the operation of equipment.

Human response to vibration of buildings is influenced by many factors (Irwin, 1981b; 1983a; 1984a; Landstrom et al., 1983). In a rural setting vibration resulting from the occassional passage of agricultural vehicles, at barely the perception threshold for sensitive humans, can raise the same level of complaint as magnitudes ten times higher in houses adjacent to urban highways or at 100 times higher for short bursts of impulsive vibration from temporary construction works or quarry operations where warning of impending events is given (British Standard, 1984; Irwin, 1978 and 1983a; ISO, 1986). In these situations associated noise is also a major influence. An extreme example involves "pogoing" (crowd en masse coordinated jumping in time to the beat of rock music) at rock concerts where vertical floor accelerations up to 8.2 m/sec^2 (27 ft/sec^2) at 2.5 Hz, or almost 1000 times average perception magnitudes at that frequency, has been recorded but was seemingly unnoticed by the frenzied fans (Irwin, 1981b and 1983a). While music often masks vibration cues, noises associated with structural motion or particular force actions can considerably lower perception thresholds of vibration as can comments by fellow occupants of a building. Subsonic sound can induce the feeling of motion in humans in the total absence of mechanical vibration.

The activity engaged in greatly influences human response to vibration (Irwin, 1979; 1981b; 1983a; Landstrom et al., 1983). This activity may range from work with electron microscopes through delicate surgical operations, skilled assembly work or decision making, routine office or factory work to the operation of power presses, task performance in flexible towers, relaxation at home or even when asleep. During sleep the influences of minor vibration intrusion and infrasound affect the quality of sleep experienced. It is generally the case that the magnitudes of vibration that people will accept are lower at night than in the day time. (British Standard, 1984; ISO, 1986).

Fear for the integrity or safety of a structure can result in adverse comments from occupants at a tenth of the normal vibration magnitude that would be likely to initiate complaints. On the other hand, knowledge of the force actions and confidence in a structure can considerably raise acceptable magnitudes of vibration as can be the case where a structure is used solely by trained personnel, as long as the vibration does not interfere with work practices or cause risk of injury.

Human response to motion in tall buildings depends on the frequencies, directions, and the modes of linear and rotational vibration present as well as the magnitudes of vibration, the duration of the events and the memory of past events, such as that of a tall building in a wind storm. The greater the period between events, the higher the vibration magnitudes before complaints begin. This argument is generally true for moderate vibration magnitudes but severe impulses such as those from a blast or earthquake, where life may be endangered, are generally found to be totally unacceptable even if there is only a single isolated occurrence.

In high-rise buildings, in contrast to some forms of traditional flexible low-rise dwellings, it would appear that occupants do not readily come to accept building motion magnitudes that cause them alarm when first experienced. Prior warning of a storm or other occurrence likely to cause motion of a tall building can raise the threshold of alarm or disturbance in the occupants.

Perception thresholds of vibration are related to the direction the vibration enters the human body. In apartment and hotel buildings a person may walk, sit, lie, or stand at or near the same location at different times of day or night and therefore assessments of satisfactory magnitudes often have to be made by considering the most stringent data for any direction. In the standing position the legs tend to flex to absorb horizontal and yaw vibratory motion and raise the comfort boundary.

Visual cues can be a major influence in the human perception of motion. In yaw motion, objects viewed at a distance may appear to move slightly and in any mode of horizontal motion relative movements of adjacent structures may be observed. False cues can also result from wind forces causing flexing of window glass and from varying light intensity patterns.

Numerous studies indicate that human response to vibration is also influenced by other environmental factors including temperature, humidity, and the purity of the air.

Field investigators have produced data on human perception of and comfort boundaries for average populations containing both sexes, most age groups, a range of backgrounds and varying states of health. In contrast to this, controlled laboratory tests have usually involved only fit young males although several investigators have conducted test series with nominally average populations (Irwin, 1981a; Irwin and Goto, 1984). Also in the laboratory situation, even in the most elaborate systems with all refinements included, relative inertial effects between body components and visual lags in simulated surroundings are different from the real world (Frank et al., 1983). Therefore the laboratory data are very useful as a check on field data but their direct application without any field correlation may be limited.

It is obvious, from a consideration of the above influences, that the subjective response of individuals to motion in tall buildings is very variable and cannot be explicitly defined. In fact even to attempt to encompass all groups within a population would require a multiplicity of reference data sets. Therefore in the later parts of this paper consideration is given only to average populations and to trained personnel in some situations.

CHOICE OF PARAMETERS

Amplitude, velocity, acceleration, jerk, pal Zeller, pal DIN and other parameters have all been put forward as the most appropriate for the assessment of human perception, comfort, annoyance, or tolerability of motion. Shocks are often assessed in terms of peak velocity for reasons concerning measurement and because constant velocity curves generally apply above about 8 Hz for structural damage and human comfort criteria for this type of event. If measurements of very low frequency vibration cannot be accurately made and recorded using available instrumentation then, in some cases and especially if secondary impulses are present at the changes in direction of the primary motion, jerk can be used as a best compromise parameter. Where visual cues regenerated by relative movement of components or structures then displacement can be a useful parameter in human response assessments. In fact some investigators have concluded that different parameters should be used for different frequency ranges and for different body axes.

Except in cases of shock motion, if measurements are taken of acceleration and frequency or if analysis of records or assessments of human response are made on this basis, then any of the other parameters mentioned above can be obtained by integration, differentiation, or by logarithmic ratios. Therefore in this paper acceleration is used to quantify perception and satisfactory magnitudes of vibration for human beings since it would appear that acceleration is the best parameter to use, it is readily measurable by available equipment, and if acceleration is adopted then compatibility is achieved with other guides. Comparisons also can be easily made with other published data.

When dealing with noise control and potential for hearing damage *A* weighted noise levels are normally employed or in some particular applications *B, C,* or *D* weightings may be appropriate. However, at low frequencies these weightings reduce significantly the noise level reading in comparison to the straight linear decibel value. Research has shown that in assessments of noise nuisance and subjective response of humans, the low frequency sound waves have a considerable influence on the findings. Therefore regarding the interaction of sound and vibration on the subjective response of humans, it is most useful to work in terms of linear decibels for the above reasons and since these can be meaningfully expressed to a frequency base.

EVENT CATEGORIES, ASSESSMENT CRITERIA, AND PROCEDURES

Comfort criteria dictate that clean air is essential in high rise buildings. Irritability is also reduced if the temperature is maintained at between about 21°C and 24°C (70°F and 75°F) and if the humidity is kept in the range 30% and 50% wherever possible. Limitation in the use of psychedelic colors can also act to damp complaint levels concerning factors such as noise and vibration. Several other factors that influence human response to motion and noise in tall buildings have been discussed above and in previous papers.

In the past, assessments of human response to vibration were generally conducted on the basis of single frequency linear vibration and occassionally on a multiaxis basis. Noise, rotational motion, multifrequency vibration, and visual cues were not included. Here a simple procedure for the assessment of the subjective response of humans to single and combined forms of translational and rotational motion in tall buildings, in the presence of noise at a full range of frequencies or when noise is not significant, is considered.

In this method, the response of the occupants of tall buildings and other tall structures to an environment in which vibration or combined noise and vibration may be present is divided into the following classifications:

a. sub-threshold influences;
b. basic threshold effects;
c. intrusion, alarm, and fear, which may be associated with minor or major adverse comment;
d. interference with activities;
e. possibility of injury or health risk.

In classification *a* the criterion is interference with work with sensitive instruments such that humans come to realize that vibratory motion is present, although it is not directly perceived by any normal human sense, body function, or component.

The criterion for the lower boundary of classification *e* is the requirement of providing restraint harness or hand holds to reduce the risk of injury to personnel caused by low frequency mechanical vibration or the necessity to provide ear protection to avoid hearing damage.

Probable human response to noise and vibration in classifications *b, c* and *d* is assessed in terms of the event category as: steady environment; impulsive events; infrequent storm action.

In general the criterion for satisfactory magnitudes of vibration in each of these categories is based on a minimum adverse comment level of 2% of the population involved. The criterion for infrequently induced low frequency vibration of tall buildings caused by wind storms is the alarm experienced by the occupants of the structures. The perception of motion may be through proprioceptive cues, the vestibular organs, or by visual cues. The influence of noise cues or continuous noise intrusion is included. Wind storm induced motion is assessed on the basis of the average acceleration caused by the peak 10 minutes of the worst wind storm in the return period under consideration (1 or 5 years or more). If infrequent events other than wind storms are considered and the duration of these events is less than 10 minutes, then those occurrences that are vividly remembered would most likely be the product of seismic action or possibly blast loading. During the peaks of wind storms, high crest factors of acceleration much in excess of the suggested satisfactory magnitudes will occur for short periods, but these higher values, briefly experienced, are not considered to make any great contribution to the memory of the storm except if the motion borders on classification *e* above. Short periods of higher acceleration that do occur during the worst 10 consecutive minutes of a storm are accounted for in the average acceleration value for the storm peak.

For structures manned by trained personnel, whose duties involve decision making or the execution of nonroutine or skilled manual operations, the criteria for infrequently induced low frequency vibration of such structures are related to the performance of those engaged in the operations.

Basic threshold effects of classification *b* for average and sensitive people apply to buildings in which precision work is carried out, other than that governed by instrumentation functions as stated for classification *a*, and in some cases to buildings caused to move by everyday wind forces or actions within the structure. Magnitudes of motion or combined noise and vibration stipulated later in the text for categories *c* and *d* should not be selected as suitable for tall buildings unless it is considered that increased complaint levels are not to be a governing factor in a design appraisal for the overall movements of the structure. They may apply to vibration of floors and other local areas within the building. Guidance in classification *e* is confined to the low frequency range as is the performance related data.

No guidance is given in this paper on short term health risk from vibration and it is emphasized that the provisions of a satisfactory environment are with respect to human response and are not concerned with structural damage, which may occur at lower magnitudes in some instances.

ASSESSMENT OF HUMAN RESPONSE TO VIBRATION AND COMBINED VIBRATION AND NOISE IN TALL BUILDINGS

Various forms of vibration at a range of intensities and combined noise and vibration have been discussed above. Vast quantities of data from field studies and laboratory tests have been gathered and used to form the following guidance for suggested upper magnitudes of whole-body vibration of humans in buildings, and for noise combined with vibration, for minimum adverse comment from those exposed to these influences. As far as is possible the recommendations encompass the various factors discussed earlier in this paper.

A. Steady Environment

Where vibration or vibration combined with noise is approximately continuous (for example from out-of-balance service motors, everyday wind or air conditioning) then in the assessment of the probable human response to this environment the following procedure can be followed:

1. Measure or calculate the frequencies and average magnitudes of the acceleration components, the noise level, and the predominant noise frequency range;
2. Where a_x and a_y are orthogonal horizontal translational acceleration components in m/sec^2 rms units (root mean square, see Fig. 1), sum these vectorially, taking account of phase, to obtain parameter a_h;

3. Evaluate,

$$a_{\alpha h} = a_h + r.a_\alpha \tag{1}$$

where a_α = yaw acceleration in radians/sec^2 rms

$$r = |1/f_\alpha - 0.57/\sqrt{f_\alpha}| + 0.2 - 0.16/f_\alpha \tag{2}$$

f_α = yaw frequency (rotation about a vertical axis) (Irwin, 1981a);

4. Obtain the noise factor N from Fig. 2 at the noise frequency f_N and the appropriate noise level dB_{linear}, and compute (Irwin, 1982 and 1983),

$$a_N = N.S \tag{3}$$

where

$$S = 0.001(1/(3.f_v) + \sqrt{f_v}) \tag{4}$$

$f_v = \Sigma f_i / n_f$ and n_f is the number of frequency components in the numerator, for example f_x, f_y, f_z and f_α and f of the same order; (5)

5. Evaluate

$$a_b = \sqrt{(a_z^2 + a_N^2 + a_{\alpha h}^2)} \tag{6}$$

6. Compare a_b from Eq. 6 with the appropriate a_s value from Fig. 3 at frequency f_v and the curve corresponding to the multiplier chosen

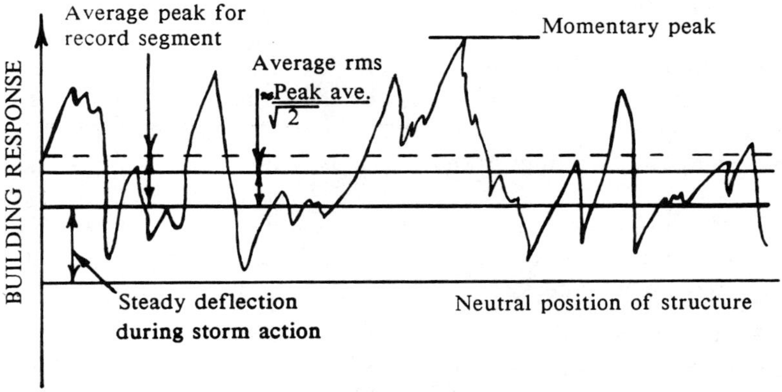

Fig. 1 Comparison of motion descriptors

from column 3 in Table 1. If a_b is less than a_s, the satisfactory magnitude, then probably the great majority of an average population would find the environment to be satisfactory.

B. Impulsive Events

For impulsive events such as impacts from service or internal transport systems and other relatively isolated occurrences, each of duration less than

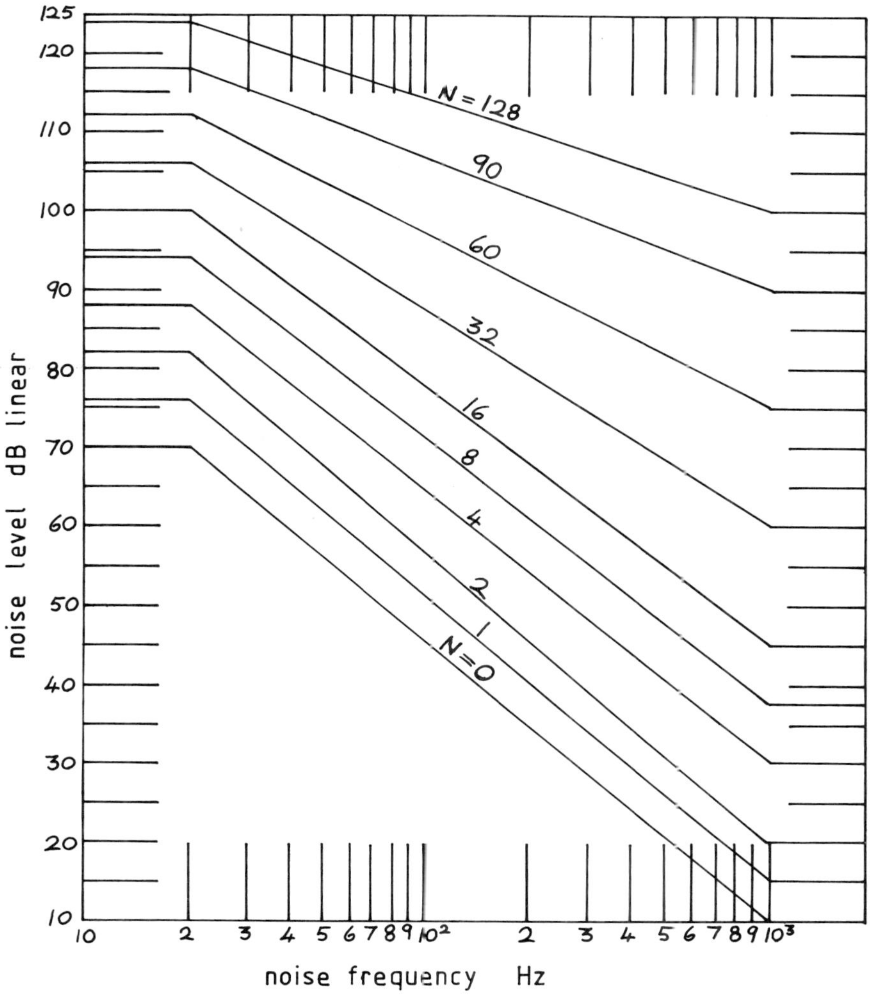

Fig. 2 Noise rating factor curves

60 sec, and where the noise and mechanical vibrations are predominantly sinusoidal in form, then follow the procedure steps 1 to 5 in A. Compare a_b at f_v with the curve value a_s in Fig. 3 when the appropriate multiplier taken from column 4 of Table 1 is unity. If the appropriate multiplier is greater than unity, then compare a_b with a_s where

$$a_s = (\text{base curve value Fig. 3 at frequency } f_v) \times (\text{column 4 factor Table 1}) \times P^{-0.5} \times T^{-d} \tag{7}$$

$d = 1.22$; $T^{-d} = 1$ for T less than 1 sec. T = event duration in sec (time rms acceleration is greater than curve 4 value in Fig. 3 at frequency f_v).

P = number of events in a working day (or 16 hr day for residences)

Note: If a_s is reduced to below that for a steady environment by this process then use a_s for the steady environment factor as in A.

When a_b is less than a_s then only minor adverse comment would normally be generated by such events, especially if prior warning is given.

If an impulsive event produces predominantly nonsinusoidal motion, with or without accompanying sound waves, then proceed as follows:

1. Evaluate V_v, the peak velocity and assess the frequency range for f_v from the form of the decay signal;
2. Obtain N from Fig. 2 as before and evaluate,

$$V_N = \text{N.S}/(f_v \cdot \pi \cdot \sqrt{2}) \tag{8}$$

3. Compute

$$V_b = \sqrt{(V_N^2 + V_v^2)} \tag{9}$$

4. Compare V_b with V_s where

$$V_s = (\text{base curve value Fig. 3 for axis concerned and frequency } f_v) \times (\text{column 4 factor Table 1}) \times P^{-0.5} \times T^{-d}/(f_v \cdot \pi \cdot \sqrt{2}) \tag{10}$$

Table 1 Satisfactory environmental multipliers above base curves

Place	Time	Steady Environment	Impulsive Events
Critical area	Any	1	1
Residential	Day	2 to 4	60 to 90*
	Night	1.4	20
Office	Any	4	128*
Workshop	Any	8	128*

*with prior warning.

C. Infrequent Storm Action

When building motion or combined motion and noise occur during infrequent storm action then the probable response of the population involved can be assessed as follows:

1. Measure or calculate the motion frequencies and the average rms acceleration components for the worst 10 consecutive minutes of the worst wind storm for the region and return period under consideration.

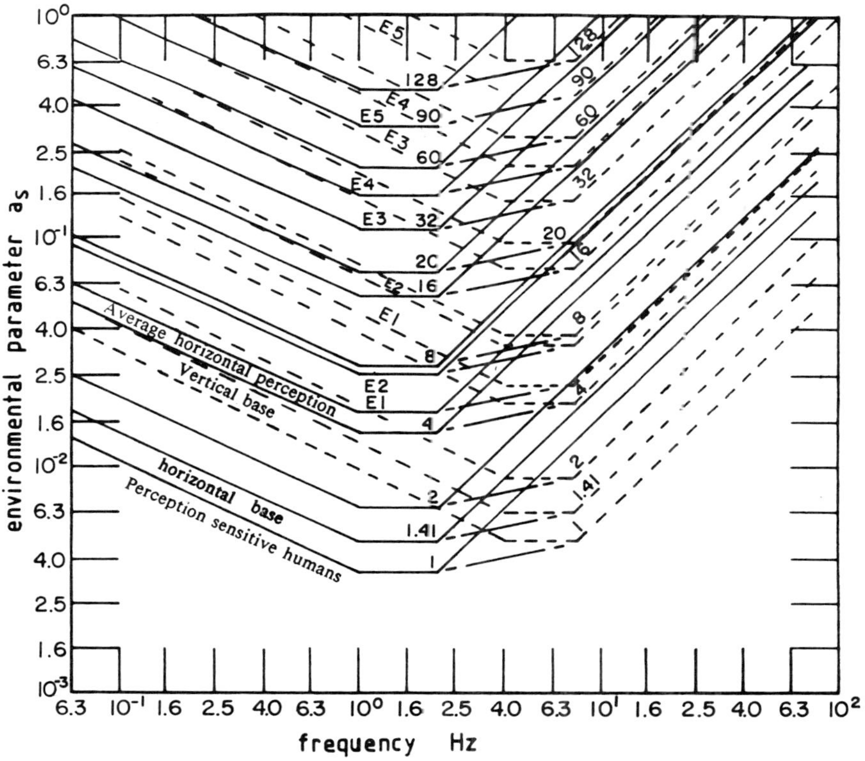

a_S = m/s² rms for linear acceleration

a_S = dimensionless parameter for combined linear & rotational motion or combined noise & vibration

—— horizontal motion predominant

- - - - vertical motion predominant

- —— - combined motion base

Fig. 3 Environment rating curves

2. Evaluate a_h as in A. 2 above, where a_x and a_y are in a narrow band or at a discrete frequency.

3. If yaw motion is present within the frequency band then evaluate $a_{\alpha h}$ as in A.3.

4. For the same part of the storm as used in 1) above, evaluate N from Fig. 2 and a_N as in A.4, and using f_v for the horizontal frequencies of significance as derived from Eq. 5 for f of the same order.

5. Compute

$$a_b = \sqrt{(a_{\alpha h} + a_N^2)} \tag{11}$$

6. Compare a_b from Eq. 11 with the appropriate a_s value from the curve in Fig. 3 at frequency f_v and corresponding to the multiplier chosen from Table 2. The curve chosen to obtain the a_s value depends on the population type (average or trained personnel), activity engaged in, the storm return period (1 or 5 years) and the level of adverse comment considered appropriate for the structure use and the economic restraints involved. The adverse comment levels are for those parts of the structure most affected by the storm-induced motion.

D. Treatment of Multiple Frequencies

When there are several discrete frequencies at which vibratory motion occurs in the frequency band 0.063 Hz to 80 Hz, then it is tentatively suggested here that these can be combined according to the following summation:

$$\Sigma(a_b/a_s) \tag{12}$$

If the sum of the ratios is less than unity, adverse comment is expected to be minimal.

Table 2 Appropriate curves and adverse comment levels for the maximum 10 minutes of the worst storm in a 1 or 5 year return period

Application	Buildings			Structures manned by trained personnel			
Curve	E1	E2		E3	E4		E5
Return Period in Years	1	5	1	1	5	1	——
Percentage Adverse Comment	2	2	12	2	2	12	Hand holds or restraint harness required

TYPICAL CASE STUDY EXAMPLES

The following represent a few of many case studies used to form the above assessment method:

a. Office building found to be satisfactory by the users. $a_z = 0.01$ m/sec^2 rms at $f_z = 4.0$ Hz; noise level of 70 dB_{linear} in the frequency range of 80 Hz. From Fig. 2, $N = 4$ and $a_N = 0.0833$ from Eq. 3. Summation by Eq. 6 yields $a_b = 0.013$. The appropriate factor from Table 1, column 3 is 4 therefore, from curve 4 in Fig. 3 at 4.0 Hz, $a_s = 0.017$ which is greater than a_b and satisfactory.

b. Multistory mill structure with a high complaint level, even if ear protectors are worn. $a_z = 0.58$ m/sec^2 rms at $f_z = 3.8$ Hz; $a_x = 0.0669$ m/sec^2 rms at $f_x = 2.9$ Hz; $a_y = 0.14$ m/sec^2 rms at $f_y = 3.2$ Hz; no rotational vibration; noise level of 105 dB_{linear} at 100 Hz. From Fig. 2, $N = 80$ and $a_N = 0.15$ from Eq. 3. $a_h = 0.155$ m/sec^2 rms and summation by Eq. 6 gives $a_b = 0.618$. Appropriate Table 1 factor is 8 and from combined base curve 8 in Fig. 3 at $f_v = 3.13$ Hz we obtain $a_s = 0.03$ which is much smaller than a_b, therefore the environment is very unsatisfactory.

c. Multistory factory building with a fairly high complaint level from occupants and fear induced in some. $a_x = 0.152$ m/sec^2 rms at $f_x = 3.3$ Hz; noise level of 85 dB_{linear} at 70 Hz range gives $N = 20$ from Fig. 2 and $a_N = 0.03825$ from Eq. 3. Summation by Eq. 6 yields $a_b = 0.157$. From Table 1 the appropriate factor is 8 and for predominantly horizontal vibration Fig. 3 at $f_v = 3.3$ Hz provides a value of $a_s = 0.09$ which is smaller than a_b and the environment is not satisfactory.

d. Tall building with an average population and subject to a five-year return storm. $a_x = 0.020$ m/sec^2 rms average for the worst 10 minutes of the storm peak and $f_x = 1.2$ Hz. $a_y = 0.005$ m/sec^2 rms at $f_y = 0.7$ Hz. $a_\alpha = 0.015$ rad/sec^2 rms at $f_\alpha = 0.8$ Hz. Therefore from Eq. 1 and $2a_{\alpha h} = 0.0206 + 0.0092 = 0.0298$. Noise level of 70 dB_{linear} at 80 Hz band results in $N = 5.5$ from Fig. 2 and $a_n = 0.00725$ from Eq. 3 at $f_v = 0.9$ Hz. Therefore from Eq. 11 $a_b = 0.0307$. The appropriate curve designation from Table 2 is E2 and from this curve in Fig. 3 at $f_v = 0.9$ Hz, $a_s = 0.03$. Very minor adverse comments were made about the building performance, which corresponds with the closeness of a_b and a_s.

e. A very similar building to that of d. and for a five-year wind storm the measured values were: $a_x = 0.017$ m/sec^2 at $f_x = 1.35$ Hz; $a_y = 0.011$ m/sec^2 rms at 0.74 Hz; $a_\alpha = 0.009$ rad/sec^2 rms at 0.64 Hz; noise level of 74 db_{linear} at 315 Hz and 82 dB_{linear} at 60 Hz. $f_v = 0.91$ Hz from Eq. 5; $a_{\alpha h} = 0.0202 + 0.0072 = 0.0274$ from Eq. 1 and 2; from Fig. 2 at 315 Hz, $N = 30$ and from Eq. 3 and 4 $a_N =$

0.0396; from Eq. 11 $a_b = 0.0482$ and, as in case study **d.** above, $a_s = 0.03$, which is less than a_b. In this building there was considerable adverse comment about the performance of the structure during storm winds.

f. A residential building subjected to 7 impulsive vibrations, each of 1.6 sec duration, in a day from adjacent site working. Only minor remarks by the building occupants. $a_x = 0.14$ m/sec^2 rms at 12.5 Hz; noise associated with the events was 106 dB_{linear} at 16 Hz. From Fig. 2, $N = 30$ and $a_N = 0.105$ from Eq. 3; $a_b = 0.175$ from Eq. 6. $a_s = 2.25 \times 10^{-2} \times 90 \times 7^{-0.5} \times 1.6^{-1.22} = 0.431$. a_s is significantly greater than a_b, which corresponds with only minor remarks made by occupants regarding the cumulative events.

g. An office building floor was subject to 11 nonsinusoidal vertical impulses per day from the impact of boxes deposited in an adjoining office. Each impulse was of average duration 0.24 sec with a peak particle velocity of 7.6×10^{-3} m/sec. The decay signal was predominantly at 16 Hz. Noise levels directly associated with the events were, on average, 83 dB_{linear} at about 120 Hz.

From Fig. 2, $N = 22$ and from Eq. 4 and 8, $V_N = 1.224 \times 10^{-3}$. From Eq. 9, $V_b = 7.701 \times 10^{-3}$. $V_s = 1 \times 10^{-2} \times 90 \times 11^{-0.5} \times 1/(\sqrt{2} \times \pi \times 16) = 3.817 \times 10^{-3}$ from Eq. 10. Significant complaints were constantly received regarding these events.

h. New type of trucks passing a row of rural cottages to collect gravel from a small quarry. 4 events per day and each of 6 sec duration. Accelerations $a_z = 0.0368$ m/sec^2 rms at $f_z = 28.6$ Hz; $a_y = 0.0177$ m/sec^2 rms at $f_y = 21.3$ Hz; $a_x = 0.0236$ m/sec^2 at $f_x = 18.2$ Hz; noise level inside cottages 64 dB_{linear} at 63 Hz (ambient noise 27.9 dBA). $N = 1.5$, $S = 0.0048$, $a_N = 0.0072$ and $a_b = 0.0477$.

$a_s = 0.014 \times 60 \times 4^{-0.5} \times 6^{-1.22} = 0.047$ (in other words lower than for continuous vibration of a house to a Table 1 factor of 4 and the 60 factor from Table 1 column 4 used since the setting is rural). Significant complaints were received from the residents about the trucks although the vibration and noise generated in the cottages by tractors passing was somewhat higher and no complaint was made about the tractors in this case.

DISCUSSION

A simple method has been presented for assessing the probable response of occupants of tall buildings to single and multiaxis motion, which can include yaw vibrations and be combined with noise to form a complex environment. A large body of data was reduced to arrive at this method. Several forms of motion and motion generating actions can be considered using the method. Average and specific populations can be assessed and a few case study examples are summarized to illustrate the use of the method.

Many of the factors that influence the response of people to noise and vibration in buildings have been referred to in this paper. Table 3 contains possible limiting criteria for vibration of buildings, other than human response factors, and which are sometimes more stringent than human requirements. Figure 4 provides data for assessing yaw motion of structures and these curves are also included to provide an indication of the influence of visual cues on human perception of motion.

The magnitudes of acceleration at which occupants of buildings begin to complain are related to perception of motion and have no bearing on either short- or long-term injury potential. In fact, very little evidence, if any, exists

Table 3 Amplitude comparison data

Projected minimum micro-chip circuit width	1 Si molecule ($\simeq 10^{-9}$ to 0.5×10^{-10}m)
Minimum present circuit width	2.54×10^{-7}m (10^{-5} in.)
Minimum disturbance for interference with interferometry (½ wavelength of visible light)	1.5×10^{-7} to 3×10^{-7}m
VLSI chip production from silicon-on-saphire wafers (sub-micron technology)	5×10^{-8}m rms at 45 Hz 1.42×10^{-7}m peak to peak 7.07×10^{-8}m amplitude
General microchip technology for greater than 1 micron accuracy	5×10^{-7}m from 0 Hz to 10 Hz (20 micro inches)
Head crash in a disk reader with a 2×10^{-6}m air cushion	2×10^{-6}m amplitude at head (allowed greater overall system amplitudes)
Perceptible to humans at 10 Hz	2.25×10^{-6}m amplitude vertical 6.75×10^{-6}m amplitude horizontal
Maximum continuous vibration to avoid interference with the operation of novibration isolated computers —range for 11 leading manufacturers	1.25×10^{-5}m to 1.5×10^{-3}m amplitude at 10 Hz (higher amplitude impulses can be tolerated)
Interference with sensitive control gear	5×10^{-5}m amplitude at 10 Hz
Switchgear in danger of tripping during seismic action	9×10^{-5}m amplitude at 10 Hz

about injuries that could be directly attributed to whole-body vibration effects other than from loss of balance resulting in a fall. It is interesting to examine Figs. 5 and 6 in this regard, since these represent motions sought after and paid for by the participants. In the case of the particular fairground ride of Fig. 5, the operators are subject to 200 second exposures, approximately 10 times per hour for 8 to 10 hours per day all season.

Care should be taken by those applying the guidance contained in this paper to carry out additional checks for damage criteria, which can be more stringent than human response limitations in some cases.

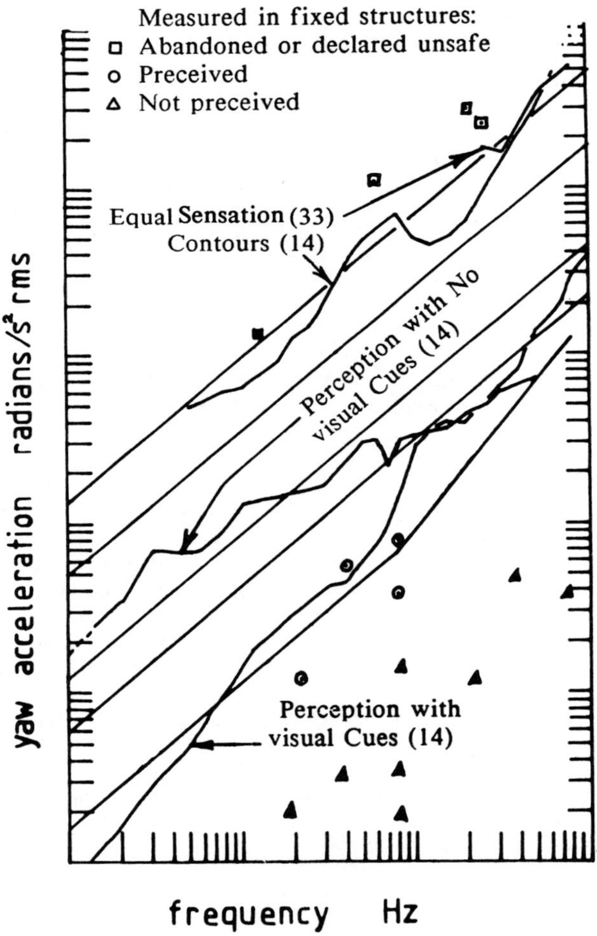

Fig. 4 Human response to yaw vibration

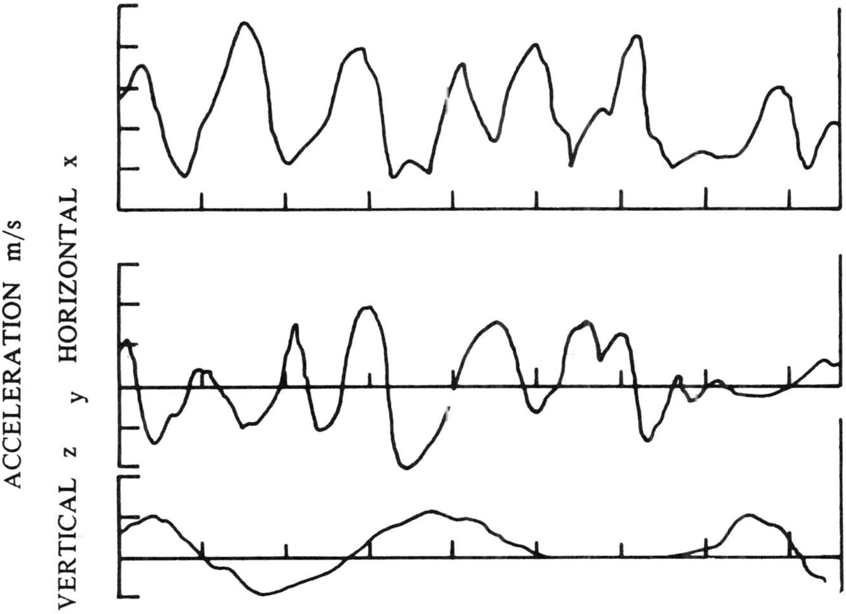

Fig. 5 Measured fairground ride motion

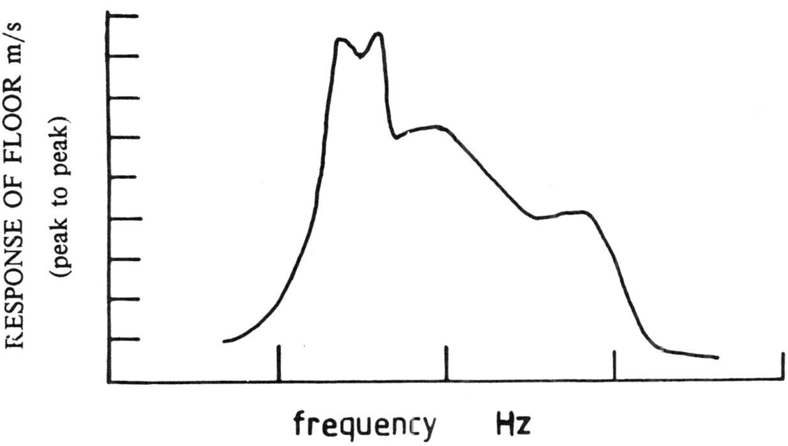

Fig. 6 Floor response to 'pogoing'

CONCLUSIONS

Factors that influence the response of humans to motion in tall buildings have been discussed. A simple method has been presented for the assessment of the probable human response to simple or complex motion in buildings and to the combined environment of motion and noise.

REFERENCES/BIBLIOGRAPHY

Blume, J. A., 1969
MOTION PERCEPTION IN THE LOW FREQUENCY RANGE, Report No. JAB-99-47, J. A. Blume and Associates Research Division, San Francisco.

British Standard, 1984
EVALUATION OF HUMAN EXPOSURE TO VIBRATION IN BUILDINGS (1 Hz to 80 Hz), British Standard BS6472, British Standards Institution, London.

Chang, F. K., 1967
WIND MOVEMENT IN TALL BUILDINGS, Civil Engineering, Vol. 1, No. 8.

Chang, F. K., 1972
PSYCHOPHYSIOLOGICAL ASPECTS OF MAN-STRUCTURE INTERACTION, Proceedings of Symposium on Planning and Design of Tall Buildings, Vol. 1a, Lehigh University, ASCE Publication.

Chen, P. W. and Robertson, L. E., 1972
HUMAN PERCEPTION THRESHOLDS OF HORIZONTAL MOTION, ASCE Journal of Structural Division, August.

Frank, L., Kennedy, R. S., Kellog, R. S. and McCauley, M. E., 1983
SIMULATOR SICKNESS: A REACTION TO A TRANSFORMED PERCEPTUAL WORLD 1 SCOPE OF THE PROBLEM, 2nd Symposium of Aviation Psychology, Ohio State University, Columbus, April.

Gibson, W. B., 1984
WORLD'S MOST TERRIFYING AMUSEMENT PARK RIDE, National Examiner, July 31.

Goto, T., 1981
HUMAN PERCEPTION AND TOLERANCE OF MOTION, Monograph of Council on Tall Buildings and Urban Habitat, Vol. PC, Planning and Environmental Criteria for Tall Buildings.

Goto, T., 1983
STUDIES OF WIND-INDUCED MOTION OF TALL BUILDINGS BASED ON OCCUPANTS' REACTION, Proceedings 6th International Conference on Wind Engineering, Session 5—Wind Loading of Tall Buildings (b), Gold Coast, Australia.

Gierke, H. E. von, 1977
GUIDELINES FOR ENVIRONMENTAL IMPACT STATEMENTS WITH RESPECT TO NOISE, Noise-Con 77 NASA Langley Research Center, Hampton, Virginia.

Hansen, R. J., Reed, J. W. and Vanmarke, E. H., 1973
HUMAN RESPONSE TO WIND-INDUCED MOTION OF BUILDINGS, ASCE Journal of Structural Division, July.

International Standards Organization, 1984
GUIDELINES FOR THE EVALUATION OF THE RESPONSE OF OCCUPANTS OF FIXED STRUCTURES, ESPECIALLY BUILDINGS AND OFF-SHORE STRUCTURES, TO LOW-FREQUENCY HORIZONTAL MOTION (0.063 to 1 Hz), International Standards Organization, ISO6897, Geneva.

International Standards Organization, 1986
EVALUATION OF HUMAN EXPOSURE TO WHOLE-BODY VIBRATION—EVALUATION OF HUMAN EXPOSURE TO CONTINUOUS AND SHOCK INDUCED VIBRATIONS IN BUILDINGS (1-80 Hz), International Standards Organization, ISO DIS2631, Part 2.

Irwin, A. W., 1975
HUMAN REACTIONS TO OSCILLATIONS OF BUILDINGS—ACCEPTABLE LIMITS, Build International, Applied Science Publishers, London.

Irwin, A. W., 1978
HUMAN RESPONSE TO DYNAMIC MOTION OF STRUCTURES, The Structural Engineer, Vol 56A, No. 9, and Vol. 58A, No. 3.

Irwin, A. W., 1979
A DIFFERENT KIND OF WALTZ, UK Conference on Human Response to Vibration, Royal Aircraft Establishment, Farnborough, UK.

Irwin, A. W., 1981a
PERCEPTION, COMFORT AND PERFORMANCE CRITERIA FOR HUMAN BEINGS EXPOSED TO WHOLE BODY PURE YAW VIBRATION AND VIBRATION CONTAINING YAW AND TRANSLATIONAL COMPONENTS, Journal of Sound and Vibration, Vol. 76, No. 4.

Irwin, A. W., 1981b
STUDY AND EVALUATION OF HUMAN RESPONSE TO PURE AND COMBINED FORMS OF LOW FREQUENCY MOTION AT VARIOUS LEVELS, International Workshop on Research Methods in Human Motion and Vibration Studies, New Orleans.

Irwin, A. W., 1982
A METHOD FOR ASSESSMENT OF PROBABLE HUMAN RESPONSE TO COMBINED NOISE AND MECHANICAL VIBRATION, Proceedings Noise and Environment, Bratislava, Czechoslovakia.

Irwin, A. W., 1983a
DIVERSITY OF HUMAN RESPONSE TO VIBRATION ENVIRONMENTS, Proceedings Conference on Human Response to Vibration, NIAE/NCAE, Silsoe, UK.

Irwin, A. W., 1983b
RELATIVE INFLUENCE OF NOISE AND WHOLE BODY VIBRATION ON THE RESPONSE OF HUMANS, Proceedings Inter-Noise 83, Vol. 11, Institute of Acoustics, Edinburgh.

Irwin, A. W., 1984a
DESIGN OF SHEAR WALL BUILDINGS, Construction Industry Research & Information Assoc., Report No. 102, London, UK.

Irwin, A. W., 1984b
INTRODUCTION TO NOISE PHYSICS, Lecture to UK Offshore Operators Association Medical Group, Aberdeen, Scotland.

Irwin, A. W. and Goto, T., 1984
HUMAN PERCEPTION, TASK PERFORMANCE AND SIMULATOR SICKNESS IN SINGLE AND MULTI-AXIS LOW FREQUENCY HORIZONTAL LINEAR AND ROTATIONAL VIBRATION, Proc. UK-HRV 84, Edinburgh, Scotland.

Khan, F. R. and Parmelee, R. A., 1971
SERVICE CRITERIA FOR TALL BUILDINGS FOR WIND LOADING, Proc. 3rd Int. Conf. on Wind Effects on Buildings and Structures, Tokyo, Japan.

Landstrom, U., Lunstrom, R. and Bystrom, M., 1983
EXPOSURE TO INFRASOUND—PERCEPTION AND CHANGES IN WAKEFULNESS, Journal of Low Frequency Noise and Vibration, Vol. 2, No. 1.

Phillips, M. H., Wood, J. H. and Docherty, J., 1984
HORIZONTAL VIBRATION OF HOUSES, Report No. 5-84/3, Central Laboratories, Ministry of Works and Development, Lower Hutt, New Zealand.

Reed, J. W., 1971
WIND-INDUCED MOTION AND HUMAN DISCOMFORT IN TALL BUILDINGS, Research Report No. R71-42, Massachusetts Institute of Technology.

Reed, J. W., Hansen, R. J. and Vanmarke, E. H., 1972
HUMAN RESPONSE TO TALL BUILDING WIND-INDUCED MOTION, Proceedings Symposium on Planning and Design of Tall Buildings, Vol. 11, Lehigh University, ASCE Publication.
Robertson, L. E., 1974
WIND ENGINEERING OF TALL BUILDINGS, Proceedings of Symposium on Tall Buildings—Planning, Design and Construction, Nashville.

Schoenberger, R. W., 1980
PSYCHOPHYSICAL COMPARISON OF VERTICAL AND ANGULAR VIBRATIONS, Aerospace Medical Association Meeting, Anaheim.
Steele, J. E., 1961
MOTION SICKNESS AND SPACIAL PERCEPTION: A THEORETICAL STUDY, Technical Report ASD-TR-61-530, National Technical Information Service, Washington, D. C.

Damping in Tall Buildings

Alan P. Jeary

The response of tall buildings to naturally occurring forces is fundamentally affected by the way in which the energy input is dissipated by the structure. In choosing the mathematical convenience of a viscous damping coefficient, early workers and researchers solved their immediate problems, but bequeathed a legacy of confusion to modern-day researchers and designers who require a more precise understanding of the process involved. This confusion is caused by the fact that (1) no physical mechanism for viscous damping in structures can be identified; and, more importantly, that (2) the damping process is in any case nonlinear.

Concern about the response of structures to earthquakes in California in the 1950s in conjunction with the electronics revolution (then in its infancy) heralded the modern techniques of measuring, with precision, the dynamic characteristics of full-scale buildings. Gradually the techniques spread to a few selected centers around the world, and a data base began to be assembled. However, the intellectual straightjacket of established mathematical methods of handling dynamic characteristics continued to confound the search for a precise definition of damping values for tall buildings. This growing problem was compounded by difficulties with equipment and data handling techniques, which sometimes allowed extremely wide confidence intervals to pass unnoticed.

In the mid-1970s, a group in England set about making measurements of the dynamic characteristics of large structures that took full advantage of the advances in microelectronics technology. The improved control achieved in shaking full-scale structures soon indicated measurement by referencing them closely to the amplitude at which the measurement was made. It was quickly realized that many of the arguments that had occurred between different authorities were merely caused by a failure to define a palpably nonlinear process in terms of one of the principal variables (amplitude).

SELECTION OF A DATA BASE

It is believed by the author that many of the arguments about appropriate damping values have been caused by the poor quality of the data on which opinions have been based. Damping measurements are made in two different circumstances. In the first, the motion of the building is caused by a random forcing function, typically the wind or an earthquake, and measurements are made by one of several alternative statistically-based techniques. In the second, the building is forced into motion by an artificial source and deterministic measurements are made. These are considered in turn:

Random Forcing

The use of statistical techniques results in fairly long record length requirements for tall buildings. Typically this length may be a few hours. It is a basic requirement of all statistically-based data analysis techniques currently used for damping estimation that stationarity of the data must exist. (This means that all statistical properties must be invariant with time). It can be appreciated that this situation never exists in the case of an earthquake, and in the case of wind engineering, it is a rare circumstance for the wind to blow continuously in magnitude and direction for some hours. It is surprising, then, that it is a rare event for an author to perform a check for stationarity on his data before making damping measurements. Accordingly all data not subjected to a check for stationarity have been excluded from the data base. In practice, this means that no data of this kind have been used.

Artificially induced motion

In this case it is necessary for the source of forcing to be well controlled. This effectively allows only those tests of tall buildings in which a very precise vibrator system is used to control shaking. In practice the precision necessary is typically one thousandth of a Hertz. When only data generated in this way are included in the data base, less than twenty buildings are left on which to work. The estimation of damping from such well-controlled tests can be performed in any one of several ways. The data forming this reduced set are reported here and have been used to arrive at a predictor of damping values.

One further restriction on the use of the data base is that sufficient structural details are reported together with the measurements to allow a correlation with damping variations to be made. This further reduces the usefulness of some of the data. However, because of the paucity of the high quality data, some use of these measurements can also be made.

HISTORICAL BACKGROUND

Early work identified three types of damping behavior, but also showed that all damping models could be treated by using the concept of *equivalent viscous damping* (Jacobsen, 1930).

The increasing concern about the fact that the viscous damping model did not reflect reality was expressed in a paper that addressed itself to differences caused by using viscous or structural damping models (Soroka, 1949).

Hudson (1965) and Jacobsen (1965) considered the bilinear model of elasto-plastic movement in buildings, the former to justify the use of an equivalent viscous damping for the response spectrum approach to earthquake engineering, and the latter to try to understand the physical mechanism involved in energy dissipation in tall buildings. Jacobsen's description of these mechanisms included the observation that the main source of energy dissipation is by slippage in joints.

Two papers presented to the Fifth World Conference on Earthquake Engineering reported measurements on full-scale buildings in which measurements at different amplitudes had shown increases of damping values with increasing amplitude of response (Hart et al., 1973; Petrovski et al., 1973).

Further progress was made when measurements were presented to show not only an increase of damping with amplitude, but also generic differences between concrete and steel buildings (Hart and Vasudevian, 1975). Their curves show concrete buildings to have approximately 20% more damping than do steel buildings vibrating at the same amplitude. In Japan, a search for correlation of damping values with lengths of shear walls in buildings showed a trend, but a very poor correlation with the measured data (Naito et al., 1956). The authors noted that "damping increases the larger the periphery of the base of the buildings and the softer the ground, that is, the shorter the wave length of the S-wave in the ground." The fact that the damping values were not related to amplitude is undoubtedly to blame for the poor correlation.

Another researcher from Japan, Nakagawa (1961), reported a correlation between damping and building dimension derived from tests on a group of seven buildings. The scatter was large, but a correlation appeared to exist despite the fact that the data were not related to amplitude. The fact that all the buildings were of a similar type of construction and were tested in the same series of tests probably means that similar vibration conditions existed in each case and the results are therefore mutually comparable. Twenty buildings were included in the series of tests.

At the First World Earthquake Conference some remarkable conclusions were presented in which a plot of the logarithm of damping against period, for several buildings, produced a remarkable negative correlation (Kawasumi and Kanai, 1956). Accordingly the authors proposed that the product of damping and period is a constant.

Not until recently has anyone proposed a damping predictor for tall buildings based on their observations (Kobayashi and Sugiyama, 1977). This predictor was actually the same form as that proposed by Kawasumi and Kanai. This lack of progress over such a long time was caused by the large scatter in the experimental results published in the interim.

The concept of a physical mechanism for damping (friction), together with a *stiction* term, which was randomly distributed with respect to amplitude, was evolved very recently (Wyatt, 1977). (Stiction represents a STuck frICTION element, in which, the early stages of an increasing applied load, no motion takes place). The integrated effect is of a response which, for practical purposes behaves as a linear viscous system at any amplitude, but in which damping increases with increasing amplitude. The benefit of the theory is that the mechanism for the energy dissipation is the physically realizable friction between elements in a building.

MECHANISM FOR DAMPING BEHAVIOR

In the mechanism proposed by Wyatt all significant energy losses are assumed to be caused by friction. Additionally stiction effects allow a gradually increasing number of participating elements with increasing amplitude which accounts for the phenomenon of damping increasing with increasing amplitude previously reported (Hart et al., 1973; Jeary and Ellis, 1981).

It has been shown that all materials show a much smaller strength than might be expected from a study of molecular forces (Griffith, 1921). He showed that this discrepancy can be explained if it is assumed that many microcracks exist throughout materials. If under the action of external forces, these microcracks lengthen, then there is a reduction in the overall strain energy. The mechanism of increasing the crack length is that the surface energy of the crack increases until sufficient energy is available to repute the material. The increase in crack length then represents an energy sink and is measured in practical terms as damping. The critical shear stress S_{cr} at which this lengthening of microcracks occurs, is given (Timoshenko, 1970) as:

$$S_{cr} = \sqrt{\frac{4ET}{l}} \tag{1}$$

where T is the surface energy of the crack and l is the length of the crack.

It is obvious that the distribution of crack lengths through a material will dictate the form of the damping variation. Equation 1 gives a rationale for the energy sink of damping, for the stiction effect, for the effect of damping increasing with increasing amplitude, and of the appearance of a low amplitude plateau region to the damping characteristic. These effects will be considered:

Damping increase with increasing amplitude. It can be assumed that the distribution of the lengths of cracks throughout a material is random and that therefore the distribution of the critical shear stress is also random. This infers that there will be an increasing number of participating cracks as amplitude increases.

The stiction effect. Since the critical value of the shear stress must be overcome before a crack lengthens, then before this value there will be an increasing force but no corresponding increment in crack length for any particular crack.

The damping effect. The energy sink of crack lengthening, or mobilization of a joint, provides the basic mechanism of damping.

The low amplitude plateau region. Joints throughout a structure can be considered in the same light as microcracks, but in this case the value of l is much greater. This implies that there will be mobilization of cracks between large dimension joints at low amplitude. Since the amount of energy dissipated in this way will be relatively large, this mechanism seems to represent, for practical purposes, a relatively constant damping value for a range of low amplitudes.

A high amplitude plateau. When all joints are mobilized and all cracks capable of acting as energy sinks in a specific amplitude have been lengthened, then no further effective increase in the damping value will occur, and accordingly, a second plateau region will be reached. Experimental work on the dynamic behavior of bridges supports this supposition (Tilley and Eyre, 1977).

Since the governing parameter to the friction mechanism is the value of the shear stress that will cause either mobilization of structural joints or the lengthening of microcracks, the form of shear stress distribution is important to consider. The distribution of strain energy throughout a building is an important consideration for the application of any predictive techniques to the problem of damping estimation. It has been shown that for the aspect ratio common in tall buildings, 80% or more of strain energy is a result of bending movement (Moore, 1983). Bending movement is often considered to decrease quadratically with height, and therefore implies that stresses which occur low down in a building will dominate the damping characteristics.

DATA REDUCTION

Using the reduced data set discussed in Table 1, a correlation between rate of increase of damping and the square root of a base dimension of the building and its attachments has been identified. This information is presented in graphical form in Fig. 1.

The basic data set is published elsewhere (Jeary, 1986). It is found that the best regression line for the data falls close to the relationship:

$$\mathrm{Log}_{10}\, \zeta_I = \frac{\sqrt{D}}{2} \tag{2}$$

The use of this relationship with data published by other workers shows good agreement with results produced in a variety of experiments by various authors (Jeary, 1986).

THE USE OF DAMPING PREDICTORS

It is apparent that this, or any other damping predictor, will produce values that vary significantly from the true damping values of real buildings in the following circumstances:

Where significant soil-structure interaction exists.
Where the amplitude of response exceeds the yield point.
Where a building has no cladding, or is composed only of a framework.
Where a positive effort to increase damping has been included in the design of the building.

The identification of a predictor for damping leads to some interesting possibilities for design. The dimensions of a building may be chosen, for

Table 1 Data relating to damping/amplitude characteristics

Rate of increase in damping (ζ_I)	Building dimension (in m)	Building name
75	15.0	South Stoneham
84	12.0	Leicester
113	17.9	Harpy (unpublished)
115	19.5	Priory Hall
212	12.0	Sutherland
220	27.0	Robert Millikan
221	12.5	Dunstan
227	17.0	South Stoneham
260	25.0	Robert Millikan
317	24.0	Leicester
318	24.0	John Russel Court
335	18.0	John Russel Court
405	24.5	Dunstan
437	22.0	Priory Hall
447	20.0	Sheffield
583	23.7	Harpy (unpublished)
1811	16.0	Exeter
2371	43.0	Sutherland
3840	50.0	Exeter

instance, to give sufficient damping at low amplitudes to satisfy serviceability requirements, while maintaining adequate values at higher amplitudes for limit state criteria.

If an unsatisfactory response of a building to naturally occurring phenomena is experienced, then the attachment of a single story unit to increase the damping value may provide a low cost solution to the problem.

It is certain that the damping predictor used here (or any others) will need further refinement, and as such they are intended to be a first step towards successful damping prediction at the design stage. Inspection of more data will be a useful exercise in order to reduce uncertainties. However, the close inspection of exceptions to the rule are of far greater potential benefit to improvement of prediction techniques.

It is obvious that it is not solely the lowest level of a building that contributes to the energy dissipation, and the study of the relative contributions of higher levels in different types of buildings will undoubtedly lead to a better understanding of the present model.

SUGGESTED SCHEME FOR DAMPING ESTIMATION

Damping values should be specified in two parts—those of low amplitude damping ζ_o and a rate of increase of damping with amplitude ζ_I.

The basic data necessary to define damping values for a building are its physical dimensions of length, breadth, and height.

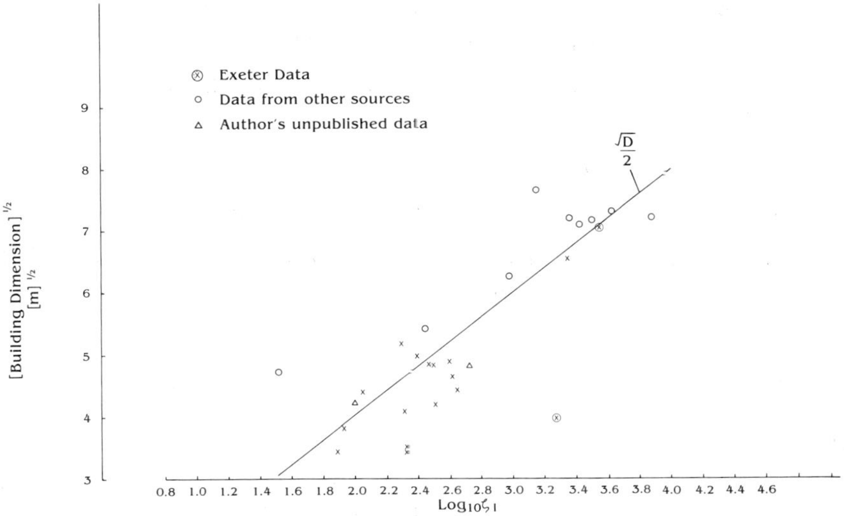

Fig. 1 Rate of change of damping vs. building dimension

The first step is to define f_o. Since this has been shown to be correlated with natural frequency (Jeary, 1981), a simple formula for the prediction of natural frequency should be used. Such a formula has been suggested previously (Ellis, 1980)

$$f_o = \frac{46}{H} \tag{3}$$

where H is the height of the building meters.

The following relationship should then be used (Jeary, 1981):

$$\zeta_o = 0.15 + 0.76\, f_o \tag{4}$$

Next the value of ζ_I can be assessed by reference to the relevant dimension of D of the building from Eq. 2. Finally, it is assumed that, for practical purposes, the low-amplitude damping plateau can be effectively reduced to a zero-amplitude value.

The absolute value of damping, at amplitude x_H, for mode j, can therefore, be estimated by the expression

$$\zeta_j = \zeta_{oj} + \zeta_{Ij} \frac{[x_H]}{H} \tag{5}$$

It is suggested, then, that the following scheme should be followed for the estimation of damping values for any building:

1. Obtain the overall dimensions of the building. These should include a ground floor plan showing attached buildings.
2. Estimate the ζ_o value of damping from the fundamental natural frequencies.
3. Estimate the ζ_I value of damping from Eq. 2.
4. Assess the amplitude at which the calculation is to operate.
5. Use Eq. 5 to calculate the damping value at the design amplitude established in the plan above.

REFERENCES/BIBLIOGRAPHY

Ellis, B. R., 1980
AN ASSESSMENT OF THE ACCURACY OF PREDICTING THE FUNDAMENTAL NATURAL FREQUENCIES OF BUILDINGS AND THE IMPLICATIONS CONCERNING DYNAMIC ANALYSIS OF STRUCTURES, Proceedings Institution of Civil Engineers, Part 2, 69, pp. 763-776.

Griffith, A. A., 1921
THEORY OF RUPTURE OF BRITTLE MATERIALS, Transactions of the Roya Society, Ser. A, Vol. 21, pp. 163-198.

Hart, G. C., Lew, M., and Di Julio, R., 1973
HIGH-RISE BUILDING RESPONSE: DAMPING AND PERIOD NON-LINEARITIES, Proceedings of the 5th World Conference on Earthquake Engineering, Rome.

Hart, G. C. and Vasudevian, R., 1975
EARTHQUAKE DESIGN OF BUILDINGS: DAMPING, Journal of the Structural Division, ASCE ST1, January.

Hudson, D. E., 1965
EQUIVALENT VISCOUS FRICTION FOR HYSTERETIC SYSTEMS WITH EARTHQUAKE-LIKE EXCITATIONS, Proceedings of the 3rd World Conference on Earthquake Engineering, Vol. 11, pp. 185-201.

Jacobsen, L. S., 1930
STEADY FORCED VIBRATION AS INFLUENCED BY DAMPING, Trans. ASME-APM-52-15, pp. 169-181.

Jacobsen, L. S., 1965
DAMPING IN COMPOSITE STRUCTURES, Proceedings of the 2nd World Conference on Earthquake Engineering, pp. 1029-1044.

Jeary, A. P., 1981
THE DYNAMIC BEHAVIOUR OF TALL BUILDINGS, PhD. thesis, University College, London, UK, March.

Jeary, A. P., 1986
DAMPING IN TALL BUILDINGS—A MECHANISM AND A PREDICTOR, Earthquake Engineering and Structural Dynamics, Vol. 14, pp. 733-750.

Jeary, A. P. and Ellis, B. R., 1981
VIBRATION TESTS ON STRUCTURES AT VARIED AMPLITUDES ASCE/EMD Conference on Dynamics of Structures, Atlanta, Georgia.

Kawasumi, H. and Kanai, K., 1956
VIBRATION IN BUILDINGS IN JAPAN, Proceedings of the 1st World Conference on Earthquake Engineering, Berkeley.

Kobayashi, H. and Sugiyama, N., 1977
VISCOUS DAMPING OF STRUCTURES RELATED TO FOUNDATION CONDITIONS, Proceedings of the 6th World Conference on Earthquake Engineering, New Delhi.

Moore, D. B., 1983
THE STRUCTURE OF PYRIMOIDAL SHELLS, PhD. thesis, Bradford University, UK.

Naito, T., Nasu, N., Takeushi, M., Kubota, G., Tanaka, Y. and Hara, M., 1956
VIBRATIONAL CHARACTERISTICS OF REINFORCED CONCRETE BUILDINGS EXISTING IN JAPAN, Proceedings of the 1st World Conference on Earthquake Engineering, Berkeley.

Nakagawa, K., 1961
VIBRATIONAL CHARACTERISTICS OF BUILDINGS, PART 2, VIBRATIONAL CHARACTERISTICS OF ACTUAL BUILDINGS DETERMINED BY VIBRATION TESTS, Waseda University Bulletin of Science and Engineering Research Laboratory Report No. 16.

Petrovski, J., Jurukovski, D. and Paskalov, T., 1973
DYNAMIC PROPERTIES OF A FOURTEEN STORY RC FRAME BUILDING FROM FULL SCALE FORCED VIBRATION STUDY AND FORMULATION OF A MATHEMATICAL MODEL, Proceedings of the 5th World Conference on Earthquake Engineering Rome.

Soroka, W. W., 1949
NOTE ON THE RELATIONS BETWEEN VISCOUS AND STRUCTURAL DAMPING COEFFICIENTS, Journal of the Aeronautical Sciences, Vol. 16, Issue 7, July.

Timoshenko, S. P. and Goodier, J. N., 1970
THEORY OF ELASTICITY, McGraw-Hill, 3rd Edition.
Tilley, G. P. and Eyre, R., 1977
DAMPING MEASUREMENTS ON STEEL AND COMPOSITE BRIDGES, Symposium on dynamic behavior of bridges, Transport and Road Research Laboratory, Crowthorne, Berkshire, May 19.

Wyatt, T. A., 1977
MECHANISMS OF DAMPING, Symposium on dynamic behavior of bridges, Transport and Road Research Laboratory, Crowthorne, Berkshire, May 19.

The Role of Damping Systems

Kenneth B. Wiesner

Structural engineers have learned from our colleagues in wind engineering and from the owners and occupants of a number of "lively" tall buildings built during the past 30 years that tall buildings respond dynamically to wind. In many cases the dynamic response is dominant, exceeding the slowly varying mean response. Taking this as a challenge, structural engineers have searched for effective ways to reduce building dynamic response in order to increase building safety, reduce occupant discomfort, and reduce non-structural damage.

Dynamic response is significantly influenced by some factors such as the site (over which we have little or no control) plus the following building-related parameters: shape and height, vibration periods, mass and mass distribution, laternal stiffness and stiffness distribution, and structural damping. We have design control, although not as much as we might like, over the building parameters of shape, height, and mass. Designing a tall building to meet a given drift limit under code-specified equivalent static wind forces can lead to serious dynamic problems. Dynamic response can be reduced to acceptable levels by designing the building to have a relatively low vibration period combined with sufficient mass. Modern lightweight construction materials and systems for structural and nonstructural elements may make it economically prohibitive to take that approach. Also some tall slender buildings are susceptible to vortex shedding. So that leaves us with structural damping. Wind induced building response is inversely proportional to the square root of total damping, which includes aerodynamic plus structural damping. If the structural damping is doubled, dynamic response is reduced by 29%, quadruple it and a 50% reduction is achieved. Most tall buildings have structural damping between 0.5% and 1.5%, but the present state of the art is such that the damping cannot be accurately estimated closer than plus/minus 30% until the building is fully constructed. Reliable, effective methods of increasing the damping by factors of 2 or more are not available if the usual construction materials and systems are used.

The last 15 years have produced some innovative approaches to increasing the effective damping of tall buildings. In 1972 each World Trade Center Tower in New York City used 10,000 passive viscoelastic dampers. In 1985 the Columbia Center office building in Seattle used 260 larger scale passive viscoelastic dampers. No information has been published concerning how much damping the devices added to these buildings. In 1975 the CN Tower in Toronto employed a passive tuned mass damper (TMD) to reduce dynamic response of the large antenna atop the structure. In 1977 and 1978 powered passive tuned mass dampers were completed in the John Hancock Tower, Boston, and the Citicorp Center office building, New York City. In 1981, a passive TMD was completed in the Sydney Tower (formerly Centrepoint Tower), Sydney, Australia. A passive TMD is currently under construction for the Bronx Whitestone Bridge, New York City.

A segment of the academic research community meanwhile has been heavily engaged in primarily theoretical research on reducing dynamic response of buildings using various methods of applying active control forces. This field is known as *structural control.* The methods include semiactive and actively controlled TMDs, actively controlled mass impact dampers, active tendon control, active pulse control, and active aerodynamic appendages. Except for the latter, these methods have application to both wind and seismic response.

In the earthquake engineering field, much progress has been made in reducing building seismic response using seismic base isolation, in which a building is supported on passive laterally flexible or sliding bearings. Only relatively low-rise buildings have been designed and built using this approach, to date. Base isolation is neither theoretically very effective nor very practical for tall buildings.

COMPARISON OF SYSTEMS

Motion reduction systems for tall buildings are compared in Table 1, which provides a qualitative comparison chart of certain characteristics of all these systems. Concerning seismic effectiveness, it should be noted that all types of TMDs and the active mass impact dampers are rated low because these systems would, in order to be effective, have to respond almost immediately (within a few seconds) to building seismic motion and would have to be designed to counteract motion in 2nd and 3rd modes of vibration in each direction as well as the fundamental mode. The floor space required for a TMD or active mass impact damper is likely to be a large part of one upper floor of a tall building. Tables 2 and 3 provide a list of primary subsystems required for each of the motion reduction systems.

POWERED PASSIVE TUNED MASS DAMPERS

The tuned mass damper system installed in the 241 m (790 ft) tall John Hancock Tower, Boston, is shown in Fig. 1. This system, which reduces both

Table 1 Systems for dynamic motion reduction

Characteristic	Visco-elastic dampers	Passive TMD	Powered passive TMD	Active TMD	Active mass impact damper	Active tendon control	Active pulse control	Active aerodyn. append.
A. Used in constructed tall buildings	Y	Y	Y	N	N	N	N	N
B. System proven reliability	Y	Y	Y	N	N	N	N	N
C. Seismic effectiveness	3	0–1	0–1	0–2	0–2	2–3	2–3	0
D. System complexity	1	2	3	3	3	2–3	2–3	3
E. System energy use	0	0	2	3	3	3	3	2–3
F. System maintenance	0	1	3	3	3	3	3	3
G. System durability	3	2	2	2	2	2	2	1–2
H. Building floor space	0–1	3	3	3	3	1–2	2	1
I. System tuning need	N	Y	Y	Y	N	N	N	N
capability	0	0–1	3	3	0	0	0	0

Key—Items C to I
0 None (0 Floor area)
1 Less (or small) (<2,000 SF Floor area)
2 Moderate (2–4,000 SF floor area)
3 More (or large) (4–10,000 SF Floor area)

transverse and torsional motion, employs a 2670 kn (300 ton) mass block at each end of an upper floor of the building. Each mass slides along a single axis. One electronic control system is used to control this dual TMD. The mass ratio of TMD to building generalized mass is 0.014 in transverse direction and 0.021 in torsion.

The TMD system installed in the 280 m (919 ft) tall Citicorp Center in New York City is shown in Figs. 2 and 3. This system, which reduces both north-

Table 2 Primary subsystems of mass damper systems

	Mass damper type			
Subsystem	Passive TMD	Powered TMD	Active TMD	Active mass impact damper
Mass	X	X	X	X
Gravity support	X	X	X	X
Springs	X	X	X	O
Damping	X	X	O	X
Overtravel restraints	X	X	X	X
Directional guides	X	X	X	X
Active force-generating	O	X	X	X
Electronic control	O	X	X	X
Moveable motion stops	O	O	O	X

Table 3 Primary subsystems of other motion reduction systems

System	Primary Subsystems
Active tendon control	Tendons Tendon guides and structural attachments Active force generating Electronic control
Active pulse control	Electronic control Active force generators Hydraulic actuators
Active aerodynamic appendages	Aerodynamic appendage Appendage support system Overtravel restraints Active force generating Electronic control
Viscoelastic dampers	Damper units Structural guides and attachments

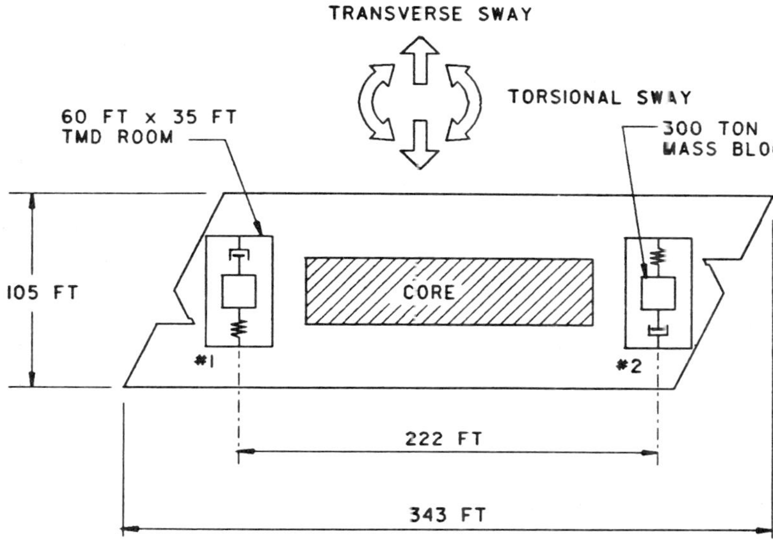

Fig. 1 Dual tuned mass damper system schematic, John Hancock Tower, Boston

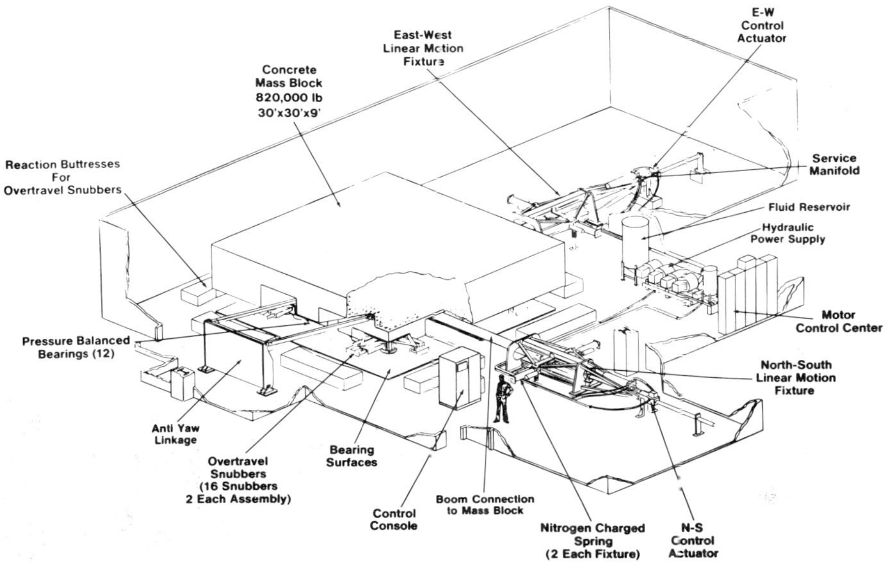

Fig. 2 Citicorp TMD system diagram

south and east-west lateral motion, has a 3650 kn (410 ton) mass block on an upper floor. The mass slides biaxially and is prevented from twisting by an antiyaw directional guide. The mass ratio of TMD to building generalized mass is 0.02 in each direction.

These powered TMD systems incorporate pneumatic springs and electronically controlled hydraulic actuators to provide the TMD damping force and additional force needed to make the TMD system behave like a theoretical passive TMD. The John Hancock Dual TMD includes an electronically controlled damping "shift" that automatically increases the internal TMD damping and thereby reduces the mass block displacements for larger building motions caused by infrequent major wind storms. Neither the Hancock nor the Citicorp TMD system is essential to the structural safety of the building. The primary purpose of each system is to increase comfort of building occupants.

ACTIVE TUNED MASS DAMPERS

Theoretical studies indicate that a properly designed active-controlled TMD system can reduce building dynamic motion significantly in comparison to a passive TMD or a powered TMD designed to be equivalent to a passive TMD. To date only passive and powered passive TMD systems have been installed in actual buildings. The classification of a system as passive,

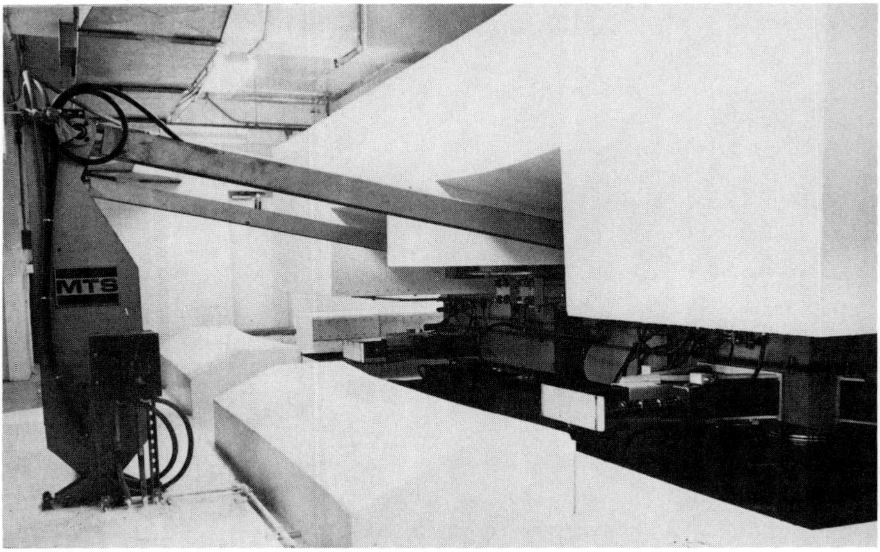

Fig. 3 Citicorp TMD system

semiactive, and so on, is based on required system inputs of information and energy. A passive system requires neither, a powered passive system requires both. A semiactive system requires information and almost no energy, and an active system requires both information and energy inputs.

ACTIVE MASS IMPACT DAMPERS

This type of damper, shown schematically in Fig. 4, is similar in some respects to a TMD. A sliding mass block slides upon but is not positively connected to the supporting building, and contacts resiliently supported motion stops when there is building dynamic motion. The active control system moves the resilient "stops" to make these impacts occur when the mass relative velocity is large, thus improving motion reduction in comparison to a passive mass impact damper.

ACTIVE TENDON CONTROL

This system requires diagonal tendons that impose lateral interstory forces on the structure. These tendon forces are timed to act counter to the dynamic building response by means of an electronic control system. Some one-fourth scale experimental work is in progress at State University of New York at

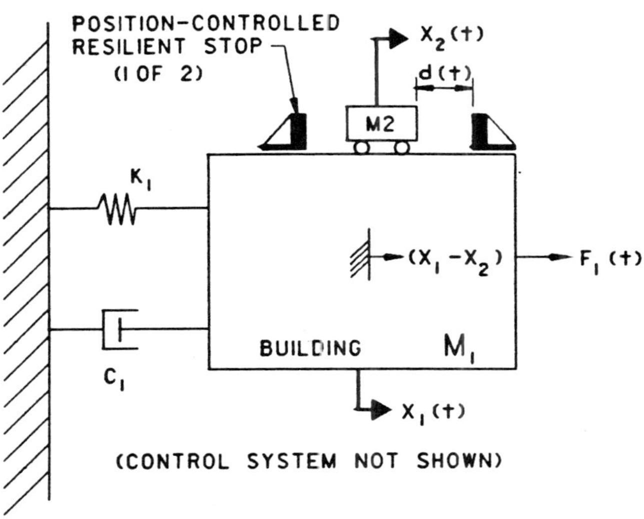

Fig. 4 Active mass impact damper-schematic with single-degree-of-freedom system

Buffalo, as shown in Fig. 5. This type of system tends to make the building dependent on active mechanical and electronic devices for lateral strength and stability because the system is most effective for a laterally flexible building.

ACTIVE PULSE CONTROL

It is possible to reduce dynamic building response significantly by means of short duration lateral forces, called pulses, applied at one or more building levels. Pulses are required once or twice per building motion cycle. One method of generating the pulses is to employ a gas pulse generator. A storage container for a pressurized gas is fitted with a metered flow nozzle and connected to the structure with a servo-controlled hydraulic actuator (Fig. 6). Research has been done at the University of Southern California.

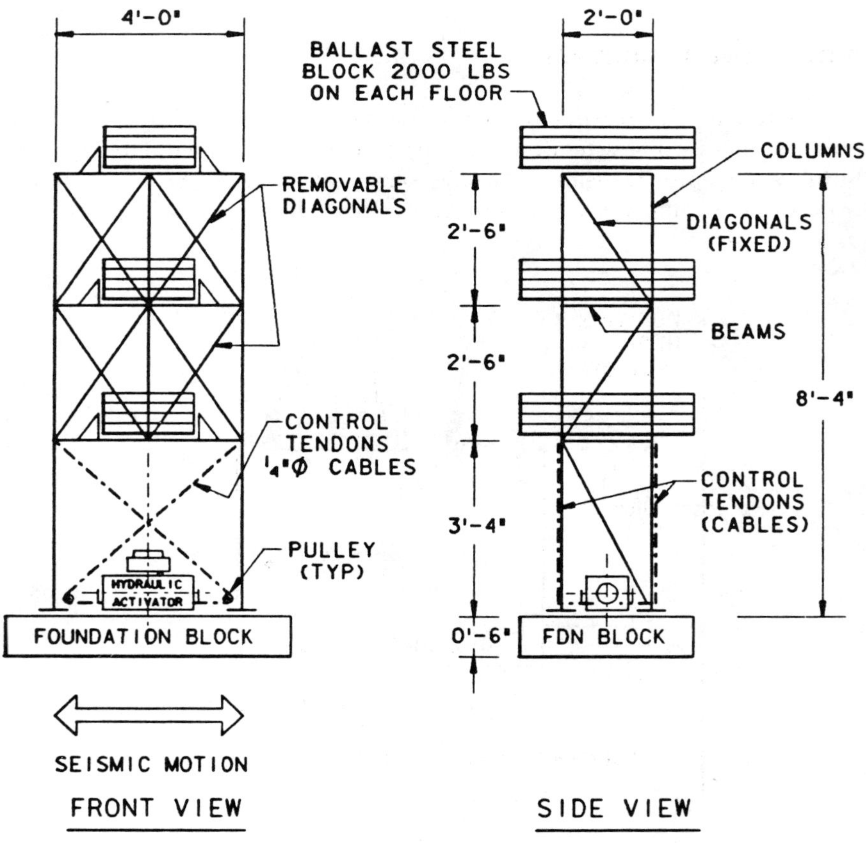

Fig. 5 Active tendon control laboratory model

ACTIVE AERODYNAMIC APPENDAGES

Familiar examples of this type of device are movable aircraft control surfaces such as wing flaps. Some theoretical work and simple small-scale wind tunnel model testing has been carried out (Fig. 7). A movable surface appendage near the top of a building is controlled actively, to deploy during the periodic building motion when building response velocity is against the wind direction and to retract when building velocity is with the wind. The main energy source providing the building motion reduction is the wind itself. However the energy to lift and retract the appendage requires a source in the building. This type of system would require the active cooperation of the architect and owner, since the appendage would have to be of the order of 2% of the building height.

VISCOELASTIC DAMPERS

Examples of viscoelastic dampers are shown in Figs. 8 and 9. The primary objective in designing viscoelastic dampers for a tall building is to reduce building accelerations for frequent storms to acceptable occupant comfort levels. The engineer must design the building for *strength* recognizing that the dampers may not be fully effective because of damper material, fabrication and erection tolerances or errors, temperature effects, and the like.

With viscoelastic dampers the building design process has to be iterative. It is desirable to increase energy dissipation by increasing damper unit strain, but this requires less building stiffness to obtain larger building displacements. The resulting higher vibration periods may lead to larger accelerations. It is helpful to try several viscoelastic loss moduli, then plot building added

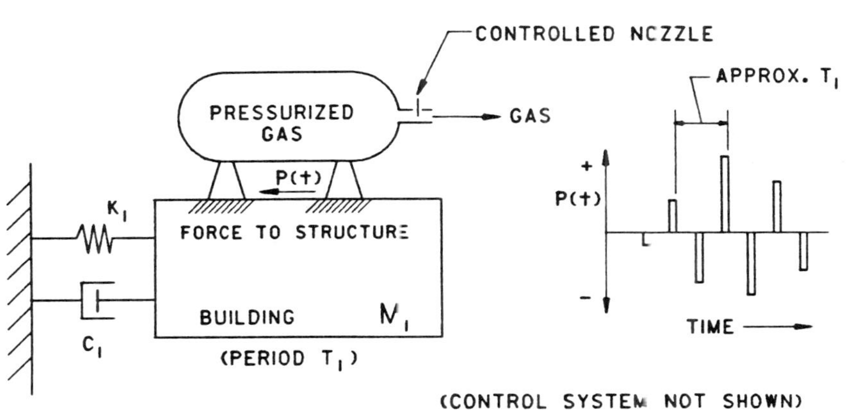

Fig. 6 Active pulse control-schematic with single-degree-of-freedom system

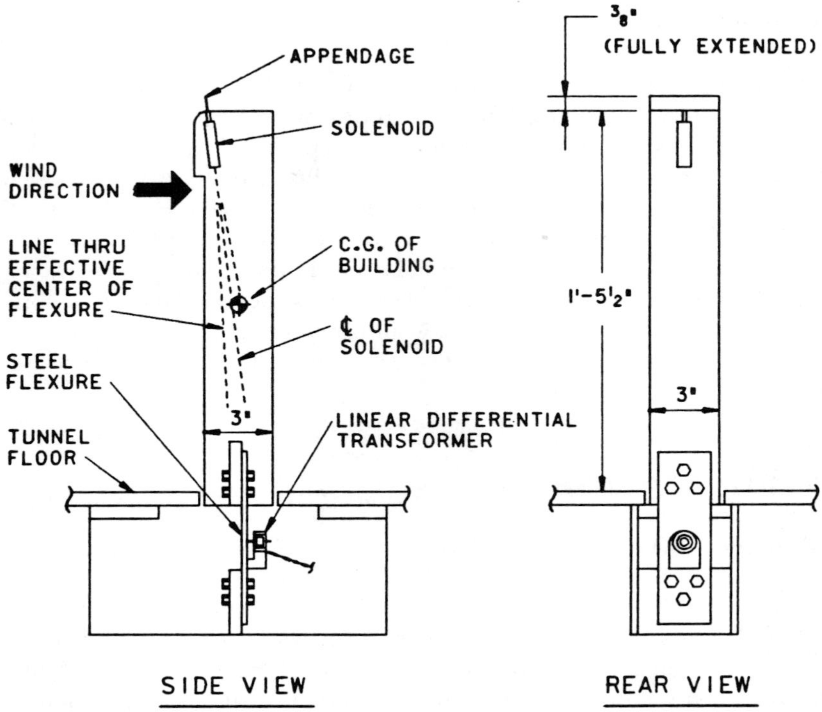

Fig. 7 Wind tunnel model of building with aerodynamic appendage

damping, building period, and acceleration versus the damper loss modulus. If results are not successful, further adjustments must be made in the structure or the dampers. The additional damping provided by viscoelastic dampers is calculated from

$$\zeta_{\text{add}} = \frac{E_d}{2E_t}, \tag{1}$$

in which E_d is the energy dissipated by the viscoelastic dampers and E_t is the total building strain energy, per cycle, including dampers. Using a single degree of freedom building representation,

$$E_t = 2\pi \left(\frac{1}{2} M^* \omega^2 \Delta^2\right) \tag{2}$$

and the required

$$E_d = 2\pi \zeta_{\text{add}} M^* \omega^2 \Delta^2. \tag{3}$$

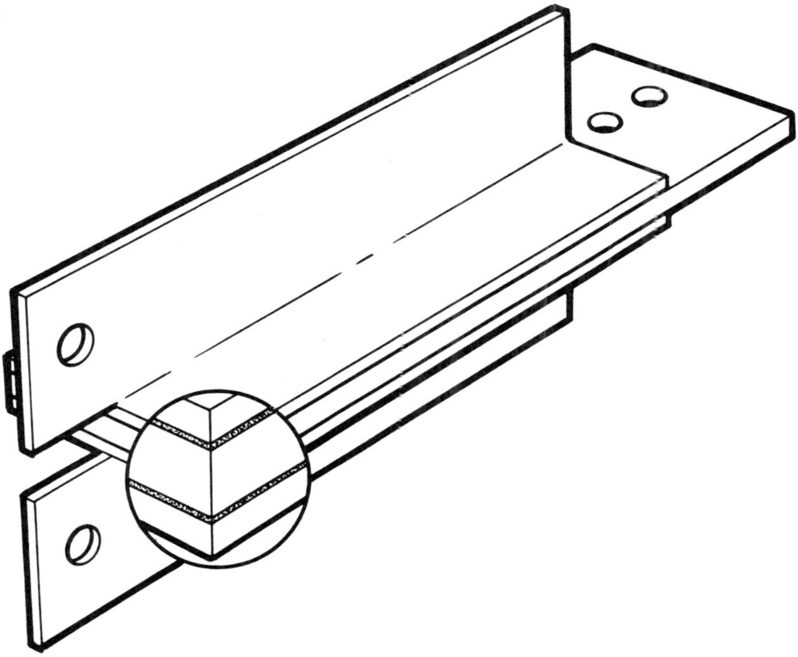

Fig. 8 Typical construction of World Trade Center and Columbia Center damper

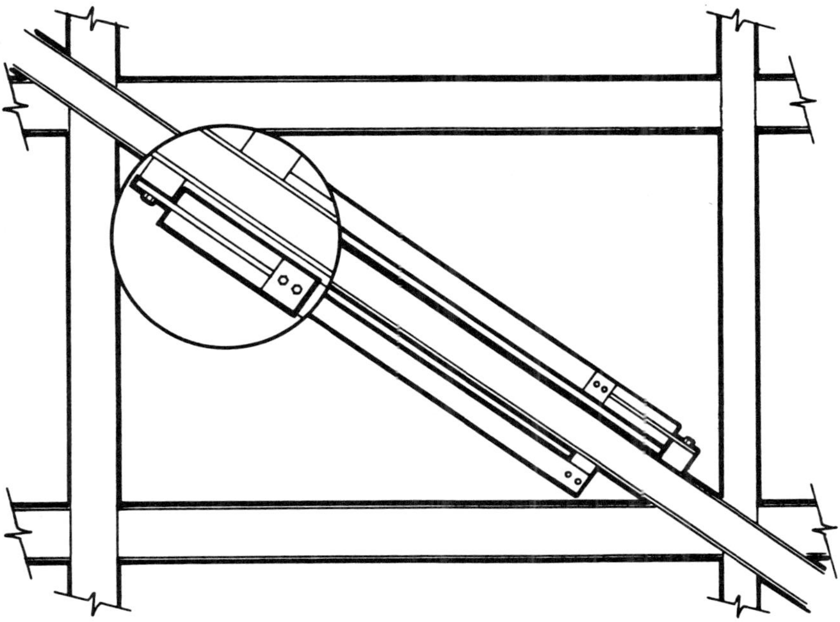

Fig. 9 Columbia Center dampers (2) in place

The provided

$$E_d = \Sigma \pi V \gamma^2 G''. \tag{4}$$

Definition of terms: $\varsigma = c/c_{cr}$ damping ratio; M^* = Building generalized mass, 1st mode; $\omega = 2\pi/T$ where T = Building period, 1st mode; Δ = 1st mode building displacement; V = Volume of a damper unit; γ = Damper shear strain; and G'' = Viscoelastic loss shear modulus. G'' varies with temperature, shear strain and vibration period, and temperature may rise with the number of motion cycles. Also, Δ and γ should be calculated for the same lateral loading, which should consist of an equivalent static wind load distributed in accordance with the inertial loading (mass × mode shape at each level). Only the dynamic part of the wind response should be used in making these energy and damping calculations.

FUTURE PROSPECTS

Some kind of added damping system or active control system should become more commonplace as we design our tallest buildings. None of the available systems can be presently considered reliable enough to be counted on fully in the strength design of the building, but all types would be appropriate for making the building more livable. The research work on active structural control has either been wholly theoretical, or when laboratory models were tested they were too simple and small-scale for extrapolation to full size. Most of the research studies assume perfect sensing and control devices. For example, in full size hardware, real time is required to physically generate the required control force or displacement. This time delay must be carefully considered in the system design and analysis.

The writer's opinion is that the most promising systems are viscoelastic damping, active TMDs, and possibly active pulse control.

Going beyond the systems already described, why not also encourage research leading to more ideal systems for increasing building damping? For example:

1. Provide highly damped permanent partition wall systems that could be installed in core areas for the full building height. Generally these would also be required by the building code to serve as fire-rated enclosures.

2. Develop a material that would significantly increase building damping after being *glued*, or applied like sprayed fireproofing, to all primary structural members. Ideally the material would also serve as the fireproofing for structural steel members.

The ideal is an added-damping system that is economical to install, durable, has low operating expense, is maintenance-free, requires little space, and is effective and reliable at all levels of building dynamic response.

REFERENCES/BIBLIOGRAPHY

Chang, J. C. H. and Soong, T. T., 1980
STRUCTURAL CONTROL USING ACTIVE TUNED MASS DAMPERS, Journal of Engineering Mechanics Division, ASCE, Vol. 106, No. EM6 December, pp. 1091-1098

Davenport, A. G. and Hill-Carroll, P., 1986
DAMPING IN TALL BUILDINGS: ITS VARIABILITY AND TREATMENT IN DESIGN, Building Motion in Wind, proceedings of a session at ASCE Convention, Seattle, Washington, April 8, pp. 42-57.

Dehghanyar, T. J., Masri, S. F., Miller, R. K., and Caughey, T. K., 1985
ON-LINE PARAMETER CONTROL OF NONLINEAR FLEXIBLE STRUCTURES, 2nd IUTAM International Symposium on Structural Control, University of Waterloo, Ontario, Canada, July.

Engineering News Record, 1976
LEAD HULA-HOOPS STABILIZE ANTENNA, Engineering News Record, Vol. 197, No. 4, July, p. 10.

Feld, L. S., 1971
SUPERSTRUCTURE FOR 1,350 FT. WORLD TRADE CENTER, Civil Engineering, Vol. 41, No. 6, June, pp. 66-70.

Hrovat, D., Barak, P., and Rabins, M., 1983
SEMI-ACTIVE VERSUS PASSIVE OR ACTIVE TUNED MASS DAMPERS FOR STRUCTURAL CONTROL, Journal of Engineering Mechanics Division, ASCE, Vol. 109, No. 3, June, pp. 691-705.

Johnson, C. R., 1981
USE OF VISCO ELASTIC DAMPERS IN REDUCING THE WIND-INDUCED MOTION OF TALL BUILDINGS, M. S. Thesis, Department of Civil Engineering, Massachusetts Institute of Technology, August.

Kaynia, A. M., Veneziano, D., and Biggs, J. M., 1981
SEISMIC EFFECTIVENESS OF TUNED MASS DAMPERS, Journal of Structural Division, ASCE, Vol. 107, No. ST8, August, pp. 1465-1484.

Keel, C. J. and Mahmoodi, P., 1986
DESIGN OF VISCOELASTIC DAMPERS FOR COLUMBIA CENTER BUILDING, Building Motion in Wind, Proceedings of a session at ASCE Convention, Seattle, Washington, April, pp. 66-81.

Klein, R. E. and Salhi, H., 1980
THE TIME-OPTIMAL CONTROL OF WIND-INDUCED STRUCTURAL VIBRATIONS USING ACTIVE APPENDAGES, Structural Control, H. H. Leipholz, ed., North-Holland Publishing Company, Amsterdam, the Netherlands, pp. 415-429.

Kwok, K. C. S., 1984
DAMPING INCREASE IN BUILDING WITH TUNED MASS DAMPER, Journal of Engineering Mechanics Division, ASCE, Vol. 110, No. 11, November, pp. 1645-1649.

Leipholz, H. H., 1980
STRUCTURAL CONTROL, North-Holland Publishing Company, Amsterdam, the Netherlands.

Lund, R., 1980
ACTIVE DAMPING OF LARGE STRUCTURES IN WINDS, Structural Control, H. H. Leipholz, ed., North-Holland Publishing Company, Amsterdam, the Netherlands, pp. 459-470.

Mahmoodi, P., 1969
STRUCTURAL DAMPERS, Journal of Structural Division, ASCE, Vol. 95, No. ST8, August, pp. 1661-1672.

Masri, S. F., 1973
RESPONSE OF THE IMPACT DAMPER TO STATIONARY RANDOM EXCITATION, Journal of Acoustic Society of America, Vol. 53, No. 1, January, pp. 200-211.

Masri, S. F., Bekey, G. A., and Caughey, T. K., 1981
OPTIMUM PULSE CONTROL OF FLEXIBLE STRUCTURES, ASME Journal of Applied Mechanics, Vol. 48, September, pp. 619-626.

Masri, S. F., Bekey, G. A., and Caughey, T. K., 1982
ON-LINE CONTROL OF NON-LINEAR FLEXIBLE STRUCTURES, ASME Journal of Applied Mechanics, Vol. 49, December, pp. 877-884.

McNamara, R. J., 1977
TUNED MASS DAMPERS FOR BUILDINGS, Journal of Structural Division, ASCE, Vol. 103, No. ST9, September, pp. 1785-1798.

Petersen, N. R., 1980
DESIGN OF LARGE SCALE TUNED MASS DAMPERS, Structural Control, H. H. Leipholz, ed., North-Holland Publishing Company, Amsterdam, the Netherlands, pp. 581-596.

Raggett, J. D., 1975
ESTIMATING DAMPING OF REAL STRUCTURES, Journal of Structural Division, ASCE, Vol. 101, No. ST9, September, pp. 1823-1835.

Roorda, J., 1975
TENDON CONTROL IN TALL STRUCTURES, Journal of Structural Division, ASCE, Vol. 101, No. ST3, March, pp. 505-521.

Sinclair, D. F., 1986
DAMPING SYSTEMS TO LIMIT THE MOTION OF TALL BUILDINGS, Building Motion in Wind, proceedings of a session at ASCE Convention, Seattle, Washington, April, pp. 58-65.

Sladek, J. R. and Klingner, R. E., 1983
EFFECT OF TUNED-MASS DAMPERS ON SEISMIC RESPONSE, Journal of Structural Engineering, ASCE, Vol. 109, No. 8, August, pp. 2004-2009.

Soong, T. T. and Skinner, G. T., 1981
EXPERIMENTAL STUDY OF ACTIVE CONTROL, Journal of Engineering Mechanics Division, ASCE, Vol. 107, No. EM6, December, pp. 1057-1067.

Soong, T. T., Reinhorn, A. M., and Yang, A. N., 1985
A STANDARDIZED MODEL FOR STRUCTURAL CONTROL EXPERIMENTS AND SOME EXPERIMENTAL RESULTS, 2nd IUTAM International Symposium on Structural Control, University of Waterloo, Ontario, Canada, July.

Vickery, B. J. and Davenport, A. G., 1970
AN INVESTIGATION OF THE BEHAVIOR IN WIND OF THE PROPOSED CENTREPOINT TOWER IN SYDNEY, AUSTRALIA, Engineering Science Research Report, BLWT 1-70, University of Western Ontario, London, Canada, February.

Wargon, A., 1983
DESIGN AND CONSTRUCTION OF SYDNEY TOWER, The Structural Engineer, Vol. 61A, No. 9, September, pp. 273-281.

Wiesner, K. B., 1979
TUNED MASS DAMPERS TO REDUCE BUILDING WIND MOTION, presented at ASCE National Convention, Boston, Massachusetts, Ajpril 2-6, ASCE Preprint No. 3510.

Yang, J. N. and Giannopoulos, F., 1978
ACTIVE TENDON CONTROL OF STRUCTURES, Journal of Engineering Mechanics Division, ASCE, Vol. 104, No. EM3, June, pp. 551-568.

Yang, J. N. and Lin, M. J., 1982
OPTIMAL CRITICAL-MODE CONTROL OF BUILDING UNDER SEISMIC LOAD, Journal of Engineering Mechanics Division, ASCE, Vol. 108, No. EM6, pp. 1167-1185.

Yang, J. N. and Samali, B., 1983
CONTROL OF TALL BUILDINGS IN ALONG-WIND MOTION, Journal of Structural Engineering, ASCE, Vol. 109, No. 1, January, pp. 50-68.

Structural Design of Tall Steel Buildings

Introductory Review

Leo Finzi

In the last three decades the structural design of tall steel buildings has undergone significant developments thanks to the use of new steel grades, new types of connections, new codes, new fabrication and construction techniques, and new structural systems. Clear evidence of this was provided by the many presentations and discussions during the workshops and the Third International Conference on Tall Buildings held in Chicago in January, 1986.

As a result of these developments, the price that we now pay for tallness is much less than before. This progress is also the fruit of increased interaction and cooperation right from the start between the architect, the mechanical engineer, the contractor, and the structural engineer. May we expect a similar kind of development during the second century of the skyscraper? I sincerely believe that we will see development, but not on the same line as we have had previously. The form and shape that skyscrapers will take in the future will have to change. If present trends are followed we will be building a one-mile-tall arrow or Christmas tree. But we should strive for a wise use of the third dimension, much as an oak spreads its branches. Architects have had a great many dreams in this direction, but engineers have always come up with designs that create terrible congestion among columns and elevator wells at the lower levels.

From whence can we expect real help when trying to increase the degrees of freedom in designing tall buildings? The first area to examine is in the loading conditions. It would be better to evaluate the effect of the actions on a tall building than to increase the strength of the structure itself. The probabilistic approach, with its greater use of information about actual loading conditions in a building, the concept of expected life of the building, the scaling of events as related to those return periods considered, give rise to a great many loading combinations. This is not over-tedious for the designer since he is now fully aided by computers, and the management of dozens of loading combinations is not nearly as costly and annoying as in the past. Among the many loading combinations, those related to wind action and earthquakes often govern the design, leading to very stiff, heavy and therefore costly structures.

The ductile behavior both of the structural details and of the structural system itself will play a great role in improving the capability of the structure to resist winds and earthquakes. From this point of view, a better knowledge of the presently disregarded interaction, involving floors and cladding with the main structure, will be a key factor.

Interactive computer-aided design and finite element approaches increasingly allow us to look at the many possible links that would generate favorable or unfavorable membrane effects in the building, causing us to consider the building more as a body than as a skeleton. It is well known that the price for tallness can be split into two parts: the one related to vertical loads and the one due to lateral actions.

A good shape, an efficient bracing system, and an adequate interaction between the main structure and its cladding and floors can keep the second at a minimum level; but what about the first? One could hope for new materials with increased elastic modulus and very modest residual stresses, but is this realistic? In this area things are not moving very quickly, as we are using, with few exceptions, the same steels that the Germans used before World War II.

Steelmakers are very prudent in changing their products. The only real innovation in the last few years has been the development of cold-rolled profiles. But such new sections are only useful for lattice transmission towers or for apartment houses of only a few floors. Nothing really new in this area can be foreseen for use in tall buildings.

I believe that an opening to a really innovative future can only come from consideration, not of a simple building the size of a typical U.S. block, but of greater agglomerations including an entire quarter of the city. A giant framework could in this case be the reference skeleton for both housing and other residential activities and for transportation systems.

Because of the great width of such macrostructures, the price for tallness could be largely reduced, and the large size of the widely spaced members would maintain slenderness ratios at low values. In a jumbo frame, moreover, one could take advantage of design concepts similar to those brilliantly adopted for long span bridges using straight or curved high tensile cables, box structures, and the composite action between steel and reinforced concrete. One must say, nevertheless, that the use of macrostructures can be adopted only through radical changes in the urban planning of the city. But for new settlements there should be no great difficulties in achieving this.

Let us now examine the unit cost of the tall building structure and determine what can be done to reduce this cost as much as possible. We must agree that the only way to obtain significant results is to go through full industrialization of the process of fabrication and construction. In Europe we have been looking at this problem for almost thirty years but with little success.

First, careful analysis of the reasons for the lack of such "break-through" development should be done, since it is now clear that standardization of structural profiles and details is necessary together with that for the height of the floors and the thickness of the floor itself. The tall building components should not be considered a masterpiece and tailored to fit the single case. Standardization is also needed for the building codes and, in particular, for the loading conditions to be considered. Changes in snow loads by simply crossing a state border are not easily understood by people with common sense.

Worldwide efforts by organizations such as ISO, AISC, ECCS, SSRC, IABSE, and others are starting to bring about change. Looking at the future is always a very difficult task and very few succeed. Therefore, please do not interpret these thoughts on the future of tall steel buildings in a prophetic sense. By looking at the recent past and listening to what was said at the Third International Conference on Tall Buildings, we can evaluate some possible trends.

Structural Standards

Geerhard Haaijer

The start of the second century of the skyscraper also marks the introduction of Load and Resistance Factor Design (LRFD) as a new structural standard for steel design in the United States. LRFD follows a worldwide trend toward Limit State Design (LSD). The aim of structural standards is to provide for reliable structures. Specification and standards committees combine the results of research, structural analyses, and engineering judgment in establishing structural standards. With LRFD it is possible to exercise much better judgment than is possible with conventional Allowable Stress Design (ASD). Whereas ASD depends on a single material factor of safety, LRFD uses several load and resistance factors that recognize the levels of uncertainty associated with different types of loading and with approximations inherent in specification formulas. Judgment is further enhanced by statistical analysis of load and strength data so that numerical values can be established for the reliability index.

The reliability of structures also requires that materials with appropriate properties be selected. The American Institute of Steel Construction, Inc. (AISC) specification lists American Society for Testing and Materials (ASTM) designations of steels that have given satisfactory service in statically loaded structures not exposed to low outdoor temperatures. For structures exposed to different ranges of temperature, guidance can be obtained from the American Association of State Highway and Transportation Officials (AASHTO) requirements for impact resistance. Finally, after the structure has been properly designed and appropriate materials selected, reliability can only be achieved if quality standards are met during fabrication and construction. As an aid in achieving the latter goal, AISC introduced a Quality Certification Program. Certified plants have the capability and commitment to produce fabricated steel of the required quality.

DESIGN

Galambos and Viest (1985) traced the history of the development of the LRFD specification in the United States. The specification is based on LSD principles and the partial factors were determined by First-Order Second-Moment (FOSM) probabilistic theory. The initial research effort (1969–1978) culminated in a preliminary draft of the LRFD specification which appeared as a series of eight papers in the September 1978 *Journal of the Structural Division* of the American Society of Civil Engineers (ASCE) and detailed the methods and the statistical data used to develop the criteria. It was evident that (1) LRFD was a possible and desirable design method, but (2) the AISC LRFD specification could not function without a common basis for load and load factor determination, which is shared by the structural design codes for all structural materials, especially reinforced concrete, masonry, aluminum, and wood. As a consequence, the action shifted to the American National Standards Institute (ANSI) Committee on Building Code Requirements for Minimum Design Loads in Buildings, which issued its load standard in 1982. This standard contains the new load factors and load combinations to go with any building material, and its basis is the FOSM method. The load factors were determined to give a target reliability index $\beta = 3.0$ for gravity load, and $\beta = 2.5$ for load combinations involving wind loads. The various material groups then were provided with a method to determine resistance factors to go with their respective design codes such that roughly the same reliability was achieved as implied by the load factors.

In the meantime various subcommittees of the AISC Specification Committee undertook to rework the LRFD draft, and the LRFD specification was approved in principle in 1981. A draft for review and trial was issued in 1983, and currently work is concluding on the final specification, including a commentary and a design manual.

Format of the LRFD Specification

The format of the design inequality contains partial factors for load effects, γ, and for resistance, ϕ:

$$\phi_k R_{nk} \geqq \gamma_D Q_D + \gamma_{Ei} Q_{ni} + \sum_{j=i}^{m} \gamma^*_{Ej} Q_{nj} \tag{1}$$

where the subscript n denotes nominal (code specified) values of the resistance R and the load effects Q, the subscript k denotes different applicable resistance limit states, the subscript D means dead load, and the subscript E defines the time-varying load effects due to occupancy, wind, snow, earthquake, and so forth. The load factors for these latter quantities count on one of the time-varying loads to have its maximum lifetime value (a 50 year life is assumed) while the others take on their arbitrary-point-in-time values. Following is an array of some load combinations to illustrate the combinatorial process:

$$1.4D \tag{2a}$$

$$1.2D + 1.6L + 0.5S \tag{2b}$$

$$1.2D + 1.6S + 0.5L \text{ OR } 0.8W \tag{2c}$$

$$1.2D + 1.3W + 0.5L + 0.5S \tag{2d}$$

D, L, S and W are dead, live, snow and wind load effects, respectively.

The specification provides the nominal resistance R_{nk} and the resistance factors ϕ_k for the various limit states appropriate to each type of member or connection. The resistance factors were determined by FOSM and they provide reliability index values from about 2.5 to 3.0 for members and 4.0 to 5.0 for connections under dead plus live loads.

Comparison with European Approach

In 1978 the European Convention for Constructional Steelwork (ECCS) developed recommendations that are the basis for EUROCODE 3. Galambos and Viest (1985) also showed the comparison shown in Table 1 of the AISC and ECCS load factors. ECCS uses a resistance factor of 1.0, whereas AISC uses 0.9 for several members including beams and tension members when the limit state is yielding. For proper comparison the AISC load factors were divided by 0.9 to account for this difference.

For columns the situation is more complicated. One column formula is

Table 1 Comparison of load factors

AISC Code 1985	ECCS Code 1978
$1.33D + 1.78L + .56S$	$1.3D + 1.5L + .75S$
$1.33D + 1.78S + .56L$	$1.3D + 1.5S + .75L$
$1.33D + 1.44W + .56(L + S)$	$1.3D + 1.5W + .75(L + S)$

used in the LRFD specification for all types of compression members. The column strength is based on the effect of residual stress and a mean value of the initial out-of-straightness of 1/1500 of the column height; end-restraint that results in an elastic effective length factor of 0.96 is assumed. The LRFD curve is above the ECCS column curve *b*, especially in the intermediate slenderness range, because the ECCS curves are based on a maximum permissible out-of-straightness of 1/1000 of the column height. It should be realized, however, that the resistance factors and the load factors are not the same for the two codes. With a resistance factor $\phi = 0.85$ and 1.0, respectively, for the AISC and the ECCS codes, it turns out that in most instances the latter is more liberal. For example, a factored design force P of 670 kN (75.3 tons) and 706 kN (79.4 tons) is obtained by the AISC and ECCS method, respectively, for the following column:

$$W12 \times 65,\ F_y = 248 \text{ MPa (36 ksi), Length} = 4.57 \text{ m (15 ft)}$$

$$P_{\text{dead}} = P \qquad P_{\text{live}} = P \qquad P_{\text{wind}} = P/2$$

The comparison would be different for heavy sections. In this instance ECCS uses a lower column curve so that the resulting design strengths will be closer. Thus, AISC gives up economy of some column designs in the interest of simplicity offered by a single column curve.

MATERIAL SELECTION

For most building applications that involve static loading and protection from low outdoor temperatures, the structural grades of steel listed by their ASTM designations in the AISC Specification have a long record of satisfactory performance. For unusual situations, the requirement of the AASHTO Specification can provide guidance in establishing toughness requirements. Material toughness can be defined as the ability to carry load or deform plastically in the presence of a notch. Rolfe and Barsom (1977) described the approach used to develop the AASHTO fracture-toughness requirements for bridge steels. An important factor in selecting toughness requirements is the strain rate encountered in the actual structure. The effect of slow loading rate compared with impact-loading rate is to shift the fracture-toughness transition to lower temperatures. For structural steels with yield points in the range 248 to 965N/mm^2 (36 to 140 ksi), the magnitude of the temperature shift from impact loading (strain rate of about 10 sec^{-1}) to slow loading (strain rate of about 10^{-5} sec^{-1}) can be approximated by the equation

$$T_s = 100 - 0.12F_y \text{ (for °C, N/mm}^2) \tag{3a}$$

$$T_s = 215 - 1.5F_y \text{ (for °F, ksi)} \tag{3b}$$

As shown in Table 2, AASHTO designates geographical zones in accordance with the average minimum temperature to which the structure may be subjected. Table 3 illustrates the impact-test requirements for several structural steels included in ASTM A709 when used in fracture-critical members of bridges in zone 2. The specific Charpy V-notch test requirements shown are for the base metal of nonredundant welded steel bridge members. The specification recognizes that the strain rates encountered in actual bridges (10^{-3} sec^{-1} in the vicinity of a notch) are closer to those for slow loading than to those for impact loading. The toughness requirements are complemented by an overall fracture-control plan that includes all factors that affect fracture behavior, including design and fabrication. This approach provides a balance between requirements and performance.

QUALITY CERTIFICATION

Peshek (1978) described the AISC Quality Certification Program that was established in recognition of the need for a comprehensive national standard in the United States for fabricator certification. The purpose of the AISC Quality Certification Program is to confirm to the construction industry that a structural steel fabricating plant has the personnel, organization, experience, procedures, knowledge, equipment, capability, and commitment to produce

Table 2 AASHTO temperature zone designation

Minimum service temperature	Temperature zone designation
0°F and above (−18°C and above)	1
−1°F to −30°F (−19°C to −34°C)	2
−31°F to −60°F (−35°C to −51°C)	3

Table 3 Base metal Charpy V-notch requirements for fracture-critical members in zone 2

ASTM designation	Thickness, in. (mm)	Minimum average energy, lbf-ft (N-m)	Testing temperature, °F(°C)
A36	to 4(102), incl	25(34)	40(4)
A588*	to 2(51), incl	25(34)	40(4)
A588*	over 2 to 4(51 to 102), incl	30(41)	40(4)
A514	to 2.5(64), incl	35(48)	0(−18)
A514	over 2.5 to 4(64 to 102), incl	45(61)	0(−18)

*If the yield point of the material exceeds 450 MPa (65 ksi), the testing temperature for the minimum average energy required shall be reduced by 15°F (8°C) for each increment or fraction of 70 MPa (10 ksi) above 450 MPa (65 ksi).

fabricated steel of the required quality for a given category of structural steelwork.

The program was developed by a group of highly qualified shop operation personnel from large, medium, and small structural steel fabricating plants throughout the United States. These individuals all had extensive experience and were fully aware of where and how problems can arise during the production process and of the steps and procedures that must be followed during fabrication to assure that the finished product meets the quality requirements of the contract.

The program was reviewed and strongly endorsed by an Independent Board of Review comprised of 17 prominent structural engineers from throughout the United States who were not associated with the steel fabricating industry but were well qualified in matters of quality requirements for reliable service of all types of steel structures.

Categories of Certification

A fabricator may apply for certification of a plant in one of the following four categories of structural steelwork:

I: Conventional Steel Structures—small public service and institutional buildings (schools, etc.), shopping centers, light manufacturing plants, miscellaneous and ornamental iron work, warehouses, sign structures, low-rise, truss beam/column structures, simple rolled beam bridges.

II: Complex Steel Building Structures—large public service and institutional buildings, heavy manufacturing plants, powerhouses (fossil, nonnuclear), metal producing/rolling facilities, crane bridge girders, bunkers and bins, stadia, auditoriums, high-rise buildings, chemical processing plants, petroleum processing plants.

III: Major Steel Bridges—All bridge structures other than simple rolled beam bridges.

MB: Metal Building Systems—Pre-engineered metal building structures.

Certification in Category II automatically includes Category I. Certification in Category III automatically includes Categories I and II. Certification in Category MB is not transferable to any other Category.

Inspection-Evaluation Procedure

An outside, experienced, professional organization, ABS Worldwide Technical Services, Inc. (a subsidiary of American Bureau of Shipping) has been retained by AISC to perform the plant inspection-evaluation in accordance with a standard check list and rating procedure established by AISC for each certification category in the program. Upon completion of this inspection-evaluation, ABS Worldwide Technical Services (commonly known as ABSTECH) will recommend to AISC that a fabricator be approved or disapproved for certification. ABSTECH's inspection-evaluation is totally independent of the fabricator's and AISC's influence, and its evaluation is not subject to review by AISC.

The number of days required for inspection-evaluation varies according to the size and complexity of the plant, but will usually require from two to five days.

Following recommendation for certification by the inspection-evaluation team, AISC will issue a certificate identifying the fabricator, the plant, and the category of certification.

FUTURE WORK

The present paper describes some aspects of codes and specifications in the United States. Committee 13 of the Council is planning to extend this work on an international basis and also cover additional standards in the United States.

REFERENCES/BIBLIOGRAPHY

Galambos, T. V. and Viest, I. M., 1985
DESIGN OF STEEL STRUCTURES WITH LOAD AND RESISTANCE FACTOR DESIGN SPECIFICATIONS, IABSE Reports, Vol. 48, Zurich.

Peshek, Jr., C., 1978
THE AISC QUALITY CERTIFICATION PROGRAM, Engineering Journal, AISC, Vol. 15, No. 3, Chicago, pp. 102-107.

Rolfe, S. T. and Barson, J. M., 1977
FRACTURE AND FATIGUE CONTROL IN STRUCTURES, Prentice-Hall, Inc., Englewood Cliffs, p. 129.

Methods of Analysis and Design

Seismic Design of Tall Steel Buildings

Egor P. Popov

Earthquake-resistant design of buildings is very complex and is perhaps the most challenging and awesome problem in structural engineering. The magnitude of a major earthquake at a given site at best can only be roughly approximated. The dynamic nature of the problem, including soil-structure interaction effects, adds to the complexity of the problem. Moreover, for reasons of economy, building frames cannot be overly conservatively designed. The strength limit state at presumed extreme loading conditions during the intended life of the structure is the dominant criterion. In this strength limit state some damage to the structure can be tolerated. However, "the overriding considerations of public safety for life, limb and property of human beings" (AISC, 1984) must be strictly adhered to. For these reasons, the steel frames, as designed, do not respond in an elastic manner during a massive earthquake and are expected to develop inelastic action in many of the members and connections. Only very recently our analytical capabilities reached a stage of development where such behavior can be corroborated reasonably well by test results and as yet has not received any significant use in practice.

What is extraordinarily important in seismic design are the engineering studies in the wake of major earthquakes. By the slow process of observation and analysis, a tremendous impact on our codes is being made continuously. In more recent years, this has been greatly augmented by coordinated experimental work in the laboratory together with analytical studies. Nevertheless, in the final analysis, field observations in the aftermath of a major destructive earthquake provide the acid test for our efforts in economically constructing earthquake-resistant structures.

The pent up desire and need for large and tall structures in climatically attractive yet seismically active areas of the world is very great. Thus, the era of tall buildings ushered in with the construction of a ten-story Home Insurance Building in Chicago in 1883, (to which two more stories were added later) remains unabated throughout the world.

LESSONS FROM HISTORY

The April 18, 1906 San Francisco great earthquake and the September 19, 1985 Mexico City major quake are used to illustrate the destruction that may be caused during such events as well as the survivability of some steel framed buildings. The Richter magnitude of the San Francisco earthquake was 8.3, and, except for the 1964 Alaska earthquake which had a magnitude of 8.4, is the strongest recorded in the U.S. The Mexico earthquake had a Richter surface wave magnitude of 8.1, and was followed the next day by and aftershock of magnitude of 7.5. However, since these magnitudes indicate only the energy release by an earthquake and are independent of the place of observation, the effect of these earthquakes on structures at a particular location should be judged differently.

The epicenter of the San Francisco earthquake was not far from the city, whereas the epicenter of the Mexico City earthquake was about 250 miles (400 km) away. Therefore, it is more appropriate to think in terms of the intensity of an earthquake at a location or site, rather than just in terms of its magnitude. The intensity of an earthquake can be expressed using the Modified Mercalli (MM) scale, which is divided into twelve intensities, with I corresponding to a barely perceptible event and XII associated with total damage (Steinbrugge, 1982). These intensity ratings based on observations of damage are not precise. Nevertheless, they are very useful and provide a reasonable basis for comparison. Using this approach, the 1906 San Francisco earthquake is considered to have had an intensity of XI, whereas the Mexico City earthquake may be said to have had intensities that varied from II to X.

The large spread in the intensities suggested above for the Mexico City requires some explanation.

> "Although staggering when viewed through the camera lens, the Mexico City disaster was a limited one. Only 250 buildings collapsed out of a total of 600,000 or more structures in the city; about one percent of the city suffered heavy damage. A great earthquake can be much worse" (Kerr, 1985, p. 633).

Thanks to the researchers in the Institute of Engineering at the National

Autonomous University of Mexico plausible reasons for severe damage and collapse of buildings can already be advanced.

In contrast to the San Francisco earthquake, the epicenter of the 1985 Mexico earthquake was located far away from Mexico City on the Pacific coast of Mexico. The recorded maximum accelerations in the epicentral region, which were 0.16 g, greatly attenuated with distance, and in one locality at some 100 km (62 miles) outside Mexico City they were only on the order of 0.03 g (Kerr, 1985). Comparable small peak accelerations occurred in the outlying districts of the city. However, in the central part of the city, the soil on a dried lake bed, on picking up resonant vibrations, amplified the peak accelerations to about 0.2 g with sustained, almost constant, periods of 2 sec (NRC/EERI, 1985). Therefore, many buildings in the 15- to 25-story range, also with natural period of about 2 sec, in the lake bed area of the city, due to resonance were either severely damaged or collapsed.

The two earthquakes selected for illustrating possible damage are different. The 1906 San Francisco earthquake, in addition to its large magnitude, is important because it provided a start in considering lateral forces in the design of steel buildings. The 1985 Mexico City earthquake dramatically demonstrated the credibility of a phenomenon, unrecognized by the codes, of sustained constant periods, together with a large acceleration (0.2 g) (NRC/EERI, 1985). The effect of such motion on buildings with the same fundamental periods is disastrous. Because of the severity and the damage this earthquake caused to modern structures, its evaluation is providing important knowledge on structural behavior. Further, it is important to note that many of the buildings in Mexico City experienced an earthquake of MM intensity VII in 1957. Therefore, some observations on the effect of a severe earthquake on new as well as repaired buildings can also be made.

Spectacular photographs of San Francisco after the earthquake and fire are shown in Figs. 1 and 2. According to Steinbrugge (1982), some 80% of the damage in the city was caused by fire which was poorly controlled because the water supply lines broke down. Both of the hotels, prominently seen in the photographs, framed in steel, are still in use today. Such good behavior of steel frames during this earthquake helped to entrench the position of steel for seismic construction.

Photographs of collapsed steel framed buildings in Mexico City are shown in Figs. 3 and 4. The building in Fig. 3 is an older structure exhibiting a typical failure in the upper stories. (A number of reinforced concrete buildings failed in the same manner.) A plausible reason for such failures is given in the NRC/EERC Report (1985): "Many structures had columns that were progressively reduced in side from the base upwards, giving the impression that the designers did their best to economize and presumably stay within building code requirements." Future studies may show that the strong column-weak girder concept was not always adhered to.

The collapsed and the damaged buildings shown in Fig. 4 are from a

Fig. 1 San Francisco downtown after 1906 earthquake and fire. St. Francis Hotel in picture still in use *(Courtesy: Karl V. Steinbrugge Collection.)*

Fig. 2 San Francisco Nob Hill after 1906 earthquake and fire. Fairmont Hotel in picture still in use *(Courtesy: Karl V. Steinbrugge Collection.)*

Fig. 3 11-story Atlas Building after 1985 Mexico City earthquake

Fig. 4 Demolition of collapsed 21-story unit of Pino Suarex complex. 1985 Mexico City earthquake

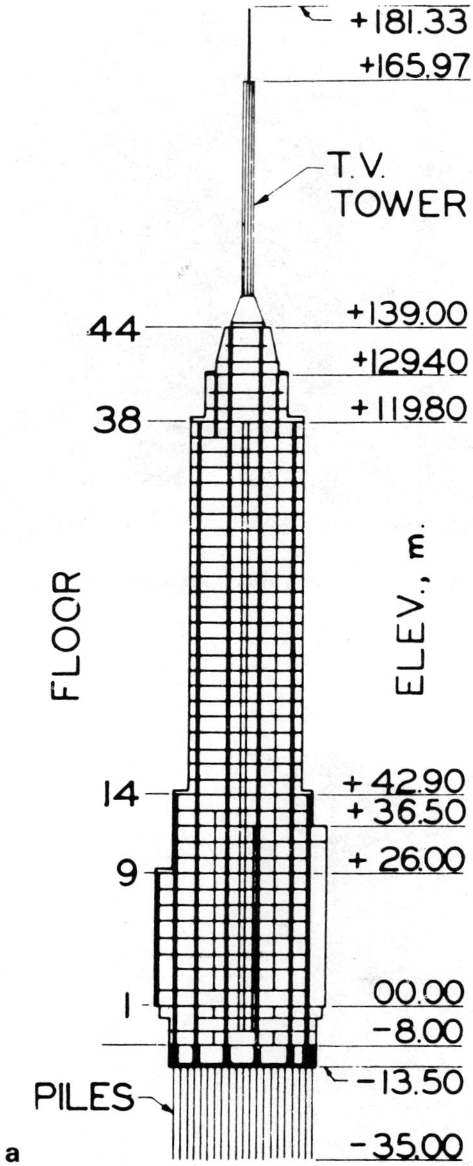

Fig. 5 Latino Americana Tower. (a) cross-section; (b) steel erection completed, 1955 *(Courtesy: Leonardo Zeevaert.)*

b

complex of two 14-story and three 21-story steel buildings. "One 21-story structural steel building and its adjacent 14-story building of similar construction were both totally destroyed" (NRC/EERC, 1985). Only future studies can clarify the reasons for this disaster.

A success story is documented in Fig. 5 and is a tribute to its designers and builders. This 44-story Latino Americana Tower did not even suffer broken windows during the 1957 Mexico City earthquake (Binder, 1962) and lost only about a dozen glass panes in the 1985 quake. The excellent behavior of this building may be attributed to the good steel skeleton, its pile foundations going down to the hardpan under the old lake bed, and its natural period, which was approximately double that of the dominant ground motion.

EVOLUTION OF CODES

The basic function of a building code is to provide *minimum* requirements for safeguarding against major failures and loss of life. Some of the user-sponsored codes, such as California's Administrative code for schools and hospitals, are concerned with minimizing damage as well as protecting the occupants. Such damage control is generally not considered in earthquake provisions of the usual building codes, and is not considered in this discussion. The main objectives in a general seismic code are based on the following philosophy:

1. A building must resist a minor shake without damage;
2. In moderate earthquakes some nonstructural damage is permissible;
3. During a major earthquake, a building must not collapse, but both some structural damage and nonstructural damage may occur.

This approach appears to have been adopted from the beginning and the seismic code provisions continue to be refined as new knowledge becomes available. San Francisco was rebuilt after the 1906 disaster under the provisions of a 30 psf (1.4 kN/m^2) wind pressure to account for either the effect of wind or earthquake. It was not until 1927 that the concept of lateral seismic forces proportional to mass was recognized in the Uniform Building Code (ICBO, 1982), and more extensively in 1933 following the Long Beach destructive earthquake (SEAOC Seismology Committee, 1976). Constant coefficients relating the mass of a building to the lateral force varied over the years. In 1943, however, the city of Los Angeles, indirectly recognizing the influence of flexibility in a building, adopted an equation for the lateral force coefficient, which depended on the number of stories. In 1948 a Joint Committee on Lateral Forces of the San Francisco Section of the American Society of Civil Engineers (ASCE) and the Structural Engineers Association of Northern California (Anderson et al., 1952) related the base shear coefficient C (base shear divided by building weight) to an estimated or calculated fundamental period T of the structure. This relationship, simply expressed as $C = K/T$, where K is a constant, is the precursor of current developments.

In 1959 the Structural Engineer Association of California (SEAOC) expanded the above approach, including the triangular distribution of lateral design forces on a structure. Then, as an outgrowth of further studies, it was recommended to establish the SEAOC Applied Technology Council (ATC) to provide the results of practical research applicable to structural engineering. The culmination of the ATC work on codes resulted in the publication in June, 1978 of *Tentative Provisions for the Development of Seismic Regulations for Buildings*, which was a cooperative effort of design professions, building code interests, and the research community (ATC, 1978). This work has been updated by the Building Seismic Safety Council (BSSC). The SEAOC Seismology Committee thoroughly revised the Recommended Lateral Force Requirements and has made this work generally available. A commentary on these recommendations is in preparation (July, 1987). Updating the current AISC provisions for seismic design of steel structures based on the compatible BSSC and SEAOC versions of the code is in progress. It is anticipated that the revised Uniform Building Code seismic requirements, closely following the SEAOC developments should become available in 1988.

Inasmuch as steel seismic codes in the United States are under periodic revisions, in the opinion of the writer, as a result of the 1985 Mexico earthquake, some findings will be used to further update the code. Very likely greater emphasis will be placed on possible effects of resonance between soil and structure. Seismic response in the soil layers and fill in the lowlands of cities requires much further study. It is also possible that in addition to the well recognized ductility requirements, a greater emphasis will emerge for the need for designing stiffer structures, that is, the deflection control of frames may become more stringent. It is likely that the importance of the torsional effects will also become more generally recognized. Redundancy in the structural systems will likely receive much greater recognition.

EVOLUTION OF STRUCTURAL STEEL FABRICATION

Along with the evolution that took place in the building codes through the years, fabrication of structural steel also changed. From the post-San Francisco-1906-earthquake era through the early 1960s, the connections were riveted. Wind bracing connections, the essential ingredient for all moment-resisting frames, were usually detailed per AISC recommendations such as those shown in Fig. 6 (AISC, 1934). It is instructive to note that in these excellent connections the steel is used in an optimum manner, either in tension along the rolling direction, or shear. A slight connection flexibility caused by bending of column flanges and attachment details is not critical. During a possible seismic event, a limited amount of flexibility in such connections is probably beneficial.

From the mid-1940s shop welding became increasingly popular, and by the mid-1950s when high-strength bolts were introduced, the fabricating industry rapidly adopted the use of welding and high-strength bolting to the

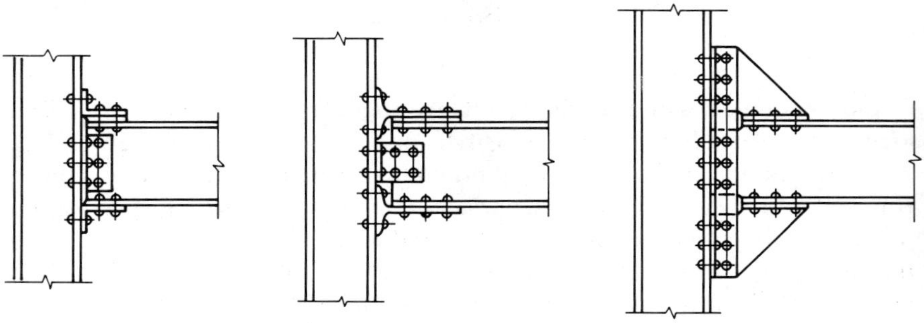

(a) (b) (c)

Fig. 6 Typical wind bracing connections circa 1930s *(After AISC, 1934.)*

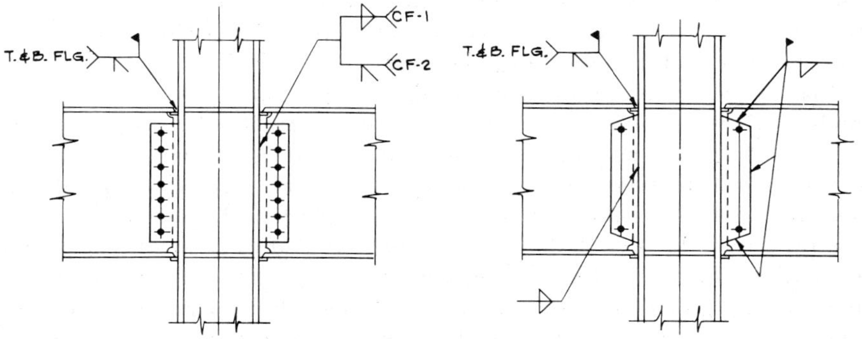

WEB BOLTED – FLANGE BUTT WELDED

(a)

FLANGE BUTT WELDED – WEB FILLET WELDED TO TAB

(b)

Fig. 7 Beam-to-column flange moment connections. (a) web bolted-flange butt welded; (b) flanges and webs welded *(Courtesy: Steel Committee of California.)*

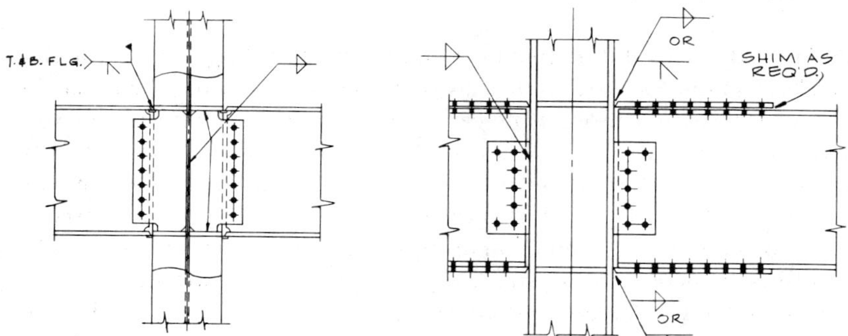

WEB BOLTED – FLANGE BUTT WELDED

(a)

WEB AND FLANGE H.S. BOLTED WITH MOMENT PLATES

(b)

Fig. 8 (a) Beam-to-column moment connection. Bolted web-flange butt welded. (b) Beam-to-column high-strength bolted connection *(Courtesy: Steel Committee of California.)*

exclusion of riveting. Typical seismic moment connections in general use today on the West Coast are shown in Figs. 7(a), 7(b), and 8(a) (Steel Committee of California, 1981). The costlier connection shown in Fig. 8(b) is now resorted to only if field bolting is necessary. All of these connections use extensive welding, and, except for ordinary erection bolts in the detail shown in Fig. 7(b), employ high-strength bolts.

Current construction of taller and more complex buildings than in the past caused the mills to roll larger sections. Thus, whereas in the 1930s the largest available column section was W14 × 426, now one can obtain W14 × 730 sections. Similarly, in addition to the earlier largest W36 × 300 beam section, one can now specify W36 × 359 sections and order custom sections up to 42 in. (1.067 m) in depth. Moreover, in addition to 36 ksi (248 MPa) steel, 50 ksi (345 Mpa) material is now generally available.

As may be noted from Figs. 7 and 8, present day fabrication of major seismic connections involves considerable amounts of welding, resulting in very rigid joints. Because of this and the use of stronger and thicker material, great care is required in specifying appropriate welding procedures in order to minimize residual stresses. In seismic design one must guard especially against the possibility of brittle fractures to which welded joints are susceptible. Experimental verification of cyclic behavior of large joints and connections simulating seismic conditions is very limited.

DESIGN OPTIONS FOR TALL STEEL BUILDINGS

Several types of steel structural systems can be used to advantage for tall buildings situated in seismic regions. In the design of such systems the basic schemes for developing lateral resistance is similar to that used for wind bracing. However, sometimes some modifications are necessary in order to achieve maximum frame ductility beyond the elastic range of behavior. For an economical seismic-resistant design, some of the joints and members are expected to behave inelastically in the event of a major earthquake. This provides a mechanism for energy dissipation, effectively damping the motion of a frame.

The importance of frame ductility can be seen by considering the diagram shown in Fig. 9. Here, a very strong earthquake is assumed to have a site acceleration of 0.33 g (compare with the 1985 Mexico City earthquake of 0.2 g). For this condition the upper curve gives the elastic response spectrum for oscillators with different periods of vibration. In this diagram the ductility ratio μ_δ is defined as the ratio of the system deflection to the deflection at yield. For the upper curve $\mu_\delta = 1$. When, because of inelastic (plastic) deformations, the system deflections increase, the values for μ_δ also increase, and the curves progressively drop down. The base shear coefficient C for which a system has to be designed dramatically changes with ductility. For example, for an elastic system with a natural period of 2 sec, the value of C corresponds to the upper arrow in the figure, whereas the lower arrow defines the value of C for a system with $\mu_\delta = 6$. The economic consequences of this are apparent.

For comparison, a dashed curve for a widely used seismic code (Uniform Building Code, 1982) based on elastic concepts is also shown in Fig. 9. This curve corresponds to the code elastic design requirements multiplied by a factor of 1.4 to approximate the limit of elastic action, that is, the beginning of yield. For the earthquake considered, the discrepancy between this curve and the upper one is very large and can be reconciled only if a structure is capable of deforming inelastically several times more than that occurring at first yield. For the assumed conditions, only buildings of over about 20 stories with a period greater than 2 sec and meeting a very severe ductility requirement of 6 would survive if designed by this code. Therefore, in seismically active regions direct use of code specified response spectra may be hazardous. For this reason, the geologically foreseeable most severe earthquake site accelerations on the order of 0.5 g have sometimes been used for design in San Francisco (Merovich and Nicoletti, 1982). On the other hand, the response spectra favor tall buildings by indicating a decreasing design base shear with increasing period, that is with increasing building height. Therefore, in some geographic locations, wind, rather than seismic requirements, controls the design of tall buildings.

The above requirements can generally be met by moment-resisting frames (MRFs), and they are by far most widely used in the seismic-resistant design of low and moderately tall buildings. MRFs possess good ductility and can be designed to control displacements at service loads. An example of framing

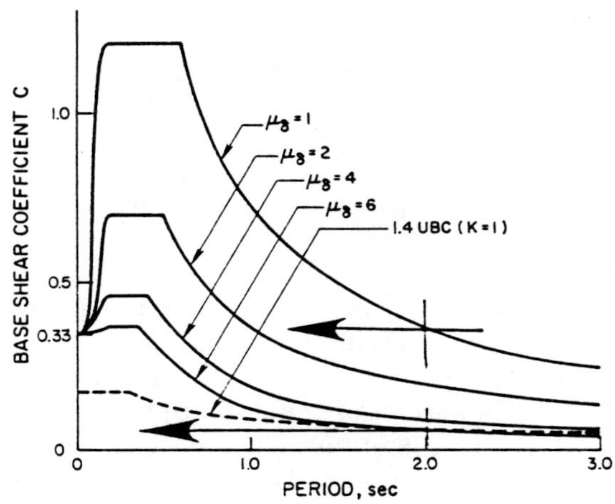

Fig. 9 Base shear coefficient curves for different ductilities for a severe earthquake

of this type is shown in Fig. 10. Connections of the type shown in Figs. 7 and 8 would normally be used for this type of framing.

The second type of framing extensively used in the past, especially along the narrow widths of buildings and elevator shafts, is shown in Fig. 11 (a). Recently, some objections (Popov and Black, 1981) developed to this particular type of concentrically braced frames (CBFs). Extensive experiments have shown (Yamanouchi et al., 1984) that during severe cycling simulating an earthquake, one of the braces at a joint may buckle or fail (Fig. 11(b)), greatly reducing the capacity of the frame. An unbalance of the vertical force components at a joint, such as C in the figure, can cause the floor beam, such as AB, to severely deflect. Fortunately, there are other bracing schemes, such as X-bracing, that can be designed to act more satisfactorily. However, it must be

Fig. 10 Example of moment-resisting framing. Fourth and Blanchard Building, Seattle *(Courtesy: United States Steel.)*

clearly recognized that under severe cyclic loading, the capacity of braces greatly deteriorates (Popov and Black, 1981). The use of CBFs in a building produces a stiff, but not particularly ductile, structure.

A bracing system that on the one hand is very stiff, and on the other, possesses good ductility for extreme cyclic loading conditions, is illustrated in Fig. 12. This type of framing is referred to as eccentrically braced framing

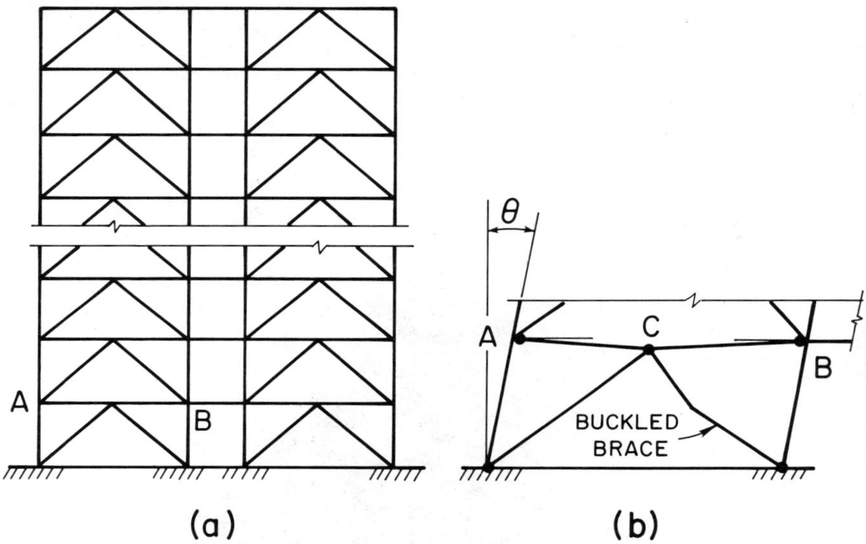

Fig. 11 (a) Example of concentrically braced framing; (b) possible brace buckling at severe loads

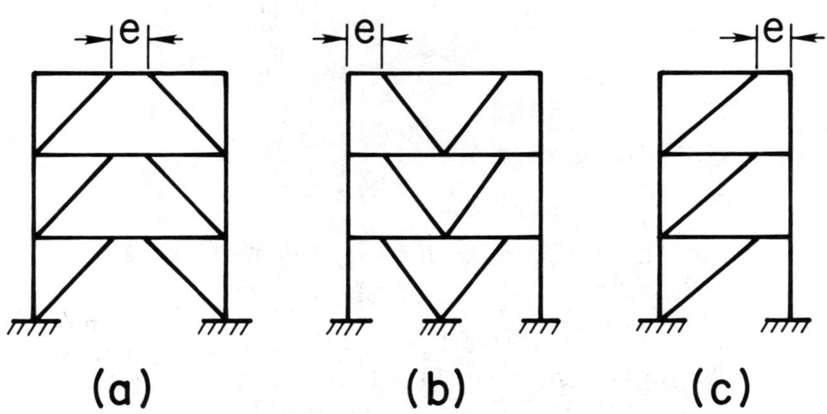

Fig. 12 Examples of eccentrically braced framing

(EBF). Although only recently introduced for seismic application (Roeder and Popov, 1978; Malley and Popov, 1984; BSSC, 1985; Kasai and Popov, 1986) it has already been adopted for many tall buildings. The distinguishing aspects of this framing consists of two features. First, flexible short beam segments or links of length e (Fig. 12) are isolated along beams. Such links transmit vertical force components in the braces mainly through shear and moment. The horizontal force components in the braces are largely resisted by the long parts of beams. Second, since the forces in the braces are limited by well predictable strength capacities of the links, the braces can be confidently designed so as not to buckle.

Eccentric bracing can be arranged in many different ways. For general use the three types shown in Fig. 12 are favored by the writer. Bracing shown in Fig. 12(a) avoids the need for field beam flange welding, and the concentric joints at columns can be bolted. The V-bracing shown in Fig. 12(b) has an advantageous kinematic mechanism in the plastic range of frame behavior. Single diagonal bracing shown in Fig. 12(c) is well suited for narrow bays. The concentric joints at columns can be bolted.

Another possible arrangement for eccentric bracing is shown in Fig. 13. If, in the spirit of the late Fazlur Khan, this bracing could be exposed, perhaps it could form a suitable motif for a garage or a building.

Several variations on the above three basic framing types have been used in the design of steel seismic-resistant buildings. One of the important variations is perimeter framing. In tall buildings it is often found advantageous to concentrate the lateral-resisting framing along the outer walls with the remainder of the framing designed to carry only gravity loading. Usually, MRFs with closely spaced columns are used in such cases. Frequently, it has been found advantageous to combine perimeter framing with EBFs or special

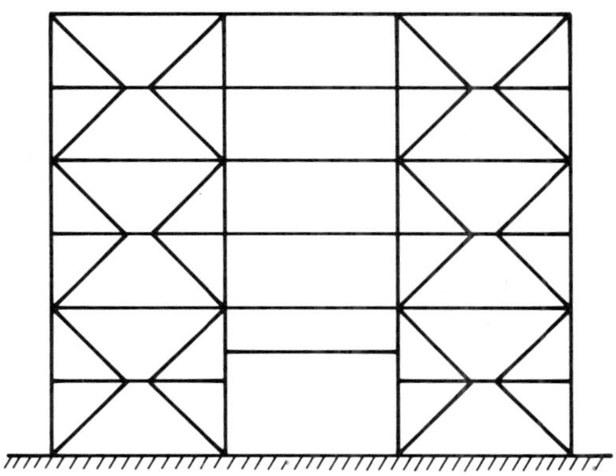

Fig. 13 Example of exposed eccentrically braced framing

reinforced concrete panels along the narrow directions of buildings. Many other combinations of the three basic types of framing are possible. Some of them will be illustrated in the next section.

DESIGN EXAMPLES

Many buildings have been constructed of considerable height in seismically active regions. A few examples of notable buildings are provided as illustrations. For low and moderate heights conventional MRFs in the two orthogonal directions are generally employed; CBFs are occasionally used along the narrow direction. For tall buildings the present trend is to use MRFs as perimeter framing. For those in the higher height range, in order to reduce drift, the use of EBFs in selected bays has become popular in the United States in recent years. Figs. 14 and 15 show examples of combined use of EBFs with MRFs. In Japan, for their tallest building to date (1976), Fig. 16, Kajima Corporation used moment-resisting perimeter framing stiffened with slitted reinforced concrete panels placed in selected bays. On the 54-story

Fig. 14 Group of four similar Embarcadero Center Buildings, San Francisco (1980). Moment-resisting frame of 45-story building in foreground is stiffened with eccentric bracing in transverse direction as shown at right *(Courtesy: URS/John A. Blume & Associates.)*

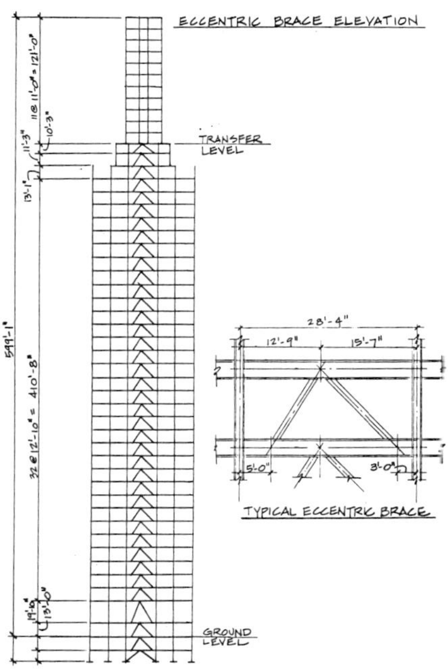

Fig. 15 345 California Street Building, San Francisco (1986). Moment-resisting perimeter framing is stiffened with eccentric bracing in narrow direction *(Courtesy. Skidmore, Owings & Merrill, San Francisco, Photo by Wes Thompson.)*

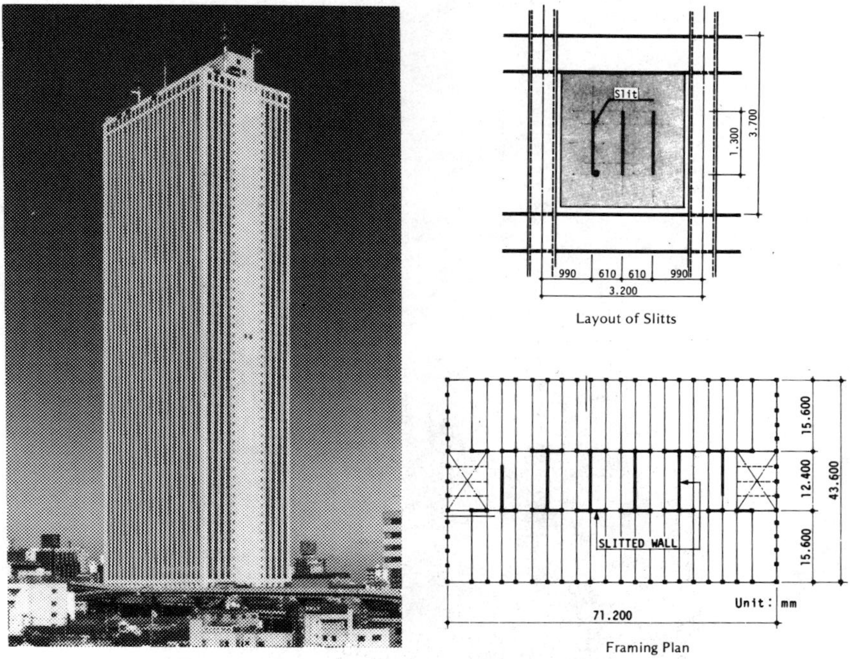

Fig. 16 60-story Ikebukuro Building, Tokyo (1976). Moment-resisting perimeter framing is stiffened with slitted walls *(Courtesy: Kajima Corporation and Muto Institute of Structural Mechanics.)*

Shinjuku Building (1979) designed by Taisei Corporation, again, in addition to a moment-resisting perimeter frame, special reinforced concrete panels with yielding elements were used.

A general view of the 44-story Latino Americana Tower discussed in connection with Fig. 5 is shown in Fig. 17. This well designed building sustained a number of major earthquakes. A well known San Francisco landmark is shown in Fig. 18. Some critical features of the design of this building were experimentally verified. An interesting structural framing for a difficult lot shown in Fig. 19 illustrates the possibilities of framing in steel.

ACKNOWLEDGEMENTS

The writer is greatly indebted to a number of individuals who generously provided some essential support material for this paper. The acknowledged illustrations were kindly provided by the following: Navin Amin and Thomas Lauck of Skidmore, Owings & Merrill, San Francisco; Joseph Nicoletti and Andrew Merovich of URS/John A. Blume; Dennis Oh of Chin and Hensolt;

Fig. 17 Latino Americana Tower, Mexico City (1955) *(Courtesy: Leonardo Zeevaert.)*

Fig. 18 Transamerica Building during construction (1972) *(Courtesy: Chin and Hensolt.)*

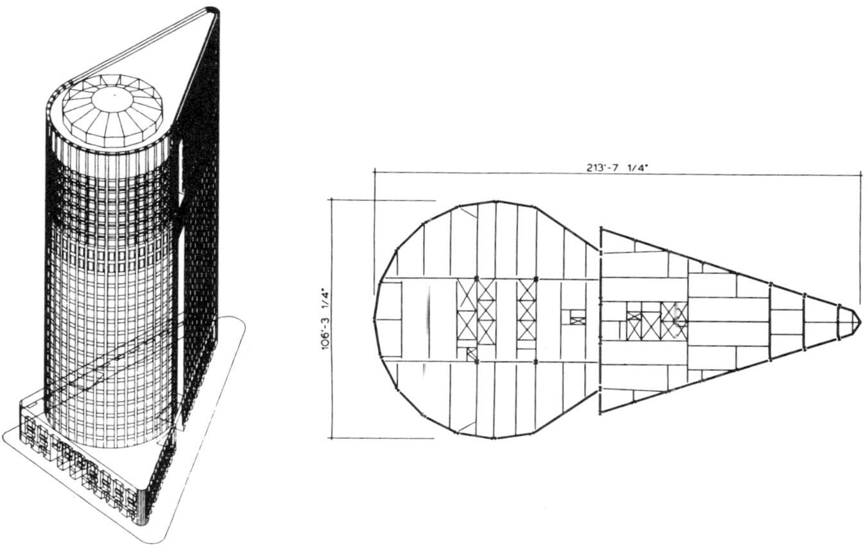

Fig. 19 388 Market Street, San Francisco. Moment-resisting perimeter framing with three special moment frames in narrow bays (1986) *(Courtesy: Skidmore, Owings & Merrill,*

Karl V. Steinbrugge; Tsunehisa Tsugawa of Kajima Corporation; and Leonardo Zeevaert. Koichi Takanashi and Atsuo Tanaka also assisted with this effort. Gail Faizel prepared several excellent illustrations, and Eric Eisman meticulously typed the manuscript and arranged the art work. The writer is also most grateful for the partial financial support provided by NSF (Grant ECE 84-18487) and AISI, which made this work possible.

REFERENCES/BIBLIOGRAPHY

AISC, 1934
STEEL CONSTRUCTION MANUAL, American Institute of Steel Construction, New York, 2nd Ed., p. 127.

AISC, 1984
SPECIFICATIONS FOR STRUCTURAL STEEL BUILDINGS, Commentary on the Proposed Load and Resistance Factor Design, Chicago, August 1, p. 18.

Anderson, A. W., et al., 1952
LATERAL FORCES OF EARTHQUAKE AND WIND, Transactions, ASCE, Vol. 117, pp. 716-780.

ATC, 1978
TENTATIVE PROVISIONS FOR THE DEVELOPMENT OF SEISMIC REGULATIONS FOR BUILDINGS, NBS Special Publication 510, U.S. Department of Commerce, Washington, DC. Applied Technology Council.

Binder, R. W., 1962
SIGNIFICANT ASPECTS OF THE MEXICAN EARTHQUAKES, May 11 and 19, Proceedings, 31st Annual Convention, Structural Engineers Association of California, pp. 96-110.

BSSC, 1985
RECOMMENDED PROVISIONS FOR THE DEVELOPMENT OF SEISMIC REGULATIONS FOR NEW BUILDINGS, Building Seismic Safety Council, Washington, DC.

ICBO, 1982
UNIFORM BUILDING CODE, International Conference of Building Officials, Whittier, CA.

ICBO, 1985
UNIFORM BUILDING CODE, International Conference of Building Officials, Whittier, CA.

Kasai, K. and Popov, E. P., 1986
GENERAL BEHAVIOR OF WF STEEL SHEAR LINK BEAMS, Journal of Structural Engineering, ASCE, Vol. 111, February.

Kerr, R. A., 1985
PREDICTABLE QUAKE DAMAGE, Science, Vol. 230, No. 4726, November p. 633.

Malley, J. O. and Popov, E. P., 1984
DESIGN OF SHEAR LINKS IN ECCENTRICALLY BRACED FRAMES, Journal of Structural Engineering, ASCE, Vol. 110, September, pp. 2275-2295.

Merovich, A. T. and Nicoletti, J. P., 1982
ECCENTRIC BRACING IN TALL BUILDINGS, Journal of the Structural Division, ASCE, Vol. 108, May, pp. 2066-2080.

NCR/EERI, 1985
IMPRESSIONS OF THE GUERRO-MECHOACAN, MEXICO EARTHQUAKE OF 19 SEPTEMBER 1985: A PRELIMINARY RECONNAISSANCE REPORT, Publication No. 85-05, National Research Council/ Earthquake Engineering Research Institute, Berkeley, CA, October, 20 pp. plus figures.

Popov, E. P. and Black, R. G., 1981
STEEL STRUTS UNDER SEVERE CYCLIC LOADING, Journal of the Structural Division, ASCE, Vol. 107, September, pp. 1857-1881.

Roeder, C. W. and Popov, E. P., 1978
ECCENTRICALLY BRACED STEEL FRAMES FOR EARTHQUAKES, Journal of the Structural Division, ASCE, Vol. 104, March, pp. 391-412.

SEAOC Seismology Committee, 1976
RECOMMENDED LATERAL FORCE REQUIREMENTS, Structural Engineers Association of California, San Francisco, CA.

Steel Committee of California, 1981
STEEL CONNECTIONS/DETAILS AND RELATIVE COSTS, El Monte, California, 24 pp.

Steinbrugge, K. V., 1982
EARTHQUAKES, VOLCANOS, AND TSUNAMIS, Scandia America Group, 280 Park Avenue, New York, NY 10017, 392 pp.

Yamanouchi, H., et al., 1984
EXPERIMENTAL RESULTS ON A K-BRACED STEEL STRUCTURE UNDER SEISMIC LOADING UTILIZING FULL-SCALE SIX-STORY TEST STRUCTURE—U.S./JAPAN COOPERATIVE RESEARCH PROGRAM, Proceedings, 1984 Annual Technical Session, Stability Under Seismic Loading, Structural Stability Research Council, San Francisco, CA.

Zeevaert, L., 1964
STRUCTURAL STEEL BUILDING FRAMES IN EARTHQUAKE ENGINEERING, Proceedings, Steel Utilization Congress, Luxembourg.

Design for Strength (Stability)

Giulio Ballio

Even though much energy has been devoted to the analysis of stability phenomena, there is still a great deal to be covered before it becomes possible to conclude that stability problems may be considered well understood and should no longer be subjected to research efforts. Different approaches are still used in solving stability problems. We study columns in a different way from that of shells or orthotropic plates. Nevertheless there is a common tendency to a convergence of studies and code formats.

In this paper the future trends and research needs are discussed in order to guarantee reliability, to unify code formats and to provide even more economical designs for compressed members, structural systems, and thin-walled elements.

We shall consider the development of our knowledge starting from the conclusions of the Third International Colloquium on Stability of Metal Structures, held in Toronto, Canada in 1983, organized by the Structural Stability Research Council (SSRC) in cooperation with European Convention for Constructional Steelwork (ECCS) and the National Council of Engineers of Romania (SSRC, 1983).

COMPRESSED MEMBERS

The research work of the past 30 years has shown without any doubt that the influence of geometrical and mechanical imperfections has a fundamendamental effect on the value of the load-carrying capacity of compressed members.

The maximum axial load changes not only with the slenderness, as codes in all part of the world recognize, but also it depends on yield limit, initial crookedness, residual stresses, shape of the section, and manufacturing procedures (Ballio and Mazzolani, 1983; Chen and Lui, 1985).

Pin-ended Columns

The approach to pin-ended columns is well known. Tests and numerical simulation were performed in Europe, Japan, and the United States. The researchers agree that the results of these studies are very well described by multiple column curves as in Fig. 1 (Zandonini, 1983).

Thus a question arises: Why are there so many objections to the introduction of multiple column curves in national and international codes? Probably we may find an answer to this question by considering the uncertainties that the research results have posed to the code makers and to the designers. It is unpleasant to admit that the reliability of a compressed bar may depend on its manufacturing process. Engineers with their deterministic feeling find it difficult to accept the fact that two columns of the same shape may have two different values of resistance depending on the factory in which they were

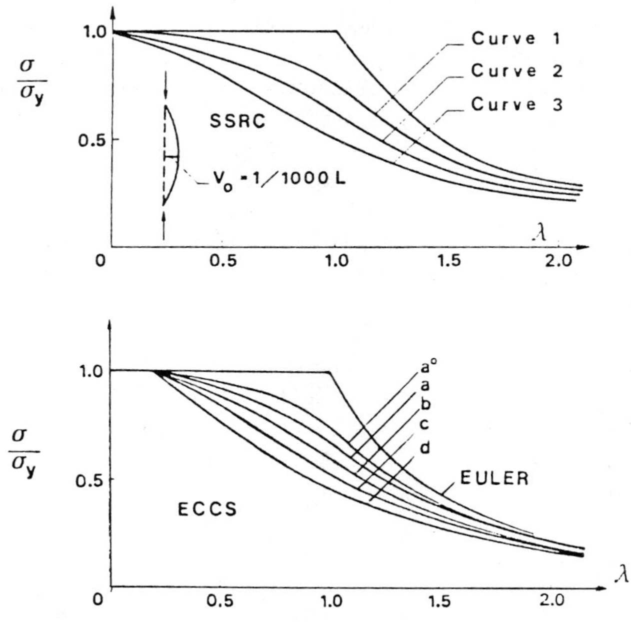

Fig. 1 SSRC and ECCS column curves

produced. For these reasons steel producers need to perform experimental and numerical analyses to verify the correct relationship between shapes and column curves. The relative effects of residual stresses in heavy sections and high strength steel bars may be especially different from those for normal sections and usual steel grades. The results obtained by the ARBED Company (Aschendorff et al., 1983) may be considered as a promising beginning. They have demonstrated that their heavy sections behave better than the predictions of the European recommendations (ECCS, 1978). Heavy sections were assigned to the lower curve "d." The experimental and numerical results show that the behavior is more favorable for buckling about both the strong and weak axes and the European Codes (Commission of the European Communities, 1984) accepted this point of view. With the same procedure it was found that I-sections in high-strength steel with a ratio of height to width greater than 1.2 gave results more favorable than those predicted by codes.

Of course the above conclusions must be confirmed by other tests before they may be accepted by codes all over the world. Nevertheless research of this type may be considered as an example of what can be done to improve reliability and economy of steel columns.

Of course, results lower than code predictions may be found. Let us consider a hollow section obtained by welding two high-strength steel cold-formed profiles (Ballio et al., 1977). The welding process caused an unfavorable pattern of residual stresses; in the range of small slenderness the resulting strength was much lower than that predicted by buckling curves (Fig. 2).

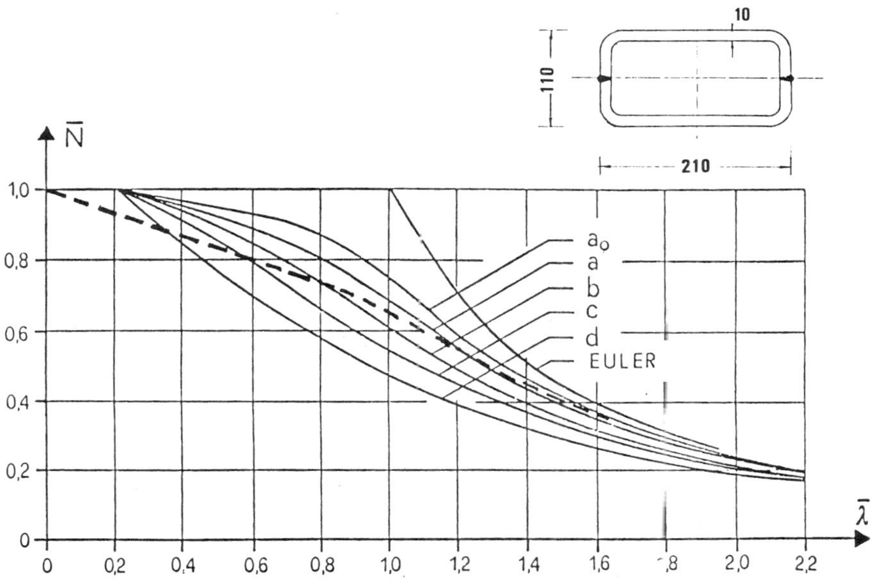

Fig. 2 Strength of tubular columns fabricated by welding of two cold-formed sections

Built-up Compressed Members

It is possible that the pressure of economy and technology will modify our method of designing built-up members. In fact the strength of built-up members is influenced by the type of connections between the cords. The shear forces acting in the connections increase exponentially with the applied axial load. Some tests and numerical simulations (Zandonini, 1985) have shown that at 80% of the ultimate load the value of the shear is only 25% of the value corresponding to the ultimate load.

I think that it is worthwhile to support research in order to allow codes to choose between the two following alternatives: (1) improve the stiffness and the strength of the connections regardless of economical aspects (welded or friction bolted connections) or, (2) use more economical connections designing for a strength lower than the loading capacity of the cords.

Perhaps codes may reconsider the practical approach often used in the past: It is economical and reliable to design built-up members for a strength of about 80% of the load capacity of the cords without giving provisions that are too severe for the quality and the tolerances of the connections.

Beams

Comprehensive reviews on studies about flexural torsional buckling are already available (Galambos, 1983; Nethercot, 1985). Our knowledge in this field benefitted from the collection and re-evaluation of several hundred buckling tests (Fukumoto and Itoh, 1983).

For symmetrical I-beams we must still perform a detailed numerical and experimental analysis of the effects of geometrical and mechanical imperfections. Residual stresses may have a detrimental effect on the carrying capacity of welded sections if compared with rolled sections. Our knowledge is more limited when considering the behavior of beams with asymmetrical sections and the interaction between flexural torsional buckling and compressed flange instability.

The restraining effects due to the rotational constraints, often acting where the loads are applied, must also be examined in detail (Lawson and Nethercot, 1985). More precisely future research considering the beam as a part of a three dimensional structure is necessary. Thus it would be possible to know the loads acting on the bracings of a beam and to remove the simplistic hypothesis that often causes underestimation of buckling loads.

Beam-Columns

A tremendous amount of work has been done in the field of beam-columns (Nethercot, 1983; Chen and Lui, 1983). Nevertheless we need to recognize the following points:

1. Research has solved some problems but a general formulation was never reached.
2. Many interaction curves have been proposed but the old equation

$$\frac{P}{P_u} + \frac{C_m M}{M_u (1\text{-}P/P_{ex})} = 1 \qquad (1)$$

   is still retained in the more recent codes, at most with some slight modifications considering biaxial bending and lateral buckling.
3. The equivalent slenderness concept may be useful only for no-sway frames; for economical design it must consider the rigidity of the connections; the sway frames equation (1) gives many uncertainties that only non-linear analysis may avoid (see the section on Structural Systems).
4. Mechanical and geometrical imperfections are not as important as for compression members; shape of the section, moment distribution, and transverse loads are decisive.
5. The interaction equation (1) is always on the safe side when the bending moment is constant; in some cases it may lead to considerable underestimation of the allowable moment (even of 100%).
6. The C_m coefficient must be defined independently from the assumed interaction curve; it must be stated on the basis of the values of bending moments corresponding to the ultimate strength of the column.
7. The C_m coefficient depends on bending moment distribution along the beam, transverse loads, type of constraints at the ends, slenderness, type of section, and axial force value (Ballio and Campanini, 1981).

Thus it is most unlikely that a new formulation will be available in the near future that is able to account for different shapes of sections, end moments, and transverse loads simultaneously.

Perhaps researchers must give up the quest; they must face reality and stop searching for a general formula. Perhaps, in time to come, it will be more useful to devote research efforts in order to develop sufficiently general, reliable, and efficient non-linear minicomputer programs that any practical engineer can use.

STRUCTURAL SYSTEMS

The efforts devoted to the formulation of the column problem and to the search for the parameters influencing its stability load absorbed the major part of the resources of our scientific community during the past 20 years. It is the author's opinion that we should devote the most of our future efforts to the study of the stability of overall structures.

Let us consider simple problems such as the restraining effects given to a column by elastic constraints, the stability of the upper chord of through bridges, and the design of the horizontal bracings stabilizing the trusses of mill buildings. Engineers still ask for theoretical solutions of the same qualitative level as those that were obtained for compressed columns. In other words we are able to evaluate, with a high degree of reliability, the maximum compressive load of a column but we design its restraints on the basis of rough hypothesis, often disregarding the influence of imperfections and of the non-linear behavior of the material. As another example let us consider steel frames. We have characteristics for braced and unbraced frames; sway and no-sway frames; rigid and semirigid connections. May we define these characteristics without ambiguity? Obviously an "unbraced sway frame with semirigid connection" does exist. Let us analyze the following statements, widely used in our literature. "In braced frames sidesway is prevented by adequate diagonal bracings or shear walls." How can a scientific measure be given to the word "adequate"? In a similar manner we maintain that in no-sway frames, second order effects can be disregarded. But how may it be proved beyond any doubt that second order effects are negligible if their order of magnitude is not evaluated? A connection may be considered as rigid if its stiffness has no significant influence on the structural behavior. Again, how may we judge the influence of joint deformation without evaluating it?

Last but not least the interest in semirigid connections is steadily increasing since the cost of the joints is an important part of the total cost of the frame.

In these past five years promising progress has been achieved. European and U.S. groups, often cooperating with each other, are developing a new philosophy for approaching these problems. Present studies concern the experimental assessment of the behavior of the most common connections and their analytical models; the influence of different types of connections on the ultimate load of frames (Fig. 3); and the statement of simplified design rules that are also applicable to semirigid connections, thus encouraging simplicity and economy of the frame.

Much work remains to be done in this field. Recently Committee 8.2 of ECCS chaired by Professor Vogel (ECCS, 1984) issued a document that may be regarded as a new and very interesting approach to the problem of system stability. The document abandons the effective length concept and gives tentative rational definitions of sway frames and of rigid joints. In detail, sway frames are called unbraced frames or braced frames in which the stiffness of the bracing systems against sidesway is less than five times the stiffness of the bare frames. Rigid joints are welded connections and friction type connections with high-strength bolts. The document recognizes that sway frames are complex systems with several independent parameters and that it is not always possible to use simple and safe formulas to calculate their ultimate limit state without loss of economy. Thus it tries to classify the methods that may be used as; elastic-plastic analysis, elastic analysis and the Merchant-Rankine approach. For each of the above methods the document states the basic hypothesis and the values of the main parameters that must be used in the structural analysis. Summarizing, the document states only the basic assumptions and the methods to be used for performing the structural analysis; it does not specify design rules. This approach is quite new and its consequences will be discussed at the end of this paper.

LOCAL STABILITY

We are assisting with a wonderful development of the use of nonstandard shapes obtained by cold-forming and by welding. The cost of money in this inflationary period of our economy discourages large stocks of hot-rolled profiles. The architects complain about the constraints imposed by predetermined shapes; they desire freedom in developing their imagination. The designers are confident of the calculation methods derived from Winter's studies (1947) and in dealing with profiles obtained from cold-forming thin sheets; that is, shapes with high b/t ratios and low values for the thickness t.

Practical engineers ask for performance of experimental research and

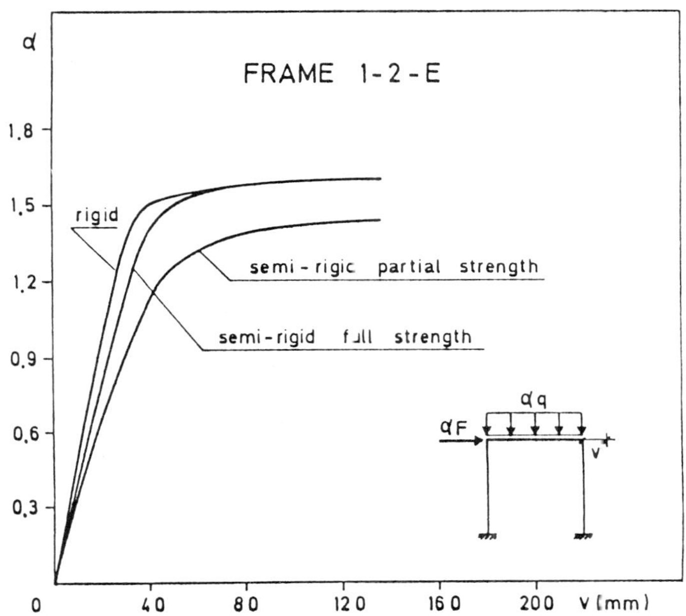

Fig. 3 Effect of different types of connections on load-carrying capacity of frame

development of calculation techniques in order to design orthotropic plates, stiffened and unstiffened webs and box girders. In this field, economy and reliability have not yet been satisfied. Last but not least, even if seismic design is beyond the scope of this paper, I wish to point out that local stability and b/t limit values are very sensitive to the loading pattern. The behavior of a section subjected to local stability phenomena is very different if loaded by monotonic or cyclic actions (Fig. 4).

CONCLUSIONS

Two conclusions may be taken from the previous points.

1. Economy and reliability requirements need further studies regarding the stability of systems. Present methods of analysis are often conservative and lead to a waste of material. Research efforts and resources must be addressed in this direction.
2. There is no hope that more sophisticated approaches in the field of stability of structural systems will be summarized in simple and practical design rules like interaction formulas or alignment charts.

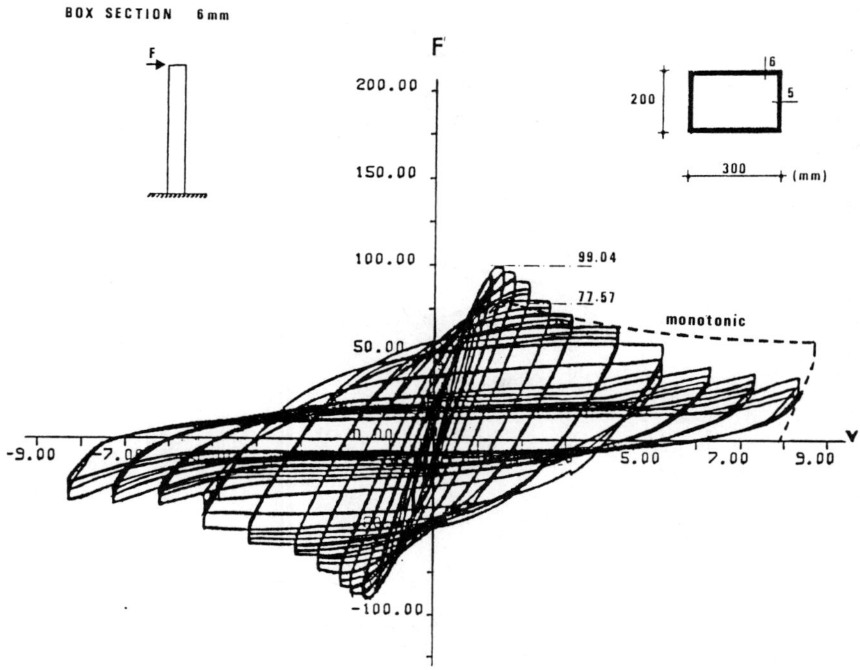

Fig. 4 Effect of local instability on hysteretic behavior of tubular beam-column

Thus we may expect that code format may change slowly in future. It is possible that codes will necessitate more precise input data (geometrical and mechanical imperfection values, construction tolerances), more severe performance requirements (serviceability limitations, maintenance critiera), and stricter procedures (quality control of the design and building process). However, codes will give engineers more freedom to select verification methods.

For this reason design will need more and more sophisticated computer programs. The validation of computerized methods of analysis is an urgent topic. Two opposite tendencies are fighting one another. Steel producers wish to spread as widely as possible general purpose computerized methods in order to allow everybody to use steel. Specialists in design of steel structures wish to keep their know-how; they prefer to appoint and check their own computer programs and they do not like validation procedures that are too severe. Once we know the winner of this competition, we shall be able to foresee the future trends in the design of steel structures.

REFERENCES/BIBLIOGRAPHY

Aschendorff, K. K., Bernard, A., Buck, O., Mang, F. and Plumier, A., 1983
OVERALL BUCKLING OF HEAVY ROLLED I-SECTION COLUMNS, Proceedings of the Third International Colloquium on Stability of Metal Structures, Toronto, pp. 37-49.

Ballio, G., Finzi, L., Setti, P., and Urbano, C., 1977
CAPACITA' DI ASTE TUBOLARI COMPRESSE IN ACCIAIO AD ELEVATO LIMITE ELASTICO, Costruzioni Metalliche, No. 2.

Ballio, G. and Campanini, G., 1981
EQUIVALENT BENDING MOMENT FOR BEAM-COLUMNS, The Journal of Constructional Steel Research, Vol. 1, No. 3.

Ballio, G. and Mazzolani, F. M., 1983
THEORY AND DESIGN OF STEEL STRUCTURES, Chapman and Hall, London and New York.

Chen, W. F. and Lui, E. M., 1983
DESIGN OF BEAM-COLUMNS IN NORTH AMERICA, Proceedings of the Third International Colloquium on Stability of Metal Structures, Toronto, pp. 253-292.

Chen, W. F. and Lui, E. M., 1985
COLUMNS WITH END RESTRAINT AND BENDING IN LOAD AND RESISTANCE DESIGN FACTOR, Engineering Journal, Vol. 22, No. 3.

Commission of the European Communities, 1984
EUROCODE NO. 3—COMMON UNIFIED RULES FOR STEEL STRUCTURES

ECCS, 1978
EUROPEAN RECOMMENDATIONS FOR STEEL CONSTRUCTION, European Convention for Constructional Steelwork.

ECCS, 1984
ULTIMATE LIMIT STATE CALCULATION OF SWAY FRAMES WITH RIGID JOINTS, ECCS Publication No. 33.

Fukumoto, Y. and Itoh, Y., 1983
EVALUATION OF BEAM STRENGTH FROM THE EXPERIMENTAL DATA-BASE APPROACH, Proceedings of the Third International Colloquium of Metal Structures, Toronto, pp. 133-150.

Galambos, T. V., 1983
A WORLD VIEW OF BEAM STABILITY RESEARCH AND DESIGN PRACTICE, Proceedings of the Third International Colloquium on Stability of Metal Structures, Toronto, pp. 113-132.

Lawson, R. M. and Nethercot, D., 1985
LATERAL STABILITY OF I BEAMS RESTRAINED BY PROFILED SHEETING, The Structural Engineering, Vol. 63B, No. 1.

Nethercot, D. A., 1983
EVALUATION OF INTERACTION EQUATIONS FOR USE IN DESIGN SPECIFICATIONS IN WESTERN EUROPE, Proceedings of the Third International Colloquium on Stability of Metal Structures, Toronto, pp. 293-314.

Nethercot, D., 1985
DESIGN OF LATERALLY UNSUPPORTED BEAMS and ELASTIC LATERAL BUCKLING OF BEAMS, Beams and Beam Columns, Applied Science Publishers Ltd.

SSRC, 1983
PROCEEDINGS OF THE THIRD INTERNATIONAL COLLOQUIUM ON STABILITY OF METAL STRUCTURES, Toronto.

Winters, G., 1947
STRENGTH OF THIN STEEL COMPRESSION FLANGES, Bulletin 35/3, Cornell University Engineering Experiment Station, Ithaca, N.Y.

Zandonini, R., 1983
RECENT DEVELOPMENTS IN THE FIELD OF STABILITY OF STEEL COMPRESSED MEMBERS, Proceedings of the Third International Colloquium on Stability of Metal Structures, Toronto, pp. 1-19.

Zandonini, R., 1985
STABILITY OF COMPACT BUILT-UP STRUTS: EXPERIMENTAL INVESTIGATION AND NUMERICAL SIMULATION, Construzioni Metalliche, No. 4.

Adaptive Expert System for Preliminary Design of Wind Bracings in Steel Skeleton Structures

Tomasz Arciszewski
Wojciech Ziarko

There is a growing belief that expert systems will revolutionize engineering design in the coming years. In the area of tall buildings, however, progress is still very limited. Research at Carnegie-Mellon University (Maher, 1984) represents the first implemented expert system for the preliminary design of tall buildings with a manually encoded set of inference rules. In a paper by Arciszewski (1986), a method of computer generation of types of wind bracing in steel skeleton structures has been proposed. The method is based on the stochastic simulation of morphological analysis. This article presents a more advanced method of development of wind bracing types for the purposes of preliminary design. The concept of an adaptive expert system has been used, and an earlier proposed solution generator became a part of the adaptive expert system under development.

In wind bracing design, two main phases can be distinguished: preliminary design and final design (Fig. 1). The subject of the preliminary design is the determination of all feasible structural alternatives or types of wind bracing and the initial selection of the most suitable one. In the final design phase the selected type of wind bracing is analyzed and the design finalized. The preliminary design of wind bracing is particularly important because there are at least several hundred known types of wind bracings (Arciszewski, 1986). The selection of the most suitable known type in a given case or the development of a new type has a decisive input as far as final structural characteristics of wind bracing are concerned. All decisions taken in preliminary design have a global character since they affect all subsequent local decisions of the final design strongly influencing the final weight, stiffness, simplicity, and such of the wind bracing. In many cases the decisions regarding the type of wind bracing are more important than the final design optimization of an assumed type as far as weight and stiffness are concerned.

As proposed in several papers (Arciszewski, 1975, 1986; Arciszewski and Pancewicz, 1976), a wind bracing scheme can be completely described by a set of qualitative and quantitative design variables. Qualitative variables of discrete character, each having a certain number of feasible states, describe one general structural form of bracing and define its type. Quantitative variables concern the dimensions of the wind bracing, its specification, and so forth and can be treated as continuous variables. A type of wind bracing is defined by a compatible combination of feasible states, when for all qualitative variables only one state is taken at a time. This approach enables formal presentation of all considered qualitative variables and their feasible states in a tabular form called a "typologic table." Such a table has been selected for the identification of different types of wind bracings to be considered in the preliminary design. A typologic table of wind bracing given by Arciszewski (1986) has been used in the proposed adaptive expert system.

As the result of extensive studies of structural shaping of wind bracing nine qualitative variables describing wind bracing were identified. These variables and their feasible states are shown in Table 1.

CONCEPT OF ADAPTIVE EXPERT SYSTEM

The adaptive expert system under development is intended to assist the designer in the development and selection of wind bracing types in the

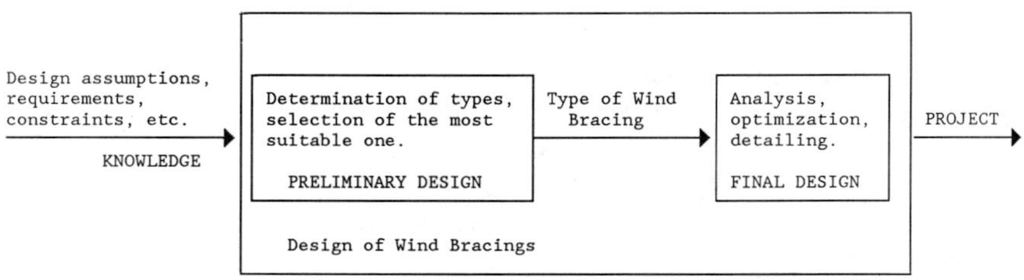

Fig. 1 Design of wind bracing

preliminary design phase. The evaluation of different bracing types in this design stage is very complex, and the application of formal multicriteria evaluation models is difficult. The evaluation and selection are usually subjective and based on the designer's experience and preferences. Often, types of wind bracing under consideration are simply evaluated as poor, good, very good, and so forth. This practice is equivalent to assigning to individual bracing types values of "structural soundness" taken from an assumed scale, say 0–100. Such an approach will ultimately be used in our system. There is also the possibility of applying the theory of generalized characteristics (Arciszewski and Pancewitz, 1976; Staniszewski, 1980) in the preliminary evaluation, but this requires extensive design studies.

The system will consist of three major components:

1. Solution Generator
2. Knowledge Base
3. Decision Rule Generator (learning component)

The Solution Generator generates random combinations of feasible states: compatible combinations, which represent types of wind bracing, and incom-

Table 1 Typologic table for wind bracing in steel skeleton structures

	Feasible States			
Qualitative variables	1	2	3	4
Static character of joints	rigid	hinged	rigid and hinged	
Number of bays entirely occupied by bracing	1	2	3	
Number of vertical trusses	0	1	2	3
Number of horizontal trusses	0	1	2	3
Number of horizontal truss systems	0	1	2	3
Material for core used	0	steel	reinforced concrete	
Number of cores	0	1	2	
Structural character of external members	0	columns and beams	cables	grid frame
Static character of bottom external joints	0	rigid	hinged	rigid and hinged

patible combinations, to be eliminated by the system. Generated types of wind bracing are presented to the designer, who evaluates them and assigns them "structural soundness" values. The Knowledge Base is used to store all these generated types of bracing, with evaluation information and decision rules. The Decision Rule Generator extracts decision rules based on the set of samples contained in the Knowledge Base.

The system features two phases of operation: learning and production. During the first phase, the system accumulates the initial information on the designer's experience and preferences by analyzing a series of training sample combinations produced by the Solution Generator. The designer analyzes all generated combinations of feasible states, identifies incompatible combinations, and assigns "structural soundness" values to all remaining combinations that represent types of bracing. This information is used to extract a set of decision rules to be used in the next phase. The system in the learning phase is illustrated schematically in Fig. 2. In the production phase the accumulated knowledge on the designer's experience and preferences is used to guide the generation and selection of the most promising solutions. These solutions are ranked in decreasing order of probability of satisfying the designer. The results of human evaluation of this group of types are also sent back to the system to refine the decision rules. The whole process is illustrated in Fig. 3.

Knowledge acquisition is a bottleneck in existing expert systems. Expert systems are normally built by interrogating human experts and encoding their knowledge in the form of decision rules. These processes are very costly

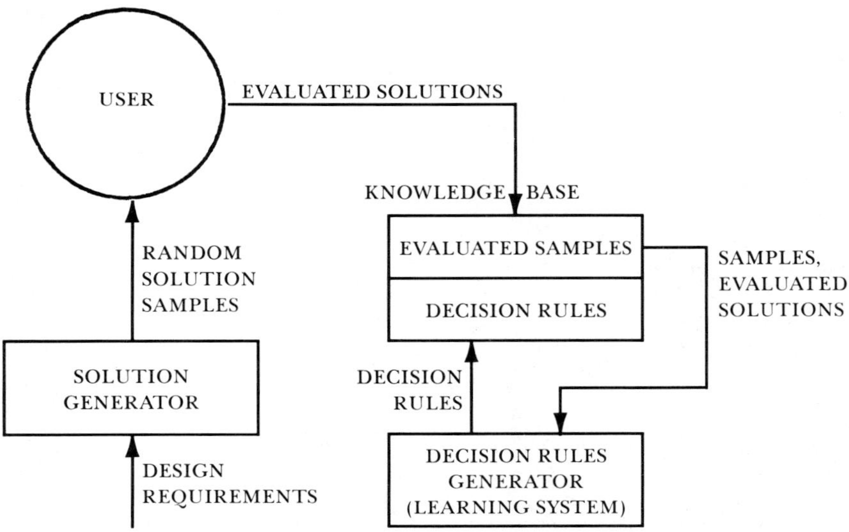

Fig. 2 Learning phase

and time-consuming, involving knowledge engineers and subject domain experts. Another approach to expert knowledge acquisition is through automatic learning, which seems to be very promising. In this approach the expert system itself is supposed to generate its decision rules, based on broadly understood observations of decisions by human experts. As indicated by Michalski et al. (1983), there are many dimensions of learning, which makes it rather unlikely that a general learning system will be built in the near future. However, encouraging results have recently been obtained in inductive learning of decision rules from examples of expert decisions represented in the form of attributes and attribute values (Quinlan, 1983). In particular, the theory of rough sets introduced by Pawlak (1982) provided a sound mathematical tool for dealing with problems associated with this kind of learning. This theory has been used to develop a general probabilistic theory of approximate classification by Wong and Ziarko, which seems to be appropriate for a certain class of adaptive expert systems (Wong and Ziarko, 1985). This class comprises systems in which decision rules are obtained through inductive learning from representative examples of expert decisions expressed by attributes and their values. The system reported in this paper falls into that category.

In the approach based on the idea of probabilistic approximate classification adopted in our system, the starting point is a population of objects *U*, called a universe, each of which can be characterized by two groups of parameters. The first group consists of syntactic parameters whose values can be easily identified. The second group is the set of semantic parameters

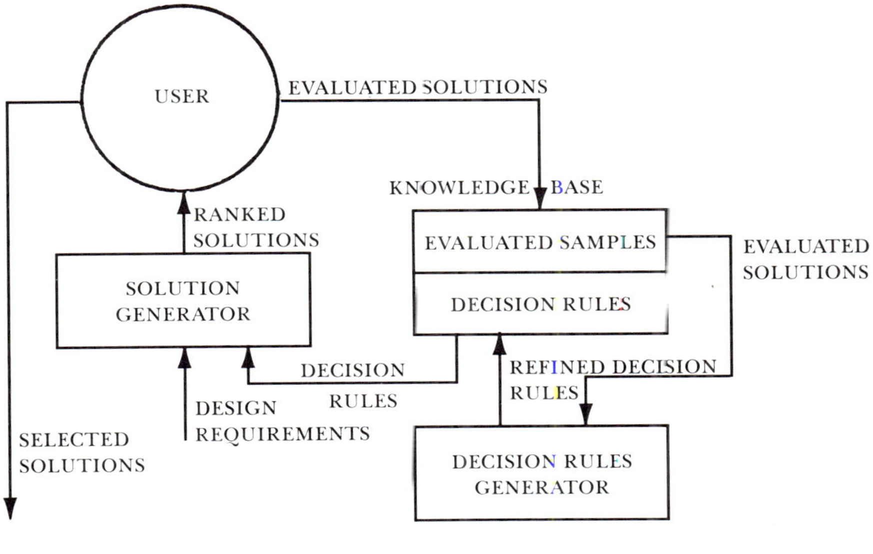

Fig. 3 Production phase

whose values must be determined based on an expert's subjective evaluation (for example, whether a particular design proposal is highly valuable). The universe of objects, for instance, can comprise all feasible solutions to a design problem. Of course, the points of this universe are not known, in general (as we never know all possible projects satisfying given requirements); we can, however, "test" our universe by randomly generating acceptable solutions. Suppose that this is the case. If so, we can look at our universe from a probabilistic point of view as a space of elementary random events Ω. Each semantic and syntactic parameter becomes a random variable in this model. To explain the idea, assume, for simplicity, that there is only one semantic parameter D and a number of syntactic parameters $C_1, C_2, \ldots, C_m$. It is known that the set of all possible values of a random variable or combination of them generates a partition of the space of random events Ω. We will call the pair (Ω, T), where T is the partition of Ω, an approximation space. Each value of the parameter D corresponds to a random event, referred to as a concept. Suppose that $T = [X_1, X_2, \ldots, X_n]$ and consider concept $Y \leqq \Omega$.

By assumption, each class of X_i of the partition T has a distinct description in the form of a combination of values of some syntactic parameters, which means that we can always classify an object in X_i based solely on values of syntactic parameters. However, in general, the description of the concept Y in terms of syntactic parameters is not known, which makes it impossible to classify an object in Y with 100% certainty. In other words, we are able to determine approximately only the value of the semantic (subjective) parameter D based on the values of syntactic parameters $C_1, C_2, \ldots, C_m$. The problem of approximate classification is fundamental to many applications of *AI*, in particular to adaptive diagnostic expert systems. This problem has been investigated in detail by Wong and Ziarko (1985). The implementation of their probabilistic algorithm forms the learning component of our system responsible for finding near optimal sets of decision rules to approximately classify generated design solutions with respect to values of some qualitative variables.

EXAMPLE OF APPLICATION

The system is still under development and implementation, but its major components are operational and have already been tested. The Solution Generator was developed a number of years ago and proved to be useful in the development of traditional and innovative solutions; its use resulted in two patent applications (Arciszewski, 1984 and 1986). The remaining parts of the system, however, particularly the Decision Rule Generator, still require extensive testing.

The first successful major test was to verify the ability of the Decision Rule Generator to identify decision rules in the area of wind bracings and to distinguish between compatible and incompatible combinations of feasible states. In other words, the ability of the system to select wind bracing types from thousands of combinations of feasible states generated by the Solution Generator has been tested.

In the learning phase 180 combinations of feasible solutions formally identified in accordance with Arciszewski (1986) were entered. One hundred sixty combinations were compatible and they represented known types of wind bracing while 20 combinations were incompatible with no apparent structural representation. All combinations were for a building, rectangular in plan, having an elevation with three bays on the narrower side. Evaluation information was in a simplified form, that is, only the compatibility of combinations of feasible states was given. Individual combinations were evaluated positive for compatible combinations representing types of wind bracing and negative for incompatible combinations. These 180 examples were used by the Decision Rules Generator to develop decision rules.

It was assumed that the Decision Rules Generator should generate three groups of decision rules differing in certainty levels. Certainty level is a probabilistic measure associated with a decision rule that indicates the probability that the considered decision rule is correct or that it will produce expected results when used.

Assumed certainty levels and numbers of generated decision rules for individual groups are given in Table 2. For example, the following decision rule in the first group was identified: if $G = 2$ and $H = 1$ then the combination of feasible states is compatible and represents a type of wind bracing. This rule can be also stated in structural terms as if there is one core and no structural members situated in exterior walls of the building, the wind bracing should be feasible. Another decision rule from the same first group is: if $G = 1$ and $H = 1$ then the combination of feasible states is incompatible and represents no wind bracing or if there is neither a core nor structural members situated in exterior walls of the building, then wind bracing is infeasible.

Decision rules from the second group are more complex. For example if $A = 1$ and $B = 1$ and $C = 1$ and $G = 2$ and $H = 1$, then the combination of feasible states should be compatible. The same rule stated in structural terms is very simple: a one-bay rigid frame can be used as a bracing.

Here are two examples of a decision rule generated by the Decision Rules Generator, this time from the third group of rules: If $B = 3$ and $G = 1$ and $H =$

Table 2 Assumed certainty levels and related numbers of generated decision rules

Group number	Certainty level	Number of generated rules
1	$\geqq 0.9$	5
2	$\geqq 0.95$	13
3	$= 1.0$	13

1 and $I = 2$ then the combination of feasible states is infeasible. This rule can be interpreted as: If you do not have external structural bracing members you cannot assume that these members are rigidly connected with the foundation. The second rule is very interesting. If $G = 2$ and $H = 1$ and $I = 1$ then the combination of feasible states is compatible, or in structural terms: If there is one core and no structural members situated in the exterior walls of the building and obviously no joints of these structural members with foundation, the wind bracing should be feasible. This is simply an extended version of the decision rules belonging to the first group, which was presented earlier in this paper.

In the production phase of the test, 86 combinations of feasible states generated by the Solution Generator were entered. The system correctly recognized the compatibility or incompatibility of individual combinations in 62% of cases. Since this was the first testing cycle, the results obtained are good and show the system's great potential.

CONCLUSIONS

The proposed adaptive expert system is still under development, and it is is not clear how efficient the final version will be. Also, it is uncertain how fast and precise the Decision Rule Generator will be, given that fact that the learning algorithm does not provide a fully optimal solution; that is, it does not minimize the number of decision rules due to the computational complexity of the problem. However, the theoretical basis of this system seems to be well prepared, and no major problems with implementation are expected.

The first test of the system showed the existence of a close relation between the assumed certainty level and the character of decision rules. When the certainty level is increased, decision rules become more reliable but also more complex and their applications may become more "local" in the sense that they will be useful only for a relatively small class of wind bracing types. Conversely, decreasing the certainty level makes decision rules less reliable but also applicable to a larger class of bracing types. An analogy can be found between certainty level in computer learning and generality level of directives in innovative design, where the character of design directives also strongly depends on the assumed generality level.

The initial experience with the system indicates that the proposed system, when implemented, will be an effective design tool to be used in the crucial preliminary phase of wind bracing design. It should be very useful in all cases where a building requires a typical wind bracings or when innovative solutions are desired.

REFERENCES/BIBLIOGRAPHY

Arciszewski, T., 1975
WIND BRACINGS IN THE FORM OF BELT TRUSS SYSTEMS IN STEEL SKELETON STRUCTURES OF TALL BUILDINGS, Ph.D. dissertation, Warsaw Technical University, Poland.

Arciszewski, T., 1986
DECISION MAKING PARAMETERS AND THEIR COMPUTER-AIDED ANALYSIS FOR WIND BRACINGS, Advances in Tall Buildings, Van Nostrand Publishing Company, New York.

Arciszewski, T. and Pancewicz, Z., 1976
AN APPROACH TO THE DESCRIPTION OF WIND BRACING CHARACTERISTICS IN SYSTEMS SKELETON STRUCTURES, Proceedings of the Wroclaw Technical University, Vol. 20, Wroclaw, Poland.

Maher, M. L., 1984
HI-RISE: A KNOWLEDGE-BASED EXPERT SYSTEM FOR THE PRELIMINARY STRUCTURAL DESIGN OF HIGH RISE BUILDINGS, Research Report, Design Research Center, Carnegie-Mellon University, Pittsburgh.

Michalski, R. S. M., Carbonell, J. G. and Mitchell, T. M., 1983
MACHINE LEARNING, Tioga Publishing Company, Palo Alto, California.

Pawlak, Z., 1982
ROUGH SETS, International Journal of Information and Computer Sciences Vol. 11, No. 5.

Staniszewski, R., 1980
CYBERNETICS OF DESIGN SYSTEMS, Ossolineum, Poland.

Quinlan, J. R., 1983
LEARNING EFFICIENT CLASSIFICATION PROCEDURES AND THEIR APPLICATION TO CHESS AND GAMES, Machine Learning: The Artificial Intelligence Approach, (eds. R. S. Michalski, J. G. Carbonell and T. M. Mitchell), Palo Alto, Tioga Press.

Wong, S. K. M. and Ziarko, W., 1985
A PROBABILISTIC MODEL OF APPROXIMATE CLASSIFICATION IN INDUCTIVE LEARNING, ISBN 0-7331-0092-3, Department of Computer Science, University of Regina.

Connection Flexibility in Steel Frames

W. F. Chen

In conventional analysis and design of steel frameworks, it is convenient to represent the beam-to-column connection as a joint point and to represent its moment-rotation behavior by idealized joint models. Two of the most commonly used idealized models representing the two extreme cases of connection behavior are the rigid-joint model and the pinned-joint model. For the rigid-joint model, it is assumed that the rotational continuity between adjoining members is fully realized. As a result of this assumption, the angle between adjacent members remains unchanged as the frame deforms and the full moment of the beam is transmitted to the column. This type of joint is known as the moment connection and is designated by the American Institute of Steel Construction (AISC) as Type 1 (Rigid) Framing Construction is the present Allowable Stress Design (ASD) Specification (AISC, 1978). For the pinned-joint model, it is assumed that the rotational continuity between adjoining members is nonexistent. Consequently, no moment is transmitted to the column by the beam. This type of joint is known as the shear connection and is designated by AISC as Type 2 (Simple Framing) Construction in the present ASD Specification.

Although these two idealized models corresponding to the two extreme cases are simple to use and easy to implement in analysis and design, their validities are not corroborated by experiments. Experiments carried out over the past decades have shown convincingly that actual joint behavior always falls in between the two extremes of fully-rigid and perfectly-pinned. In addition, the moment-rotational deformation behavior of the connections is usually nonlinear and always irreversible for almost the entire range of rotations. This type of joint is known as the semi-rigid connection and is designated by the AISC ASD specification as Type 3 (Semi-rigid Framing) Construction (AISC, 1978). Almost all beam-to-column connections in steel building construction are of the semirigid type, but structural analysis and design based on Type 3 Construction is seldomly used in practice because of the difficulty in assessing the strength and deformational behavior of such connections.

With the advent of computers, much effort has already been devoted to analyses of frames with semirigid connections (Chen, 1987 and to be published). Numerical analyses of frames that implement actual connection behavior in the solution procedures have made it possible for engineers and designers to get a better understanding of the response characteristics of real frames (Goto and Chen, 1987). Connection flexibility not only affects the behavior of beams and columns of the frame (Lui and Chen, 1983), but it has a definitive influence on the overall strength and stability of the structure.

Recognizing the importance of connection flexibility, and with the recent adoption of the limit states design philosophy as an alternative design format, the AISC limit states design code, referred to as the Load and Resistance Factor Design (LRFD) specification (1986), specifically designates two types of construction in its provision: Type FR (Fully-Restrained) Construction, which acknowledges the use of moment connection as a good assumption for the analysis of a rigid framing type of construction, and Type PR (Partially Restrained) Construction, which encourages the use of actual connection stiffness that always exists in an actual beam-to-column connection for the analysis and design of this type of steel frame. The use of Type PR Construction means that the designer should now take full advantage of the inherent rotational stiffness of a semi-rigid connection and include this beneficial effect explicitly in addition to member behavior in his analysis and design processes, instead of simply assuming the rotational continuity between adjoining members through semirigid connections to be nonexistent (a pinned-joint).

The stability analysis of flexibly-connected frames requires connection modeling. Since connection moment-rotation behavior is usually nonlinear, the inclusion of the connection as a structural element in a limit state analysis requires the use of nonlinear structural theory. With the advent of computer technology, great advances have been made in computer-aided analysis and design of structures. Currently, first- and second-order elastic analyses of structures can conveniently be made for nearly all types of structures (Timoshenko and Gere, 1961). Analysis of structures loaded into the inelastic range can also be made for certain types of structures (Chen and Atsuta, 1976, 1977). The continued development in computer hardware and software has made it possible for engineers and designers to predict structural behavior rather accurately. The advancement in structural analysis techniques coupled with the increased understanding of structural behavior has made it possible for engineers to adopt the limit state design philosophy. A limit state is defined as a condition at which a structural member or its component ceases to perform its intended function under normal conditions (serviceability limit state) or failure under severe conditions (ultimate limit state). Load and Resistance Factor Design (LRFD) is based on the limit state philosophy and thus it represents a more rational approach to the design of structures (Chen and Lui, 1987a).

In view of this development, it seems inevitable that the influences of connection flexibility on the behavior and design of steel frames must be reckoned with by designers in a limit state analysis and design procedure. Herein, the incorporation of this flexibility factor in column and frame analyses and in the proposed design procedures are summarized and discussed.

THE LITERATURE ON THE STATE OF THE ART

No attempt will be made here to review the vast literature on connection restraint characteristics, nonlinear analysis of steel frames with flexible connections, and design of flexibly-connected frames. These advances have been made or will be reported by the following publications:

1. ASCE Proceedings of a Session on "Connection Flexibility and Steel Frames" edited by W. F. Chen, October 24, 1985, 122 pp.
2. Special Issue of the Journal of Constructional Steel Research on "Steel Beam-To-Column Building Connections," edited by W. F. Chen, to be published.
3. Special Issue of the Journal of Constructional Steel Research on "Flexibility of Connections in Steel Frames," edited by W. F. Chen, 1987.
4. Special Reports on "Connection Restraint Characteristics," Task Group 25, Structural Stability Research Council, Washington, D.C. Annual Meeting, 1986.
5. SSRC Guide to the Stability Design Criteria for Metal Structures, 4th edition, T. V. Galambos, editor, John Wiley, New York, 1987.
6. Proceedings of a Workshop on "Steel Research Needs for Buildings," edited by C. Culver, N. Iwankiw, and A. Kuentz, Workshop sponsored by NBS, NSF, AISC, AISI, and MBMA, Gaithersburg, Md., May, 1985.
7. State-of-the-art Papers in Solid Mechanics Archives on "Steel Beam-to-Column Moment Connections" by W. F. Chen and E. M. Lui, 1986 and 1987b.

These papers and proceedings provide a valuable source of information on the current state of the art, and help identify research needed to fill the gaps in present knowledge and engineering practice.

CONNECTION RESTRAINT CHARACTERISTICS

The analyses and design of Type FR (fully-restrained) and Type PR (partially-restrained) frames differ in that for Type PR Construction, the effect of connection flexibility must be taken into account. Since a connection is a highly statical indeterminate element, a rigorous analytical study of its behavior is quite a formidable task. In view of this, a special Task Group (TG-25) of the Structural Stability Research Council was set up with the aim of investigating connection behavior theoretically and experimentally.

The behavior of a connection is best described by its moment-rotation relationship. Since the moment-rotation relationship of most connections is nonlinear almost from the start of loading, the analysis of structures including the effect of connection flexibility is inherently a nonlinear problem. To simplify the analysis technique, a number of simplified models have been proposed.

Connection Modeling

Figure 1 shows two simple linear models. The first model (Rathbun, 1936) utilizes the initial stiffness R_{ki} of the connection to represent the behavior of

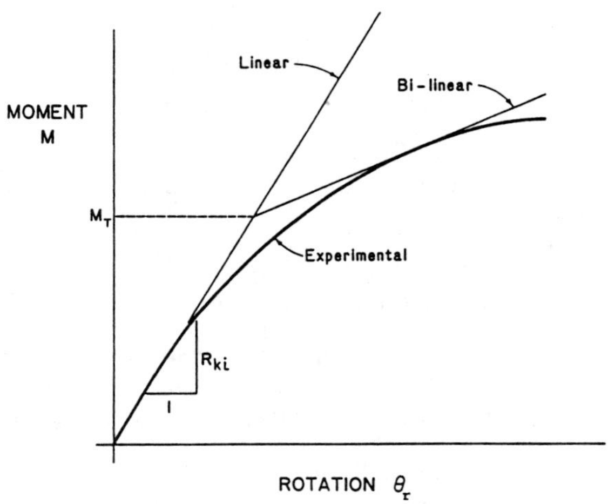

Fig. 1 Linear M-θ_r models

the connection for the entire range of loading. As can be seen, the validity of this linear model deteriorates as the moment increases. To get a better representation of the connection stiffness, a bilinear model (Romstad and Subramanian, 1970) was used. In the bilinear model, the initial slope of the moment-rotational line was replaced by a shallower line at a certain transition moment M_T. A direct extension of the bilinear model is the piecewise linear model (Razzaq, 1983) in which the nonlinear moment-rotation (M-θ_r) curve of the connection is represented by a series of straight line segments. Although the linear, bilinear, or piecewise linear models are easy to implement, the inaccuracies and sudden jump in stiffness inherent in these models make them undesirable for use in a limit state analysis routine.

To this end, Frye and Morris (1975) proposed a polynomial model in which a polynomial is used to represent the connection M-θ_r behavior (Fig. 2). However, there is a major drawback in this model. Since the nature of a polynomial is to peak and trough within a certain range, the stiffness of the connection (as represented by the first derivation of the polynomial) may be negative, which is physically unjustifiable. To overcome this, Jones et al. (1982) use a cubic B-spline curve-fitting technique to improve the polynomial model (Fig. 2). In the cubic B-spline model, a cubic polynomial is used to fit segments of a curve. Continuity between the first and second derivatives of each segment of curve are enforced. Although the cubic B-spline model gives a good representation of the connection behavior and circumvents the problem of negative stiffness, a large number of data points are necessary for the curve-fitting process. To overcome this, the power model proposed by Colson and Louveau (1983) and the exponential model proposed by Lui (1985) can be used.

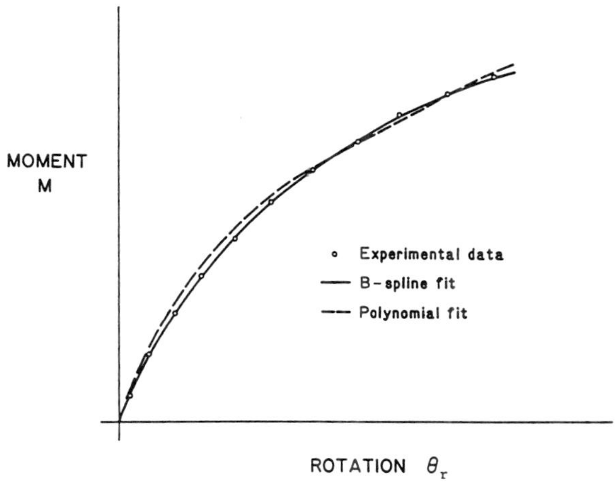

Fig. 2 B-spline and polynomial curve fit models

In the power model (Colson and Louveau, 1983) a power function is used to represent the connection $M\text{-}\theta_r$ behavior. It has the form

$$\theta_r = \frac{M}{R_{ki}} \frac{1}{1-(\frac{M}{M_{cu}})a} \tag{1}$$

where (refer to Fig. 3)

R_{ki} = initial connection stiffness
M_{cu} = ultimate moment capacity of the connection
a = a parameter to account for the curvature of the $M\text{-}\theta$ relationships.

In the exponential model (Lui, 1985), the connection $M\text{-}\theta_r$ behavior is represented by an exponential function of the form

$$M = \sum_{j=1}^{n} C_j (1 - e^{-\frac{\theta_r}{2j\alpha}}) + M_o + R_{kf}\theta_r \tag{2}$$

where

M_o = initial moment
R_{kf} = final or strain-hardening connection stiffness
α = scaling factor
C_j = connection model parameters

The connection model parameters are merely curve-fitting constants that can be obtained by using an optimization technique.

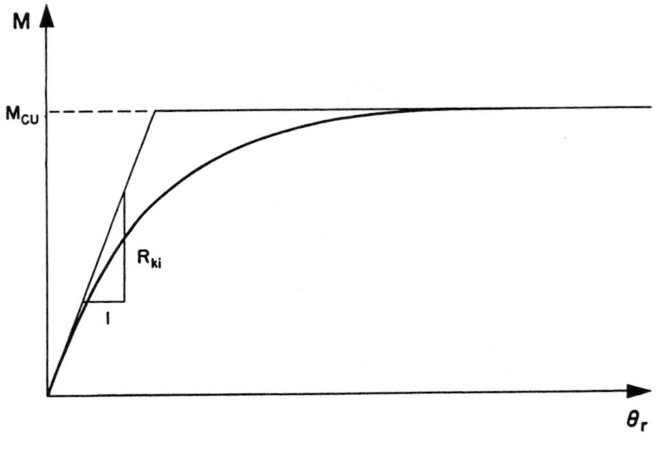

Fig. 3 Connection moment-rotation idealization used in the power model

To demonstrate the validity of the exponential model, two experimentally obtained moment-rotation curves are curve-fitted with Eq. 2 using four curve-fitting constants and ten sets of data from each curve. The results are shown in Figs. 4 and 5 respectively. The connection used in Fig. 4 was a double web angle connection tested by Lewitt, Chesson and Munse (1969). The connection used in Fig. 5 was a T-stub connection tested by Rathbun (1936). As can be seen, the exponential model gives an excellent representation of the test curves.

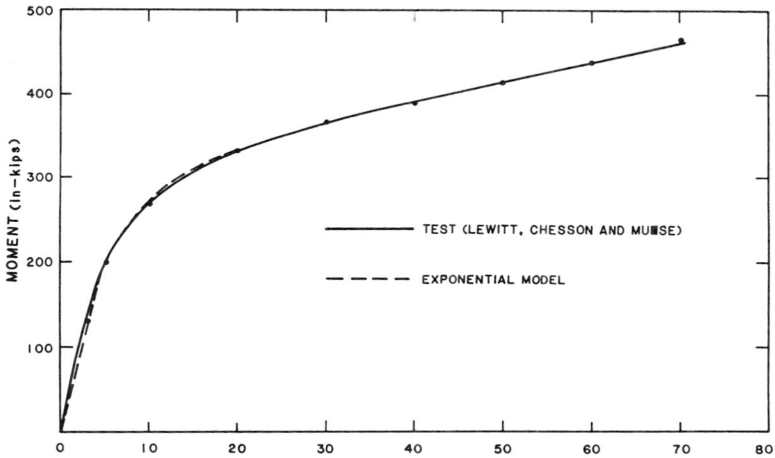

Fig. 4 Comparison of exponential connection model with test by Lewitt, Chesson and Munse

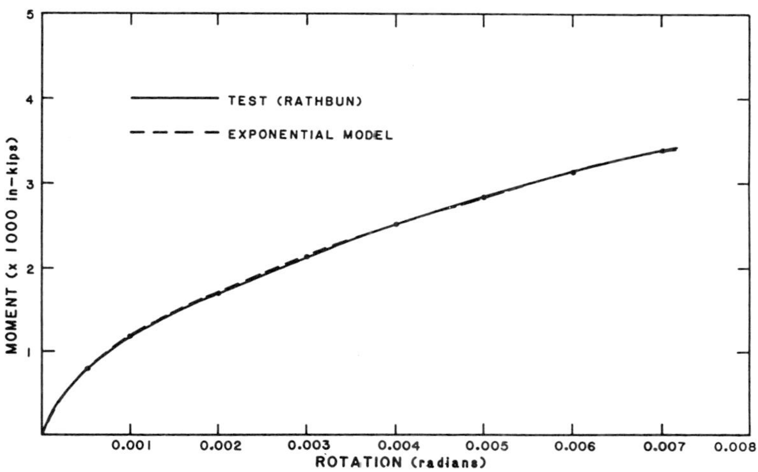

Fig. 5 Comparison of exponential connection model with test by Rathbun

Research Need

As mentioned earlier, because of the complex geometries and stress distributions of most connections, most $M\text{-}\theta_r$ curves are nonlinear and thus almost all existing $M\text{-}\theta_r$ curves available today are obtained from experiments (Kishi and Chen, 1986). Since most of these experiments were performed on connections that have become obsolete, it is essential that additional analytical and experimental investigations on commonly-used connections be conducted in view of the advancement made in the limit state approach to analysis and design of steel structures (Chen and Lui, 1985).

BEHAVIOR OF COLUMNS WITH RIGID AND SEMIRIGID CONNECTIONS

For columns in frames, another important phenomenon that engineers should be aware of is the moment transfer mechanism between the beams and columns. One commonly posed question is: How can a beam restrain a column if at the same time it is inducing moment to the column? Whether a beam restrains or induces moment to the column depends on a number of factors. Some of the important ones are (1) the rigidity of the connection, (2) the relative stiffness of the beam and column, and (3) the load patterns and load sequences on the frame.

MOMENT TRANSFER—RIGID CONNECTION

To study the moment transfer mechanism between the beam and the column, it is advantageous to look at the behavior of some simple subassemblages analyzed by Sugimoto (1983). Figure 6 shows a T-shaped subassemblage consisting of two beams and a column rigidly connected to one another. A concentrated load Q equal to half the yield load of the beam is applied to the midspan of each beam. An axial load P is then applied to an imperfect column with the influence of residual stress as well as with an initial out-of-straightness of 0.001L. The moment distribution of the joint is plotted as P increases. By assuming that the beams behave elastically for the entire range of loading and the joint is rigidly connected, the moment at the joint is considered to consist of three parts.

1. Bending moment M_1 caused by load Q with joint fixed.
2. Bending moment M_2 caused by joint translation as column buckles.
3. Bending moment M_3 caused by joint rotation.

These three components of bending moments are shown schematically in Fig. 7.

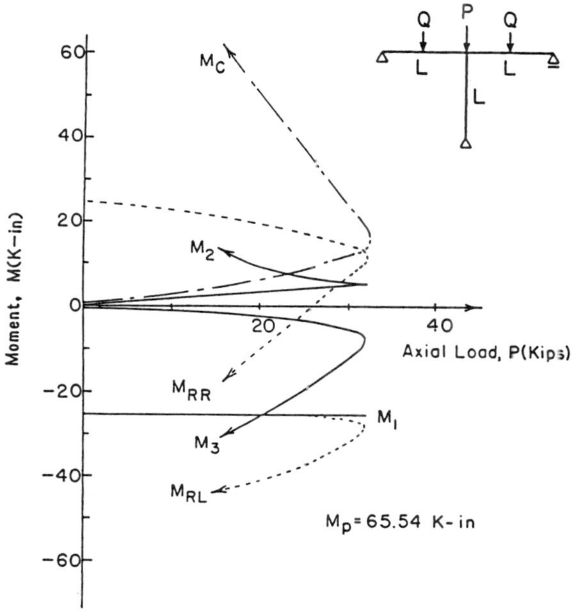

Fig. 6 Moment distributions at joint of subassemblage *(Sugimoto, 1983)*

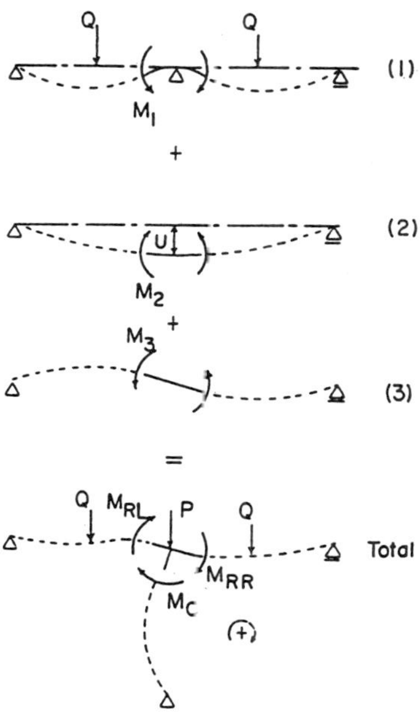

Fig. 7 Components of bending moments at joint of subassemblage

The bending moments in the left beam M_{RL} and in the right beam M_{RR} can be expressed as follows:

$$M_{RL} = M_1 + M_2 - M_3 \tag{3}$$

$$M_{RR} = M_1 - M_2 - M_3 \tag{4}$$

The bending moment in the column M_C can be obtained by considering joint equilibrium.

$$M_C = -(M_{RL} + M_{RR}) = 2M_3 \tag{5}$$

The variation of these bending moments with the axial load P is plotted in Fig. 6 and it can be seen that the moment due to the buckling of the column is not negligible. Not only does it reduce the moment of the left beam, but, together with the moment arising from joint rotation, it restrains the column during the final stage of loading. The moment of the right beam M_{RR}, at first inducing moment to the column, decreases gradually and at $P = (115.65 \text{ kn})$ (26 kips) reverses sign and becomes a restraining moment to the column. On the other hand, the moment of the left beam M_{RL} is always negative and thus always restains the column.

Moment Transfer—Flexible Connection

If the connections are not rigid, the moment transfer mechanisms between the beams and the columns are more complicated because of the loading/unloading characteristic of the connections. Figure 8 demonstrates this characteristic schematically. For this subassemblage (Fig. 8a), the beams are connected to the column by semirigid connections. Beam loads w_L, w_R are first applied to simulate the dead load of the structure. Fig. 8b shows the directions of moments acting on the left- and right-hand side of the joint of the subassemblage. The corresponding M-θ_r curves for the left and right connections are also shown. The left connection will follow curve OA' and the right connections will follow curve OA''. The moment acting on the column will be M_{1L} on the left side of the joint and M_{1R} on the right side of the joint.

Now, a column axial load P is applied to the subassemblage to simulate the live load. Under the action of P, the column will shorten and bend as shown by the dashed line in Fig. 8a. The moment induced caused by shortening of the column is shown in Fig. 8c. Note that the directions of moments on both sides of the column are opposite to that of Fig. 8b. Therefore, unloading of the connections will occur. As a result, the M-θ_r curve of the left connection will follow path $A'B'$ and that of the right connection will follow path $A''B''$. The slopes of $A'B'$ and $A''B''$ are parallel to the initial slopes to the corresponding M-θ_r curves. In addition to column shortening, there is bending deformation in the column. As a result of bending, the joint will rotate. If rotation is in the direction as shown in Fig. 8a, the direction of the induced moment will be that as shown in Fig. 8d. The induced moment to the left of the joint has the same direction as that of Fig. 8b, but the direction of the induced moment to the right of the joint has opposite direction to that of Fig. 8b. In other words, the connection to the left of the column will load while the connection to the right of the column will unload as a result of joint rotation.

Since the two columns' deformations, shortening, and bending occur simultaneously as P is applied, the phenomenon depicted in Figs. 8c and 8d are concurrent events. Consequently, the connection on the left-hand side of the column may follow path $A'B'$ or $A'C'$ (that is unload or load) depending on whether M_{2L} is greater or smaller than M_{3L}. On the other hand, the connection on the right-hand side of the column will always unload and so it will always exhibit a restraining effect on the column.

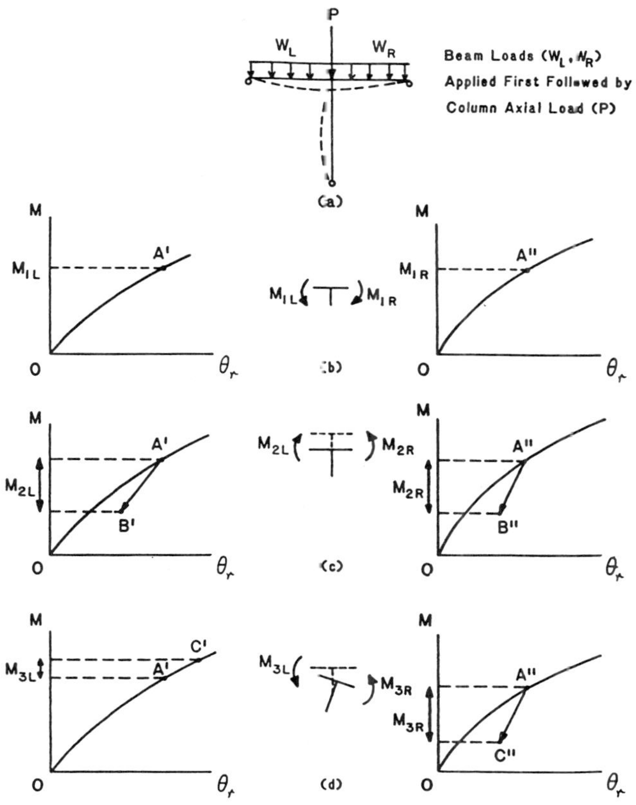

Fig. 8 Schematic representation of moment transfer mechanism of a flexibly-connected subassemblage

To study the behavior of flexibly-connected frames, recourse to numerical methods is inevitable because of the inherent nonlinear nature of the problem. To give the reader an insight into the restraint characteristic between members of flexibly-connected frames, the behavior of the following I- and T-subassemblages will be briefly discussed. Detailed analyses of these subassemblages are given elsewhere (Lui, 1985).

I-Shaped Subassemblage Study

Connections with Tangent Stiffness Restraint. Figure 9 shows the geometry and loading conditions of an I-shaped subassemblage. In load sequence 1, the column and beams of this subassemblage are preloaded with concentrated forces of 1.11 kn (0.25 kips). All beam loads are applied at midspans of the beams. The horizontal column loads are applied to simulate the out-of-straightness of the column. In load sequence 2, in addition to the column load of 5*P*, midspan beam loads of *P* are applied in the upper right beam and lower left beam so that single curvature bending of the column will result. The subassemblage was analyzed twice, first with rigid connections

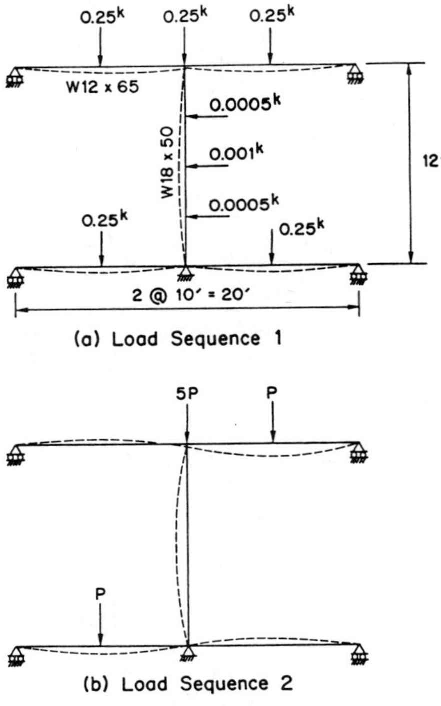

Fig. 9 I-shaped subassemblage—connections with tangent stiffness restraint

and then with flexible connections with the connection moment-rotation behavior shown in Fig. 10.

Figure 11 shows the distribution of joint moments for the upper joint of the rigidly-connected and flexibly-connected cases. In both cases, the moment of the left beam M_{BL} and the moment of the column M_C act opposite to the moment of the right beam M_{BR}. Nevertheless, their relative magnitudes are quite different. In Fig. 12, the moment ratios M_{BL}/M_{BR} and M_C/M_{BR} for both cases are plotted. For the rigidly-connected case, the percentage of M_{BR} shared by M_{BL} and M_C do not differ much because the relative stiffness of the beam and column are quite comparable. For the flexibly-connected case, the percentage of M_{BR} shared by M_{BL} and M_C differ significantly because of the presence of the connection, and the apparent stiffness of the beam is reduced. For the entire range of loadings, no unloading of the connection was detected in the analysis. As a result, the tangent stiffness governed the behavior of all the connections for the whole range of loadings.

If we examine the directions of the column end moments it can be seen that they are enhancing and not restraining the buckling of the column (Fig. 13). For both cases, the analysis was terminated when a plastic hinge formed at the mid-height of each column.

Connections with Initial Stiffness Restraint. Figure 14 shows a subassemblage geometrically identical to the one in Fig. 9. The loadings in load sequence 1

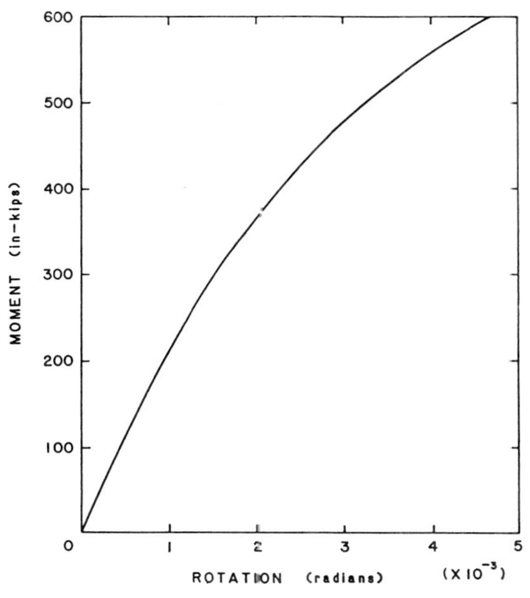

Fig. 10 Connection moment-rotation behavior used for the I-shaped subassemblage

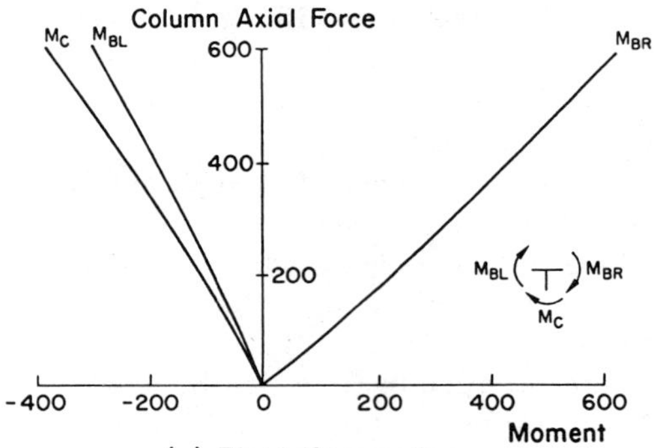

(a) Rigid Connection

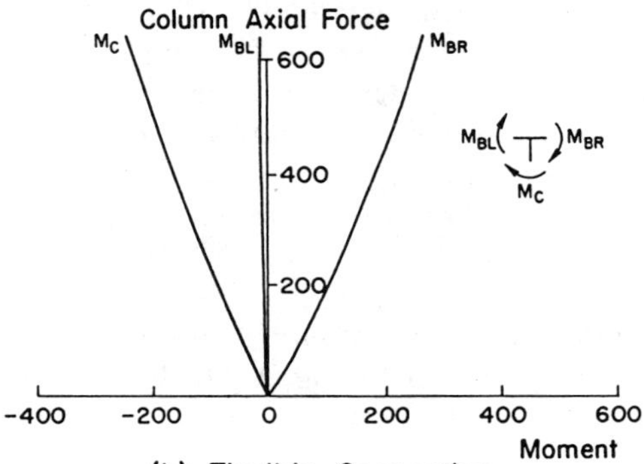

(b) Flexible Connection

Fig. 11 Distribution of upper joint moments

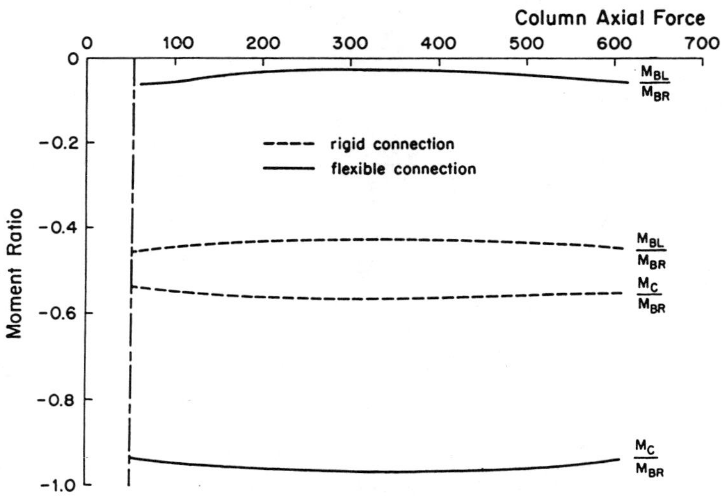

Fig. 12 Moment ratios of rigidly-connected and flexibly-connected I-shaped subassemblage

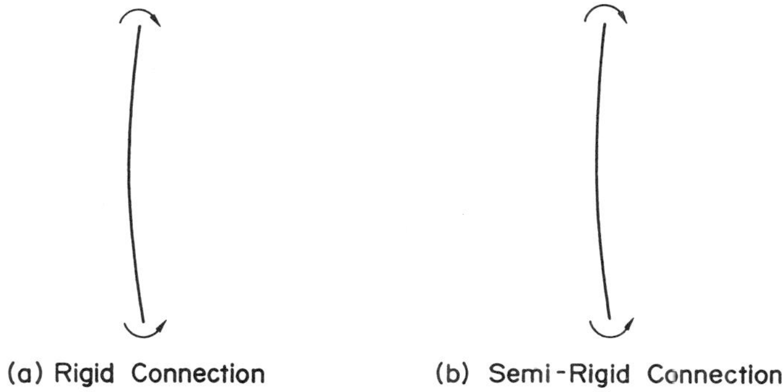

Fig. 13 Column end moments

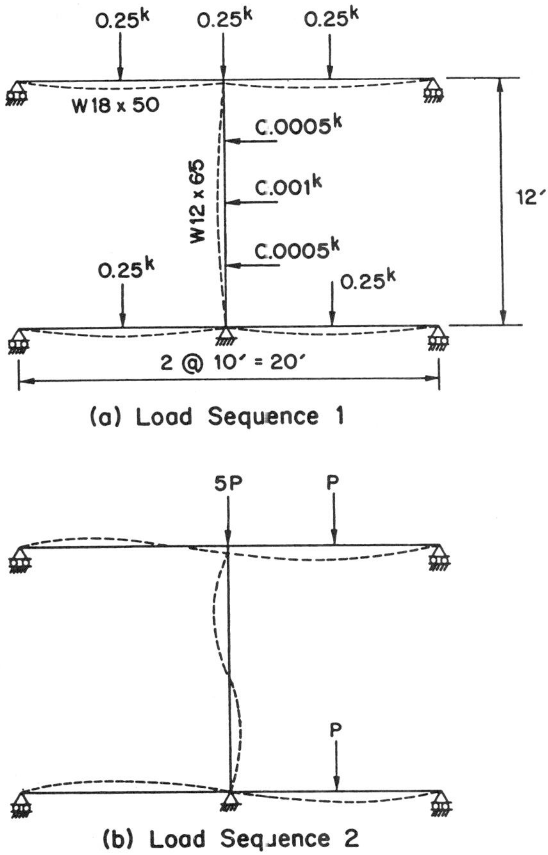

Fig. 14 I-shaped subassemblage—connections with initial stiffness restraint

are also identical. However, the loadings in load sequence 2 are altered so that double curvature bending of the column will result. The distribution of joint moments of the upper joint for two cases (rigid connection and flexible connection with the moment-rotation behavior shown in Fig. 10) are shown in Fig. 15. During the analysis, unloading occurred in the connection that is attached to the right beam at the upper joint. Consequently, the initial stiffness rather than the tangent stiffness was used for this connection. Note that the directions of the column end-moments are such that the columns in both the rigidly-connected and flexibly-connected subassemblages are restrained

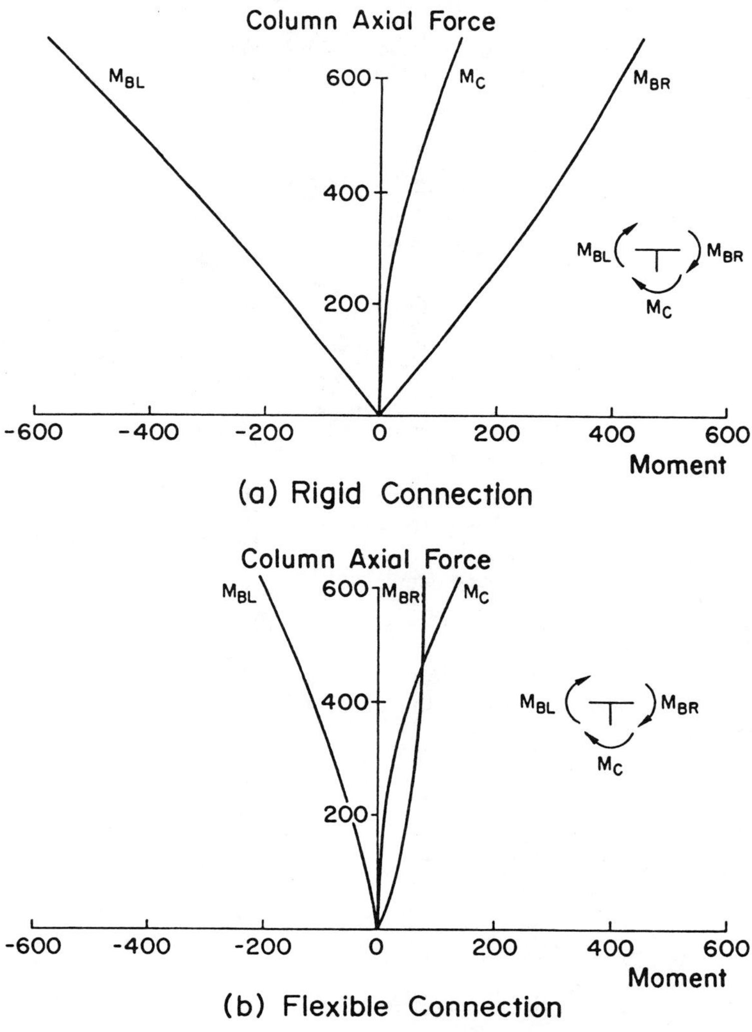

Fig. 15 Distribution of upper joint moments

against the buckling (Fig. 16). Analysis was terminated when plastic hinges developed in the column at the lower and upper joints of the subassemblages for both the rigidly-connected and flexibly-connected cases.

Connections with No Restraint. If the midspan beam loads are present in all the four beams in the second load sequence (Fig. 17b), the behavior of the I-shaped subassemblage can be studied by examining Fig. 18 in which the distribution of joint moments of the upper joint are plotted. As can be seen, although the magnitudes of beam moments for the rigidly-connected case are much larger than those for the flexibly-connected (Fig. 10) case, the column end moments for both cases are almost zero, which means the beams neither induce moment to the column nor restrain it against buckling. Consequently, the column behaves like a centrally loaded column and the ultimate load is the tangent modulus load.

T-Shaped Subassemblage Study

To study the behavior of flexibly-connected frames further, the subassemblages shown in Figs. 19 and 20 are analyzed with the two load sequences applied as shown. The differences between the subassemblage of Fig. 19 and that of Fig. 20 is that rigid connections are assumed in the subassemblage of Fig. 19 and flexible connections, with a moment-rotation behavior of Fig. 21, are used in the subassemblage of Fig. 20. The loadings for the two subassemblages are identical. The two beams are first loaded at midspan with a 22.24-kn (5-kip) concentrated load in load sequence 1. The column is also loaded with a 22–24-kn (5-kip) concentrated load in load sequence 1. The column loads (vertical at the joint and horizontal at quarter-points of the columns) are then applied monotonically in load sequence 2 until a plastic hinge forms in the column.

(a) Rigid Connection (b) Semi-Rigid Connection

Fig. 16 Column end moments

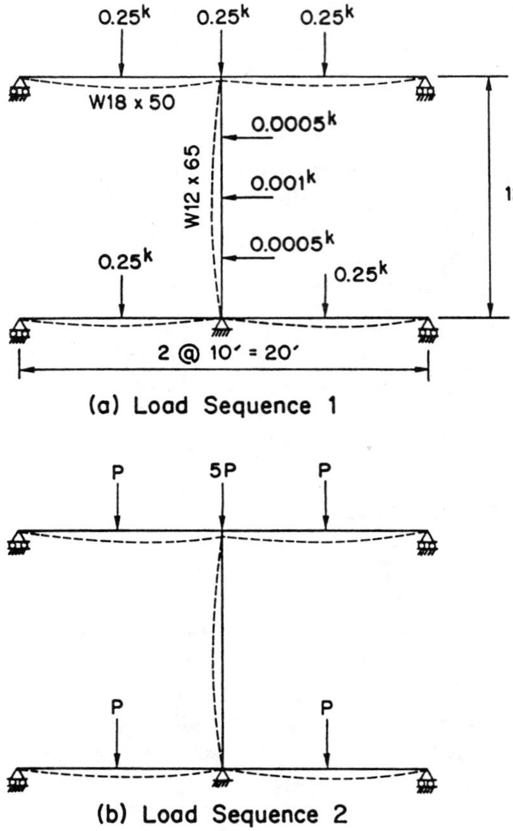

Fig. 17 I-shaped subassemblage—connections with no restraint

The distribution of joint moments for the rigidly-connected and flexibly-connected subassemblages are shown in Fig. 22a and 22b respectively. The following observations can be made from the plots:

1. The column is restrained against buckling even though the beams are preloaded. For the rigidly-connected subassemblage, restraint is offered by the left-hand beam until at $P = 169.03$ kn (38 kips) the right-hand beam starts to provide the restraint. At $P = 240.20$ kn (54 kips), restraint is offered solely by the right-hand beam. For the flexibly-connected subassemblage both beams provide the restraint to the column.
2. The restraining effect is more pronounced for rigid connections than for flexible connections.

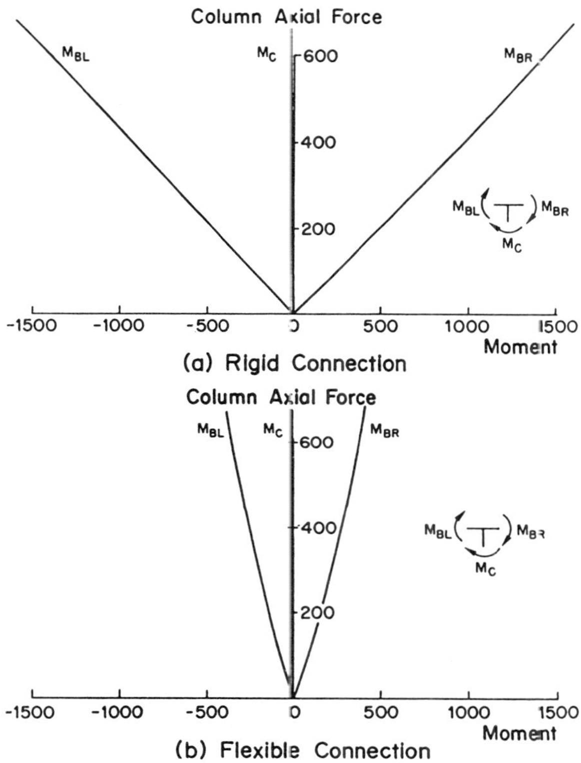

Fig. 18 Distribution of upper joint moments

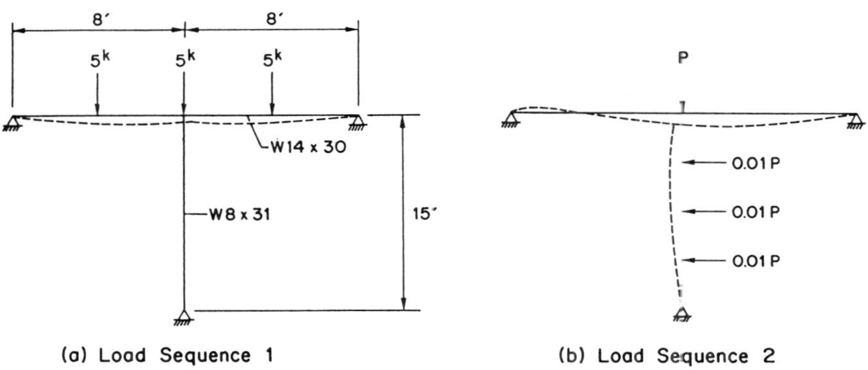

Fig. 19 Rigidly-connected T-shaped subassemblage

The difference in moment distribution around the joint is apparent in Fig. 22. Of particular interest is the direction of M_{BL}. For the flexibly-connected subassemblage, M_{BL} is always negative whereas for the rigidly-connected subassemblage M_{BL} is only negative at low values of P but becomes positive at high values of P. The reason for this can be explained by reference to Fig. 23 in which the beam end moments at the joint are decomposed. At the end of

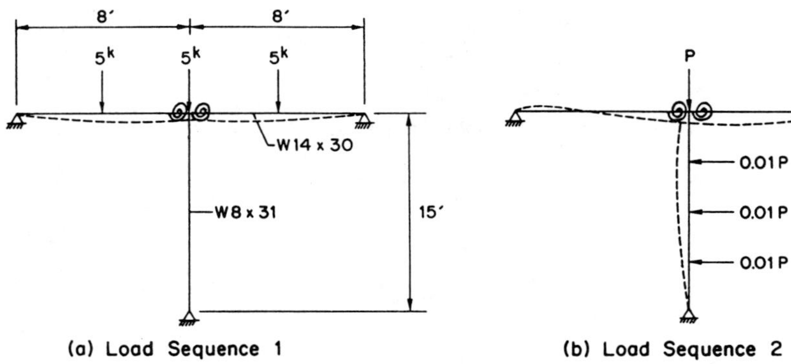

Fig. 20 Flexibly-connected T-shaped subassemblage

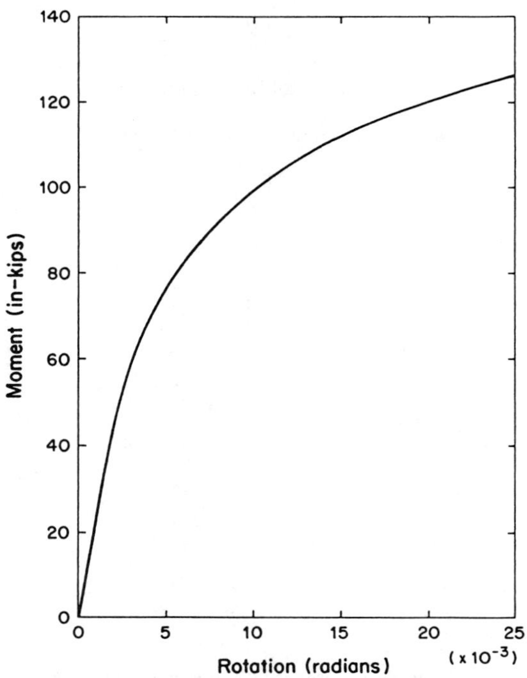

Fig. 21 Connection moment-rotation behavior used for the T-shaped subassemblage

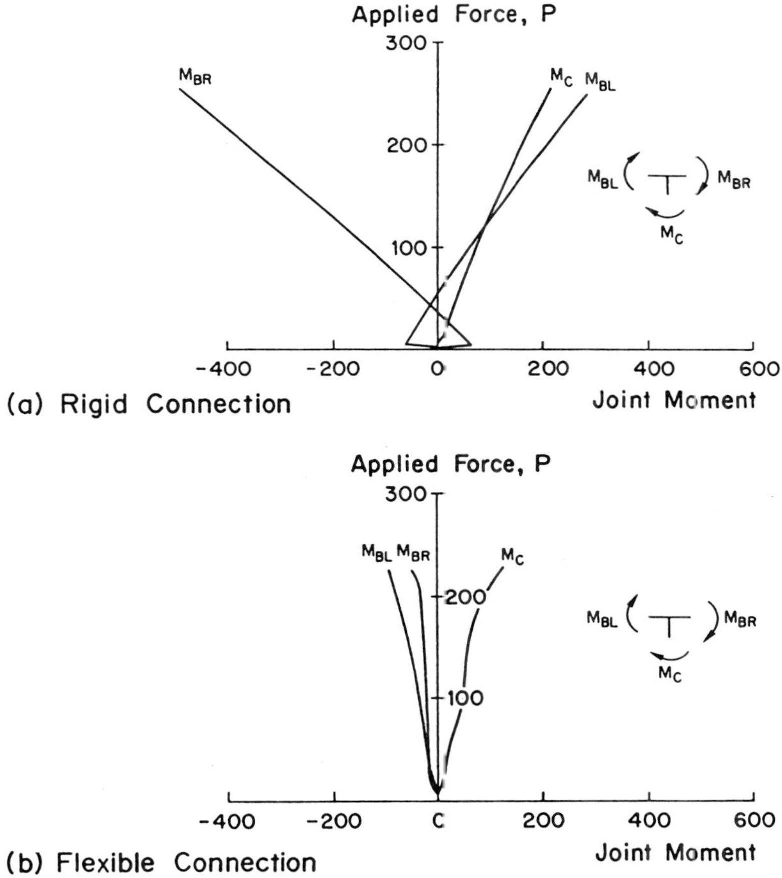

Fig. 22 Applied force vs. Joint moment relationships

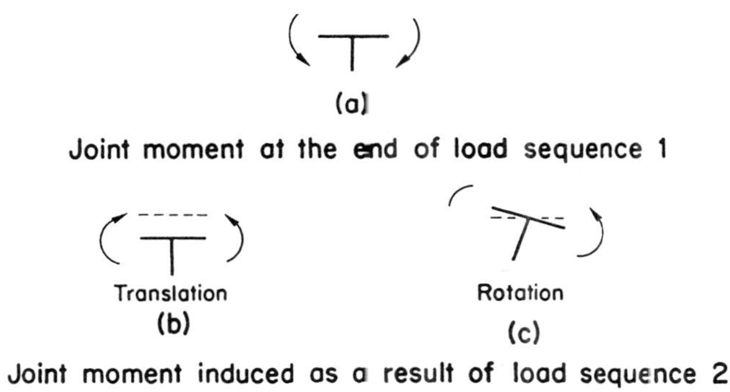

Fig. 23 Decomposition of joint moments

load sequence 1, M_{BL} is negative (that is counterclockwise, Fig. 23a). However, as load sequence 2 commences, the induced moment M_{BL} may be positive (that is clockwise, Fig. 23b) as a result of joint translation, or negative (that is counterclockwise, Fig. 23c) as a result of joint rotation. Whether the final value of M_{BL} is positive or negative depends on whether joint translation or joint rotation dominates. In Fig. 24, the magnitude of joint translation and joint rotation as a function of the applied force P are plotted. Although the magnitude of joint translation for both the rigidly-connected and flexibly-connected subassemblages are comparable, the joint rotation of the flexibly-connected subassemblage is significantly larger than that of the rigidly-connected subassemblage. As a result, the moment induced as a result of joint rotation will outweigh that of joint translation and hence the final value of M_{BL} for the flexibly-connected frame is negative.

As for the right-hand beam, regardless of whether joint translation or joint rotation dominates, the induced M_{BR} is almost always negative. As a result, this beam, except at the initial loading stage for the rigidly-connected subassemblage, will always provide restraint to the column regardless of whether the connection is rigid or flexible. It should be mentioned that

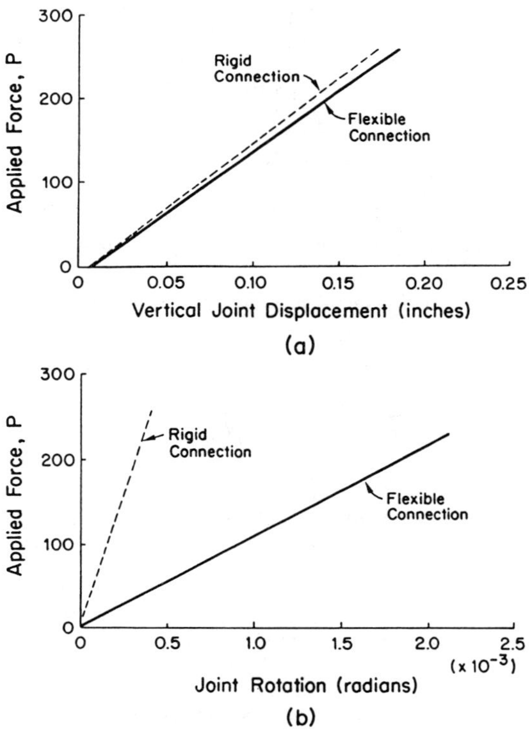

Fig. 24 Joint displacement and rotation of the T-shaped subassemblage

unloading occurs at the connection of the right-hand beam to the column because the direction of moment at this location is opposite for load sequence 1 and load sequence 2 (See Fig. 20).

A more detailed analysis and discussion of the behavior of subassemblages with flexible connections are given elsewhere (Lui, 1985).

DESIGN OF COLUMNS WITH RIGID AND SEMI-RIGID CONNECTIONS

For design purposes, if the connections are rigid, one can just perform a first-order analysis on a trial frame. With the end moments and axial force known for each member, an interaction formula (see for example, Chen and Lui, 1985) can be used to check the trial sizes of the members. However, if the connections are not rigid, then care must be exercised in using the interaction equations that are contained in the LRFD Specification (AISC, 1986), because the maximum moment in the member as determined by the equation $M_u^* = B_1 M_{nt} + B_2 M_{lt}$ [Eq. (H1-2) in the LRFD Specification] will no longer be valid since the moment amplification factors B_1 and B_2 are only defined for rigidly-connected frames (Chen and Lui, 1987a).

One plausible solution is to use computer-aided analysis and design in which a second-order analysis is performed on the flexibly-connected frame to determine the maximum moment including the P = δ and P = Δ effects directly (Goto and Chen, 1987). However, in lieu of such analysis, simplified design methods (Lui, 1985; Disque, 1975) can be used. By assuming that

1. the nonlinear moment-rotation behavior of the connection can be idealized as shown in Fig. 3,
2. the connection attains its ultimate moment capacity M_{CU} under gravity loads only, and
3. the connection behaves elastically with the stiffness R_{ki} when unloading

the moments acting on the column can be expressed in terms of the connection ultimate moment capacity M_{CU}.

Columns in a Nonsway Frame

For example, for an exterior column in a nonsway frame (Fig. 25a), the column can be designed according to Fig. 25b. For an interior column in a nonsway frame (Fig. 26a), the column can be designed according to Fig. 26b. Note that in Fig. 26b, it is assumed that the columns above and below have the same stiffness as the column to be designed. If the columns have different stiffnesses, then instead of using the factor 1/2, a factor reflecting the difference

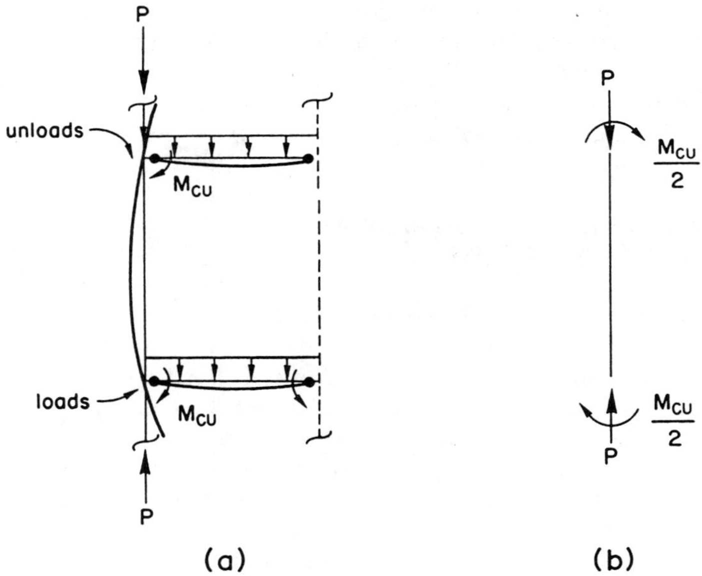

Fig. 25 Design of exterior columns in a nonsway frame

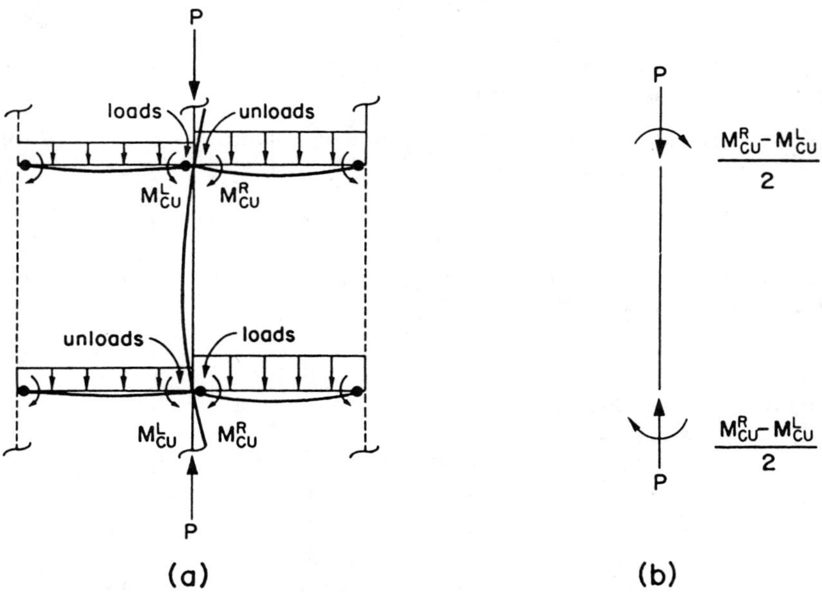

Fig. 26 Design of interior columns in a nonsway frame

in column stiffnesses can be used. For instance if two columns meeting at a joint have stiffnesses k_1 and k_2 respectively for the upper and lower columns, then it can be assumed that $k_1/(k_1 + k_2)$ of the beam moment will go to the upper column and $k_2/(k_1 + k_2)$ of the beam moment will go to the lower column.

Column in a Sway-Permitted Frame

For columns in a sway-permitted frame (Fig. 27), the windward exterior column and the interior column can be designed according to Fig. 27b and 27c respectively. Again, the factor 1/2 is used for distributing the beam moment indicating that equal column stiffnesses are assumed.

It should be pointed out that because of unloading, which some connections experience, the actual moment acting on the column may be greater or smaller than the moments indicated in the figures depending on the loading patterns on the beams. Therefore, the above procedure is just an approximate design procedure. A connection will unload when the moment induced as a result of column bending or frame sway acts opposite to that induced from beam gravity loads.

Additional Studies

Additional studies on the role of connections in affecting the strength and stiffness of frames have been reported by Ackroyd (1979), Moncarz and

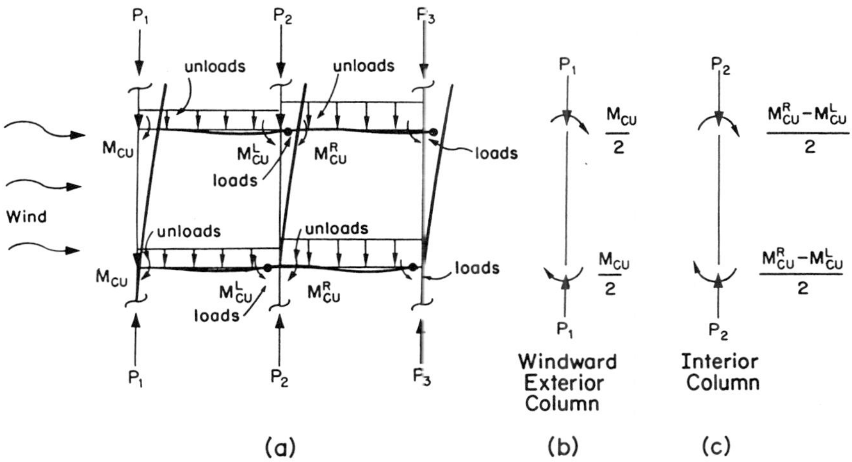

Fig. 27 Design of exterior and interior columns in a sway permitted frame

Gerstle (1981), Simitses and Giri (1982), Simitses and Vlahinos (1982), and Simitses et al. (1984). Ackroyd's work shows that an increase in connection stiffness does not always result in an increase in frame strength. For long-span frames with only a few stories, where the effects of lateral loads are small compared with those of gravity loads, an increase in connection stiffness may cause a decrease in frame strength. A parameter that can be used as an index was defined to determine whether provision of overstiff connections will be beneficial or detrimental.

In addition to connection flexibility, another important factor that will affect the limit state behavior of a frame is panel zone deformation. The study of panel zone deformation on frame behavior has been reported by Fielding and Huang (1971), Fielding and Chen (1973), Becker (1975), Kato (1982) and Krawinkler (1978). In these studies, attention was given to the modeling of the shear deformation of the panel zone. The important results of these studies were summarized by Chen and Lui (1984). Recently, a finite element model of the panel zone, which can represent all modes (extension, shear and bending) of deformation of the panel zone, has been reported (Lui, 1985). Generally speaking, the strength and stiffness of frames will be reduced if the effect of panel zone deformation is taken into account in the analysis procedures. The analysis of nonlinear elastic-plastic frames including the effects of connection flexibility and panel zone deformation on the behavior of steel columns and frames is now underway at Purdue University and has already produced some interesting results (Chen, 1987 and to be published). These results will hopefully shed much more light on this important subject.

REFERENCES/BIBLIOGRAPHY

Ackroyd, M. H., 1979
NONLINEAR INELASTIC STABILITY OF FLEXIBILITY-CONNECTED PLANE STEEL FRAME, Ph.D. Dissertation, Department of Civil, Environmental and Architectural Engineering, University of Colorado, Boulder, Colorado.

AISC, 1978
SPECIFICATION FOR THE DESIGN, FABRICATION AND ERECTION OF STRUCTURAL STEEL FOR BUILDINGS, American Institute of Steel Construction, Chicago, Illinois.

AISC, 1986
LOAD AND RESISTANCE FACTOR DESIGN SPECIFICATION FOR STRUCTURAL STEEL BUILDINGS, American Institute of Steel Construction, Chicago, Illinois.

Becker, R., 1975
PANEL ZONE EFFECT ON THE STRENGTH AND STIFFNESS OF STEEL RIGID FRAMES, AISC Engineering Journal, Vol. 12, No. 1, 1st Quarter, Chicago, pp. 19-29.

Chen, W. F., Ed., 1987
FLEXIBILITY OF CONNECTIONS IN STEEL FRAMES, Journal of Constructional Steel Research, Elsevier Applied Science, Vol. 7, July, United Kingdom.

Chen, W. F., Ed., to be published
STEEL BEAM-TO-COLUMN BUILDING CONNECTIONS, Journal of Constructional Steel Research, Special issue, Elsevier Applied Science, United Kingdom.

Chen, W. F. and Atsuta, T., 1976
THEORY OF BEAM-COLUMNS: VOL. 1—IN-PLANE BEHAVIOR AND DESIGN, McGraw-Hill, New York.
Chen, W. F. and Atsuta, T., 1977
THEORY OF BEAM-COLUMNS: VOL. 2—SPACE BEHAVIOR AND DESIGN, McGraw-Hill, New York.
Chen, W. F. and Lui, E. M., 1984
EFFECTS OF CONNECTION FLEXIBILITY AND PANEL ZONE SHEAR DEFORMATION ON THE BEHAVIOR OF STEEL FRAMES, Seminar on Tall Structures and Use of Prestressed Concrete in Hydraulic Structures, New Delhi, India, May, pp. 155-176.
Chen, W. F. and Lui, E. M., 1985
COLUMNS WITH END RESTRAINT AND BENDING IN LOAD AND RESISTANCE FACTOR DESIGN, AISC Engineering Journal, Vol. 22, No. 3, 3rd Quarter, Chicago, pp. 105-131.
Chen, W. F. and Lui, E. M., 1986
STEEL BEAM-TO-COLUMN MOMENT CONNECTIONS—PART I: FLANGE MOMENT CONNECTIONS, Vol. II, Issue 4, December, pp. 257-316, Solid Mechanics Archives, Oxford University Press, Oxford, England.
Chen, W. F. and Lui, E. M., 1987a
STRUCTURAL STABILITY: THEORY AND IMPLEMENTATION, Elsevier, New York.
Chen, W. F. and Lui, E. M., 1987b
STEEL BEAM-TO-COLUMN MOMENT CONNECTIONS—PART II: WEB MOMENT CONNECTIONS, Vol. 12, Issue 1, March, pp. 327-378, Solid Mechanics Archives, Oxford University Press, Oxford, England.
Colson, A. and Louveau, J. M., 1983
CONNECTIONS INCIDENCE ON THE INELASTIC BEHAVIOR OF STEEL STRUCTURES, Euromech Colloquium 174, October.

Disque, R. O., 1975
DIRECTIONAL MOMENT CONNECTIONS—A PROPOSED DESIGN METHOD FOR UNBRACED STEEL FRAMES, AISC Engineering Journal, Vol. 12, No. 1, 1st Quarter, Chicago, pp. 14-18.

Fielding, D. J. and Huang, J. S., 1971
SHEAR IN BEAM-TO-COLUMN CONNECTIONS, The Welding Journal, Vol. 50, July.
Fielding, D. J. and Chen, W. F., 1973
FRAME ANALYSIS AND CONNECTION DEFORMATION, Journal of Structural Division, ASCE, Vol. 99, No. ST1, pp. 1-18.
Frye, M. J. and Morris, G. A., 1975
ANALYSIS OF FLEXIBILITY CONNECTED STEEL FRAMES, Canadian Journal of Civil Engineers, Vol. 2, No. 3, Canada, pp. 280-291.

Goto, Y. and Chen, W. F., 1987
SECOND-ORDER ELASTIC ANALYSIS FOR FRAME DESIGN, Journal of Structural Engineering, ASCE, Vol. 113, No. 7, July, pp. 1501-1519.

Jones, S. W., Kirby, P. A., and Nethercot, D. A., 1982
COLUMNS WITH SEMI-RIGID JOINTS. Journal of the Structural Division. ASCE, Vol. 108, No. ST2, February, pp. 361-372.

Kato, B., 1982
BEAM-TO-COLUMN CONNECTION RESEARCH IN JAPAN, Journal of the Structural Division, ASCE, Vol. 108, No. ST2, February, pp. 343-360.
Kishi, N. and Chen, W. F., 1986
DATA BASE ON STEEL BEAM-TO-COLUMN CONNECTIONS, Structural Engineering Report No. CE-STR-86-26, School of Civil Engineering, Purdue University, two volumes, 653 pp.
Krawinkler, H., 1978
SHEAR IN BEAM-COLUMN JOINTS IN SEISMIC DESIGN OF STEEL FRAMES, AISC Engineering Journal, 3rd Quarter, pp. 82-91.

Lewitt, C. S., Chesson, E., and Munse, W. H., 1969
RESTRAINT CHARACTERISTICS OF FLEXIBLE RIVETED AND BOLTED BEAM-TO-COLUMN CONNECTIONS, Engineering Experiment Station Bulletin No. 500, University of Illinois at Urbana-Champaign, Illinois, January.

Lui, E. M., 1985
EFFECTS OF CONNECTION FLEXIBILITY AND PANEL ZONE DEFORMATION ON THE BEHAVIOR OF PLANE STEEL FRAMES, Ph.D. Dissertation, School of Civil Engineering, Purdue University, West Lafayette, Indiana.

Lui, E. M. and Chen, W. F., 1983
END RESTRAINT AND COLUMN DESIGN USING LRFD, AISC Engineering Journal, Vol. 20, No. 1, 1st Quarter, pp. 29-39.

Moncarz, P. D. and Gerstle, K. H., 1981
STEEL FRAMES WITH NON-LINEAR CONNECTIONS, Journal of the Structural Division, ASCE, Vol. 107, No. ST8, August, pp. 1427-1441.

Rathbun, J. C., 1936
ELASTIC PROPERTIES OF RIVETED CONNECTIONS, Transactions of the American Society of Civil Engineers, Vol. 101, pp. 524-563.

Razzaq, Z., 1983
END RESTRAINT EFFECT ON STEEL COLUMN STRENGTH, Journal of the Structural Division, ASCE, Vol. 109, No. ST2, pp. 314-334.

Romstad, K. M. and Subramanian, C. V., 1970
ANALYSIS OF FRAMES WITH PARTIAL CONNECTION RIGIDITY, Journal of the Structural Division, ASCE, Vol. 96, No. ST11, pp. 2283-2300.

Simitses, G. J. and Giri, J., 1982
NON-LINEAR ANALYSIS OF UNBRACED FRAMES OF VARIABLE GEOMETRY, International Journal of Nonlinear Mechanics, Vol. 17, No. 1, pp. 47-61.

Simitses, G. J. and Vlahinos, A. S., 1982
STABILITY ANALYSIS OF A SEMI-RIGIDLY CONNECTED SIMPLE FRAME, Journal of Constructional Steel Research, Vol. 2, No. 3, September, pp. 29-32.

Simitses, G. J., Swisshelm, J. D., and Vlahinos, A. S., 1984
FLEXIBILITY-JOINTED UNBRACED PORTAL FRAMES, Journal of Constructional Steel Research, Vol. 4, pp. 27-44.

Sugimoto, H., 1983
STUDY OF OFFSHORE STRUCTURAL MEMBERS AND FRAMES, Ph.D. Dissertation, School of Civil Engineering, Purdue University, West Lafayette, Indiana.

Timoshenko, S. P. and Gere, J. M., 1961
THEORY OF ELASTIC STABILITY, 2nd Edition, Engineering Societies Monographs, McGraw-Hill, New York.

Load and Resistance Factor Design
(Limit States Design)

Serviceability Limits for Tall Buildings

Gerhard Sedlacek
Stefan Bild

Although serviceability criteria are seldom clearly defined in design codes, they are very important for designing structures against vibration. Two examples demonstrate how vibration criteria may control the structural design. The first example concerns a steel structure on top of a concrete building, which was modified after erection to reduce vibration problems. The second example is of an exhibition hall, for which computer simulations of the dynamic behavior were used to demonstrate the suitability of the proposed composite deck.

MEANING OF SERVICEABILITY CHECKS

Allowable stress design methods usually contain hidden serviceability criteria, since compliance with the stress limits imposes strain limits on the material and therefore provides deformation and stiffness limits.

The new limit states design method distinguishes between ultimate limit states and serviceability limit states. Ultimate limit states usually include physical fracture and collapse caused by instability or unbounded deformations. Serviceability limits are often related to sensory perceptions, such as those of visual deflections or cracks, or accelerations experienced by humans and are usually defined by stiffness and deformation criteria, as shown in Fig. 1.

Whereas codes describe very precise ultimate limit states and analysis models, it is remarkable that the criteria for most serviceability phenomena are often ill-defined. Often these criteria are agreed on in individual cases by the contract partners, and so codes usually give only general recommendations. For example, the recommendations in the monograph of the Council on Tall Buildings and Urban Habitat (1979) are quite vague, and in general they only give deformation limits related to satisfactory structures. For instance, the requirements for the maximum lateral displacement D of a tall building of height H vary between $H/1000$ and $H/200$, and appear to be somewhat arbitrary.

The lack of precise serviceability criteria in design codes is in obvious contrast to the practical importance of these criteria. The owner assumes that sufficient load capacity will be provided as a matter of course, and focuses his

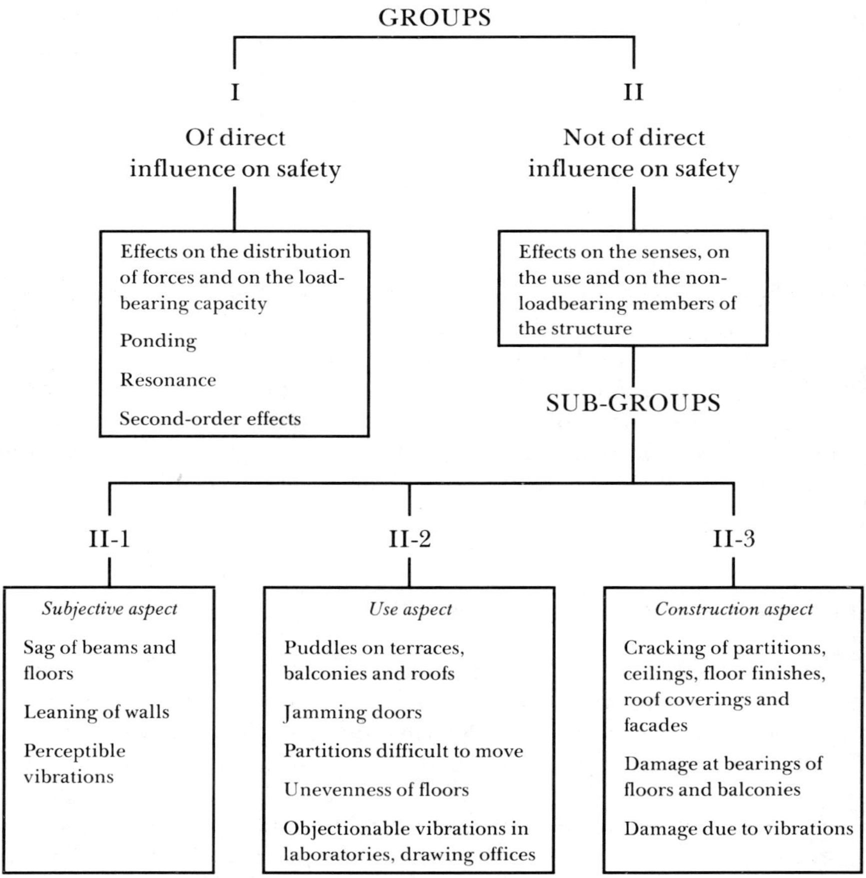

Fig. 1 Effects of deformation *(Steelbuilding Association, 1980)*

interest on the service performance. Defects and complaints made by owners seldom refer to inadequate load capacity but very often to violations of serviceability conditions. The following two examples of floor designs illustrate the importance of serviceability checks.

SERVICEABILITY CHECK AFTER BUILDING ERECTION

Cantieni (1985) reports a case in Zürich, Switzerland, where the composite floor shown in Fig. 2 was repaired following the measurement and assessment of its service performance. The two-story steel frame was erected on top of a concrete building and was to be loaded only at the external walls and the internal columns. Thus the columns of the steel frame were elastically supported by the load distributing girder system shown in Fig. 3. Any kind of vibration of the intermediate floor beams of profiles IPE 330 spaced at 1.55 m (5.09 ft) was considered to be unsatisfactory. The following measurements were made:

1. Determination of the floor's natural frequencies following impulsive excitation by a falling mass.
2. Evaluation of the maximum floor response to a man skipping with frequencies of approximately 2, 3 and 4 Hz.
3. Measurement of the behavior under service conditions simulated by three persons moving in step.

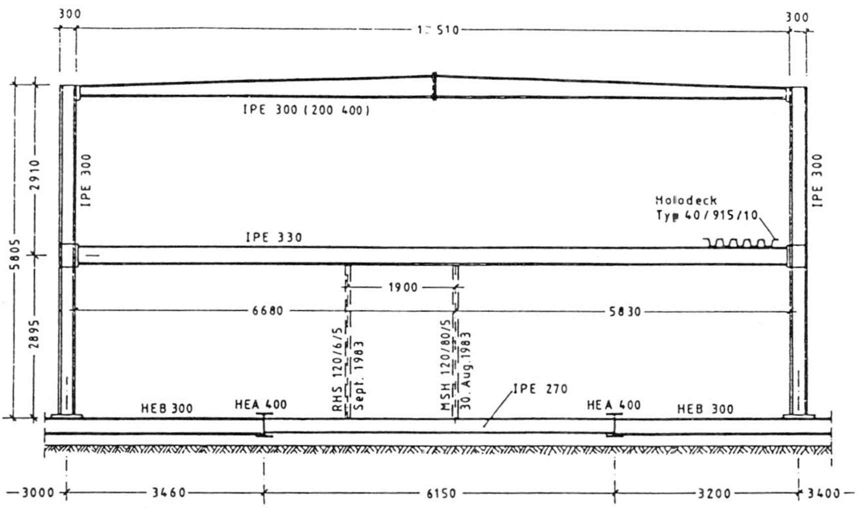

Fig. 2 Framed steel structure built on a concrete building in Zurich, Switzerland

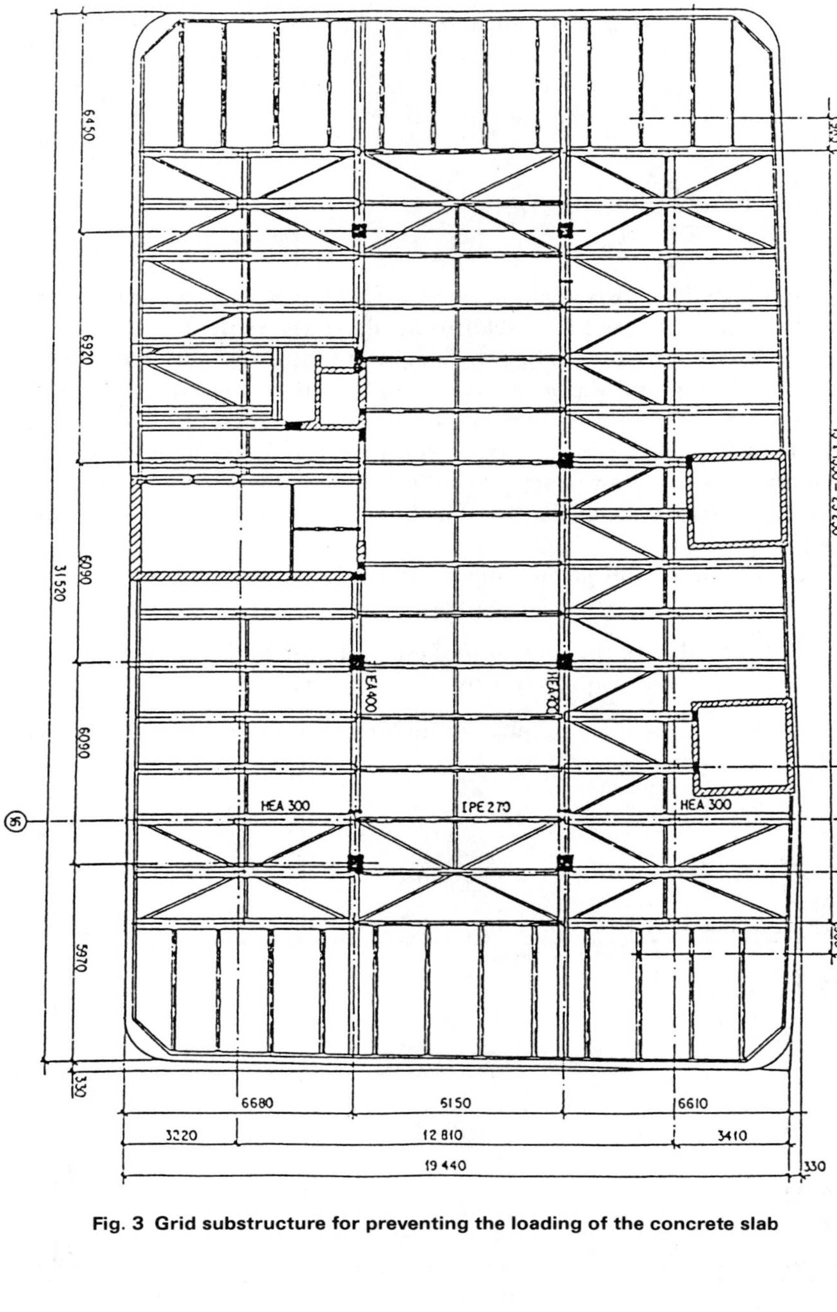

Fig. 3 Grid substructure for preventing the loading of the concrete slab

Fortunately, measurements and assessments were able to be made of both the original (and unsatisfactory) structure and of the structure after stiffening by intermediate columns.

The acceleration-time histories were measured from which the frequency spectra of the root mean square accelerations were derived. These were then compared with comfort limits given by the ISO (code 2631, 1976), the German VDI (Recommendation 2057, 1979), the German code (DIN 4150, 1975), and the British Standard (BS 6472, 1984), as shown in Fig. 4.

The limit curves of these codes appear to be almost identical, and their maximum sensitivity lies between frequencies of 4 Hz and 8 Hz. However, concerning the admissibility the limit curves are quite different being most restrictive in the proposed German code DIN 4150 (1975).

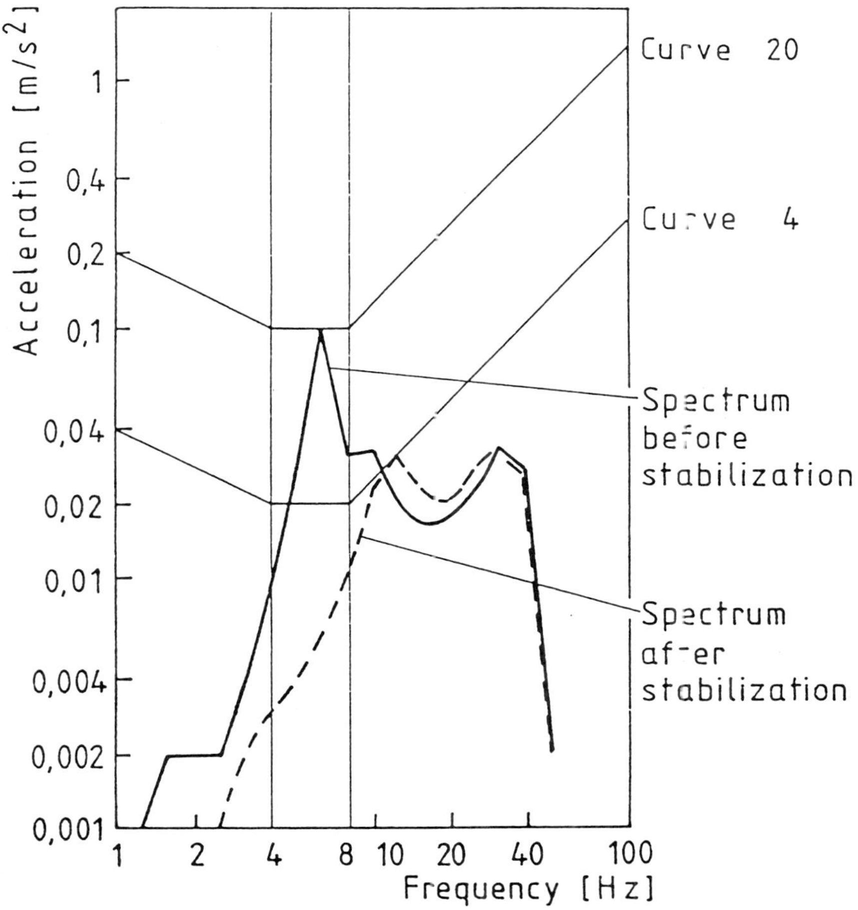

Fig. 4 **Human comfort curves vs. root mean square values for the acceleration and spectra before and after stabilization**

The assessment method of the BS 6472 (1984) is developed in the most general manner. Not only must the limit curve be satisfied at all points, but in addition a procedure similar to the determination of the decibel (A) values in acoustics must be used. By introducing a basic curve for the variation of the acceptable acceleration with the frequency, an effective dimensionless root mean square value is calculated from

$$M_f = \sqrt{\sum_i \left(\frac{T_{pi}}{B_{ci}}\right)^2} \tag{1}$$

in which T_p is the measured acceleration and B_c is the corresponding value of the basic curve.

In the original state before stiffening, M_f was 22.1, indicating that the structure should certainly be modified. After modifications M_f had reduced to 7.1. This is almost twice the lower bound of the maximum permissible value ($4 < M_f < 8$), and so the assessment method of the BS 6472 (1984) indicates that further modifications might be made. No modifications would be required if the M_f value was less than 4. The measurements showed that the spectra measured after stiffening did not exceed the Limit Curve 4 of BS 6472 (1984) for people moving in office buildings at any location.

The purpose of the structural modifications is to shift the building's natural frequencies out of the critical range where significant responses can be expected. This can be done most effectively by changing the design span as indicated by the frequency formula

$$\omega = 2\pi f = \left(\frac{\pi}{1}\right)^2 \cdot \sqrt{\frac{EI}{m}} \tag{2}$$

Other methods could be to change the end restraint or else to increase the amount of damping when the natural and excitation frequencies are too close.

SERVICEABILITY CHECK DURING A BUILDING'S DESIGN

When it is difficult to predict the building's vibration behavior, then it is advisable to make provisions for the future installation of vibration controls should the performance prove unacceptable. However, the major effort should be to increase the range of predictions that lead to reliable structures.

An example of such a prediction can be seen in Fig. 5. It shows an exhibition hall in Berlin, West Germany, which was planned with a composite floor structure. The contract design was based on the results of an analytical simulation of the floor's vibration behavior under the actions of fork-lift trucks and dances. After the building's completion, vibration tests were carried out, which confirmed the simulation results and demonstrated the adequacy of the building.

In detail, the composite floor structure consists of two spans of 15 m each. It is supported by a girder in the center and built-in to the columns at the ends. The natural frequency calculated was in the range of those frequencies that were measured when a man skipped on the floor, as shown in Fig. 6. Cantieni (1985) considers that lowest natural frequencies of approximately 5 Hz are still critical and recommends that natural frequencies be not less than 8 Hz to 10 Hz. Nevertheless, the floor proved to be satisfactory.

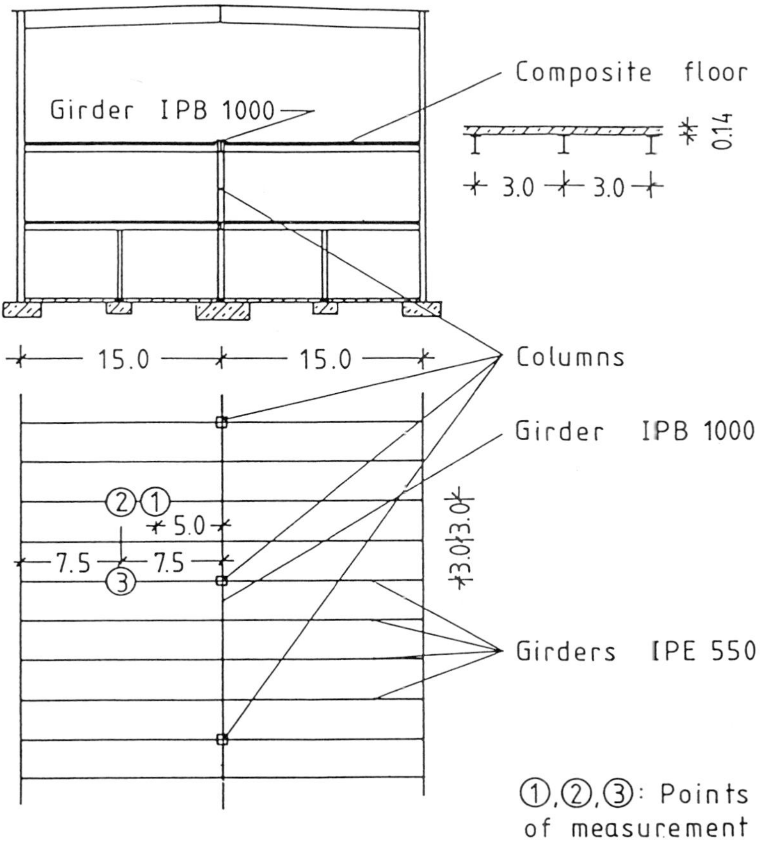

Fig. 5 Exhibition hall with a composite deck, Berlin, West Germany

	Natural frequency [Hz]	At intermediate columns		Between columns		
Calculated	Dead load only Girder load : 11.6 kN/m (794.9 lb/ft)	6.49	9.34	5.80	6.49	Not rem.
Calculated	Dead load and live load Girder load : 23.2 kN/m (1589.7 lb/ft)	4.60	6.60	4.10	4.60	Remeasured
Measured	Dead load and live load Girder load : ~23.6 kN/m (1617.1 lb/ft)	5.03	11–12	4.90	5.00	Remeasured

Fig. 6 Comparison of calculated and measured frequencies

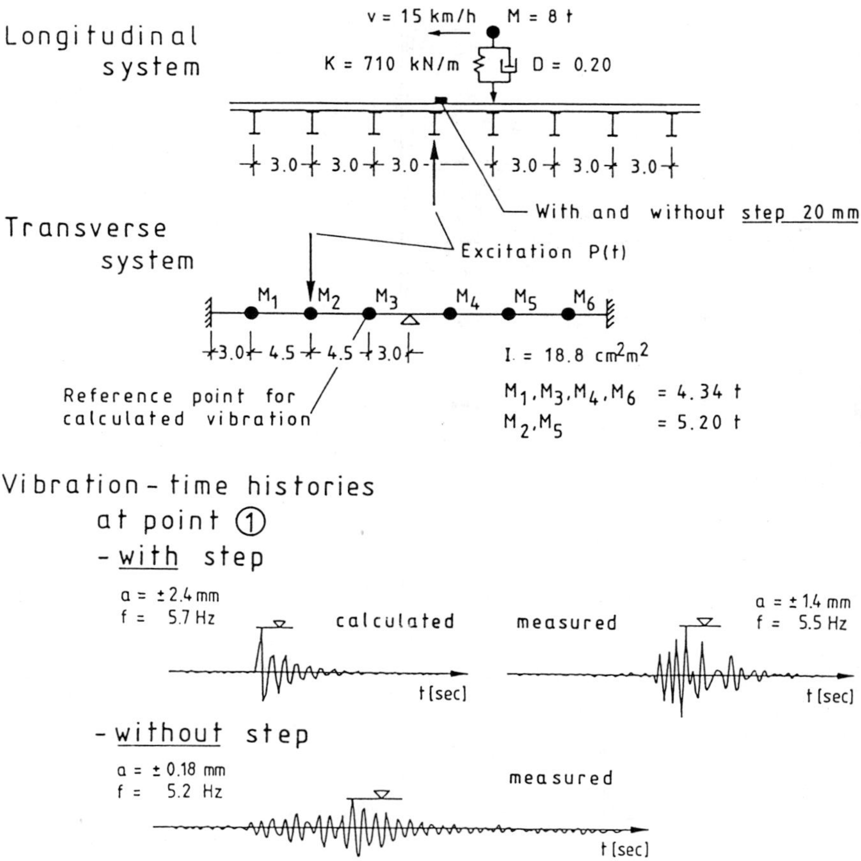

Fig. 7 Modeling of the composite deck and the vehicle as mass-spring-dashpot element system

For the computer simulation, the floor was modelled as a system consisting of several masses and springs with an overall damping factor of 4% as shown in Fig. 7. The vehicle was simulated by a spring-mass-dashpot system and directed over a step 20 mm (0.8 in.) high at a speed of 15 km/hr (9.32 miles/hr). The vibration amplitudes and frequencies were assessed according to the VDI (Recommendation 2057, 1979) and the DIN (4150, 1975) using the range of admissible perception magnitudes shown in Fig. 8.

Measurements made after the completion of the building gave actual damping factors between 3.7% and 6.8%, and confirmed the damping value assumed. The predicted vibration amplitudes were shown to be satisfactory. For the analytical simulation of dancing, the assumed load-time histories (see Fig. 9) produced building responses below the admissible perception magnitudes. The measured vibration behavior of the composite floor was acceptable and gave strong support for the analytical checks.

Recent proposals of EUROCODE 3 (1983) require natural frequencies of not less than 5 Hz and deflections caused by live load of not greater than 10 mm (0.4 in.) for those floors that may be loaded by rhythmic skipping. These conditions are met by the actual system, which had a natural frequency of 5 Hz and a maximum deflection caused by live load of 5 mm (0.2 in.).

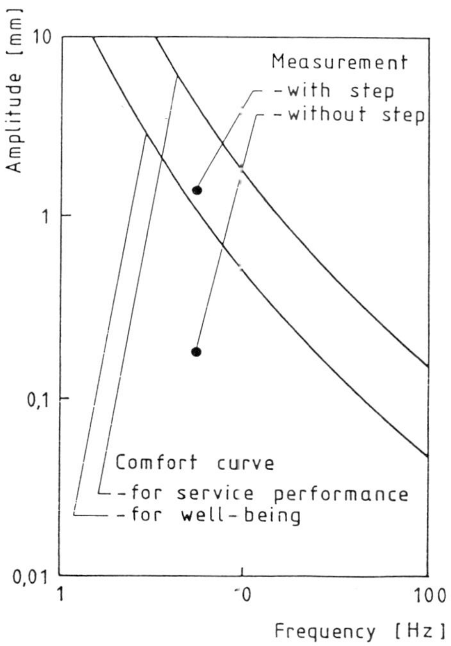

Fig. 8 Comparison of the maximum amplitudes with the comfort curves vs. frequencies

FUTURE DEVELOPMENT

In the future, the basis of the procedure should be improved and standardized comfort limits should be established. Computer simulations of the behavior of floor structures under moving masses are practicable if the floors can be described by mass-spring-dashpot systems. It should also be possible to derive simple design rules similar to those contained in the EUROCODE 3 (1983) that might help predict and assess the vibration behavior of structures.

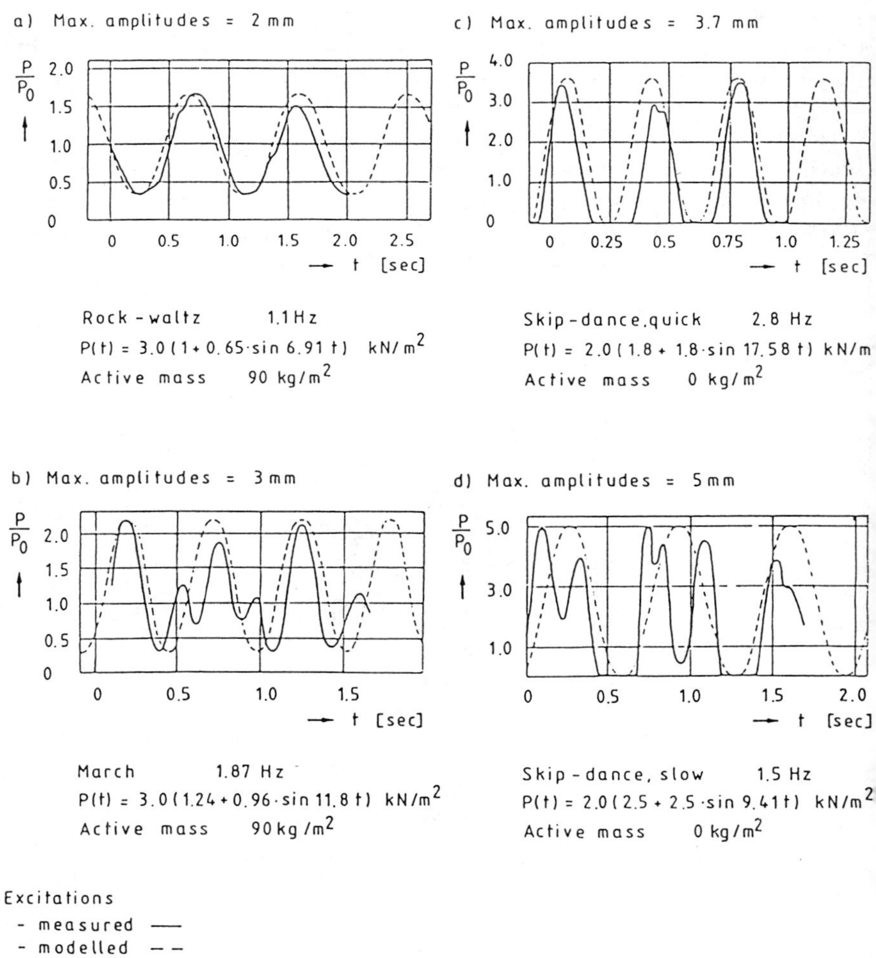

Fig. 9 Simulation of dancing performance

REFERENCES/BIBLIOGRAPHY

British Standards Institution, 1984
BS 6472, GUIDE TO EVALUATION OF HUMAN EXPOSURE TO VIBRATIONS IN BUILDINGS, London, UK.

Cantieni, R., 1985
RESEARCH ON FLOOR VIBRATIONS OF AN OFFICE BUILDING (Untersuchungen von Deckenschwingungen bei einem Burogebaude), Report No. 116, EMPA, Zurich, Switzerland.

Commission of the European Communities, 1983
EUROCODE 3, COMMON UNIFIED CODE OF PRACTICE FOR STEEL STRUCTURES, Draft, Brussels, Belgium.

Council on Tall Buildings and Urban Habitat, Group SB, 1979
STRUCTURAL DESIGN OF TALL BUILDINGS, Volume SB of Monograph on Planning and Design of Tall Buildings, ASCE, New York, USA.

Deutsches Institut fur Normung, 1975
SHOCK AND VIBRATION IN BUILDING STRUCTURES (Erschutterungen im Bauwesen), Draft of DIN 4150 Deutsches Institut fur Normung, Berlin, West Germany.

International Standards Organization, 1976
ISO, Code 2631, Oslo, Norway.

Pfluger, A., 1970
VIBRATIONS OF PEDESTRIAN BRIDGES (Schwingungen von Fusswegbrucken), Report No. 15, Institut fur Statik der TU Hannover, Hannover, West Germany.

Sedlacek, G., 1984
ASPECTS OF SERVICEABILITY OF STEEL STRUCTURES (Aspekte der Gebrauchstuchtigkeit von Stahlbauten), Stahlbau, 53, pp. 305-310.

Steelbuilding Association, 1980
DEFORMATION REQUIREMENTS FOR BUILDINGS, Rotterdam Concrete Association, Zoetermeer, The Netherlands, Technical Translation, Ottawa, Canada.

Talin, A. and Ellingwood, B., 1985
WIND-INDUCED MOTION OF TALL BUILDINGS, Engineering Structures, Vol. 7, October, pp. 245-252.

Verein Deutscher Ingenieure, 1979
VDI, RECOMMENDATION 2057, PARTS 1-3, Dusseldorf, West Germany.

Structural Design of Tall Concrete and Masonry Buildings

Introductory Review

Ignacio Martin

The second century of concrete skyscrapers will bring a new concrete with new mixtures and changes in the nature of cement that will permit ultra-high-strength and the decrease or elimination of creep and shrinkage.

Construction methods will change dramatically, but these changes will have to be adapted to the socio-economic characters of the location of each project.

Design methods will continue to be more satisfactory and new building codes will be changed to reflect computer technology.

Extensive research will be required not only in materials research, but research applied to construction methods. Concrete construction must be industrialized in the true sense of the word.

Concrete strength will continue to increase. High-strength concrete of 100 mPa (14 ksi) is commercially feasible today, and 138 mPa (20 ksi) concrete could be available in the year 2000. Concrete strength of 210 mPa (30 ksi) can be attained with silica fume concrete, but there are warnings about the exposure behavior of this type of concrete in high temperature. Mile-high concrete buildings are proposed using 100 mPa (14 ksi) concrete. The predicted elastic, creep, and shrinkage shortening of this building would be 3.5 m (11 ft), which would have to be compensated for.

A National Science Foundation funded U.S. research program on masonry, which will be coordinated with a Japan program, aims at safer, more efficient and taller masonry buildings. The research includes bond of reinforcement to grout and the possibility of using limit state design for masonry buildings.

There is a need for research on high-strength concrete for shear bond and the effect of confinement. There is a further need to have the same factors of safety for steel and concrete building codes, in order to facilitate the use of mixed construction.

Stone and monumental materials are coming back and are restoring the stiffness that tall buildings used to have. On the other hand, buildings with an aspect ratio of 1 to 16 are being proposed with a period of vibration of 14 sec. A study of the mile high concrete building shows a period of 25 sec. "Post modernism" in architecture may well bring back the use of precast concrete in facades.

Special attention will be paid to the problem of durability of concrete and in particular to the corrosion of reinforcement of parking structures and buildings near the ocean. Also important are problems such as chemical attack and acid rain and the need for low permeability of concrete.

The building codes of the future will be written for the use of computers which will enhance the applicability of three-dimensional analysis. Building codes must deal more effectively with the problem of torsional instability. More knowledge is needed on the mathematical modelling of concrete structures especially regarding stiffness and behavior with time.

The use of precast and prestressed construction in modern concrete tall buildings is due for an increase. It is not possible to use precast concrete efficiently and at the same time fast track construction.

Also more innovations in the construction process will have to be introduced.

There is a continuing need to prevent failures in the planning and design stages. More owners and developers are getting involved at that time and in the decision making process and must be selective in choosing structural engineers and contractors involved in the construction of megastructures. They must allow the time and means for the design and the planning of construction of tall buildings. In the future owners and developers will have to share the responsibility in construction failures for the decision they make.

Concrete tall buildings in seismic zones require ductility. This ductility must be present in the structure, but should not be abused in building codes to improperly reduce the equivalent seismic loads.

The Mexico City earthquake will teach us important lessons such as the importance of soil conditions of three-dimensional analysis to consider torsional effects, and the adequate structural system selection. A new generation of analysis of performances of buildings will be born: instead of explaining why buildings fail we will try to obtain answers from the study of the buildings that performed well.

Before this century is over we will witness dramatic changes in the construction of tall concrete buildings that will push the imagination of developers, designers, and builders to limits unknown today.

Building Codes

James G. MacGregor

In an essay on "Standardization and its Abuse," Prof. Hardy Cross (1952, p. 141) wrote:

> In the field of structural design the effort to get intelligence through standardization has been carried pretty far In reinforced concrete, for example, it has been necessary to set up elaborate standards. Out of this work came a narrowly circumscribed standardization of procedure, which is called "the theory of reinforced concrete" and to which unfortunate students are exposed. Few will question that the standardized theory of reinforced concrete is perhaps as complicated a bit of nonsense as has been conceived by the human mind. It does, however, work pretty well as a check on undiscriminating unintelligence.
>
> In engineering there is no attempt to standardize unless there is some reason for it. Some, however, wish to standardize where there is no real advantage and so fasten for a long time upon the profession as complex assembly line that has characteristics of a cartoon.

This paper reviews current trends in the design of tall concrete buildings and how these are affected by current code statements.

THE ROLE OF CODES IN BUILDING DESIGN

The primary reason for building codes is to protect the health and safety of the public. If this definition is accepted, adequate strength is essential, serviceability is outside the legal sphere of the code unless it affects health and safety, and durability is only important if it affects the safety of the structure. Increasingly, however, codes for concrete structures are including clauses on serviceability (deflections, crack control) and durability since designers want standardized ways of dealing with these problems.

APPLICATION OF CURRENT CODES TO TALL CONCRETE BUILDINGS

Current building codes for concrete structures pay very little attention to particular problems encountered in tall buildings for two reasons. Our building codes have evolved over an 80-year period, but very tall concrete buildings are a recent phenomenon. Second, and more important, only 2% of all buildings built in the United States in 1983 exceeded 15 stories. For nonresidential buildings, only 6% exceeded 15 stories. As a result, the problems of tall building design per se have not received a great deal of attention in code committees. Several areas requiring more thought and guidance follow.

Compatibility of Codes with Computers

A code such as the ACI Building Code is the product of decades of work. Each edition incorporates much from the previous edition with incremental changes in a few areas. The assumptions for structural analysis of beams, frames, and two way slabs in that code were all formulated prior to the computer era and envisioned moment distribution as the primary method of analysis. These assumptions and design methods need a careful re-examination in view of the needs and abilities of modern analyses.

Modern analyses have made it possible to model member behavior more precisely—or have they? Collins et al. (1985) describe the results of a competition in which 43 engineers from 13 countries used state-of-the-art analyses to predict the test results of four reinforced concrete panels subjected to combined shear and axial forces. The predicted failure stresses ranged from 27% to 280% of the measured failure stresses. The standard deviation of the computed values averaged about 30% of the measured values. This suggests that considerable care should be taken before the results of an exotic analysis are adopted for design purposes.

Modern analyses have made it possible to consider large numbers of loading cases for complex frames. Engineers need guidance in this field. How many load cases are good enough? How many iterations should be carried out with updated member properties or updated cracking scenarios? Beeby and Taylor (1978) touch on that question:

> . . . the tendency of some engineers to do as much calculation as they can in the design stage. This design approach, used uncritically, leads to the dangerous tendency of which many of us may be accused: if it is possible to carry out calculations for a particular aspect of behavior, then we do; if no method currently exists, then calculations for this aspect of design are deemed unnecessary.
>
> More simply, we put forward our aphorism for bad engineering practice: I can, therefore I must—I cannot, therefore I need not . . . (p. 210).

Irregular Floor Plans and Structural Systems

During the decade from 1965 to 1975, tall buildings tended to have consistent structural systems throughout their height. Frequently these consisted of frames and shear walls or tube-in-tube types of structures with similar floor plans on all floors. During the past decade, the advent of multistory atria and multiuse buildings has resulted in buildings with significant changes in structural framing at one or more points in the height. Thus, an atrium space may have columns of varying heights. A multiuse building may have one column spacing in the retail space and atrium space, a different column spacing in office floors, and a different spacing in the residential floors. In such a building, major transfer girders or trusses are needed at changes in the structural system.

The design of the deep beams, which serve as transfer girders, is poorly handled in most codes. Recently, design models based on equilibrium plasticity solutions have been proposed in a number of countries and these show considerable promise.

Mixed Structural Systems

In North America and Europe, structural systems involving combinations of structural steel and reinforced concrete are frequently used. The most common of these are:

1. Concrete cores, sometimes slipformed, with a structural steel frame and composite floors.
2. Concrete cores and a concrete external tube, with floors consisting of composite beams, stub-girders or trusses spanning from the core to the exterior.
3. Structures similar to (2) with a light exterior steel frame for erection purposes. This frame is eventually incorporated into an exterior composite tube system to resist lateral loads.

In North America the concrete code writing community has not responded to these developments and the current design rules have been derived largely by the steel design community. There is a danger that the unique properties of concrete are idealized excessively in this treatment.

A more pressing concern, however, is that the design of the concrete structure, the steel structure, and the foundation all be carried out with a consistent design philosophy using consistent load factors. Moves are afoot for the American Institute of Steel Construction to adopt a limit states design format similar to the one used by the American Concrete Institute Code and for both codes to move toward the same or very similar load factors. In Canada the same load factors are used for limit states design of concrete, steel, and timber structures.

High Strength Concrete Structures

Since 1972 more than 20 buildings have been built in Chicago alone with columns having design compressive strengths of 62 MPa (9000 psi) or higher.

Since high-strength concrete is a relatively new material, its structural properties are not adequately recognized in current building codes. In particular, it has a more brittle stress-strain curve that necessitates a modification of the compression stress block. The modulus of elasticity and the shear strength of high-strength concrete beams both appear to increase more slowly with concrete strength than current code relationships indicate. Finally, lateral confinement appears to be less effective in enhancing the strength of high-strength concrete than it is for normal concrete. Since the floors in buildings with high-strength concrete columns generally are made from considerably lower-strength concrete, special provisions must be made to transfer the column loads through the floors. Various codes treat this in different ways.

Very Slender Buildings

The desire to provide plaza areas or to use small pieces of property has resulted in tall slender buildings with height-to-width ratios exceeding eight in some cases. At the same time owners and architects are insisting on more and more open spaces at the ground level of buildings where stiffness is most needed. As a result, stability is a continuing problem in the design of tall buildings.

Increasingly, second-order analyses are used to determine stability effects in tall buildings. The assumptions in such analyses must be documented more carefully, particularly for concrete structures. Questions such as what member properties are to be used, what is the effect of creep, what loading cases should be considered, and so on all need informed discussion. How the results of such analyses are combined with the moments resulting from checkerboard gravity loads needs more study. The 1984 Canadian code (Canadian Standards Association, 1984) attempts to discuss this latter point.

Although torsional instability and the coupling of torsional and orthogonal modes can be a problem in some tall buildings, this question is not specifically mentioned in any current building design code nor are there established ways of considering it in design.

Durability of Tall Buildings

In Australia great concern is currently being expressed about the corrosion damage to facades and other parts of buildings near the ocean. In Canada and the United States, corrosion damage to parking garages in the basements of high-rise buildings has reached epidemic proportions. In these countries, codes are being changed to reflect the concept that durability is essential to good structural performance.

SUMMARY

This paper has reviewed several areas where current codes need improvement in dealing with tall concrete buildings. In recent years the ponderous code writing machinery has lagged behind the state of the art in the design of tall concrete buildings.

REFERENCES/BIBLIOGRAPHY

Beeby, A. W. and Taylor, H. P. J., 1978
THE USE OF SIMPLIFIED METHODS IN CP110—IS RIGOUR NECESSARY?, The Structural Engineer, Vol. 156A, No. 8, August, pp. 209-215.

Canadian Standards Association, 1984
DESIGN OF CONCRETE STRUCTURES FOR BUILDINGS, CSA Standard CAN 3-A23.3-M84, 281 pp.

Collins, M. P., Vecchio, F. J. and Mehlhorn, G., 1985
AN INTERNATIONAL COMPETITION TO PREDICT THE RESPONSE OF REINFORCED CONCRETE PANELS, Canadian Journal of Civil Engineering, Vol. 12, No. 3, September, pp. 624-644.

Cross, H., 1952
ENGINEERS AND IVORY TOWERS, McGraw-Hill, New York, p. 141.

Selection of Structural Systems

Concrete Buildings—A Mile High

Joseph P. Colaco

The number and height of tall buildings in general and of tall concrete buildings in particular has increased dramatically in the last 40 years. In the last 25 years there have been dramatic advancements in the technology of construction of tall concrete buildings with the advent of new forming systems such as slip-forming, flying forms, gang-forms, and so forth. Also, the development of ultimate strength design, the development of structural lightweight concrete, the development of high strength concrete, the use of admixtures (in particular superplasticizers), and concrete pumping techniques have given concrete a great boost for tall structures.

Prior to 1965, concrete buildings in the 40-story range such as the Brunswick Building in Chicago were designed and built. The evolution of structural systems particular to concrete construction, notably by the late Dr. Fazlur Khan, gave rise to the potential for taller concrete structures. In 1968, One Shell Plaza in Houston, a 50-story lightweight concrete building was designed and constructed and became the tallest concrete building in the world. In Hong Kong, the Connaught Center project with its exterior slip form concrete walls and round windows gave rise to the development of an extremely efficient "tubular" structure where the axial forces flow around the windows and minimize bending stresses. The slip forming of the exterior wall has the added advantage of speed of construction. In the 1970s Water Tower Place in Chicago was built and to this day holds the record as the world's tallest concrete building at 263 m (864 ft) in height. (The tallest concrete structure, however, is the CN Tower in Toronto which is 457 m (1500 ft) tall.) In 1977 the 75-story, 305 m (1000 ft) tall Texas Commerce Plaza in Houston was constructed. This building is the tallest exterior composite building in the world and has two unique features; first, all the concrete in the project was pumped and secondly, self-jacking exterior forms were used for the construction of the exterior composite frame. The pumping of the concrete to 305 m (1000 ft) stands today as a record for the tallest height of pumping of concrete with a single stage pump. The self-jacking exterior forms enabled the construction to proceed at a very rapid pace, and 72 floors of the building were constructed in 11 months due to this combination of techniques.

What are the advantages of concrete for tall building construction?

Tall buildings in nonseismic areas are governed not so much by strength considerations but by performance characteristics under wind loads. The most important considerations are the sway of the building under wind loads as well as motion perception that could affect occupant comfort. This consideration is particularly critical in steel buildings, which have a very high strength-to-weight ratio as compared to concrete. In tall structures the two most effective methods of obtaining better performance under wind loads is to increase the mass and to increase the damping. Whereas a steel structure generally has a mass in the range of 96 to 128 kg/m^3 (6 to 8 lb/ft^3), a comparable concrete building will have a mass in the range of 160 to 240 kg/m^3 (10 to 15 lb/ft^3). With regard to damping, it is commonly acknowledged that a steel structure has an approximate damping value of 1% of critical damping whereas a concrete structure has a damping value ranging from 1%–2% of critical damping. In general therefore, concrete structures inherently perform much more satisfactorily under wind loads than steel structures because of both a higher mass and a higher damping.

PROPOSED BUILDING SCHEME

In the 1950s, Frank Lloyd Wright proposed a mile-high building in Chicago. Chicago also currently boasts the tallest office building in the world, namely the Sears Tower at 442 m (1450 ft) in height. In view of this, it seemed that it would be interesting to develop a concrete building a mile high and to study its implications.

In order to obtain a good aspect ratio (ratio of height/width) for the building, a structure having a base dimension of approximately 152 m (500 ft) square was desirable. Fig. 1 shows a floor plan of the building. The basic arrangement of the building is in modules of 30.5 m (100 ft) square (similar to the modular arrangement of the Sears Tower). The exterior of the building is shown in Fig. 2. Taking advantage of the ability to shape concrete into any reasonable form and with the desire to develop true "tubular" behavior, it was decided to utilize diamond shaped windows as shown in Fig. 3, which give rise to a natural "truss" appearance on the outside of the building. This "trussed megatube" is an extremely rigid element and functions primarily to tie the building together for wind loads and secondarily, also to make the exterior walls rigid against other forces and movements. With regard to the interior floor plan, columns are spaced at 6.1 m (20 ft) on centers along the modular lines and are run continuously from top to bottom without any transfers as shown in Figs. 4 through 8. This gives rise to 30.5 m (100 ft) square column-free open spaces in the building, which would be adequate to take care of most occupancy needs. Tall buildings require a great deal of vertical transportation and as the elevators drop off, the modules of the structure are

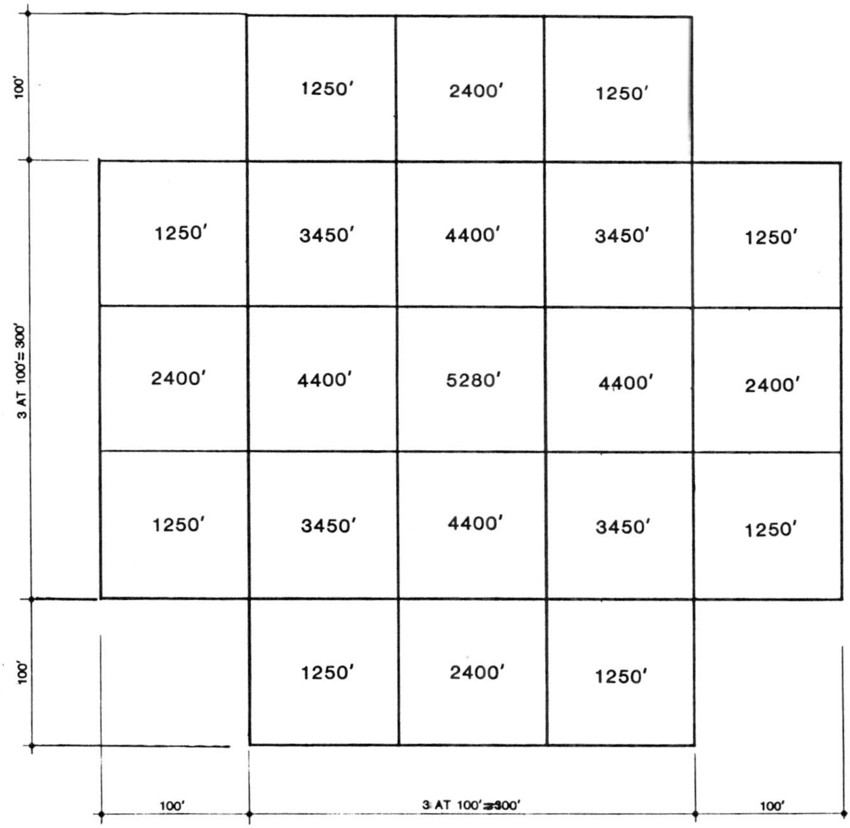

Fig. 1 Key plan

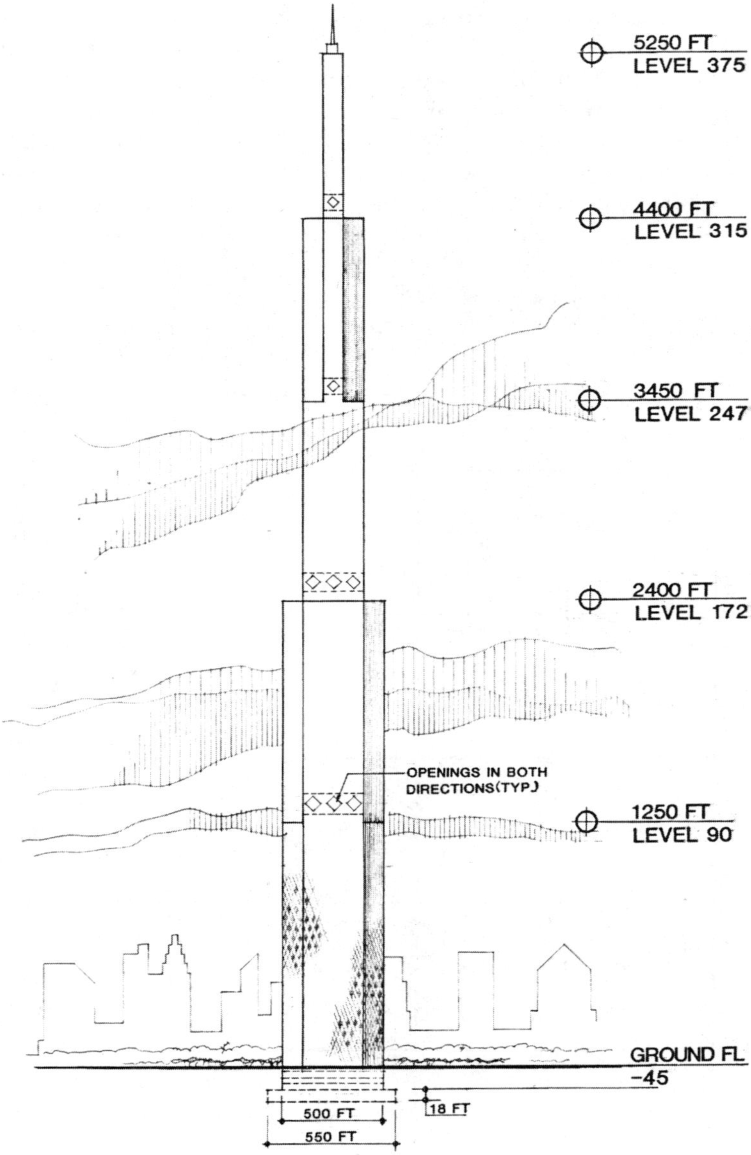

Fig. 2 Elevation

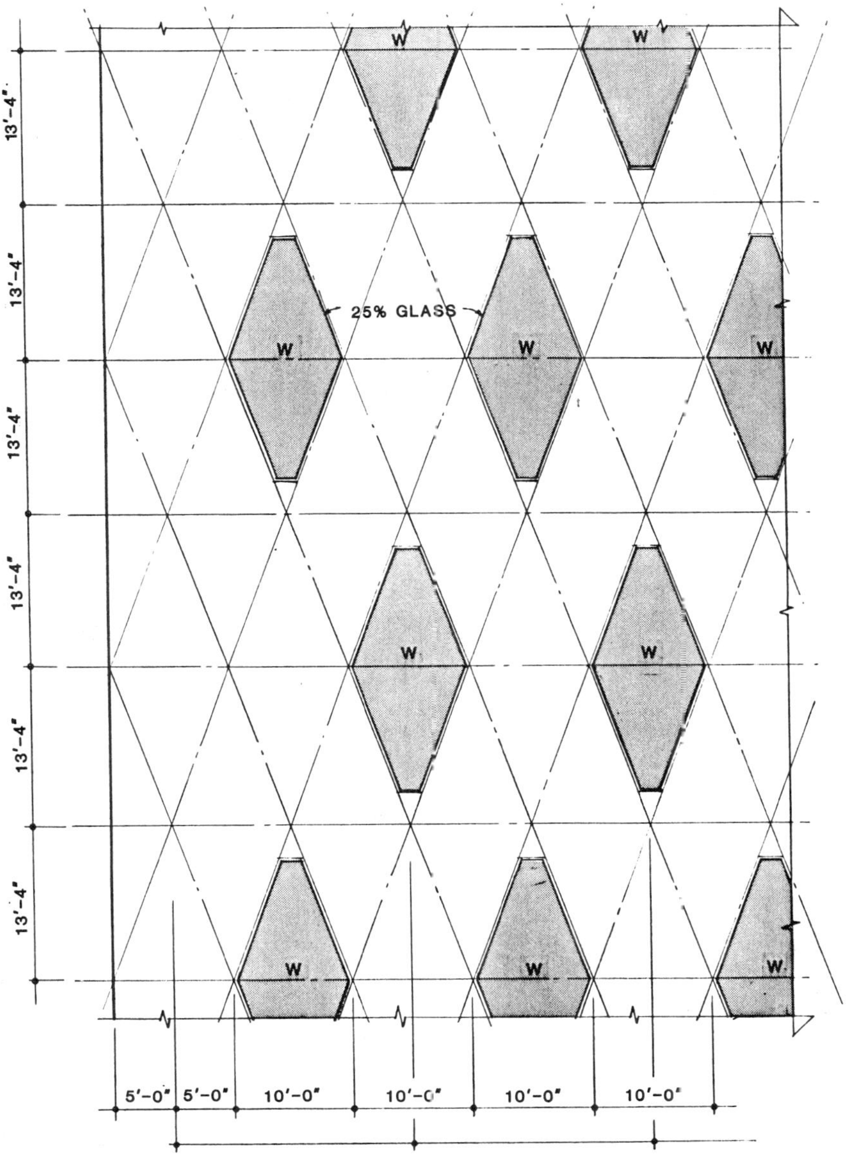

Fig. 3 Trussed megatube with "diamond" shaped windows

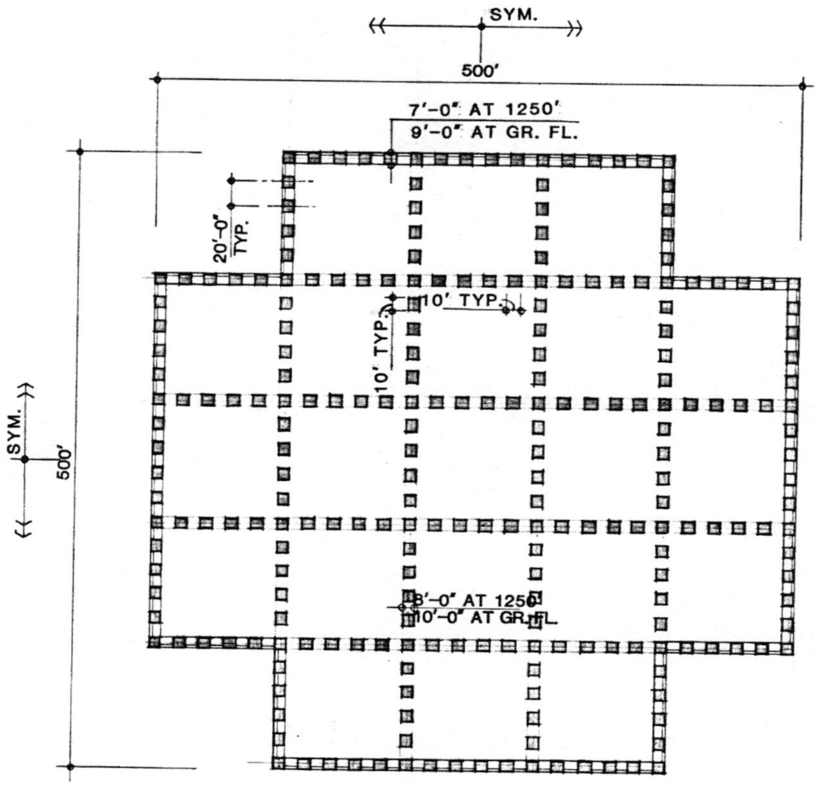

NOTE : ALL COLUMNS AT 20'-0" O.C.

Fig. 4 Floor plan from base to 1250 ft

dropped off to give rise to a tapered appearance on the sky line. All of the drop-offs occur in modular fashion so once again no transfers are required at any point. The basic building blocks and drop-offs occur at 381 m (1250 ft), 732 m (2400 ft), 1052 m (3450 ft), 1295 m (4250 ft), and then on to the top of the building, at 1609 m (5280 ft).

Two types of floor framing systems were considered for this building. From an intellectual standpoint the desired floor framing system is a *super-waffle* floor system with waffle ribs at 6.1 m (20 ft) centers in each direction (see Fig. 4). A 140 mm (5½ in.) thick slab will complete the floor assembly. In order to minimize depths, the waffle ribs are 690 mm (2 ft 3 in.) deep at mid-span and approximately 1.07 m (3 ft 6 in.) deep at the columns. The floor slabs could be cast in lightweight concrete to minimize some of the dead load coming down the structure. An advantage to the waffle floor slab is that the monolithic

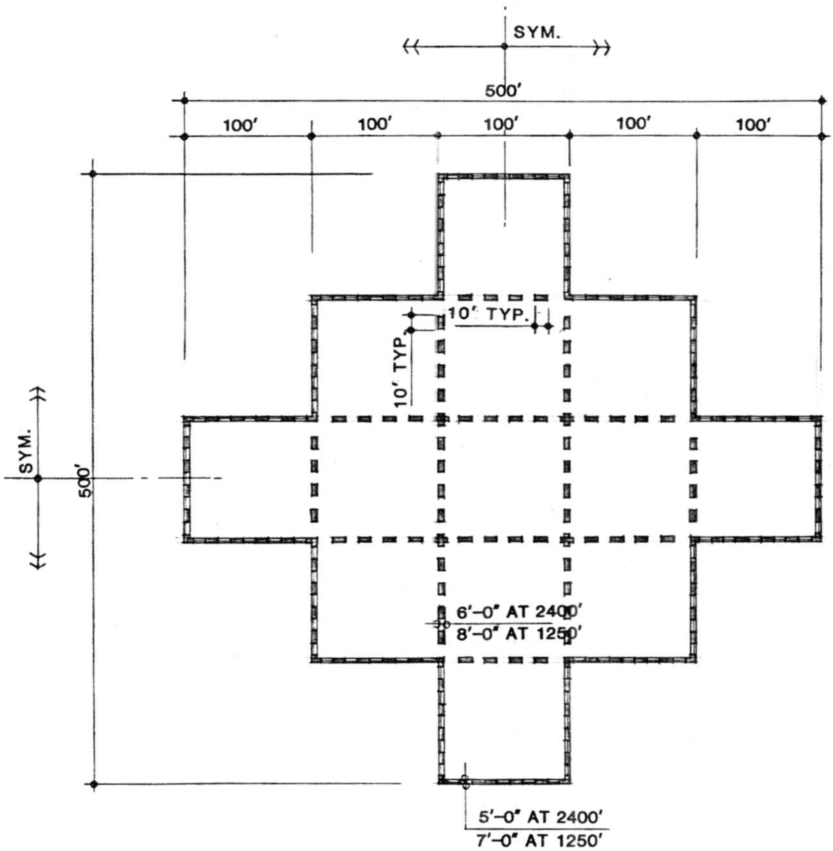

Fig. 5 Floor plan from 1250-2400 ft

nature of the structure enables gravity loads to be distributed very well. The major drawback of this kind of system is the large amount of formwork required that will slow down the construction process. An alternate system would be a precast structural slab such as a filigree slab. Figure 9 shows an elevation of the interior columns. In order to obtain better efficiency as a bundled tube and also to provide restraint against building movements, it is proposed to provide a "diagonal" tie through the interior columns.

WIND ENGINEERING

In the design of a tall building the basic wind forces to which the structure is subjected are generally determined by extensive testing of a model of the

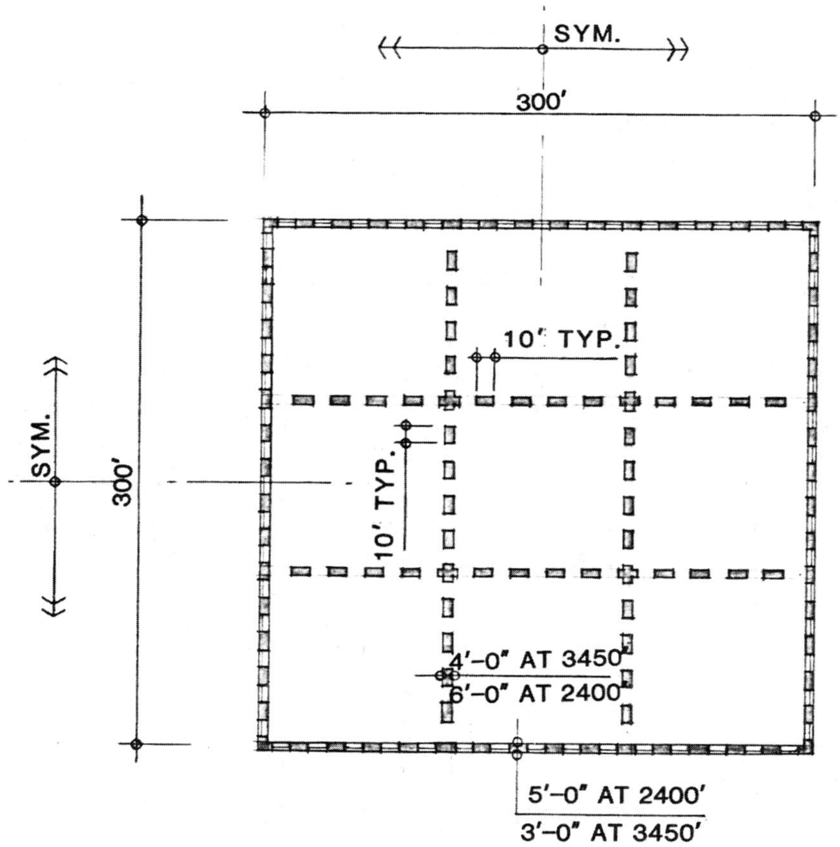

Fig. 6 Floor plan from 2400-3450 ft

structure in a boundary-layer wind tunnel. The wind climate is evaluated at the particular site. For this exercise, the provisions of the ANSI Code were used. Since the building is likely to be much taller than its neighbors, Exposure C was considered to be appropriate. The design was based on a downtown site utilizing a 145 km/hr (90 mph) basic wind velocity at 9 m (30 ft) and importance factor equal to 1.0. The shape factor was obtained from consultation with Dr. Isyumov of the University of Western Ontario and for preliminary purposes the value of 0.9 is used for this shape of building. Utilizing these values a basic wind profile was obtained for this structure as shown in Fig. 10. Wind pressures increase gradually from the bottom to top of the structure. Using these numbers, the wind shear at the base of the building

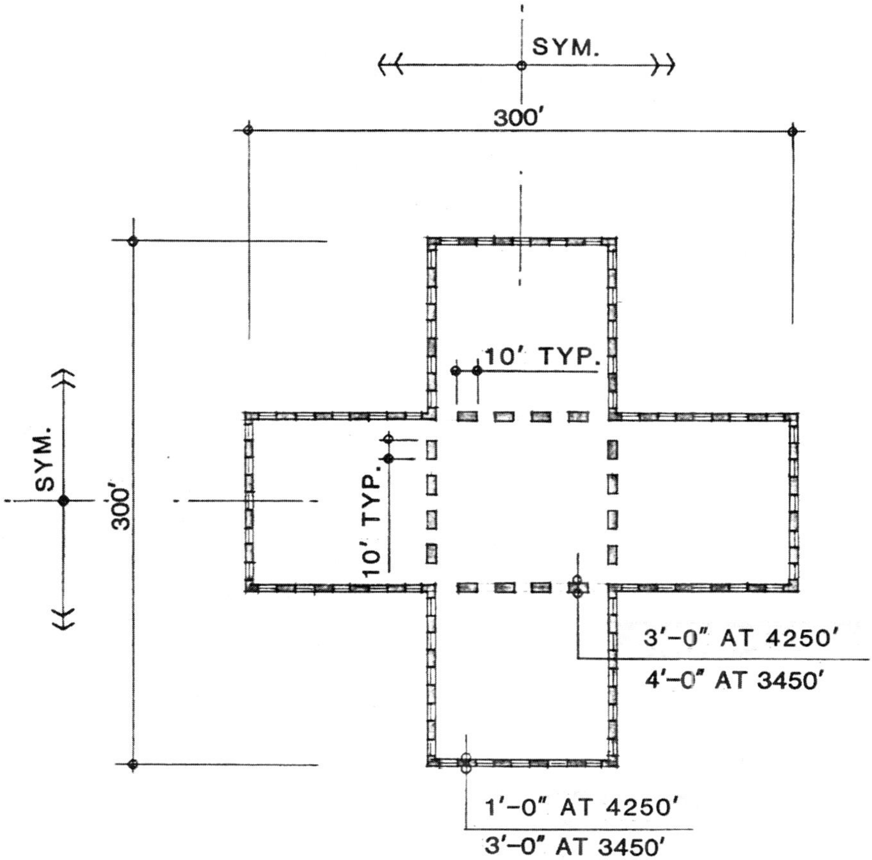

Fig. 7 Floor plan from 3450-4250 ft

is approximately 420 MN (95,000 kips) and the base overturning movement is 312,000 MN-m (230,000,000 kip-ft). Based on the preliminary sizes of the structure shown in Figs. 4 through 8, the sway of the building under the above wind loads is approximately 2.6 m (8 ft 6 in.) (heighted divided by 621). The maximum wind stress in the exterior wall at the base is approximately 5.69 MPa (825 psi) whereas the gravity stress under working loads is 42 MPa (6100 psi). Utilizing the Canadian Building Code, the gust response factor G came out to be 1.02. Figure 11 summarizes the wind engineering criteria. The fundamental period of the building is approximately 25 seconds. The building density is 400 kg/m^3 (25 lb/ft^3), which is substantially higher than in any other kind of construction. For damping value of 2% of critical, the Canadian

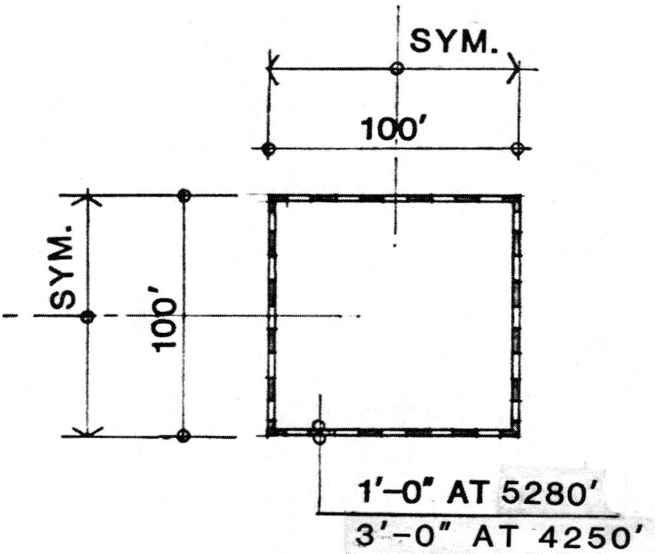

Fig. 8 Floor plan from 4250 to 5280 ft

Code Analysis indicates acceleration of the top of the building of approximately 0.04 g. Experience has shown that the above analysis overestimates the acceleration as compared to wind tunnel results by about 25%–30%. Even then, this number is a little higher than desirable. Since the mass of the building is so high, it is unlikely that tuned-mass dampers or an active damping system will be viable. Hence, other means such as openings through the building as shown in Fig. 2 could be used to minimize the accelerations. Obviously, a site specific wind tunnel test will have to be used to obtain the expected accelerations.

FOUNDATIONS

Because of the dia-grid arrangement of the main building structure, the most desirable foundation system for this kind of structure is a mat foundation. The foundation loads are in the range of 2.20 MPa (46 kips/ft^2) so that a minimum allowable soil bearing of 2.4 MPa (50 kips/ft^2) is necessary. This bearing capacity is readily available in several major metropolitan areas such as New York, Chicago, Dallas, Phoenix, and others. A mat foundation 168 m × 168 m (550 ft × 550 ft) with a thickness of 5.5 m (18 ft), will suffice to carry the building. The mat can be thinned out in the center part of each module to reduce the concrete volume.

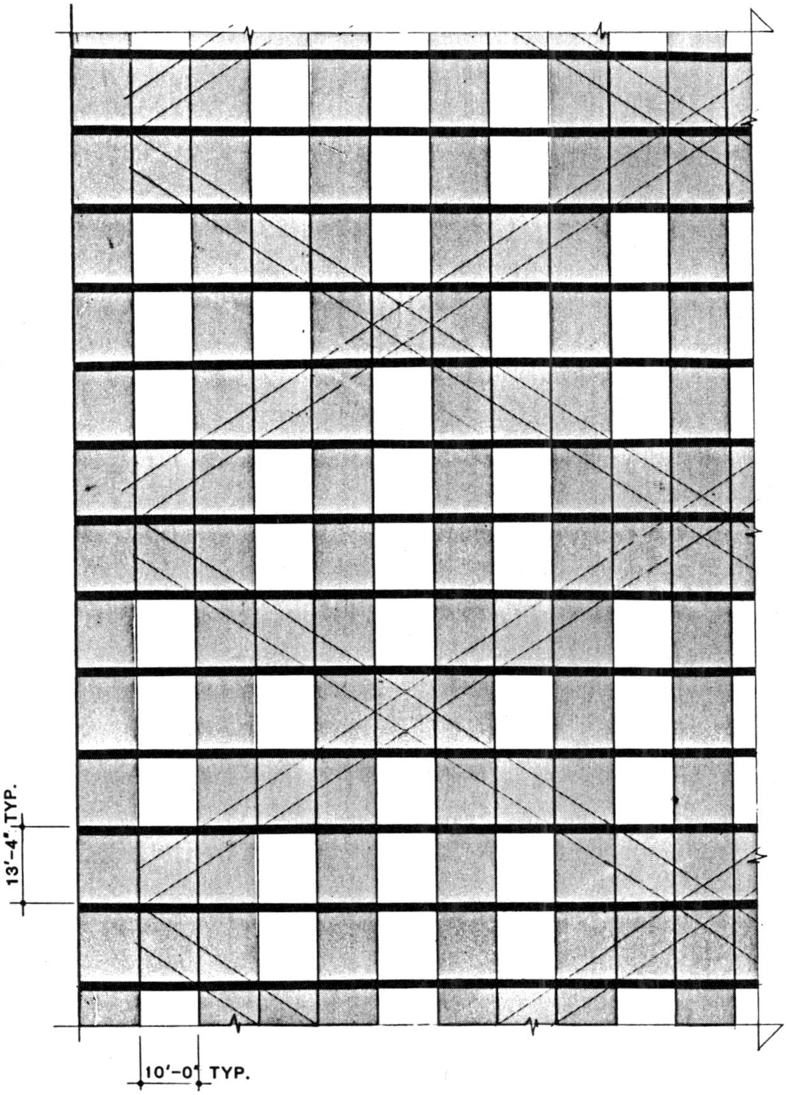

Fig. 9 Elevation of internal concrete columns

ARCHITECTURAL CONSIDERATIONS

It was felt that architectural exposed concrete should be utilized both as a cost savings device and because of the relatively good results with previous structures. One of the main draw backs of this approach from a structural standpoint is that the exterior concrete structure is exposed to temperature variations. If the building were to be built in the Sunbelt, the mean low winter temperatures are in the range of −7° C (20° F). If one assumes that that the interior of the building will be maintained at 21° C (70° F), then temperature graphs can be developed as shown in Fig. 12 to study the average temperature of the exterior elements under the worst winter condition. A study of one exterior column indicates that under these conditions the average temperature of the concrete columns at the lower levels would be 9° C (49° F), thereby having 12° C (21 °F) temperature differential between the exterior and the interior. For the same considerations in northern climates, the temperature differential jumps to 19 °C (35° F). The interior arrangement of the concrete column with diagonals has the ability to resist these thermal movements. More detailed analyses can be conducted to study the forces in the cross-walls and the resulting exterior wall movements.

CONSTRUCTION CONSIDERATIONS

Recent analysis has indicated that a cost effective way to design concrete columns is to provide for reinforcing in the range of 1% to 2% and utilize as

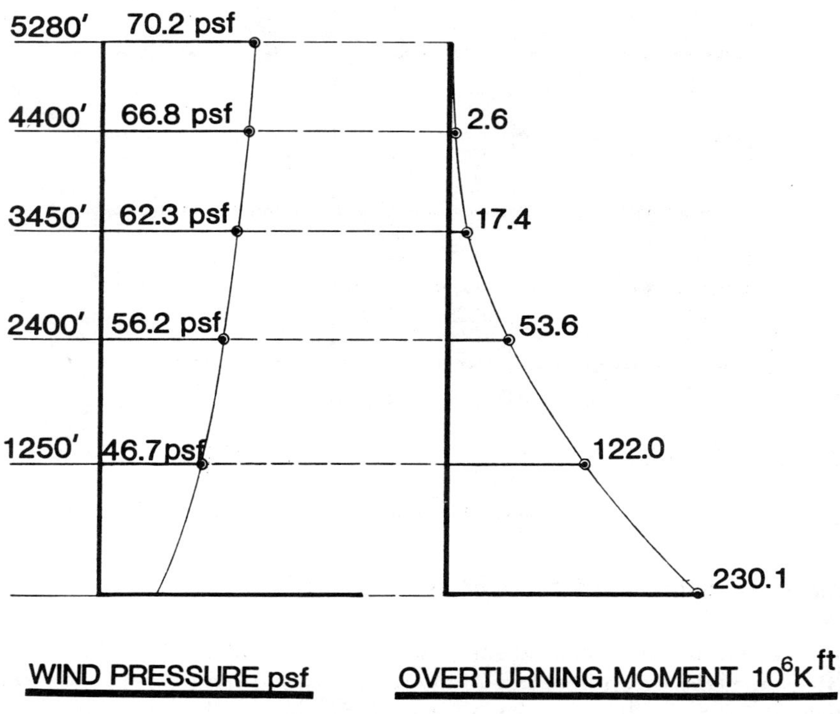

Fig. 10 Wind pressures & overturning moment

high a concrete strength as possible, the basic philosophy utilized in this building. The thickness of the walls and columns throughout the structure are based on the above criteria. The maximum concrete strength required at the base is 97 MPa (14,000 psi), which is readily available in Chicago, Dallas and most metropolitan areas at this time. Reinforcing is kept to a minimum and the detailing will be simplified especially at splices in the reinforcing.

Discussions with a contractor indicate several techniques that could be used for the rapid construction of this project. It is considered very essential that a batch plant be established on the job site so that the large concrete volumes needed have an ensured delivery. Since several basement levels are envisioned for this building it is felt that the batch plant could be in the basement.

1. V_{30} = 90 M. P. H.: Exposure Zone "C"
2. Velocity Profile:

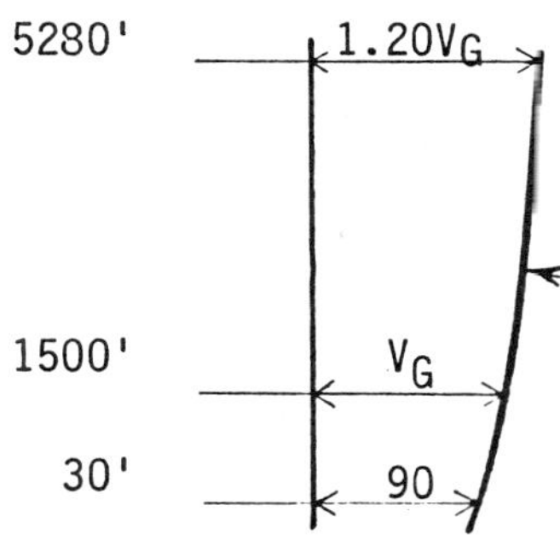

$$V_h = V_G \left(\frac{h}{1500}\right)^{2/7}$$

3. Drag Coefficient = 0.9
4. Damping = 2% of Critical
5. Gust Factor $\overline{G}$ = 1.02
6. Acceleration $a_d \propto \dfrac{1}{\sqrt[4]{KM\beta^2}}$

   $a_L \propto \dfrac{1}{\sqrt[4]{KM\beta}}$

7. Vortex Shedding $V_{cr} = \dfrac{f_o D}{0.12}$

Fig. 11 Wind engineering criteria

Since pumping to a height of 305 m (1000 ft) has been successfully achieved, the construction technique envisioned is to provide hoisting equipment to lift the concrete using a regular materials' hoist to a point 305 m (1000 ft) below the point of deposition. For the last 305 m (1000 ft), the concrete will be pumped. It is envisioned that the form work for the columns and walls will be self-jacking insulated forms. All the materials and personnel hoists will be on

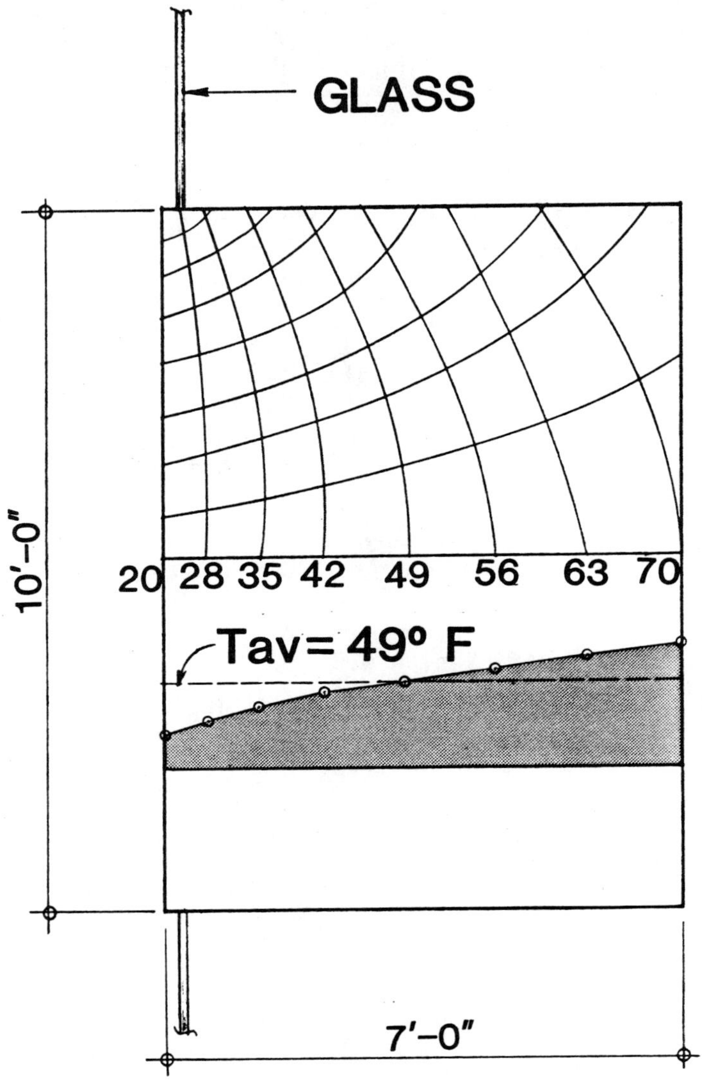

Fig. 12 Temperature variation through exterior wall

the inside of the building to provide for protection against weather. Since very large wall thicknesses are used, techniques will have to be devised (such as insulation) to gradually dispose of the heat of hydration.

CONCLUSIONS

A mile-high building in concrete is definitely feasible from a technical standpoint. Concrete has many advantages for tall buildings and with careful planning, most of the disadvantages can be mitigated.

ACKNOWLEDGMENTS

The author would like to acknowledge the help of many professionals who have contributed their expertise. They are: Mr. Harwood Taylor and Mr. Victor Lundy of Taylor/Lundy/HKS Architects, Mr. Sam Edwards of Elevator Service Professionals, Dr. Nick Isyumov of the University of Western Ontario, Mr. Jack Buckley of I. A. Naman and Associates (Consulting Engineers) and Mr. Dick Corry of Turner Construction. Their helpful suggestions are greatly appreciated.

Methods of Analysis in the Design of Tall Concrete and Masonry Buildings

Alex Coull

INTRODUCTION

The purpose of an analysis is to check the adequacy of a design, from the points of view of safety and serviceability of the structure, when subjected to specified gravitational and lateral loads. The performance of the structure must be appraised, and checks made on whether stress levels, lateral deflections, stability, dynamic behavior, and the like fall within prescribed limits of established design criteria.

During the 1960s and early 1970s, the evolution of new structural forms for tall buildings, coupled with a worldwide trend toward high-rise construction, provided great stimulus for the development of appropriate methods of analysis. Simultaneously, the increasing availability and power of digital computers allowed corresponding developments in numerical techniques for the solution of engineering problems; in particular, the great advances in matrix techniques of structural analysis were as relevant to tall buildings as to other structural forms.

The initial stimulus has now diminished. Indeed, in some countries high-rise residential buildings have been rejected on sociological grounds, possibly as a result of an over-emphasized acceptance of this mode of construction as a panacea for the post-Second World War housing shortage. Much of the fundamental research has been completed, and both accurate and approximate analytical methods are available for virtually all the identifiable regular forms of tall buildings. As a result, there appear to have been few radically new developments in the elastic analysis of tall concrete and masonry buildings in the past few years, and this has been reflected in a lessening of the number of new publications. The period has been essentially one of consolidation, with the initial research work being digested and introduced as a working tool into the design office through the availability of general purpose computer programs based on the stiffness method of analysis. However, specific techniques have continued to be refined and advanced, and gaps in the field of knowledge are gradually being filled while the search continues for improved accuracy and efficiency. The continuing rapid evolution of computer systems and the introduction of the new series of micro- and mini-computers has led to some reappraisal of structural analysis programs. There has also been a greater awareness of the need to consider soil/structural interaction and more realistic foundation conditions.

Although it is well appreciated that concrete and masonry behave in a nonlinear manner, a rigorous linear elastic analysis still remains the most important check on a design. Techniques do exist for the prediction of nonlinear behavior, but they are not yet sufficiently well-developed or economically feasible for undertaking a realistic detailed analysis of a highly indeterminate tall building structure. In strength design, any set of forces in equilibrium with the applied loads are acceptable, provided there is sufficient ductility to accommodate the implied resulting plastic deformations. Consequently, the elastic analysis has the merit that, in addition to predicting the initial quasi-elastic working load response, it provides a condition of static equilibrium, and hence, even if cross-sections are dimensioned on ultimate conditions, a safe design.

The power of the modern digital computer, allied with sophisticated and versatile programs based on the highly developed stiffness method, using either line or surface finite elements, means that, in principle at least, it is possible to analyse accurately virtually any complex elastic structure. The only constraints are the capacity of the computer available, time, and cost. The results are of course no more accurate than the initial input of applied loads and structural properties, and precise calculations do not necessarily imply a high degree of accuracy in predicting real structural behavior. Consequently, it is rarely economic, or necessary, to analyze as accurately, or exactly, as possible the large three-dimensional structural assemblies encountered in tall buildings. Some simplified mathematical model is normally adopted from which it is judged that reasonable results may be obtained within the capacity of the computing facilities available. The main problem then lies in devising the simplest model that will reproduce with sufficient accuracy the main structural actions of the building, and development work is still under way in this field. The first requirement in modeling is to recognize the dominant modes of action and interaction, so that the major components that contribute to these can be incorporated in the idealized structure. An appreciation of the load-deformation behavior of the different independent structural assemblies involved, which can frequently be assessed by simplified continuum techniques that allow the form of response to be evaluated through the medium of one or more simple relative stiffness parameters (Stafford Smith et al., 1981) may be helpful in the development of a suitable model to simulate the resistance to lateral forces.

The general classification of the structure and the most appropriate techniques for analysis were presented in the Monograph, Vol. CB (Council on Tall Buildings, 1978), and these are still applicable in general terms. Specific developments since then were discussed in the Monograph Update (Coull and Stafford Smith, 1983).

The literature devoted to the analysis of tall concrete and masonry structures is now very extensive. For example, a bibliography on shear wall structures (Singh and Schwaighofer, 1976) covering the period 1928–1975 contains over a thousand entries. Consequently, it is not possible in a short paper to cover in detail all the techniques available for analysis, particularly since many have been devised for specific situations. The objective is rather to provide a broad overview of the current position and to attempt to emphasize the most important techniques available for the practical analysis of high-rise structures. The main emphasis is on the determination of the response of the structure to lateral forces, since this is generally the most difficult to predict but most rewarding for innovative structural design. Methods of elastic analysis form the dominant theme, since these are likely to remain for the immediate future the main analytical tools of the practicing engineer.

ROLE OF ANALYSIS IN DESIGN PROCESS

Architectural and functional requirements tend to dictate the possible locations and maximum desirable sizes of the load-bearing elements. However, once the functional layout has been decided, the structural design process generally follows a well defined iterative procedure. After a structural model has been devised, preliminary calculations of member sizes are usually based on the gravity loading augmented by an arbitrary increment to take account of wind forces. The cross-sectional areas of vertical members will be determined from the accumulated loading from their associated tributary areas, with a reduction for multiple floor levels. The initial sizes of horizontal beams are normally based on codified mid-span and end moments, or on some very simple approximate method of analysis. A check is then made on the maximum lateral deflection, again by means of some simple technique, and adjustments made to provide increased lateral stiffness if the deflection is considered excessive. In the process, it must be borne in mind that increasing the stiffness of a component will lead to that component picking up an increased share of the load. The procedure is repeated for different possible load combinations, allowing an envelope of worst combinations of forces to be obtained for the different members throughout the structure.

Modifications will invariably be made to the layout of the building as the design evolves, and the various stages will be repeated until a satisfactory solution is reached. A rigorous final analysis, using a more refined analytical model, will be made to check on member strengths and deflections including the influence of second order geometrical (P-Δ) effects. A dynamic analysis may be required if there are possibilities of vibration or comfort criteria being exceeded, and any possible deleterious effects of differential movements due to creep, shrinkage, or temperature differentials, checked.

The iterative process will thus call upon different levels of structural analysis, from the crude and approximate techniques of the early stages where changes and modifications occur regularly to the sophisticated final check.

In the design process, an appreciation of the structural behavior of individual and combined components is essential for the development of an appropriate, efficient, and economic load resisting system that minimizes the penalty for resisting parasitic lateral loads and simultaneously satisfies the basic serviceability requirements.

ASSUMPTIONS IN ANALYSIS

Certain basic assumptions are usually made in the modeling and analysis of tall building structures. The most common and important are:

1. All structural materials behave linearly elastically.
2. Only the primary structural components are active in resisting applied loads. Secondary components such as cladding, partitions, and staircases are non-load-bearing.
3. Plane sections before bending remain plane after bending for columns, beams, and wall elements. The cross-section rotational effects caused by the finite depths of the walls may then be significant and should be included.
4. Unless the building is very long and slender, or unless there are large openings in plan, the floor slabs are so stiff in their own planes that they undergo only rigid body displacements. At any floor level, the entire building will then be subjected to three components of horizontal movement only, two orthogonal displacements and a rotation.

The relatively thin floor slabs are frequently assumed to be perfectly flexible out of plane, particularly if the joints with the vertical elements have little rotational stiffness. In that case, they transmit horizontal forces between the vertical elements and distribute gravitational forces to the vertical components. However, with bending-resistant connections, their effect in coupling vertical elements such as shear walls or open-box cores may be significant and should be taken into account in assessing the stiffness of such assemblies.

5. Only the major stiffnesses of component elements are normally included. Such actions as out-of-plane bending and torsion of slender walls and frames may justifiably be neglected.

   Only the most significant deformations are included. Generally, shearing deformations of slender walls or beams, axial strains in beams (especially as a result of assumption (4)) and possibly axial deformations in columns, provided the building is not too tall, are neglected.

6. The effects of cracking in concrete members subjected to bending or tensile stresses may be allowed for by a reduction in cross-sectional properties or stiffness. This is, however, difficult to estimate accurately.

7. A small amount of rotational movement at the base of a tall building can have a significant effect on its overall stiffness. Consequently, it is important to estimate if any relative vertical or rotational movements may occur at the foundation level, and include these in the analytical model, perhaps through the device of effective foundation "springs."

DISCRETE AND CONTINUOUS APPROACHES

Methods used in the analysis of tall building structures may be conveniently divided into two main groups, the discrete and the continuum approaches.

Continuum Approach

In the continuum approach, each horizontal element (beam or slab) connecting two vertical elements is "smeared" over the story height involved to form an effective continuous connection of equivalent stiffness throughout the height. If, for example, vertical elements in a building are constrained to deflect equally in a horizontal plane by the surrounding floor slabs, the action of the latter may be simulated by a series of axially rigid pin-ended links that transmit only horizontal forces between elements. These links may then be replaced by an equivalent continuous axially-rigid pin-ended medium (Fig. 1), which constrains the elements to deflect equally throughout the height of the building and allows a continuous distribution of axial forces between them.

Similarly, in the uniform coupled shear walls of Fig. 2, each connecting beam of flexural rigidity EI_b may be smeared over the story height h to give a continuous connection of flexural rigidity EI_b/h per unit height. The discrete structure is transformed into a continuous one over the height of the building, consisting of two walls and a central connecting shear-resisting medium. The discrete set of shear forces, moments, and axial forces in the connecting beams may then be replaced by corresponding distributions of forces (shear force/unit height, etc.) throughout the structure. The two-dimensional structure is thereby replaced by an equivalent one-dimensional system, in which the internal forces and displacements are continuous functions of, and vary only with, the height coordinate. In that case, conditions of compatibility and equilibrium enable the behavior of the structure to be expressed in the form of a low order ordinary linear differential equation in one of the internal stress-resultants, or the lateral deflection, yielding a general contin-

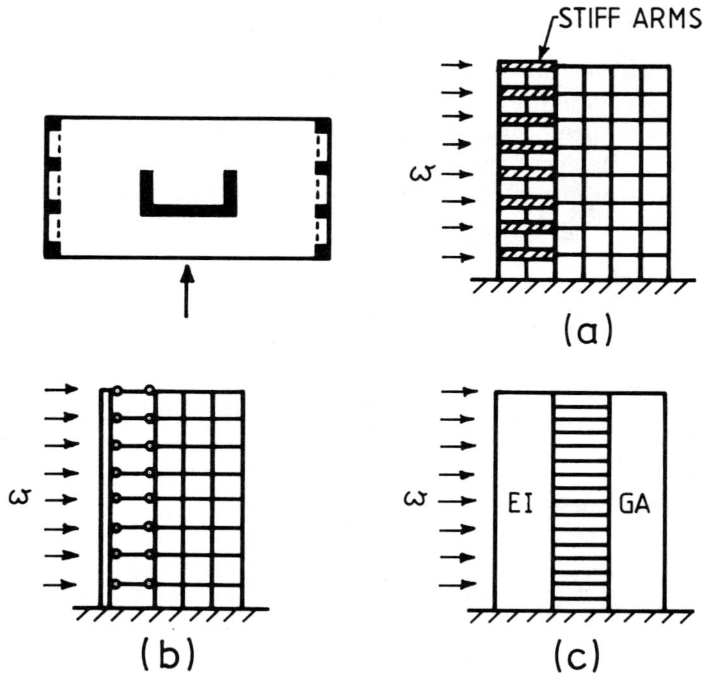

Fig. 1 Planar representations of idealized wall-frame structure: (a) wide-column model for wall; (b) linked wall-frame model; (c) continuum model, with frame modeled by equivalent shear cantilever

uous closed-form solution. Once the solution for the equivalent continuous structure is achieved, the forces in the real discrete system may be determined by integrating the continuous force distributions over the levels concerned.

A similar approach may be used for the components of a three-dimensional structure, with the horizontal elements being replaced by equivalent pin-ended or moment-resisting continuous connecting media. All actions may then be expressed as continuous functions of the height coordinate only; the forces and displacements may be expressed in terms of the three displacement components, two translations and a rotation, at each floor level, which are also variables of the height coordinate only.

The technique may be used for both plane and spatial structures, which are essentially regular in form throughout the height. It is possible to handle structures with a limited number of discontinuities in stiffness by obtaining solutions for each uniform segment and matching them at the junction by such techniques as the transfer matrix method. However, this tends to be laborious and to destroy the essential simplicity of the technique and the closed-form solutions that exist for the entire structure.

It has the merit that the solution is independent of the number of stories involved, and, in fact, the accuracy increases as the number of stories increases. It is the only feasible hand calculation approach for tall building structures.

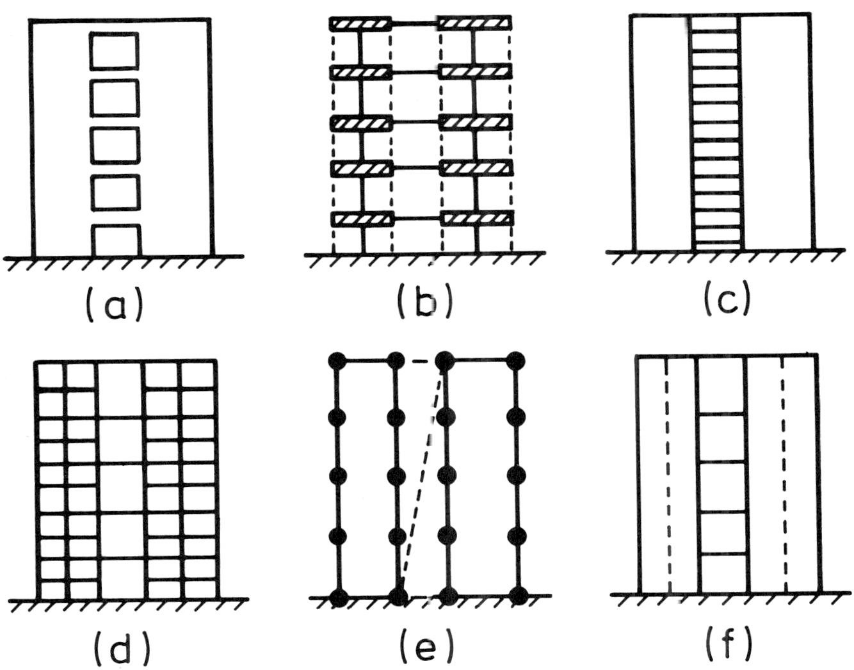

Fig. 2 Coupled wall structure (a) represented by: (b) wide-column frame analogy, (c) continuous medium, (d) finite elements, (e) higher order finite elements, (f) finite strips

Its major limitation lies in its inability to deal with irregular or discontinuous structures and with other than simple boundary conditions. The analysis requires the solution of a governing differential equation for each independent connecting medium; consequently, with a more complex structure involving a number of different forms of vertical shear-resisting connections, the solution of a set of simultaneous differential equations is required. Although the integration can be carried out numerically on a computer, standard programs for this are not normally available to the structural engineer.

Its main use has been in the analysis of uniform plane and spatial coupled shear walls, core structures, and in certain classes of uniform or symmetrical three-dimensional assemblies such as cross-wall structures. If, as discussed later, a rigidly jointed frame is modeled by an equivalent continuous system, or shear cantilever, the method can be used for the analysis of wall-frame structures. Similar models may be used to represent frame elements by panels of equivalent axial and shearing rigidity in more complex structures such as framed or bundled tubes.

The closed form of solution obtained, in conjunction with the very limited number of parameters required to describe the structural behavior (Stafford Smith et al., 1981), lends itself to the production of simple design curves (Coull and Choudhury, 1967). These curves can be very useful in design offices for giving rapid assessments of structural behavior, particularly in the early design stages, and for assessing qualitatively the relative merits of different systems. It has been shown (Stafford Smith et al., 1981) that, by expressing their behavior by the continuum approach, three common components, the uniform braced frame, rigid frame, and coupled wall structures belong to a family of cantilevers whose mode of lateral deflection can be defined by two parameters based on their bending and shearing characteristics. The relative values of the bending and shear parameters give an indication of the form of deformation that predominates, and can give both guidance on structural modeling and on the mechanism of resisting lateral forces.

Since it gives formulae that express explicitly the horizontal load-deflection relationship, it can be used to provide a matrix of flexibility influence coefficients for the component concerned. These may then be used in a more extensive three-dimensional analysis to give relatively quickly and accurately the distributions of lateral forces between the assemblies forming the complete building. As the method is based on a continuous structure, the equations of horizontal equilibrium and compatibility may be set up for any suitable set of reference heights, independent of the story levels. The engineer can decide independently the number of reference levels to employ to produce results of acceptable accuracy. It has been found that if the structure is reasonably uniform throughout the height, a relatively small number of reference levels, say no more than about 4-10, will yield results of sufficient accuracy for many practical applications. The orders of the matrices involved in the calculation will then be very small also.

However, as far as is known, no general computer packages have ever been produced for the use of continuum methods, and individuals have produced their own programs for the solution of specific structural forms.

As noted in succeeding sections, similar assumptions regarding the replacement of a discrete set of beams or a frame by an equivalent vertical panel element have been used in recent years in conjunction with the finite strip method, or the finite element method using higher order elements that cover the entire height of the building for the solution of two- and three-dimensional structures.

Discrete Methods

As the name implies, the discrete approach to analysis involves the modeling of the structure by a more realistic assembly of story-height and bay-width components that comprise the structural model. These consist generally of vertical and horizontal line or surface elements connected together at the joints or junctions.

Beams and columns may be represented by line elements joining node points. Unless severe discontinuities are present, plane walls may generally be represented also by columns positioned along their centroidal axes. However, if they are connected to other members, the finite width of the wall should be incorporated at each floor level by a stiff horizontal arm connecting the centroidal axis to the external fibers. This device effectively ensures that plane sections remain plane as the wall deflects, and ensures that the effects of the wall rotations are correctly included at the junction with an external connection (Fig. 1a). Walls with discontinuities and spatial walls may be modeled by an equivalent frame or by surface elements. The surface components may be subdivided into finite elements or finite strips for analytical purposes.

The modeling yields a three-dimensional structural assembly that consists of a series of line or plate elements connected at a number of nodes. The analysis of such systems is most conveniently achieved by the stiffness matrix method (McGuire and Gallagher, 1979), whose highly formalized and disciplined approach makes it ideally suited for digital computation, and allows general purpose programs to be developed. The main power of the method springs from the ability of computers to assemble and solve large numbers of equations quickly and economically. The method has been extensively developed over the past three decades and now forms a major component of undergraduate courses in structural analysis. Consequently, it is unnecessary to describe the method in any detail, although it is relevant to consider how it is generally employed in the analysis of tall buildings.

The stiffness method of analysis of linear elastic structures essentially involves four basic stages (McGuire and Gallagher, 1979).

The first stage requires the evaluation of the local stiffness matrix for each element; this is an array of coefficients that define the relationships between the nodal forces and nodal displacements of the element. In a three-dimensional frame system, for example, there will be six components of displacement (three translations and three rotations relative to an orthogonal set of axes) and six corresponding forces and moments at each node; in a two-dimensional or plane system, there will be three components of displacement (two translations and one rotation) and three corresponding forces at each node. The methods of obtaining the stiffness coefficients for either line or plate elements are well established and may be evaluated explicitly, or numerically in the computer program, using standard techniques.

In the second stage, the stiffness matrix for the entire structure is assembled from the individual member stiffness matrices. This is accomplished by setting up the equations of translational and rotational equilibrium at each node, and by stipulating that the members meeting at a node have the same corresponding displacements. When this global stiffness matrix has been assembled, the known displacement boundary conditions can be imposed by setting displacements at rigid supports zero, or by adding stiffness coefficients to simulate elastic supports or some associated supporting structure.

The third stage of the analysis involves setting up the equations of equilibrium that relate the externally applied forces to the internal forces at the nodes. This gives the fundamental matrix equation relating the external forces P to the unknown nodal displacements Δ through the global stiffness matrix K,

$$P = K\Delta \tag{1}$$

The solution of the resulting set of equations yields the complete set of nodal displacements Δ.

In the final stage, the nodal forces acting on the individual elements may be determined by multiplying the local stiffness matrices by the nodal displacements concerned. The effects of the nodal forces may then be combined with those obtained by analyzing the element with applied loads and fixed supports to obtain the total effect on the element.

Although these form the essential four stages in the formulation, they may be organized in different ways for different elemental forms and loads. The approach is essentially the same for different structural forms, the main differences lying in the stiffness matrices of the components involved and the method of including the load terms.

Exact analyses are possible for linear beam or column elements, and their local stiffness matrices are readily determined in explicit form. Using standard beam theory, for example, the stiffness matrix of the most basic uniform plane beam element ij of Fig. 3 may readily be shown to be given by the relationship (McGuire and Gallagher, 1979),

$$\begin{bmatrix} Fxi \\ Fyi \\ Mzi \\ \hline Fxj \\ Fyj \\ Mzj \end{bmatrix} = \left[\begin{array}{ccc|ccc} r & 0 & 0 & -r & 0 & 0 \\ 0 & 6k & 3kL & 0 & -6k & 3kL \\ 0 & 3kL & 2kL^2 & 0 & -3kL & kL^2 \\ \hline -r & 0 & 0 & r & 0 & 0 \\ 0 & -6k & -3kL & 0 & 6k & -3kL \\ 0 & 3kL & kL^2 & 0 & -3kL & 2kL^2 \end{array}\right] \begin{bmatrix} \delta xi \\ \delta yi \\ \theta zi \\ \hline \delta xj \\ \delta yj \\ \theta zj \end{bmatrix}$$

or

$$\mathrm{P}ij = \mathrm{K}ij\ \Delta ij \tag{2}$$

where $r = {}^{EA}/L$, $k = {}^{2EI}/L^3$, Fx, Fy, and Mz are the end forces and moments, δx, δy and θz are the corresponding end displacements and rotations, and E, A, I and L are respectively the elastic modulus, cross-sectional area, second moment of area, and span of the beam. The square symmetrical local stiffness matrix Kij is of order 6 × 6.

The effects of shearing deformations may be added to the terms of Eq. 2, and, if required, the stiffness matrix can be readily transformed into other systems of axes.

A corresponding 12 × 12 stiffness matrix exists for a line element in space (McGuire and Gallagher, 1979), relating the six end forces (two shears, axial force, two bending moments, and a twisting moment) to the corresponding displacements referred to three orthogonal coordinate axes.

The orders of the element stiffness matrices involved may be reduced by suppressing or neglecting certain actions. For example, the axial deformation of beams may be assumed zero because of the high in-plane stiffness of

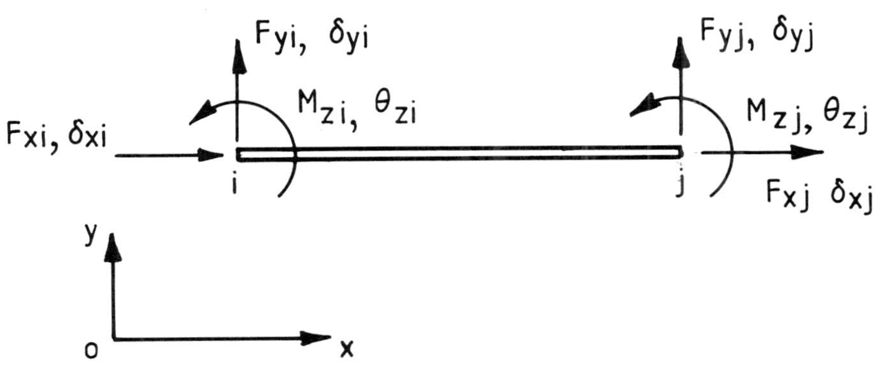

Fig. 3 Typical beam element

the associated slabs, or the torsional stiffnesses of frame or wall elements may be neglected.

The equilibrium conditions at the nodes may then be set up by the addition of all equations of the same form as Eq. 2, yielding an overall structure stiffness matrix of the same order as the number of degrees of freedom used to define the displacements of the structure. The analysis thus requires the solution of a set of simultaneous algebraic equations to give all nodal displacement components. The internal stress resultants then follow.

The finite element method may be considered as a generalized displacement method for the analysis of plane and spatial surface structures such as walls, cores, and floor slabs. In this technique, the surface concerned is divided into a series of elements, generally rectangular or triangular, or possibly quadrilateral in shape, by some chosen mesh. The elements are assumed to be interconnected at a discrete number of nodes situated on their boundaries (Fig. 2d). The displacements of these nodal points are the basic unknown parameters, as in the earlier discrete frame structural analysis.

Unlike line elements, exact solutions for plate elements in plane stress or flexure do not exist, and the element stiffness matrix must be formulated by choosing a set of functions to define uniquely the pattern of displacements within each finite element in terms of the prescribed nodal displacements. This then defines uniquely the state of strain, and, hence, from the constitutive relationships, the state of stress, within the elements in terms of the nodal displacements. A system of nodal forces equilibrating the boundary stresses and any distributed loads may then be determined, using the principle of virtual work or the principle of minimum potential energy to give an element stiffness matrix of the same form as Eq. 2, the order depending on the number of nodes and degrees of freedom involved (Gallagher, 1975). The stiffness matrix for the complete structure can then be formed by adding the individual element stiffnesses, and the solution achieved by the same procedure as discussed earlier.

In selecting a displacement function, the aim should be to choose a function that will enable the element to deform in a reasonably similar manner to the deformation that occurs in the corresponding region of the continuum, including the strain-free rigid body movements. The displacement compatibility along the inter-element boundaries must also be considered. Although a choice of displacement functions exists, best results will be achieved if the function satisfies certain established criteria (Zienkiewicz, 1977). A wide variety of elements and their stiffness matrices for both bending and in-plane actions have been developed for general analysis and may be used for tall building structures. If local deformations are small, so that the bending and membrane actions in a flat element are uncoupled, the stiffness matrix for a flat shell element, which includes both types of actions, can be built up from the constituent bending and plane stress stiffnesses in a relatively straightforward manner.

The elements used may be simple, with only nodes at the corners, and the displacements expressed in terms of the most elementary degrees of freedom possible, or may be more complex multi-noded elements with higher order degrees of freedom included. (Figs. 2d and 2e). It may then be necessary to choose between a larger number of low order elements or a smaller number of higher order elements to model the structural behavior. With higher order plane-stress elements for example, it may be possible to use only one or two elements to model a wall over the entire height of the building (Cheung, 1983).

The technique has the advantage that, by using combinations of rectangular and triangular elements, or quadrilateral elements, it is possible to grade the mesh used. A fine mesh may be used in regions of particular interest, where high stress gradients occur, with a very coarse mesh in regions of low or uniform stresses.

Commercial programs are now widely available for linear elastic analysis by the stiffness method, using both line and surface elements, and the packages include a wide variety of elements. These programs will differ in both their facilities and abilities, and in selecting one the engineer must consider the range of elements, amount of data preparation required, and range of problems that can be readily handled.

In the analysis, it is important to ensure that displacements are compatible. Problems will, arise, for example, when beams modeled by line elements with three degrees of freedom at each node frame into walls modeled by simple plane stress elements with only two translational degrees of freedom at each node. Special devices must be employed to ensure rotational compatibility at the interface, such as special finite elements with an additional rotational degree of freedom at each node (Sisodiya, et al., 1972; MacLeod, 1969) or by using an auxiliary beam connected across the interior elements to give the necessary rotational resistance to the exterior beam.

A relatively recent addition to the methods available for the analysis of tall buildings is the finite strip technique, which has particular advantages for structures that are essentially uniform in one direction. The method may be regarded as a combination of the primary features of the classical Levy solution for plate structures and those of the finite element approach (Cheung, 1983).

In this method, the vertical elements in the structure are divided into a series of strip elements in the vertical direction (Fig. 2f). The displacements are defined in terms of simple polynomials in the horizontal direction and continuously differentiable functions in the vertical direction that satisfy the boundary conditions at top and bottom. Once the displacement functions are known, the stiffness and load matrices may be determined by standard finite element procedures. The overall structure stiffness matrix may then be formed and a solution obtained as before.

By virtue of the continuous nature of the solution in the vertical direction, allied to the formulation in terms of stiffness matrices for each element, the method may be regarded in a sense as lying between the discrete and continuum approaches.

The technique can deal with discontinuities throughout the height and can offer significant savings in computer time over the finite element method. However, it does not yet appear to have been widely adopted in design offices, and general computer programs are not generally available other than in institutions where developments have taken place (Cheung, 1984).

ANALYSIS OF THREE-DIMENSIONAL STRUCTURAL SYSTEMS

When devising a suitable model to simulate the behavior of the structure, consideration must be given initially to the modes of interaction between the different components. The resistance to lateral forces will depend to a considerable degree on whether the horizontal components are capable of transmitting bending and shearing forces in addition to axial forces.

If the slabs are considered to be either very flexible out of plane, or the connections with the vertical elements are not moment-resistant, they transmit only horizontal axial forces and serve to distribute lateral wind forces between the vertical elements. The building generally contains elements and components that deform in different configurations. For example, under the action of lateral forces, walls deflect in a flexural mode and rigid frames in a predominantly shearing mode. Consequently, if the different elements are constrained to act together by the action of the floor slabs, horizontal interactive forces must be mobilized through the slabs to force the different elements to take up the same configuration (Fig. 4). A redistribution of lateral forces, or shears, then takes place throughout the height. Since only horizontal forces

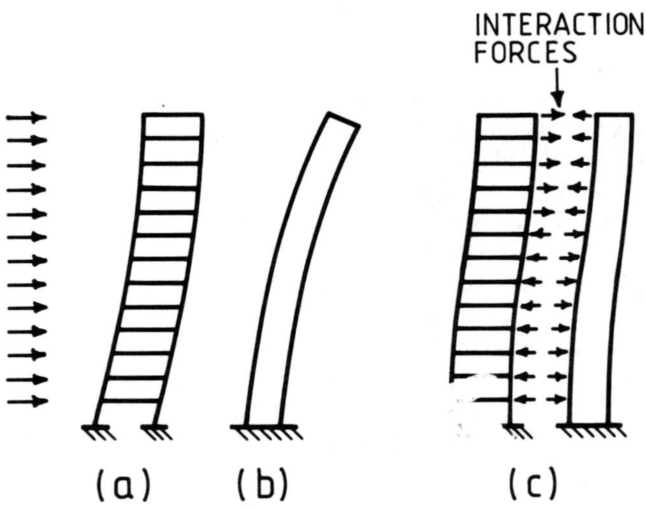

Fig. 4 Interaction between frame (deforming in shear mode) and wall (deforming in flexural mode) constrained to act together by floor slabs

are transmitted between the components. the stiffness of the building is given by the aggregate response provided by the independent actions of the components (Fig. 5a). The wind moment M is then resisted by the sum of the moments in the individual components.

On the other hand, if the horizontal connections are capable of resisting moments and can transmit vertical shear forces as well as axial forces, they will resist the free bending of the vertical elements. Vertical axial forces and resisting moments will be induced in the vertical elements, which will contribute significantly to the moment of resistance and lateral stiffness of the system. The wind moment is now resisted by the sum of the moments in the individual components, together with moment of resistance of the axial forces, N, (Fig. 5b). The lateral stiffness will then be greater than the aggregate of the independent components. This mode of interaction increases greatly the complexity of the analysis, but must be included in view of its importance.

Any twisting moments on the structure will be resisted largely by differential shearing actions in the frames and walls, and, apart from torsionally stiff elements such as cores, the torsional stiffnesses of individual structural components may usually be ignored. The actions involved in resisting bending and twisting are then essentially similar in form, although in the latter case the resistance is proportional to the product of the stiffness of the component and its distance from the center of twist.

A three-dimensional analysis will thus involve essentially an assessment of the horizontal shear interaction between components, which within themselves involve vertical shear interaction. It is then useful to be able to consider independently the behavior of the latter.

Apart from some special cases of uniform and repetitive structures, a

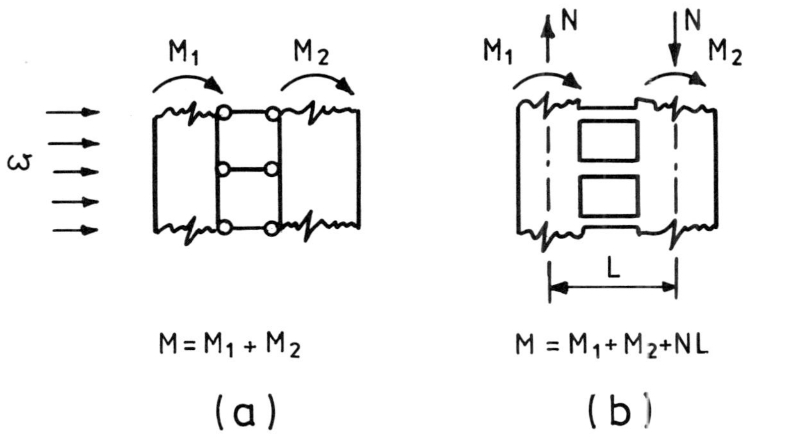

Fig. 5 Interactions between components that are (a) linked together (horizontal shear interaction) and (b) connected by moment-resistant elements (vertical and horizontal shear interaction)

generally asymmetric and nonuniform complex three-dimensional assembly of different structural elements must be analyzed by a three-dimensional stiffness approach. The equilibrium equations must be expressed in terms of the complete three-dimensional stiffness matrix in the general form of Eq. 1 and solved to give the components of displacement at the node points. However, in a complete analysis of this form involving all vertical and horizontal elements, the number of degrees of freedom involved would normally render the analysis prohibitive in both computing time and cost. The problem must be reduced as far as possible to a practicable size by making use of the assumptions discussed earlier, particularly in neglecting the axial and torsional stiffnesses of beams, out-of-plane bending of walls and frames and the torsional stiffnesses of slender elements, to reduce the number of degrees of freedom in the model to be analyzed. In particular, the assumption that floor slabs are rigid in their own plane implies that the horizontal deflection of any element in the building may be expressed in terms of three components only, two orthogonal translations of, and a rotation about, some particular axis in the cross-section.

This may be achieved in a standard program by such devices as including in the model a fictitious horizontal cross-bracing system of very stiff pin-ended links connecting the vertical elements to constrain the cross-section to deform horizontally as a rigid body. This approach does however complicate the analysis by the inclusion of a number of additional members and connected nodes. A better approach is to express the horizontal displacements of all nodes in terms of three datum nodal displacements corresponding to the three rigid body actions. Such a "master-slave" facility to include the horizontal internodal constraints is a very useful provision in a computer program to be used for tall building structures (Stafford Smith, 1985; MacLeod, 1977). In the analysis, consideration must be given to the numbering of the nodes involved in order to reduce as far as possible the bandwidth of the resulting equations for computational purposes.

The desirable features of general purpose structural analysis programs for general applicability, as well as consideration of ease of input, output, and potential errors, have been documented in the literature (Allwood and Robins, 1983).

When analysing very large structural systems, it may prove necessary to use the substructuring method (McGuire and Gallagher, 1979) or to lump a number of stories together to form large modules in the overall analysis (Stafford Smith, 1985). This method is most successful if the stories involved are identical in stiffness. To judge the accuracy of this lumping procedure, it may be necessary to analyze a segment of the structure to ensure that the larger modules do represent accurately the structural behavior of their components.

Typical approaches to the analysis of three-dimensional structures are described by Weaver, et al. (1971), Gluck and Kalev (1972) and Nair (1975).

In special circumstances, with uniform and repetitive structures, it may be possible to analyze a three-dimensional structure by the continuum approach, in which the vertical elements will be connected by continuous media, which carry either horizontal forces only, or both horizontal and vertical shear forces.

Setting up the equations of equilibrium between external and internal moments involves the stiffnesses of the different independent load-resisting components, which are connected by linking members. The linking members allow the redistribution of horizontal forces to ensure that all independent components deflect in unison as a result of the rigidity of the floor slabs. The analysis requires the setting up of horizontal compatibility equations, whereby all displacements are expressed in terms of the three rigid body displacements, and of the vertical compatibility conditions between the vertical elements, which are connected by bending resistant members, and thus involve the more complex vertical shear interactions. An analysis will then involve as unknowns, either explicitly or implicitly, the three horizontal displacement functions and an axial displacement function for each coupled system.

The problem reduces to the solution of a set of simultaneous linear differential equations in the unknown functions. Standard procedures may be used for the numerical integration of the equations, but general purpose programs are not available for structural analysis by this technique. Solutions have largely been derived for systems of uniform spatial coupled wall and frame structures (Rosman, 1971; Danay, et al., 1974; Stamato and Mancini, 1973).

Considerable simplifications are possible if the structure contains parallel systems of repetitive components and both exact and approximate solutions are relatively straightforward if only horizontal shear interaction occurs between vertical components (Coull and Khachatoorian, 1982; Coull and Mohammed, 1983).

It should be stressed that the presence of vertical shear connections through coupling beams or slabs greatly complicates this type of analysis by adding an additional unknown function for each independent coupling system.

The amount of computation may be reduced considerably if advantage can be taken of conditions of symmetry and skew-symmetry in the structure. If a structure is symmetric in plan and is subjected to symmetrical loading, only one half of the structure need be analyzed, subjected to half the loading. If the structure is in addition symmetric about an axis perpendicular to the plane of loading, it will deform in a skew-symmetric mode about that axis, and only one-quarter of the structure need be analyzed. Appropriate releases or constraints must be imposed along the lines of symmetry and skew-symmetry (McGuire and Gallagher, 1979).

Structures that are symmetrical in plan and subjected to loading that is symmetrical about the axis of symmetry will not twist, and can be analyzed as a two-dimensional plane structure. In the representative structure of Fig. 6, for example, the different components are assembled in one plane and connected by horizontal rigid pin-ended links to simulate the floor slabs and produce a common horizontal deflection. Any beams must be assumed to be axially rigid for the same reason. The loads may then be applied to the nodes of any convenient bent, and the analysis achieved by a plane frame stiffness program.

Uniform structures with few vertical shear interactions may be analyzed directly by the continuum approach.

The horizontal forces will be distributed to the various components in proportion to their stiffnesses, which will be expressed in discrete matrix form or possibly in continuum terms.

The equation of equilibrium for applied horizontal forces will be of the general form,

$$\begin{aligned} H &= H_1 + H_2 + H_3 + \ldots = \Sigma H_i \\ &= K_1\delta + K_2\delta + K_3\delta + \ldots = \Sigma K_i\delta \end{aligned} \tag{3}$$

where H is the vector of applied horizontal loads, H_i is the vector of forces on component i, of stiffness K_i, and δ is the vector of horizontal deflections. The solution for δ then leads to the distribution of forces on the different components.

If a building subjected to bending contains groups of identical components, such as frames, walls, or coupled walls, they may be combined by adding their component stiffnesses to form a single composite unit as the forces on

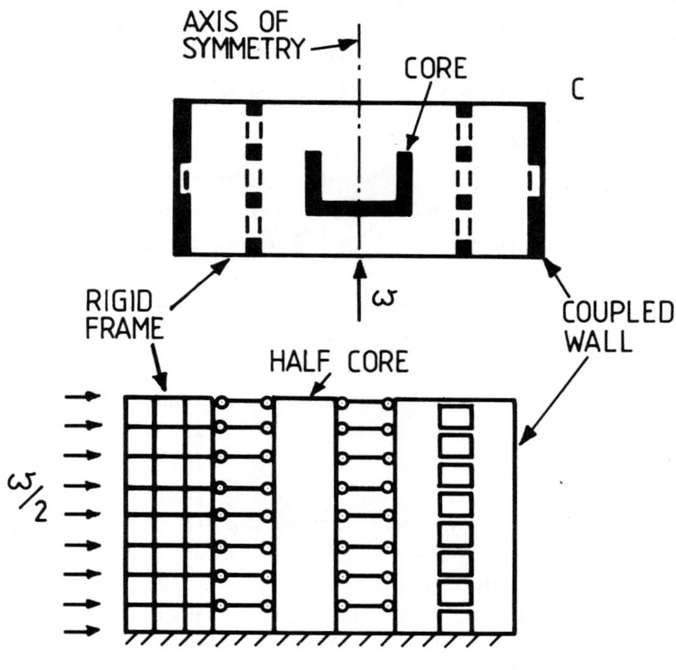

Fig. 6 Representation of three-dimensional symmetric structure by equivalent planar system

each will be identical. In the extreme case of a symmetrical structure with parallel identical assemblies of walls or fames, all will behave identically and the analysis of one, subjected to its proportion of the load, will suffice.

If the properties are uniform throughout the height, a solution will be possible by the continuum method by replacing the connecting links with an equivalent continuous medium. Simple closed form solutions are possible if the structure contains identical coupled walls or not more than two different types of coupled walls. If more than two different types exist, the solution becomes complicated by the necessity to solve numerically the resulting set of simultaneous linear differential equations.

If the symmetrical structure is subjected to eccentric loads, the load may be replaced by a concentric transverse load through, and a twisting moment T about, the central axis. The former can be treated as described earlier.

Under the action of torsional loading, the structure will rotate about a vertical axis through the plane of symmetry, and the horizontal deflection of any component will be proportional to its distance from the axis. The forces on identical components will be proportional to the deflection, and the resultant resisting twisting moment will consequently be proportional to the square of the distance from the central axis. The three-dimensional structure may again be replaced by an equivalent two-dimensional linked system in some convenient datum plane, at a distance l_n say from the central axis. In the equivalent structure, the stiffness matrix K_i of each bent at the distance l_i from the axis must be replaced in the datum plane by an effective stiffness matrix equal to $K_i\ (l_i\ /\ l_n)^2$. After transforming all bents into the datum plane, the plane structure is subjected to a set of horizontal forces of magnitude $T/2l_n$, and an analysis again performed by the stiffness method. Torsionally stiff components such as cores may be replaced by effective flexural and shear elements. After the analysis has been completed, the equivalent loads and deflections in the real three-dimensional structure may be determined by transforming the results back to the original bent positions.

Using the same procedure, single closed-form solutions may again be achieved for uniform structures containing walls, frames, and coupled shear walls by the continuum method.

In the special case of a shear wall building, the resistance to lateral forces is provided entirely by uncoupled structural walls.

If the walls are proportionate, that is, the flexural rigidities of all walls bear a constant ratio to each other throughout the height, they will deflect with similar configurations. There will then be no horizontal interactive forces between them and no redistribution of loads. In that case the forces acting on the walls may be obtained from the equations of equilibrium, in an analogous manner to a traditional "rivet group analysis."

If the walls are nonproportionate, the deflected configurations will be dissimilar, and the interactive forces developed in the stiff floor slabs will produce a redistribution of lateral loads. If no twisting occurs, the walls may be linked in plane, as before. If twisting occurs, a three-dimensional analysis is required, with the walls being constrained by a series of pin-ended horizontal stiff beams or by using the master-slave technique.

The general overall analysis may be carried out to produce directly all member forces and deflections. However, it is often convenient to carry out a preliminary distribution of the lateral load to the various bents by some approximate estimation of their stiffnesses. One possibility that has been used is to distribute the loads to the various bents in proportion to their top stiffness or deflection. This in itself may require some effort, but the results may be reasonably accurate provided that the load-deformation characteristics of the bents are similar in form. The more complex the system, in terms of the number of elements and assemblies with radically different load-deformation characteristics, the more in error are approximate techniques likely to be. This is particularly the case if twisting as well as bending occurs.

In the early design stages of structures, which consist only of combinations of shear walls, rigid frames or braced frames, a much simplified analysis is possible if all components are represented by single cantilever elements of equivalent bending and shearing rigidities (Stafford Smith et al., 1981), which are positioned at their respective centers of resistance.

The constraining actions of the rigid floor slabs may again be incorporated in a standard program by means of the master-slave technique or by a system of rigid pin-ended beams connecting the nodes at all floor levels.

Alternatively, if the components are of uniform stiffness throughout the height, an analytical solution may be achieved using the continuum approach to set up the conditions of horizontal equilibrium and compatibility throughout the height.

A particular form of three-dimensional symmetrical structure, which may be modeled very simply by an equivalent plane system, is the framed tube, shown in plan in Fig. 7a. In its simplest form, this is essentially a perforated box comprising four orthogonal frame panels of closely spaced columns connected by spandrel beams around the perimeter at each floor level, which is designed to resist all lateral forces.

The frames normal to the wind direction are constrained by the floors to deform as flanges in the same mode as the side frames, and can play a significant part in resisting wind moments. These normal frames are subjected mainly to axial forces and the side frames to the primary shearing actions involved in resisting the lateral forces. The primary action is complicated by the flexibility of the spandrel beams allowing a shear lag, which increases the stresses in the corner columns and reduces those in the inner columns, of the normal frame. The major interactions between the two types of frame are the vertical shear transfer forces at the corners. By recognizing the dominant modes of action of the panels, an accurate assessment of the structural behavior may be achieved using a plane frame analysis.

Because of the double symmetry about the two central axes, only one quarter of the structure need be considered (Fig 7b). The side and normal panels may then be assumed to lie in the side plane, with one quarter of the applied load applied to the side panel. The shear transfer at the corners may be effected by connecting the side and normal panels in parallel by very short fictitious connecting beams, which allow only shears to be transmitted between the two (Rutenberg, 1974). This may be achieved by incorporating a hinged release in the fictitious members, or alternatively by assigning to these members high shear stiffnesses and very low bending stiffnesses. The former approach is simpler to include in a standard plane frame analysis program. The device allows the condition of vertical compatibility to be maintained at the corner, and enables the shear lag to be determined. It does, however, implicitly assume that torsional effects are negligible in the girders.

This basic model does not include the out-of-plane bending of the columns in the normal frames, which may be significant in the lower levels, especially if bending occurs about the weaker axis. This effect may be included in the

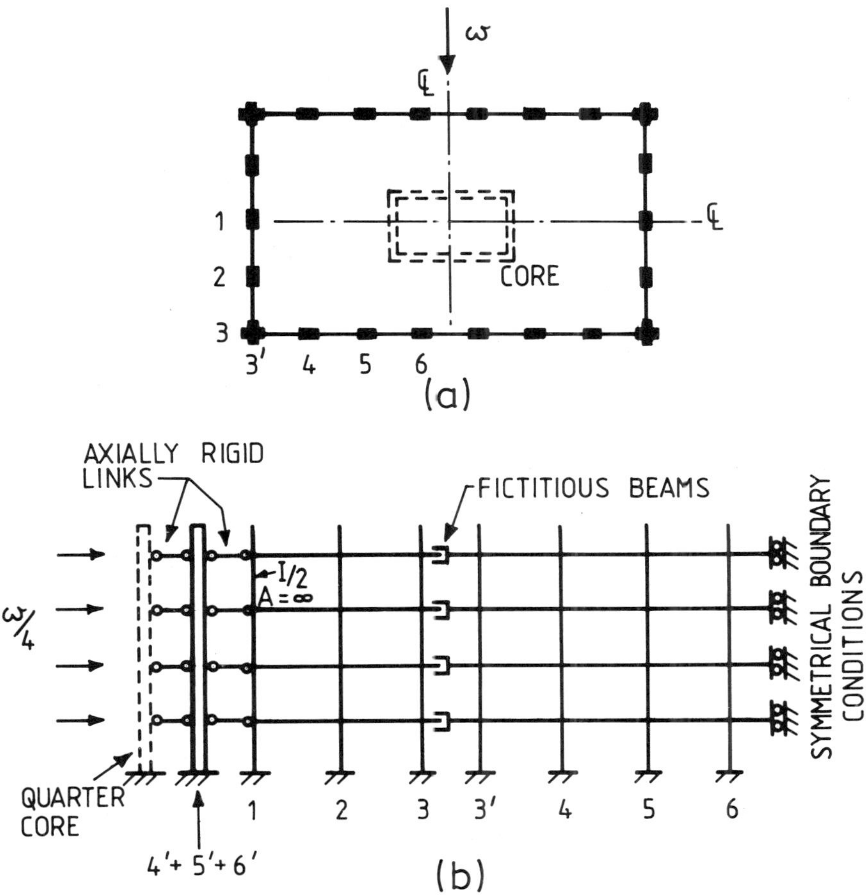

Fig. 7 Representation of three-dimensional framed tube structure by equivalent plane frame

model by adding an additional equivalent column whose flexural rigidity is equal to the sum of the flexural rigidities of the normal columns about the axis of bending, which is pin-ended to the existing system to constrain the column to have the same horizontal deflection as the side panel members.

If the structure contains an internal core that may be modeled as a line element, it may also be included by adding a further column of the required flexural rigidity, pin-connected in series to the existing system (Fig. 7b).

Similar techniques may be employed to include torsional behavior caused by eccentric loading (Rutenberg, 1974).

The example illustrates how a recognition of the dominant modes of action in the structural model may be employed to produce a sufficiently accurate analysis with a minimum amount of computation.

In the building, gravity loads will be transferred by the floor slabs to the vertical elements from their appropriate tributary areas in association with the form of construction employed. The effects of any relative vertical deformation that arises due to nonuniform axial stress levels must then be examined, using either the stiffness matrix method or some simpler technique such as moment distribution. The effects of relative deformations arising as a result of creep and shrinkage should also be considered (Fintel, 1974).

In a column, the strains caused by creep and shrinkage effects depend upon the percentage of reinforcement and the volume-to-surface ratio as well as on the basic material properties. The differential movements are cumulative over the height of the building, and the cumulative distortions in the upper levels may cause damage to nonstructural elements. The movements are functions of both sequence and time of construction. It is difficult to make accurate predictions of these effects, but a practical design procedure exists (Fintel and Khan, 1968). This procedure allows the amount of movement in the columns and shear walls to be estimated, based on a knowledge of loading history, member size and reinforcement, and the environment concerned. It is then possible to assess the magnitudes of the resulting member forces caused by both elastic and inelastic shortening of the vertical elements.

The basic discrete and continuum analytical techniques have been described in the context of the complete building, but they are equally applicable to the determination of the structural behavior of the two basic components that provide the lateral stiffness of a tall building, the rigid frame and the shear wall.

CONCLUSIONS

A review has been presented of the more important practical techniques for the analysis of tall concrete and masonry structures. Both discrete and continuum methods have been described for the overall analysis of three-dimensional structures, and for the determination of the characteristics of their individual components. Particular attention has been devoted to the effects of lateral loads.

REFERENCE/BIBLIOGRAPHY

Allwood, R. J. and Robins, P. J., 1983
COMPUTER APPLICATIONS 1: USE OF LARGE COMPUTERS, Chapter 21, Handbook of Structural Concrete, McGraw-Hill, New York.

Cheung, Y. K., 1983
TALL BUILDINGS 2, Chapter 38, Handbook of Structural Concrete, McGraw-Hill, New York.

Cheung, Y. K., 1984
COMPUTER ANALYSIS OF TALL BUILDINGS, Proceedings of Third International Conference on Tall Buildings, Hong Kong and Guangzhou, December, pp. 8-15.

Coull, A. and Choudhury, J. R., 1967
STRESSES AND DEFLECTIONS IN COUPLED SHEAR WALLS, Journal ACI, Vol. 64, pp. 65-72.

Coull, A. and Khachatoorian, H., 1982
DISTRIBUTION OF LATERAL FORCES IN STRUCTURES CONSISTING OF CORES, COUPLED SHEAR WALLS AND RIGIDLY-JOINTED FRAMES, Proceedings of ICE, London, Vol. 73, Part 2, December, pp. 731-745.

Coull, A. and Mohammed, T. H., 1983
SIMPLIFIED ANALYSIS OF LATERAL LOAD DISTRIBUTION IN STRUCTURES CONSISTING OF FRAMES, COUPLED SHEAR WALLS AND CORES, Structural Engineer, London, Vol. 61B, March, pp. 1-8.

Coull, A. and Stafford Smith, B., 1983
RECENT DEVELOPMENTS IN ELASTIC ANALYSIS OF TALL CONCRETE BUILDINGS, Developments in Tall Buildings 1983, Council on Tall Buildings and Urban Habitat, pp. 569-581.

Council on Tall Buildings and Urban Habitat, 1978
ELASTIC ANALYSIS, Structural Design of Tall Concrete and Masonry Buildings, Vol. CB, Chapter CB-5, Monograph on Planning and Design of Tall Buildings.

Danay, A., Gellert, M., and Gluck, J., 1974
THE AXIAL STRAIN EFFECTS ON LOAD DISTRIBUTION IN NON-SYMMETRIC TIER BUILDINGS, Building Science, Vol. 9, pp. 29-38.

Fintel, M., 1974
MULTISTORY STRUCTURES, Chapter 10, Handbook of Concrete Engineering, Van Nostrand Reinhold, New York.

Fintel, M. and Khan, F. R., 1968
EFFECTS OF COLUMN CREEP AND SHRINKAGE IN TALL STRUCTURES—PREDICTION OF INELASTIC COLUMN SHORTENING, Journal of ACI, Vol. 66, December, pp. 957-967.

Gallagher, R. H., 1975
FINITE ELEMENT ANALYSIS: FUNDAMENTALS, Prentice-Hall, New Jersey.

Gluck, J. and Kalev, I., 1972
COMPUTER METHOD FOR ANALYSIS OF MULTI-STORY STRUCTURES, Journal of Computers and Structures, Vol. 2, pp. 897-913.

MacLeod, I. A. 1969
NEW RECTANGULAR FINITE ELEMENT FOR SHEAR WALL ANALYSIS, Journal of Structural Division, Proceedings ASCE, Vol. 95, No. ST3, March, pp. 399-409.

MacLeod, I. A., 1977
STRUCTURAL ANALYSIS OF WALL SYSTEMS, Structural Engineer, London, Vol. 55, pp. 487-494.

McGuire, W. and Gallagher, R. H., 1979
MATRIX STRUCTURAL ANALYSIS, Wiley, New York.

Nair, R. S., 1975
LINEAR STRUCTURAL ANALYSIS OF MULTISTORY BUILDINGS, Journal of Structural Division, Proceedings of ASCE, Vol. 101, No. ST3, March, pp. 551-565.

Rosman, R., 1971
STATICS OF NON-SYMMETRIC SHEAR WALL STRUCTURES, Proceedings of ICE, London, Paper 7393S., Vol. 48, pp. 211-244.
Rutenberg, A., 1974
ANALYSIS OF TUBE STRUCTURES USING PLANE FRAME PROGRAMS, Proceedings of Regional Conference on Tall Buildings, Bangkok, Thailand, pp. 397-413.

Singh, G. and Schwaighofer, J., 1976
A BIBLIOGRAPHY OF SHEAR WALLS 1928-1976, University of Toronto, Department of Civil Engineering, Publication No. 76-02, May.
Sisodiya, R. G., Cheung, Y. K., and Ghali, A., 1972
NEW FINITE ELEMENTS WITH APPLICATION TO BOX GIRDER BRIDGES, Proceedings ICE, London, Vol. 49, Supplement Paper No. 7479S, pp. 207-225.
Stafford Smith, B., 1985
MODELING OF HIGH-RISE STRUCTURES FOR ANALYSIS BY STANDARD COMPUTER PROGRAMS, Paper presented to ASCE, New York Metropolitan Section, Structures Group, April.
Stafford Smith, B., Kuster, M., and Hoendercamp, J. C. D., 1981
A GENERALIZED APPROACH TO THE DEFLECTION ANALYSIS OF BRACED FRAME, RIGID FRAME AND COUPLED WALL STRUCTURES, Canadian Journal of Civil Engineering, Vol. 8, pp. 230-240.
Stamato, M. C. and Mancini, E., 1973
THREE-DIMENSIONAL INTERACTION OF WALLS AND FRAMES, Journal of Structural Division, Proceedings of ASCE, Vol. 99, No. ST12, December, pp. 2375-2390.

Weaver, W., Brandow, G. E., and Manning, T., 1971
TIER BUILDINGS WITH SHEAR CORES, BRACING AND SETBACKS, Journal of Computers and Structures, Vol. 1, pp. 57-84.

Zienkiewicz, O. C., 1977
THE FINITE ELEMENT METHOD, McGraw-Hill, Maidenhead, England.

Precast and Prestressed Concrete Directions in Australia and Southeast Asia

W. L. Meinhardt

This paper outlines, with reference to recent structures, special uses of precast and prestressed concrete in tall building design and construction with the objective of stimulating the development of new structural systems to answer the needs of *post modern* architecture in high-rise buildings. The emphasis is on examples of the use of precast concrete elements and prestressing as techniques to achieve particular results as distinct from the substitution of prefabricated concrete for insitu concrete or prestressed concrete in lieu of reinforced concrete.

INTRODUCTION

Economic circumstances vary greatly in different countries and, accordingly, the development of the construction industry varies considerably from one country to another. Contrasting the highly industrialized countries with the threshold and developing countries, most very tall buildings in North America for example have been designed and constructed using structural steel and lightweight fire protection while tall buildings in Australia and Southeast Asia have, with few exceptions, been designed and constructed using reinforced, prestressed, and precast concrete.

Great advances have been made in the analysis, the computing techniques, and the design of structures and also in the quality and properties of building materials. However, construction systems for reinforced concrete structures have remained largely unchanged.

In its simplest form, a concrete structure is traditionally constructed by erecting extensive falsework and formwork, followed by placing the reinforcement and then pouring the wet concrete, and in the case of a prestressed concrete structure, followed by post-tensioning. Modern systems and highly industrialized procedures for handling concrete construction have been developed, but any contribution or improvement to the existing "art of building" must be capable of achieving substantial commercial gains. To this end the use of precast concrete and of prestressing is looked at as a means of introducing industrialized systems into the "art of building."

Precast concrete, however, is not a natural ally of high-rise construction because of the weight and difficulty of handling components and it is doubtful that it will ever feature in the forefront of buildings claiming record breaking heights.

While most tall buildings are structural steel or insitu reinforced or prestressed concrete, *apartment buildings*, due to the uniformity of building modules, can lend themselves to the use of an entirely precast panel systems approach. There is nothing new about panel construction in itself; however, the construction of 15,000 high-rise units in Singapore is reviewed here because the engineering of the project constitutes a total building system from factory to complete project.

For *commercial* high-rise structures, precast concrete is not a direct alternative structural system to structural steel or reinforced concrete, but prefabricated concrete elements can constitute or complement a systems approach to design and construction. The construction of a 52-floor reinforced concrete commercial structure is described using precast elements that provide an attractive facade and in addition the formwork and falsework for the insitu concrete structure.

In the area of *performance* and *serviceability* of structures, the use of partial prestressing of columns in the 65-floor Rialto Towers Building, Melbourne, is reviewed. Partial prestressing is particularly interesting in building construction and, in the opinion of the author, far too rarely used.

INDUSTRIALIZED PREFABRICATED CONCRETE BUILDING

In 1981 the Housing and Development Board, Singapore, awarded a turnkey contract to White Industries Australia for the construction of 125 buildings at 31 sites around the island (Fig. 1). The contract included the engineering of the "system" and the construction of a major precasting factory. Four 12-story buildings, each containing 132 apartments, are presently being produced and erected every 28 days; or one 12-story building every seven days; demonstrating the optimization of a total precast building system to dwelling construction.

The system is designed on the principle that "Todays production is erected tomorrow." Components are handled direct from mold to site, requiring only a minimum number of units to be stockpiled and minimization of double handling.

The precast components include bathrooms, rubbish chutes and elevator wells, which are produced as complete box units (Figs. 2 and 3), access corridors, floor planks, internal and external walls, staircases, and roof panels. The floor system is prestressed, hollow-core planking extruded in 180 m (590 ft) lengths and then cut to design length. Vertical load-carrying elements consist of nominally reinforced load-bearing walls, three column portal frames with integral nonstructural infill panels, and the room size box units for bathrooms. Elevator shafts are story-height rectangular tubes. Long span floors, up to 9 m (30 ft), and nonstructural partitions, which can be moved at a later stage, allow alterations to be made to suit living style.

A description of the engineering design for normal and abnormal loads is given by Hely and Taylor (1983).

White Industries appointed the author's firm to provide full consulting engineering on the ground in Singapore for the precasting factory. The system starts from the precasting factory. It is essential that this facility be tuned precisely to provide the prefabricated concrete components to match the transportation, which must in turn match the erection program, which hands over four 12-story buildings every 28 days.

Fig. 1 Singapore housing apartments

Each day's production requires only one casting from each mold in the 24-hour cycle. Accelerated steam curing commences three hours after concreting and follows a cycle lasting six hours. Molds are then left to cool, and stripping takes place 15 hours following concreting.

Review

The major advantage of a factory-based system is that it can be operated by a relatively unskilled workforce, which is a particular advantage in remote and underdeveloped countries. After a short training period, workmen can

Fig. 2 Complete volumetric units ready for erection

Fig. 3 Precasting factory showing volumetric molds

produce components of a consistent high quality provided adequate supervision is maintained.

The industrialized prefabricated method has proved to be 30% faster than conventional methods on similar sites under similar climatic conditions. It should be noted that Singapore has an average rainfall in excess of 2300 mm (90 in.) with some 180 wet days per year.

The critical activity is the finalization of the outline dimensions of the concrete products and associated connection details, prior to mold manufacture, and takes place during the design activity.

A disadvantage is the capital intensive nature in the early stages.

SYSTEM BUILDING WITH PRECAST ELEMENTS

A system of construction used by the author's firm some ten years ago was short-lived because it was followed by the almost universal acceptance of the total glass curtain wall. However, the emergence of post modern architecture heralds various articulated structures having in many cases some form of masonry facade.

Nauru House, Melbourne, Australia is a 52-story reinforced concrete structure having a total precast concrete facade forming the architectural finish (Fig. 4). The structure is 177 m (581 ft) high and the design is based on the tube-in-tube concept. The structure of floor slabs, 16 exterior columns and the spandrel beams is constructed using the progressive strength system, with the exposed aggregate precast panels providing permanent formwork to the columns and spandrel (Fig. 5).

The column panels are generally 125 mm (5 in.) thick. Although their cross-sectional area is not considered in the design of the columns for vertical dead and live load plus bending, it is considered in determining both the immediate (elastic) and long term (creep) column shortening. The thickness of the panels is also considered in the investigation of the effect of various temperature changes across the column.

Spandrel panels are rationalized down to two basic types. Each spandrel beam is divided into three separate precast sections to satisfy size limitations and the visual arrangement of joints. The three sections forming each spandrel are cast at the same time to maintain aggregate and color uniformity. The reinforcement is fixed to each individual panel prior to lifting it into place. Splice bars are placed after the units are correctly located. Because of the large clear span of the spandrel beams, the precast units are temporarily supported on a truss that spans from column to column.

The columns are also rationalized down to two basic types. The columns taper throughout their height, and to facilitate the casting of the tapered sections, specially hinged forms to the molds were designed, enabling the variable dimensioned sections to be cast in the same mold.

Fig. 4 Nauru House—total precast facade

The construction sequence model (Fig. 6) shows the precast column units erected ahead of wet construction to provide the outside formwork.

Review

The system proved successful in achieving time savings and economies. The building system, however, is sensitive to the timely supply of the prefabricated units, which become the critical activity as they constitute formwork and falsework.

PRESTRESSING FOR SERVICEABILITY

The object of this review is to introduce "partial" prestressing in the sense of provision of just sufficient active force at the most effective position to achieve a specific result. In the Rialto Towers, partial prestressing has been used to induce pseudo gravity loads.

Currently the tallest office building in Australia and the Southern Hemisphere, the 65-story Rialto Tower extends 243 m (797 ft) above basement level

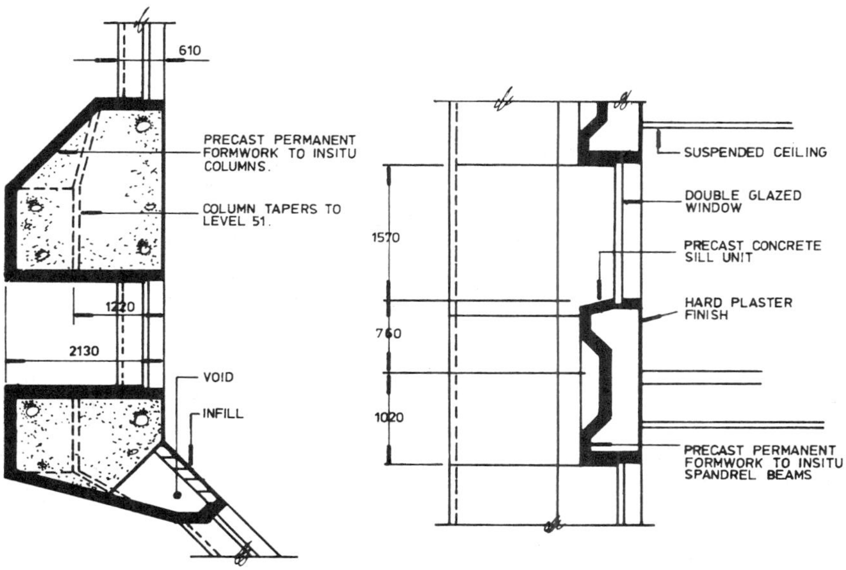

Fig. 5 System details

and is the second tallest reinforced concrete building in the world. Total floor space provided by the building is 144,000 m^2 (1,550,000 ft^2) with basement parking for 800 cars (Fig. 7).

Tower A rises 65 stories and Tower B 48 stories above basement level. The construction of two integrated towers with substantially different heights using reinforced concrete gives rise to a number of specific problems that are insignificant in a steel framed tower. The most obvious of these is concrete creep and shrinkage. It is estimated that total nonelastic shortening of the 65-story tower would be in the order of 150 to 200 mm (6 to 8 in.). Provided allowances are made in the attachment of nonload-bearing elements such as elevator rails and the facade, the magnitude of this nonelastic deformation is not highly significant.

However, as Tower A and Tower B form an integrated structure, a differential in the order of 40 mm (1.6 in.) is estimated between adjacent columns at level 41 (Tower B roof) due to effects of the additional 17 stories of Tower A. The distance between these columns is only 4 m (13 ft) and clearly such movements cannot be accommodated in the construction. Jointing of the towers is neither feasible nor practical and the provision of a *belt* at this level unsuitable to the architecture as well as inducing long-term out-of-plumb at the top of Tower A.

To actively control the behavior of the structure, prestressing cables are introduced from level one to level 38 and stage stressed as Tower A construction proceeds (Fig. 8). All columns below level 38 are subject to the same

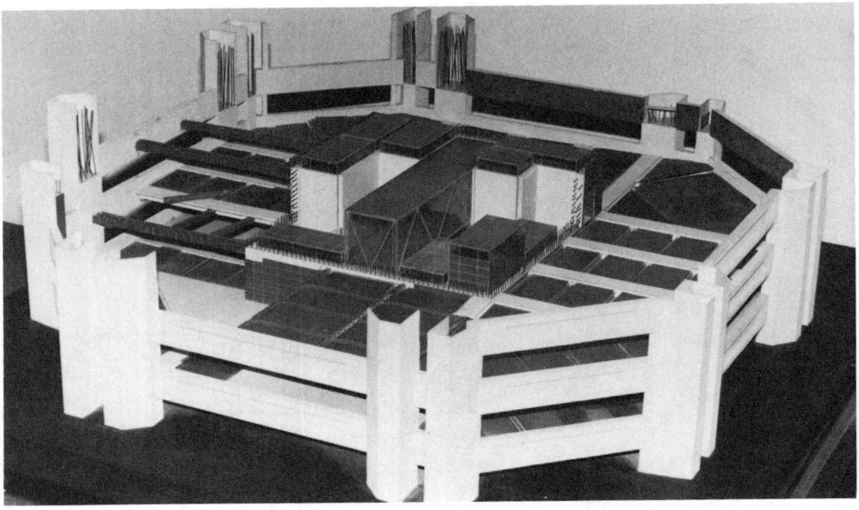

Fig. 6 Structural system

Fig. 7 Rialto Towers Building, Melbourne, Australia

loadings at the same time and therefore elastic and nonelastic shortening values are relatively consistent for the lifetime of the building.

It could be said that the designer has played a confidence trick on Tower B, making the structure believe itself to be 17 stories taller. Levels since taken on adjacent columns on Tower A and B have confirmed the behavior of the structure to be closely related to that estimated.

FUTURE DIRECTIONS

New modernism is bringing more hybrid and articulated buildings. Yet another hurdle is the growing acceptance of multiuse buildings. When the

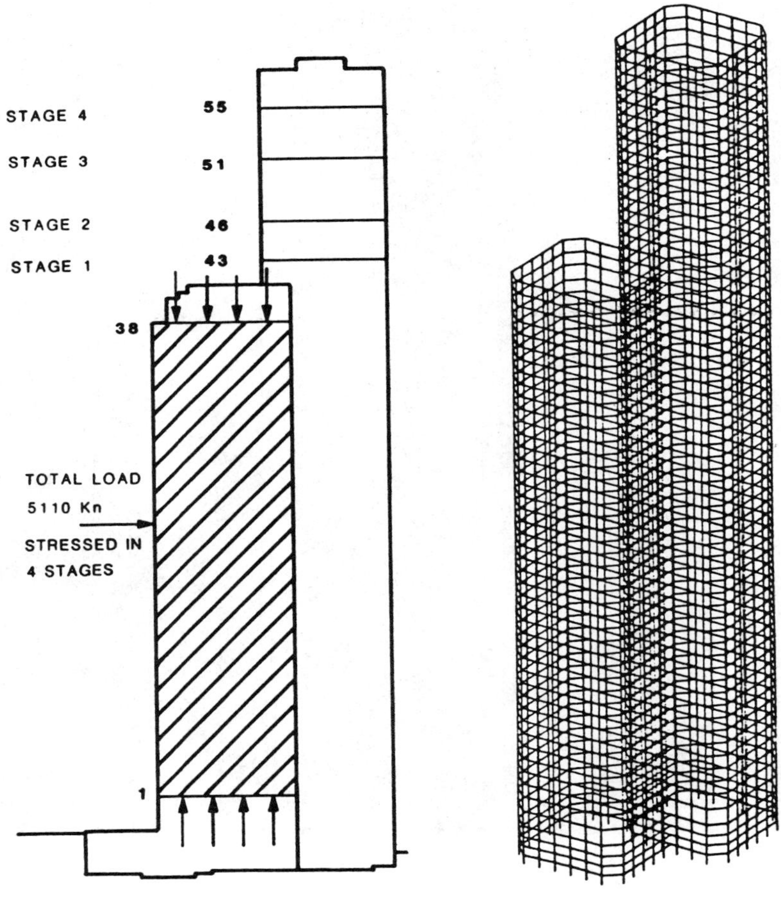

Fig. 8 Rialto Towers. Tower B prestressing

different uses are enclosed in one envelope and the building also takes on a hybrid and articulated form, the structural engineer is forced to rethink the systems necessary to resist the horizontal and vertical loads involved.

There is a tendency to withdraw into the core or inner tube of the building when the exterior shape does not fit into a standard solution or becomes difficult. While a concrete core can provide stiffness, the principle is not an economical solution. It is well proven that the most economical way to resist horizontal loads is by using the outside skin of the building.

Fortunately at the same time that *post modernism* has introduced structural complexity, the computer has provided the facility to analyze configurations that would not previously have been possible. It is no longer necessary to classify or think of structures in terms of tubes, bundled tubes, super frames, and so on. It is possible to analyze the forces in the three dimensional envelope of the articulated structure.

Referring again to "partial prestressing" as opposed to full prestressing, the term was originally used to describe a structural concrete member in which part of the section is stressed in tension under service conditions. This definition is restrictive in that it suggests that tensile stress is always involved in a partially prestressed situation. This term may be extended to cover any situation of mixed steel, where normal reinforcement is present in a considerable quantity, that is in more than just minimum or distributive steel amounts and just sufficient prestressing force is applied to achieve a particular desired objective. The review of Rialto demonstrates the use of "partial prestress" to induce pseudo gravity loads, so as to match creep characteristics.

It follows that we can provide just sufficient active force at the desired position and actively orchestrate all the forces to achieve a desired situation rather than acting passively to the design action effects of loading.

Looking now at the use of both precast concrete and prestressing in building structures, it appears that the very considerable experience and successes in bridge design, using post-tensioned precast segmental elements has been largely ignored when designing building structures. The most positive method of utilizing the full cross-section of precast elements and totally integrating them into a structure is post-tensioning.

The author's firm has adopted this design approach on a high-rise residential structure (Figs. 9 & 10). The facade consists of prestressed precast concrete elements that are post-tensioned to form vierendeel girders spanning the full face of the structure. The post-tensioned facade is supported on four corner columns and provides vertical support to the edge of the floor slab and the full lateral support to the structure.

There is a challenge to examine further the application of "partial prestressing" or small amounts of prestress to control crack width in horizontal facade elements, to integrate precast elements, and to improve long-term durability of finishes. Over and above everything that has preceded, when dealing with precast and prestressed concrete the first essentials are

1. A sound engineering concept,
2. intelligent design calculation that takes into account both normal and abnormal load conditions, and
3. an appreciation of building.

Fig. 9 High-rise residential building with prestressed precast facade

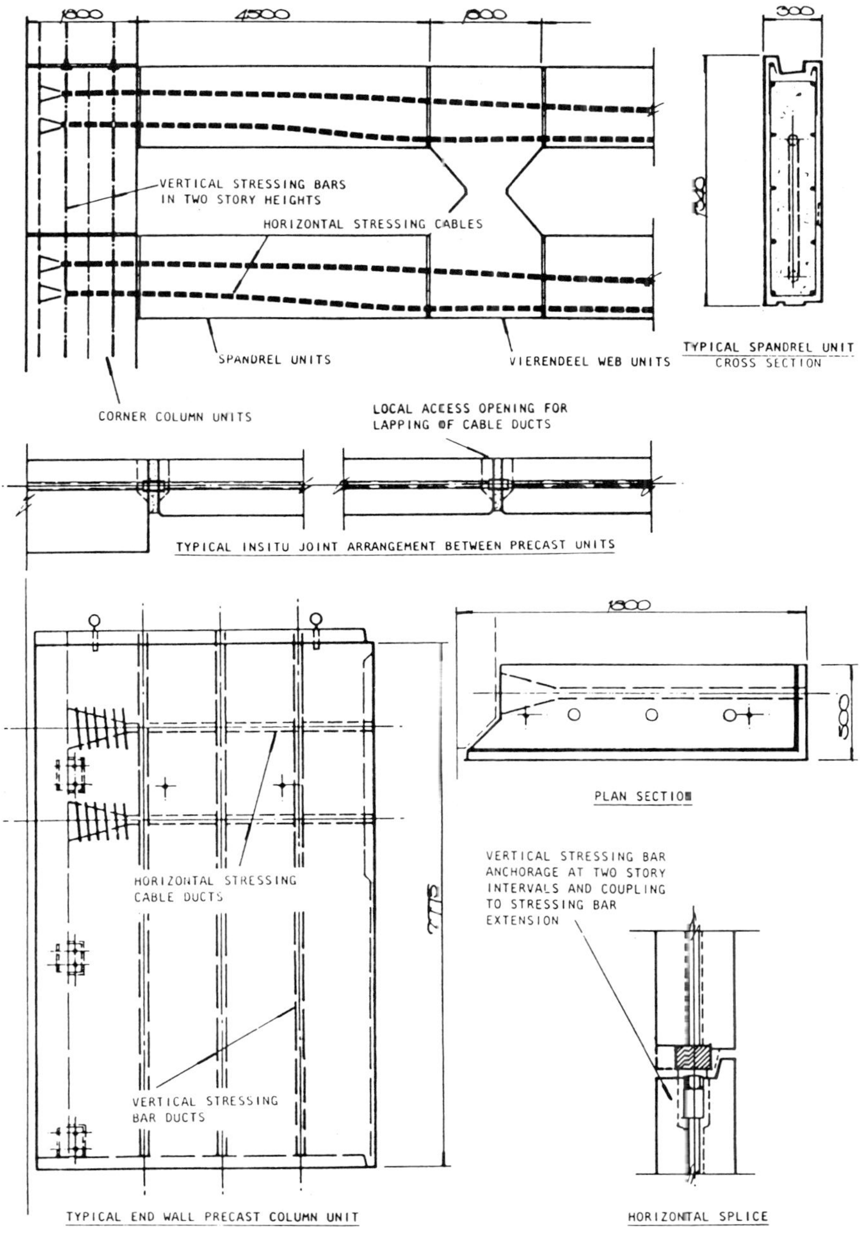

Fig. 10 Design approach for post-tensioned facade

With the advent of computers there is a tendency to believe that more sophisticated design methods yield better structures. Complex calculation will not compensate for poor engineering judgment.

REFERENCES/BIBLIOGRAPHY

Bennett, E. W., 1984
PARTIAL PRESTRESSING—A HISTORICAL OVERVIEW, P.C.I. Journal, September-October.

Buggeling, A. S. G., 1985
CONCRETE STRUCTURES—A DESIGN CHALLENGE, P.C.I. Journal, March-April.

Hely, B. G. and Taylor, D. C., 1983
INDUSTRIALIZED HOUSING PROJECT—SINGAPORE, Concrete Institute of Australia Conference.

Hose, R., 1984
THE RIALTO PROJECT—CARPARK AND OFFICE BUILDING, The Association of Consulting Structural Engineers N.S.W. Australia, Modern Structural Engineering.

Rossiter, S. G., 1975
NAURU HOUSE MELBOURNE—DESIGN PHILOSOPHY, Meinhardt Group, Technical Bulletin.

Deterioration of Concrete

James R. Clifton

Concrete is usually a durable material whose usefulness equals or exceeds its intended service life. Premature deterioration of a concrete building or its components can occur, however, if the concrete is not of adequate quality, the building and the concrete were not properly designed for the service environment, or the environment was not as anticipated or changed during the service life of the building. In many cases the premature deterioration of concrete can be prevented by selecting the constituents of concrete based on an understanding of the major deterioration processes of concrete. As our knowledge of the relationships between concrete constituents and durability improves, fewer incidents of premature deterioration will occur if the knowledge is properly disseminated.

NATURE OF CONCRETE

The durability of concrete is largely controlled by the physical and chemical nature of its constituents and their compatibilities. Hardened concrete consists predominantly of mineral aggregates and hardened cement paste. Distributed between and within these constituents is a considerable amount of void space, a major factor in controlling the durability of concrete. With normal weight concrete, most of the void space is located in the hardened cement matrix. In concrete technology, space in the hardened cement matrix is divided into (in sequence of decreasing size) entrapped air, air voids or bubbles, and capillaries and gel pores. The void space associated with capillaries and gel pores is considered to form the porosity of the hardened cement paste. The minimum porosity of a completely hydrated cement matrix is around 28%. The permeability of portland cement paste increases with porosity (Fig. 1), with the cement paste of durable concrete usually having porosities below 40% (Wood, 1968) and with the extent that the pores are interconnected. In addition, some aggregates can have an appreciable amount of porosity that can affect the durability of concrete, especially its freeze-thaw resistance. Regarding the chemical properties of concrete, the hydration of portland cement yields alkaline products that are susceptible to attack by acid precipitation.

CAUSES OF CONCRETE DETERIORATION

The major causes of deterioration of concrete in a building, excluding structural misuse (for example, overloading and fatigue), include:

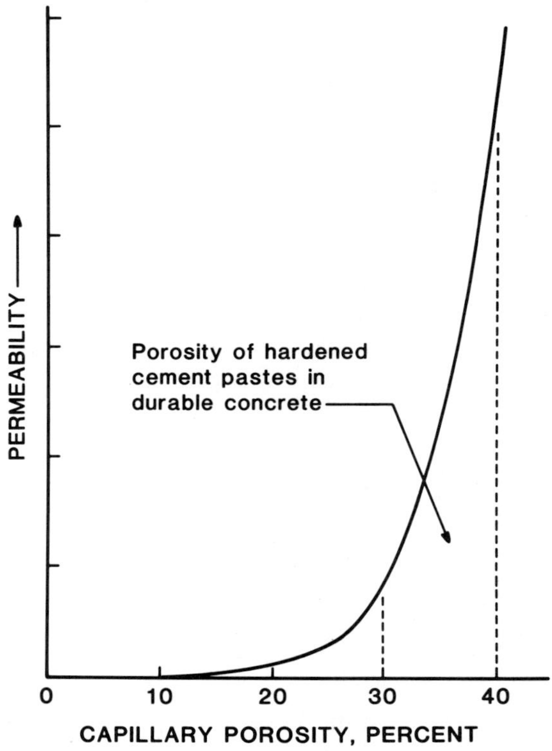

Fig. 1 Permeability of portland cement paste

1. freezing and thawing while in a water saturated condition
2. dissolution of soluble salts by running water
3. reactions between alkalies in the cement and certain aggregates (primarily of siliceous or calcareous mineralogy)
4. corrosion of reinforcing steel
5. chemical attack (for example, acid deposition containing sulfates)

This list is clearly not complete and a concrete could be subject to more than one type of distress.

Note that all of the listed causes of concrete deterioration involve the diffusion of water or an aqueous solution into concrete. Through either physical or a combination of chemical and physical processes, the build-up of internal tensile stresses in concrete often can result in cracking of the concrete. The simplest but often the most difficult way to prevent deterioration is to keep the concrete dry. Because of the difficulty in keeping concrete dry, other means of preventing concrete deterioration have been developed based on an understanding of the mechanisms involved in the deterioration processes.

Freezing and thawing damage is believed to be caused by the expansion of water in the void space of concrete, resulting in internal stresses exceeding the tensile strength of concrete. The common method of preventing freeze-thaw damage is to deliberately add closely-spaced but unconnected air bubbles in the concrete to accommodate the expanding water as it freezes.

The major salts in concrete are, at best, slightly soluble, with calcium hydroxide usually being the most soluble major reaction product of portland cement. Calcium hydroxide can be converted into an essentially insoluble silica gel by adding a reactive pozzolan such as fly ash or silica fumes to the concrete mix.

Some form of reaction occurs between almost all aggregate and the hydration products of portland cement, especially the soluble alkalies. Often these reactions are not of any consequence. However, certain siliceous and calcareous aggregates can react with alkalies to form products that expand by imbibing water. In some cases, the expansion results in the cracking of concrete and its failure (Wood, 1968). The most practical means of preventing alkali-aggregate problems is to select unreactive aggregates or a low-alkali cement. In addition, the alkali content in concrete can be reduced by replacing part of the cement with a pozzolan.

The corrosion of steel in concrete is usually a slow process because concrete provides an alkaline environment (pH of around 12.5), which results in the passivation of steel. The pH of concrete can be reduced by carbonation, however, this process is slow in quality concrete. Most corrosion problems are attributed to the corrosive effects of chloride ions. Chloride ions have the ability to depassivate steel, which can result in the rapid corrosion of steel reinforcement. In addition to chloride ions, both moisture and oxygen must be present for corrosion to take place. Common sources of chloride ions are deicing salts and concrete admixtures, especially calcium chloride, which is used to accelerate the hardening of concrete in cold-weather construction. Several methods have been developed to retard the corrosion rate of steel reinforcement (ACI, 1985) including restricting the amount of chloride ions in concrete constituents, coating steel reinforcement with epoxy coatings, cathodic protection of the reinforcement, and the use of low permeability concrete.

A recently publicized type of chemical attack of construction materials is acid deposition, which includes acid rain and dry depositions containing sulfate ions. Acid depositions are able to attack concrete because concrete is chemically a basic material, having a pH of about 12.5. A common constituent of air pollution is sulfur dioxide, which often is converted to sulfuric acid. Sulfuric acid can attack concrete by two main processes: by a typical acid-base reaction involving hydroxide ions in the pore solution of concrete, and by reaction of sulfate ions with hydrated calcium aluminates in hardened cement paste to form an expansive product known as ettringite (Wood, 1968).

EXPERT SYSTEMS

Based on field experiences and laboratory research, many of the factors causing premature deterioration of concrete are sufficiently understood that few incidents of premature deterioration should be occurring. However, such incidents are all too commonplace. A major reason for many of the cases are a lack of knowledge on the part of individuals responsible for the design, construction, and maintenance of concrete structures, even though such knowledge is possessed by concrete experts. Expert systems, also known as knowledge-based systems, appear to be an effective means for transferring the knowledge of the durability of concrete possessed by experts to decision makers. Expert systems are considered to be computer programs that use knowledge and inference procedures to solve problems that are difficult enough to require significant human expertise for their solution (Feigenbaum, 1981). The knowledge of most current expert systems consist of facts and heuristics.

DURCON (DURable CONcrete) is a prototype expert system being developed at the National Bureau of Standards to give recommendations on the selection of constituents for durable concrete (Clifton, 1985). DURCON is being developed to demonstrate the application of expert systems to improving the process of selecting construction materials. Development of DURCON is considered feasible because much specialized and heuristic knowledge exists on relations between the design of concrete mixtures, including constituents, and the durability of concrete. Four major deterioration problems are being covered; freeze-thaw, sulfate attack, corrosion of reinforcing steel, and cement-aggregate reactions. The factual knowledge-base is based on the American

Concrete Institute Guide to Durable Concrete (ACI, 1977). Heuristic knowledge is being obtained from experts on the durability of concrete.

DURCON asks questions of the user relative to the specific concrete structure being considered, such as the type of structure, type of cement and concrete that will be used, information concerning properties of the aggregate, and information on exposure conditions. Then depending on the durability problems likely to be encountered, recommendations on concrete mix design and properties of the hardened concrete are given. For example, if freeze-thaw durability could be a problem, DURCON will give recommendations on the water to cement ratio of the concrete, amount of entrained air in the concrete, the compressive strength of the concrete at 28-days, and the selection of acceptable aggregates.

SUMMARY

The major causes of the deterioration of concrete in a building are associated with the diffusion of water or an aqueous solution into concrete, and through either physical or chemical processes, the build-up of internal stresses in concrete that can result in concrete cracking. Knowledge exists on means of preventing the premature deterioration of concrete and improvement in the means of disseminating this knowledge will result in fewer incidents of premature failures. Expert systems appear to be an effective means for transferring knowledge from experts to individuals responsible for the design, construction, and maintenance of concrete structures.

REFERENCES/BIBLIOGRAPHY

ACI, 1977
GUIDE TO DURABLE CONCRETE, ACI201.2R-77, Journal of the American Concrete Institute, December.

ACI, 1985
CORROSION OF METALS IN CONCRETE, ACI222R-85, Journal of the American Concrete Institute, January.

Clifton, J., Oltikar, B., and Johnson, S., 1985
DEVELOPMENT OF DURCON, AN EXPERT SYSTEM FOR DURABLE CONCRETE: Part 1, NBSIR 85-3186, National Bureau of Standards.

Feigenbaum, E., 1981
EXPERT SYSTEMS IN THE 1980's, in Machine Intelligence, A. Bond ed., Pergamon Press, Elmstead, New York.

Wood, H., 1968
DURABILITY OF CONCRETE CONSTRUCTION, ACI, Monograph No. 4, Detroit.

U.S. Coordinated Program for Masonry Building Research

James L. Noland

Load-bearing masonry buildings have been built in the United States for many years—nearly since the time European settlers first arrived. Masonry buildings are a significant percentage of buildings built from that time up to the 1930s and are generally unreinforced.

Reinforced concrete and structural steel buildings gradually became a larger and larger part of the construction market. Their use was and is encouraged by the great amount of research and development done and by the gradual improvement in design methods and codes.

In contrast, the use of masonry as the primary structural system declined in comparison to the use of steel and concrete possibly because of the perception that masonry design technology was not at the same level as that for steel and concrete and because of the perception that masonry structures perform poorly in earthquakes.

While some U.S. research in structural masonry was done earlier, it began to increase in the mid-to-late 1960s and continues to increase. Masonry building code improvement has also recently been emphasized as exemplified by the 1985 Uniform Building Code and by the work of the joint American Society of Civil Engineers/American Concrete Institute (ASCE/ACI) Committee 530 on masonry. This author believes that some of the impetus for renewed interest in the structural use of masonry is because of its architectural flexibility and that it can be competitively economical for certain types of building plans. The introduction of reinforcement has been shown to provide the ductility required for adequate performance in earthquake conditions.

However, the design methodology is still basically a working stress approach based on the assumption of linear-elastic material behavior although the latest Uniform Building Code (UBC) does permit strength methods for certain applications. Although many reinforced masonry structures have been built and have performed successfully, economic considerations and a better ability to predict ultimate behavior would make structural masonry a more viable alternative as a building system, particularly for earthquake conditions. A need was recognized for complete strength-method design techniques for reinforced masonry based on adequate experimental data to bring masonry structural technology to a level consistent with modern needs.

RESEARCH TEAM ORGANIZATION

A team of masonry researchers known as the Technical Coordinating Committee for Masonry Research (TCCMAR) was formed to identify the research necessary to organize so that the research can be done in an orderly fashion. The organization of TCCMAR and the basic research areas are depicted in Fig. 1.

Funding is provided by The National Science Foundation (Dr. A. J. Eggen-

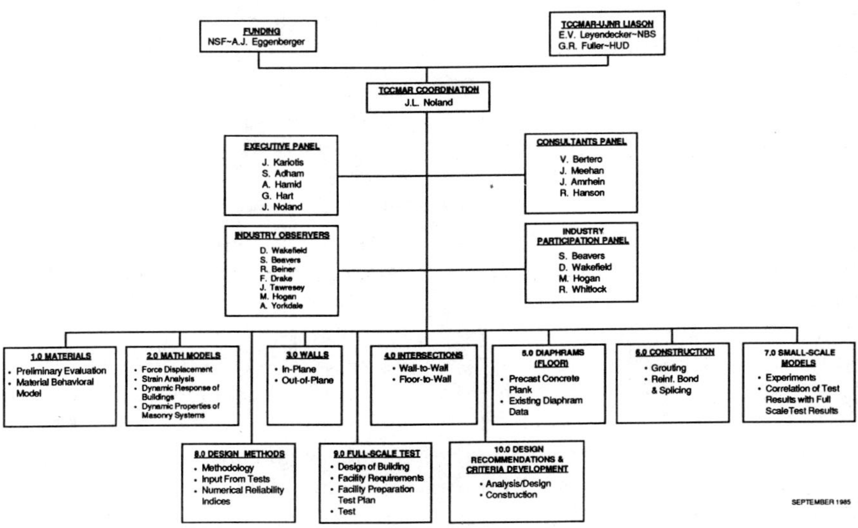

Fig. 1 TCCMAR Organization (U.S.)

berger). Liason with the U.S.-Japan Cooperative Program in Natural Resources (UJNR) Committee on Wind and Seismic Effects is done by G. R. Fuller of the U.S. Dept. of Housing and Urban Development (HUD) and Dr. H. S. Lew of the National Bureau of Standards (NBS). A single coordinator (J. Noland) is responsible for the overall operation of the U.S. program with the advice and assistance of the Executive Panel. The Consultants Panel is composed of individuals highly qualified to provide basic guidance on the course and conduct of the program. Two were intimately associated with previous U.S.-Japan joint research programs.

A panel of Industry Observers was formed to provide industry perspective and to relay information about the program to the industry. The purpose of the Industry Participation Panel is to define ways in which the industry can assist the program and arrange for that assistance.

RESEARCH PLAN

The U.S. masonry research program consists of a group of coordinated research tasks beginning with tasks that address basic material behavior, progressing to tasks addressing behavior of masonry structural components, for example walls, and finally to experiments on a full-size segment of a masonry building. Development of analytical methods for stress and strain analysis and system behavior will proceed in parallel with experimental tasks. The final task will be the preparation of recommendations for design procedures and building code provisions.

The research tasks identified by TCCMAR fell into 10 categories as shown in Fig. 1. The following is a more detailed list of research tasks in each category and their objectives.

Category	*Task*	*Title-Purpose*
1.0	1.1	Preliminary Material Studies: To establish the range of continuity of masonry behavior to provide a basis for selection of the type or types of masonry to be used. To establish standardized materials test procedures for all the experimental tasks.
1.0	1.2	Materials Models: To evaluate K1, K2, and K3 for the flexural stress-block. To determine uniaxial and biaxial material properties for analytical models (Task 2.1 and 2.2) including post-peak behavior. To evaluate nonisotropic behavior.
2.0	2.1	Force-Displacement Models for Masonry Components: To develop force-displacement mathematical models that accurately characterize reinforced masonry components

Category	*Task*	*Title-Purpose*
		under cyclic loading to permit pretest predictions of experimental results. To develop models suitable for parameter studies and models suitable for design engineering.
2.0	2.2	Strain Analysis Model for Masonry Components: To develop a strain model for reinforced masonry components in conjunction with Task 2.1 to enable regions of large strain to be identified, thus assisting in experimental instrumentation planning. To develop a simplified model to be used to provide data for strength design rules and in-plane shear design procedures.
2.0	2.3	Dynamic Response of Masonry Buildings: To develop a generalized dynamic response model to predict interstory displacements using specified time histories. To correlate force-displacement models and to investigate force-displacement characteristics of structural components in the near-elastic and inelastic displacement range. To provide data for building test planning.
2.0	2.4	Dynamic Properties of Masonry Systems: To develop consistent, unified, rationale for seismic design of masonry buildings considering elastic and inelastic response of masonry buildings and of the soil/structure interaction, and related to seismic hazard zones.
3.0	3.1(a)	Response of Reinforced Masonry Story-Height Walls to Fully Reversed In-Plane Lateral Loads: To establish the behavior of story-height walls subjected to small and large amplitude reversals of in-plane lateral deflection, axial force, and bending moments, considering aspect ratios, reinforcement ratios and patterns, and the effect of openings.
3.0	3.1(b)	Development of a Sequential Displacement Analytical and Experimental Methodology for the Response of Multi-Story Walls to In-Plane Loads: To develop a reliable methodology for investigating, through integrated analytical and experimental studies, the in-plane behavior of multistory reinforced hollow unit masonry walls. The methodology will be the basis of studying the response of a full-scale masonry research building.
3.0	3.2(a)	Response of Reinforced Masonry Walls to Out-of-Plane Static Loads: To verify the behavior of flexural models developed using material models, to evaluate the influence of mortar joint spacing, unit properties, reinforce-

Category	*Task*	*Title-Purpose*
		ment ratios and grouting upon wall behavior. To provide stiffness data for correlation with dynamic wall test results (Task 3.2(b)).
3.0	3.2(b)	Response of Reinforced Masonry Walls to Out-of-Plane Dynamic Excitation To determine effects of slenderness, reinforcement amounts and ratios, vertical load and grouting on dynamic response, to verify mathematical response models, to develop design coefficients for equivalent static load methods.
4.0	4.1, 4.2	Wall-to-Wall Intersections and Floor-to-Wall Intersections of Masonry Buildings: To determine the effectiveness of intersection details to connect masonry wall components, to construct a nonphenomenological analytical model of intersection behavior.
5.0	5.1	Concrete Plank Diaphragm Characteristics: To investigate experimentally concrete plank floor diaphragms to determine modes of failure and stiffness characteristics including yielding capacity in terms of distortion as needed for masonry building models.
5.0	5.2	Assembly of Existing Diaphragm Data: To assemble extensive existing experimental data on various types of floor diaphragms, to reduce to a form required for static and dynamic analysis models.
6.0	6.1	Grouting Procedures for Hollow Unit Masonry: To identify methods of grouting hollow unit masonry such that the cavity is solidly filled and reinforcement is completely bonded.
6.0	6.2	Reinforcement Bond and Splices in Grouted Hollow Unit Masonry: To develop data and behavioral models on the bond strength and slip characteristics of deformed bars in grouted hollow unit masonry, to develop data and behavioral models on the bond strength and slip characteristics of deformed bar lap splices in grouted hollow unit masonry as needed for building modeling.
7.0	7.1	Small Scale Models: To experimentally evaluate the use of small-scale modeling for reinforced hollow unit masonry walls by correlating test results with those of full-scale walls of the same configuration. To determine if tests of small-scale specimens can reveal basic characteristics and failure modes of full-scale masonry specimens.

Category	*Task*	*Title-Purpose*
8.0	8.1	Limit State Design Methodology for Reinforced Masonry: To select an appropriate limit state design methodology for masonry. To select and document a procedure to compute numerical values for strength reduction factors. To review program experimental research tasks to assure that statistical benefits are maximized and proper limit states are investigated.
8.0	8.2	Numerical Reliability Indices: To develop numerical values of statistically-based strength reduction (that is, Ø) factors using program, experimentally developed data, other applicable data, and judgment.
9.0	9.2	Design of Reinforced Masonry Research Building, Phase I: To develop the preliminary designs of the potential research buildings which reflect a significant portion of modern U.S. masonry construction. To estimate interstory displacements using methods developed in Category 2 tasks and the associated load magnitudes and distributions. To select a single configuration in consultation with TCCMAR that will be used as a basis for defining equipment and other laboratory facilities in Task 9.2.
9.0	9.2	Facility Preparation: Define, acquire, install, and check-out equipment required for experiments on a full-scale masonry research building.
9.0	9.3	Full-Scale Masonry Research Building Test Plan: To develop a detailed and comprehensive plan for conducting static load-reversal tests on a full-scale reinforced masonry research building.
9.0	9.4	Full-Scale Test: To conduct experiments on a full-scale reinforced masonry research building in accordance with the test plan and acquiring data indicated. To observe building response and adjust test procedures and data measurements as required to establish building behavior.
10.0	10.1	Design Recommendations and Criteria Development: To develop and document recommendations for the design of reinforced masonry buildings subject to seismic excitation in a manner conducive to design office utilization. To develop and document corresponding recommendations for masonry structural code provisions.

Category	*Task*	*Title-Purpose*
11.0	11.1	Coordination: To fully coordinate the U.S. research tasks to enhance data transfer among researchers and timely completion of tasks. To schedule and organize TCCMAR and Executive Panel meetings. To establish additional program policies as the need arises. To stimulate release of progress reports and dissemination of results. To coordinate with industry for the purposes of informing industry and arranging industry support. To interface with NSF and UJNR on overall funding and policy matters.

The names, affiliations, and research task arrangements of current TCCMAR members are as follows:

Task	*Researcher*	*Institution*
1.1	R. Atkinson	Atkinson-Noland & Assoc., Boulder
1.2(a)	A. Hamid	Drexel Univ.
1.2(b)	R. Brown	Clemson Univ.
2.1	R. Englekirk	AKEH Joint Venture
2.2	G. Hart	AKEH Joint Venture
2.3	J. Kariotis	AKEH Joint Venture
2.4	R. Ewing	AKEH Joint Venture
3.1(a)	J. Noland/B. Shing	Univ. of Colo.-Boulder
3.1(b)	G. Hegemier/F. Seible	Univ. of Calif.-San Diego
3.2(a)	A. Hamid	Drexel Univ.
3.2(b1)	S. Adham	Agbabian Assoc.
3.2(b2)	R. Mayes	Computech
4.1	G. Hegemier	Univ. of Calif.-San Diego
4.2	F. Seible	Univ. of Calif.-San Diego
5.1	M. Porter	Iowa State University
5.2	M. Porter	Iowa State University
6.1	L. Tulin	Univ. of Colo.-Boulder
6.2	L. Tulin	Univ. of Colo.-Boulder
7.1	A. Hamid	Drexel Univ.
8.1	G. Hart	Univ. of Calif.-Los Angeles
8.2	TCCMAR*	---
9.1	J. Kariotis/A. Johnson	Barnes-Kariotis Joint Venture
9.2	G. Hegemier	Univ. of Calif.-San Diego
9.3	TCCMAR*	---
9.4	TCCMAR*	Univ. of Calif.-San Diego
10.1	TCCMAR*	---
11.1	J. Noland	Atkinson-Noland & Assoc.

*Task to be done by TCCMAR/U.S. or subgroup thereof.

TASK COORDINATION

The research tasks are interdependent, that is, results from a given task may be required for the execution of others and vice-versa. Analytical tasks will generally require interaction with experimental tasks on a fairly continuous basis so that analytical model development may incorporate data as they are obtained. The needs of the analytical tasks will in turn serve to define, in part, the manner in which experimental tasks are designed and conducted and the data to be obtained.

The anticipated intra-task interaction is depicted generally in Fig. 2.

SCHEDULE

The schedule for tasks comprising the U.S. program is shown in Fig. 3. The total time required to complete the program is estimated to be approximately 6 years.

REVIEW OF 1985 PROGRESS

Task 1.1, Preliminary Studies, was completed in June 1985. Work was begun on Tasks 1.2(a and b), 2.1, 2.2, 2.3, 3.1(a), 3.2(a), 5.1, 5.2, 6.2, 8.1 and 9.1 in the Fall, 1985; therefore results are not yet available.

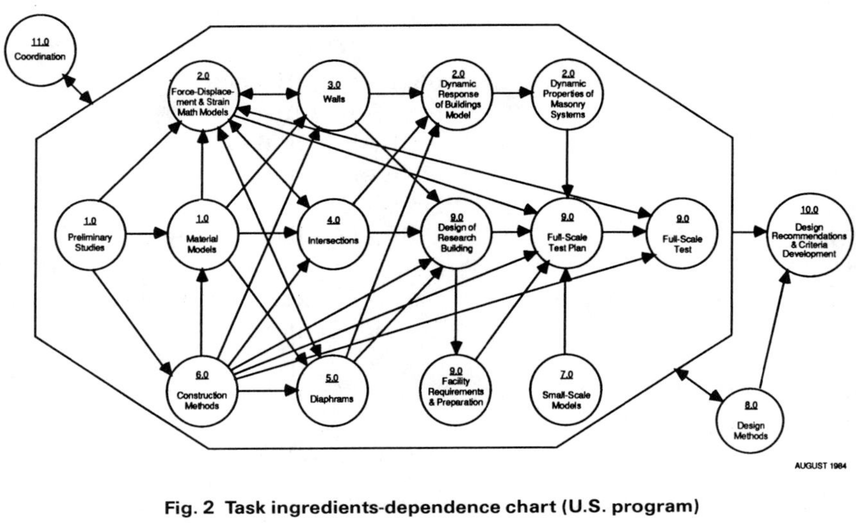

Fig. 2 Task ingredients-dependence chart (U.S. program)

CONCLUSIONS

The U.S. program was designed to provide a strength design methodology for reinforced masonry buildings based on an integrated experimental and analytical program of specific research tasks. Study of the task descriptions will reveal that only issues believed to be essential to this development are addressed. Budget limitations also restrict the number of parameters that can be evaluated in these tasks. However, the work will provide a unified framework and reference point so that future research, as funds become available, will be able to build effectively on the knowledge developed in this program.

ACKNOWLEDGMENT

The support of the National Science Foundation is greatly acknowledged. The contribution of hollow concrete units by the Concrete Masonry Association of California-Nevada and of hollow clay units by the Western States Clay Products Association for specimen construction is also acknowledged.

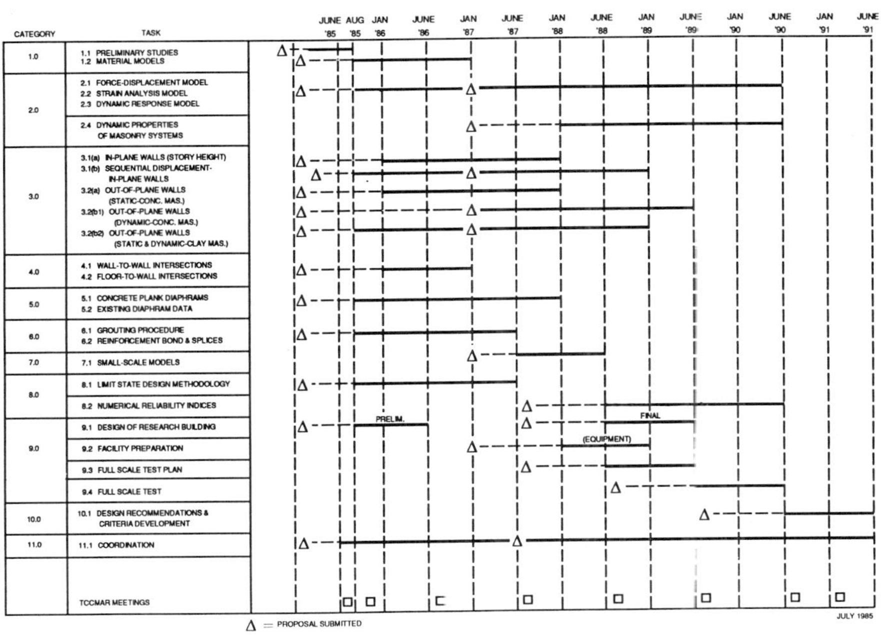

Fig. 3 Task schedule (U.S. program)

Superstructure of OUB Centre, Singapore

S. Sasaki
M. Suko
Y. Atsukawa

The high-rise tower of OUB Centre at Raffles Place, Singapore, when completed, may well be up for the title of world's tallest building outside of North America. As is usual with such high-rise buildings, the structural system of this tower was greatly affected by the unique architectural parameters. The steel skeleton was finally chosen employing the composite system efficiently at the lower part of the structure as a result of the practical study of construction costs. The result of the space frame analysis of this asymmetric and slender structure is briefly discussed together with some topics from the structural design through the erection work.

OUTLINE OF OUB CENTRE

This prestigious building will be the new headquarters of the Overseas Union Bank, Singapore, (Fig. 1). The architectural design was developed by Kenzo Tange Associates, Japan. The original structural design by Bylander Meinhardt Partnership (BMP), Australia, of the main skeleton was purely of steel. However, at the time of a tender call for the steel structure, the alterna-

Fig. 1 Final stage of steel erection on OUB Centre

tive design proposed by Nippon Kokan K. K. (NKK), Japan, which was based on the steel structural system with the composite system at the lower part of the structure, was adopted in this project after a study of the total construction cost and of the structural efficiency. The structure consists of a high-rise office tower and a podium for commerical purposes. Visually, the tower is composed of two different size and height right-angled triangular towers. The smaller one is 30 × 30 m (98 × 98 ft) and rises to the 49th floor. The other, 40 × 40 m (131 × 131 ft), rises to the 64th floor. These architectural parameters governed the structural planning such as the layout of columns, girders, and beams. The tower has a four-story basement and a typical story height is 4 m (13 ft). The total floor area of the tower is 75,000 m^2 (807,000 ft^2), and the structural steel used is about 11,000 tons (10 million kg). The podium, which is built of concrete, is separated from the tower by the expansion joints mainly because of the assumed differential settlement of foundation piles.

DESIGN CRITERIA

Because of the high slenderness ratios of the tower, much attention had been paid to the stiffness from the beginning. Primary design criteria established for this building are as follows:

1. Wind load was determined referring to the result of the wind tunnel studies conducted at Monash University, Australia, and the record of wind data at Singapore. The dynamic pressure at the top of the tower is 1.11 kPa (23 psf).
2. The standard deviation of the horizontal acceleration for a one year return period is kept less than 10 milli-g for the comfort of the occupants.
3. The horizontal deflection of the top occupied floor is kept less than 0.0015 of the height.

Other specific remarks are

Design calculation is according to BS 449, CP114 and CP117.
Imposed loads are according to Singapore Building Regulation.
Grade of steel: Primary member—G50 and G55
Secondary member—G43
Grade of concrete: Slab—30 mpa
Wall—30 and 40 mpa

CHOICE OF COMPOSITE SYSTEM

In the original design, a steel skeleton with concrete floor slabs was chosen. To increase the stiffness of this tall building to the adequate level, a concept of a mega-frame system was used. A megaframe is a large frame enclosing a whole building with another structural subsystem within it. The interaction between the megaframe and the intermediate subsystems increases the stiffness and the strength of the whole structure by extending the lever of the couple in either direction. The conceptual sketch for this megaframe system is shown in Fig. 2.

In the alternative design, a steel skeleton employing a composite system at the lower part of the structure was chosen. The primary reason that the client accepted this alternative design was that the total construction cost was greatly reduced irrespective of the increase of the concrete work. The steel walls required to maintain the stiffness of the tower in the original design were replaced by the thick RC bearing walls below the 25th floor and by the RC shear walls above that floor. (For layout of the RC walls, see Figs. 3 and 4.) Box girders 8 m (26 ft) deep and 1 m (3 ft) wide placed at mechanical floors were changed to heavy trusses for simplicity of construction.

The following remarks describe advantages of the alternative design.

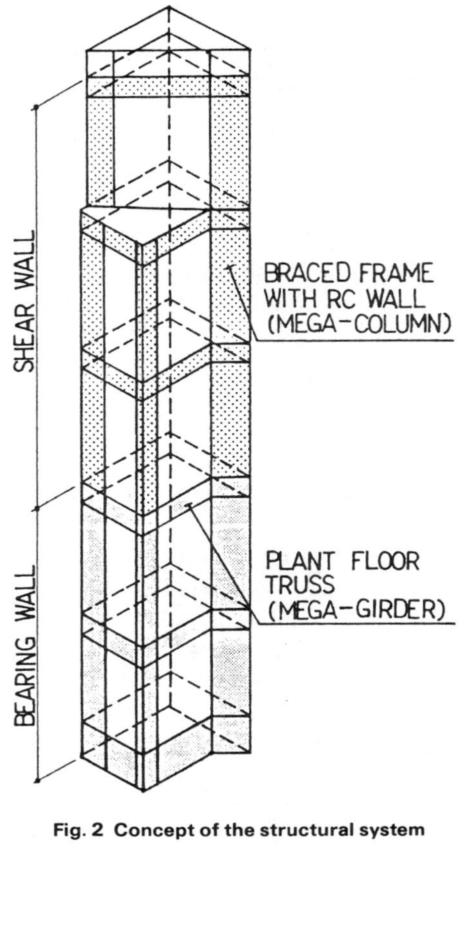

Fig. 2 Concept of the structural system

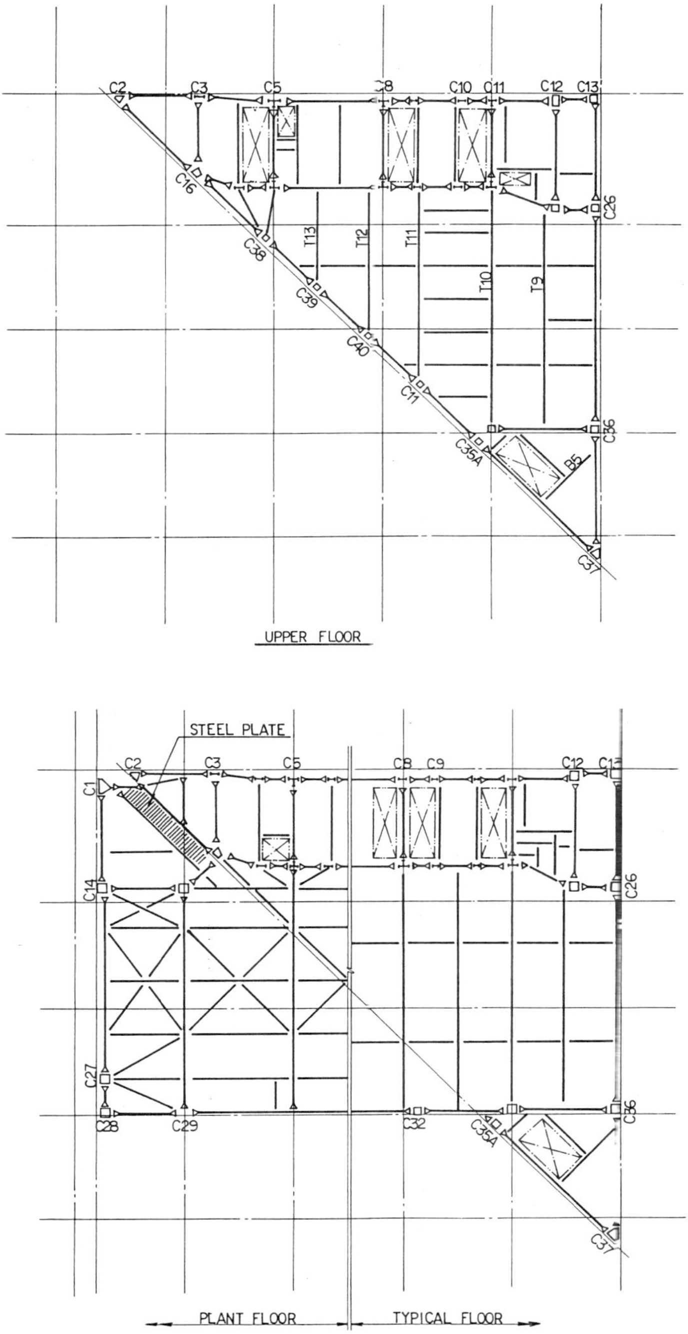

Fig. 3 Typical framing plans

Stiffness

Adequate stiffness of the tower is maintained by placing RC walls as composite elements, reducing the steel weight by about 40%. Choice of compact column sections and simple truss details also contributed to this reduction. In a megaframe system, the stiffness of megacolumns, which are composed of steel columns, bracings, girders, and RC walls, is very important to control the total stiffness. RC walls are quite adequate to combine these elements and to control the stiffness of megacolumns by changing wall thicknesses. Increasing the stiffness of the structure only by using steel elements such as bracings and steel plate walls is less efficient from a structural and economical point of view.

Overturning moment

Because the concrete walls work as counter weights against the overturning moment, uplift forces on the peripheral columns caused by wind loads expected during and after the construction was diminished. As a result, anchor bolts to the foundation and the size of weldment for the column field joints were much reduced.

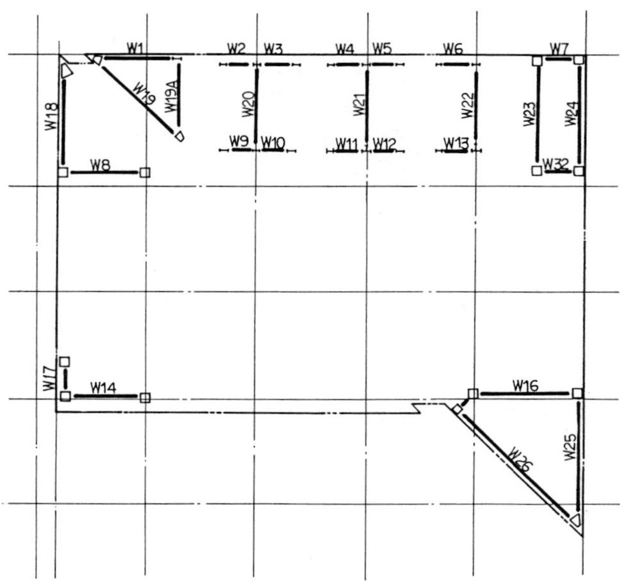

Fig. 4 Layout of structural walls

Construction speed

Since the steel skeleton was designed strong enough to withstand the dead load and wind load during the construction time, the speed of construction of this composite structure remains the same as other orthodox steel buildings. The erection of the tower was completed in 300 days and on schedule.

Wall

All of the RC walls are required not only structurally but also architecturally. RC walls function as partitions, fire walls, and sound insulations.

Weight of the tower

As the RC walls are thick only at the lower part of the tower, the total weight of the tower is close to that of a pure steel structure, which would result in reducing the required beam capacity of the foundation.

Figure 5 shows the connection details of thick RC walls and steel columns. The shear studs welded to the steel column flanges transfer the force between steel columns and RC walls. In addition to this, the steel girders and bracing built into the RC walls help in transferring forces as if they were shear studs.

Two concrete floor slabs at and just above mechanical floors are designed to be strong enough to integrate a megaframe structure as horizontal diaphragms.

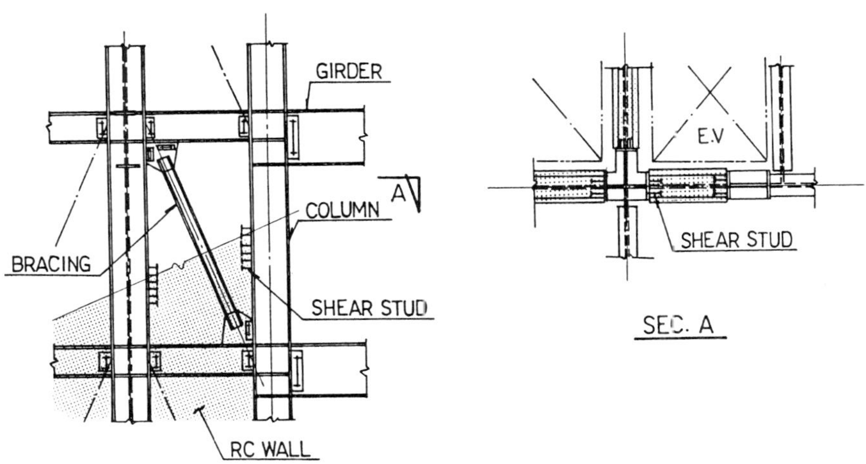

Fig. 5 Details around structural walls

Some parts of these floors are relatively highly stressed and reinforced with steel plates and bracing. The bracing elements at these floors also work as diaphragms during the steel erection. Figures 6 and 7 show the details of a mechanical floor truss. Part "A" works as a panel zone in a megaframe and is reinforced with steel diagonals and RC walls. Shear studs are placed on the two truss chords, top and bottom. The 20 m (65 ft) long floor trusses give rise to a wide column-free office space of 20 × 40 m (65 × 130 ft). The 950 mm (37 in.) deep floor truss is designed as a composite beam with 150 mm (6 in.) thick by 4320 mm (170 in.) wide slab.

ANALYSIS

The three-dimensional structural analysis of the tower of OUB Centre was conducted using the sophisticated computer programs, SAP VI, developed at the University of Southern California. To grasp the total structural behavior of such an asymmetric high-rise building, three-dimensional analysis is indispensable today. The number of modeled nodes are about 4500, and the number of elements are about 20,000. For the analytic model, see Fig. 8.

In the coding work of this large scale problem, the pre-processor and post-processor programs written in-house were very helpful to reduce errors

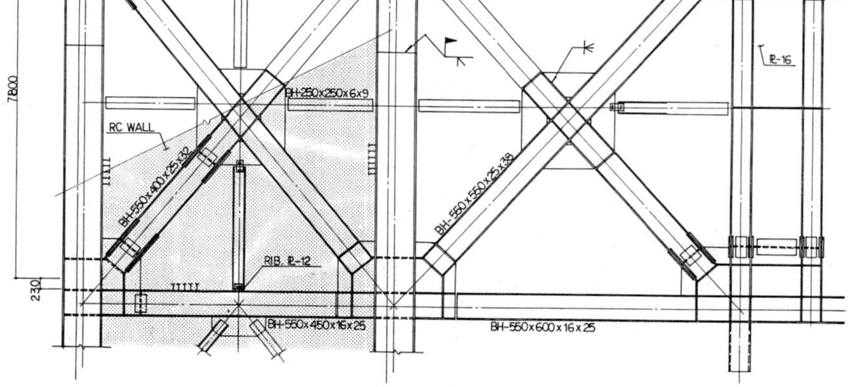

Fig. 6 Details of a plant floor truss

and to improve the design efficiency. The former program has a function of auto-generating node and element data. The latter helps to investigate the result of analysis in a shorter time by supplying visual information such as working force diagrams and deformation patterns on a display or in print out.

The sizes of columns and girders chosen based on the calculated working force by the tributary area method were checked against several combina-

Fig. 7 Outrigger trusses

tions of loadings using the in-house program. Through the several trials, the final member sizes were determined.

SPECIAL PROBLEMS

Occupant Comfort

A dynamic analysis was conducted using a five-mass structural system with 15 degrees of freedom. The stiffness properties and rotational properties of the complete structure were calculated referring to the result of the static structural analysis. The result of the calculation is as follows:

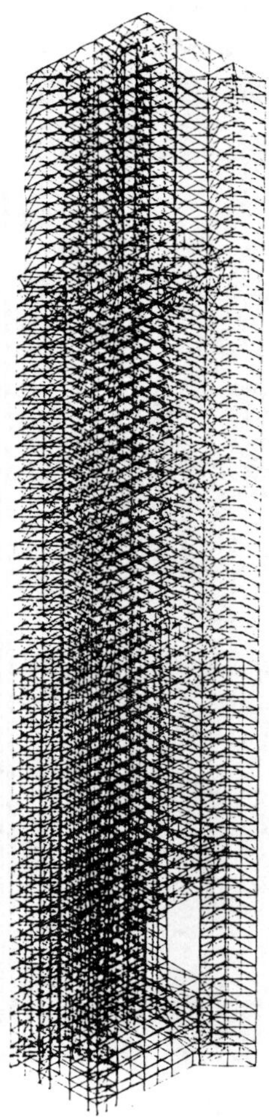

Fig. 8 Plot out of a three-dimensional analytic model

Dominant modes

mode	frequency	natural period	direction
1	0.1378	7.256	y-y
2	0.1654	6.045	x-x
3	0.2069	4.833	rotation

Based on this result, the standard deviation of the horizontal acceleration of the top occupied floor (62F) during the wind-induced vibration is calculated to check occupant comfort (Fig. 9). The damping ratio assumed is 2% in each direction. The calculation result is as follows:

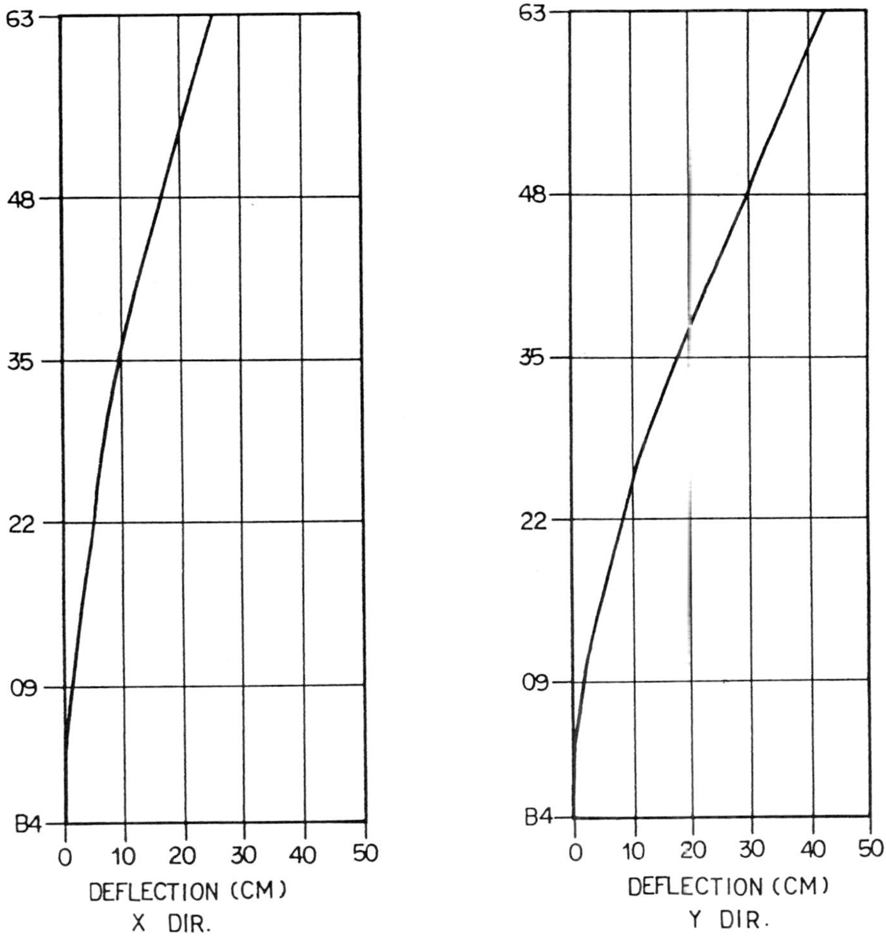

Fig. 9 Deflection of the tower against wind loadings

direction	acceleration
y-y	5.7 mg
x-x	5.4 mg

Trapezoidal box column

Built-up H and box sections were selected for the columns. The unique architectural floor plan, which is composed of two different size triangles, required the development of trapezoidal box columns at the vertices of these triangles. For details, see Fig. 10. Fabrication procedure of these sections is quite similar to that of ordinary box sections. Seam welds of skin plates were carried out by a submerged arc welding with tandem electrodes ensuring stable quality with high efficiency. Butt welding of diaphragm plates to skin plates (which could not be done before placing a closing skin plate) was performed by consumable-nozzle electroslag welding.

Level floor

It is almost impossible to attain a theoretical level floor in a high-rise building because of the differential shortening of vertical elements, the foundation movement, and the fabrication tolerances. The tower of OUB Centre has a composite system for the lower 40% of the building and a steel structural system at the top. These RC walls were distributed, and this arrangement minimizes the differential shortening among columns. To attain a level floor in the acceptable range, steel column segments were fabricated some millimeters longer depending on the location. To adjust the column elevations, which were different from the calculation, shim plates were placed at the field joints of the columns.

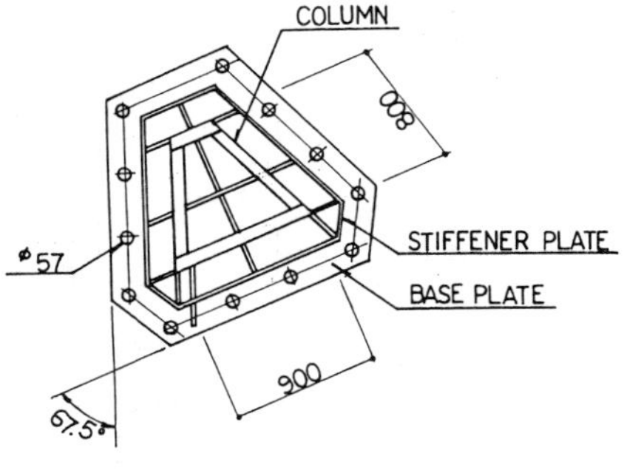

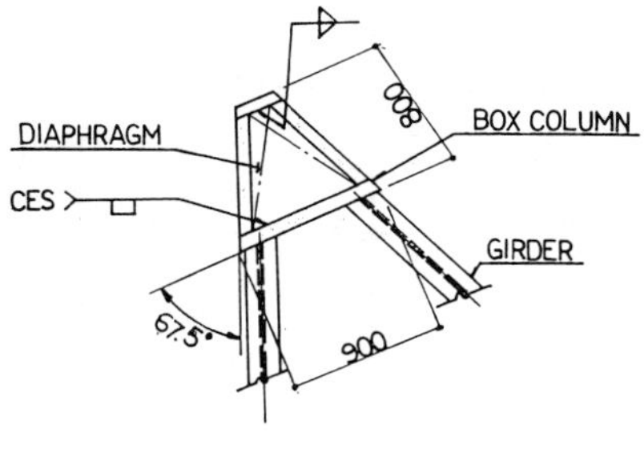

Fig. 10 Detail of a trapezoidal box column

Structural safety during construction

It was requested and confirmed through the check analysis of the several models assumed for typical construction sequences that 26 stories of structural steel frame were allowed to proceed above the completed concrete walls and slabs. Horizontal rigidity of the steel frames was maintained by the floor deck plates installed and the temporary steel bracing.

The Skyscraper's Base: Architecture, Landscape and Use in the Dallas Arts District

Stuart O. Dawson

Too much attention is given to the skyline, to net-to-gross, and to unique fenestration. There is too much "leadership" by the construction manager, too much attention to security and travertine, too much affection for the automobile, and much too *little* concern for people on the street! If we believe in the future of our cities, we should understand what works in our great cities and be willing to eliminate what doesn't work. "Landscape," no matter how good or how much, can soften but not solve what's bad. Cities need "great streets." Great streets can only exist through a complicated joint venture of people and policy. In addition to good urban design and landscape, the developer and the architect must either volunteer or be required to say "hello" to the *public*—the users of the street!

Holly Whyte warns, "We are losing our former good streets to blank walls" (Whyte, 1983) (Fig. 1). Pedestrians enter self-contained shopping arcades from underground parking, tunnels, and overhead skyways (Fig. 2). Shopping center development standards are being reapplied in our cities, which is all right for the suburbs, but unhealthy for the downtown. The city should be different from the suburban mall, yet most efforts are being made to make

Fig. 1 "We are losing our former good streets to blank walls."—*William H. Whyte*

Fig. 2 Overhead skyways keep people off the streets

them the same. The architectural and developer focus of the late 1950s and 1960s is haunting us today.

The Dallas Arts District is an excellent example of a current attempt to create a great street—a people place for the citizens of Dallas.

CONTEXT

The Dallas Arts District, located at the edge of downtown Dallas, was proposed in the 1970s by Carr Lynch Associates of Cambridge, who recommended that the City of Dallas create a concentration of arts activities near the central business district (Fig. 3). In 1982, the land owners within the emerging 60-acre Arts District decided to create a consortium to plan and manage the area. This was a major step for Dallas, a city not known for holistic planning and design. Land owners included the City of Dallas, the Art Museum, the Symphony, Tri-Land Development, Trammell Crow Company, Tishman Realty, Young Gentek Company, and Luedtke, Aldredge & Pendleton. They decided that a coordinator without conflict of interest should head up the effort with an unbiased and objective point of view. Dr. Philip O'B.

Fig. 3 The Dallas Arts District (outlined) is located on the edge of downtown

Montgomery, Jr. was chosen for this task. It was then decided to conduct a design competition for the district. While first prize money was not significant, the selected consultant was assured a continuing role.

COMPETITION

Sasaki Associates, with its extensive experience in the planning and design of urban environments, decided to enter the competition. Recognizing that, along with esthetic considerations, the potential client would respond best to a strong market and economic plan, the firm asked Halcyon, Inc. of Hartford, Connecticut to be part of the team.

From the start it was clear that Dallas needed a "great street" and that Flora Street, which was then a rather bleak stretch of pavement through the center of the District (Fig. 4), could be turned into a magnificent urban space (Fig. 5). Trees, shrubs, and paving alone could not create a great street, however; an energetic retail program at the edges of the right-of-way would enhance the existing institutional uses within the district and would help assure a lively

Fig. 4 Construction photo shows Flora Street beginning to take form in front of the new LTV Center

and intimate street. To give a distinctive pedestrian-scale identity to the street, which is some 610 m (2000 ft) long and 30 m (100 ft) wide, retail activity was grouped around three themes: Museum Crossing, Concert Lights, and Fountain Plaza (Fig. 6).

In their presentation to the consortium, Sasaki Associates and Halycon stressed good design with a strong marketing foundation. The competition plan proposed that major works of art, both permanent and temporary, be placed throughout the district; that a variety of regularly scheduled and impromptu entertainment be programmed; and that the life of the district should extend well into the evening hours with programmed activity, transparent shops, restaurants, and active institutional festivities. The "Sasaki Plan" was selected.

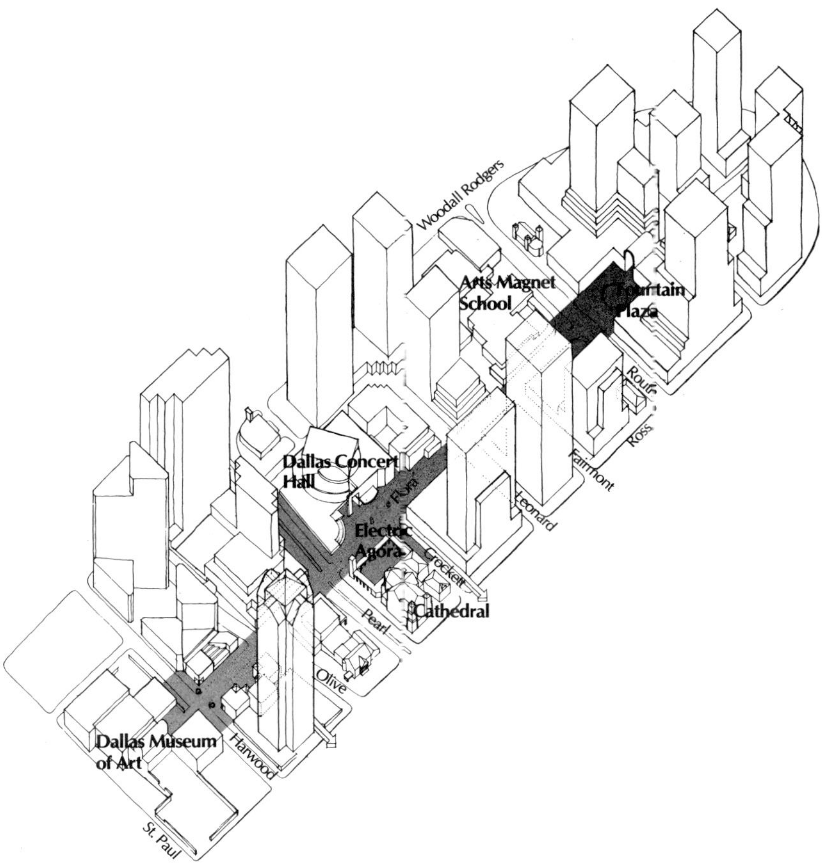

Fig. 5 Flora Street is the unifying element of the Arts District

THE PLAN

The final plan was developed based upon the ideas generated from the competition. The plan (Fig. 7) deals with specific issues such as a pedestrian and transit network, vehicular circulation and underground parking, landscape character and night lighting, and a cost-benefit calculation. The final plan made very effort to be as financially detailed as it was design detailed. It is estimated that $200 million of public money is needed for streets, garages, and institutions, which in turn will generate almost $2 billion of private investment over a 20-year development period.

Three legal documents were created for presentation to City Council: the Dallas Arts District Plan Development District Ordinance; the Dallas Arts District Cost Sharing for Streetscape Improvements; and the Dallas Arts District Plan for Maintenance and Management. The City Council would

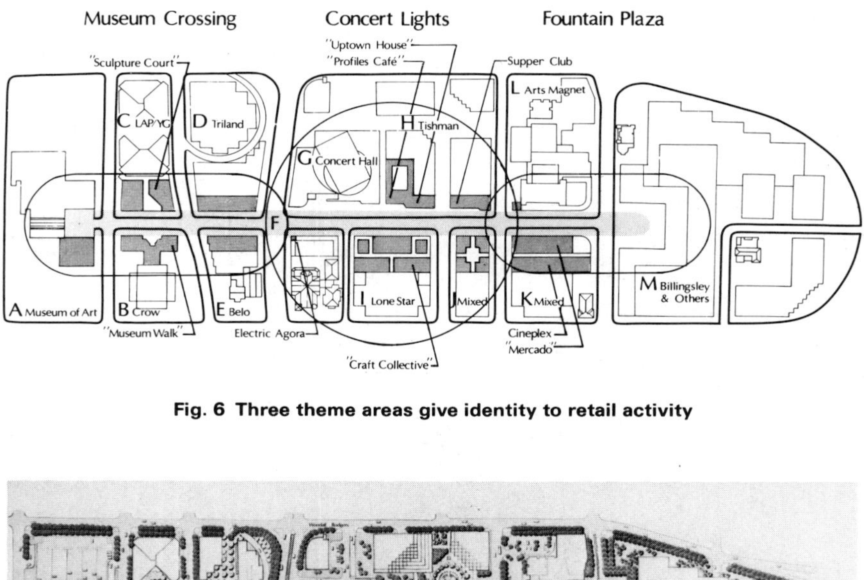

Fig. 6 Three theme areas give identity to retail activity

Fig. 7 Flora Street is planted with triple rows of trees; side streets are also landscaped to give the district a special "green" character

not approve an ordinance until issues of cost sharing, immediate and long-term maintenance, and management were resolved. The ordinance was passed in February, 1983.

THE DESIGN

Much of the legal ordinance is design-oriented. For instance, only a maximum of 25% of any block may be used for office space. And up to 50% of any block can be used for park space (Fig. 8). Three rows of trees are required on either side of Flora Street and graphics throughout the district must fit within the cohesive whole. The edge of the district will consist of a double row of trees around the entire 60 acres, with portals or gateways some 8 m (25 ft) high announcing and celebrating the Arts District.

In cross-section the 30-m (100-ft) Flora Street right-of-way may be flanked by a maximum of two 15-m (50-ft) high facades; those facades must be at least 50% glass (Fig. 9). The setback line then ascends at 45° to a horizontal dimension 15 m (50 ft) inside the 30-m right-of-way. Above that point, the building can be as high as the FAR formula allows. This terraced setback

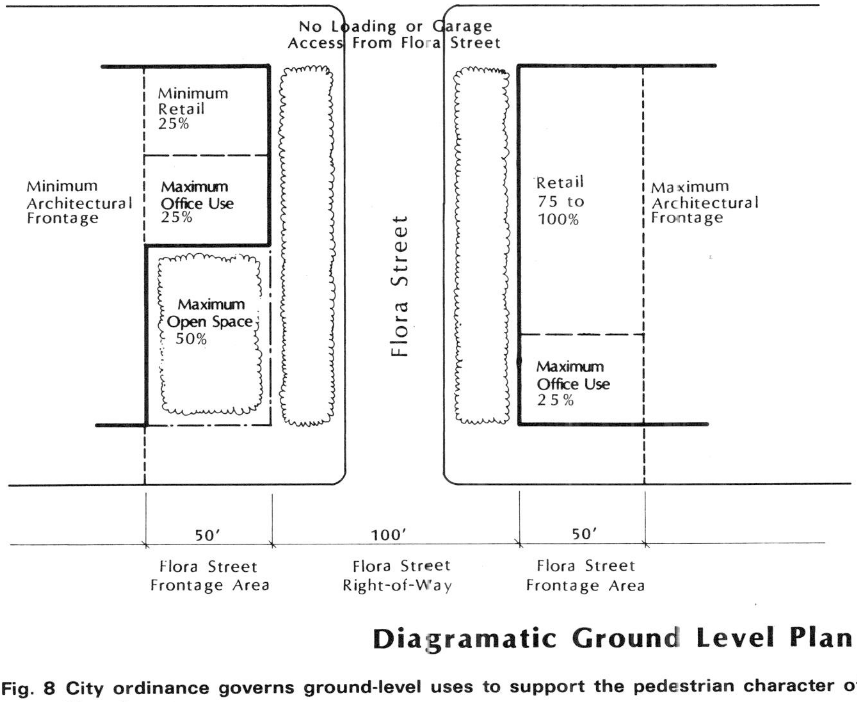

Fig. 8 City ordinance governs ground-level uses to support the pedestrian character of Flora Street

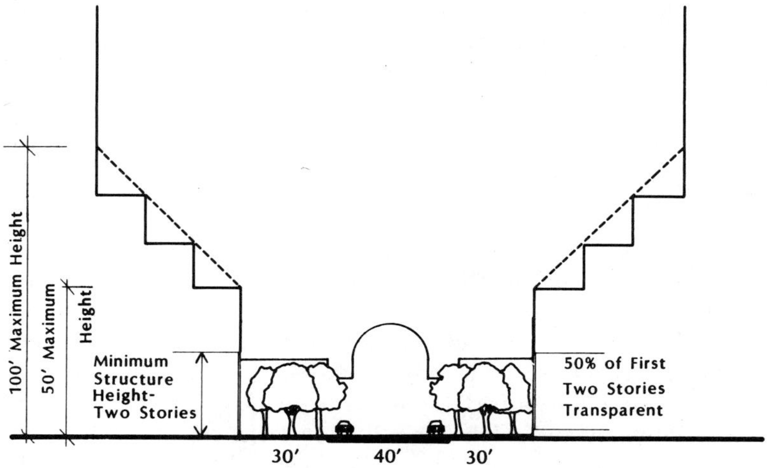

Fig. 9 Height, setback, and transparency of buildings on Flora Street are also governed by ordinance

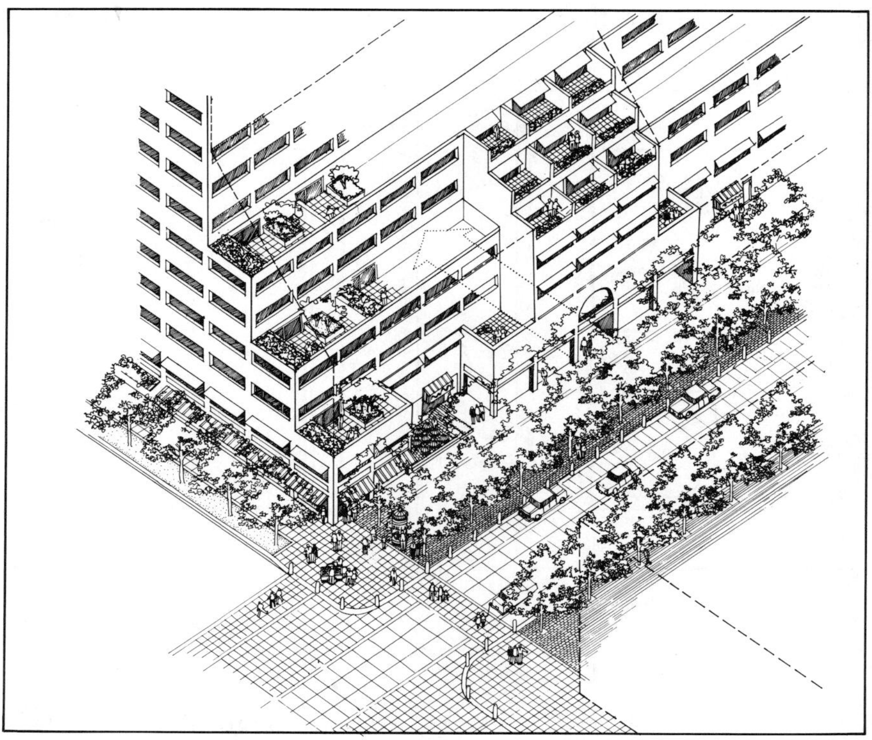

Fig. 10 Design guidelines focus on building details such as materials, facade crenelation, terraces, lighting, and graphics

provides the added benefits of greater sun penetration (up to 15 m more sunlight on Flora Street) and wind dampening protection against building facade wind shear.

No curb cuts are allowed and uses such as parking garages, bus stations, banks, and savings and loan institutions are not permitted.

Design guidelines are included within the ordinance and focus on such details as facade crenelation, retail one-half level below and one-half level above street level, awnings, canopies, terraces, regional materials such as stone and earth-tone stucco concrete, graphics, lighting, and extended retail around corners on adjacent streets (Fig. 10).

The new Concert Hall designed by I. M. Pei is under construction and four new buildings are on the boards. The City of Dallas was recently able to purchase the 8-acre "Borden tract" in the heart of the district (Fig. 11). This purchase will enable the City to provide uses within the district that the private sector would not be able to finance, such as a new opera house, gallery space, studios, practice rooms, below-market rate housing, arts-related retail, outdoor performance areas, and special gardens.

FLORA STREET

The Flora Street design is rather simple (Fig. 12). There are six rows of trees; light fixtures which are also flagpoles and a location for informational, directional, and regulatory graphics; and street furniture including bollards, fountains, kiosks, drinking fountains, pots, benches, trash receptacles, and telephone kiosks. The materials are elegant and simple in nature. To allow easy flow of pedestrians back and forth across the street, no curb has been built so that Flora Street, which is paved in brick, is essentially a pedestrian

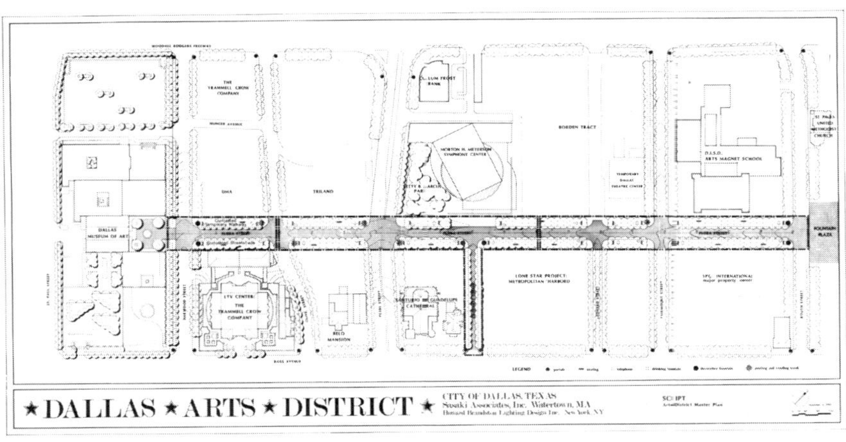

Fig. 11 Land ownership in the Arts District represents public and private interests

plaza that is open to vehicular traffic but can be closed in part or in total at will. Since it does not lead anywhere and provides no access to parking facilities, Flora Street becomes the ultimate "great street" that Dallas so sorely needs.

In the Arts District as in other urban settings, the landscape should be

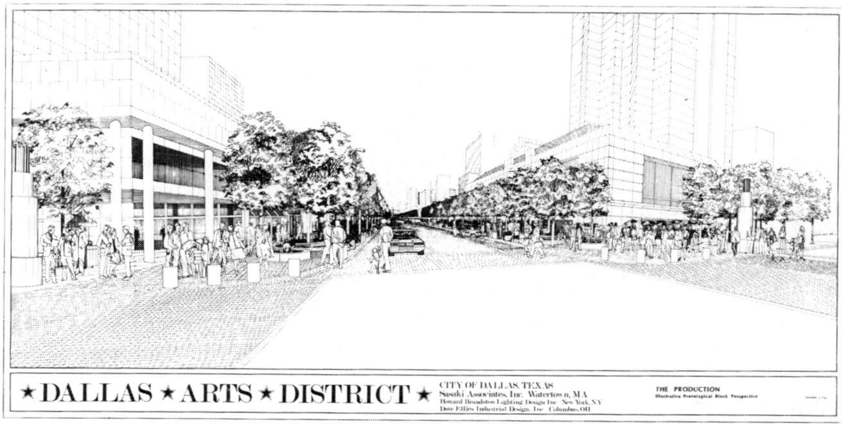

Fig. 12 Artist's rendering of the view of Flora Street from the Art Museum

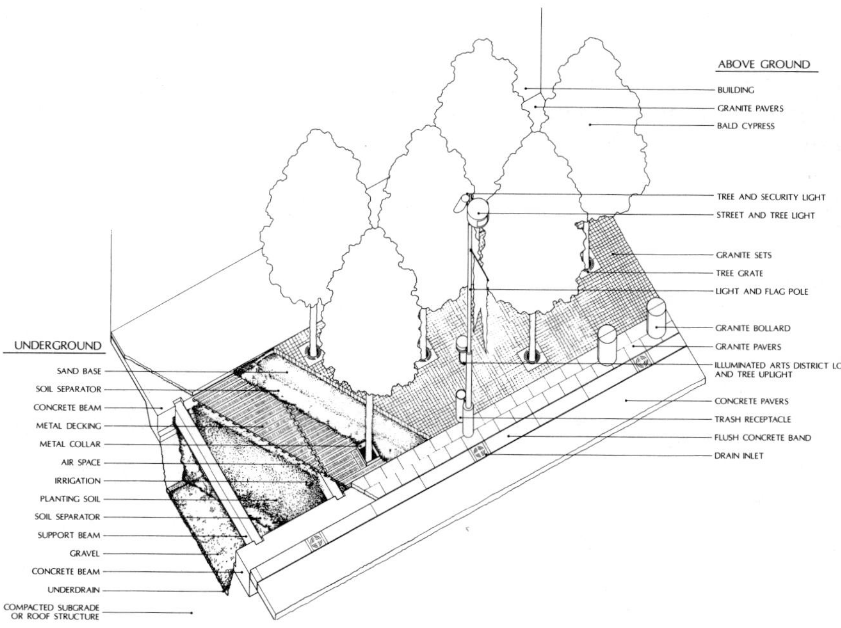

Fig. 13 Planting requirements assure healthy growing conditions for street trees

frosting on the cake. The real issues regarding urban vitality and beauty relate to ground-level uses and building facade transparency. Trees in the city are indisputably important, but all too often they are planted in poor soil conditions or in poor climatic conditions. Hybrid species, which have little chance of survival even in the native countryside, are often used.

To assure the success of the trees in the district, very rigid planting requirements were developed, including a continuous tree pit with at least 2 m (6 ft) of soil, provisions for drainage and irrigation, and a structure for the planting bed that supports a system of galvanized joists covered with pavers set in sand (Fig. 13). As a result, no pedestrian traffic touches the root zone and the air space of almost 305 mm (12 in.) between the earth and the paving surface creates an insulation layer, allowing the roots to stay cooler in the summer and warmer in the winter. This construction avoids one of the biggest problems for urban street trees, which is not water or air so much as it is temperature and pedestrian compaction of the root zone. The paving system also allows easy addition or removal of street furniture or trees.

Pink and green granite pavers used for walking surfaces are being applied directly to a concrete slab for stability, and precast concrete unit pavers are being used in the street areas with broad cast-in-place concrete headers for lane divisions and stop bands (Fig. 14). Pedestrian crossings are identified with green granite pavers. All site metal is white to reflect the architectural metals in both the new Art Museum and the new Concert Hall.

Fig. 14 Granite paving distinguishes pedestrian and vehicular areas along Flora Street

OPERATIONS

The Arts District is well underway and moving forward at a rapid pace, although completion may take up to 20 years. Three associations have been formed to manage the district: a Management Association, which is responsible for operating the District; the Arts District Foundation, which is responsible for enriching the District; and the Arts District Friends, whose purpose is public support including funding and volunteers.

ACTIVITIES

Even though the Arts District is only 20% complete, the level of maintenance and amount of activity suggests a very successful future indeed (Fig. 15). As the result of a design competition, a new logo now appears on the portals, light poles, banners, stationery, and souvenirs. The activities that have taken place over the last year include the Mirage 500 Fair, the largest art fair in the state; folk dancing in the streets; mimes, clowns, and vendors; lunchtime concerts; Friday night jazz concerts; and the Smithsonian-sponsored "American Anthem" exhibit. At last, there is a good reason to stay downtown after 5:00 pm (Fig. 16).

While the jury is still out on the long-term success of the district, the approaches taken there are likely to work not only within the Arts District but may become the foundation for an ordinance applied to the entire 1000-acre Central Business District.

Fig. 15 Area employees and visitors enjoy the completed first block of Flora Street

A quote from Holly Whyte best sums up what the Arts District is all about and where other cities should head if they intend to provide a successful urban experience and lifestyle. "The street is the City's river of life. It is the unifying place. The way to a successful open space is a fine relationship with the street. The best public places are those in which you don't know where the street ends and the place begins" (Whyte, 1983).

REFERENCES/BIBLIOGRAPHY

Christian Science Monitor, 1983
DALLAS FINDS ITS $2.6 BILLION ARTS DISTRICT IS GOOD FOR BUSINESS, Christian Science Monitor, November 7.

Dawson, S., 1983
DALLAS ARTS DISTRICT, Urban Design International, Purchase, New York.

Horizon Magazine, 1984
DALLAS: ARTS IN THE HEART OF TEXAS, Horizon Magazine, May.

Urban Land Institute, 1985
CULTURAL FACILITIES IN MIXED-USE DISTRICTS, Urban Land Institute, Case Study 9, Dallas Arts District, Washington, DC.

Whyte, W. H., 1983
QUOTATIONS BY WILLIAM H. WHYTE, from a speech entitled "The Blank Wall," given at the "What Makes A City?" conference, April 29, 1983, sponsored by the Dallas Institute of Humanities and Culture, Dallas, Texas.

Fig. 16 A temporary canopy marks the edge of Flora Street opposite the completed first block

High-Rise Building Data Base

Since the inception of the Council, various surveys have been made concerning the location, number of stories, height, material, and use of tall buildings around the world. The first report on these surveys was published in the Council's Proceedings of the First International Conference in 1972. *Tall Building Systems and Concepts* (Volume SC) brought that information up to date with a detailed survey in 1980.

Changes to some of the data were reflected in *Developments in Tall Buildings—1983,* with further updating in *Advances in Tall Buildings* (January, 1986), *High-Rise Buildings: Recent Progress* (November, 1986), and *Tall Buildings of the World* (February, 1987). This present volume provides the newer information more recently received.

The original survey was based mainly on information collected from individuals in the major cities of the world. The main criterion for the selection of a city was generally its population. Another criterion was the availability of a Council member or other contact who might provide the needed information.

This Appendix updates Table 1. Because the data came from so many sources, complete accuracy cannot be guaranteed. Buildings change names or new ones break ground, and this information is sometimes slow in reaching the Council headquarters. In this sense the survey keeps its nature as a "living document."

Additions and corrections to the information presented here are welcomed, and should be brought to the attention of Council Headquarters at Lehigh University.

Table 1: World's Tallest Buildings. This is a list of the world's 100 tallest buildings. The information presented here includes the city in which each building is located, the year of completion or "under construction" (UC) and expected year of completion, number of stories, height in meters and feet, structural material, and use of each building.

Table 1 100 Tallest buildings in the world

Building	City	Year Comp	No. of Stories	Height Meters	Height Feet	Material	Use
Sears Tower	Chicago	1974	110	443	1454	Steel	Office
World Trade Center—North	New York	1972	110	417	1368	Steel	Office
World Trade Center—South	New York	1973	110	415	1362	Steel	Office
Empire State	New York	1931	102	381	1250	Steel	Office
Bank of China	Hong Kong	UC88	72	368	1209	Mixed	Office
Amoco	Chicago	1973	80	346	1136	Steel	Office
John Hancock	Chicago	1968	100	344	1127	Steel	Multiple
Chrysler Bldg.	New York	1930	77	319	1046	Steel	Office
Library Square Tower	Los Angeles	UC90	75	310	1017	Steel	Office
Texas Commerce Plaza	Houston	1981	75	305	1002	Mixed	Office
Allied Bank Plaza	Houston	1983	71	296	970	Steel	Office
1 Wacker Drive	Chicago	UC90	80	295	967	Concrete	Office
311 Wacker Drive	Chicago	UC90	65	292	959	Concrete	Office
Columbia Center	Seattle	1985	76	291	955	Mixed	Office
American Intl. Bldg.	New York	1931	66	290	950	Steel	Office
First Bank Tower	Toronto	1975	72	285	935	Steel	Office
40 Wall Tower	New York	1966	71	283	927	Steel	Office
Interfirst Plaza Tower	Dallas	1985	70	281	921	Mixed	Office
Citicorp Center	New York	1977	59	280	919	Steel	Multiple
Overseas Union Bank	Singapore	UC86	63	280	919	Steel	Office
Scotia Plaza	Toronto	UC88	68	275	901	Mixed	Office
Transco Tower	Houston	1983	64	275	901	Steel	Office
900 N. Michigan	Chicago	UC89	67	267	875	Mixed	Multiple
AT&T Corp. Center	Chicago	UC88	64	267	875	Mixed	Office
Mellon Bank	Philadelphia	UC88	56	265	870	Mixed	Office
Water Tower Place	Chicago	1976	74	262	859	Concrete	Multiple
United California Bank	Los Angeles	1974	62	262	858	Steel	Office
Transamerica Pyramid	San Francisco	1972	48	260	853	Steel	Office
RCA Rockefeller Center	New York	1933	70	259	850	Steel	Office
First National Bank	Chicago	1969	60	259	850	Steel	Office
U.S. Steel	Pittsburgh	1970	64	256	841	Steel	Office
One Liberty Place	Philadelphia	UC87	60	256	840	Steel	Office
Atlantic Center	Atlanta	UC88	50	250	820	Mixed	Office
Cityspire	New York	UC87	72	248	815	Mixed	Multiple
One Chase Manhattan Plaza	New York	1961	60	248	813	Steel	Office
Pan American	New York	1963	59	246	808	Steel	Office
Momentum Place	Dallas	UC87	60	244	797	Steel	Office
Rialto Center	Melbourne	UC86	70	243	794	Concrete	Office
Woolworth Bldg.	New York	1913	57	242	792	Steel	Office
1 Palac Kultury I Nauki	Warsaw	1955	42	241	790	Mixed	Office
John Hancock Tower	Boston	1973	60	241	790	Steel	Office

Building	City	Year Comp	No. of Stories	Height Meters	Height Feet	Material	Use
World Wide Plaza	New York	UC89	53	250	788	Steel	Office
M.L.C. Centre	Sydney	1976	70	240	786	Concrete	Office
Commerce Court West	Toronto	1974	52	239	784	Steel	Office
Republic Bank Center	Houston	1983	56	238	780	Steel	Office
Bank of America	San Francisco	1969	52	237	778	Steel	Office
Office Towers	Caracas	1985	60	237	778	Mixed	Office
3 First National Plaza	Chicago	1981	58	236	775	Mixed	Office
I D S Center	Minneapolis	1972	57	235	772	Mixed	Office
Norwest Center	Minneapolis	UC88	57	235	770	Mixed	Office
Singapore Treasury Bldg.	Singapore	1986	52	235	770	Mixed	Office
One Penn Plaza	New York	1972	50	234	766	Steel	Office
Korea Ins. Company	Seoul	1985	63	233	764	Steel	Office
Tun Abdul Razak	Penang	1985	61	232	761	Concrete	Office
Equitable Tower W.	New York	1985	51	230	755	Steel	Office
Maine Montparnasse	Paris	1973	64	229	751	Mixed	Office
Prudential Center	Boston	1964	52	229	750	Steel	Office
Federal Reserve Bldg.	Boston	1975	32	229	750	Steel	Office
Exxon	New York	1971	54	229	750	Steel	Office
First International Plaza	Houston	1981	55	228	748	Mixed	Office
Republic Plaza	Denver	1983	56	227	745	Mixed	Office
Morgan Bank	New York	UC88	50	227	745	Steel	Office
Security Pacific National Bank	Los Angeles	1974	55	226	743	Steel	Office
One Liberty Plaza (U.S. Steel)	New York	1972	54	226	743	Steel	Office
Ikebukuro Tower (Sunshine 60)	Tokyo	1978	60	226	742	Steel	Office
Raffles City Hotel	Singapore	1986	70	226	741	Concrete	Hotel
20 Exchange Place (Citibank)	New York	1931	55	226	741	Steel	Office
Crocker Center	Los Angeles	1983	55	225	740	Steel	Multiple
Renaissance 1	Detroit	1977	73	225	739	Concrete	Hotel
World Financial Center	New York	1985	51	225	739	Steel	Office
Toronto Dominion Bank Tower	Toronto	1967	56	224	736	Steel	Office
Two Union Square	Seattle	UC88	56	224	736	Mixed	Multiple
1600 Smith	Houston	1984	55	223	732	Steel	Office
Southeast Financial Center	Miami	1984	53	222	730	Mixed	Office
One Astor Plaza	New York	1972	54	222	730	Mixed	Office
Olympia Center	Chicago	1981	63	222	728	Concrete	Multiple
One Mellon Bank Center	Pittsburgh	1983	54	222	727	Steel	Office
Gulf Tower	Houston	1982	52	221	726	Mixed	Office
9 West 57th St.	New York	1974	50	221	725	Steel	Office
Peachtree Center Plaza Hotel	Atlanta	1975	71	220	723	Concrete	Hotel
Carlton Centre	Johannesburg	1973	50	220	722	Concrete	Office
Texas Commerce Tower	Dallas	UN87	56	219	720	Concrete	Office
Metropolitan Tower	New York	1986	66	219	716	Concrete	Multiple
One Shell Plaza	Houston	1971	50	218	714	Concrete	Office
Petro-Canada I	Calgary	1983	52	217	710	Mixed	Office
First International Bldg.	Dallas	1973	56	216	710	Steel	Office
Hopewell Centre	Hong Kong	1981	65	216	709	Concrete	Office
Shinjuku Center	Tokyo	1979	54	216	709	Steel	Office
Terminal Tower	Cleveland	1930	52	216	708	Steel	Office
Union Carbide Bldg.	New York	1960	52	215	707	Steel	Office
General Motors	New York	1968	50	214	705	Steel	Office

Building	City	Year Comp	No. of Stories	Height Meters	Height Feet	Material	Use
Petrolaos Mexicanos	Mexico City	1984	52	214	702	Steel	Office
American Fletcher Center	Indianapolis	UC89	60	214	704	Steel	Office
Metropolitan Life	New York	1909	50	213	700	Steel	Office
Atlantic Richfield Plaza A	Los Angeles	1972	52	212	699	Steel	Office
Atlantic Richfield Plaza B	Los Angeles	1972	52	212	699	Steel	Office
One Shell Square	New Orleans	1972	51	212	697	Steel	Office
500 Fifth Avenue	New York	1931	58	212	697	Steel	Office
IBM Bldg.	Chicago	1973	52	212	695	Steel	Office
Four Allen Center	Houston	1983	50	212	695	Steel	Office

Nomenclature

GLOSSARY

Accelerograph. An instrument that measures and records the time-history of ground acceleration at a point during an earthquake.

Actions. Load and imposed or constrained deformations that cause stresses and deformations in structural members.

Agora. An open space in an ancient Greek town, used as a market and general meeting place. The forum had the same function in Roman architecture.

Air-Conditioning. Artificial ventilation with air at a controlled temperature and humidity.

Alkai-aggregate reaction. Chemical reaction in mortar or concrete between alkalies (sodium and potassium) from portland cement or other sources and certain constituents of some aggregates; under certain conditions, deleterious expansion of the concrete or mortar may result.

Allowable load. The ultimate load divided by factor of safety.

Allowable stress. Maximum permissible stress used in design of members of a structure and based on a factor of safety against rupture or yielding of any type.

Allowable stress design or working stress design. A method of proportioning structures such that the computed elastic stress does not exceed a specified limiting stress.

Arcade. Strictly, a line of arches carried on columns, either free-standing or attached to a wall. The term was applied during the nineteenth century to glass-roofed shopping areas modeled on the Crystal Palace, and is now being used for passages with shops on one or both sides, irrespective of construction.

Art deco. A decorative style stimulated by the Paris Exposition International des Arts Decoratifs et Industrielles Modernes of 1925, widely used in the architecture of the 1930s, characterized by sharp angular or zigzag surface forums and ornaments.

Aspect ratio. In any rectangular configuration, the ratio of the longer dimension to the shorter dimension.

Atrium. The main courtyard of a Roman house. Later, a courtyard of a building or a group of buildings. In contemporary architecture, an open multifloor space within a building.

Augercast pile. A pile with cylindrical shape produced by augercasting method.

Beam. A structural member, usually horizontal, subjected primarily to flexure, on which the weight of a floor, partition, or other beam is carried.

Beam-and-slab floor. A reinforced concrete floor system in which the floor slab is supported by beams of reinforced concrete.

Beam-column. A beam that transmits an axial load along its longitudinal axis in addition to end moments or transverse loads, or both.

Bearing wall. A wall or partition that supports the portion of the building above it in addition to its own weight.

Block. A concrete masonry unit.

Bond. Adhesion and grip of concrete or mortar to reinforcement or to other surfaces against which it is placed; also, the arrangement of units in masonry and brickwork so that vertical joints are discontinuous.

Bond strength. Resistance to separation of mortar and concrete from reinforcing steel and other materials with which they are in contact.

Bracing. The ties and struts used for supporting and strengthening a frame.

Brittle fracture. Abrupt cleavage with little or no prior ductile deformation.

Buckling. A general term descriptive of the failure of a compression member or compression flange of a member due to loss of its original straight or nearly straight shape.

Building automation. Computerized systems for automatic operation of building.

Building codes. Regulations that establish standards for construction and housing, the former for assuring structural strength, reasonable safety from fire, and proper plumbing, electrical, and heating installations; the latter prescribing the minimum conditions under which a building, or parts of it, may be lawfully occupied as a dwelling, and usually include regulations to prevent overcrowding, for basic sanitary facilities, light and ventilation, maintenance, and heating.

Building services. Mechanical and electrical systems, vertical and horizontal transportation, plumbing and fire protection.

Building standard. A document defining minimum standards for design.

Building tiedown. A pile underneath the building foundation and connected to the basement mat, used to resist hydrostatic uplift pressure on a building.

Bundled tube. A structural system in which structural framed tubes are arranged or bundled together in such a way that common walls of contiguous tubes are combined into single walls, thereby forcing total compatibility of stresses at the interface of such contiguous tubes. In a bundled tube, individual tube elements may be terminated at any appropriate level.

CADD system. Computer aided designing and drafting system

Caisson. A reinforced concrete pier usually excavated by auger that may have a shell to laterally support the soil (a) a deep foundation member used to deliver structural loads to the founding material, installed by excavating, (b) a large box used to advance an excavation while men work within it.

Camber. A slight, usually upward, curvature of structural member or form to improve appearance or to compensate for anticipated deflections.

Campanile. A tall, detached bell tower.

Cantilever. In construction, a portion of a floor or deck extending beyond the vertical column, an overhang.

Cast iron. Iron with a total carbon content between 1.8% and 4.5%.

Cast-in-place. Mortar or concrete that is deposited in the place where it is required to harden as part of the structure, as opposed to precast concrete.

Chilled water. In air conditioning, refrigerated water that is piped to coils for air cooling.

Claustrophobia. An abnormal fear of being in an enclosed place.

Code. Building code, a legal document providing design criteria for buildings in a particular jurisdiction.

Coefficient of variation. The ratio of the standard deviation to the mean of a random variable.

Cofferdam. A watertight enclosure placed under water and pumped dry to allow for construction or repairs.

Colonial architecture. A style derived from British Georgian architecture. It is applied to American and Australian architecture of the eighteenth and early nineteenth centuries that follows the classical tradition, and to modern revivals.

Column. A structural member whose prime function is to carry loads parallel to its longitudinal axis.

Column, long, slender. A column whose axial load capacity is reduced because of its slenderness, due to moments induced by deflection of the column.

Column, short. A column whose axial load capacity need not be reduced because of its slenderness.

Column curve. A curve expressing the relationship between axial column strength and slenderness ratio.

Composite construction. A type of construction made up of different materials (e.g., concrete and structural steel) or of members produced by different methods (e.g., cast-in-place concrete and precast concrete).

Compressive strength. The measured maximum resistance of a concrete or mortar specimen to axial loading; expressed in stress, or the specified resistance used in design calculations; in United States expressed in lb per sq in. (psi) and designated f'_c.

Concentrically-braced frame. A frame in which the resistance to lateral load or frame instability is provided by diagonal, *K*, or other auxiliary system of bracing.

Concrete. A composite material that consists essentially of a binding medium within which are embedded particles or fragments of aggregate; in normal portland cement concrete, the binder is a mixture of portland cement and water and the aggregate is sand and gravel.

Concrete, lightweight. Structural concrete made with lightweight aggregate; the unit density usually is in the range of 1440 kg/m^3 to 1850 kg/m^3 (90 lb/ft^3 to 115 lb/ft^3).

Concrete, normal-weight. Concrete having a unit density of approximately 2400 kg/m^3 (150 lb/ft^3) made with aggregates of normal weight.

Condominium. Individual ownership in multiunit housing.

Confinement. Concrete contained by various structural elements or by closely spaced special transverse reinforcement that restrains the concrete in directions perpendicular to the applied stress.

Construction joint. The surface where two successive placements of concrete meet.

Construction management. Method of placing construction supervision in the hands of a qualified manager.

Context. A circumstance that contains facts that reflect the current state of a problem.

Continuous medium. An analogy for a regular system of discrete connecting members in a structure. The analogy consists of a continuous connecting medium whose bending and shear properties are those of the discrete members distributed uniformly over the system.

Core. An assembly or group of shear walls, usually joined together to form an open or partially closed box structure, often used as a bracing element for a building structure. The core often includes elevator, stairs, mechanical shaft and toilets.

Core space ratio. Ratio of core space to office space.

Cornice. In classical architecture, the projecting section of an entablature.

Coupled frame-shear wall structure. A shear wall and frame structure in which the shear walls and the frame are coupled together for analysis purposes to determine the stiffness that the coupled structure provides against lateral movement under lateral loads.

Coupled shear-walls. Shear walls that are connected together by bending resistant members.

Coupled walls. Shear walls that are connected together by bending resistant members.

Coupling beams. Beams connecting two sections of a shear wall.

Creep. The slow time-dependent change in dimensions of concrete under a sustained load, primarily in the direction in which the load acts. It is a dimensionless quantity having units of strain.

Critical Path Method. A project networking and planning technique with the deterministic mode.

Cul-de-sac. A dead-end street, blind alley.

Cupola. A small dome, rising above a roof.

Curtain wall. Outside covering, usually of glass, on a building supported by steel or concrete structure.

Curvature. The angle change per unit length of a loaded structural member.

Cyclic behavior. Structural response to repeated, reversed, or alternating load.

Damper. A device that dissipates vibrational energy, thus decreasing the amplitude of motion of a vibrating structure.

Damping. The resistance of a structure to displacement from an externally applied disturbance that is subsequently removed. The resistance may be provided by internal frictional resistance of the molecules of the structure, the drag effects of the surrounding medium, or contact with or connection to an adjacent structure.

Damping coefficient. Coefficient by which velocity is multiplied to obtain the corresponding damping force.

Dead load. The actual weight of the structural elements. (This is a gravity load.)

Decentralization. The redistribution of population and industry from urban areas to outlying areas.

Decision rules. Statements that define actions to be taken when a given sequence of conditions are found to be true.

Deep beam. A beam in which the elastic distribution of stresses is nonlinear and in which there are significant compressive forces between the loads and reactions.

Deflection. The horizontal or vertical movement of a point on a structure due to loads, creep, shrinkage, temperature changes, or settlements.

Density. A measure related to the extent of the volume, area, or length of a Cartesian spatial system, Can be either avolumetric, planar or areal, or linear density, respectively.

Depreciation. A percentage of the value of an improvement deducted each year for wear and tear.

Design spectrum. A diagram used for determining the design force for seismic loading. It is usually derived from the average of the response spectra of several recorded earthquakes, but reduced to account for the effective structural damping resulting from inelastic member behavior.

Development length. The length of reinforcement embedment required to develop by bond the design strength of the reinforcement at a critical section.

Diagonal tension crack. Crack in reinforced concrete member caused primarily by a shear force.

Diaphragm. A thin platelike structural element, whose in-plane rigidity may be used to stiffen the structure; for example, a floor slab that maintains the cross-sectional shape of the building.

Drift. Lateral displacement due to lateral force.

Drift index. Ratio of the height of the building to the horizontal deflection of the top of the building.

Drift limit. The maximum possible drift, usually expressed in terms of a drift index.

Drying shrinkage. The shrinkage that takes place due to the loss of water in hardened cement paste.

Ductility. That property of a material by virtue of which the material may undergo large permanent deformation without rupture.

Ductility factor. The ratio of the total deformation of a frame to the elastic limit deformation of the frame.

Durability of concrete. Ability of concrete to resist weathering action, chemical attack, abrasion, or any other process of deterioration.

Dynamic knowledge. Knowledge in a domain is extensive and exists in many disciplines.

Dynamic stabilization. The process of preventing the buckling or excessive displacements of a structure when repeated, reversed, or alternating loads are applied.

Early occupancy. The lower part of a building occupied by its tenants while the upper part of the building is still under construction.

Eccentrically-braced frame. Frame in which the center lines of braces are offset from the points of intersection of the center lines of beams and columns.

Eclectic. Choosing what appears to be the best from diverse sources, systems or styles.

Ecological corridor. Large area on the ground floor to accommodate shops, offices, exhibition areas and growing transportation needs.

Economizer cycle. Use of outside air for cooling in winter.

Edge beam. A stiffening beam at the edge of a slab.

Effective length. The equivalent length of a member which, in the Euler formula for a hinged end column, results in the same elastic critical loads as for the member or other compression element under consideration at its theoretical critical load.

Elastic analysis. A method of analysis that satisfies equilibrium and compatibility based on a linear stress-strain relationship.

Elastic deformation. The change in dimension of concrete occurring instantaneously at the time of load application.

Elastic model. A model made of a material different from that of the structure, and therefore capable of being used only within the elastic range.

Electroslag. Welding by feeding electrode weld wire into an enclosed chamber.

Electro-thermal cooling. Method of providing local cooling by direct use of electrical energy.

Elevator core. The hoistway in which elevators move.

Elevatoring. Provision of vertical transportation system.

Energy dissipation. The process of reducing kinetic energy by converting it into other forms of energy (e.g., work done by friction, plastic strain energy).

Environmental loads. Loads on a structure due to wind, snow, earthquake, or temperature.

Equivalent frame. A single-bay substitute frame used to replace one or more stories of a multibay, multistory building for the purpose of simplifying analysis.

Esthetic testing. A computer methodology that allows one to view a project from different angles in three dimensional space.

Ettringite. A mineral, high sulfate calcium sulfoaluminate; occurring naturally or formed by sulfate attack on mortar and concrete.

Expansion joint. A separation between adjoining parts of a concrete structure that is provided to allow small relative movements, such as those caused by thermal changes, to occur independently.

Expert system. A knowledge-based system (interactive computer program) that is particularly oriented to incorporate judgement, experience, rules of thumb, intuition, and other expertise to provide knowledgeable advice about a variety of tasks. (See Knowledge-based system.)

Explanation module. An knowledge-based system function that provides explanations of the inferences.

Express elevators. Elevators that operate non-stop to sky lobbies or observation floors.

Facade. The face, and especially the principal elevation of a building.

Facility management. Activities concerned with the design, construction, maintenance and management of the physical environment as it relates to people and work processes.

Facility planning. A plan developed as a facility specification that meets an organization's strategic plan and maintains a high level of productivity.

Fact. A truth known by actual experience or observation (as used in knowledge-based or expert systems).

Factor of safety. The ratio of the ultimate strength (or yield point) of a material to the working stress assumed in design (stress factor of safety); or the ratio of the ultimate load, moment, or shear of a structural member to the working load, moment, or shear respectively, assumed in design (load factor of safety).

Factored loads. The specified, working, or characteristic loads multiplied by appropriate load factors.

Failure. A condition where a limit state is reached. This may or may not involve collapse or other catastrophic occurrences.

Fast track. A phasing technique between design and construction whereby the construction starts before design is complete.

Fastest-mile wind speed. Highest wind velocity in one day, measured as the average of the velocities recorded during the time it takes a horizontal column of air one mile long to pass the anemometer.

Fenestration. The windows and openings in a building envelope.

Finial. A decorative detail at the uppermost extremity of a pinnacle or gable, or an ornamental cap for a spire.

Finite element. A single element in an analogy developed for stress analysis, in which a continuous structure is subdivided into a number of discrete linear, two-dimensional or three-dimensional segments.

Finite element model. Mathematical approximation of a structure obtained by discretizing it into a collection of geometric subdomains, each of which exhibits that same load-deformation characteristics as the material comprising the structure according to a presumed form for the displacement field.

First-order analysis. An analysis in which equilibrium is formulated on the undeflected structure and which, as a result, ignores the interaction of vertical loads and lateral deflections.

Flat plate. A flat slab without column capitals or drop panels (See Flat slab).

Flat slab. A concrete slab reinforced in two or more directions, generally without beams or girders to transfer the loads to supporting members.

Flexural crack. Crack in a reinforced concrete member caused by bending of the member.

Flexural-torsional buckling. A mode of column buckling where both twisting and transverse displacements of cross-sections occur.

Floor area ratio. A specified ratio of permissible total floor space to lot area, in which the inducement to reduce lot coverage is an important component. The basic ratio is frequently modified by providing "bonus" or "premium" floor space for such aspects as arcades, setbacks, and plazas.

Floor populations. Number of potential elevator passengers per floor.

Fly ash. The finely divided residue resulting from the combustion of ground or powdered coal and which is transported from the firebox through the boiler by flue gases.

Flying buttress. A type of projecting structure that is built against a wall to resist horizontal reaction and is suspended in the air.

Flying form. Large and unitized sections of forming systems, mechanically lifted and moved through the air; frequently include supporting trusses, beams, or scaffolding units.

Forcing function. A function that describes the spatial and time variation of the forces acting on a structure.

Formwork. The total system of support for freshly placed concrete, including the mold or sheathing that contacts the concrete, as well as all supporting members, hardware, and necessary bracing. Falsework and Shuttering are also used with essentially the same meaning.

Foundation. The material or materials through which the load of a structure is transmitted to the earth.

Frame. An assembly of structural members. Most frames are rectangular, that is they consist of horizontal and vertical members.

Freeze-thaw cycle. Frost action that causes deterioration of hardened concrete, depending not only on characteristics of the concrete but also on specific environmental conditions. Cracking, spalling, and scaling are common forms of frost damage.

Fresco. Painting on a wet lime plaster with pigments mixed with water. (The term "fresco painting" is sometimes used incorrectly for painting on dry plastered walls and ceilings.)

Fundamental period of vibration. Time necessary to complete one cycle of motion.

Gang form. Prefabricated concrete form panels joined to make a much larger unit, up to 30 × 50 feet for convenience in erecting, stripping and reusing; usually braced with walers, strongbacks or special lifting hardware.

Girder. A main horizontal supporting member or beam in a structure.

Gothic architecture. A style characterized structurally by pointed arches, ribbed vaults, and flying buttresses. It originated in France in the late twelfth century.

Hammering. See Pounding.

Hardpan. An extremely dense hard layer of soil, boulder clay or gravel, difficult to excavate.

Heating, ventilating, and air conditioning. Includes the mechanical systems for heating, cooling, humidity control, and movement of air in buildings.

Heuristics. An exploratory problem-solving technique that utilizes self-educating approach to improve performance.

High range water-reducing admixture. See Superplasticizer.

High tech. High technology, advanced technology, normally incorporating advanced computer systems.

High-strength concrete. Concrete with a 28-day compressive strength of 6000 psi or higher.

High-strength steel. Concrete reinforcing bars generally having a minimum yield strength above 414 MPa (60,000 psi, 4220 kgf/cm^2).

Hybrid construction. The frame construction composed of different structural building materials, such as concrete and steel.

Hydraulic elevator. An elevator raised by a hydraulic ram underneath it. It requires a high-pressure water supply. (Some hydraulic lifts are still in use.)

Inductive learning. The accumulation of new general knowledge by compiling the facts and actions of specific examples.

Inelastic action. Deformation of a material that does not disappear on removal of the force that produced it.

Inference mechanism. A procedure that manipulates the context by using a knowledge-based system.

Infill wall. Non-load-carrying wall inserted within a frame of a building.

Infiltration. Air moisture or dust flowing inward as through walls or cracks.

Instability. A condition reached during buckling under increasing load in a compression member, element, or frame at which the capacity for resistance to additional load is exhausted and continued deformation results in a decrease of load-resisting capacity.

Intelligent buildings. Buildings that provide more effective life safety, security, energy conservation, communication capability (through shared data systems) and environmental design through better use of high technology planning and design, and means such as microprocessors, better materials, and better construction technologies.

International style. Style stemming from the Bauhaus; characterized by purity, functionalism, and impersonality.

Interstory drift. The relative horizontal displacement of two successive floors of a particular building.

Inverted base plate. A steel plate that transfers the load of the column to the larger area underneath the plate.

Jerk. Rate of change of acceleration.

Joint. The interface between adjoining or abutting elements. In monolithic construction, the concrete between the ends of the columns and beams entering the joint.

Joist. One of a series of closely spaced horizontal structural members interacting with or supporting a deck.

Knowledge acquisition module. An interface between an expert and an expert system.

Knowledge base. The knowledge specific to a domain of the problem to be solved.

Knowledge-based system. Computer program that uses expert knowledge to attain high levels of performance in a problem area; knowledge (i.e. facts and rules) represented symbolically and the system mimics the analytical processes of human professionals. (See also Expert systems.)

Knowledge engineering. The art of designing and building expert systems and other knowledge-based programs.

Knowledge representation. A combination of data structures and interpretive procedure.

Land use. A comprehensive term classifying particular areas or plots by the manner in which they are being utilized, and their relationships with adjacent uses and integral components (transit systems, etc.).

Lateral stability system. A system that resists lateral loads, such as wind and horizontal earthquake load.

Limit states. Condition in which a structure or a part thereof ceases to fulfill one of its functions or to satisfy the conditions for which it was designed. Limit states can be classified in two categories: (1) ultimate limit states, corresponding to the load-carrying capacity of the structure—safety is usually related to these types of limit state; and (2) serviceability limit states, related to the criteria governing normal use of the structure.

Limit states design. A design process that involves identification of all potential modes of failure (limits states) and maintaining an acceptable level of safety against their occurrence. The safety level is usually established on a probabilistic basis.

Load and resistance factor design. A design method in which, at a chosen limit state, load effects and resistances are separately multiplied by factors that account for the inherent uncertainties in the determination of these quantities.

Load-bearing shear wall. A wall which in its own plane carries both vertical forces caused by gravity and lateral forces caused by transverse loads.

Load combinations. Loads likely to act simultaneously.

Load effects. Moment, shears, and axial forces in a member due to loads or other actions.

Load factors. Factors applied to a load to express probability of not being exceeded; safety factors.

Local stability. The absence of buckling of compression elements that make up a cross-section.

Logistical control. A control method dealing with the procurement, maintenance, and movement of material and personnel.

Longitudinal. In the direction of the longer plan dimension.

LVDT. A device called linear variable differential transducer, which measures the movement of extensometer plunger by electrical field changes. It can be hooked onto a deep benchmark to read building settlements, or to the end of a fixed invar tape to measure horizontal movements.

Machine room. Also referred to as "mechanical space." Areas in a building reserved for mechanical equipment.

Market-rate rental. A building constructed for rental purposes rather than for sale.

Masonry. Construction composed of shaped or molded units, usually small enough to be handled by one man and composed of stone, ceramic brick or tile, concrete, glass, adobe, or the like.

Master plan. A plan, usually graphic and drawn on a small scale but often supplemented by written material, that depicts all the elements of a project or scheme.

Material failure. A column failure occurring when the forces and moments at the failure cross section reach the capacity of that cross section.

Maximum load (ultimate load). Plastic limit load or stability limit load, as defined (also the maximum load-carrying capacity of a structure under test).

Mean. The arithmetic average of a group of values.

Mechanical room. A room used to house the bulk of the HVAC equipment.

Megastructure. A structure, usually large, providing for multiple use, combining living, working, and service functions within the whole.

Mercalli scale. A measure of the local intensity of an earthquake. It is based solely on observations of damage and other phenomena.

Minaret. A tall, slender tower on a mosque with one or more projecting balconies from which a muezzin summons the people to prayer.

Missile impact. The effect of suddenly applied force on a body or subject.

Model analysis. Analysis of structure by means of measurements on an accurately scaled model.

Modernism. A style of architecture that came into existence in the 1920s, and had its classical period in the 1930s and the late 1940s.

Modularity. System of standardized units selected for ease of construction.

Module. A unit of least dimension, usually the window mullion spacing or some multiple of it.

Moment magnification. The increase in moment in a member resulting from the increase in the eccentricity of the applied axial load within the member due to deflections.

Moment redistribution. Moment transfer from section to section within a structure due to successive formation of cracking and plastic hinges.

Moment-resisting frame. An integrated system of structural elements possessing continuity and hence capable of resisting bending forces. (These frames usually develop minor axial forces.)

Monitored time frame. A schedule of deliveries that is able to place construction materials on time and at the precise place of work.

Monolithic structure. Structure that is in one piece, not able to be dismantled into parts.

Mortar. A mixture of cement paste, lime, and sand, used to bond together masonry elements. The proportions are expressed as a ratio (cement: lime:sand).

Mullions. The horizontal or vertical members of a window wall or curtain wall system that are normally attached to the floor slab or beams, and support the glass and/or elements of a window wall.

Multiple column curves. A series of curves, each with selected applications, dependent upon material, manufacture, fabrication, and cross-sectional properties of the columns for which they are applicable.

Multiple-use. A building that contains more than one type of occupancy. For example, a building with apartments on the upper floors and business establishments on the lower floors is a multiuse building. Also referred to as "mixed-use."

Natural frequency of a building. Number of cycles per second when the structure is vibrating in its most easily excited mode.

Neo-Gothic. An architectural movement that flourished particularly in the nineteenth century. Also called Gothic Revival.

Network. Relationships among people, often reflected by friendships. Network analysis is a method for analyzing relationships among people.

Network theory. Techniques concerned with the analysis of sequences of activities that can be represented as a network. More than one form of representation is possible. The critical path method represents an important application of network theory.

Node. A real or hypothetical joint between two or more structural elements, used for reference in structural analysis.

Nominal load effect. Calculated using a nominal load; the nominal load frequently used is defined with reference to a probability level; e.g., 50-year mean recurrence interval wind speed used in calculating wind load.

Nominal resistance. Calculated using nominal material and cross-sectional properties and a rationally developed formula based on an analytical and/or experimental model of limit state behavior.

Nomogram. A diagram used for the evaluation of an equation.

Normal modes. Characteristic deflected shapes exhibited by a structure as it vibrates. These shapes depend on the distribution of mass and stiffness along the structure.

Occupancy sensors. Illumination control system that senses when a person is within the system space.

One-way construction. A structural system where the arrangement of the steel reinforcement is intended to resist stresses due to bending in one direction only.

Open section. A thin-walled structural member whose cross section does not form a closed contour.

Optimization. The process, either by mathematical or other procedures, of defining the particular state of a system, which realizes the greatest benefit or which is the best compromise between opposing tendencies.

P-Δ effect. Increase in moment in a frame or member resulting from the vertical loads acting through the lateral displacement.

P-Δ moments. Moments resulting from the lateral displacement of the vertical loads acting on a frame.

Pan-joist. A horizontal structural concrete member usually supported by beams constructed by aid of prefabricated form units (pans).

Panel building system. A method of construction using precast wall and slab panels that are connected together at the building site to form the finished building structure.

Panel zone deformation. Distortion of the web of an I-shaped column in the region defined by the top and bottom flanges of the attached beam(s).

Partial factor of safety. A safety factor used to account for the variability of one aspect of structural safety.

Partial prestressing. Prestressing applied to a member in which flexural tensile stresses will be produced under working load.

Pediment. In classical architecture, a low-pitched gable above a portico.

Perimeter moment frame. A lateral load-resisting system consisting mainly of frames around building surfaces.

Permanent form. Any form that remains in place after the concrete has developed its design strength. The form may or may not become an integral part of the structure.

Pile. A long slender column of timber, concrete, or steel embedded in the foundation.

Plane frame. A framed structure that can be idealized as a two-dimensional structure.

Plastic hinge. A yielded zone that forms in a structural member when the plastic moment is attained. The beam rotates as if hinged, except that it is restrained by the plastic moment, M_p.

Plinth. In classical architecture, the projecting base of a wall or column pedestal, moulded or chamfered at the top. The term is now used for a slight widening at the base of a wall or column.

Plinth area. The ground level of a building.

Podium. Lower floors of a tall building from which the upper portion steps back, the podium serving as a foundation.

Population density. Refers to the number of people in a specified area of space. This is sometimes referred to as "social density" as well.

Post-tensioning. A method of prestressing reinforced concrete in which tendons are tensioned after the concrete has hardened.

Pounding. Contact of a structure with an adjacent one during oscillation in an earthquake.

Pozzolan. A siliceous or siliceous and aluminous material, which in itself possesses little or no cementitious value but will, in finely divided form and in the presence of moisture, chemically react with calcium hydroxide at ordinary temperatures to form compounds possessing cementitious properties.

Precast concrete. Concrete cast elsewhere than its final position in the structure.

Prestressed concrete. Concrete in which internal stresses of such magnitude and distribution are introduced that the tensile stresses resulting from the service loads are counteracted to a desired degree; in reinforced concrete the prestress is commonly introduced by tensioning the tendons.

Pretensioning. A method of prestressing reinforced concrete in which the tendons are tensioned between fixed abutments before the concrete is placed and are released after the concrete has hardened and bonded to the tendons.

Private knowledge. A type of knowledge that may not be available through books, but rather is acquired over years of experience and is thus in the hands of human experts.

Probabilistic. A quantity is said to be probabilistic if its value varies from occurrence to occurrence. The opposite of deterministic.

Probabilistic behavior. A design method that explicitly utilizes probability theory in the safety checking process.

Probability of failure. The probability that the limit state is exceeded or violated.

Progressive collapse. A situation in which the failure of one element causes failure of adjacent elements, which spreads progressively through a large part of the structure. In general, the final amount of damage is many times that expected from the initial failure itself.

Public knowledge. A knowledge medium such as published definitions, facts, and theories associated with the field or domain in question.

Quality control. A formalized system of procedures and controls used to determine as-produced quality or acceptance quality.

Radiation. Energy transmitted by electromagnetic waves.

Reinforced concrete, masonry. Concrete or masonry containing reinforcement and designed on the assumption that the two materials act together in resisting forces.

Reinforcement. Metal bars, wires, or other slender members that are embedded in concrete in such a manner that the metal and the concrete act together in resisting forces.

Reinforcement ratio. Ratio of the effective area of the reinforcement to the effective area of the concrete at any section of a structural member.

Release. In structural analysis, the removal of a structural continuity at a section in a member that renders the member incapable of transmitting that particular force component across the section. In pretensioning, the time when the reaction between the tendons and the prestressing bed is removed.

Reliability index. A computed quantity defining the relative reliability of a structure or structural element.

Reshoring. The construction operation in which the original shoring or posting is removed and replaced in such a manner as to avoid damage to the partially cured concrete.

Residual stress. The stresses that remain in an unloaded member after it has been formed into a finished product. (Examples of such stresses include, but are not limited to, those induced by cold bending, cooling after rolling, or welding.)

Resistance factor. A partial safety factor to account for the probability of understrength of materials or structural members.

Resonance. Vibrations of large amplitude in a member or structure that occur when the natural frequency of the member or structure is close to the frequency of the applied force causing the vibration.

Resource planning. A plan centralizing the production and supply of essential building materials.

Response spectrum. A diagram of the maximum response vs. the natural frequency of vibration for a group of single-degree-of-freedom structures subjected to base motions from an earthquake.

Return period. Average length of time between occurrences of wind, snow, and so forth of a given magnitude.

Reversed loading. A loading condition in which the load is first applied in one direction, then completely removed and reapplied in the opposite direction.

Richter magnitude. Logarithm of the amplitude of motion recorded by a specific type of seismograph located on firm ground 100 km from the epicenter. The Richter magnitude is related to the energy released during the earthquake.

Rigid frame. A structure made up of beam and column members joined together at their intersections in such a manner that there is no relative rotation between the intersection members at a joint under applied load or deformation.

Rock coring technique. A process through which nonmagnetic oriented core is produced by a modified drill rig holding the rock barrel in a fixed position. Markings are scribed into the rock as it moves up into the core barrel.

Romanesque architecture. The style current in Europe from about the ninth century until the advent of Gothic. It includes both the Anglo-Saxon and the Norman architecture of England.

Rule. An "if-then" statement. The action that should be taken once certain conditions are satisfied.

S wave. An elastic wave consisting of shear deformation in the medium.

Sandwich wall. A wall with more than one layer.

Second-order analysis. An analysis that includes the effect on deflections and moments of the lateral displacement of the vertical loads on the structure.

Second-order deflections, moments. Deflections or moments calculated by a second-order analysis.

Security. Control of access and movement within buildings. Detection of emergency situations.

Seismometer. Instrument similar to an accelerograph for measuring ground velocity.

Seismic. Pertaining to earthquakes.

Semi-rigid connection. A connection designed to permit some rotation, either in steel or in reinforced concrete construction. It is intermediate between a rigid joint and a pin joint.

Sensor signal processing. A systematical series of actions to handle the signals detected by sensors.

Service loads. Loads anticipated during the normal life of the building.

Service run. A service system such as vertical riser ducts providing water and sewerage service, electric service, HVAC system, and so forth.

Serviceability limit state. A condition occurring in the service life of the building in which the structure becomes unfit for its intended use due to excessive deflections, cracking, vibrations, damage to contents.

Service-based economy. An economical system in a country with net new employment concentrated in office rather than factory settings.

Setback. The withdrawal of the face of a building to a line some distance from the boundary of the property or from the street.

Shear. An internal force tangential to the plane on which it acts.

Shear reinforcement. Reinforcement designed to resist shear or diagonal tension stresses; dowels are not considered to be shear reinforcement.

Shear stud. A short mild-steel rod with flattened head, welded to a steel member, to transfer shear force between steel and surrounding concrete.

Shear wall. A structural wall which, through in-plane shear forces, resists lateral forces resulting from wind, earthquake, or other transverse loads.

Shear wall, coupled. Two shear walls connected by beams or slabs that permit the two walls to act together as a load-resisting system.

Shear wall-frame structure. A structure in which shear walls and frames interact in resisting lateral loads.

Shore. Temporary vertical support for formwork and fresh concrete or for recently built structures that have not developed full design strength. Also called prop, tom, post, strut.

Shrinkage. Volume decrease caused by drying and chemical changes; a function of time but not of stress due to external load or thermal expansion. It is generally expressed in terms of a linear strain.

Shuttle elevator. An express elevator between two or among three landings to transport pedestrian traffic from the street lobby to a sky lobby above, where a transfer is made to a bank or banks of local elevators.

Sidesway. The lateral movement of a structure under the action of lateral loads, unsymmetrical vertical loads, or unsymmetrical properties of the structure.

Silica fume. A by-product of the induction arc furnaces in the silicon metal and ferrosilicon alloy industries. Tiny spherical particles two orders of magnitude finer than normal portland cement particles.

Sill. The horizontal member below a door or window opening.

Single tee. A precast, one-way slab member consisting of a wide top flange and a single web.

Site. The area on which a facility or other improvement is constructed.

Skewed. Not parallel or perpendicular.

Skin. Cladding of a building.

Sky lobby. A major lobby above the street to permit transfer from a bank of express shuttle elevators to a bank or banks of local elevators.

Skylight. A window placed in a flat or sloping roof.

Skyline. The outline of a group of buildings seen against the sky.

Skyscraper. The name originally applied in the United States to tall, multistory buildings built entirely with an iron skeleton, later with steel or reinforced concrete, or both.

Skyway. An elevated highway.

Slab. A flat, usually horizontal or nearly so, molded layer of plain or reinforced concrete usually of uniform thickness, but sometimes of variable thickness; as the flat section of floor or roof either on the ground or supported by beams, columns, or other framework. (See also Flat slab.)

Slab type or high-rise building. A building in the shape of a vertical slab standing on the ground on its short dimension.

Slenderness ratio. This term may refer to the geometric slenderness ratio or the mechanical slenderness ratio, generally the latter.

Slip form. A form that moves, usually continuously, during placing of the concrete. Movement may be either horizontal or vertical. Slip forming is like an extrusion process with the forms acting as moving dies to shape the concrete.

Slurry or diaphragm wall technique. A concrete wall designed to retain soil and water, keeping it away from an excavation.

Smart building. (See Intelligent building).

Soffit. The underside of a subordinate part or member of a building, such as a beam, stairway, arch, and the like.

Soft story. A story in a building that is significantly weaker than the other stories.

Software. A term used in contrast to hardware to refer to all programs that can be used on a particular computer system.

Space frame. A three-dimensional framework as contrasted to a plane frame.

Space management. A management procedure designed to maintain an organization's strategic plan and productivity goals as they relate to the physical environment at a minimum cost.

Spandrel. A beam spanning between columns on the exterior of a building.

Spandrel beam. A floor level beam in the face of a building, usually supporting the edges of the floor slabs.

Spirally reinforced column. A column in which the vertical bars are enveloped by spiral reinforcement, that is, closely spaced continuous hooping.

Splice. Connection of one reinforcing bar to another by overlapping, welding, mechanical end connectors, or other means.

Sprinkler system. A system of pipes affixed to the ceiling or roof of a building. Valves open at predetermined temperatures to release water for extinguishing fires.

Stability. The capacity of a member or structure to recover from displacement induced by an applied force or disturbance.

Staggered wall beam system. A structural system for a building with frames in one direction and frames braced in the other direction by use of story-deep beams staggered in location at alternate frames on every other floor of the building.

Stair pressurization. Increasing the air pressure in stair wells (usually with fan systems) to provide refuge area from fire and smoke.

Standard deviation. A quantity used to measure the dispersion of a set of values equal to the root-mean-square deviation of the values from their mean value.

Stiffness. The resistance to deformation of a member of structure measured by the ratio of the applied force to the corresponding displacement.

Stiffness matrix. An ordered set of quantities arranged in a rectangular array with values representing the coefficients of the correction rotation or displacement terms in the linear set of equations formulated for a structure under load or applied disturbance by the displacement method of analysis.

Stochastic process. A method that is probabilistic in nature as opposed to deterministic.

Strand. A prestressing tendon composed of a number of wires most of which are twisted about a center wire or core.

Strength design. A method of proportioning structures or members to have failure capacities equal to or greater than the elastically computed moments, shears, and axial forces corresponding to a specified multiple of the working loads and assuming a nonlinear distribution of flexural stresses.

Strength limit state. The condition of a structural element in which it reaches its maximum load carrying capacity for a given loading.

Strong-column-weak-beam concept. A design approach where columns and girders are relatively sized so that beams attain their plastic capacity prior to the attached columns reaching their plastic capacity.

Structural damping. A property inherent in structures that causes vibrational energy to be dissipated. It is an aggregate of several factors including friction.

Subassemblage. A truncated portion of a structural frame.

Subgrade. The soil beneath the foundation structure of the building.

Substructure. A part of a structure that may be analyzed apart from the remainder of the structure.

Subsystems. Parts of a larger system, but also being systems in themselves.

Suburb. A town, village, or other community just outside a city or larger town.

Superplasticizer. A concrete admixture consisting of long-chain, high-molecular-weight anionic surfactants with a large number of polar groups in a hydrocarbon chain; it is able to reduce three to four times the water in a given concrete mixture compared to normal water-reducing admixtures.

Suspended maneuvering system. A helicopter-hung fire-fighting platform fully equipped to tackle the fire and rescue the occupants.

Suspended system. A structure supported from above so as to allow its members to develop tensile stress.

Systems methodology. A computer method that not only can do mathematical calculation but also can assist engineers to performing their tasks by access into knowledge-based and intelligence-based systems.

T-beams. A beam composed of a stem and a flange in the form of a T.

Temperature reinforcement. Reinforcement designed to carry stresses resulting from temperature changes; also the minimum reinforcement for areas of members not subjected to primary stresses or necessarily to temperature stresses.

Tendon. A steel element such as a wire, cable, bar, rod, or strand used to impart prestress to concrete when the element is tensioned.

Thermal conductivity. Rate of transfer of heat along a body by construction.

Thermodynamics. Study of the relation between heat and energy.

Time-dependent volume changes. The combined effect of creep, shrinkage, and temperature change.

Tolerance. The permitted variation from a given dimension or quantity.

Topographic data. Data on maps and charts describing the physical features of an area, such as mountains or rivers.

Torsion. The stress caused when one end of an object is twisted in one direction and the other end is held motionless or twisted in the opposite direction.

Transfer girder. A large beam (normally situated at first or second story levels) used to redistribute vertical forces from the closely spaced columns above to the wider spaced ground floor columns.

Transfer structure. Structure that supports heavy concentrations of loads with limited movement.

Transfer truss. Fabricated structural member that supports load and usually contains the escalator or moving walk equipment.

Translation. A rigid body linear movement, in any specific direction, of the cross section of the building.

Transverse. In the direction of the shorter plan dimension.

Transverse load. Horizontal lateral force applied at right angles to the vertical axis of the building.

Tube. A structure with continuous perimeter frame designed to act in a manner similar to that of a hollow cylinder.

Tube system. A building in which columns are closely spaced around the perimeter of the building and interconnected by stiff spandrel beams to optimize stiffness against lateral loads.

Tube-in-tube system. A building with an inner core tube system and an exterior perimeter tube system.

Tuned mass damper. A large mass mounted on a structure using springs. The natural frequency of the mass-spring system is adjusted (tuned) so that the mass responds in the opposite direction to the structure's motion and produces a restoring force on the structure.

Two-way construction. A structural system intended to resist stresses caused by bending in two directions. The reinforcement is placed at right angles to each other.

U tube. A structure with continuous perimeter frame designed to act in a manner similar to that of a hollow cylinder.

Ultra high-strength concrete. Concrete with a 28-day compressive strength substantially higher than 10,000 psi.

Ultimate limit state. A structure or structural element reaches an ultimate limit state when its useful life is terminated by collapse, fracture, overturning, instability, and so forth.

Ultimate strength design. See Strength design.

Unbraced frame. A structural frame that can buckle in a sidesway mode.

Urban. Communities with 2500 or more persons.

Urban core. The central business district or downtown of an urban area.

Urban design. To make plans for the structures, facilities, and services of a city (or a significant portion of it) that will meet the human activity needs thereof, including industrial, governmental, economic, social, cultural, and esthetic. To fashion a city.

User interface module. Function that provides an interface between users and computerized knowledge system.

Utility. Service related to a building operation for which the consumer pays. Commercial energy may be in the form of electricity, oil, distillate, coal, gas, propane, wood, solid waste, waste heat. Other services include a sewer, garbage removal, water supply, and telephone.

Variable air volume (VAV) systems. Constant temperature HVAC system. Differentiates air flow or volume rate.

Vault. An arched masonry or concrete roof.

Ventilation. Systems, either mechanical or structural, for providing fresh air to a building.

Vierendeel action. Using a planar rectangular grid of members working in flexure to act as a truss for longer spans for loads in that plane.

Viscous damping coefficient. See Damping coefficient.

Waffle slab. A type of two-way concrete-joist floor construction.

Web. A vertical plate, or its equivalent, that joins the top and bottom flanges in a beam.

Working load. The load for which a structure is designed under normal service and conditions.

Working stress design. A method of proportioning structures or members for prescribed working loads at stresses well below the ultimate, and assuming linear distribution of flexural stresses.

Wrought iron. Iron with a low carbon content, less than mild steel.

Wythe. Each continuous vertical section of a wall one masonry unit in thickness.

Yaw motion. Twisting, or torsional motion.

Yield point. The point during increasing stress at which the proportion of stress to strain becomes substantially less than it has been at smaller values of stress, below which the stress-strain curve may be assumed to be linear, and above which the curve is usually nonlinear.

Yield strength. The stress corresponding to the yield point.

Zoning. The legal regulation of the use of land and buildings, in which the density of population and the height, bulk, and spacing of structures is also specified.

SYMBOLS

A = tributary area
A_c = concrete area
A_{eo} = equivalent orifice area
a_s = environmental parameter
C = flow coefficient
C_d = discharge coefficient
$C_{p\,code}$ = peak pressure coefficient from code
$C_{p\,model}$ = peak pressure coefficient from model test
C_{pe} = external pressure coefficient
C_{pi} = internal pressure coefficient
E = energy
E_d = energy dissipated by dampers
E_{MR} = machine room heat release
E_t = total building strain energy
EI_b = flexural rigidity of beam
F = forces at end of members
H = applied horizontal force vector
h = story height
K = stiffness
$\underset{\sim}{K}$ = global stiffness matrix
$\underset{\sim}{K}_i$ = stiffness matrix of bent i
k = flexural stiffness of member = $2EI/L^3$
L = length of member
L_{10} = sound pressure level reached or exceeded for 10% of the time
M = moment at end of member
m = mass
N = axial force
P = pressure difference
$\underset{\sim}{P}$ = external force vector
p = peak pressure
Q = air flow
r = extensional stiffness of member = EA/L
T = twisting moment
V = volume of damper unit
V_{30} = wind velocity 30 ft above ground
V_g = wind velocity at reference height (1500 ft)
V_h = design mean wind speed
V_h = design max. gust wind speed
V_h = wind velocity at h ft above ground
β = damping coefficient
γ = damper shear strain
γ = air density
Δ = first mode building displacement

$\underset{\sim}{\Delta}$ = nodal displacement vector
δ = displacement of member end
ϕ = cross spectral density
ω = frequency
ℓ = distance
ρ = air density
ρ = reinforcement ratio
σ = standard deviation
θ = rotation of member end
ζ = damping ratio

ABBREVIATIONS

ACI—American Concrete Institute
ACRS—Accelerated Cost Recovery System
ANSI—American National Standards Institute
BOMA—Building Owners and Managers Association
BRE—Building Research Establishment
CBD—Central Business District
CEB—European Concrete Committee
CIB—International Council for Building Research, Studies, and Documentation
CPF—Central Provident Fund
CPF—Compulsory Provident Fund
CPM—Critical Path Method
CSA—Canadian Standards Association
ECI—Equatorial Comfort Index
ENSU—Essential Maintenance Service Unit
FAR—Floor Area Ratio
FSI—Floor Space Index
FT—Fully Tempered
HDB—Housing and Development Board
HS—Heat-Strengthened
HUD—U.S. Department of Housing and Urban Development
HVAC—Heating, Ventilating and Air Conditioning
IG—Insulating Glass
INTEGRO—Open-Elements Load-Bearing Construction System for Public and Industrial Buildings
ISIR—International Symposium on Industrial Robots
JPDP—Japan Power Demonstration Reactor
JRR—Japan Research Reactor
LVDT—Linear Variable Differential Transducer
MITI—Ministry of International Trade and Industry
MRT—Mass Rapid Transit
NASA—National Aeronautic and Space Administration
NATM—New Austrian Tunneling Method

NBS—National Bureau of Standards
NFPA—National Fire Protection Association
NSF—National Science Foundation
NWS—Neighborhood Watch Scheme
PDA—Pile Driving Analyzer
PERT—Project Evaluation and Review Technique
RC—Resident's Committees
RERC—Real Estate Research Corporation
SIT—Singapore Improvement Trust
SMS—Suspended Manuevering System
SOM—Skidmore, Owings & Merrill
SSR—Shimizu Site Robot
TCCMAR—Technical Coordinating Committee for Masonry Research
TDR—Transferable Development Rights
TMD—Tuned Mass Damper
TP—Trip Time
UJNR—U.S.-Japan Cooperation Program on Natural Resources
VAV—Variable Air Volume
VPM—Vertical Production Method
WASCOR—Waseda Construction Robot

UNITS

In the table below are given conversion factors for commonly used units. The numerical values have been rounded off to the values shown. The British (Imperial) System of units is the same as the American System except where noted. Le Système International d'Unités (abbreviated "SI") is the name formally given in 1960 to the system of units partly derived from, and replacing, the old metric system.

SI	American	Old Metric
	Length	
1 mm	0.03937 in.	1 mm
1 m	3.28083 ft	1 m
	1.093613 yd	
1 km	0.62137 mile	1 km
	Area	
1 mm^2	0.00155 $in.^2$	1 mm^2
1 m^2	10.76392 ft^2	1 m^2
	1.19599 yd^2	
1 km^2	247.1043 acres	1 km^2
1 hectare	2.471 acres[(1)]	1 hectare

SI	American	Old Metric
Volume		
1 cm^3	0.061023 $in.^3$	1 cc
		1 ml
1 m^3	35.3147 ft^3	1 m^3
	1.30795 yd^3	
	264.172 gal[2] liquid	
Velocity		
1 m/sec	3.28084 ft/sec	1 m/sec
1 km/hr	0.62137 miles/hr	1 km/hr
Acceleration		
1 m/sec^2	3.28084 ft/sec^2	1 m/sec^2
Mass		
1 g	0.035274 oz	1 g
1 kg	2.2046216 lb[3]	1 kg
Density		
1 kg/m^3	0.062428 lb/ft^3	1 kg/m^3
Force, Weight		
1 N	0.224809 lbf	0.101972 kgf
1 kN	0.1124045 tons[4]	
1 MN	224.809 kips	
1 kN/m	0.06853 kips/ft	
1 kN/m^2	20.9 lbf/ft^2	
Torque, Bending Moment		
1 N-m	0.73756 lbf-ft	0.101972 kgf-m
1 kN-m	0.73756 kip-ft	101.972 kgf-m
Pressure, Stress		
1 N/m^2 = 1 Pa	0.000145038 psi	0.101972 kgf/m^2
1 kN/m^2 = 1 kPa	20.8855 psf	
1 MN/m^2 = 1 MPa	0.145038 ksi	
Viscosity (Dynamic)		
1 $N\text{-}sec/m^2$	0.0208854 $lbf\text{-}sec/ft^2$	0.101972 $kgf\text{-}sec/m^2$
Viscosity (Kinematic)		
1 m^2/sec	10.7639 ft^2/sec	1 m^2/sec
Energy, Work		
1 J = 1 N-m	0.737562 lbf-ft	0.00027778 w-hr
1 MJ	0.37251 hp-hr	0.27778 kw-hr

SI	American	Old Metric
	Power	
1 W = 1 J/sec	0.737562 lbf ft/sec	1 w
1 kW	1.34102 hp	1 kw
	Temperature	
K = 273.15 + °C	°F = (°C × 1.8) + 32	°C = (°F − 32)/1.8
K = 273.15 + 5/9(°F − 32)		
K = 273.15 + 5/9(°R − 491.69)		

(1) Hectare as an alternative for km^2 is restricted to land and water areas.
(2) 1 m^3 = 219.9693 Imperial gallons.
(3) 1 kg = 0.068522 slugs.
(4) 1 American ton = 2000 lb. 1kN = 0.1003612 Imperial ton. 1 Imperial ton = 2240 lb.

References/Bibliography

The citations that follow include all publications referred to or cited in the articles, and a bibliography for further reading. The material is arranged alphabetically by author, followed by the year of publication. Since the citation in the text is to author and year, there will be instances in which reference is made to two different articles published in the same year by the same author. In those instances it has been necessary to affix letters to the year to provide proper identification.

Where articles are published in a language other than English, the translation of the title is given first, followed by the title in the original language.

Additional bibliographies are available through the Council.

ACI, 1977
GUIDE TO DURABLE CONCRETE, ACI201.2R-77, Journal of the American Concrete Institute, December.

ACI, 1985
CORROSION OF METALS IN CONCRETE, ACI222R-85, Journal of the American Concrete Institute, January.

AISC, 1934
STEEL CONSTRUCTION MANUAL, American Institute of Steel Construction, New York, 2nd Ed., p. 127.

AISC, 1978
SPECIFICATION FOR THE DESIGN, FABRICATION AND ERECTION OF STRUCTURAL STEEL FOR BUILDINGS, American Institute of Steel Construction, Chicago, Illinois.

AISC, 1984
SPECIFICATIONS FOR STRUCTURAL STEEL BUILDINGS, Commentary on the Proposed Load and Resistance Factor Design, Chicago, August 1, p. 18.

AISC, 1986
LOAD AND RESISTANCE FACTOR DESIGN SPECIFICATION FOR STRUCTURAL STEEL BUILDINGS, American Institute of Steel Construction, Chicago, Illinois.

ATC, 1978
TENTATIVE PROVISIONS FOR THE DEVELOPMENT OF SEISMIC REGULATIONS FOR BUILDINGS, NBS Special Publication 510, U.S. Department of Commerce, Washington, DC, Applied Technology Council.

Ackroyd, M. H., 1979
NONLINEAR INELASTIC STABILITY OF FLEXIBILITY—CONNECTED PLANE STEEL FRAME, Ph.D. Dissertation, Department of Civil, Environmental and Architectural Engineering, University of Colorado, Boulder, Colorado.

Akiner, V. T., 1986
KNOWLEDGE ENGINEERING IN OBJECT AND SPACE MODELING, in Expert Systems in Civil Engineering (Kostem, C. N. and Maher, M. L., eds.), ASCE, New York, pp. 204–218.

Albers, J., 1961
DESPITE STRAIGHT LINE, Yale University Press, New Haven, Conn.

Allen, D. E. and Rainer, J. H., 1976
VIBRATION CRITERIA FOR LONG SPAN FLOORS, Canadian Journal of Civil Engineering, Vol. 3, No. 2, June.

Allwood, R. J. and Robins, P. J., 1983
COMPUTER APPLICATIONS 1: USE OF LARGE COMPUTERS, Chapter 21, Handbook of Structural Concrete, McGraw-Hill, New York.

Alves, M. and Berica, A., 1984
CINFAB—AN INTEGRATED STRUCTURAL STEEL FABRICATION SYSTEM, Seminar—Modern Structural Engineering, The Assn. Cons. Struct. Engrs. NSW, Sydney, Australia, August 15.

Amarel, S., 1978
BASIC THEMES AND PROBLEMS IN CURRENT AL RESEARCH, in Ceilsielske, V. B. (editor), Proceedings of the Fourth Annual AIM Workshop, held at Rutgers University, June, New Holland Publishers, New York.

American National Standards Institute, 1982
MINIMUM DESIGN LOADS FOR BUILDINGS AND OTHER STRUCTURES, ANSI A58.1-1982, New York, NY.

Anderson, A. W., et al., 1952
LATERAL FORCES OF EARTHQUAKE AND WIND, Transactions, ASCE, Vol. 117, pp. 716–780.

Anderson, S. D. and Woodhead, R. W., 1980
PROJECT MANPOWER MANAGEMENT: MANAGEMENT PROCESSES IN CONSTRUCTION PRACTICE, John Wiley and Sons, Inc., New York, NY.

Apgar, W., 1985
HOUSING FUTURES FORECAST MODEL, Joint Center for Housing Studies of MIT and Harvard University, Cambridge.

Appleyard, D., 1980
LIVABLE STREETS, University of California Press, Berkeley, California.

Arciszewski, T., 1975
WIND BRACINGS IN THE FORM OF BELT TRUSS SYSTEMS IN STEEL SKELETON STRUCTURES OF TALL BUILDINGS, Ph.D. dissertation, Warsaw Technical University, Poland.

Arciszewski, T., 1986
DECISION MAKING PARAMETERS AND THEIR COMPUTER-AIDED ANALYSIS FOR WIND BRACINGS, Advances in Tall Buildings, Van Nostrand Reinhold Company, New York.

Arciszewski, T. and Pancewicz, Z., 1976
AN APPROACH TO THE DESCRIPTION OF WIND BRACING CHARACTERISTICS IN SYSTEMS SKELETON STRUCTURES, Proceedings of the Wroclaw Technical University, Vol. 20, Wroclaw, Poland.

Armer, G. S. T., 1983
THE STABILITY OF STRUCTURES, Building Research and Practice, July/August, pp. 216–221.

Aschendorff, K. K., Bernard, A., Buck, O., Mang, F., and Plumier, A., 1983
OVERALL BUCKLING OF HEAVY ROLLED I-SECTION COLUMNS, Proceedings of the Third International Colloquium on Stability on Metal Structures, Toronto, pp. 37–49.

BSSC, 1985
RECOMMENDED PROVISIONS FOR THE DEVELOPMENT OF SEISMIC REGULATIONS FOR NEW BUILDINGS, Building Seismic Safety Council, Washington, DC.

Badger, D. D., 1865
ILLUSTRATIONS OF IRON ARCHITECTURE, MADE BY THE ARCHITECTURAL IRON WORKS OF THE CITY OF NEW YORK, Architectural Iron Works, New York.

Ballio, G., Finzi, L., Setti, P., and Urbano, C., 1977
CAPACITA' DI ASTE TUBOLARI COMPRESSE IN ACCIAIO AD ELEVATO LIMITE ELASTICO, Costruzioni Metalliche, No. 2.

Ballio, G. and Campanini, G., 1981
EQUIVALENT BENDING MOMENT FOR BEAM-COLUMNS, The Journal of Constructional Steel Research, Vol. 1, No. 3.

Ballio, G. and Mazzolani, F. M., 1983
THEORY AND DESIGN OF STEEL STRUCTURES, Chapman and Hall, London and New York.

Bannister, T. C., 1951
THE FIRST IRON FRAMED BUILDINGS, Architectural Review No. 107.

Bannister, T. C., 1956
BOGARDUS REVISITED. PART I: THE IRON FRONTS, Journal of the Society of Architectural Historians, Vol. 15, No. 4, December, pp. 12–22.

Bannister, T. C., 1957
BOGARDUS REVISITED. PART II: THE TOWERS, Journal of the Society of Architectural Historians, Vol. 16, No. 1, March, pp. 11–19.

Barnes, S., 1984
DICE DESIGN INTERFACE FOR CIVIL ENGINEERING, Master's thesis, Carnegie-Mellon University, September.
Barr, A., Cohen, P., and Feigenbaum, E. A., 1981
THE HANDBOOK OF ARTIFICIAL INTELLIGENCE, William Kaufmann, Palo Alto, Vol. 1, pp. 143–216.
Barrie, D. S., ed., 1981
DIRECTIONS IN MANAGING CONSTRUCTION: A CRITICAL LOOK AT PRESENT AND FUTURE INDUSTRIAL PRACTICES, PROBLEMS AND POLICIES, John Wiley and Sons, Inc., New York, NY.
Bate, S. C. C., 1975
REPORT ON THE FAILURE OF ROOF BEAMS AT SIR JOHN CASS'S FOUNDATION AND RED COAT CHURCH OF ENGLAND SECONDARY SCHOOL STEPNEY, Building Research Establishment Current Paper CP58/74, June.
Beason, W. L., 1980
A FAILURE PREDUCTION MODEL FOR WINDOW GLASS, Inst. for Disaster Research, Texas Tech. University, Lublock.
Becker, R., 1975
PANEL ZONE EFFECT ON THE STRENGTH AND STIFFNESS OF STEEL RIGID FRAMES, AISC Engineering Journal, Vol. 12, No. 1, 1st Quarter, Chicago, pp. 19–29.
Beeby, A. W. and Taylor, H. P. J., 1978
THE USE OF SIMPLIFIED METHODS IN CP110—IS RIGOUR NECESSARY?, The Structural Engineer, Vol. 156A, No. 8, August, pp. 209–215.
Bennett, E. W., 1984
PARTIAL PRESTRESSING—A HISTORICAL OVERVIEW, P.C.I. Journal, September-October.
Bennett, J. S. and Engelmore, R. S., 1979
SACON: A KNOWLEDGE-BASED CONSULTANT FOR STRUCTURAL ANALYSIS. In Proceedings of Sixth International Joint Conference on Artificial Intelligence (IJCAI), William Kaufman, Inc., Los Altos, August 20–23, pages 47–49.
Binder, R. W., 1962
SIGNIFICANT ASPECTS OF THE MEXICAN EARTHQUAKES, May 11 and 19, Proceedings, 31st Annual Convention, Structural Engineers Association of California, pp. 96–110.
Bixby, W. H., 1895
WIND PRESSURES IN ENGINEERING CONSTRUCTION, Engineering News, 33:11, March, 175–184.
Blume, J. A., 1969
MOTION PERCEPTION IN THE LOW PREQUENCY RANGE, Report No. JAB-99-47, J. A. Blume and Associates Research Division, San Francisco.
Bogardus, J., 1856
CAST IRON BUILDINGS: THEIR CONSTRUCTION AND ADVANTAGES, J. W. Harrison, New York.
Bonissone, P. P., 1982
OUTLINE OF THE DESIGN AND IMPLEMENTATION OF A DIESEL ELECTRICAL ENGINE TROUBLESHOOTING AID. Technical Conference of the BCS SGES, U. K.

British Standards Institution, 1978
SMOKE CONTROL IN PROTECTED ESCAPE ROUTES USING PRESSURIZATION, BS 5588, Part 4, British Standards Institution, London.

British Standards Institution, 1984
EVALUATION OF HUMAN EXPOSURE TO VIBRATION IN BUILDINGS (1 Hz to 80 Hz), British Standard BS6472, British Standards Institution, London.

British Standards Institution, 1978–1981
CODE OF PRACTICE FOR THE DESIGN OF BUILDINGS AND STRUCTURES FOR AGRICULTURE, British Standards Institution, BS5502, London.

Brooks, M. A., 1912
REMINISCENCES OF THE EARLY DAYS OF FIREPROOF BUILDING CONSTRUCTION IN NEW YORK CITY, Engineering News, 68:22, November 28, pp. 986–987.

Brown, W. G., 1974
A PRACTICABLE FORMULATION FOR THE STRENGTH OF GLASS AND ITS SPECIAL APPLICATION TO LARGE PLATES, Nat. Res. Council of Canada, Publ. No. NRC 14372, Ottowa, November.

Bruegmann, R., 1978
CENTRAL HEATING AND FORCED VENTILATION: ORIGINS AND EFFECTS ON ARCHITECTURAL DESIGN, Journal of the Society of Architectural Historians, 37:3, pp. 143–160.

Buggeling, A. S. G., 1985
CONCRETE STRUCTURES—A DESIGN CHALLENGE, P.C.I Journal, March-April.

Burgess, E. W., 1925
THE GROWTH OF THE CITY: The City, University of Chicago Press, Chicago, IL, pp. 47–62.

CIB, 1983
A CONCEPTIONAL APPROACH TOWARDS A PROBABILITY BASED DESIGN GUIDE ON STRUCTURAL FIRE SAFETY, CIB/W14, Fire Safety Journal, Vol. 6, No. 1.

CIB 84 Singapore, 1984
HIGHRISE CONSTRUCTION TECHNIQUES AND MANAGEMENT FOR THE 1990's, conference proceedings, 23–25 February.

Calderone, I., 1985
DIRECT INTEGRATION OF TIME LOAD HISTORY FOR GLASS DESIGN, thesis submitted for M. Eng. Sc. degree, Monash University, November.

Canadian Standards Association, 1984
DESIGN OF CONCRETE STRUCTURES FOR BUILDINGS, CSA Standard CAN 3-A23.3-M84, 281 pp.

Cantienai, R., 1985
RESEARCH ON FLOOR VIBRATIONS OF AN OFFICE BUILDING (Untersuchungen von Deckenschwingungen bei einem Burogebaude), Report No. 116, EMPA, Zurich, Switzerland.

Chang, F. K., 1967
WIND MOVEMENT IN TALL BUILDINGS, Civil Engineering, Vol. 1, No. 8.

Chang, F. K., 1972
PSYCHOPHYSIOLOGICAL ASPECTS OF MAN-STRUCTURE INTERACTION, Proceedings of Symposium on Planning and Design of Tall Buildings, Vol. 1a, Lehigh University, ASCE Publication.

Chang, J. C. H. and Soong, T. T., 1980
STRUCTURAL CONTROL USING ACTIVE TUNED MASS DAMPERS, Journal of Engineering Mechanics Division, ASCE, Vol. 106, No. EM6, December, pp. 1091–1098.

Chen, W. F., ed., 1977
FLEXIBILITY OF CONNECTIONS IN STEEL FRAMES, Journal of Constructional Steel Research, Elsevier Applied Science, Vol. 7, July, United Kingdom.

Chen, W. F., ed., in press
STEEL BEAM-TO-COLUMN BUILDING CONNECTIONS, Journal of Constructional Steel Research, Special issue, Elsevier Applied Science, United Kingdom.

Chen, P. W. and Robertson, L. E., 1972
HUMAN PERCEPTION THRESHOLDS OF HORIZONTAL MOTION, ASCE Journal of Structural Division, August.

Chen, W. F. and Atsuta, T., 1976
THEORY OF BEAM-COLUMNS: VOL. 1—IN-PLANE BEHAVIOR AND DESIGN, McGraw-Hill, New York.

Chen, W. F. and Atsuta, T., 1977
THEORY OF BEAM-COLUMNS: VOL. 2—SPACE BEHAVIOR AND DESIGN, McGraw-Hill, New York.

Chen, W. F. and Lui, E. M., 1983
DESIGN OF BEAM-COLUMNS IN NORTH AMERICA, Proceedings of the Third International Colloquium on Stability of Metal Structures, Toronto, pp. 253–292.

Chen, W. F. and Lui, E. M., 1984
EFFECTS OF CONNECTION FLEXIBILITY AND PANEL ZONE SHEAR DEFORMATION ON THE BEHAVIOR OF STEEL FRAMES, Seminar on Tall Structures and Use of Prestressed Concrete in Hydraulic Structures, New Delhi, India, May, pp. 155–176.

Chen, W. F. and Lui, E. M., 1985
COLUMNS WITH END RESTRAINT AND BENDING IN LOAD AND RESISTANCE FACTOR DESIGN, AISC Engineering Journal, Vol. 22, No. 3, 3rd Quarter, Chicago, pp. 105–131.

Chen, W. F. and Lui, E. M., 1985
COLUMNS WITH END RESTRAINT AND BENDING IN LOAD AND RESISTANCE DESIGN FACTOR, Engineering Journal, Vol. 22, No. 3.

Chen, W. F. and Lui, E. M., 1986
STEEL BEAM-TO-COLUMN MOMENT CONNECTIONS—PART I: FLANGE MOMENT CONNECTIONS, Vol. II, Issue 4, December, pp. 257–316, Solid Mechanics Archives, Oxford University Press, Oxford, England.

Chen, W. F. and Lui, E. M., 1987a
STRUCTURAL STABILITY: THEORY AND IMPLEMENTATION, Elsevier, New York.

Chen, W. F. and Lui, E. M., 1987b
STEEL BEAM-TO-COLUMN MOMENT CONNECTIONS—PART II: WEB MOMENT CONNECTIONS, Vol. 12, Issue 1, March, pp. 327–378, Solid Mechanics Archives, Oxford University Press, Oxford, England.

Cheung, C. K. and Melbourne, W. H., 1984
CLADDING PRESSURES ON RECTANGULAR HIGH RISE BUILDINGS, workshop on Wind Engineering and Industrial Aerodynamics, C.S.I.R.O. Division of Building Research, Australia, pp. 108–114.

Cheung, Y. K., 1983
TALL BUILDINGS 2, Chapter 38, Handbook of Structural Concrete, McGraw-Hill, New York.

Cheung, Y. K., 1984
COMPUTER ANALYSIS OF TALL BUILDINGS, Proceedings of Third International Conference on Tall Buildings, Hong Kong and Guangzhou, December, pp. 8–15.

Christian Science Monitor, 1983
DALLAS FINDS ITS $2.6 BILLION ARTS DISTRICT IS GOOD FOR BUSINESS, Christian Science Monitor, November 7.

Chua, B. H., 1981
HDB'S INTEREST IN SOCIAL MANAGEMENT, a talk given at the Social Awareness Seminar, JTC Auditorium, October 3 and 4, p. 3.

Civil Engineering, 1985
ENGINEERS SHOULD TAKE RESPONSIBILITY FOR DETAILS, Civil Engineering, Vol. 55, No. 12, December, p. 12.

Cizek, P., 1985
OPEN PREFAB LOADBEARING SYSTEM, Stavebnicka rocenka, ALFA, Bratislava.

Clawson, W. C. and Darwin, D., 1982
STRENGTH OF COMPOSITE BEAMS AT WEB OPENINGS, Journal of the Structural Division, ACSE, Vol. 108, No. ST3, Paper No. 16939, March, pp. 623–641.

Clifton, J., Oltikar, B., and Johnson, S., 1985
DEVELOPMENT OF DURCON, AN EXPERT SYSTEM FOR DURABLE CONCRETE: Part 1, NBSIR 85-3186, National Bureau of Standards.

Collins, M. P., Vecchio, F. J. and Mehlhorn, G., 1985
AN INTERNATIONAL COMPETITION TO PREDICT THE RESPONSE OF REINFORCED CONCRETE PANELS, Canadian Journal of Civil Engineering, Vol. 12, No. 3, September, pp. 624–644.

Colson, A. and Louveau, J. M., 1983
CONNECTIONS INCIDENCE ON THE INELASTIC BEHAVIOR OF STEEL STRUCTURES, Euromech Colloquium 174, October.

Comite Euro-International du Beton, 1985a
CEB MODEL CODE FOR SEISMIC DESIGN OF CONCRETE STRUCTURES, CEB Bulletin No. 165.

Comite Euro-International du Beton, 1985b
CEB DRAFT GUIDE FOR THE DESIGN OF PRECAST WALL CONNECTIONS, CEB Bulletin No. 169.

Commission of the European Communities, 1983
EUROCODE 3, COMMON UNIFIED CODE OF PRACTICE FOR STEEL STRUCTURES, Draft, Brussels, Belgium.
Commission of the European Communities, 1984
EUROCODE NO. 3—COMMON UNIFIED RULES FOR STEEL STRUCTURES.
Condit, C., 1964
THE CHICAGO SCHOOL OF ARCHITECTURE, The University of Chicago Press, Chicago, Il.
Condit, C. W., 1974
THE WIND BRACING OF BUILDINGS, Scientific American, 230:2, February, pp. 92–105.
Conlin, W. F., Sr., 1972
ECONOMICS OF HIGH-RISE BUILDINGS, Proceedings of the International Conference on Planning and Design of Tall Buildings, Vol. 1(a), Lehigh University, Bethlehem, PA, August, pp. 119–132.
Cooper, L., 1983
A CONCEPT FOR ESTIMATING SAFE AVAILABLE EGRESS TIME IN FIRES, Fire Safety Journal, Vol. 5.
Coull, A. and Choudhury, J. R., 1967
STRESSES AND DEFLECTIONS IN COUPLED SHEAR WALLS, Journal ACI, Vol. 64, pp. 65–72.
Coull, A. and Khachatoorian, H., 1982
DISTRIBUTION OF LATERAL FORCES IN STRUCTURES CONSISTING OF CORES, COUPLED SHEAR WALLS AND RIGIDLY-JOINTED FRAMES, Proceedings of ICE, London, Vol. 73, Part 2, December, pp. 731–745.
Coull, A. and Mohammed, T. H., 1983
SIMPLIFIED ANALYSIS OF LATERAL LOAD DISTRIBUTION IN STRUCTURES CONSISTING OF FRAMES, COUPLED SHEAR WALLS AND CORES, Structural Engineer, London, Vol. 61B, March, pp. 1–8.
Coull, A. and Stafford Smith, B., 1983
RECENT DEVELOPMENTS IN ELASTIC ANALYSIS OF TALL CONCRETE BUILDINGS, Developments in Tall Buildings 1983, Council on Tall Buildings and Urban Habitat, Hutchinson Ross Publishing Co., Stroudsburg, pp. 569–581.
Council on Tall Buildings, Beedle, L. S., 1978–1981
PLANNING AND DESIGN OF TALL BUILDINGS, A Monograph in 5 volumes, ASCE, New York.
Council on Tall Buildings and Urban Habitat, 1978
ELASTIC ANALYSIS, Structural Design of Tall Concrete and Masonry Buildings, Vol. CB, Chapter CB-5, Monograph on Planning and Design of Tall Buildings, ASCE, New York.
Council on Tall Buildings and Urban Habitat, Group CB, 1978
STRUCTURAL DESIGN OF TALL CONCRETE AND MASONRY BUILDINGS, Monograph on Planning and Design of Tall Buildings, Volume CB, ASCE, New York.
Council on Tall Buildings and Urban Habitat, Group SB, 1979
STRUCTURAL DESIGN OF TALL BUILDINGS, Volume SB of Monograph on Planning and Design of Tall Buildings, ASCE, New York.

Council on Tall Buildings and Urban Habitat, Group CL, 1980
CRITERIA AND LOADING, Vol. CL, Monograph on the Planning and Design of Tall Buildings, ASCE, New York.
Council on Tall Buildings, Committee 37, 1981
SOCIAL EFFECTS OF THE ENVIRONMENT, Chapter PC-3, Vol. PC of Monograph on Planning and Design of Tall Buildings, ASCE, New York.
Council on Tall Buildings, Group PC, 1981
PLANNING AND ENVIRONMENTAL CRITERIA FOR TALL BUILDINGS, Volume PC of the Monograph on the Planning and Design of Tall Buildings, ASCE, New York.
Coyle, D. C., 1929
MUSHROOM SKYSCRAPERS, American Architect, pp. 829–832.
Coyle, D. C., 1931
MEASURING THE BEHAVIOUR OF TALL BUILDINGS, Engineering News Record, pp. 310–313.
Cross, H., 1952
ENGINEERS AND IVORY TOWERS, McGraw-Hill, New York. p. 141.

Dalgliesh, W. A., 1979
ASSESSMENT OF WIND LOADS FOR GLAZING DESIGN IAHR/IUTAM Symposium on Practical Experience with Flow Induced Vibrations, Springer-Verlag, Berlin, pp. 696–708.
Danay, A., Gellert, M., and Gluck, J., 1974
THE AXIAL STRAIN EFFECTS ON LOAD DISTRIBUTION IN NON-SYMMETRIC TIER BUILDINGS, Building Science, Vol. 9, pp. 29–38.
Davenport, A. G., 1961
THE SPECTRUM OF HORIZONTAL GUSTINESS NEAR THE GROUND IN HIGH WINDS, Quarterly Journal, Royal Meteorological Society, Vol. 87, April, pp. 194–211.
Davenport, A. G., 1966
THE TREATMENT OF WIND LOADING ON TALL BUILDINGS, Proceedings of Symposium of Tall Buildings, University of Southampton, Pergamon Press, Inc., New York.
Davenport, A. G., 1967
GUST LOADING FACTORS, Journal of the Structural Division, ASCE, Vol. 93 (ST3), pp. 11–34.
Davenport, A. G., 1977
WIND ENGINEERING—ANCIENT AND MODERN—THE RELATIONSHIP OF WIND ENGINEERING RESEARCH TO DESIGN, Proceedings of the Sixth Canadian Congress of Applied Mechanics, May 29–June 3, Vancouver, B.C.
Davenport, A. G., Isyumov, N., and Jandali, T., 1971
A STUDY OF WIND EFFECTS FOR THE SEARS PROJECT, The University of Western Ontario, Engineering Science Research Report, BLWT-5-1971.
Davenport, A. G. and Surry, D., 1983
THE ESTIMATION OF INTERNAL PRESSURES DUE TO WIND WITH APPLICATION TO CLADDING PRESSURES AND INFILTRATION, Presented at the ASCE Structural Engineering Conference, Houston, Texas.

Davenport, A. G. and Hill-Carroll, P., 1986
DAMPING IN TALL BUILDINGS: ITS VARIABILITY AND TREATMENT IN DESIGN, Building Motion in Wind, proceedings of a session at ASCE Convention, Seattle, Washington, April 8, pp. 42–57.

Davis, K., 1973
CITIES: THEIR ORIGIN, GROWTH AND HUMAN IMPACT (Readings from Scientific American), W. H. Freeman and Co., San Francisco.

Davis, R. et al., 1981
THE DIPMETER ADVISOR: INTERPRETATION OF GEOLOGIC SIGNALS. In Proceedings Seventh IJCAI, pages 846–849, William Kaufman, Inc., Los Altos, Ca.

Dawson, S., 1983
DALLAS ARTS DISTRICT, Urban Design International, Purchase, New York.

Dehghanyar, T. J., Masri, S. F., Miller, R. K., and Caughey, T. K., 1985
ON-LINE PARAMETER CONTROL OF NONLINEAR FLEXIBLE STRUCTURES, 2nd IUTAM International Symposium on Structural Control, University of Waterloo, Ontario, Canada, July.

Del Valle C., E., 1980
SOME LESSONS FROM THE MARCH 14, 1979 EARTHQUAKE IN MEXICO CITY, Proceedings of the Seventh World Conference on Earthquake Engineering, Volume 4, Istanbul, Turkey.

Del Valle C., E., 1984
EARTHQUAKE DAMAGE TO NON-STRUCTURAL ELEMENTS, Proceedings of the Eighth World Conference on Earthquake Engineering, Vol. V, San Francisco, California.

Deutsches Institut fur Normung, 1975
SHOCK AND VIBRATION IN BUILDING STRUCTURES (Erschutterungen im Bauwesen), Draft of DIN 4150 Deutsches Institut fur Normung, Berlin, West Germany.

Disque, R. O., 1975
DIRECTIONAL MOMENT CONNECTIONS—A PROPOSED DESIGN METHOD FOR UNBRACED STEEL FRAMES, AISC Engineering Journal, Vol. 12, No. 1, 1st Quarter, Chicago, pp. 14–18.

Downs, A., 1981
NEIGHBORHOODS AND URBAN DEVELOPMENT, The Brookings Institution, Washington, DC.

Dryden, H. L. and Hill, G. C., 1933
WIND PRESSURE ON A MODEL OF THE EMPIRE STATE BUILDING, Bureau of Standards Journal of Research, Vol. II, pp. 493–523.

Dubos, R., 1981
CELEBRATIONS OF LIFE, McGraw-Hill Book Company, New York, St. Louis, San Fancisco, Hamburg, Toronto, Mexico.

Duda, R. O., et al., 1979
A COMPUTER-BASED CONSULTANT FOR MINERAL EXPLORATION. Final Report SRI Project 6415 SRI International edition.

Duda, R. and Gasching, J. G., 1981
KNOWLEDGE-BASED EXPERT SYSTEMS COME OF AGE. BYTE 6(9):238–279, September.

Duda, R. O. and Shortliffe, E. H., 1983
EXPERT SYSTEMS RESEARCH. Science 220:261–268, April.

Dunn, J. and Fry, J., 1966
FIRES FOUGHT WITH FIVE OR MORE JETS, Fire Research Technical Paper No. 16, London, HMSO.

Dutt, A. J., 1984
ENVIRONMENTAL EFFECT OF WIND FOR NATURAL VENTILATION IN APARTMENT BLOCK IN SINGAPORE, International Journal of Housing Science, Vol. 8, No. 3, July, pp. 259–267.

ECCS, 1978
EUROPEAN RECOMMENDATIONS FOR STEEL CONSTRUCTION, European Convention for Constructional Steelwork.

ECCS, 1984
ULTIMATE LIMIT STATE CALCULATION OF SWAY FRAMES WITH RIGID JOINTS, ECCS Publication No. 33.

Eiffel, G., 1885
PROJECT FOR A TOWER IN IRON OF 300 M HEIGHT. MEMOIRS OF THE CIVIL ENGINEERING SOCIETY (Project D'une Tour Eufer de 300m De Nauteur. Memoires De La Societe des Ingenieurs Civils), Paris, pp. 345-370.

Eiffel, G., 1900
SCIENTIFIC EXPERIMENTS CONDUCTED ON THE 300 M TOWER FROM 1889 TO 1900 (Travaux Scientifiques Executes a La Tour De Trois Cents Metres de 1889 A 1900), L. Maretheux, Paris France.

Ellingwood, B., Galambos, T. V., MacGregor, J. G., and Cornell, C. A., 1980
DEVELOPMENT OF A PROBABILITY BASED LOAD CRITERION FOR AMERICAN NATIONAL STANDARD A58, NBS Special Publication 57, June.

Ellis, B. R., 1980
AN ASSESSMENT OF THE ACCURACY OF PREDICTING THE FUNDAMENTAL NATURAL FREQUENCIES OF BUILDINGS AND THE IMPLICATIONS CONCERNING DYNAMIC ANALYSIS OF STRUCTURES, Proceedings Institution of Civil Engineers, Part 2, 69, pp. 763–776.

Ellis, E. P., 1953
THERMAL COMFORT IN WARM AND HUMID ATMOSPHERES, Journal of Hygiene, Vol. 51.

Engineering News Record, 1976
LEAD HULA-HOOPS STABILIZE ANTENNA, Engineering News Record, Vol. 197, No. 4, July, p. 10.

Engineering News Record, 1983
JAPAN TAKES EARLY LEAD IN ROBOTICS, Engineering News Record, July 21, USA.

Engineering Times, 1985
ENGINEERS BLAMED IN SKYWALK DEATHS, Engineering Times, Vol. 12, No. 12, December, p. 1.

Fagan, L. M., Kunz, J. C., Feigenbaum, E. A., and Osborn, J. J., 1979
REPRESENTATION OF DYNAMIC CLINICAL KNOWLEDGE: MEASUREMENT INTERPRETATION IN THE INTENSIVE CARE UNIT, In Proceedings Fifth IJCAI, pages 1014–1019, William Kaufman, Inc., Los Altos, Ca.

Feigenbaum, E., 1981
EXPERT SYSTEMS IN THE 1980's, in Machine Intelligence, A. Bond ed., Pergamon Press, Elmstead, N.Y.
Feigenbaum, E. A., 1981
EXPERT SYSTEMS IN THE 1980s, in Bond, A., ed., Infotec State of the Art Report, Pergamon Press, Elmstead, N.Y.
Feld, L. S., 1971
SUPERSTRUCTURE FOR 1,350 FT. WORLD TRADE CENTER, Civil Engineering, Vol. 41, No. 6, June, pp. 66–70.
Fenves, S. J. and Rehak, D. R., 1984
ROLE OF EXPERT SYSTEM IN CONSTRUCTION ROBOTICS, Proceedings of Workshop Conference on Robotics in Construction, Carnegie-Mellon University, Pittsburgh, PA.
Fenves, S. J. and Garrett, J. H., Jr., 1985
STANDARDS REPRESENTATION AND PROCESSING. In IABSE-ECCS Symposium on Steel in Buildings, pages 107–114, ETH, Hongger, Berg, CH. Zurich.
Ferguson, E., 1976
A HISTORICAL SKETCH OF CENTRAL HEATING, 1800–1860, Charles E. Peterson, ed., Building Early America, Chilton Book Co., Radnor, Pa., pp. 165–185.
Ferriss, H., 1986
METROPOLIS, exhibition in the Whitney Museum of American Art at Equitable Center, New York, June 6–July 30.
Fielding, D. J. and Huang, J. S., 1971
SHEAR IN BEAM-TO-COLUMN CONNECTIONS, The Welding Journal, Vol. 50, July.
Fielding, D. J. and Chen, W. F., 1973
FRAME ANALYSIS AND CONNECTION DEFORMATION, Journal of Structural Division, ASCE, Vol. 99, No. ST1, pp. 1–18.
Fintel, M., 1974
MULTISTORY STRUCTURES, Chapter 10, Handbook of Concrete Engineering, Van Nostrand Reinhold, New York.
Fintel, M. and Khan, F. R., 1968
EFFECTS OF COLUMN CREEP AND SHRINKAGE IN TALL STRUCTURES —PREDICTION OF INELASTIC COLUMN SHORTENING, Journal of ACI, Vol. 66, December, pp. 957–967.
Firey, W., 1945
SENTIMENT AND SYMBOLISM AS ECOLOGICAL VARIABLES, American Sociological Review, Vol. 10.
Firkins, A., 1984
CITY BUILDINGS—THE STEEL SOLUTION, (Presented at International Conference on Steel Structures, Singapore, March 7–9), C1-Premier Ltd., Singapore.
Flemming, U., Coyne, R., Glavin, T., and Rychener, M., 1986
A GENERATIVE EXPERT SYSTEM FOR THE DESIGN OF BUILDING LAYOUTS, Proceedings of the First International Conference on Applications of A1 in Engineering, April, Computational Mechanics Publications, Southampton.
Francisco, M., 1977
HISTORY OF THE SKYSCRAPER, De Capo Press, New York, New York.

Frank, L., Kennedy, R. S., Kellog, R. S., and McCauley, M. E., 1983
SIMULATOR SICKNESS: A REACTION TO A TRANSFORMED PERCEPTUAL WORLD 1 SCOPE OF THE PROBLEM, 2nd Symposium of Aviation Psychology, Ohio State University, Columbus, April.

Frein, J. P., ed., 1980
HANDBOOK OF CONSTRUCTION MANAGEMENT AND ORGANIZATION, second edition, Van Nostrand Reinhold Company, New York, NY.

Freitag, J. K., 1901
Architectural Engineering, 1895/1901/1906/1911/1912, John Wiley and Sons, New York.

Freudenthal, A., 1947
THE SAFETY OF STRUCTURES, Trans., ASCE, Vol. 112.

Frye, M. J. and Morris, G. A., 1975
ANALYSIS OF FLEXIBILITY CONNECTED STEEL FRAMES, Canadian Journal of Civil Engineers, Vol. 2, No. 3, Canada, pp. 280–291.

Fukumoto, Y. and Itoh, Y., 1983
EVALUATION OF BEAM STRENGTH FROM THE EXPERIMENTAL DATABASE APPROACH, Proceedings of the Third International Colloquium of Metal Structures, Toronto, pp. 133–150.

GICC, 1984
SLOPED GLAZING, proposed change to the Uniform Building Code, 1984 Code Change Proposals, International Conference of Building Officials, Whittier, CA.

Galambos, T. V., 1983
A WORLD VIEW OF BEAM STABILITY RESEARCH AND DESIGN PRACTICE, Proceedings of the Third International Colloquium on Stability of Metal Structures, Toronto, pp. 113–132.

Galambos, T. V. and Ravindra, M. K., 1978
LOAD AND RESISTANCE FACTOR DESIGN, Journal of the Structural Division, ASCE, No. ST9, September.

Galambos, T. V. and Viest, I. M., 1985
DESIGN OF STEEL STRUCTURES WITH LOAD AND RESISTANCE FACTOR DESIGN SPECIFICATIONS, IABSE Reports Vol. 48, Zurich.

Gallagher, R. H., 1975
FINITE ELEMENT ANALYSIS: FUNDAMENTALS, Prentice-Hall, New Jersey.

Gatton, T. M., 1984
ROBOTIC ASSEMBLY FOR MOBILIZATION CONSTRUCTION, Proceedings of Workshop Conference on Robotics in Construction, Carnegie-Mellon University, Pittsburgh, PA.

Gertis, K., 1975
TALL BUILDINGS AND THERMAL INSULATION (Hochhäuser und Sinnovoller Warmeschutz), Proceedings of German Conference on Tall Buildings, Mainz, October 2–4, Deutsche Gruppe der Internationalen Vereinigung fur Brückenbau und Hochbau (IVBH), Wiesbaden/Köln, German Federal Republic, pp. 257–271.

Gertis, K., 1980
NON-STEADY HEAT AND MOISTURE TRANSFER PROBLEMS IN BUILDING PHYSICS, 11th Congress—Vienna, Introductory Report, Theme VIC, (Proceedings of Congress held in Vienna, Austria, August 31–September 5), IABSE, Zurich, Switzerland, pp. 127–132.

Gibson, W. B., 1984
WORLD'S MOST TERRIFYING AMUSEMENT PARK RIDE, National Examiner, July 31.

Giedion, S., 1928
BUILDING IN FRANCE. IRON, REINFORCED CONCRETE (Bauen in Frankenreich. Eisen, Eisenbeton), Second Edition, Klunkhardt & Biermann, Leipzig.

Giedion, S., 1967
SPACE, TIME AND ARCHITECTURE: THE GROWTH OF A NEW TRADITION, Harvard University Press, Cambridge, Massachusetts.

Gierke, H. E. von, 1977
GUIDELINES FOR ENVIRONMENTAL IMPACT STATEMENTS WITH RESPECT TO NOISE, Noise-Con 77 NASA Langley Res. Center, Hampton, Virginia.

Gluck, J. and Kalev, I., 1972
COMPUTER METHOD FOR ANALYSIS OF MULTI-STORY STRUCTURES, Journal of Computers and Structures, Vol. 2, pp. 897–913.

Godfrey, R. S., 1982
BUILDING SYSTEMS COST GUIDE, second edition, Robert S. Means Company, Inc., Kingston, MA.

Goh, K. S., 1956
"HOUSING" URBAN INCOMES AND HOUSING; A Report on the Social Survey of Singapore, Singapore Government Printers, pp. 61–68.

Goldberger, P., 1981
THE SKYSCRAPER, Alfred A. Knopf, Publisher, New York, New York, Created by Media Pojects, Inc.

Goldberger, P., 1985a
THE PROSPECT OF BIGGER TOWERS CAST A SHADOW, New York Times, December 29.

Goldberger, P., 1985b
IS TRUMP'S LATEST PROPOSAL JUST A CASTLE IN THE AIR?, New York Times, December 22.

Goldman, J., 1980
THE EMPIRE STATE BUILDING, St. Martin's Press, New York.

Good, K. O. and Carolina, J., 1980
CONSTRUCTION FOR PROFIT, Reston Publishing Company, Reston, Virginia.

Goto, K., Fukuda, T., Tanba, T., and Otsubo, Y., 1983
SELF CLIMBING INSPECTION MACHINE FOR EXTERNAL WALL, Robot No. 38, Japan Industrial Robot Association (JIRA), Tokyo.

Goto, T., 1981
HUMAN PERCEPTION AND TOLERANCE OF MOTION, Monograph of Council on Tall Buildings and Urban Habitat, Vol. PC, Planning and Environmental Criteria for Tall Buildings.

Goto, T., 1983
STUDIES OF WIND-INDUCED MOTION OF TALL BUILDINGS BASED ON OCCUPANTS' REACTION, Proceedings 6th International Conference on Wind Engineering, Session 5—Wind Loading of Tall Buildings (b), Gold Coast, Australia.

Goto, Y. and Chen, W. F., 1987
SECOND-ORDER ELASTIC ANALYSIS FOR FRAME DESIGN, Journal of Structural Engineering, ASCE, Vol. 113, No. 7, July, pp. 1501–1519.

Government of Maharashtra, 1974
FIRE PRECAUTIONS IN HIGH RISE BUILDINGS IN BOMBAY, Urban Development and Public Health Department, Bombay.

Government of Maharashtra, 1984
DEVELOPMENT CONTROL RULES FOR GREATER BOMBAY, as amended up to 1984, Urban Development and Public Health Department, Bombay.

Grannatt, M. H., III, 1975
THE ECONOMIC IMPACT OF AN OFFICE BUILDING HEIGHT RESTRICTION, Doctoral Dissertation, Lehigh University, April, pp. 43–89.

Griffith, A. A., 1921
THEORY OF RUPTURE OF BRITTLE MATERIALS, Transactions of the Royal Society, Ser. A, Vol. 21, pp. 163–198.

Gropius, W., 1946
DESIGN, Vol. 47, No. 8, Living Architecture (or International Style), April.

HMSO, 1965
COLLAPSE OF FLATS AT RONAN POINT, CANNING TOWN, Report of the Inquiry, HMSO, London.

HMSO, 1984
THE ASSESSMENT OF WIND LOADS, BRE Digest 119, Garston, UK.

Halpin, D. W. and Woodhead, R. W., 1980
CONSTRUCTION MANAGEMENT, John Wiley and Sons, Inc., New York, NY.

Hamilton, H. B., 1941
THE USE OF CAST IRON IN BUILDING, Transactions of The Newcomen Society, Vol. 21.

Hansen, R. J., Reed, J. W., and Vanmarke, E. H., 1973
HUMAN RESPONSE TO WIND-INDUCED MOTION OF BUILDINGS, ASCE Journal of Structural Division, July.

Harris, C. and Ullman, E., 1945
THE NATURE OF CITIES, Annals of the American Academy of Political and Social Science, Vol. 252.

Hart, G. C., Lew, M., and Di Julio, R., 1973
HIGH-RISE BUILDING RESPONSE: DAMPING AND PERIOD NONLINEARITIES, Proceedings of the 5th World Conference on Earthquake Engineering, Rome.

Hart, G. C. and Vasudevian, R., 1975
EARTHQUAKE DESIGN OF BUILDINGS: DAMPING, Journal of the Structural Division, ASCE ST1, January.

Haryott, R. B. and Glover, M. J., 1984
DEVELOPMENTS IN MULTI STOREY BUILDINGS. TRENDS IN THE DESIGN OF STEEL CONSTRUCTION, Nat. Struct. Steel Conference "New Developments in Steel Construction Part 1," Br. Const. Steel Assn., London, December 11–12.

Hasegawa, Y., 1981
ROBOTIZATION OF REINFORCED CONCRETE BUILDING CONSTRUCTION, Proceedings of 11th International Symposium for Industrial Robots (ISIR), JIRA, Tokyo.

Hasegawa, Y., 1983
ROBOTIZATION OF CONSTRUCTION WORK, Robot No. 38, Japan Industrial Robot Association (JIRA), Tokyo.

Hasegawa, Y., 1984
ROBOTIZATION OF CONSTRUCTION WORK, Proceedings of Workshop Conference on Robotics in Construction, Carnegie-Mellon University, Pittsburgh, PA.

Hasegawa, Y. and Sugimoto, N., 1982
INDUSTRIAL SAFETY AND ROBOTS, Proceedings of 12th ISIR, Paris, ISF Publications, England.

Hasegawa, Y. and Tamaki, K., 1985
MODULES FOR CONSTRUCTION ROBOTICS SYSTEMS, Proceedings of 2nd Workshop Conference on Robotics in Construction, Carnegie-Mellon University, Pittsburgh, PA.

Hasegawa, Y. and Tamaki, K., 1985
ROBOT MODULE APPLICATION FOR COMPLICATED CONSTRUCTION SYSTEM, Proceedings of 15th ISIR, Tokyo.

Hayes-Roth, F., Waterman, D. A., and Lenat, D. B., 1983
AN OVERVIEW OF EXPERT SYSTEMS in Hayes-Roth, F., Waterman, D. A., and Lenat, D. B. (eds.), Building Expert Systems, Addison-Wesley, Reading, Mass.

Hayes-Roth, F., Waterman, D., and Lenat, D., 1983
BUILDING EXPERT SYSTEMS. Addison-Wesley, Reading, Ma.

Hely, B. G. and Taylor, D. G., 1983
INDUSTRIALIZED HOUSING PROJECT—SINGAPORE, Concrete Institute of Australia Conference.

Henkel, D. J., Philips, A. B., and Gulager-Nielsen, E., 1984
NATIONAL BANK HOUSE, MELBOURNE—FOUNDATION DESIGN AND PERFORMANCE, Proceedings 4th Australian & New Zealand Conference on Geomechanics, Perth, Australia, May 14–18, pp. 300–304.

Hinkley, P., 1975
WORK BY THE FIRE RESEARCH STATION ON THE CONTROL OF SMOKE IN SHOPPING CENTRES, Building Research Establishment CP83/75, Borehamwood.

Hitchcock, H. R. and P. Johnson, 1932
THE INTERNATIONAL STYLE: ARCHITECTURE SINCE 1922, W. W. Norton & Co.

Holmes, J. D., 1984
WIND ACTION ON GLASS AND BROWN'S INTEGRAL, workshop on Wind Engineering and Industrial Aerodynamics, C.S.I.R.O. Division of Building Research, Australia, pp. 35–38.

Home Office, 1980
FUTURE FIRE POLICY: A CONSULTATIVE DOCUMENT, Home Office, Scottish Home and Health Department, London, HMSO.

Horizon Magazine, 1984
DALLAS: ARTS IN THE HEART OF TEXAS, Horizon Magazine, May.

Hose, R., 1984
THE RIALTO PROJECT—CARPARK AND OFFICE BUILDING, The Association of Consulting Structural Engineers N.S.W. Australia, Modern Structural Engineering.

Housing and Development Board, 1970
FIRST DECADE IN PUBLIC HOUSING 1960–1969, Housing and Development Board, Singapore.

Housing and Development Board, 1982
OUR HOME, A Housing and Development Board Publication, June.

Housing and Development Board, 1984/85
ANNUAL REPORT, Housing and Development Board, Singapore.

Housing and Development Board, 1985
DESIGNED FOR LIVING—PUBLIC HOUSING ARCHITECTURE IN SINGAPORE, Housing and Development Board, Singapore.

Hoyt, H., 1939
THE STRUCTURE AND GROWTH OF RESIDENTIAL NEIGHBORHOODS IN AMERICAN CITIES, U.S. Federal Housing Administration, U.S. Government Printing Office, Washington, DC.

Hrovat, D., Barak, P., and Rabins, M., 1983
SEMI-ACTIVE VERSUS PASSIVE OR ACTIVE TUNED MASS DAMPERS FOR STRUCTURAL CONTROL, Journal of Engineering Mechanics Division, ASCE, Vol. 109, No. 3, June, pp. 691–705.

Hruska, E., 1985
TOWARD THE CREATION OF URBAN ENVIRONMENT, (in Slovak: K tvorbe urbanistickeho prostredia), Zvaz slovenskych architektov, Bratislava, Czechoslovakia.

Hudson, D. E., 1965
EQUIVALENT VISCOUS FRICTION FOR HYSTERETIC SYSTEMS WITH EARTHQUAKE-LIKE EXCITATIONS, Proceedings of the 3rd World Conference on earthquake engineering, Vol. 11, pp. 185–201.

Hutsch, V., 1980
THE MUNICH GLASS PALACE 1854–1931 (Der Munchner Glaspalast 1854–1931), Munchen.

ICBO, 1982
UNIFORM BUILDING CODE, International Conference of Building Officials, Whittier, CA.

ICBO, 1985
UNIFORM BUILDING CODE, International Conference of Building Officials, Whittier, CA.

INTEGRO, 1986a
CATALOGUE OF PROJECTS AND BUILDINGS, ZIPP. Bratislava.

INTEGRO, 1986b
Proceedings of Conference in Prague.

International Association for Earthquake Engineering, 1984
EARTHQUAKE RESISTANT REGULATIONS. A WORLD LIST, Tokyo, Japan.

International Standards Organization, 1976
ISO, Code 2631, Oslo, Norway.

International Standards Organization, 1984
GUIDELINES FOR THE EVALUATION OF THE RESPONSE OF OCCUPANTS OF FIXED STRUCTURES, ESPECIALLY BUILDINGS AND OFF-SHORE STRUCTURES, TO LOW-FREQUENCY HORIZONTAL MOTION (0,063 to 1 Hz), International Standards Organization, ISO6897, Geneva.

International Standards Organization, 1986
EVALUATION OF HUMAN EXPOSURE TO WHOLE-BODY VIBRATION—EVALUATION OF HUMAN EXPOSURE TO CONTINUOUS AND SHOCK INDUCED VIBRATIONS IN BUILDINGS (1-80 Hz), International Standards Organization, ISO DIS2631, Part 2.

Irwin, A. W., 1975
HUMAN REACTIONS TO OSCILLATIONS OF BUILDINGS—ACCEPTABLE LIMITS, Build International, Applied Science Publishers, London.

Irwin, A. W., 1978
HUMAN RESPONSE TO DYNAMIC MOTION OF STRUCTURES, The Structural Engineer, Vol. 56A, No. 9, and Vol. 58A, No. 3.

Irwin, A. W., 1979
A DIFFERENT KIND OF WALTZ, UK Conference on Human Response to Vibration, Royal Aircraft Establishment, Farnborough, UK.

Irwin, A. W., 1981a
PERCEPTION, COMFORT AND PERFORMANCE CRITERIA FOR HUMAN BEINGS EXPOSED TO WHOLE BODY PURE YAW VIBRATION AND VIBRATION CONTAINING YAW AND TRANSLATIONAL COMPONENTS, Journal of Sound and Vibration, Vol. 76, No. 4.

Irwin, A. W., 1981b
STUDY AND EVALUATION OF HUMAN RESPONSE TO PURE AND COMBINED FORMS OF LOW FREQUENCY MOTION AT VARIOUS LEVELS, International Workshop on Research Methods in Human Motion and Vibration Studies, New Orleans.

Irwin, A. W., 1982
A METHOD FOR ASSESSMENT OF PROBABLE HUMAN RESPONSE TO COMBINED NOISE AND MECHANICAL VIBRATION, Proceedings Noise and Environment, Bratislava, Czechoslovakia.

Irwin, A. W., 1983a
DIVERSITY OF HUMAN RESPONSE TO VIBRATION ENVIRONMENTS, Proceedings Conference on Human Response to Vibration, NIAE/NCAE, Silsoe, UK.

Irwin, A. W., 1983b
RELATIVE INFLUENCE OF NOISE AND WHOLE BODY VIBRATION ON THE RESPONSE OF HUMANS, Proceedings Inter-Noise 83, Vol. 11, Institute of Acoustics, Edinburgh.

Irwin, A. W., 1984a
DESIGN OF SHEAR WALL BUILDINGS, Construction Industry Research & Information Assoc., Report No. 102, London, UK.

Irwin, A. W., 1984b
INTRODUCTION TO NOISE PHYSICS, Lecture to UK Offshore Operators Association Medical Group, Aberdeen, Scotland.

Irwin, A. W. and Goto, T., 1984
HUMAN PERCEPTION, TASK PERFORMANCE AND SIMULATOR SICKNESS IN SINGLE AND MULTI-AXIS LOW FREQUENCY HORIZONTAL LINEAR AND ROTATIONAL VIBRATION, Proc. UK-HRV 84, Edinburgh, Scotland.

Ishizaki, H., 1983
WIND PROFILES, TURBULENCE INTENSITIES AND GUST FACTORS FOR DESIGN IN TYPHOON-PRONE REGIONS, Journal of Wind Engineering and Industrial Aerodynamics, Vol. 13, Nos. 1-3, December, pp. 55–66.

Isyumov, N., Davenport, A. G., and Monbaliu, J., 1984
CN TOWER, TORONTO: MODEL AND FULL SCALE RESPONSE TO WIND, International Association for Bridge and Structural Engineering (IABSE) 12th Congress, Session VI, September 3–7, Vancouver, B.C.

Jacobs, J., 1984
PRINCIPLES OF ECONOMIC LIFE, Random House, New York.

Jacobsen, L. S., 1930
STEADY FORCED VIBRATION AS INFLUENCED BY DAMPING, Trans. ASME-APM-52-15, pp. 169–181.

Jacobsen, L. S., 1965
DAMPING IN COMPOSITE STRUCTURES, Proceedings of the 2nd World Conference on Earthquake Engineering, pp. 1029–1044.

Jeary, A. P., 1981
THE DYNAMIC BEHAVIOUR OF TALL BUILDINGS, Ph.D. thesis, University College, London, UK, March.

Jeary, A. P., 1986
DAMPING IN TALL BUILDINGS—A MECHANISM AND A PREDICTOR, Earthquake Engineering and Structural Dynamics, Vol. 14, pp. 733–750.

Jeary, A. P. and Ellis, B. R., 1981
VIBRATION TESTS ON STRUCTURES AT VARIED AMPLITUDES, ASCE/EMD Conference on Dynamics of Structures, Atlanta, Georgia.

Johnson, C. R., 1981
USE OF VISCO ELASTIC DAMPERS IN REDUCING THE WIND-INDUCED MOTION OF TALL BUILDINGS, M.S. Thesis, Department of Civil Engineering, Massachusetts Institute of Technology, August.

Joint Committee on Tall Buildings, 1973
PLANNING AND DESIGN OF TALL BUILDINGS (Proceedings of ASCE-IABSE International Conference held at Lehigh University, August, 1972), ASCE, New York (5 volumes).

Jones, S. W., Kirby, P. A., and Nethercot, D. A., 1982
COLUMNS WITH SEMI-RIGID JOINTS, Journal of the Structural Division, ASCE, Vol. 108, No. ST2, February, pp. 361–372.

Kangari, R., 1985
ROBOTICS FEASIBILITY IN THE CONSTRUCTION INDUSTRY, Proceedings of 2nd Workshop Conference on Robotics in Construction, Carnegie-Mellon University, Pittsburgh, PA.

Kano, N., 1985
CONSTRUCTION PLANNING BY A ROBOT, Proceedings of 2nd Workshop Conference on Robotics in Construction, Carnegie-Mellon University, Pittsburgh, PA.

Kareem, A., 1985
HURRICANE ALICIA: ONE YEAR LATER, Proceedings of a Specialty Conference (Galveston, Texas, August 16–17, 1984), ASCE, New York.

Kareem, A. and Stevens, J. G., 1985
WINDOW GLASS PERFORMANCE AND ANALYSIS IN HURRICANE ALICIA, Proceedings, Specialty Conference, Hurricane Alicia: One Year Later (Galveston, Texas, August 16–17, 1984), ASCE, New York, pp. 178–186.

Kasai, K. and Popov, E. P., 1986
GENERAL BEHAVIOR OF WF STEEL SHEAR LINK BEAMS, Journal of Structural Engineering, ASCE, Vol. 111, February.

Kateley, R. and Lachman, M., 1985
INVESTMENT OPPORTUNITIES IN APARTMENTS, John Hancock Realty, Boston.

Kato, B., 1982
BEAM-TO-COLUMN CONNECTION RESEARCH IN JAPAN, Journal of the Structural Division, ASCE, Vol. 108, No. ST2, February, pp. 343–360.

Kawasumi, H. and Kanai, K., 1956
VIBRATION IN BUILDINGS IN JAPAN, Proceedings of the 1st World Conference on Earthquake Engineering, Berkeley.

Kaynia, A. M., Veneziano, D., and Biggs, J. M., 1981
SEISMIC EFFECTIVENESS OF TUNED MASS DAMPERS, Journal of Structural Division, ASCE, Vol. 107, No. ST8, August, pp. 1465–1484.

Keel, C. J. and Mahmoodi, P., 1986
DESIGN OF VISCOELASTIC DAMPERS FOR COLUMBIA CENTER BUILDING, Building Motion in Wind, Proceedings of a session at ASCE convention, Seattle, Washington, April 8, pp. 66–81.

Kendik, E., 1985
ASSESSMENT OF ESCAPE ROUTES IN BUILDINGS AND A DESIGN METHOD FOR CALCULATING PEDESTRIAN MOVEMENT, Paper at SFPE 35th Anniversary Engineering Seminar, Chicago, May.

Kerr, R. A., 1985
PREDICTABLE QUAKE DAMAGE, Science, Vol. 230, No. 4726, November, p. 633.

Khan, F., 1974
A CRISIS IN DESIGN—THE NEW ROLE OF THE STRUCTURAL ENGINEER, Proceedings of the Conferences on Tall Buildings (Kuala Lumpur, Malaysia, December), Institution of Civil Engineers, Kuala Lumpur, Malaysia.

Khan, F. R. and Parmelee, R. A., 1971
SERVICE CRITERIA FOR TALL BUILDINGS FOR WIND LOADING, Proc. 3rd Int. Conf. on Wind Effects on Buildings and Structures, Tokyo, Japan.

Kishi, N. and Chen, W. F., 1986
DATA BASE ON STEEL BEAM-TO-COLUMN CONNECTIONS, Structural Engineering Report No. CE-STR-86-26, School of Civil Engineering, Purdue University, 2 volumes, 653 pp.

Klein, R. E. and Salhi, H., 1980
THE TIME-OPTIMAL CONTROL OF WIND-INDUCED STRUCTURAL VIBRATIONS USING ACTIVE APPENDAGES, Structural Control, H. H. Leipholz, ed., North-Holland Publishing Company, Amsterdam, the Netherlands, pp. 415–429.

Klote, J. and Fothergill, J., 1983
DESIGN OF SMOKE CONTROL SYSTEMS FOR BUILDINGS, ASHRAE, September.

Kobayashi, H. and Sugiyama, N., 1977
VISCOUS DAMPING OF STRUCTURES RELATED TO FOUNDATION CONDITIONS, Proceedings of the 6th World Conference on Earthquake Engineering, New Delhi.

Kobayashi, K., 1985
DEVELOPMENT OF ADVANCED SYSTEM FOR CONSTRUCTION TECHNOLOGIES WITH PROPER USE OF ELECTRONICS, Proceedings of 15th ISIR, Tokyo.

Krawinkler, H., 1978
SHEAR IN BEAM-COLUMN JOINTS IN SEISMIC DESIGN OF STEEL FRAMES, AISC Engineering Journal, 3rd Quarter, pp. 82–91.

Kwok, K. C. S., 1984
DAMPING INCREASE IN BUILDING WITH TUNED MASS DAMPER, Journal of Engineering Mechanics Division, ASCE, Vol. 110, No. 11, November, pp. 1645–1649.

Lachman, M. and Miller, R., 1985
DOWNTOWN HOUSING—WHERE THE ACTION IS, Journal of Real Estate Development, Vol. 1, #1, pp. 15–27.

Landstrom, U., Lunstrom, R., and Bystrom, M., 1983
EXPOSURE TO INFRASOUND—PERCEPTION AND CHANGES IN WAKEFULNESS, Journal of Low Frequency Noise and Vibration, Vol. 2, No. 1.

Lansdown, J., 1982
EXPERT SYSTEMS, THEIR IMPACT ON THE CONSTRUCTION INDUSTRY, Report in RIBA Conference Fund, London.

Larson, G. E., 1981
FIRE, EARTH AND WIND: TECHNICAL SOURCES OF THE CHICAGO SKYSCRAPER, Inland Architect, 25:7, September, pp. 20–29.

Law, M., 1985a
FIRE PROTECTION IN TERMINAL BUILDINGS, Paper in Symposium on Building Services for Airports, Gatwick, CIBSE, November 6–7.

Law, M., 1985b
TRANSLATION OF RESEARCH INTO PRACTICE: BUILDING DESIGN, Paper in First International Symposium on Fire Safety Science, Gaithersburg, October 9–11.

Lawson, R. M. and Nethercot, D., 1985
LATERAL STABILITY OF I BEAMS RESTRAINED BY PROFILED SHEETING, The Structural Engineering, Vol. 63B, No. 1.

Leipholz, H. H., 1980
STRUCTURAL CONTROL, North-Holland Publishing Company, Amsterdam, the Netherlands.

Lewicki, B., 1966
BUILDING WITH LARGE PREFABRICATES, Elsevier Publishing Company, Amsterdam.

Lewicki, B. and coll., 1982
PROGETTAZIONE DI EDIFICI MULTIPIANO INDUSTRIALIZZATI, ITEC editrice, Milano.

Lewicki, B., Cholewicki, A., and Makulski, W., 1983
LARGE-PANEL BUILDING: BEHAVIOUR IN PARTIAL DAMAGE, Building Research & Practice, July/August.

Lewitt, C. S., Chesson, E., and Munse, W. H., 1969
RESTRAINT CHARACTERISTICS OF FLEXIBLE RIVETED AND BOLTED BEAM-TO-COLUMN CONNECTIONS, Engineering Experiment Station Bulletin No. 500, University of Illinois at Urban-Champaign, Illinois, January.

Lie, T. T., 1972
FIRE AND BUILDINGS, Applied Science Publishers, London.

Lim, B. P. and Rao, K. R., 1977
ENVIRONMENTAL CONTROL OF BUILDING, Journal of the Singapore National Academy of Science, Vol. 6, No. 1, p. 78 ff.

Lim, B. P., Rao, K. R., and Rao, S. P., 1980a
AIR-CONDITIONING IN SINGAPORE—IS IT NECESSARY?, Proceedings of the Conference on Energy Conservation and Management in Buildings, Institute of Engineering, Singapore, March, pp. 108–138.

Lim, B. P., et al., 1980b
DEVELOPING NOISE CRITERIA FOR SINGAPORE, Seminar on Noise: Problems and Control, The Science Council of Singapore et al., January, Singapore, pp. 48–74.

Lim, W. S. W., 1983
PUBLIC HOUSING AND COMMUNITY DEVELOPMENT: THE SINGAPORE EXPERIENCE, MIMAR 7, Architecture in Development.

Lin, M., 1985
BEAUTY AND THE BANK, The New Republic, December 23.

Lin, T. Y. and Stotesbury, S. D., 1981
STRUCTURAL CONCEPTS AND SYSTEMS FOR ARCHITECTS AND ENGINEERS. John Wiley and Sons, New York.

Liu, T.-K., 1979
PUBLIC HOUSING—THE SINGAPORE EXPERIENCE, paper prepared for the 17th IFAWPCA Convention, Singapore, 23–28 September.

Liu, T.-K., 1983
HOUSING POLICIES AND LIFE STYLE, Paper presented at the 1983 Singapore Professional Convention.

Liu, T.-K., 1984
HIGH-RISE, HIGH-DENSITY LIVING, Singapore Professional Centre, SPC Convention 1983, Selected Papers, August, pp. 10–24.

Ludgin, M. K. and Masoti, L. H., 1985
DOWNTOWN DEVELOPMENT, CHICAGO: 1979–1984, Evanston, IL; Center for Urban Affairs and Policy Research, Northwestern Univ.

Ludgin, M. K. and Masoti, L. H., 1986
DOWNTOWN DEVELOPMENT, CHICAGO: 1985–1986, joint publication of Northwestern University Center for Urban Affairs and the City of Chicago, Department of Planning.
Lui, E. M., 1985
EFFECTS OF CONNECTION FLEXIBILITY AND PANEL ZONE DEFORMATION ON THE BEHAVIOR OF PLANE STEEL FRAMES, Ph.D. Dissertation, School of Civil Engineering, Purdue University, West Lafayette, Indiana.
Lui, E. M. and Chen, W. F., 1983
END RESTRAINT AND COLUMN DESIGN USING LRFD, AISC Engineering Journal, Vol. 20, No. 1, 1st Quarter, pp. 29–39.
Lund, R., 1980
ACTIVE DAMPING OF LARGE STRUCTURES IN WINDS, Structural Control, H. H. Leipholz, ed., North-Holland Publishing Company Amsterdam, the Netherlands, pp. 459–470.

Mackenzie, I. M., 1980
THE CBA CENTRE, SYDNEY, Steel Construction Journal, Australia, Steel Const., Sydney, Vol. 14, No. 1.
Mackenzie, I. M., 1981
THE CBA CENTRE, SYDNEY, The Arup Journal, London, Vol. 16, No. 4, December.
MacLeod, I. A., 1969
NEW RECTANGULAR FINITE ELEMENT FOR SHEAR WALL ANALYSIS, Journal of Structural Division, Proceedings ASCE, Vol. 95, No. ST3, March, pp. 399–409.
MacLeod, I. A., 1977
STRUCTURAL ANALYSIS OF WALL SYSTEMS, Structural Engineer, London, Vol. 55, pp. 487–494.
Maher, M. L., 1984
HI-RISE: A KNOWLEDGE-BASED EXPERT SYSTEM FOR THE PRELIMINARY STRUCTURAL DESIGN OF HIGH RISE BUILDINGS, Research Report, Design Research Center, Carnegie-Mellon University, Pittsburgh.
Maher, M. L. and Fenves, S. J., 1984
HI-RISE: AN EXPERT SYSTEM FOR THE PRELIMINARY STRUCTURAL DESIGN OF HIGH RISE BUILDINGS, in Knowledge Engineering in Computer-Aided Design, North-Holland, Netherlands.
Mahmoodi, P., 1969
STRUCTURAL DAMPERS, Journal of Structural Division, ASCE, Vol. 95, No. ST8, August, pp. 1661–1672.
Malley, J. O. and Popov, E. P., 1984
DESIGN OF SHEAR LINKS IN ECCENTRICALLY BRACED FRAMES, Journal of Structural Engineering, ASCE, Vol. 110, September, pp. 2275–2295.
Marans, R. and Spreckelmeyer, K. F., 1981
EVALUATING BUILT ENVIRONMENTS: A BEHAVIORAL APPROACH, Survey Research Center and Architectural Research Laboratory, University of Michigan, Ann Arbor.

Marrey, B. and Chemetov, P., 1976
FAMILIARLY UNKNOWN . . . ARCHITECTURE, PARIS 1848–1914 (familierement inconnues . . . Architectures, Paris 1848–1914), Catalogue, n.d., n.p. Paris, Bellamy & Martet.

Masri, S. F., 1973
RESPONSE OF THE IMPACT DAMPER TO STATIONARY RANDOM EXCITATION, Journal of Acoustic Society of America, Vol. 53, No. 1, January, pp. 200–211.

Masri, S. F., Bekey, G. A., and Caughey, T. K., 1981
OPTIMUM PULSE CONTROL OF FLEXIBLE STRUCTURES, ASME Journal of Applied Mechanics, Vol. 48, September, pp. 619–626.

Masri, S. F., Bekey, G. A., and Caughey, T. K., 1982
ON-LINE CONTROL OF NON-LINEAR FLEXIBLE STRUCTURES, ASME Journal of Applied Mechanics, Vol. 49, December, pp. 877–884.

Matsubara, S., 1983
FLOOR CLEANING ROBOT, Robot No. 38, JIRA, Tokyo.

McDermott, J., 1980
R1: A RULE-BASED CONFIGURER OF COMPUTER SYSTEMS. Technical Report CMU-CS-80-119, Carnegie-Mellon University, Pittsburgh, Pa.

McGraw-Hill Information Systems Company, 1986
DODGE BUILDING COST CALCULATOR AND VALUATION GUIDE, Quarterly, McGraw-Hill Information Systems Company, New York, NY.

McGuire, W. and Gallagher, R. H., 1979
MATRIX STRUCTURAL ANALYSIS, Wiley, New York.

McNamara, R. J., 1977
TUNED MASS DAMPERS FOR BUILDINGS, Journal of Structural Division, ASCE, Vol. 103, No. ST9, September, pp. 1785–1798.

Mehrtens, G. C., 1908
LECTURES IN ENGINEERING SCIENCE, Part II. Iron Bridge Construction, Vol. I (Vorlesungen Uber Ingenieur-Wissenschaften, II, Teil, Eisenbruckenbau, Bd. 1), Wilhelm Engelmann, Leipzig.

Mehta, J. B., 1977
AMENITIES IN TALLER BUILDINGS IN DEVELOPING COUNTRIES, 2001, Urban Space for Life and Work (Proceedings of Conference held in Paris, 1977), Vol. II, CTICM, p. 63–66.

Mehta, J. B., 1978
HIGH RISE BUILDINGS, published by the author.

Melbourne, W. H., 1979
TURBULENCE EFFECTS ON MAXIMUM SURFACE PRESSURES—A MECHANISM AND POSSIBILITY OF REDUCTION, Proc. 5th Int. Conf. on Wind Engineering, Fort Collins, Pergamon Press, pp. 541–552.

Melchers, R. E. and Harrington, M. V., 1984
HUMAN ERROR IN STRUCTURAL RELIABILITY—I. INVESTIGATION OF TYPICAL DESIGN TASKS, Report No. 2, Civil Engineering, Monash University.

Melinek, S. and Booth, S., 1975
AN ANALYSIS OF EVACUATION TIMES AND THE MOVEMENT OF CROWDS IN BUILDINGS, Building Research, Establishment CP96/75, Borehamwood.

Merovich, A. T. and Nicoletti, J. P., 1982
ECCENTRIC BRACING IN TALL BUILDINGS, Journal of the Structural Division, ASCE, Vol. 108, May, pp. 2066–2080.

Michael, D. and Anderson, J. M. D., 1967
SUSPENDED FRAME BUILDINGS AND IN PARTICULAR THE STANDARD BANK CENTRE, Arup Journal, May, pp, 10–19.

Michalski, R. S. M., Carbonell, J. G., and Mitchell, T. M., 1983
MACHINE LEARNING, Tioga Publishing Company, Palo Alto, Calif.

Michelson, W. (1975)
BEHAVIORAL RESEARCH METHODS IN ENVIRONMENTAL DESIGN, Dowden, Hutchinson & Ross, Inc., Stroudsburg, Pa.

Ministry of Housing and Local Government, 1968
REPORT OF THE INQUIRY INTO THE COLLAPSE AT RONAN POINT, London.

Minor, J. E., 1984
WINDOW GLASS DESIGN, Workshop Notes, Fourth Canadian Workshop on Wind Engineering (Toronto, November 19–20, 1984), Canadian Wind Engineering Association, National Research Council of Canada, Ottawa, pp. 195–212.

Minor, J. E., 1985
WINDOW GLASS PERFORMANCE AND HURRICANE EFFECTS, Proceedings, Specialty Conference, Hurricane Alicia: One Year Later (Galveston, Texas, August 16–17, 1984), ASCE, New York, pp. 151–167.

Mocak, V., 1979
SECURITY PROVISIONS IN CONCENTRATED HOUSING DEVELOPMENT, Proceedings of International Conference on Housing Planning, Financing and Construction, pp. 650–661, Florida International University, Pergamon Press, New York, Oxford, Toronto, Sydney Frankfurt, Paris.

Mocak, V., 1983
SKYLOBBIES: THE EMERGING NEW INFRASTRUCTURE ELEMENTS, Lecture at Fifth International Conference on Urban Design, Washington, DC, October 26–29.

Mocak, V., 1984
SKYSCRAPERS AT WATERFRONT LANDS, Lecture at School of Architecture, Pratt Institute, New York, August 1.

Mocak, V., 1986
ZONING OF SKYSCRAPERS AT WATERFRONT LANDS, Lecture at Skidmore, Owings & Merrill, New York, January 15.

Modjeski, R., Webster, G. S., and Ball, L. A., 1927
THE BRIDGE OVER THE DELAWARE RIVER CONNECTING PHILADELPHIA, PA. AND CAMDEN, N.J., Final Report of the Board of Engineers to the Delaware River Bridge Joint Commission of the States of Pennsylvania and New Jersey, n.p.

Moncarz, P. D. and Gerstle, K. H., 1981
STEEL FRAMES WITH NON-LINEAR CONNECTIONS, Journal of the Structural Division, ASCE, Vol. 107, No. ST8, August, pp. 1427–1441.

Moore, D. B., 1983
THE STRUCTURE OF PYRIMOIDAL SHELLS, Ph.D. thesis, Bradford University, UK.

Moore, J. F. A., 1982
DISCUSSION ON PAPER BY ROSE AND BURBAGE, The Structural Engineer, V60A, No. 8, August.

Morgan, H. and Chandler, S., 1981
FIRE SIZES AND SPRINKLER EFFECTIVENESS IN SHOPPING COMPLEXES AND RETAIL PREMISES, Fire Surveyor, October.

Morgan, H. and Hansell, G., 1985
FIRE SIZES AND SPRINKLER EFFECTIVENESS IN OFFICES—IMPLICATIONS FOR SMOKE CONTROL DESIGN, Fire Safety Journal, Vol. 8, No. 3.

Mori, M., Nakamura, T., Matsushita, H., and Tase, Y., 1983
SOME EXAMPLES OF ROBOTIZATION IN OUR COMPANY, Robot No. 38, JIRA, Tokyo.

Morris, Tasker and Company, 1858
PRICE LIST, Morris, Tasker and Company, Philadelphia, Pa.

Morris, Tasker and Company, 1861
ILLUSTRATED CATALOGUE, Fourth Edition, Morris, Tasker and Company, Philadelphia, Pa.

Mullarkey, P. M., 1983
DEVELOPMENT OF AN EXPERT SYSTEM FOR INTERPRETATION OF GEOTECHNICAL CHARACTERIZATION DATA FROM CONE PENETROMETERS. Ph.D. thesis proposal, Department of Civil Engineering, Carnegie-Mellon University, November.

Murai, T., Aogyagi, H., and Kawamura, T., 1983
CONCRETE DISTRIBUTING ROBOT, Robot No. 38, JIRA, Tokyo.

NBS, 1978
TENTATIVE PROVISIONS FOR THE DEVELOPMENT OF SEISMIC REGULATIONS FOR BUILDING, NBS Special Publication 510, June, National Bureau of Standards.

NCR/EERI, 1985
IMPRESSIONS OF THE GUERRO-MECHOACAN, MEXICO EARTHQUAKE OF 19 SEPTEMBER 1985: A PRELIMINARY RECONNAISSANCE REPORT, Publication No. 85-05, National Research Council/Earthquake Engineering Research Institute, Berkeley, CA, October, 20 pp. plus figures.

Nair, R. S., 1975a
LINEAR STRUCTURAL ANALYSIS OF MULTISTORY BUILDINGS, Journal of Structural Division, Proceedings of ASCE, Vol. 101, No. ST3, March, pp. 551–565.

Nair, R. S., 1975b
OVERALL ELASTIC STABILITY OF MULTISTORY BUILDINGS, Journal of the Structural Division, ASCE, December.

Nair, R. S., 1983
A SIMPLE METHOD OF OVERALL STABILITY ANALYSIS FOR MULTISTORY BUILDINGS, Developments in Tall Buildings 1983, Council on Tall Buildings and Urban Habitat, Hutchinson Ross Publishing Co.

Naito, T., Nasu, N., Takeushi, M., Kubota, G., Tanaka, Y., and Hara, M., 1956
VIBRATIONAL CHARACTERISTICS OF REINFORCED CONCRETE BUILDINGS EXISTING IN JAPAN, Proceedings of the 1st World Conference on Earthquake Engineering, Berkeley.

Nakagawa, K., 1961
VIBRATIONAL CHARACTERISTICS OF BUILDINGS, PART 2, VIBRATIONAL CHARACTERISTICS OF ACTUAL BUILDINGS DETERMINED BY VIBRATION TESTS, Waseda University Bulletin of Science and Engineering Research Laboratory Report No. 16.

National Fire Protection Association, 1981
LIFE SAFETY CODE, National Fire Protection Association.

Nau, D. S., 1983
EXPERT COMPUTER SYSTEMS, Computer 16:63–85, February.

Nethercot, D. A., 1983
EVALUATION OF INTERACTION EQUATIONS FOR USE IN DESIGN SPECIFICATIONS IN WESTERN EUROPE, Proceedings of the Third International Colloquium on Stability of Metal Structures, Toronto, pp. 293–314.

Nethercot, D., 1985
DESIGN OF LATERALLY UNSUPPORTED BEAMS and ELASTIC LATERAL BUCKLING OF BEAMS, Beams and Beam Columns, Applied Science Publishers Ltd.

New York, City of, 1984
NEW YORK CITY ZONING RESOLUTION, Department of City Planning, New York.

New York Planning Commission, 1982
MIDTOWN ZONING, New York Planning Commission, Department of City Planning, Library of Congress, Catalogue Card #82-81294.

Newberry, C. W. and Eaton, K., 1974
WIND LOADING HANDBOOK, Department of the Environment, Building Research Establishment, London, H.M.S.O.

Newell, J. E., 1979
BUILDER'S GUIDE TO CONSTRUCTION FINANCING, Craftsman Book Company, Solana Beach, CA.

Nillson, N. J., 1980
PRINCIPLES OF ARTIFICIAL INTELLIGENCE, Tioga Publishing Company, Palo Alto, California.

Nof, S. Y., 1985
HANDBOOK OF INDUSTRIAL ROBOTICS, John Wiley and Sons, New York.

Nutt, J. G., 1984
WIND AND THE DESIGN OF HIGH RISE BUILDINGS, Seminar—Modern Structural Engineering, The Association of Construction Engineers of NSW, Sydney, Australia, August 15.

Nutt, J. G. and Haworth, D. P., 1976
CAPITAL TOWER, The Arup Journal, No. 3, Vol. 11, October, Ove Arup Partnerships, London.

Otis Elevator Company, 1953
THE FIRST HUNDRED YEARS, Otis Elevator Company, New York.

Ove Arup & Partners, 1984
GROSVENOR PLACE, SYDNEY REPORT ON TRANSFER SYSTEM PART 1—INTERACTION WITH SURROUNDING STRUCTURE, October.

Ove Arup & Partners, 1984
GROSVENOR PLACE, SYDNEY REPORT ON TRANSFER SYSTEM PART 2—ASSESSMENT OF TRANSFER FRAMES, November.

Parkinson, B. and Hirst, J., 1981
TSUEN WAN RESIDENTIAL BLOCKS, Arup Journal, April, pp. 2–6.

Pascoe, G. W., 1984
STRUCTURAL STEEL FLOOR SYSTEMS, Seminar—Floor Systems in Building Construction, Institute of Engineers, Australia, Victoria Branch, Melbourne, September 14.

Pauls, J., 1977
MOVEMENT OF PEOPLE IN BUILDING EVACUATIONS, Human Response to Tall Buildings, Chapter 21, P. J. Conway, ed., Dowen, Hutchinson and Ross, Stroudsburg, Pennsylvania, Community Development Series, Vol. 34.

Paulson, Jr., B. C., 1984
THE POTENTIAL FOR ROBOTICS IN CONSTRUCTION, Technical paper of Manufacturing Engineers, Robots West Conference, Anaheim, CA.

Pawlak, Z., 1982
ROUGH SETS, International Journal of Information and Computer Sciences, Vol. 11, No. 5.

Perrow, C., 1984
NORMAL ACCIDENTS, Basic Books, Inc., New York.

Peshek, Jr., C., 1978
THE AISC QUALITY CERTIFICATION PROGRAM, Engineering Journal, AISC, Vol. 15, No. 3, Chicago, pp. 102–107.

Petak, W. and Hart, G. C., 1980
DAMAGE AND DECISION MAKING IN WIND ENGINEERING, Proceedings, Fifth International Conference on Wind Engineering (Ft. Collins, CO, July, 1979), Pergamon Press, Oxford, pp. 61–74.

Peters, T. F., 1981
THE DEVELOPMENT OF LONG-SPAN BRIDGE BUILDING, 3rd ed., Zurich: ETH.

Petersen, N. R., 1980
DESIGN OF LARGE SCALE TUNED MASS DAMPERS, Structural Control, H. H. Leipholz, ed., North-Holland Publishing Company, Amsterdam, the Netherlands, pp. 581–596.

Petrovski, J., Jurukovski, D. and Paskalov, T., 1973
DYNAMIC PROPERTIES OF A FOURTEEN STOREY RC FRAME BUILDING FROM FULL SCALE FORCED VIBRATION STUDY AND FORMULATION OF A MATHEMATICAL MODEL, Proceedings of the 5th World Conference on Earthquake Engineering Rome.

Pfluger, A., 1970
VIBRATIONS OF PEDESTRIAN BRIDGES (Schwingungen von Fussweg-brucken), Report No. 15, Institut fur Statik der TU Hannover, Hannover, West Germany.

Phillips, M. H., Wood, J. H., and Docherty, J., 1984
HORIZONTAL VIBRATION OF HOUSES, Report No. 5-84/3, Central Laboratories, Ministry of Works and Development, Lower Hutt, New Zealand.

Popov, E. P. and Black, R. G., 1981
STEEL STRUTS UNDER SEVERE CYCLIC LOADING, Journal of the Structural Division, ASCE, Vol. 107, September, pp. 1857–1881.

Popov, E., Takahashi, K., and Roeder, C. W., 1976
STRUCTURAL STEEL BRACING SYSTEMS. BEHAVIOR UNDER CYCLIC LOADING, Earthquake Engineering Research Center, Report UCB/EERC-76/17, University of California, Berkeley.

Port Authority of New York and New Jersey, 1984
JOURNAL SQUARE TRANSPORTATION CENTER CONCOURSE CEILING COLLAPSE, Special Investigation Report, June.

Prince, J., et al., 1985
INFORMES IPS-10A, IPS-10B, IPS-10C, IPS-10D, IPS-10E, GAA-1A, GAA-1B and GAA-IC, Instituto de Ingenieria, UNAM, September–October.

Purandare, D. D., 1982
FIGHTING FIRE IN THE SKY, The Illustrated Weekly, of India, June.

Quimby, H. H., et al., 1892–1893
WIND BRACING IN HIGH BUILDINGS, Engineering Record 26:25 (19 Nov. 1892), p. 394; 27:5 (31 Dec. 1892), p. 99; 27:7 (14 Jan. 1893), p. 138; 27:8 (21 Jan. 1893), pp. 161–162; 27:9 (28 Jan. 1893), p. 180; 27:13 (25 Feb. 1893), p. 260; 27:15 (11 Mar. 1893), pp. 298–299; 27:16 (18 Mar. 1893), p. 320.

Quinlan, J. R., 1980
FUNDAMENTALS OF THE KNOWLEDGE ENGINEERING PROBLEMS, Technical Report No. 149, Basser Department of Computer Science, The University of Sydney.

Quinlan, J. R., 1983
LEARNING EFFICIENT CLASSIFICATION PROCEDURES AND THEIR APPLICATION TO CHESS AND GAMES, Machine Learning: The Artificial Intelligence Approach, (eds. R. S. Michalski, J. G. Carbonell and T. M. Mitchell), Palo Alto, Tioga Press.

R. S. Means Company, Inc., 1986
BUILDING CONSTRUCTION COST DATA (Annual), R. S. Means Company, Inc., Kingston, MA.

Rader, M., 1985
POTENTIAL FOR ROBOTIZATION AND AUTOMATION IN GERMAN CIVIL ENGINEERING AND CONSTRUCTION INDUSTRY, Proceedings of 2nd Workshop Conference on Robotics in Construction, Carnegie-Mellon University, Pittsburgh, PA.

Raggett, J. D., 1975
ESTIMATING DAMPING OF REAL STRUCTURES, Journal of Structural Division, ASCE, Vol. 101, No. ST9, September, pp. 1823–1835.

Randall, F. A., 1949
HISTORY OF THE DEVELOPMENT OF BUILDING CONSTRUCTION IN CHICAGO, University of Illinois Press.

Rathbun, J. C., 1936
ELASTIC PROPERTIES OF RIVETED CONNECTIONS, Transactions of the American Society of Civil Engineers, Vol. 101, pp. 524–563.

Rathbun, J. C., 1940
WIND FORCES ON A TALL BUILDING, Transactions ASCE, Vol. 105, pp. 1–41.

Razzaq, Z., 1983
END RESTRAINT EFFECT ON STEEL COLUMN STRENGTH, Journal of the Structural Division, ASCE, Vol. 109, No. ST2, pp. 314–334.

Rea, D., Shah, H. C., and Bouwkamp, J., 1971
DYNAMIC BEHAVIOR OF A HIGH RISE DIAGONALLY BRACED STEEL BUILDING, Earthquake Engineering Research Center, Report UCB/EERC-71/5, University of California, Berkeley.

Real Estate Research Corporation, 1985
EMERGING TRENDS IN REAL ESTATE, Equitable Real Estate, New York.

Reed, J. W., 1971
WIND-INDUCED MOTION AND HUMAN DISCOMFORT IN TALL BUILDINGS, Research Report No. R71-42, Massachusetts Institute of Technology.

Reed, J. W., Hansen, R. J., and Vanmarke, E. H., 1972
HUMAN RESPONSE TO TALL BUILDING WIND-INDUCED MOTION, Proceedings Symposium on Planning and Design of Tall Buildings, Vol. 11, Lehigh University, ASCE Publication.

Rich, E., 1983
ARTIFICIAL INTELLIGENCE, McGraw-Hill, New York.

Robertson, L. E., 1973a
DESIGN CRITERIA FOR VERY TALL BUILDINGS, Proceedings of the Australian and New Zealand Conference on the Planning and Design of Tall Buildings, August 14–17.

Robertson, L. E., 1973b
LIMITATIONS ON SWAYING MOTION OF TALL BUILDINGS IMPOSED BY HUMAN RESPONSE FACTORS, Proceedings of the Australian and New Zealand Conference on the Planning and Design of Tall Buildings, Sydney, Australia, August, pp. 171–180.

Robertson, L. E., 1974
WIND ENGINEERING OF TALL BUILDINGS, Proceedings of Symposium on Tall Buildings—Planning, Design and Construction, Nashville.

Roeder, C. W. and Popov, E. P., 1978
ECCENTRICALLY BRACED STEEL FRAMES FOR EARTHQUAKES, Journal of the Structural Division, ASCE, Vol. 104, March, pp. 391–412.

Rogers, F., 1977
FIRE LOSSES AND THE EFFECT OF SPRINKLER PROTECTION OF BUILDINGS IN A VARIETY OF INDUSTRIES AND TRADES, Building Research Establishment CP9/77, Borehamwood.

Rolfe, S. T. and Barson, J. M., 1977
FRACTURE AND FATIGUE CONTROL IN STRUCTURES, Prentice-Hall, Inc., Englewood Cliffs, N.J., p. 129.

Romstad, K. M. and Subramanian, C. V., 1970
ANALYSIS OF FRAMES WITH PARTIAL CONNECTION RIGIDITY, Journal of the Structural Division, ASCE, Vol. 96, No. ST11, pp. 2283–2300.

Roorda, J., 1975
TENDON CONTROL IN TALL STRUCTURES, Journal of Structural Division, ASCE, Vol. 101, No. ST3, March, pp. 505–521.

Rosenblueth, E., Meli, R., and Resendiz, D., 1985
EL TEMBLOR DEL 19 DE SEPTEIEMBRE DE 1985 Y SUS EFECTOS EN LAS CONSTRUCCIONES DE LA CIUDAD DE MEXICO, Informe preliminar del Instituto de Ingenieria de la Universidad Nacional Autonoma de Mexico, September.

Rosman, R., 1971
STATICS OF NON-SYMMETRIC SHEAR WALL STRUCTURES, Proceedings of ICE, London, Paper 7393S., Vol. 48, pp. 211–244.

Rossiter, S. G., 1975
NAURU HOUSE MELBOURNE—DESIGN PHILOSOPHY, Meinhardt Group, Technical Bulletin.

Royer, K., 1981
THE CONSTRUCTION MANAGER IN THE 80s, Prentice-Hall, Inc., Englewood Cliffs, NJ.

Rutenberg, A., 1974
ANALYSIS OF TUBE STRUCTURES USING PLANE FRAME PROGRAMS, Proceedings of Regional Conference on Tall Buildings, Bangkok, Thailand, pp. 397–413.

Rutstein, R., 1979
THE ESTIMATION OF THE FIRE HAZARD IN DIFFERENT OCCUPANCIES, Fire Surveyor, April.

SEAOC Seismology Committee, 1976
RECOMMENDED LATERAL FORCE REQUIREMENTS, Structural Engineers Association of California, San Francisco, CA.

SSRC, 1983
PROCEEDINGS OF THE THIRD INTERNATIONAL COLLOQUIUM ON STABILITY OF METAL STRUCTURES, Toronto.

Saito, M., Tanaka, N., Arai, K., and Banno K., 1985
THE DEVELOPMENT OF A MOBILE ROBOT FOR CONCRETE SLAB FINISHING, Proceedings of 15th ISIR, Vol. 1, Tokyo.

San Francisco Department of City Planning, 1981
GUIDING DOWNTOWN DEVELOPMENT, Department of City Planning, City and County of San Francisco, May.

Sander, D. M., 1974
FORTRAN IV PROGRAM TO CALCULATE AIR INFILTRATION IN BUILDINGS, National Research Council of Canada, Division of Building Research, Computer Program No. 37, Ottawa.

Sangrey, D. A., 1985
QUALITY AND RELIABILITY AS MOTIVATIONS FOR CONSTRUCTION ROBOTICS, Proceedings of 2nd Workshop Conference on Robotics in Construction, Carnegie-Mellon University, Pittsburgh, PA.

Sangrey, D. A. and Warszawski, A., 1984
CONSTRAINTS ON THE DEVELOPMENT OF ROBOTS FOR CONSTRUCTION, Proceedings of Workshop Conference on Robotics in Construction, Carnegie-Mellon University, Pittsburgh, PA.

Saxon, R., 1983
ATRIUM BUILDINGS: DEVELOPMENT AND DESIGN, Architectural Press, London.

Schoenberger, R. W., 1980
PSYCHOPHYSICAL COMPARISON OF VERTICAL AND ANGULAR VIBRATIONS, Aerospace Medical Association Meeting, Anaheim.

Scully, V., 1985
BUILDINGS WITHOUT SOULS, The New York Times Magazine, September 8.

Sedlacek, G., 1984
ASPECTS OF SERVICEABILITY OF STEEL STRUCTURES (Aspekte der Gebrauchstuchtigkeit von Stahlbauten), Stahlbau, 53, pp. 305–310.

Shaw, C. Y., 1979
A METHOD FOR PREDICTING AIR INFILTRATION RATES FOR A TALL BUILDING SURROUNDED BY LOWER STRUCTURES OF UNIFORM HEIGHT, ASHRAE Transactions, Vol. 85, Part 1, pp. 72–84. Also reprint of National Research Council of Canada, Division of Building Research, NRCC 18029.

Shaw, C. Y., Sander, D. M., and Tamura, G. T., 1973
AIR LEAKAGE MEASUREMENTS OF THE EXTERIOR WALLS OF TALL BUILDINGS, ASHRAE Transactions, Vol. 79, Part 2, pp. 40–48. Also reprint of National Research Council of Canada, Division of Building Research, NRCC 13951.

Shortliffe, E. H., 1976
COMPUTER-BASED MEDICAL CONSULTATONS:MYCIN, American Elsevier, New York.

Simitses, G. J. and Giri, J., 1982
NON-LINEAR ANALYSIS OF UNBRACED FRAMES OF VARIABLE GEOMETRY, International Journal of Nonlinear Mechanics, Vol. 17, No. 1, pp. 47–61.

Simitses, G. J. and Vlahinos, A. S., 1982
STABILITY ANALYSIS OF A SEMI-RIGIDLY CONNECTED SIMPLE FRAME, Journal of Constructional Steel Research, Vol. 2, No. 3, September, pp. 29–32.

Simitses, G. J., Swisshelm, J. D., and Vlahinos, A. S., 1984
FLEXIBILITY-JOINTED UNBRACED PORTAL FRAMES, Journal of Constructional Steel Research, Vol. 4, pp. 27–44.

Sinclair, D. F., 1986
DAMPING SYSTEMS TO LIMIT THE MOTION OF TALL BUILDINGS, Building Motion in Wind, proceedings of a session at ASCE Convention, Seattle, Washington, April 8, pp. 58–65.

Singh, G. and Schwaighofer, J., 1976
A BIBLIOGRAPHY ON SHEAR WALLS 1928-1976, University of Toronto, Department of Civil Engineering, Publication No. 76-02, May.

Sisodiya, R. G., Cheung, Y. K., and Ghali, A., 1972
NEW FINITE ELEMENTS WITH APPLICATION TO BOX GIRDER BRIDGES, Proceedings ICE, London, Vol. 49, Supplement Paper No. 7479S, pp. 207–225.

Skempton, A. W., 1858–1860
THE BOAT STORE, SHEERNESS (1858–1860), AND ITS PLACE IN STRUCTURAL HISTORY, Transactions of the Newcomen Society, Vol. XXXII, pp. 57–58.

Skempton, A. W., 1959
EVOLUTION OF THE STEEL FRAME BUILDING, Guilds Engineer, Vol. X, pp. 37–51.

Skempton, A. W., 1959–1960
THE BOAT STORE, SHEERNESS (1858–1860), AND ITS PLACE IN STRUCTURAL HISTORY, Transactions Newcomen Society, Vol. 32.

Skempton, A. W., 1960
THE BOAT STORE, SHEERNESS (1858–1860), paper, read at the Science Museum, London on February 3.

Skempton, A. W. and Johnson, H. R., 1962
THE FIRST IRON FRAMES, Architectural Review No. 119.

Skibniewski, M. and Hendrickson, C. T., 1985
EVALUATION METHOD FOR ROBOTICS IMPLEMENTATION: APPLICATION TO CONCRETE FORM CLEANING, Proceedings of 2nd Workshop Conference on Robotics in Construction, Carnegie-Mellon University, Pittsburgh, PA.

Sladek, J. R. and Klingner, R. E., 1983
EFFECT OF TUNED MASS DAMPERS ON SEISMIC RESPONSE, Journal of Structural Engineering, ASCE, Vol. 109, No. 8, August, pp. 2004–2009.

Soong, T. T. and Skinner, G. T., 1981
EXPERIMENTAL STUDY OF ACTIVE CONTROL, Journal of Engineering Mechanics Division, ASCE, Vol. 107, No. EM6, December, pp. 1057–1067.

Soong, T. T., Reinhorn, A. M., and Yang, A. N., 1985
A STANDARDIZED MODEL FOR STRUCTURAL CONTROL EXPERIMENTS AND SOME EXPERIMENTAL RESULTS, 2nd IUTAM International Symposium on Structural Control, University of Waterloo, Ontario, Canada, July.

Soroka, W. W., 1949
NOTE ON THE RELATIONS BETWEEN VISCOUS AND STRUCTURAL DAMPING COEFFICIENTS, Journal of the Aeronautical Sciences, Vol. 16, Issue 7, July.

Spurr, D., 1930
WIND BRACING—THE IMPORTANCE OF RIGIDITY IN HIGH TOWERS, McGraw-Hill, New York.

Sriram, D., 1983
KNOWLEDGE-BASED APPROACH TO INTEGRATED STRUCTURAL DESIGN. Unpublished Ph.D. Thesis Proposal, Department of Civil Engineering, Carnegie-Mellon University.

Sriram, D., Maher, M. L., and Fenves, S. J., 1985
KNOWLEDGE-BASED EXPERT SYSTEMS IN STRUCTURAL DESIGN, Computers and Structures, Vol. 20, No. 1-3, pp. 1–9.

Stafford Smith, B., 1985
MODELING OF HIGH-RISE STRUCTURES FOR ANALYSIS BY STANDARD COMPUTER PROGRAMS, Paper presented to ASCE, New York Metropolitan Section, Structures Group, April.

Stafford Smith, B., Kuster, M., and Hoendercamp, J. C. D., 1981
A GENERALIZED APPROACH TO THE DEFLECTION ANALYSIS OF BRACED FRAME, RIGID FRAME AND COUPLED WALL STRUCTURES, Canadian Journal of Civil Engineering, Vol. 8, pp. 230–240.

Stamato, M. C. and Mancini, E., 1973
THREE-DIMENSIONAL INTERACTION OF WALLS AND FRAMES, Journal of Structural Division, Proceedings of ASCE, Vol. 99, No. ST12, December, pp. 2375–2390.
Staniszewski, R., 1980
CYBERNETICS OF DESIGN SYSTEMS, Ossolineum, Poland.
Starrett, Col. W. A., 1928
SKYSCRAPERS AND THE MEN WHO BUILD THEM, C. Scribner's, New York/London.
Steel Committee of California, 1981
STEEL CONNECTIONS/DETAILS AND RELATIVE COSTS, El Monte, California, 24 pp.
Steele, J. E., 1961
MOTION SICKNESS AND SPACIAL PERCEPTION: A THEORETICAL STUDY, Technical Report ASD-TR-61-530, National Technical Information Service, Washington, D.C.
Steelbuilding Association, 1980
DEFORMATION REQUIREMENTS FOR BUILDINGS, Rotterdam Concrete Association, Zoetermeer, The Netherlands, Technical Translation, Ottawa, Canada.
Stefik, M. and Martin, N., 1977
A REVIEW OF KNOWLEDGE-BASED PROBLEM SOLVING AS A BASIS FOR A GENETICS EXPERIMENT DESIGNING SYSTEM. Technical Report STAN-CS-77-596, Computer Science Department, Stanford University, March.
Stefik, M., 1981
PLANNING WITH CONSTRAINTS (MOLGEN 1), Artificial Intelligence 16:111–140.
Steinbrugge, K. V., 1982
EARTHQUAKES, VOLCANOS, AND TSUNAMIS, Scandia America Group, 280 Park Avenue, New York, NY 10017, 392 pp.
Steiner, H. M., 1980
PUBLIC AND PRIVATE INVESTMENTS: SOCIOECONOMIC ANALYSIS, John Wiley, New York, pp. x to 414.
Steyert, R. D., 1972
THE ECONOMICS OF HIGH RISE APARTMENT BUILDINGS, Proceedings of the International Conference on Planning and Design of Tall Buildings, Lehigh University, Bethlehem, PA, August, pp. 103 to 118.
Stussi, F., 1943/1944
AN UNKNOWN REPORT BY NAVIER (Un rapport inconnu de Navier) in Publications of the International Association of Bridge and Structural Engineering, Vol. VII, 1943/1944, Zurich, pp. 1–13.
Sugimoto, H., 1983
STUDY OF OFFSHORE STRUCTURAL MEMBERS AND FRAMES, Ph.D. Dissertation, School of Civil Engineering, Purdue University, West Lafayette, Indiana.
Swanson, V. E., 1975
THE ECONOMICS OF HIGH RISE OFFICE BUILDINGS, Pan Pacific Tall Building Conference Proceedings. Paper read before the Regional Conference under the auspices of the Joint Committee on Tall Buildings, University of Hawaii, Honolulu, Hawaii, January, pp. 97–110.

Talin, A. and Ellingwood, B., 1985
WIND-INDUCED MOTION OF TALL BUILDINGS, Engineering Structures, Vol. 7, October, pp. 245–252.

Tanaka, H., 1985
HUMAN IMPLICATIONS OF ROBOTIZATION IN THE WORKSITE: THE JAPANESE EXPERIENCE, Robotics, Vol. 1, No. 3, Elsevier Science Publishers B. V., Netherlands.

Taranath, B. S., 1982
COMPOSITE DESIGN OF FIRST CITY TOWER, HOUSTON, TEXAS, The Structural Engineer, London, Vol. 60A, No. 9, September.

Tassios, T. P. and Tsoukoutas, S. G., 1983
BEHAVIOR OF LARGE PANEL CONNECTIONS, Building Research & Practice, July/August.

Terai, T., 1986
THE CONSTRUCTION INDUSTRY IN THE ADVANCED INFORMATION SOCIETY, Science & Technology in Japan, Vol. 5, No. 17, Tokyo.

The Washington Post, 1984
THE WASHINGTON POST, Washington, DC, 14 October.

The Washington Post, 1985a
THE WASHINGTON POST, Washington, DC, 23 February.

The Washington Post, 1985b
THE WASHINGTON POST, Washington, DC, 1 December.

Thomas, P., et al., 1963
INVESTIGATIONS INTO THE FLOW OF HOT GASES IN ROOF VENTING, Fire Research Technical Paper No. 7, London, HMSO.

Thompson, P., 1976
THE O.C.B.C. CENTRE, Arup Journal, October, pp. 3–11.

Thompson, P. J. and Ainsworth, 1985
COMPOSITE BEAMS WITH WEB PENETRATIONS, GROSVENOR PLACE, SYDNEY (Third Conference on Steel Developments held in Melbourne, Australia, May 1985), Australian Institute of Steel Construction.

Thompson, P. J., O'Hea, R. S., and Bergin, R. J., 1976
THE OCBC CENTRE SINGAPORE, The Arup Journal, Ove Arup Partnerships, London, No. 3, Vol. 11, October.

Tilley, G. P. and Eyre, R., 1977
DAMPING MEASUREMENTS ON STEEL AND COMPOSITE BRIDGES, Symposium on dynamic behavior of bridges, Transport and Road Research Laboratory, Crowthorne, Berkshire, May 19.

Timoshenko, S. P. and Gere, J. M., 1961
THEORY OF ELASTIC STABILITY, 2nd Edition, Engineering Societies Monographs, McGraw-Hill, New York.

Timoshenko, S. P. and Goodier, J. N., 1970
THEORY OF ELASTICITY, McGraw-Hill, 3rd Edition.

Tschanz, T. and Davenport, A. G., 1983
THE BASE BALANCE TECHNIQUE FOR THE DETERMINATION OF DYNAMIC WIND LOADS, Sixth International Conference on Wind Engineering, Gold Coast, Australia, March 21–25.

Tucker, J. B., 1985
SKYSCRAPERS: AIMING FOR 200 STORIES, High Technology, January.

Turak, T., 1966
WILLIAM LE BARON JENNEY: 19TH CENTURY ARCHITECT, Ph.D. Dissertation, University of Michigan.

Ueno, T. and Yoshida, T., 1983
ROBOTIZATION OF SPRAY WORK FOR FIREPROOFING STEEL STRUCTURE, Robot No. 38, JIRA, Tokyo.
Urban Land Institute, 1985
CULTURAL FACILITIES IN MIXED-USE DISTRICTS, Urban Land Institute, Case Study 9, Dallas Arts District, Washington, DC.

van Eyck, A., 1981
1981 ANNUAL DISCOURSE for the Royal Institute of British Architects, London.
Verein Deutscher Ingenieure, 1979
VDI—RECOMMENDATION 2057, PARTS 1–3, Dusseldorf, West Germany.
Vernez-Moudon, A., 1983
CITY FORM AND TALL BUILDINGS: CATHEDRALS, PALAZZI, TALL DOWNTOWNS, AND TALL CITIES, Developments in Tall Buildings 1983, Council on Tall Buildings and Urban Habitat, Hutchinson Ross Publishing Company, Stroudsburg, PA.
Vickery, B. J. and Davenport, A. G., 1970
AN INVESTIGATION OF THE BEHAVIOR OF WIND ON THE PROPOSED CENTREPOINT TOWER, IN SYDNEY, AUSTRALIA, Engineering Science Research Report, BLWT 1-70, University of Wesyern Ontario, London, Canada, February.
Vogel, R. M., 1961
ELEVATOR SYSTEMS OF THE EIFFEL TOWER, 1889, United States National Museum Bulletin 228, Washington, Smithsonian Institution.
Vogel, R. M., 1976
BUILDING IN AN AGE OF STEAM, Charles E. Peterson, ed., Building Early America, Chilton Book Co., Radnor, Pa., pp. 119–134.
von Dyck, W., 1912
GEORG VON REICHENBACH. DEUTSCHES MUSEUM BIOGRAPHIES AND DOCUMENTS (George von Reichenbach. Deutsches Museum Lebensbeschreibungen und Urkunden), Deutsches Museum, Munich.

WSN 321-75, 1976
RECOMMENDATIONS FOR DESIGN OF LARGE PANEL BUILDINGS—IN RUSSIA (Instrukciya po proyektirovaniyu panielych zdanly), Moscow.
Walker, G. R., 1984
INTERACTION OF WINDOW GLASS AND WIND, workshop on Wind Engineering and Industrial Aerodynamics, C.S.I.R.O. Division of Building Research, Australia, pp. 32–34.
Wargon, A., 1983
DESIGN AND CONSTRUCTION OF SYDNEY TOWER, The Structural Engineer, Vol. 61A, No. 9, September, pp. 273–281.

Warszawski, A., 1984
APPLICATION OF ROBOTICS TO BUILDING CONSTRUCTION, Proceedings of Workshop Conference on Robotics in Construction, Carnegie-Mellon University, Pittsburgh, PA.

Warszawski, A., 1984
ECONOMIC EVALUATION OF ROBOTICS IN BUILDING, Proceedings of Workshop Conference on Robotics in Construction, Carnegie-Mellon University, Pittsburgh, PA.

Warszawski, A., 1985
ECONOMIC IMPLICATIONS OF ROBOTICS IN BUILDING, Building and Environment, Vol. 20, No. 2, Great Britian.

Weaver, W., Brandow, G. E., and Manning, T., 1971
TIER BUILDINGS WITH SHEAR CORES, BRACING AND SETBACKS, Journal of Computers and Structures, Vol. 1 pp. 57–84.

Webb, C. G., 1960
THERMAL DISCOMFORT IN AN EQUATORIAL CLIMATE, Journal of the Institute of Heating and Ventilation Engineering, January.

Webster, J. C., 1959
THE SKYSCRAPER: LOGICAL AND HISTORICAL CONSIDERATIONS, Journal of the Society of Architectural Historians, 18:4, December, pp. 126–139.

Wiesner, K. B., 1979
TUNED MASS DAMPERS TO REDUCE BUILDING WIND MOTION, presented at ASCE National Convention, Boston, Massachusetts, April 2–6, ASCE Preprint No. 3510.

Whyte, W. H., 1980
THE SOCIAL LIFE OF SMALL OPEN SPACES, The Conservative Foundation, Washington, DC.

Whyte, W. H., 1983
QUOTATIONS BY WILLIAM H. WHYTE, from a speech entitled "The Blank Wall," given at the "What Makes A City?" conference, April 29, 1983, sponsored by the Dallas Institute of Humanities and Culture, Dallas, Texas.

Wilsher, Peter and Richter, R., 1974
THE EXPLODING CITIES, Quadrangle, The New York Times Book Co., New York.

Wight, P. B., 1876
THE FIRE QUESTION, American Architect and Building News, v. 1 (17 June 1876), pp. 195–197; (24 June 1876), pp. 203–205; (1 July 1876), pp. 211–212.

Wight, P. B., 1878
THE FIRE QUESTION, American Architect and Building News, v. 3 (2 Mar. 1878), p. 76.

Winograd, T., 1971
PROCEDURES AS REPRESENTATION FOR DATA IN A COMPUTER PROGRAM FOR UNDERSTANDING NATURAL LANGUAGE, Technical Report AI TR-17, MIT, Cambridge, MA.

Winters, G., 1947
STRENGTH OF THIN STEEL COMPRESSION FLANGES, Bulletin 35/3, Cornell University Engineering Experiment Station, Ithaca, N.Y.

Wong, S. K. M. and Ziarko, W., 1985
A PROBABILISTIC MODEL OF APPROXIMATE CLASSIFICATION IN INDUCTIVE LEARNING, ISBN 0-7331-0092-3, Department of Computer Science, University of Regina.
Wood, H., 1968
DURABILITY OF CONCRETE CONSTRUCTION, ACI, Monograph No. 4, Detroit.
Wyatt, T. A., 1977
MECHANISMS OF DAMPING, Symposium on dynamic behavior of bridges, Transport and Road Research Laboratory, Crowthorne, Berkshire, May 19.

Yamamoto, K., 1985
CURRENT CONDITIONS AND PROSPECTS OF RESEARCH AND DEVELOPMENT ON THE MOST ADVANCED ROBOTICS TECHNOLOGY IN JAPAN, Proceedings of 1985 ICAR, Tokyo.
Yamanouchi, H., et al., 1984
EXPERIMENTAL RESULTS ON A K-BRACED STEEL STRUCTURE UNDER SEISMIC LOADING UTILIZING FULL-SCALE SIX-STORY TEST STRUCTURE—U.S./JAPAN COOPERATIVE RESEARCH PROGRAM, Proceedings, 1984 Annual Technical Session, Stability Under Seismic Loading, Structural Stability Research Council, San Francisco, CA.
Yang, J. N. and Giannopoulos, F., 1978
ACTIVE TENDON CONTROL OF STRUCTURES, Journal of Engineering Mechanics Division, ASCE, Vol. 104, No. EM3, June, pp. 551–568.
Yang, J. N. and Lin, M. J., 1982
OPTIMAL CRITICAL-MODE CONTROL OF BUILDING UNDER SEISMIC LOAD, Journal of Engineering Mechanics Division, ASCE, Vol. 108, No. EM6, pp. 1167–1185.
Yang, J. N. and Lin, Y. K., 1981
ALONG-WIND MOTION OF A MULTI-STORY BUILDING, Journal of Engineering Mechanics Division, ASCE, Vol. 107, No. EM2, April, pp. 295–307.
Yang, J. N. and Samali, B., 1983
CONTROL OF TALL BUILDINGS IN ALONG-WIND MOTION, Journal of Structural Engineering, ASCE, Vol. 109, No. 1, January, pp. 50–68.
Yeh, S. H. K., 1975
PUBLIC HOUSING IN SINGAPORE, Singapore University Press.
Yoshida, T., Ueno, T., Nonaka, M., and Yamazaki, S., 1984
DEVELOPMENT OF SPRAY ROBOT FOR FIREPROOF COVER WORK, Proceedings of Workshop Conference on Robotics in Construction, Carnegie-Mellon University, Pittsburgh, PA.
Yoshida, T. and Ueno, T., 1985
DEVELOPMENT OF A SPRAY ROBOT FOR FIREPROOF TREATMENT, Shimizu Technical Bulletin, No. 4, Shimizu Construction Co., Tokyo.
You, P. S. and Lim, C. Y., 1984
SINGAPORE: TWENTY-FIVE YEARS OF DEVELOPMENT, Nan Yan Xing Shou Liahne Zaobao, Singapore, p. 323.

Zandonini, R., 1983
RECENT DEVELOPMENTS IN THE FIELD OF STABILITY OF STEEL COMPRESSED MEMBERS, Proceedings of the Third International Colloquium on Stability of Metal Structures, Toronto, pp. 1–19.

Zandonini, R., 1985
STABILITY OF COMPACT BUILT-UP STRUTS: EXPERIMENTAL INVESTIGATION AND NUMERICAL SIMULATION, Construzioni Metalliche, No. 4.

Zeevaert, L., 1964
STRUCTURAL STEEL BUILDING FRAMES IN EARTHQUAKE ENGINEERING, Proceedings, Steel Utilization Congress, Luxembourg.

Zienkiewicz, O. C., 1977
THE FINITE ELEMENT METHOD, McGraw-Hill, Maidenhead, England.

Zunz, G. J., Heydenrych, R. A., and Michael, D., 1971
THE STANDARD BANK CENTRE, JOHANNESBURG, Proc. I.C.E., February, Pamphlet 7346.

Zunz, G. J., Glover, M. J., and Fitzpatrick, A. J., 1985
THE STRUCTURE OF THE NEW HEADQUARTERS FOR THE HONG KONG & SHANGHAI BANKING CORPORATION, HONG KONG, The Structural Engineer, September, Vol. 63A, NO. 9, pp. 255–284.

Indexes

Building Index

The following index enables the reader to identify the page number on which a particular building is mentioned. Numbers in italics designate pages with illustrations. Names of cities appear in the subject index.

1 Liberty Plaza (U.S. Steel), New York, 1005
1 Mellon Bank Center, Pittsburgh, 1005
1 Palac Kultury 1 Nauki, Warsaw, 1004
2 Union Square, Seattle, 1006
3 First National Plaza, Chicago, 1004
5 Fremont Plaza, San Francisco, 264
9 West 57th Street, New York, 1005
16 Cook Street, Liverpool, England, *197*
20 Exchange Place (Citibank), New York, 1005
30 South Wacker, Chicago, 37
33 West Monroe, Chicago, 37
40 Wall Tower, New York, 1004
42nd Street Development, New York, *428*
55 Water Street, New York, 1006
71 Stevenson Street, San Francisco, 265
100 First Street, San Francisco, 274
101 California Street Building, San Francisco, 274
123 North Wacker Drive, Chicago, *261*
190 South LaSalle Street, Chicago, 5, *9*
311 Wacker Drive, Chicago, 1006
333 Wacker Drive, Chicago, *157*
345 California Street, San Francisco, *831*
383 Madison Avenue, New York, *154, 155*
388 Market Street, San Francisco, *833*
444 Market Street, San Francisco, 264
500 Fifth Avenue, New York, 1005
525 Market Street, San Francisco, 274
525 Mission Street, San Francisco, 274
712 Fifth Avenue, New York, *161*
860–880 North Lake Shore Drive, Chicago, 36
900 North Michigan, Chicago, 34, 353, *354,* 359, 1004
1600 Smith, Houston, 1005

AT&T Corporate Headquarters, New York, 5, *7,* 108
AT&T Corporate Center, Chicago, 1006
Alexandra House, Hong Kong, *221*
Allegheny International, Pittsburgh, *152, 153*
Allied Bank Plaza, Houston, 1004
American Fletcher Center, Indianapolis, 1006
American International Building, New York, 1004
Amoco Building, Chicago, 37, *231,* 597, 1004
Andraus Building, Sao Paulo, 325
Armley Mill, Leeds, UK, 191
Associate Center, Chicago, 37
Astor House, New York, 21
Atlantic Center, Atlanta, 1006
Atlantic Richfield Plaza A, Los Angeles, 1005
Atlantic Richfield Plaza B, Los Angeles, 1005
Atlas Building, Mexico City, *819*
Auditorium Building, Chicago, 35, *40*

Bab Zowayla, Cairo, 118, *127*
Bank of America, San Francisco, 274, 328, 1004
Bank of China, Hong Kong, 184, 1004
Bank of the Southwest (proposed), Houston, 149, 723
Barclay Square, Paris, 124

Benyon and Marshall Flax Mill, England, 13
Boat Store, Sheerness, England, *195,* 198
Brunswick Building, Chicago, 905
Bukit Merah, Singapore, *460*
Business Men's Assurance Building, Kansas City, 122

CN Communications Tower, Toronto, 705, 722, 790, 906
Calico Mill, Liverpool, 13
Canary Wharf, London, 123, *146*
Capital Building, Lincoln, *129*
Capital National Bank Plaza, Houston, 1006
Capital Tower, Melbourne, *374*
Carlton Centre, Johannesburg, 1005
Carson Pirie Scott, Chicago, 36, *43*
Castle Foregate Flax Mill, Shrewsbury, UK, *190*
Centrepoint Tower (now Sydney Tower), Sydney, 790
Chanin Building, New York, 1006
Chartres Cathedral, Paris, 118, *126*
Chemical Bank Trust Building, New York, 1006
Chicago Board of Trade Building, Chicago, 36, 109, 120, *135*
China Resources Centre, Hong Kong, 221, *225*
Chrysler Building, New York, 34, 104, *106,* 160, 183, 1004
Citicorp Center, New York, 108, 279, 790, 792, *793, 794,* 1004
City Mutual Headquarters, Sydney, *178, 179*
Cityspire, New York, 1004
Civic Center, Chicago, 121
Civic Opera, Chicago, 36
Cleveland Arcade, Cleveland, 279, *282*
Columbia Center, Seattle, 790, *799,* 1004
Columbus Circle, New York, 121
Commerce Court West, Toronto, 1004
Concert Hall, Dallas, 997
Connaught Centre, Hong Kong, 209, *211,* 905
Continental Illinois Bank, Chicago, 36
Cooper Union, New York, 16, 21
Crocker National Bank, San Francisco, 274
Crocker Center, Los Angeles, 1005
Crown Zellerbach Building, San Francisco, 274
Crystal Palace, London, 14, 194, 195, 198

Daily News Building, Chicago, 36, *45*
Darlington Building, New York, 30
Dock and Canal development, Chicago, 229
Du Pont Asia Pacific Limited, Hong Kong, *214*

Edinburgh Tower, Hong Kong, 219, *223*
Eiffel Tower, Paris, 119, *129,* 705
Embarcadero Center, San Francisco, *245, 251,* 279, *280, 830*
Embarcadero One, San Francisco, 274
Embarcadero Two, San Francisco, 274
Embarcadero Three, San Francisco, 274
Embarcadero Four, San Francisco, 274
Empire State Building, New York, 34, 101, *102,* 104, 160, 183, 320, 327, 597, 646, 705, *714, 715,* 1004
Entex Building, Houston, *399*
Equitable Life Assurance Building, New York, 22
Equitable Tower West, New York, 1005
Esso Plaza, Calgary, 409
Etoiles, Paris, 124
Exchange Square, Hong Kong, 210, *212*
Exposition Building, Chicago, 17
Exxon, New York, 1005

The Fair Store, Chicago, 30
Fagus Factory, Germany, 198
Fairmont Hotel, San Francisco, *818*
Far East Finance Centre, Hong Kong, 216
Federal Center, Chicago, 121
Federal Reserve Building, Boston, 1005
Ferry Building, San Francisco, 34
Fifth Avenue Hotel, New York, 22
First Bank Tower, Toronto, 1004
First Canadian Place, Toronto, 409, *410*
First International Building, Dallas, 1005

First International Plaza, Houston, 1005
First National Bank, Chicago, 37, *48,* 121, 1004
First and Market Street, San Francisco, 265
Ford Foundation, New York, 285
Four Allen Center, Houston, 1005

Galleria, Houston, 333
Gardener Building, Glasgow, UK, 193, *194*
Garisenda and Awinelli Towers, Bologna, *28*
General Motors, New York, 1005
Glaspalast, Bavaria, West Germany, 194, 198
Gopala Towers, Dehli, 324
Grand Central Station, New York, 120
Grosvenor Place, Sydney, *166, 167, 168, 342, 344,* 346, *348,* 349, 377, *379*
Gulf Tower, Houston, 1005
Gulf-Western Building, New York, 1006

Haughwout Building, New York, 18
Home Insurance Building, Chicago, 11, *12,* 18, 29, 34, 35, 229, 422, 646, 816
Hong Kong Club and Office Building, Hong Kong, *172*
Hong Kong and Shanghai Bank Building, Hong Kong, 56, 209, *210,* 377, *378*
Hopewell Centre, Hong Kong, 1005
Houston Center, Houston, 82, *83*
Humana Building, Louisville, 110
Humboldt Building, San Francisco, 274
Hyatt Regency Hotel, Kansas City, 401

IBM Building, Chicago, 1005
IBM Building, New York, 279
IDS Center, Minneapolis, *57,* 108, *281,* 1004
Ikebukuro Building. (*See* Sunshine 60)
Inland Steel Building, Chicago, 36, *47,* 121, *137*
Interfirst Plaza Tower, Dallas, 1004
International Place, Boston, 5, *10*

Joelma Building, Sao Paulo, 325
John Hancock Center, Chicago, 34, 37, *50,* 121, *231, 232,* 1004
John Hancock Tower, Boston, 790, 791, *793,* 1004
John Hancock Tower, New Orleans, 122

Kendall Building, Chicago, 17, 22
Korea Insurance Company, Seoul, 1005

La Defense, Paris, 107
Lake Point Tower, Chicago, 37, *231*
Lamac Commercial Center, Bratislava, *503*
Latino Americana, Mexico City, 661, *820,* 821, 832, *833*
Legislative Assembly, Maharashtra, *480*
Lever House, New York 111
Levi's Plaza, San Francisco, 264
Liberty Square, Los Angeles, 1006
Life Insurance Corporation Building, Madras, 317
Luk Yeung Sun Cheun, Hong Kong, 37[illegible] *380*

McGraw-Hill Building, New York, 4
M.L.C. Centre, Sydney, *173, 174, 175,* 100[illegible]
Maine Montparnasse, Paris, 1005
Manufacturer's Hanover, Chicago, *260*
Marina City, Chicago, 37
Marquette Building, Chicago, 36
Marshall Field Wholesale Store, Chicago, 35
Mellon Bank, Philadelphia, 1006
Menier Chocolate Factory, Noisel-sur-Marne, 29
Metropolitan Life Tower, New York, 3[illegible] 104, 1005
Mies Van Der Rohe apartments, Chicago, *231*
Minneapolis City Center, Minneapoli[illegible] 1006
Momentum Place, Dallas, 1004
Monadnock Building, Chicago, 114, 7[illegible]
Morgan Bank, New York, 1006
Municipal Building, Chicago, 114

National Australia Bank House, Sydney, *344*
National Place, Washington, D.C., 184, *187*
Nauru House, Melbourne, 949, *950*
New World Centre, Hong Kong, 213, *214*
New York Coliseum, New York, 81, 121, *141*
Newport City, New Jersey, 84
Nikko Hotel, Chicago, 229
North Loop Redevelopment Project, Chicago, 230
Norwest Center, Minneapolis, 1005

Overseas-Chinese Banking Corp. (OCBC), Singapore, *345, 346, 347,* 373, *376*
OUB Centre, Singapore, 975, *976*
Office Towers, Caracas, Venezuela, 1006
Olympia Center, Chicago, 409, 1005
Omni International, Atlanta, 82
One Astor Plaza, New York, 1005
One Chase Manhattan Plaza, New York, 1004
One Houston Center, Houston, 1006
One Liberty Place, Philadelphia, 1004
One Magnificent Mile, Chicago, 37
One Penn Plaza, New York, 1005
One Shell Plaza, Houston, 121, *138,* 905, 1005
One Shell Square, New Orleans, 1005
One Wacker Drive, Chicago, 1006
Onterie Center, Chicago, 122, *467*
Overseas Union Bank, Singapore, 1004

P.P.G. Place, Pittsburgh, 5, *8*
Palmolive Building, Chicago, 36
Pan American Building, New York, *136,* 320, 1004
Panorama Building, Bombay, *481*
Park Row Building, New York, 18
Parliament, English Houses, London, 22
Peachtree Center Plaza, Atlanta, 1005
Peninsula Centre, Hong Kong, *216*
Penn Mutual Tower, Philadelphia, 184, *185, 186*
Pennzoil Place, Houston, *330*
Perimeter Center, Atlanta, 122
Petro-Canada 1, Calgary, 1005
Petrolaos Mexicanos, Mexico City, 1005
Philadelphia City Hall, Philadelphia, 184
Phelan Building, San Francisco, *247*
Pino Suarex complex, Mexico City, *819*
Portland Building, Oregon, 110
Presidential Towers, Chicago, *471*
Price Tower, Bartlesville, 120, *133*
Prudential Building, Chicago, *231*
Prudential Center, Boston, 1005
Pruitt-Igoe, St. Louis, 55

Quadrangle Store, Sheerness, *190*

RCA Building, New York, 107
Raffles City Hotel, Singapore, 1005
Railway Exchange Building, New York, 37
Reliance Building, Chicago, 36, *42,* 119, *129*
Renaissance 1, Detroit, 1005
Renaissance Center, Detroit, 82
Republic Bank, Dallas, 122, 146
Republic Bank Tower, Houston, 108, *330,* 1004
Republic Plaza, Denver, 1005
Rialto Center, Melbourne, 946, 951, *953,* 1004
River City project, Chicago, *258*
Riverside Centre, Brisbane, *170, 171*
Rockefeller Center, New York, 107, 119, *131,* 154, 160, *279,* 1004
Ronan Point, London, 485, 645, 743, *744*
Rookery, The, Chicago, 35, 37, *41*
Russ Building, San Francisco, 274

Sabah Foundation Headquarters, East Malaysia, *374*
St. Anne's Church, Liverpool, 13
St. Francis Hotel, San Francisco, *818*
St. Ouen Docks, Paris, 14
San Gimignano, Italy, 118, *125*
San Marco, Venice, *128*
Santa Fe Center, New York, 38
Sayner Hutte, West Germany, 191, *192*
Schrannenhalle, Munich, 194
Scotia Plaza, Toronto, 1004

Seagram Building, New York, 114, 264, 285
Sears Tower, Chicago, *31,* 34, 37, *51,* 101, 121, 183, *231,* 597, 646, 705, 906, 1004
Second Leiter Building, New York, 38
Security Pacific National Bank, Los Angeles, 1005
Shell Building, San Francisco, 274
Shinjuka Center, Tokyo, 1005
Singapore Treasury Building, Singapore, 1005
Singer Building, New York, 103
Southeast Financial Center, Miami, 1005
Southern Bell Building, Atlanta, 1006
South Ferry Plaza, New York, *150, 151*
Stalinallee, East Berlin, 169
Standard Bank, Johannesburg, 373, *375*
Standard Oil Building. (*See* Amoco Building)
State Bank, Sydney, *344*
State of Illinois Center, Chicago, 37
Stock Exchange Building, Boston, 409
Sunning Plaza, Hong Kong, *219*
Sunshine 60, Tokyo, *832,* 1005
Sydney Tower (formerly Centrepoint Tower), Sydney, 790

Tacoma Building, Chicago, 13, 15, 198
Taj-Continental Building, Singapore, *480*
Taj-Mahal Hotel, Singapore, *480*
Tao Payoh Town Center, Singapore, *458*
Television City, New York, 34, 54, 112
Terminal Tower, Cleveland, 1005
Texas Commerce Plaza, Houston, 906, 1004, 1006
Three Allen Center, Houston, 1006
Three First National, Chicago, 37
Times Square cluster (concept), New York, *291*
Times Square Center, New York, 427, *431*
Toronto Dominion Bank Tower, Toronto, 1005
Tower of Babel, 124
Transamerica Building, San Francisco, 258, 274, 328, *833,* 1004
Transco Tower, Houston, 108, 109, 327, 329, *331, 334*
Transco Tower 2, Houston, 1004
Tribune Tower, Chicago, 4, 36, *44,* 114, 120, *232*
Trump Tower, New York, 279
Tsimshatsui East, Hong Kong, 213
Tun Abdul Razak, Penang, 1005

U.S. Steel, Pittsburgh, 1004
Union Carbide Building, New York, 1005
United California Bank, Los Angeles, 1004
United Nations Secretariat Building, New York, 198
United States Gypsum Building, Chicago, 37, *49*

Victoria Tower, London, 22

Washington/Montgomery Tower, San Francisco, 265
Water Tower Place, Chicago, 34, 37, 232, 905, 1004
Westpac Centre (formerly CBA Centre), Sydney, *343*
Woolworth Building, New York, 34, 104, *105,* 111, 1004
World Financial Center, New York, 109, 409, *411,* 417, 421, 1005
World Trade Center, New York, 104, 433, 705, 719, 790
World Trade Center (North), 597, 1004
World Trade Center (South), 597, 1004
World Wide Plaza, New York, 1006
World's Tallest Building (study), New York, 289
Wrigley Tower, Chicago, 36

Xerox Center, Chicago, 37

ZOD Ruzinov, Bratislava, 501

Name Index

The following list cites the page numbers on which the indicated names are mentioned. It includes authors and others named in the text, as well as names of organizations. Names of codes, standards and specifications are found in the subject index, as are the names of cities.

Names followed by years refer to citations in the text that are included in the appendix, "References/Bibliography." Consult this appendix for additional authors and citations that may not have been mentioned in the text.

AASHTO, 810
ABSTECH, 813
AISC, 340, 805
AKEH Joint Venture, 971
ANSI Code, 912
ARBED, 839
ATC (1978), 823
Ackroyd, M. H., 882
Ackroyd, M. H. (1979), 881
Adham, S., 971
Adler and Sullivan, 15
Agbabian Associates, 971
Agency of Industrial Science and Technology, 517
Akiner, V. T., 591
Akiner, V. T. (1986), 592
Albers, J. (1961), 177
Althans, L., 191
Alves M., and Berica, A. (1984), 344
Amarel, S. (1978), 573
American Society of Civil Engineers. (*See also* AISC), 965
American Bureau of Shipping, 813
American Concrete Institute, 962, 965
American Institute of Steel Construction, 902
Amin, N., and Lauck, T., 832
Anderson et al. (1952), 822
Apgar, W. (1985), 465
Appleyard, D. (1980), 82
Arciszewski, T., 847
Arciszewski, T. (1975), 848
Arciszewski, T. (1984), 852
Arciszewski, T. (1986), 842, 847, 848
Arciszewski, T., and Pancewitz, Z. (1976), 849, 848
Armer, G. S. T., 646, 741
Aschendorff et al. (1983), 839
Atkinson, R., 971
Atkinson-Noland and Associates, 971
Atsikawa, Y., 975

Badger, D., 14, 193, 198
Bage, C., 13, 190
Baldwin, C., 18
Ballio, G., 837
Ballio, G., and F. Mazzolani, (1983), 838
Ballio et al. (1977), 839
Bannister, T. C. (1951), 189
Barnes, E. L., 109
Barnes, S. (1984), 576
Barnes-Kariotis Joint Venture, 971
Barr et al. (1981), 592
Beason, W. L. (1980), 686
Becker, R. (1975), 882
Beeby, A., and H. Taylor (1978), 900
Beedle, L. S., 603
Bello, A. A., 657
Benesch, A. and Company, 360
Bennett, J., and R. Engelmore (1979), 585
Bild, S., 885

Binder, R. W. (1962), 821
Bogardus, J., 14, 193, 198
Bombay Fire Brigade, 324
Bombay Municipal Corporation, 322
Bonissone, P. P. (1982), 574
Borda, J. C., 14
Borja, G., 657
Bosshardt, H., 621
Boulton, M., 19
Boundary Layer Wind Tunnel Laboratory, 705
Bowen, J., 330, 333
Boyd, D., 103, 309
Boyer, R., 273
Bramah, J., 20
Breuer, M., 176
Brown, R., 971
Brown, W. G. (1974), 686
Brubaker, C. W., 33
Bruegel, P., 124
Buckley, J., 919
Building Owners and Managers Association, 557
Building Research Station, 756
Building Seismic Safety Council, 823
Burgee, J., 108, 330, 427
Burgess, E. W. (1925), 79
Burke, E. M., 433
Burnham, D., 197, 198
Burnham and Root, 30, 35
Burnham, D. H., and Company, 36, 42
Bylander Meinhardt Partnership, 975

Calderone, I. (1985), 686
Cantieni, R. (1985), 887
Cantor, I. W., 427
Carnegie-Mellon University, 519, 847
Carr Lynch Associates, 991
Cermak and Plate, 720
Chen, W. F., 857, 859
Chen, W. (1987), 858
Chen, W. F. (1987), 882
Chen, W., and Atsuta, T. (1976, 1977) 858
Chen, W., and E. Lui (1983), 840
Chen, W., and E. Lui (1984), 882
Chen, W., and E. Lui (1985), 838, 864
Chen, W., and E. Lui (1986, 1987a, 1987b), 859
Chen, W., and E. Lui (1987a), 859, 879
Cheung, C., and Melbourne, W. (1984), 686, 687
Cheung, Y. (1983), 933
Cheung, Y. (1984), 934
Chin and Hensolt, 832, 833
Chua, B. H. (1981), 455
Civil Engineers Associated Group, 657
Cizek, P., 493
Clawson, W. C., and Darwin, D. (1982), 344
Clemson University, 971
Clifton, J. R., 959
Club of Rome, 184
Cobb, H., 122
Cochrane, T. L., 16
Colaco, J. P., 905
Collins et al. (1985), 900
Colson, A. and Louveau, J. (1983), 861, 862
Comite Euro-International du Beton (1985a), 489
Computech, 971
Concrete Masonry Association of California and Nevada, 973
Condit, C., 11, 30, 35
Construction Industry Development Board, 449
Cooper, L. (1983), 731
Cooperman, D., 53
Corry, D., 919
Corry, H. R., 427
Cosentini Associates, 427
Coull, A., 921
Coull, A., and J. Choudhury (1967), 928
Coull, A. and Khachatoorian, H. (1982), 937
Coull, A., and T. Mohammed (1983), 937
Coull, A., and Smith, S. (1983), 923
Council on Tall Buildings, 27, 55, 600, 603, 923, 1003
Council on Tall Buildings (1978), 489
Council on Tall Buildings (1978–1981), 594, 597
Council on Tall Buildings (1981), 27, 310
Council on Tall Buildings (1987), 487
Coyle, C. (1929), 709
Coycle, C. (1983), 709

Cramer-Klett, 194
Cross, H., 899
Cubitt, W., 195
Culver, C., N. Iwankiw, and A. Kuentz, 859

Deutsches Institut fur Normung 4150 (1975), 889
Dalgliesh, W. A., 647, 689
Dalgliesh, W. A. (1979), 686
Danay et al. (1974), 937
Davenport, A. G., 645, 705
Davenport, A. G. (1961), 700
Davenport, A. G. (1966), 700
Davenport, A., and D. Surry (1983), 691
David, E. (1985), 643
Davis, R. (1981), 574
Davy, H., 20
Dawson, S. (1983), 989
Del Valle, E. (1980), 652, 649
Del Valle, E., 649, 657
Desaguliers, J. T., 21
Diaz, M., 657
Disque, R. O. (1975), 879
Downs, A. (1981), 82
Drexel University, 971
Dryden, H., and G. Hill (1933), 710
Dubos, R., 293
Dubos, R. (1981), 292
Duda, R. (1981), 574
Duda, R., and J. Gasching (1981), 572
Duda, R., and E. Shortliffe (1983), 572
Dunn, J., and J. Fry (1966), 728
Dutt, A. J. (1984), 72

Eads, J. B., 16
Edison, T., 21
Edwards, S., 919
Eiffel, G., 30, 122, 706
Elevator Service Professionals, 919
Ellingwood et al. (1980), 735
Ellis, E. P. (1953), 70
Ellis, P., 29, 197, 198
Engineering News Record, 710
Engineering Times (1985), 401
Englekirk, R., 971
Environmental Simulation Laboratory, 236, 240

Epstein, A., 37
Escamilla, Segio, 657
Evans, O., 19
Ewing, R., 971

Fagan et al. (1979), 574
Fairbairn, W., 14, 27
Farquhar, G. S., 557
Feigenbaum, E. A. (1981), 593
Fenves, S. J., 571
Ferris, H., 119, 130
Fielding, D., and W. Chen (1973), 882
Fielding, D., and J. Huang (1971), 882
Fintel, M. (1974), 942
Fintel, M., and F. Khan (1968), 942
Finzi, L., 803
Firey, W. (1945), 80
Firkins, A. (1984), 340
Flagg, E., 103
Fleming, L. J., 16
Fleming et al. (1986), 586
Foote, F. L., 183
Fontaine, H., 14
Foster Associates, 56, 210
Foster, N., 103
Fox, C., 194
Fox, K. P., 108, 109, 360
Freitag, J. K. (1901), 30
Freudenthal, A. (1947), 735
Frye, M., and G. Morris (1975), 861
Fujikawa and Johnson, 37
Fukumoto, Y., and Y. Itoh (1983), 840
Fuller, B., 38
Fuller, G. R., 967

Galambos, T. V., 859
Galambos, T. V. (1983), 840
Galambos, T. V., and M. Ravindra (1978), 735
Galambos, T. V., and L. Viest (1985), 808, 809
Gallagher, R. (1975), 932
Gary, B., 273
Gauthey, E. M., 26
Gensler, M. A., 201
George, Warren, Inc., 439
Gertis, K. (1975), 595
Gibson, W. B. (1984), 760

Giedion, S., and W. Starrett, 30
Gilbert, C., 4, 111, 153, 159
Gilman, A., 22
Glass Research and Testing Laboratory, 406
Gluck, J., and I. Kalev (1972), 936
Goh, K. S. (1956), 443
Goldberg, B., 37
Goldberger, P., 101
Goldberger, P. (1985a), 54
Goldberger, P. (1985b), 57
Goodhue, B., 119, 124
Goto, Y., and W. Chen (1987), 879, 858
Graham, B. J., 117
Gramme, T. Z., 14
Graves, M., 110
Greene, G. T., 195
Griffith, A. (1921), 782
Gropius, W., 120, 198
Gropius, W. (1946), 169, 170
Gumi, O., 548

Haaijer, G., 807
Halcyon, Inc., 992
Hale, W. E., 18
Hamid, A., 971
Hamilton, H. B. (1941), 189
Harbour, A., 201
Harkness, J. E., 553
Harris, C., and E. Ullman (1945), 79
Hart, G., 971
Hart et al. (1973), 781
Hartung, G., 189
Haryott and Glover (1984), 340
Hasegawa, Y., and K. Tamki (1985), 511
Hasegawa, Y., 516
Haviland, J., 184
Hayes-Roth et al. (1983), 572, 593
Hegemier, G., 971
Heidegger, M., 155
Heller and Leake, 267
Heller, J., 263
Hely, B., and D. Taylor (1983), 947
Hines, G., 121, 327
Hinkley, P. (1975), 729
Hodgkinson, E., 14
Holabird and Root, 36, 14
Hollander, E. L., 229
Holmes, J. D. (1984), 686
Hood, R., 4, 114
Hooke, R., 14
Housing and Development Board, 946
Housing and Development Board (1970), 443
Housing and Development Board (1984–1985), 448
Housing Committee of Singapore (1947), 59
Hoyt, H. (1939), 79
Hoyt, P., 563
Hudson, D. (1965), 781
Hutsch, V., 193
Hyde, H. B., 22

IABSE, 805
ISO, 805
International Facilities Management Association, 553
Iowa State University, 971
Irwin, A. W., 759
Irwin, A. W. (1978), 761
Irwin, A. W. (1979), 761
Irwin, A. W. (1981a), 762
Irwin, A. W. (1981b), 760, 761
Irwin, A. W. (1983a), 760, 761
Irwin, A. W. (1984a), 760
Irwin, A. W., and T. Goto (1984), 762
Isyumov, N., 912, 919
Iyengar, H., 121

Jacobsen, L. (1930), 781
Jahn, H., 54, 109, 113
Japan Atomic Energy Research Institute, 519
Jeary, A. P., 647, 779
Jeary, A. P. (1981), 786
Jeary, A. P. (1986), 784
Jeary, A., and Ellis, B., 782
Jencks, C., 109
Jenkin and Huntington, 427
Jenney, W. L., 30, 35, 29, 422
Jervis, J. B., 20
Johnson, A., 971
Johnson, G. H., 17
Johnson, G., and B. Kreischer, 22
Johnson, P., 5, 330, 427

Johnson, P., and J. Burgee, 108
Johnson, P., and H. R. Hitchcock (1932), 169
Joint Committee on Tall Buildings (1973), 735
Jones et al. (1982), 861

KKMC, 299
Kahn, L., 159
Kajima Corp., 514, 525, 542, 830, 832
Kaplan/McLaughlin/Diaz architects, 277
Kareem, A., 398
Kareem, A. (1985), 402
Kareem, A., and J. Stevens (1985), 398
Kariotis, J., 971
Kasai, K., and E. Popov (1986), 829
Kateley, R., 465
Kateley, R., and M. Lachman (1985), 468
Kato, B. (1982), 882
Kawasumi, H., and K. Kanai (1956), 782
Kendall, E., 22
Kenzo Tange Associates, 975
Kerr, R. A. (1985), 816, 817
Khan, F. R., 26, 121, 829, 905
Khan, F. R. (1974), 595
Kishi, N., and W. Chen (1986), 864
Klote, J., and J. Fothergill (1983), 730
Kobayashi, K. (1985), 516
Kobayashi, H., and N. Sugiyama, 782
Kobe Steel Ltd., 514
Komuten, T., 547, 550
Kozai, 340
Krawinkler, H. (1978), 882
Kreischer, B., 17
Krier, L., 151
Kwiecinski, M. W., 361

Lam, P. K., 443
Landstrom et al. (1983), 760, 761
Lansdown, J. (1982), 592
Larson, G. E. (1981), 22
Latrobe, B., 13, 20
Laves, G. L., 192
Law, M., 646, 727
Law, M. (1985b), 732
LeBeun, N., 104
Le Corbusier, 111, 150, 177, 198
LeFrak, S., 84
Lew, H. S., 967
Lewicki, B., 483
Lewicki, B. (1966), 484
Lewicki et al. (1982), 486, 487
Lewicki, B., Cholewicki, A., and Makulski, W. (1983), 486
Lewitt, C., Chesson, E., and Munse, W. (1969), 863
Lie, T. T. (1972), 730
Lim, B. P., 1, 59
Lim, B. P., and K. R. Rao (1977), 71
Lim, B. P., K. R. Rao, and S. P. Rao (1980a), 72
Lim, W. S. W. (1983), 454
Lin, M. (1985), 56
Lincoln, A., 335
Loebl, Scholssman, Bennett and Dart, 37
Lollini, T., 275
Luedtke, Aldredge and Pendleton, 991
Lui, E. M. (1985), 861, 862, 868, 879, 882
Lui, E. M., and W. Chen (1983), 858
Lundy, V., 919

MacGregor, James G., 899
MacLeod, I. (1969), 933
Maher, M. L., 571
Maher, M. (1984), 847
Maher, M., and S. Fenves (1984), 573
Malley, J., and E. Popov (1984), 829
Manus, C., 273
Marans, R., and K. Spreckelmeyer (1981), 55
Marks, D., 271, 273
Marrey, B., and P. Chemetov (1976), 29
Martin, I., 679, 897
McDermott, J. (1980), 574
McGuire, W., and R. Gallagher (1979), 929, 937, 930, 936
McConnel, R., 193
McKim, Mead and White, 114
Mehrtens, G. C. (1908), 26
Mehta, J. B., 317, 473
Mehta, J. B. (1978), 322
Meinhardt, W. L., 647, 681, 945
Melbourne, W. H. (1979), 682
Melchers, R. E., and M. V. Harrington (1984), 736
Meli, R., 667

Melinek, S., and S. Booth (1975), 731
Merovich, A. T., 832
Merovich, A., and J. P. Nicoletti (1982), 826
Michalski et al. (1983), 851
Michelson, W. (1975), 55
Mighty Jack (Shimizu), 540
Ministry of Housing and Local Government (1986), 485
Ministry of International Trade and Industry, 516
Minor, J. E., 397, 405, 647
Minor, J. E. (1984), 399
Miradrain, 441
Mitchell/Giurgola Architects, 184
Mocak, V., 287
Modjeski, R., 27
Modjeski, R., G. S. Webster, and L. A. Ball (1927), 27
Molgen, 574
Monash University, 686
Moncarz, P., and K. Gerstle (1981), 881
Montgomery, P. O'B., 991
Moore, D. B. (1983), 783
Moore, J. (1982), 755
Morgan, H., and S. Chandler (1981), 728
Morgan, H., and G. Hansell (1985), 729
Morris, Tasker and Company, 19
Mosena, D. R., 257
Moser, M., 209
Mayne, J. R., 756
Mullarkey, P. M. (1983), 586
Monoz, G., 657
Murdock, W., 13, 20
Murphy, C. F., Associates, 36, 37
Muto Institute, 832
Mycin, 574

NASA, 613
Nair, R. S., 353, 360
Nair, R. S. (1975), 936
Nakagawa, K. (1961), 781
Naman, I. A., 607
Naman, I. A., and Associates, 919
Nash, J., 26
Nason, J., 19
National Science Foundation, 406, 897, 966
Nau, D. S. (1983), 572
Nethercot, D. (1983), 840
Nethercot, D. (1985), 840
Newberry, C., and K. Eaton (1974), 691
Newman, J. H., 422, 635
Nicoletti, J., 832
Nillson, N. J. (1980), 573
Noland, 965
Norris, J., 409
Nutt, J. G., 339

O'Brien, J. J., 297
Oh, D., 832
Ohbayashi-Gumi Ltd., 515
Olympia and York, 409
Oppenheim, I. J., 519
Otis, E. G., 17, 34
Overseas Union Bank, 975

Paine, C., 27
Park Tower Realtors, 429
Pascoe, G. W. (1984), 341
Pauls, J. (1977), 731
Paulson, B. C. (1984), 507, 534
Pawlak, Z. (1982), 851
Pawlowski, A. Z., 361
Peck, R., 442
Pedersen, W., 149
Pei, I. M., 103, 184, 216, 997
Pelli, C., 108, 109
Perkins, J., 19
Perkins and Will, 37, 360
Perrow, C., 737
Petak, W., and G. C. Hart (1980), 398
Peters, T. F., 25, 189
Peters, T. F. (1981), 27
Peterson, F. A., 16
Petrovski et al. (1973), 781
Pettit, H. M., 114
Planning and Environmental Criteria, 594, 601
Popov, E. P., 815
Popov, E. P., and R. Black, (1981), 827
Port Authority of New York and New Jersey (1984), 735
Porter, M., 971
Post, G. B., 22, 103
Progressive Architecture, 383

Project Management Institute, 299
Project Software and Development, Inc., 299
Prospector, 474
Purdue University, 882

Quimby, H. H., 15
Quimby et al. (1892–93), 15
Quinlan, J. R. (1980), 593
Quinlan, J. R. (1983), 851

Randall, F. A. (1949), 29, 30
Rankine, J., 337
Rathbun, J. C., 711, 713, 714, 863
Rathbun, J. C. (1936), 860
Rathbun, J. C. (1940), 710
Razzaq, Z. (1983), 861
Real Estate Research Corporation (1985), 465
Real Estate Research Corporation, 466
Reiich, R., 422
Rich, E. (1983), 573
Richardson, H. H., 35
Rigg, G., 605
Ritchie, A., 3
Robertson, L. E., 184
Roche, K., 109
Rodriguez, L. B., 657
Roeder, C., and E. Popov (1978), 829
Rogers, F. (1977), 728
Rosenblueth, E. (1960), 598
Rosman, R. (1971), 937
Rouse, W., 81
Rowe, C., 150, 151, 158
Ruchelman, L. I. 79
Rutenberg, A. (1974), 941, 942

Saarinen, E., 120
Samali, B., 697
Sander, D. M. (1974), 690, 691
Sasaki Associates and Halycon, 993
Sasaki, S., 975
Saulnier, J. (1817–1881), 29
Savery, T., 20
Saxon, R. (1983), 731
Schinkel, K. F., 191
Schipporeit-Heinrich, 37
Schlossman, L., and J. Bennett, 36
Schroeder, J., 621, 629
Schultz, C. N., 156, 158
Schuyler, M., 103
Scruton, 720
Scully, V. (1985), 82
Sedlacek, G., 885
Seible, F., 971
Seidler, H., 165
Shaw, C. Y. (1979), 689
Shaw et al. (1973), 689
Shimizu Construction Co., Ltd., 514
Shing, B., 971
Shortliffe, E. (1976), 574
Shute, L. V., 427
Simitses, G., and J. Giri (1982), 882
Simitses, G., and A. Vlahinos (1982), 882
Simitses et al. (1984), 882
Singh, G., and J. Schwaighofer, (1976), 923
Sisodiya et al., (1972), 933
Skempton, A. W. (1959–1960), 29
Skempton, A. W. (1960), 195
Skempton, A. W., and H. R. Johnson (1962), 189
Skidmore, Owings & Merrill, 36, 108, 213, 264, 723, 831, 832, 833
Smeaton, J., 14
Smith, W. S., 16
Smith et al. (1981), 923, 928
Smith, S. (1985), 936
Societe des Ingenieurs, 706
Speer, A., 169
Sprague, F. J., 18
Sriram et al. (1985), 572
Stamato, M., and E. Mancini (1973), 937
Staniszewski, R. (1980), 849
Starrett brothers, 198
Starrett, W. A. (1928), 29
State University of New York/Buffalo, 795
Steel Committee of California, 824
Steel Committee of California (1981), 825
Steele, J. (1961), 760
Stefik, M. (1981), 574
Stefik, M. and Martin (1977), 573
Steinbrugge, K. V., 818
Steinbrugge, K. V. (1982), 817
Steiner, H. M., 87

Stephenson, R., 27, 192
Stockdale, R., 563
Stone, E. D., 37
Strauss, J., 27
Stroubel, C., 198
Structural Design of Tall Concrete and Masonry Buildings, 594
Structural Design of Tall Steel Buildings, 594
Structural Engineers Association of California (1976), 822
Structural Engineers Association of California, 37, 823
Structural Engineering Association of Northern California, 822
Structural Stability Research Council (SSRC), 805, 837, 859
Structural Stability Research Council (1983), 837
Strutt, W., 13, 19, 189
Sturdevant, B. F., 21
Stussi, F. (1943–1944), 27
Sugimoto, H. (1983), 864, 865
Suko, M., 975
Sullivan, L., 4, 6, 35, 36, 40, 108, 153, 198
Sullivan, L. (1924), 159, 164
Susaki Associates, 992
Suzuki, S., 507
Swan, J., 21
Swason, V. E., 90

Taisei Corporation, 514, 832
Takenaka Komuten Co., Ltd., 515
Tall Building Criteria and Loading, 594
Tall Building Systems and Concepts, 594
Tasker, M., 20
Tassios, T. P., and S. G. Tsoukoutas, 489
Taylor, H., 919
Taylor/Lundy/HKS Architects, 919
Technical Coordinating Committee for Masonry Research, 966
Texas Tech University, 406
Thomas et al. (1963), 730
Thompson and Ainsworth (1985), 344
Thompson et al. (1976), 345
Tilley, G., and R. Eyre (1977), 783
Times Square Redevelopment Corporation, 429
Timoshenko, S., and J. Gere (1961), 858
Tishman Realty, 991
Tishman Research Company, 422
Tokyo Electric Power Co., Inc., 515
Trammell Crow Company, 991
Tri Land Development, 991
Trump, D., 84, 422
Tschanz, T., 724
Tulin, L., 971
Turak, T. (1966), 30
Turkstra, C., 646, 735
Turner Construction, 919

U.S. Housing and Urban Development, 967
U.S. Japan Cooperative Program in Natural Resources, 967
Ueno, T., 507
Uniform Building Code (1982), 826
United States Steel Corporation, 827
University of California/Berkely, 234
University of California/Los Angeles, 971
University of California/San Diego, 971
University of Colorado/Boulder, 971
University of Singapore, 72
University of Southern California, 796, 982
University of Western Ontario, 912
Urban Investment and Development Company, 360
Urban Land Institute, 230

Van Alen, 114
van der Rohe, M., 36, 37, 107, 120, 134, 180, 198
van Eyck, A. (1981), 169
Vasudevian, R. (1975), 781
Vazquez, A., 657
Verin Deutscher Ingenieure VDI Recommendation 2057 (1979), 889
Vituvius, 16
Voit, 194
von Dyck (1912), 26
von Reichenbach, G., 26
von Siemens, E. W., 18

Walker, G. R., 404
Walker, G. R. (1984), 686
Waseda University, 511
Washington Post (1984), 89
Washington Post (1985a), 89
Washington Post (1985b), 89
Watt, J., 19
Weaver et al. (1971), 936
Webb, C. G. (1960), 70
Wenzelberger, J. P., 87
Werder, L., 194, 195
Western States Clay Product Association, 973
White Industries Australia, 946
Whyte, H., 989, 990, 1001
Whyte, W. (1980), 82
Wiebeking, F., 26
Wiesner, K. B., 789
Wight, P. B. (1876, 1878), 16
Williams, G., 233
Winograd, T. (1971), 592
Wise, C., 367
Wong, S., and W. Ziarko (1985), 851
Wood, H. (1968), 960, 962
Wren, C., 124
Wright, F. L., 108, 111, 120, 422, 906
Wyatt, S., 26
Wyatt, T. A. (1977), 782

Yamamoto, K. (1985), 516
Yamanouchi et al. (1984), 827
Yang, J. N., 703
Yang, J., and Y. Lin, 699
Yang, J., and Y. Lin (1981), 700
Yeh, S. H. (1975), 61
Yoshida et al. (1984), 507, 533
Young Gentek Company, 991
Young, D., 195

Zandonini, R. (1983), 838
Zandonini, R. (1985), 840
Zeevaert, L., 820, 833
Zeevaert, L. (1975), 598
Zetlin, L., 383
Ziarko, W., 847
Zienkiewicz, O. (1977), 932
Zunz, J., 367

Subject Index

Names of individuals and organizations are found in the Name Index. Names of buildings are found in the Building Index. Names of cities are in this Subject Index, as are names of codes, specifications, and standards. Italicized numbers designate pages with figures or photographs.

AASHTO requirements, 807
Abbreviations, 1033
Abnormal loads, 485
Accelerations, 718
Acceleration type responses, 697
Accelerographs, 665
Access roads, 418
ACI Building Code, 900, 902
Acoustical privacy, 213
Adaptability, 294
Adaptive expert system, 848
Adhocism, 176
Aerodynamic appendages, 797
Aeroelastic models, 719, 722
Aeronautical wind tunnels, 719
After-tax analysis, 93
Aged, 1
Agora, 278
Agoraphobia, 292
Air conditioning, 608
Air infiltration, 689, 694
Air rights, 293
Air supply system, 614
AISC, 815, 823, 824, 857, 858, 879
AISC Code (1985), 809
AISC LRFD Specification, 808
AISC quality certification program, 811
AISC Specification, 807, 810
Alternative to high-rises, 184
Amenities, 277, 319, 333
Amenity building, *458*
American cities, 156, 317
Amplification, 670
Amplitude of excitation, 723
Analysis and design, 815
Analysis model, 968
ANSI Code, 808
ANSI load factors, 735
Apartments, 62, 452
Appliances, 322
Application of robots, 510
Application of systems methodology, 553
Applications of expert systems, 584
Aqueduct, 22
Arcades, 123
Architects, 119
Architectural concept, 189
Architectural considerations, 915
Architectural design quality, 455
Architectural directions, 165
Architectural heritage, 259
Architecture, 101, 149
Arrow diagramming ,311
Articulated housing blocks, 463
Aseismic provisions, 488
ASHRAE Standard, 639
Asian cities, 318
Asian countries, 453
Assessment criteria, 763
Assumptions in analysis, 924
ASTM A709, 811
ASTM Designations, 807
Atrium, 108, 730
Augercast piles, 435
Australian Wind Loading Code, 687
Automatic concrete sprayer, 523
Automatic learning, 851

Back-office buildings, 268
Baroque Rome, 152
Base balance measurements, 714
Base balance technique, 716
Basement waterproofing, 439
Base moment coefficients, 714
Base shear coefficient, 826
Basic threshold effects, 764
Beam cracks, 653
Beam-columns, 840
Beams, 840
Beam-to-column connection, 857
Bearing walls, 978
Beaux arts style, 235
Benching, *251*
Bilinear model, 861
Bombay, 317, 324, 473
Boundary Layer Wind Tunnel, 711, 716, 720, 722
Bracing system, 828
Bratislava, 503
Brick cladding, 394
Brittle behavior, 675
Brittle fractures, 825
BS 6472, 761, 889, 890
BSI, 730, 755
BSSC, 829
Buckling, 874
Buckling loads, 840
Building automation, 619
Building codes, 476, 477
Building forms, 239, *253*
Building managers, 560
Building resonance, 707
Building service systems, 207, 605
Building sway, 709
Building regulations, 642
Built-up compressed members, 840
Bulk controls, 238
Bulk diagram, *252*
Bundled tube, 27, 121

CADD, 563
Cairo cityscape, 126
Caisson, 16
Calcutta, 317, 473
Calgary, 409
Canada, 409
Canadian Building Code, 913
Canadian Code, 903
Canadian Code Analysis, 913
Carnegie steel tables, 198
Case history of Transco Tower, 329
Case study, 771
Casual design components, 55
CEB model code, 489
Ceiling heights, 333
Cellular floors, 423
Central Provident Fund, 63, 448
Central business districts, 54
Certainty levels, 853
Certification, 812
Changing views, 103
Charpy V-notch requirements, 811
Chicago, 33, 101, 229
Chicago perspective, 257
Chicago school, 111, 197, 198
Chicago style, 4
Children, 76, 481
Cladding, 397
Cladding elements, 681
Cladding pressure, 686
Classicism, 176
Classification of structural systems, 674
Classification of structure, 923
Claustrophobia, 292
Clean room testing robot, 527
Cluster, 287
Code formats, 837
Code evolution, 822
Codes, 899
Coefficients, 717
Comfort level, 611
Commentary on structural standards, 899
Commercial lease tax, 123
Communal facilities, 61
Communications, 423, 620
Community development, 445
Competition, 635
Complex steel building structures, 812
Composite construction, 339
Composite high-rise, 339
Composite system, 975
Composition, 156
Compressed members, 837

Computer age, 421, 614
Computer-aided analysis, 858
Computer-aided design, 122, 804
Computer control, 620
Computer floors, 424
Computer revolution, 638
Computerized facilities management, 560
Computerized management, 559
Computerized methods, 845
Computerized system, 560
Computerized scheduling, 298
Computers and codes, 900
Concentrated keys, 489
Concentrically braced framing, 828
Conceptual framework, 591, 600
Concrete, advantages of, 906
Concrete and masonry building, 897
Concrete placing crane, 528
Concrete pumping, 918
Concrete robot, 528
Concrete sprayer, 544
Concrete strength, 897
Condensation, 615
Condominium office, 322
Connection flexibility, 867, 858, 859
Connection modeling, 860
Connections, 857
Constructability critics, 586
Construction, 409
Construction costs, 435, 441, 448
Construction failures, 389
Construction industry, 448
Construction investment, 229
Construction know-how, 302
Construction management, 306
Construction methods, 897
Construction productivity, 635
Construction robot, 507, 510
Construction scheduling, 312, 421
Construction systems, 409, 427
Construction workers, 449
Consultation services, 201
Context, 573
Context tree, 582
Contextualism, 176, 257
Continuum approach, 925, 937
Core, 162
Core-hanger, 361
Cornices, 108, 156, 235
Corporate image, 328
Cost aspects, 349, 805
Costs and benefits, 90
Coupled systems, 737
Coupled wall structure, 927
Coupling beams, 489
CPM, 298
CPM scheduler, 300
Creep, 952
Creep, shrinkage and temperature effects, 959
Crime rates, 67
Criteria and loading, 645
Critical environment, 516
Critical path, 298, 309
Cross-power spectral density, 698
Culture, 189
Cumulative damage criterion, 686
Curtain wall, 18, 30, 197
Cutting robot, 524
Cyclone Tracy, *405*

Dallas Arts District, 989
Damage analysis, 652
Damage classification, 568
Damage control, 822
Damaged buildings, 671
Damaged building characteristics, 671
Damaged columns, 653
Damage evaluation, 657, 659
Damage statistics in Mexico City, 657
Damage zonification, 658
Damping behavior, 782
Damping estimation, 785
Damping predictors, 784
Damping systems, 647, 655, 680, 779, 789, 792
Dancing performance, 894
Decision-makers, 636, 641
Decision rule generator, 849
Deep beams, 901
Deep knowledge, 570
Deep spandrel beams, 652
Deflection, 707, 713, 716
Demographic factors, 468
Depreciation, 88

Design, 574
Design for safety, 748
Design loads, 753
Design modifications, 412
Design of facilities, 204
Design options, 825
Design optimization, 848
Design philosophy, 902
Design pressures, 687
Design process, 923
Design quality, 328
Design spectrum, 660
Detection, 728
Developers, 84, 329, 416, 638, 898
Developing countries, 317
Developing tall buildings, 327
Development and investment, 327
Development guidelines, 84
Development plan, 257, 476
Development regulations, 271
Development rights, 236
Development site, *320*
Diagnosis, 574, 641
Diagonal steel bracing, 649, 655
Diagramming, 311
Diaphragm wall, 433, 488
DIN 4150, 893
Discontinuous structures, 928
Discrete methods, 929
Discretionary time, 642
Displacement function, 932
Displacement response, 697
Distributed leakage, 690
Downstream, 682
Downtown belt courses, *242*
Downtown development, 233
Downtown housing, 470
Downtown Plan, 233, *248*
Downtown renewal, 230
Downtown skyline, *255*
Drag direction, 721
Drainage system, 440
Drilling robot, 523
Drive system, 621
Ductile behavior, 675, 804
Ductility reduction factors, 679
Durability, 755, 903, 959
Durcon, 962
Dust concentration, 537
Dynamic analysis, 984
Dynamic building movements, 717
Dynamic characteristics, 779
Dynamic displacements, 718
Dynamic knowledge, 591
Dynamic pressures, 681
Dynamic properties, 968
Dynamic response, 661, 789, 968
Dynamic sway, 718
Dynamic wind response, 646

Early occupancy, 416
Earthquake characteristics, 657
Earthquake of Bucharest, 487
Earthquake loading, 646
Earthquake intensity, 488
Earthquake loading and response, 649
Earthquake-resistant design, 815
Earthquake response, 485
Earthquakes, 597, 779
Earthquakes and panel systems, 487
Eccentric bracing, 828, 829
ECCS, 805, 809, 842
ECCS Code, 809, 839
ECCS column curve, 810, 838
Ecological character, 80
Ecological factors, 79, 80, 84
Economic factors, 468
Economical design, 837, 841
Economics, 1, 87
Economizer cycle, 610
Economy, 202
Effect of infill walls, 675
Effective management, 555
Efficiency, 202
Eiffel Tower movements, 708
Elastic analysis, 921
Elastic constraints, 842
Elastic system, 826
Elderly, 470
Electro-thermal cooling, 613
Elevator configurations, 629
Elevator consumption, 622
Elevator core, 630
Elevators, 17, 412
Elevators, climbing, 414
Elevators, double-deck, 623, 629

Elevators, express, 333, 623
Elevators, local, 289
Elevators, quadruple-deck, 631
Elliptical movement, 707
Employee convenience, 332
Energy, 638
Energy consumption of elevators, 621
Energy dissipation, 825
Engineering services, 613
Entablatures, 156
Enterprise zone, 123
Envelope potential, 273
Environmental change, 745
Environmental context, 454
Environmental controls, 329
Environmental criteria, 2
Environmental factors, 59, 82
Environmental simulation, *248*
Epicenter, 816
Epoxy resins, 653
Equatorial comfort index, 70, *71*
Equivalent constant pressures, 686
Equivalent slenderness, 841
Escape, 731
Esthetic consideration, 163
Esthetic criteria, 293
Esthetic statements, 110
Esthetic testing, 563
Eurocode 3, 809, 893, 894
Europen towns, 361
Evaluation, 578, 582
Evolution of skyscrapers, 422
Existing buildings, 658
Expert systems, 571, 593, 847, 962
Explanation module, 573
Exponential connection model, 863
Exponential model, 861
Exterior column, 881

Facade, 152
Facilities management, 553, 557
Facilities planning, 553, 554
Failure of columns, 675
Failure paths, 746
Family life-cycle, 56
Family size, 69
FAR (Floor Area Ratio), 238, 259, 271, 293, 317, 477, 475
FAR envelope, 272
Fast tracking, 350
Fears, 292
Feasible states, 854
Feedback control, 19
Final design, 847
Finance and economics, 446
Financial factors, 470
Financing low-income housing, 63
Financing scheme, 449
Finishing materials, 414
Finish-to-finish relationships, 311
Finite element method, 804, 932
Finite strip technique, 933
Fioratura, 153
Fire, 727
Fire engineering, 727
Firefighting, 324
Fire safety, 618, 646, 727
Fire spread, 728
First International Conference, 1003
First-order analysis, 879
First vibrational mode, 597
First World Earthquake Conference, 781
Flab slabs, 677
Flats. (*See* Apartments)
Flexibility of connections, 859, 866, 869, 870, 872
Flexibly-connected frames, 858, 868
Flexural torsional buckling, 840
Floor area ratio. (*See* FAR)
Floor loading, 754
Floor populations, 624
Floor slabs, 496
Floor space standards, 454
Floor structures, 342
Flow resistance, 690
Flying buttress, 14
Footprint, *274*
Foundation cofferdams 439
Foundation design, 433
Foundation failure, 677
Foundations, 433, 914
Fourier transforms, 698
Fracture-control plan, 811
Frame ductility, 825
FR (Fully-Restrained) construction, 858, 860

Framed systems, 661
Frame-wall interaction, 934
Friction type connections, 842
Full scale behavior, 709, 723
Full scale experiments, 709
Functional considerations, 162
Functional systems, 578
Future demand, 468
Future of skyscrapers, 3

Galleria/Post Oak, 331
Gas explosions, 485, 744
General Seismic Code, 822
Genius Loci, 156
Geological aspects, 658
Geometrical imperfection, 837
Glass breakage, 397
Glass panels, 648
Global character, 848
Global competition, 637
Gold coast, *231*
Golden age, 4
Gothic, 36, 118
Gravity system, 581
Greco-Roman urban form, 277
Grid, 576
Growth of tall buildings, 327
Growth limitation, 267
Growth management, 233
Gust factor, 683
Gust wind speed, 687

Habitability, 18
Handicapped, 1, 208
Haphazard storage, 411
HDB, 59
Heating, ventilating and air conditioning (HVAC), 605, 607
Heat release, 627
Heat transfer, 617
Heavy sections, 839
Height and bulk control, 236
Height limits, 238, 239, 259
Height map, *249*
Height/Bulk regulation, 271
Heights between floors, 423
Helicopters, 325, 423
Heuristic knowledge, 570, 593
High-density living, 445
High-rise, 571
High-rise building data base, 1003
High-rise core, 236
High-rise development, 317, 480
High-rise housing, 337
High-rise living, 59, 65
High-rise residents, 465
High-strength bolted connection, 824
High-strength concrete, 902
High-tech era, 421
High-tech tall buildings, 635
Historic view, *241*
History of tall buildings, 11, 117, 189
Hoisting facilities, 412, 415
Home ownership, 452
Hong Kong, 209, *210*, 293, 479
Horizontal deflection, 977
Horizontal joints, 489
Household servants, 322
Housing, 56, 443, 465, 473
Housing and Development Board, 59
Housing estates, 60, 61
Housing in India, 473
Housing markets, 259
Housing needs of the masses, 479
Housing program, 59
Housing shortage, 474
Houston, 149, 335
Human comfort criteria, 762
Human error, 736
Human perception, 760
Human scale, 258, 455
Hurricane Alicia, 402
Hybrid construction, 353
Hysteretic behavior, 844

IBO Uniform Building Code, 395, 822
Ice storage systems, 613
Ideal form, 159
Impact of expert systems, 587
Impact of technology, 474
Impacts, 275
Imperfections, 837, 841
Impulsive events, 767
In-plan symmetry, 676
Inclined tall buildings, 388
Inclusivism, 176

India, 317, 473
Indoor spaces, 278
Industrial development, 63, 118
Industrialized building, 483
Industrialized methods, 451
Inefficient handling, 411
Inference mechanism, 573, 601
Inference rules, 847
Infiltration, 615
Information age, 636
Initial stiffness restraint, 871
Injuries, 416
Innovative future, 804
Innovative solutions, 854
Inspection-evaluation procedure, 813
Insulation, 617
Integrated systems, 639
Integrative applications, *586*
Integro, 493
Intelligent buildings, 381, 422, 639
Intensities, 816
Interaction curves, 841
Interaction system, 361
Interaction with service, 372
Interaction with soil, 372
Interior column, 881
Interior design, 201
Interior planning systems, 206
Interior spaces, 283
Internal pressure coefficients, 691
Internal pressures, 689
Internal rate of return method, 90
International style, 4, 107, 169, 288
Interpretation, 574
Interstory displacements, 970
Interstory drift, 486
Invention, 25
Iron columns, 13
Iron framing, 13
Irritability, 763
ISO, 791, 805
ISO Code 2631, 889

Japan, 507
Jerk, 762
Jet-cutting robot, 545
Joint models, 857
Joints in external walls, 483
Joint stiffness, 487
Journal square failure, 736

Knee bracing, 14, 15
Knowledge acquisition 573, 593, 850
Knowledge-base, 570, 849, 850
Knowledge-based systems, 591
Knowledge engineering, 592
Knowledge representation, 592
Kobe Steel, 544

Labor management, 449
Labor productivity, 509
Labor shortage, 509
Labyrinth, 154
Landscaping, 63, 989, 284
Land use, 80, 445
Land values, 475
Large panel system, 744
Larger sections, 825
Late modernism, 109, 176
Lateral force coefficient, 822
Lateral seismic forces, 822
Lateral stability system, 368, 371
Leakage paths, 689
Level floor, 986
Life-course effects, 56
Life safety, 208
Life span, 591
Lifestyle factors, 469
Lift direction, 721
Light, 319
Lighthouse, 15
Lighting, 20, 619, 638
Lighthouse, 15
Lighting, 20, 619, 638
Lightweight wall panelling, 684
Limit state analysis, 859
Limit states design, 750, 807, 885. *See also* LRFD
Linear elastic analysis, 92, 933
Linear model, 861
Linkage and boundary, 158
Liveliness, 470
Load and resistance factor design (LFRD), 807, 885
Load-bearing system, 498
Load duration, 685

Load factors, 808, 809
Load patterns, 864
Load sequences, 864, 871
Loading data, 753
Loading dock, 415
Loading/unloading characteristic, 866
Loads, 753
Local stability, 843
Long duration motion, 658
Long response time, 684
Long-term reduction, 685
Los Angeles, 81
Lotus 1-2-3, 89
LRFD, 645, 735, 750. *See also* Limit states design
LRFD Specifications, 808, 879
LVDT, 437

Machine room heat release, 625
Macrostructures, 804
Madras, 317
Maintenance, 66, 474
Major tenant, 330
Management, 311
Management information systems, 299
Manhattan, 84, 466
Market/Hawkers' Center, 459
Market-rate projects, 465
Masonry structures, 965
Massive masonry curtain, 684
Mass produced product, 638
Mass-spring-dashpot element system, 892
Material properties, 752
Material selection, 810
Materials management, 449
Materials models, 967
Maximum accelerations, 817
Maximum amplitudes, 893
Mean density ratio, 720
Mean response, 721
Measurement techniques, 752
Mechanical imperfection, 837
Mechanization program, 450
Mega-frames, 723, 978
Megastructures, 6, 121
Melbourne, 165
Metal deck, 423
Metal formwork, 450
Methods of analysis, 921
Mexico City, 654, 667, 668, 670, 816
Mexico City Code, 654
Mexico City earthquakes, 598, 649, 679, 816, 898
Microcracks, 782
Micrometeorological wind tunnel, 720
Microwave relays, 335
Middle income housing, 466, 478
Mile-high building, 905, 906
Mixed systems, 353, 901
Mixed use, 467
Mixed-use habitats, 56
Mixed-use skyscraper, 37
Modelling assistants, 585
Modern architecture, 150, 180
Modernism, 4, 54
Modern movement, 177
Modified Mercalli (MM) Scale, 816
Moisture, 615
Moment coefficients, 716
Moment connection, 824, 857, 859
Moment-resisting frames, 826, 827, 830
Moment transfer, 864
Monitoring, 574
Monitoring devices, 442
Moratorium, 81
Motion characteristics, 658
Motion descriptors, 766
Motion perception and tolerance, 759
Motion reduction systems, 790, 792
Motion sickness, 760
Movement, 292
Moving workers, 412
MSCS, 299
Mullion design, 616
Multiple column curves, 838
Multipurpose travelling vehicle, 528
Multiracial society, 67
Multitier systems, 372
Multistory factories, 61

Nariman point, 320
National Bureau of Standards, 55, 489, 967
National Science Foundation, 406, 897, 966

Natural frequencies, 703, 887, 893
Natural light, 319
Natural period, 669, 707
Negative float, *314*
Neighborhood, 82
Neo-Gothic, 119
Net pressures, 691
Network format, *300*
Networking, 299
New Delhi, 317, 324, 473
New developments, 922
New towns, 60, 445
New York, 81
New York Zoning Ordinance, 265
NFPA, 731
Night lighting, 994
No-growth forces, 267
Noise control, 72, 763, 773
Nolli plan of Rome, 155
Nomenclature, 1007
Nominal resistance, 809
Nonlinear analysis, 841
Nonmagnetic, 439
Nonsway frame, 879, 880
Nonsymmetric walls, 675
Normal modes, 697
NRC/EERC, 821
Nuclear family, 70
Nuclear power, 517

Occupancy sensors, 639
Occupant comfort, 984
Office core, 257
Office growth, 259
Office market, 328
On-site access, 418
On-site medical facilities, 417
Opening area, 693
Open joint concept, 484
Open space, 237, 273, 277, 318, 322, 476, 482
Operating costs, 562
Operation, 611
Order of construction, 429
Oriented core, 439
Outdoor room, 461
Out-of-straightness, 810
Overcrowded cities, 293
Overload, 677
Overturning moment, 930
Owners, 309, 635, 742, 898
Ownership, 466

P-Δ effect, 677
Palazzo, 157
Panel venting, 685
Panel zone deformation 882
Parameter selection, 578
Parking, 332
Partial factors, 751
Partial prestressing, 951
Peak accelerations, 817
Peak coefficients, 683
Peak static deflections, 717
Pedestrian access, 281
Pedestrians, 82, 276, 234, 285
Pediments, 156
Pendulum, 709
Percentage of damage, 672
Perimeter framing, 829
Periods, 654
Periods of soils, 680
PERT, 298
Philadelphia, 80
Philosophy of tall buildings, 3
Piecewise linear, 861
Pile driving analyzer, 435
Piling, 15
Pin-ended columns, 833
Pinned-joint model, 857
Pioneer buildings, 35
Piping network, 569
Plank diaphragm, 969
Planning, 417, 574
Planning flexibility, 213
Planning module, 205
Planting bed, 999
Plastic hinge, 869
Plaza, 36, 111
Plinths, 156, 476
Plumb bob, 711, 713
Plumbing, 19, 619
Pluralism, 176
Pneumatic caisson, 16
Polish tall buildings, 362
Polynomial model, 861

Population density, 293
Porosity, 960
Portland cement, 961
Post-colonial building, 13
Post-modern classicism, 176
Post-modernism, 5, 108, 176, 287, 945
Post-tensioning, 955
Pounding, 654, 676
Power model, 862
Power requirements, 625
Prague, 504
Precast concrete, 946
Precast elements, 483, 949
Precast facade, 956
Precasting factory, 948
Precast members, 493
Precast roof, 747
Precedence diagramming, 311
Preconstruction study, 427
Prefabricated concrete systems, 483
Prefabricated tall buildings, 338, 483, 493 946
Preferred story, *66*
PR (partially-restrained) construction, 858
PR frames, 860
Preliminary analysis, 578
Preliminary design, 847
Premature deterioration, 959
Pre-planning, 410
Present worth method, 90
Preservation controls, 234, 236
Pressure coefficients, 687, 711
Prestressed concrete, 945
Prestressed concrete floor slabs, 496
Prestressed facade, 956
Prewiring, 423
Private knowledge, 593
Probabilistic approach, 804
Probabilities of failure, 745
Probabilities of occurence, 748
Probability distribution, 723
Product liability, 643
Production phase, 851
Productivity, 509, 640
Progressive collapse, 485, 488
Project management, 297, 309
Project management software, 299
Prototype, 575
Public attitudes, 278, 335
Public housing, 443
Public participation, 455
Public spaces, 275, 277, 283
Pulse control, 796
Punching failure, 677

Qualitative variables, 848
Quality assurance, 735, 738
Quality certification, 808, 811
Quality of life, 54, 277, 455
Quantitative testing, 563

Race for height, 34
Rain screen system, 647
Raised floors, 424
Random fluctuation, 698
Random forcing, 780
Ratio of the height to width, 720
Real estate developers, 328
Reasons to build tall, 327
References, 1037
Regulators, 642
Regulatory action, 642
Reinforced concrete panels, 832
Reinforced masonry, 966
Reinforcing bar arranging robot, 525
Reliability, 807, 837, 970
Rental market, 163
Renters, 466
Reporting system, 310
Resale, 64
Rescue operations, 517
Research, 53, 731
Research and development, 613
Research projections, 55
Resettlement, 445
Residential downtowns, 470
Residents, 455
Residents committees, 68
Residual stresses, 839
Resistance factors, 809
Resonance, 661
Response spectra, 659, 660, 826
Response to vibration, 760
Restraining effect, 840, 874
Retirement housing, 470

Retrofitting, 649
Revivalist architecture, 180
Rigid-connected subassemblage, 874
Rigid connection, 870, 872
Rigid joints, 842
Rigid-joint model, 857
Rigidly-connected, 869
Risk factor, 643
Robots and tall buildings, 507
Rock coring, 439
Romance of tall buildings, 329
Romanesque, 36
Romantic modernism, 109
Root mean square moment, 717
Rotational vibration, 761
Rubbish, 416
Rule base, 601

SACON, 585
Safety level, 508, 735, 741, 742, 743, 749
Sandwich panel wall, 484
San Francisco, 235, *244*, *254*, 257, *264*, 818
San Francisco Downtown Plan, 233, 263, 277
San Francisco earthquake, 235, 816
SAPVI, 982
Saudi Arabia, 299
Scenic view, 319
Scheduling, 309, 421
Scheduling programs, 310
Seating, 284
Secondary stability system, 486
Second century, 37
Second-order analyses, 902
Sectorization, 662
Security, 66, 284, 618, 748
Segregation, 69
Seismic design, 680, 815
Seismicity aspects, 657
Seismic motion, 791
Seismic regulations, 674
Seismic-resistant design, 825
Seismic waves, 658
Semantic parameters, 851
Semi-rigid connection, 842, 858, 871, 879
Semi-rigid framing, 858
Seniors, 470
Sense of community, 82
Sense of place, 455
Sensitivity analyses, 87
Serviceability, 885, 951
Service performance, 887
Services, 640
Service staff, 322
Setbacks, *253*, 264, *268*
Shear connection, 677, 857
Shearing deformations, 931
Shear interaction, 935
Shear layer, 682
Shear transfer, 941
Shear wall building, 489, 939
Shimizu site robot-2, 520
Shopping arcades, 989
Shopping center, 989
Short columns, 676
Shrinkage, 952
Shuttles, 289, 623
Sidewalks, 284
Simplified analysis, 940
Singapore, 59, *62*, 443, 456, 479
Single-family home, 470
Site assemblage, 271
Skyline, 80, 234
Sky lobbies, 287, 288, 623, 629
Skyscrapers, 11, 707, 1043
Skyscraper construction, 329
Skyscraper development, 81
Skywalk, 332
Skyways, 56
Slab-finishing robot, 522, 542
Slenderness, 160, 902, 977
Slitted walls, 832
Slurry wall, 433
Small opening, 693
Smart buildings, 422, 427
Smoke extract, 730
Smoke spread, 729
Social and cultural factors, 481
Social disutility, 55
Social effects, 53
Social interaction, 67, 107, 481
Social research, 53, 55
Socio-political influences, 79
Soft ground floor, 676
Soft soil zone, 659
Soft-site analysis, 271

Soil-structure interaction, 815
Solar access and shadowing, 238, 239, 271, 293
Solar fan, 239, *270*
Solution generator, 849, 853
Sound pressure, 76
Space management, 554
Space planning, *203*
Space requirements, 629
Space-saving elevators, 629
Spandrels, 37
Spectral density, 700
Sprinklers, 728
SSRC column curves, 838
SSRC guide, 859
Stability, 837
Stability analysis, 858
Stability of overall structures, 842
Stacked structures, 386
Stair pressurization, 618
Standard deviations, 703
Start-to-start relationship, 311
Static pressure model, 722
Statistical analysis, 807
Statistical model, 723
Statistics of damaged buildings, 662
Status symbol, 319
Steel fabrication, 823
Steel frames, 857
Steel seismic codes, 823
Steel vs concrete, 421
Steel/concrete interface, 358
Stiffness, 847, 929, 980
Stochastic process, 698
Storage area, 415
Storm action, 769
Strain rates, 811
Stratigraphic zoning, 658
Street effects, 82, 84
Streetscape, 234, 455
Strength design, 973
Strength of glass, 647
Structural behavior, 659
Structural control, 575, 790
Structural design, 897
Structural design of tall steel buildings, 803
Structural failure types, 675
Structural modifications, 890
Structural safety, 735, 741, 987
Structural stability, 727
Structural standards, 339, 595, 599, 807, 841, 901
Stud welding robot, 525
Subassemblages, 873
Submittal, 314
Subsoil zonation, 671
Substructuring method, 936
Suburban high-rise, *318*
Supertall buildings, 6, 112, 646, 705, 706
Super-waffle, 910
Surface knowledge, 570
Survey, 1003
Suspended maneuvering system, 325
Suspended systems, 364
Sway frames, 841
Sway-permitted frame, 881
Sydney, *167*
Symbolic attachments, 84
Symbols, 117, 1032
Syntactic parameters, 851
System design, 608
System failure, 737
System selection, 579
System stability, 842
Systems and concepts, 337
Systems and subsystems, 312

Tall building clusters, 54
Tallest buildings, 1004
Tangent stiffness restraint, 868
Technical evolution, 11
Technology, 451
Telecommunications, 328
Temporal linkage, 157
Tenants, 208, 314, 465, 561
Tendon control, 796
Tension cables, 383
Tension loading, 651
Tension nets, 383
Terrorist attacks, 618
Testing laboratories, 216
Thermal comfort, 70
Thermal deformation, 484
Thermal expansion, 292
Thermal insulation, 484

Third International Colloquium on Stability, 837
Third International Conference, 27, 101, 229, 803
Third World, 473
Three-dimensional systems, 934
Threshold of sensation, 709
Tie beams, 488
Tiedowns, 435, 438
Tilted, 388
Time lost, 411
Time to build, 640
Time-varying loads, 809
Torsional effects, 823
Torsional loading, 939
Torsional stiffnesses, 935
Total environment, 445
Toughness requirements, 810
Town center landscape, 460
Town centers, 62, 445
Toxicity, 643
Traditional city, 151
Traffic, 76, 430
Transfer structures, 344, 367
Transferable development rights, 236, 259, 273
Transit network, 994
Transportation, 1, 475
Transportation of materials, 412
Transverse partition walls, 650
Trapezoidal column, 986
Tributary area, 983
Tube-in-tube, 949
Tube structure, 121
Tubular columns, 839
Tubular construction, 26
Tubular structure, 25
Tuned mass damper, 722, 790
Tunnel construction, 523
Turbulent boundary layer, 681, 719, 722
Types of damage, 661
Typical floor, 300
Typical trip, 621

Ultimate limit states, 885
Unbraced sway frame, 842
Uncertainties, 838
Underground parking, 994
Underseas operations, 517
Uniform Building Code, 398, 823, 965
United States, 335
Upper West Side, 84
Upstream, 681
Urban architecture, 149
Urban behavioral patterns, 285
Urban condition, 149
Urban design, 233
Urban design corridors, 259
Urban development, 165
Urban ecology, 79
Urban environment, 184, 455
Urban fabric, 257
Urban places, 279
Urban planning and design, 229, 275
Urban responsibilities, 158
User interface, 573
User-sponsored codes, 822
Utility cost, 613

Vacancy rates, 328
Value engineering, 207
Variable Air Volume (VAV), 608
Variable wind pressures, 710
Vaulted, 15
Vehicular circulation, 994
Ventilation, 21, 72, 319, 595
Vertical and horizontal transportation, 621
Vertical joints, 488
Vertical production method, 299, 300
Vertical shear connections, 937
Vertical style, 36
Vibration, 710, 885
Vibration and noise, 765
Vibration criteria, 885
Vibration periods, 651, 665
Views, 235
Viscoelastic dampers, 790, 797
Viscous damping, 781
Visual cues, 762
Visual impact study, 567
Visual linkage, 156, 235
Vortex shedding, 723
VPM, 300

Wall inspection, 546
Wall Street district, *276*

Wall tile detector, 525, 546, 547
Welding procedures, 825
Welding robot, 545
Wind bracing, 14, 15, 824, 847
Wind engineering, 705, 911
Wind-excited tall buildings, 697
Wind-excited vibrations, 697
Wind force, 705
Wind load, 754, 977
Wind loading and wind effects, 681
Wind load spectra, 700
Wind pressure, 681, 683, 707
Wind resisting structure, 341
Wind speed, 684, 710
Wind storms, 764
Wind tunnels, 72, 258, 681, 798
Wind turbulence, 698
Wind velocity coefficients, *73*
Window-mullion spacings, 206
Wrought iron, 14

Zoning, 80, 112, 119, 167, 183, 234, 264, 293